Electrical Wiring

Residential

Based on the 1990
NATIONAL ELECTRICAL CODE®

Electrical Wiring

Residential

Tenth Edition

Ray C. Mullin

DELMAR PUBLISHERS INC.®

NOTICE TO THE READER

Publisher does not warrant or guarantee any of the products described herein or perform an independent analysis in connection with any of the product information contained herein. Publisher does not assume, and expressly disclaims, any obligation to obtain and include information other than that provided to it by the manufacturer.

The reader is expressly warned to consider and adopt all safety precautions that might be indicated by the activities described herein and to avoid all potential hazards. By following the instructions contained herein, the reader willingly assumes all risks in connection with such instructions.

The publisher makes no representations or warranties of any kind, including but not limited to, the warranties of fitness for particular purpose or merchantability, nor are any such representations implied with respect to the material set forth herein, and the publisher takes no responsibility with respect to such material. The publisher shall not be liable for any special, consequential or exemplary damages resulting, in whole or in part, from the readers' use of, or reliance upon, this material.

Cover photo courtesy of Cooper Lighting, Cooper Industries

Delmar staff
Senior Administrative Editor: David Anthony
Developmental Editor: Marjorie A. Bruce
Editing Manager: Barbara A. Christie
Production Supervisor: Larry Main
Design Coordinator: Susan C. Mathews

For information, address Delmar Publishers Inc.
2 Computer Drive West, Box 15-015
Albany, New York 12212

Printed in the United States of America
Published simultaneously in Canada
by Nelson Canada,
a division of The Thomson Corporation

10 9 8 7 6 5 4 3 2

Library of Congress Cataloging in Publication Data

Mullin, Ray C.
Electrical wiring, residential, 10th ed.

Includes index.
1. Electrical wiring, Interior. I. Title.
TK3285.M84 1990 621.319'24 89-25859
ISBN 0-8273-3490-7 (pbk.)
ISBN 0-8273-3491-5 (instructor's guide)

CONTENTS

PREFACE

INTRODUCTION

The tenth edition of ELECTRICAL WIRING – RESIDENTIAL is based on the 1990 *National Electrical Code®* *(NEC®)*. The *NEC®* is used as the basic standard for the layout and construction of electrical systems. Thorough explanations are provided of Code changes as they relate to residential wiring. To gain the greatest benefit from this text, the learner must use the *National Electrical Code®* on a continuing basis.

State and local electrical codes may contain modifications of the *National Electrical Code®* to meet local requirements. The instructor is encouraged to furnish students with any variations from the *NEC®* as they affect this residential installation in a specific area.

This text takes the learner through all of the necessary minimum requirements as set forth in the *National Electrical Code®* for residential occupancies. Although Code minimums are covered in detail, the learner will find much information above and beyond the minimum requirements.

THE ELECTRICAL TRADE

Electricity is safe when it is handled properly. If electrical installations are made in an unsafe manner, great potential exists for damage to property as a result of fire. But more importantly, electrical installations that do not meet the minimum requirements of applicable codes and standards can and do result in personal injury and/or death.

Most communities, cities, counties, and states have regulations that adopt by reference the *National Electrical Code®*. These regulations also specify any exceptions to the Code to cover local conditions.

Most building codes and standards contain definitions for the various levels of competency of workers in the electrical industry.

Apprentice shall mean a person who is required to be registered as such under Section XYZ, who is in compliance with the provisions of this article, and who is working at the trade in the employment of a registered electrical contractor and is under the direct supervision of a licensed master electrician, journeyman electrician, or residential wireman.

Residential Wireman shall mean a person having the necessary qualifications, training, experience, and technical knowledge to wire for and install electrical apparatus and equipment for wiring one-, two-, three-, and four-family dwellings. A residential wireman is sometimes referred to as a *Class B Electrician*.

National Electrical Code® and *NEC®* are registered trademarks of the National Fire Protection Association, Inc., Quincy, MA 02269.

Journeyman Electrician shall mean a person having the necessary qualifications, training, experience, and technical knowledge to wire for, install, and repair electrical apparatus and equipment for light, heat, power and other purposes, in accordance with standard rules and regulations governing such work.

Master Electrician means a person having the necessary qualifications, training, experience, and technical knowledge to properly plan, lay out, and supervise the installation and repair of wiring apparatus and equipment for electric light, heat, power and other purposes, in accordance with standard codes and regulations governing such work, such as the *National Electrical Code*.®

Electrical Contractor means any person, firm, copartnership, corporation, association, or combination thereof who undertakes or offers to undertake for another the planning, laying out, supervising and installing, or the making of additions, alterations, and repairs in the installation of wiring apparatus and equipment for electrical light, heat, and power.

THE TENTH EDITION

Continuing in the tradition of previous editions, this text thoroughly explains how Code changes affect wiring installations. New and revised illustrations supplement the explanations to ensure that electricians understand the new Code requirements. New photos reflect the latest wiring materials and components available on the market. Revised review questions test student understanding of the new content. New tables that summarize Code requirements offer a quick reference tool for students. Another reference aid is the tables reprinted directly from the 1990 edition of the *National Electrical Code*.® The extensive revisions for the tenth edition make ELECTRICAL WIRING – RESIDENTIAL the most up-to-date and well-organized guide to house wiring.

An all-new residence serves as the basis for the wiring schematics, cable layouts, and discussions provided in the text. Each unit dealing with a specific type of wiring is referenced to the appropriate plan sheet. All wiring systems are described in detail — lighting, appliance, heating, service entrance, and so on.

The house selected for this edition is scaled for current construction practices and costs. Note, however, that the wiring, lighting fixtures, appliances, number of outlets, number of circuits, and track lighting are not all commonly found in a home of this size. The wiring may incorporate more features than are absolutely necessary. This was done to present as many features and Code issues as possible to give the learner more experience in wiring a residence.

This text does *not* focus on such basic skills as drilling a hole, splicing two wires together, stripping insulations from wire, or fishing a wire through walls. These skills are certainly necessary for good workmanship, but it is assumed that the learner (electrical apprentice or journeyman electrician) has mastered the mechanical skills on the job.

This text *does* focus on the technical skills required to perform electrical installations. It covers such topics as calculating conductor sizes, calculating voltage drop, determining appliance circuit requirements, sizing service, connecting electric appliances, grounding service and equipment, installing recessed fixtures, and much more. These are critical skills that can make the difference between an installation that "meets Code" and one that does not. The electrician must understand the reasons for following

Code regulations to achieve an installation that is essentially free from hazard to life and property.

CHANGES FOR THIS EDITION

The following list highlights the changes that have been made in the tenth edition. Only the major changes are listed here. A separate list follows of the major changes in the text required by the 1990 *National Electrical Code*®

- Added second color to the text and art to highlight critical information and 1990 Code changes.
- New residence plans at the back of the text reflect a current house design scaled for today's construction practices and costs.
- All information that is revised due to 1990 Code changes is identified at the beginning of the revised text by the symbol ►. At the end of the revised text, the symbol ◄ appears.
- New unit 31 — SMART HOUSE, a project of the National Association of Home Builders to develop and demonstrate systems for an all-electric residence
- Many new and revised circuit diagrams, cable layouts, illustrations, and calculation examples
- New content on transient voltage surge suppressors; isolated ground receptacles; appliance leakage circuit interrupters; immersion detection circuit interrupters; lighting (general...accent...task...bathroom...track...lay-in...roughing-in); "infinite heat" control switches for electric ranges
- New — separate unit on GFCI protection, replacing existing receptacles, testing, need for GFCI protection on construction sites, shock hazard, and cautions if they are subjected to short circuits
- Greatly expanded coverage of grounding: grounding electrode conductors, equipment grounding conductors, UFER grounding method, bonding, and panel requirements for the termination of equipment grounding conductors; water heater circuits, calculations and connections; electric furnace load computations
- Expanded coverage of nonmetallic boxes and wiring methods; conduit wiring; flexible metal and flexible nonmetallic conduit; use of BX and flexible metal conduit for "fixture whips"; color coding of conductors; recessed incandescent and fluorescent lighting fixtures; circuit loading guidelines; paddle fan installations; conduit fill with examples; conductor ampacity correction factors and derating factors; hydromassage tubs, spas and hot tubs; remote control, signaling, and power-limited circuits to clarify mixing power conductors, Class 1 conductors and Class 2 conductors in same enclosures, compartments and raceways
- New — *Quick Chek Box Size Selector Guide*
- Clarification of Code and UL requirements for the termination of 60°C, 75°C, and 90°C conductors; termination identification for copper and aluminum conductors
- Explanation of watts versus volt-amperes and the effect on load calculations
- Separate unit on fire and smoke detectors, *NEC*® and NFPA requirements

NEC®-RELATED CHANGES

The requirements of the 1990 *National Electrical Code*® resulted in the following topic changes:

- New requirements for
  - protection of Type NMC cable where it is run parallel to framing members
  - neutral conductors for services and feeders in residential installations
  - use of Type NM cable for multistory buildings
  - grounding and bonding for submersible water pumps and well casings
  - use of flexible cord connections for dishwashers, food waste disposers, and trash compactors
  - limitation of Class 2 wiring systems
  - identification of main bonding screws in service equipment
  - identification of series rated (series connected) systems
  - identification of circuits in panelboards — directory must be filled out
  - spacing of receptacle outlets in kitchens where these outlets serve counter tops
  - installing raceways as complete systems before pulling in wires
  - separate circuit for central heating (electric, gas, or oil)
  - liquid immersion detection circuit interrupter protection for hand-held hair dryers and freestanding hydromassage units
  - overcurrent protection on nonmotor-operated appliances rated at not over 13.3 amperes
- Completely revised swimming pool section with revised drawings and revised plan
- New rules for number of receptacle outlets and GFCI requirements for basements, garages, counter tops, islands, peninsulas, outdoors, feedthrough capabilities, exclusions for certain conditions such as sump pumps and dedicated spaces for appliances
- New ampacity tables for service-entrance and feeder conductors for residential installations
- New 1,000-volt limitation for residential wiring
- New discussion of underground wiring
- Expanded coverage of various classes of low-voltage conductors
- Other changes: box fill calculations; wire size designation change from MCM to kcmil; requirements for lighting fixtures in closets; no-niche swimming pool lighting fixtures; making surface extensions from concealed wiring; installation of lighting fixtures and wiring in trees; bonding of gas pipes to grounding system (no longer permitted as grounding electrode)

SUPPLEMENTS

An Instructor's Guide is available and consists of the following information: selected references (texts and audio-visuals), answers to unit-end reviews, a final examination covering the content of the entire text, and answers to the final examination.

New to the tenth edition is a package of blank floor plans. These plans enable learners to design their own electrical layouts to meet *National Electrical Code*® requirements. The instructor may also require students to prepare cable layouts and complete wiring diagrams based upon the electrical layouts.

ABOUT THE AUTHOR

This text was written by Ray C. Mullin, former electrical circuit instructor for the Electrical Trades, Wisconsin Schools of Vocational, Technical and Adult Education. Mr. Mullin, a former member of the International Brotherhood of Electrical Workers, is presently a member of the International Association of Electrical Inspectors, the Institute of Electrical and Electronic Engineers, and the National Fire Protection Association, Electrical Section.

Mr. Mullin completed his apprenticeship training and has worked as a journeyman and supervisor. He has taught both day and night electrical apprentice and journeyman courses and has conducted engineering seminars.

Mr. Mullin presents his knowledge and experience in this text in a clear-cut manner that is easy to understand. This presentation will help learners to understand fully the essential content required to pass the residential licensing examinations and to perform residential wiring that meets Code.

Mr. Mullin is the coauthor of *Electrical Wiring — Commercial*, 7th Edition and has contributed technical material for *Electrical Grounding* by Ronald O'Riley.

He serves on the Executive Board of the Western Section, International Association of Electrical Inspectors. He also serves on their National Electrical Code Committee and on their Code Clearing Committee. He also serves on the Electrical Commission in his hometown.

Mr. Mullin conducts many technical Code workshops and seminars at state Chapter and Section meetings of the International Association of Electrical Inspectors and serves on their Code panels.

Mr. Mullin is currently Director, Technical Liaison, and Code Coordinator for a large electrical manufacturer. He has developed extensive technical literature for use by his company's field engineering personnel.

THE DELMAR 1990 ELECTRICAL WIRING SERIES

In addition to *Electrical Wiring — Residential, 10th Edition*, Delmar is pleased to recommend the following texts for use in your electrical trades program. All have been thoroughly updated to meet 1990 *NEC®* standards, and all are *guaranteed* to provide a solid basis for a *comprehensive* program of practical, trade-oriented learning.

- **Electrical Wiring — Commercial, 7th Edition**
  Ray C. Mullin and *Robert L. Smith*
  An outstanding, *step-by-step* guide to wiring of a commercial building. With full-sized building plans, *plus* the latest in wiring, lighting methods, U.L. references, and load and conductor sizing calculations. Ideal for courses emphasizing 1990 Code interpretation, electrical theory, and hands-on blueprint reading!

- **Electrical Wiring — Industrial, 7th Edition**
  Robert L. Smith and *Stephen L. Herman*
  The source for the latest standards for layout and construction of electrical circuits required in an industrial installation. Its hallmark *"systems approach"* encourages students to work with the text, the *1990 Code Book*, and *real-life industrial blueprints* (included in the back of the text)!

- **Electrical Grounding, 2nd Edition**
  Ronald P. O'Riley
  A *"one-of-a-kind"* text that begins with simple situations and a review of electrical theory before advancing to the most complex and specific applications of modern grounding technology. With examples, equations, introductions, and quotes from the *1990 NEC®* in *every* chapter!
- **Interpreting the National Electrical Code®, 2nd Edition**
  Truman Surbrook
  Field-tested, based on the *1990 NEC®*, and organized by subject (not by Code articles) for easy reference! This workbook-style manual will teach your students how to 1) read and interpret the meaning of the Code, and 2) use the actual Code book to quickly locate practical information about *all* types of wiring installations.
- **Interpreting the Changes in the 1990 National Electrical Code®**
  Truman Surbrook
  A quick reference guide that lets you "zero in" on the differences between the 1987 and the 1990 *NEC®*. A priceless, convenient resource for *any* electrical program!

ADDITIONAL RELATED TITLES

Electricity 1: Devices, Circuits, and Materials, 4th Edition
Thomas Kubala

Electricity 2: Devices, Circuits, and Materials, 4th Edition
Thomas Kubala

Electricity 3: Motors and Generators, Controls, Transformers, 4th Edition
Walter Alerich

Electricity 4: Motors, Controls, Alternators, 4th Edition
Walter Alerich

Alternating Current Fundamentals, 3rd Edition
John Duff and *Stephen Herman*

Direct Current Fundamentals, 3rd Edition
Orla Loper and *Edgar Tedsen*

Practical Problems in Mathematics for Electricians, 4th Edition
Crawford Garrard and *Stephen Herman*

Industrial Motor Control, 2nd Edition
Stephen Herman and *Walter Alerich*

Electronics for Industrial Electricians, 2nd Edition
Stephen Herman

Industrial Electricity, 2nd Edition
Nadon, Gelmine, and *McLaughlin*

Technician's Guide to Programmable Controllers, 2nd Edition
Richard Cox

Electric Motor Control, 4th Edition
Walter Alerich

Electrical Control for Machines, 3rd Edition
Kenneth Rexford

Electrical Systems in Buildings
S. David Hughes

Electricity and Controls for Heating, Ventilating, and Air Conditioning
Stephen Herman and *Bennie Sparkman*

Electricity for Refrigeration, Heating, and Air Conditioning, 3rd Edition
Russell Smith

ACKNOWLEDGMENTS

Helen J. Mullin, my wife...editing, typing, encouragement

The following individuals provided recommendations for changes and additions for the new edition.

Professor Robert L. Smith
University of Illinois
Urbana, IL

Dr. Truman Surbrook
Michigan State University
East Lansing, MI

Charles Trout
Electrical Contractor and Code Instructor
Chicago, IL

Joel Rencsok
Electrical Plans Engineer for the city of Phoenix
Phoenix, AZ

H. Brooke Stauffer
Manager, Codes and Standards Development
SMART HOUSE, L.P.
Upper Marlboro, MD

The following individuals reviewed portions or all of the revised text. Their critical analyses and recommendations helped the author revise and verify the extensive changes for this edition.

Carl L. Everett
Oakland Community College
Auburn Hills, MI 48057

Robert F. Kurr, CDE
Florida Electrical Apprenticeship and Training, Inc.
Winter Park, FL 32789

Rodney Stanley
Industrial Education Technology
Morehead State University
Morehead, KY 40351

Wellington Hiester
Technical College of the Low Country
Beaufort, SC 29902

Richard Curbello
San Jacinto College
Pasadena, TX 77505

Richard Loyd
R and N Associates
Code Consultant
Perryville, AR

Phil Cox
NEMA Field Representative
Little Rock, AR

Peter Van Putten
Instructor and Retired Electrical Inspector
Byron Center, MI

Mel Sanders
Electrical Engineer
Electrical Inspector, Code Consultant, and Code Instructor

Appreciation is expressed to Charles Williams, National Electrical Contractors Association, Bethesda, MD, for developing the cross-reference between text content and 1990 Code changes provided on the inside covers of the text.

Special appreciation is expressed to Charles W. Talcott and Home Planners, Inc. for the basic plans upon which the residence plans found at the back of the text are based.

The author wishes to thank the following companies for their contributions of data, illustrations, and technical information:

AFC/A Nortek Co.
American Home Lighting Institute
Anchor Electric Division, Sola Basic Industries
Appleton Electric Co.
Arrow-Hart, Inc.
BRK Electronics, A Division of Pittway Corporation
Brazos Technologies, Inc.
Bussmann Division, Cooper Industries
Carlon
Chromalox
Electri-Flex Co.
General Electric Co.
Halo Lighting Division, Cooper Industries
Heyco Molded Products, Inc.
Honeywell
Hubbell Incorporated, Wiring Devices Division
International Association of Electrical Inspectors
Juno
KEPTEL
The Kohler Co.
Leviton Manufacturing Co., Inc.
Midwest Electric Products, Inc.
Milbank Manufacturing Co.
Moe Light Division, Thomas Industries
NuTone Inc.
Pass & Seymour, Inc.
Progress Lighting
Rheem Manufacturing Co.
Seatek Co., Inc.
Sierra Electric Division, Sola Basic Industries
SMART HOUSE, L.P.
Square D Co.
Superior Electric Co.
THERM-O-DISC
Thomas & Betts Corporation
Winegard Company
Wiremold Co.
Wolberg Electrical Supply Co., Inc.
Woodhead Industries, Inc.

General Information for Electrical Installations

OBJECTIVES

After studying this unit, the student will be able to

- explain how electrical wiring information is conveyed to the electrician at the construction or installation site.
- demonstrate how the specifications are used in estimating cost and in making electrical installations.
- explain why symbols and notations are used on electrical drawings.
- list the agencies that are responsible for establishing electrical standards and insuring that materials meet the standards.
- discuss the metric system of measurement.
- begin to refer to the *National Electrical Code*®

THE WORKING DRAWINGS

The architect uses a set of working drawings or plans to make the necessary instructions available to the skilled crafts which are to build the structure shown in the plans. The sizes, quantities, and locations of the materials required and the construction features of the structural members are shown at a glance. These details of construction must be studied and interpreted by each skilled construction craft — masons, carpenters, electricians, and others — before the actual work is started.

The electrician must be able to: (1) convert the two-dimensional plans into an actual electrical installation, and (2) visualize the many different views of the plans and coordinate them into a three-dimensional picture, as shown in figure 1-1.

The ability to visualize an accurate three-dimensional picture requires a thorough knowledge of blueprint reading. Since all of the skilled trades use a common set of plans, the electrician must be able to interpret the lines and symbols which refer to the electrical installation and also those which are used by the other construction trades. The electrician must know the structural makeup of the building and the construction materials to be used.

SPECIFICATIONS

Working drawings are usually complex because of the amount of information which must be included. To prevent confusing detail, it is standard practice to include with each set of plans a set of detailed written specifications prepared by the architect.

These specifications provide general information to be used by all trades involved in the construction. In addition, specialized information is given for the individual trades. The specifications include

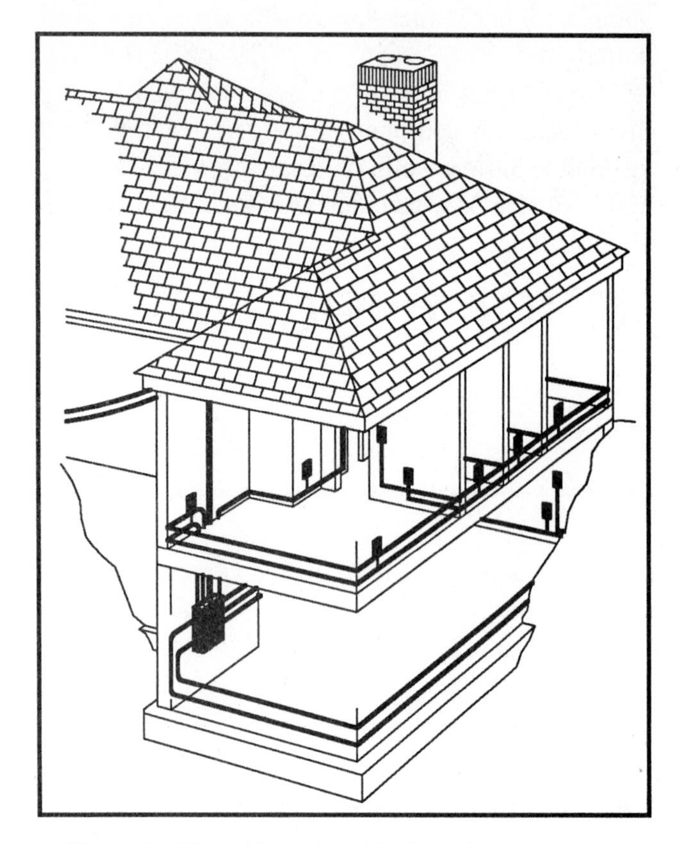

Fig. 1-1 Three-dimensional view of house wiring.

information on the sizes, the type, and the desired quality of the standard parts which are to be used in the structure.

Typical specifications include a section on "General Clauses and Conditions," which is applicable to all trades involved in the construction. This section is followed by detailed requirements for the various trades — excavating, masonry, carpentry, plumbing, heating, electrical work, painting, and others.

The plan drawings for the residence used as an example for this text are included in the back of the text. The specifications for the electrical work indicated on the plans are given in the Appendix.

In the electrical specifications, the listing of standard electrical parts and supplies frequently includes the manufacturers' names and the catalog numbers of the specified items. Such information insures that these items will be of the correct size, type, and electrical rating, and that the quality will meet a certain standard. To allow for the possibility that the contractor will not always be able to obtain the specified item, the phrase "or equivalent" is usually added after the manufacturer's name and catalog number.

The specifications are also useful to the electrical contractor in that all of the items needed for a specific job are grouped together and the type or size of each item is indicated. This information allows the contractor to prepare an accurate cost estimate without having to find all of the data in the plans.

SYMBOLS AND NOTATIONS

The architect uses symbols and notations to simplify the drawing and presentation of information concerning electrical devices, appliances, and equipment. For example, an electric range outlet looks like this:

Most symbols have a standard interpretation throughout the country as adopted by the American National Standards Institute (ANSI). Symbols are described in detail in unit 2.

A notation will generally be found on the plans (blueprints) next to a specific symbol calling attention to a variation, type, size, quantity, or other necessary information. In reality, a symbol might be considered to be a notation because symbols do represent a system that "represents words, phrases, numbers, quantities, etc." as defined in the dictionary.

Another method of using notations to avoid cluttering up a blueprint is to provide a system of symbols that refer to a specific table. For example, the written sentences on plans could be included in a table referred to by a notation. Figure 2-9 is an example of how this could be done. The special symbols that refer to the table would have been shown on the actual plan.

NATIONAL ELECTRICAL CODE® (NEC)

Because of the ever-present danger of fire or shock hazard through some failure of the electrical system, the electrician and the electrical contractor must use approved materials and must perform all work in accordance with recognized standards. The *National Electrical Code®* is the basic standard which governs electrical work. The purpose of the Code is to provide information considered necessary for the safeguarding of people and property against electrical hazards. It states that the installation must be essentially hazard-free, but that such an installation is not necessarily efficient, convenient, or adequate, *Section 90-1(b)*. (Note: All *National Electrical Code®* section references used throughout this text are printed in italics.) It is the electrician's responsibility to insure that the installation meets these criteria. In addition to the *National Electrical Code®*, the electrician must also consider local and state codes. The purpose and scope of the *National Electrical Code®* are discussed in *Article 90* of the Code book and should be studied by the student at this time.

The first *National Electrical Code®* was published in 1897. It is revised and updated every three years by changing, editing, adding technical material, adding sections, adding articles, adding tables, and so on, so as to be as up-to-date as possible relative to electrical installations. For example, the Code added requirements pertaining to circuits supplying computers and similar electronic equipment.

Another prime example is that we have included a chapter on a "SMART HOUSE," which relates to the requirements of *Article 780, Closed Loop and Programmed Power Distribution*.

As materials, equipment, and technologies change, the members of the Code-making panel solicit comments and proposals from individuals in the electrical industry or others interested in electrical safety. The panel members then meet to act upon the many proposals for changes to the Code. Final action (voting) on these proposals is taken at the National Fire Protection Association's annual meeting.

The Code-making panel consists of individuals representing over 65 organizations such as the International Association of Electrical Inspectors, the Institute of Electrical and Electronic Engineers, the National Electrical Contractors Association, the Edison Electric Institute, Underwriters Laboratories, Consumer Product Safety Commission, International Brotherhood of Electrical Workers, and many others.

Copies of the *National Electrical Code®* may be ordered from

National Fire Protection Association
Batterymarch Park
Quincy, Massachusetts 02269

or

International Association of Electrical Inspectors
930 Busse Highway
Park Ridge, Illinois 60068-2398

AMERICAN NATIONAL STANDARDS INSTITUTE

The American National Standards Institute (ANSI) is an organization that coordinates the efforts and results of the various standards-developing organizations, such as those mentioned in previous paragraphs. Through this process, ANSI approves standards which then become recognized as American national standards. One will find much similarity between the technical information found in ANSI standards, the Underwriters standards, the International Electronic and Electrical Engineers standards, and the *National Electrical Code®*.

Code Definitions

The electrical industry uses many words (terms) that are unique to the electrical trade. These terms need clear definitions to enable the electrician to understand completely the meaning intended by the Code.

Article 100 of the *National Electrical Code®* is a "dictionary" of these terms. A few of the terms defined in the Code follow.

Ampacity: The current in amperes a conductor can carry continuously under the conditions of use without exceeding its temperature rating.

Approved: Acceptable to the authority having jurisdiction.

Authority Having Jurisdiction: An organization, office, or individual responsible for "approving" equipment, an installation, or a procedure, *Section 90-4*.

Dwelling Unit: One or more rooms for the use of one or more persons as a housekeeping unit with space for eating, living, and sleeping, and permanent provisions for cooking and sanitation. (Note: The terms *dwelling* and *residence* are used interchangeably throughout this text.)

Fine Print Notes (FPN): Fine Print Notes (FPNs) are found throughout the Code. FPNs are "explanatory" in nature, in that they make reference to other Sections of the Code. FPNs also define things where further description is necessary.

Identified: (As applied to equipment.) Recognizable as suitable for a specific purpose, function, use, environment, or application, where described in a particular Code requirement. Suitability of use, marked on or provided with the equipment, may include labeling or listing.

Labeled: Equipment or materials to which has been attached a label, symbol, or other identifying mark of an organization acceptable to the authority having jurisdiction and concerned with product evaluation, that maintains periodic inspection of production of labeled equipment or materials and by whose labeling the manufacturer indicates compliance with appropriate standards or performance in a specified manner.

Listed: Equipment or materials included in a list published by an organization acceptable to the authority having jurisdiction and concerned with product evaluation, that maintains periodic inspection of production of listed equipment or materials and whose listing states either that the equipment or material meets appropriate standards or has been tested and found suitable for use in a specified manner.

Shall: Indicates a mandatory requirement. Examples are as follows:

shall be	compulsory; mandatory; a requirement; must be.
shall have	the same as "shall be."
shall not	not allowed; not permitted to be done; must not be; against the Code.
shall be permitted ..	is allowed: may be done; not against the Code.

One of the most far-reaching *NEC*® rules is *Section 110-3(b)*. This section states that the use and installation of listed or labeled equipment must conform to any instructions included in the listing or labeling. This means that an entire electrical system and all of the system's electrical equipment must be *installed* and *used* in accordance with the *National Electrical Code*® and the numerous standards against which the electrical equipment has been tested. (See Underwriters Laboratories, page 5.)

In the past, programs such as *Adequate Wiring*, *House Power*, *Live Better Electrically*, *Bronze Medallion*, and *Gold Medallion* were instituted to supplement established Code standards. The purpose of these programs was to promote the installation of efficient, convenient, and useful home wiring systems. After a period of time, as the recommendations of these programs were gradually written into the *National Electrical Code*®, the programs were phased out.

CODE USE OF METRIC (SI) MEASUREMENTS

The 1981 *National Electrical Code*® introduced the use of metric measurements in addition to English measurements. The metric system is known as the *International System of Units (SI)*.

Metric measurements appear in the Code as follows:

- in the Code paragraphs, the approximate metric measurement appears in parentheses following the English measurement.
- in the Code tables, a footnote shows the SI conversion factors.

A metric measurement is not shown for conduit size, box size, wire size, horsepower designation for motors, and other "trade sizes" that do not reflect actual measurements.

Guide to Metric Usage

In the metric system, the units increase or decrease in multiples of 10, 100, 1 000, and so on. For instance, one megawatt (1 000 000 watts) is 1 000 times greater than one kilowatt (1 000 watts).

By assigning a name to a measurement, such as a *watt*, the name becomes the unit. Adding a prefix to the unit, such as *kilo*, forms the new name *kilowatt*, meaning 1 000 watts. Refer to figure 1-2 for prefixes used in the metric system.

The prefixes used most commonly are *centi*, *kilo*, and *milli*. Consider that the basic unit is a meter (one). Therefore, a centimeter is 0.01 meter, a kilometer is 1 000 meters, and a millimeter is 0.001 meter.

Some common measurements of length and equivalents are shown in figure 1-3.

mega	1 000 000	(one million)
kilo	1 000	(one thousand)
hecto	100	(one hundred)
deka	10	(ten)
the unit	1	(one)
deci	0.1	(one-tenth) (1/10)
centi	0.01	(one-hundredth) (1/100)
milli	0.001	(one-thousandth) (1/1 000)
micro	0.000 001	(one-millionth) (1/1 000 000)
nano	0.000 000 001	(one-billionth) (1/1 000 000 000)

Fig. 1-2 Metric prefixes and their values.

one inch	=	2.54	centimeters
	=	25.4	millimeters
	=	0.025 4	meter
one foot	=	12	inches
	=	0.304 8	meter
	=	30.48	centimeters
	=	304.8	millimeters
one yard	=	3	feet
	=	36	inches
	=	0.914 4	meter
	=	914.4	millimeters
one meter	=	100	centimeters
	=	1 000	millimeters
	=	1.093	yards
	=	3.281	feet
	=	39.370	inches

Fig. 1-3 Some common measurements of length and their equivalents.

inches (in) × 0.025 4	= meters (m)
inches (in) × 0.254	= decimeters (dm)
inches (in) × 2.54	= centimeters (cm)
centimeters (cm) × 0.393 7	= inches (in)
inches (in) × 25.4	= millimeters (mm)
millimeters (mm) × 0.039 37	= inches (in)
feet (ft) × 0.304 8	= meters (m)
meters (m) × 3.280 8	= feet (ft)
square inches (in^2) × 6.452	= square centimeters (cm^2)
square centimeters (cm^2) × 0.155	= square inches (in^2)
square feet (ft^2) × 0.093	= square meters (m^2)
square meters (m^2) × 10.764	= square feet (ft^2)
square yards (yd^2) × 0.836 1	= square meters (m^2)
square meters (m^2) × 1.196	= square yards (yd^2)
kilometers (km) × 1 000	= meters (m)
kilometers (km) × 0.621	= miles (mi)
miles (mi) × 1.609	= kilometers (km)

Fig. 1-4 Useful conversions (English/SI - SI/English) and their abbreviations.

Electricians will find it useful to refer to the conversion factors and their abbreviations shown in figure 1-4.

UNDERWRITERS LABORATORIES INC. (UL)

Underwriters Laboratories (UL), founded in 1894, is a highly qualified, nationally recognized testing laboratory. UL develops standards and performs tests to these standards. Most reputable manufacturers of electrical equipment submit their products to the Underwriters Laboratories where the equipment is subjected to numerous tests. These tests determine if the products can perform safely under normal and abnormal conditions to meet published standards. After UL determines that a product complies with the specific standard, a manufacturing firm is then permitted to *label* its product with the UL logo. The products are then *listed* in a UL directory.

It must be noted that UL does not *approve* any product. Rather, UL *lists* those products that conform to its safety standards.

Useful UL publications are:

Electrical Construction Materials Directory (Green Book)

Electrical Appliances and Utilization Equipment Directory (Orange Book)

Hazardous Location Equipment Directory (Red Book)

General Information for Electrical Construction, Hazardous Location, and Electric Heating and Air Conditioning Equipment (White Book)

Recognized Component Directory (Yellow Book)

Many inspection authorities continually refer to these books in addition to the *National Electrical Code*®.

If the answer to a question cannot be found readily in the *National Electrical Code*®, then it generally can be found in the listed UL publications.

The Green, Orange, Red, and Yellow Books give the names of manufacturers, as well as the manufacturers' identification numbers. The White Book does not include this information, but it does show the technical requirements for listing a product in all of the categories that are shown in the Green, Orange, Red, and Yellow Books. It is extremely useful for an electrician and/or electrical inspector to refer to when looking for the specific requirements about a certain product.

The White Book is an excellent companion to the *National Electrical Code*.® Copies of Underwriters Laboratories Inc. publications may be obtained from

Underwriters Laboratories Inc.
333 Pfingsten Road
Northbrook, IL 60062

ETL Testing Laboratories, Inc. is another nationally recognized testing laboratory. They provide a labeling, listing and follow-up service for the safety testing of electrical products to nationally recognized safety standards or specifically designated requirements of jurisdictional authorities. Information can be obtained by writing to

ETL Testing Laboratories, Inc.
Main Office
Industrial Park
Cortland, NY 13045

REVIEW

Note: Refer to the *National Electrical Code*® or the plans where necessary.

1. What is the purpose of specifications? ____________________

2. In what additional way are the specifications particularly useful to the electrical contractor? ____________________

3. What is done to prevent a plan from becoming confusing because of too much detail? ____________________

4. Name three requirements contained in the specifications regarding material.
   a. __________ c. __________
   b. __________

5. The specifications state that all work shall be done ____________________

6. What phrase is used when a substitution is permitted for a specific item? ____________________

7. What is the purpose of an electrical symbol? ____________________

8. What is a notation? ____________________

9. Where are notations found? ____________________

10. List at least 12 electrical notations found on the plans for this residence. Refer to the plans at the back of the text. ______

11. What three parties must be satisfied with the completed electrical installation?

    a. ______ b. ______ c. ______

12. What Code sets standards for electrical installation work? ______

13. What authority enforces the standards set by the Code? ______

14. Does the Code provide minimum or maximum standards? ______

15. What do the letters *UL* signify? ______

16. What section of the Code states that all listed or labeled equipment shall be used or installed in accordance with any instructions included in the listing or labeling?

17. When the words "shall be" appear in a Code reference, they mean that it (must) (may) be done. (Underline the correct word.)

18. What is the purpose of the *National Electrical Code*®?

19. Does compliance with the Code always result in an electrical installation that is adequate, safe, and efficient? Why? ______

20. Name two nationally recognized testing laboratories. ______

21. Does Underwriters Laboratories approve products? What does UL do?

Electrical Symbols and Outlets

OBJECTIVES

After studying this unit, the student will be able to

- identify and explain the electrical outlet symbols used in the plans of the single-family dwelling.
- discuss the types of outlets, boxes, fixtures, and switches used in the residence.
- explain the methods of mounting the various electrical devices used in the residence.
- understand the meaning of the terms *receptacle outlet* and *lighting outlet.*
- understand the preferred way to position receptacles in wall boxes.
- know how to position wall boxes in relation to finished wall surfaces.
- know how to make surface extensions from concealed wiring methods.
- understand how to determine the number of wires permitted in a given size box.
- discuss interchangeable wiring devices.

ELECTRICAL SYMBOLS

Electrical symbols used on an architectural plan show the location and type of electrical device required. A typical electrical installation as taken from a plan is shown in figure 2-1.

Fig. 2-1 Use of electrical symbols and notations on a floor plan.

The *National Electrical Code®* describes an *outlet* as "a point on a wiring system where current is taken to supply utilization equipment."

A receptacle outlet is "an outlet where one or more receptacles are installed," figure 2-2.

A lighting outlet is "an outlet intended for the direct connection of a lampholder, a lighting fixture, or a pendant cord terminating in a lampholder," figure 2-3.

A toggle switch is *not* an outlet.

The term *outlet* is used broadly by electricians to include noncurrent-consuming switches and similar control devices in a wiring system when estimating the cost of the installation. Each type of outlet is represented on the plans as a symbol.

The *NEC®* defines a device as "a unit of an electrical system which is intended to carry but not utilize electric energy." In figure 2-1 the outlets are shown by the symbols ⊖ and -○-. The standard electrical symbols are shown in figures 2-4 through 2-9 (pages 10–14).

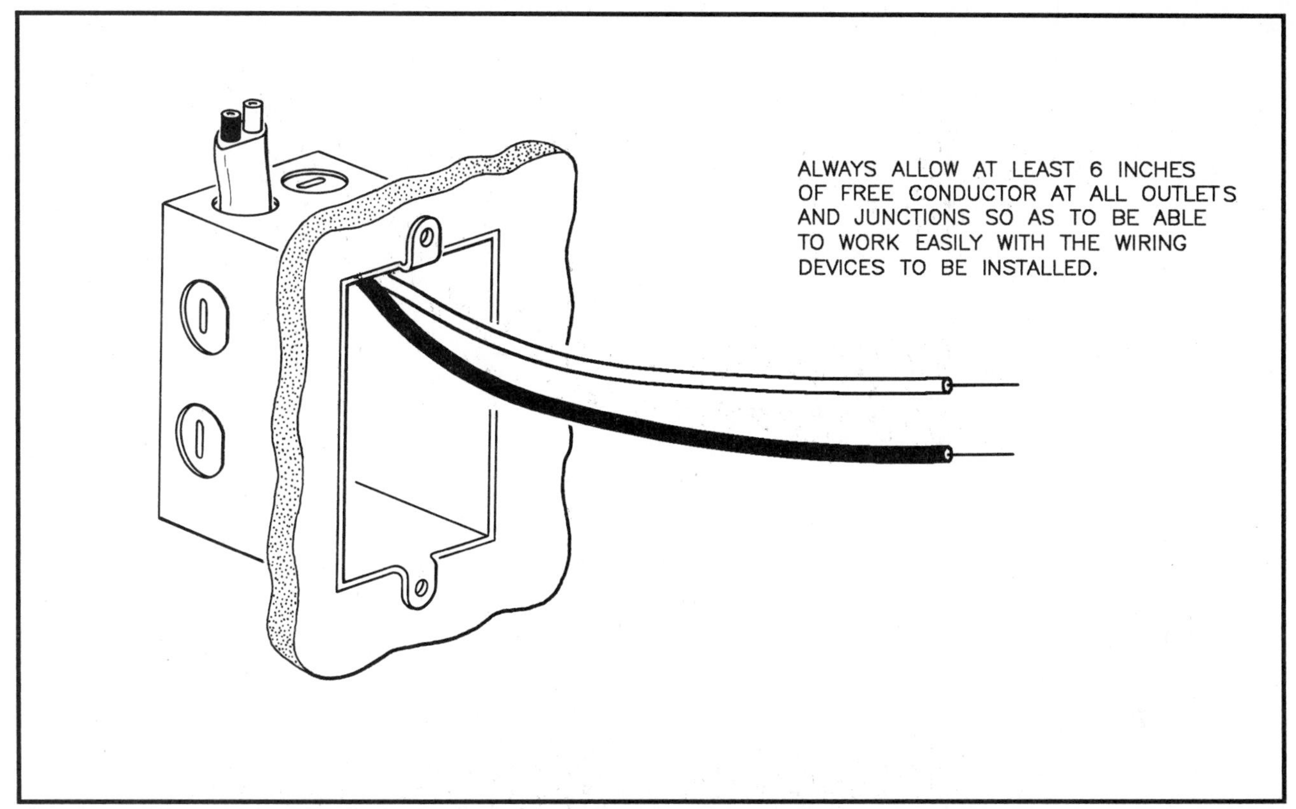

Fig. 2-2 When a receptacle is connected to the wires, the outlet is called a *receptacle outlet*. For ease in working with wiring devices, the Code in *Section 300-14* requires that at least 6 inches of free conductor be provided.

The dash lines in figure 2-1 run from the outlet to the switch or switches which control the outlet. These lines are usually curved so that they cannot be mistaken for invisible edge lines. Outlets shown on the plan without curved dash lines are independent outlets and have no switch control.

A study of the plans for the single-family dwelling shows that many different electrical symbols are used to represent the electrical devices and equipment used in the building.

In drawing electrical plans, most architects, designers, and electrical engineers use symbols approved by the American National Standards Institute (ANSI) wherever possible. However, plans may contain symbols that are not found in these standards. When such unlisted (nonstandard) symbols are used, the electrician must refer to a legend which interprets these symbols. The legend may be included on the plans or in the specifications. In many instances, a notation on the plan will clarify the meaning of the symbol.

Figures 2-4 through 2-9 list the standard, approved electrical symbols and their meanings. Many of these symbols can be found on the accompanying plans of the residence. Note in these figures that several symbols have the same shape. However, differences in the interior presentation indicate that the meanings of the symbols are different. For example, different meanings are shown in figure 2-10 (page 14) for the outlet symbol. A good practice to follow in studying symbols is to learn the basic forms first and then add the supplemental information to obtain different meanings.

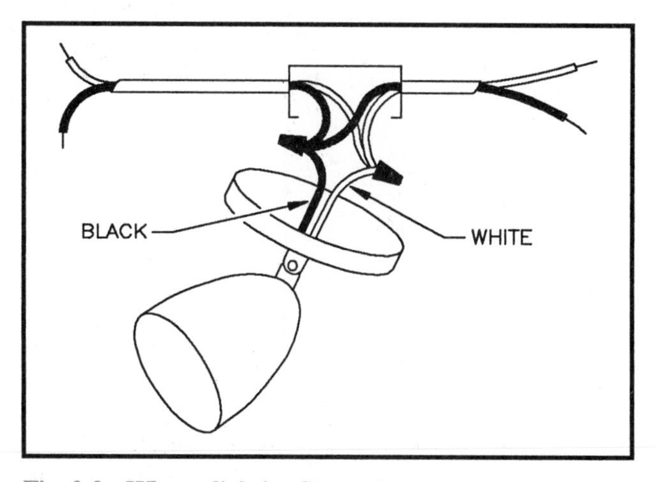

Fig. 2-3 When a lighting fixture is connected to the wires, the outlet is called a *lighting outlet*. The Code requires that at least 6 inches of free conductor be provided.

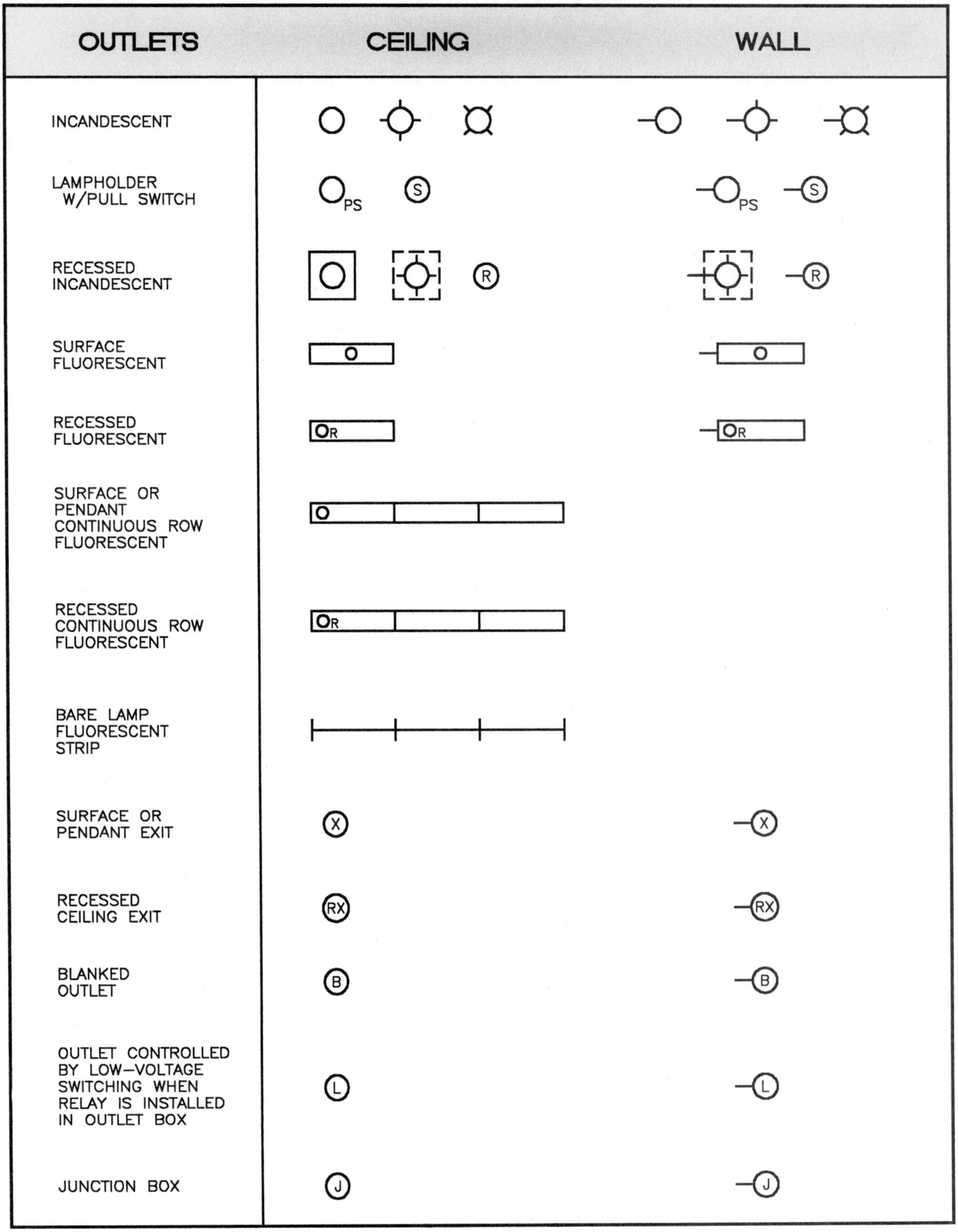

OUTLETS	CEILING	WALL
INCANDESCENT		
LAMPHOLDER W/PULL SWITCH	PS S	PS S
RECESSED INCANDESCENT	R	R
SURFACE FLUORESCENT	O	O
RECESSED FLUORESCENT	OR	OR
SURFACE OR PENDANT CONTINUOUS ROW FLUORESCENT	O	
RECESSED CONTINUOUS ROW FLUORESCENT	OR	
BARE LAMP FLUORESCENT STRIP		
SURFACE OR PENDANT EXIT	X	X
RECESSED CEILING EXIT	RX	RX
BLANKED OUTLET	B	B
OUTLET CONTROLLED BY LOW-VOLTAGE SWITCHING WHEN RELAY IS INSTALLED IN OUTLET BOX	L	L
JUNCTION BOX	J	J

Fig. 2-4 Lighting outlet symbols.

RECEPTACLE OUTLETS	
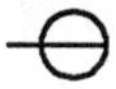	SINGLE RECEPTACLE OUTLET
	DUPLEX RECEPTACLE OUTLET
 IG	INSULATED (ISOLATED) GROUND RECEPTACLE OUTLET
	TRIPLEX RECEPTACLE OUTLET
	DUPLEX RECEPTACLE OUTLET, SPLIT-CIRCUIT
	TRIPLEX RECEPTACLE OUTLET, SPLIT-CIRCUIT
 WP	WEATHERPROOF RECEPTACLE OUTLET
 GFCI	GROUND-FAULT CIRCUIT INTERRUPTER RECEPTACLE OUTLET
 DW	SPECIAL-PURPOSE OUTLET (SUBSCRIPT LETTERS INDICATE SPECIAL VARIATIONS: DW = DISHWASWER. ALSO a, b, c, d, ETC. ARE LETTERS KEYED TO EXPLANATION ON DWGS. OR IN SPECIFICATIONS).
 R	RANGE OUTLET
 D	CLOTHES DRYER OUTLET
F	FAN OUTLET
C	CLOCK OUTLET
	FLOOR OUTLET
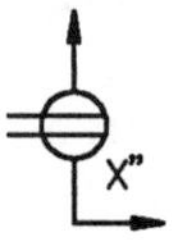 X"	MULTIOUTLET ASSEMBLY; ARROW SHOWS LIMIT OF INSTALLATION. APPROPRIATE SYMBOL INDICATES TYPE OF OUTLET. SPACING OF OUTLETS INDICATED BY "X" INCHES.
	FLOOR SINGLE RECEPTACLE OUTLET
	FLOOR DUPLEX RECEPTACLE OUTLET
	FLOOR SPECIAL-PURPOSE OUTLET

Fig. 2-5 Receptacle outlet symbols.

FIXTURES AND OUTLETS

Architects often include in the specifications a certain amount of money for the purchase of electrical fixtures. The electrical contractor includes this amount in the bid, and the choice of fixtures is then left to the homeowner. If the owner selects fixtures whose total cost exceeds the fixture allowance, the owner is expected to pay the difference between the actual cost and the specification allowance. If the fixtures are not selected before the roughing-in stage of wiring the house, the electrician usually installs outlet boxes having standard fixture mounting studs.

Most modern surface-mount lighting fixtures can be fastened to a fixture stud in the box (figure 2-11 (page 15), top row, picture A) or an outlet box or plaster ring using appropriate #8-32 metal screws and mounting strap furnished with the fixture (figure 2-11, third row, raised plaster cover; outlet boxes.)

A box must be installed at each outlet or switch location, *Section 300-15*. There are exceptions to this rule, but, in general, they relate to special manufactured wiring systems where the "box" is an integral part of the system. For standard wiring methods, such as cable or conduit, a box is usually required.

Be careful when roughing in boxes for fixtures. UL states in its Green and White Books that it has not investigated hanging ceiling fixtures to

- nonmetallic device (switch) boxes.
- nonmetallic device (switch) plaster rings.
- metallic device (switch) boxes.
- metallic device (switch) plaster rings.
- any nonmetallic box, unless specifically marked on the box or carton for use as a fixture support, or to support other equipment, or to accommodate heat-producing equipment.

Therefore, unless the box, device plaster ring, or carton is marked to indicate that it has been listed by UL for the support of fixtures, do not use where fixtures are to be hung. Refer to figure 2-12 (page 16).

SWITCH OUTLETS

S	SINGLE–POLE SWITCH
S_2	DOUBLE–POLE SWITCH
S_3	THREE–WAY SWITCH
S_4	FOUR–WAY SWITCH
S_D	DOOR SWITCH
S_{DS}	DIMMER SWITCH
S_K	KEY SWITCH
S_L	LOW–VOLTAGE SWITCH
S_{LM}	LOW–VOLTAGE MASTER SWITCH
S_P	SWITCH WITH PILOT LAMP
S_R	VARIABLE–SPEED SWITCH
S_T	TIME SWITCH
S_{WP}	WEATHERPROOF SWITCH

Fig. 2-6 Switch outlet symbols.

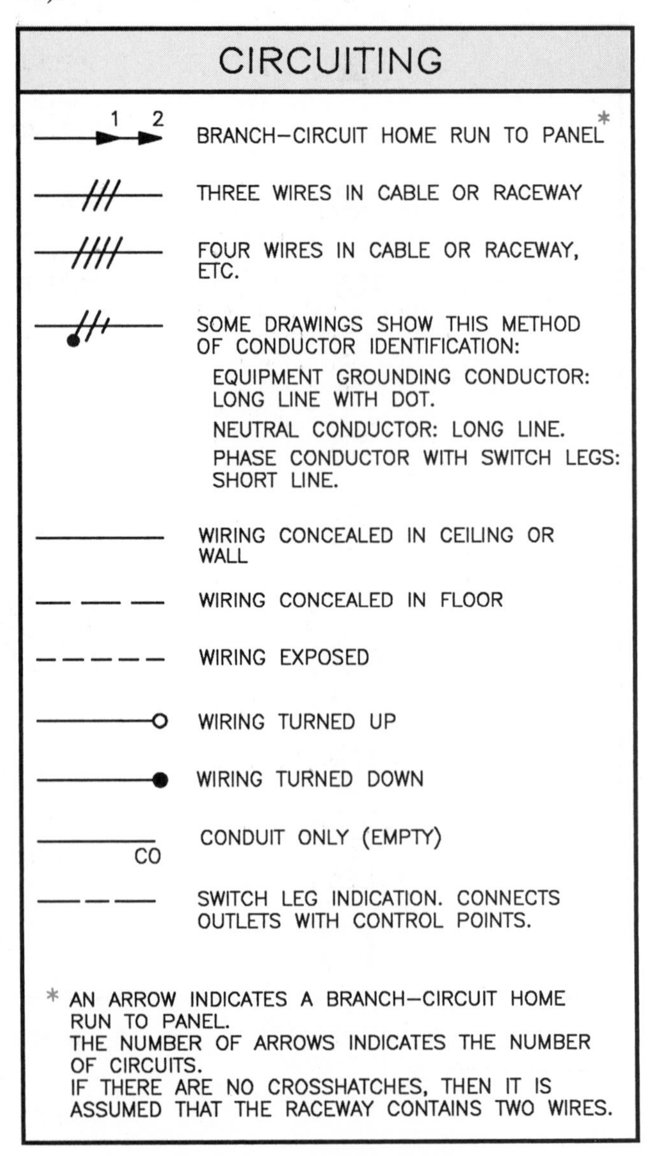

Fig. 2-7 Circuiting symbols.

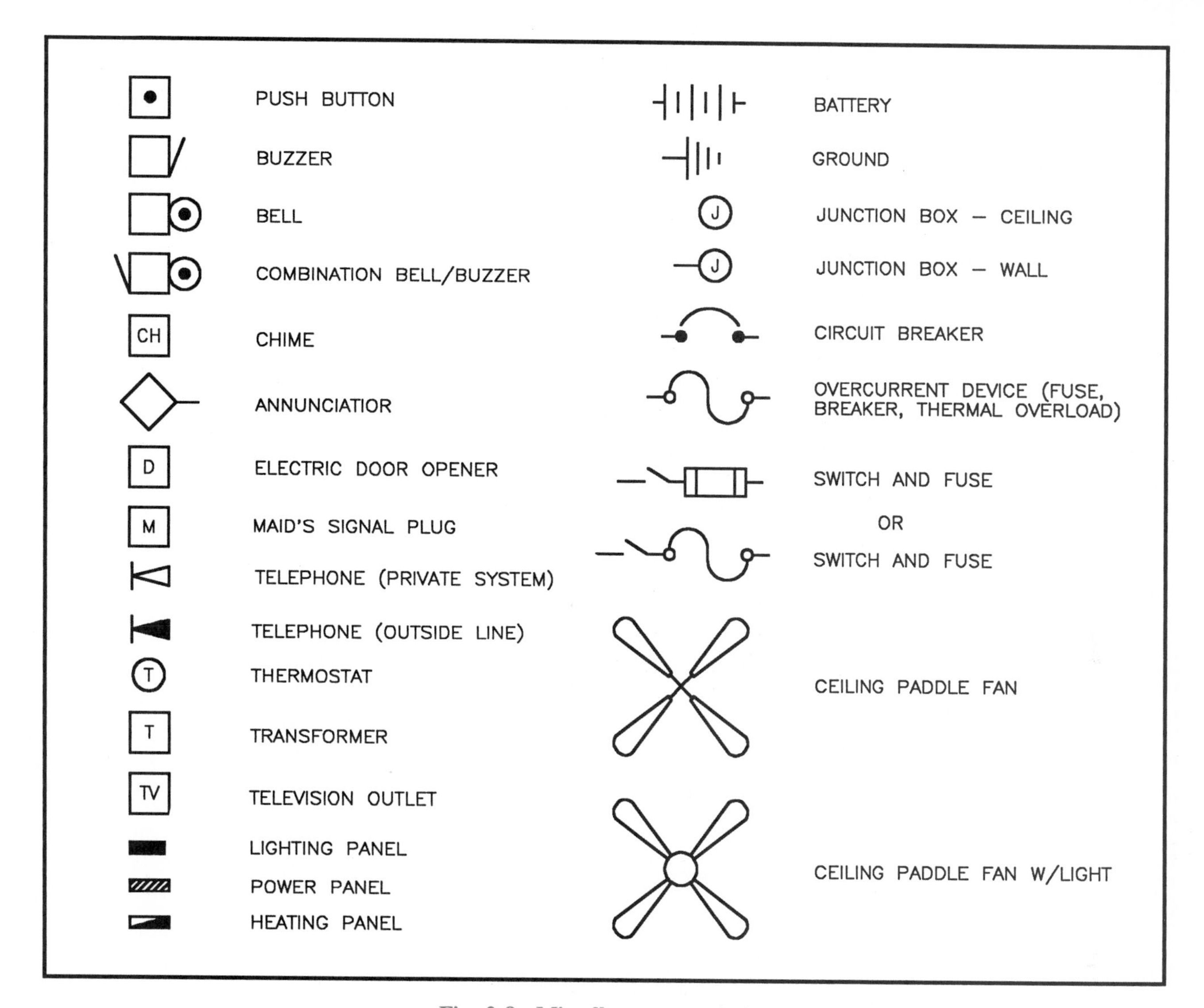

Fig. 2-8 Miscellaneous symbols.

Be careful when installing a ceiling outlet box for the purpose of supporting a fan. See unit 9 for detailed discussion of paddle fans and their installation.

If the owner selects fixtures prior to construction, the architect can specify these fixtures in the plans and/or specifications. Thus, the electrician is provided with advance information on any special framing, recessing, or mounting requirements for the fixtures. This information *must* be provided in the case of recessed fixtures, which require a specific wall or ceiling opening.

Many types of lighting fixtures are presently available. Figure 2-11 shows several typical lighting fixtures that may be found in a dwelling unit. Also shown are the electrical symbols used on plans to designate these fixtures and the type of outlet boxes or switch boxes on which the lighting fixtures can be mounted. A standard receptacle outlet is shown as well. The switch boxes shown here are made of steel. Switch boxes may also be made of plastic, as shown in figure 2-12. Other types of outlets will be covered in later units.

FLUSH SWITCHES

Some of the standard symbols for various types of switches are shown in figure 2-13 (page 17). Typical connection diagrams are also given. Any sectional switch box or 4-inch square box with a side mounting bracket and raised switch cover can be used to install these switches. Refer to figure 2-11.

JUNCTION BOXES AND SWITCH (DEVICE) BOXES *(ARTICLE 370)*

Junction boxes are sometimes placed in a circuit for convenience in joining two or more cables or conduits. All conductors entering a junction box are joined to other conductors entering the same box to form proper hookups so that the circuit will operate in the manner intended.

All electrical installations must conform to the *National Electrical Code*® standards requiring that junction boxes be installed in such a manner that the wiring contained in them shall be accessible without removing any part of the building. In house wiring, this requirement limits the use of junction boxes to unimproved basements, garages, and open attic spaces because flush blank covers exposed to view detract from the appearance of a room. Of course, an outlet box such as the one installed for the front hall ceiling fixture is really a junction box because it contains splices. Removing the fixture makes the box accessible, thereby meeting Code requirements. Refer to figures 2-14 (page 17) and 2-15 (page 18).

Section 300-15 requires that a box or fitting be installed wherever splices, switches, outlets, junction points, or pull points are required (figure 2-14). However, there are instances where a change is made from one wiring method to another, in which case a box is not required. This is permitted by *Exception No. 9* to *Section 300-15(b)*. Note that the fitting where the change is made must be accessible after installation, figure 2-16 (page 19).

SYMBOL	NOTATION
1	PLUGMOLD ENTIRE LENGTH OF WORKBENCH. OUTLETS 18" O.C. INSTALL 48" TO CENTER FROM FLOOR. GFCI PROTECTED.
2	TRACK LIGHTING. PROVIDE 5 LAMPHOLDERS.
3	TWO 40 WATT RAPID START FLUORESCENT LAMPS IN VALANCE. CONTROL WITH DIMMER SWITCH.

Fig. 2-9 Example of how certain *notations* might be added to a symbol when the symbol itself does not fully explain its meaning. The architect or engineer has a choice of explaining fully the meaning directly on the plan if there is sufficient room; if insufficient room, then a *notation* could be used.

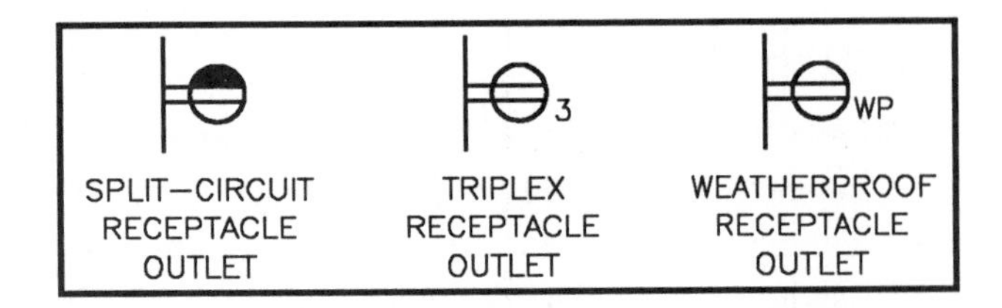

Fig. 2-10 Variations in significance of outlet symbols.

NONMETALLIC OUTLET AND DEVICE BOXES

Section 370-3 of the Code permits nonmetallic outlet and device boxes to be installed where the wiring method is nonmetallic sheathed cable or nonmetallic raceway. Nonmetallic shall not be used with armored cable or metallic raceways.

The house wiring system usually is formed by a number of specific circuits. Each circuit consists of a continuous run of cable from outlet to outlet or from box to box. The residence plans show many branch circuits for general lighting, appliances, electric heating, and other requirements. The specific Code rules for each of these circuits are covered in later units.

GANGED SWITCH (DEVICE) BOXES

A flush switch or convenience outlet for residential use fits into a standard 2″ × 3″ sectional switch box (sometimes called a *device box*). When two or more switches (or outlets) are located at the same point, the switch boxes are ganged or fastened together to provide the required mounts, figure 2-17 (page 19).

Three switch boxes can be ganged together by removing and discarding one side from both the first and third switch boxes and both sides from the second (center) switch box. The boxes are then

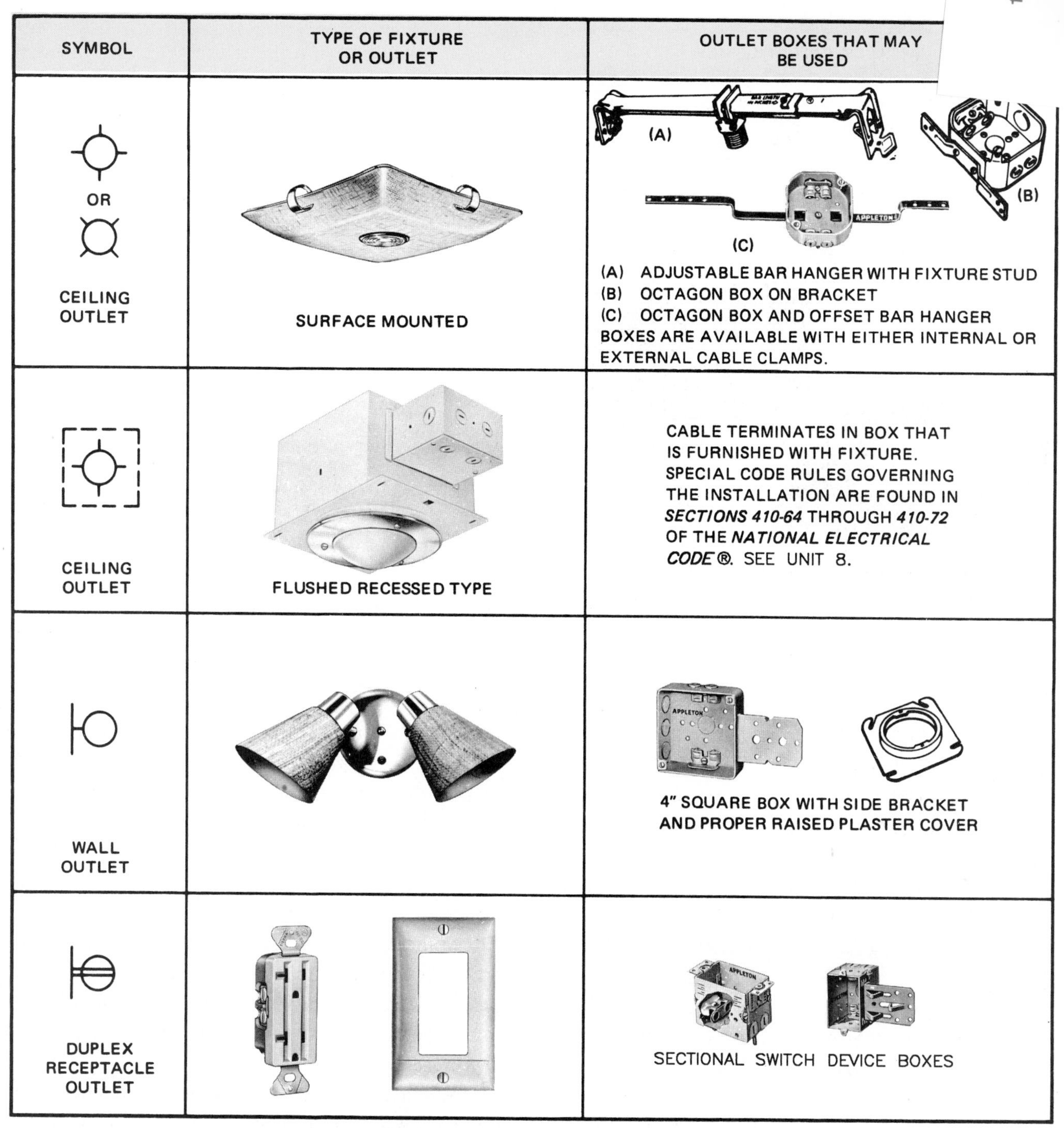

SYMBOL	TYPE OF FIXTURE OR OUTLET	OUTLET BOXES THAT MAY BE USED
OR CEILING OUTLET	SURFACE MOUNTED	(A) (B) (C) (A) ADJUSTABLE BAR HANGER WITH FIXTURE STUD (B) OCTAGON BOX ON BRACKET (C) OCTAGON BOX AND OFFSET BAR HANGER BOXES ARE AVAILABLE WITH EITHER INTERNAL OR EXTERNAL CABLE CLAMPS.
CEILING OUTLET	FLUSHED RECESSED TYPE	CABLE TERMINATES IN BOX THAT IS FURNISHED WITH FIXTURE. SPECIAL CODE RULES GOVERNING THE INSTALLATION ARE FOUND IN *SECTIONS 410-64* THROUGH *410-72* OF THE *NATIONAL ELECTRICAL CODE®*. SEE UNIT 8.
WALL OUTLET		4" SQUARE BOX WITH SIDE BRACKET AND PROPER RAISED PLASTER COVER
DUPLEX RECEPTACLE OUTLET		SECTIONAL SWITCH DEVICE BOXES

Fig. 2-11 Types of fixtures and outlets.

joined together as shown in figure 2-17. After the switches are installed, the gang is trimmed with a gang plate having the required number of switch handle or receptacle outlet openings. These plates are called two-gang wall plates, three-gang wall plates, and so on, depending upon the number of openings.

The dimensions of a standard sectional switch box (2″ × 3″) are the dimensions of the opening of the box. The depth of the box may vary from 1 1/2 inches to 3 1/2 inches, depending upon the require-ments of the building construction and the number of conductors and devices to be installed. *NEC® Article 370* covers outlet, switch, and junction

METALLIC SWITCH (DEVICE)

ANY NONMETALLIC BOX

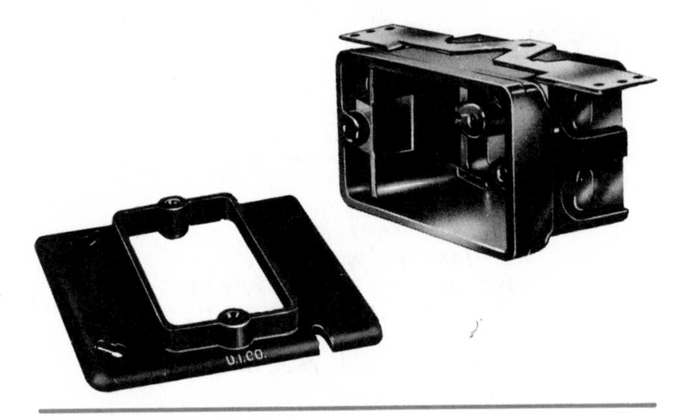

Fig. 2-12 Supporting (hanging) a ceiling fixture from these types of boxes is not permitted unless specifically marked on the box or carton (UL requirement).

boxes. See figure 2-18 (page 20) for a complete listing of box dimensions.

Section 370-10 of the Code states that boxes must be mounted so that they will be set back not more than 1/4 inch (6.35 mm) when the boxes are mounted in noncombustible walls or ceilings made of concrete, tile, or similar materials. When the wall or ceiling construction is of combustible material (wood), the box must be set flush with the surface, figure 2-19 (page 21). These requirements are meant to prevent the spread of fire if a short circuit occurs within the box.

Ganged sectional switch (device) boxes can be installed using a pair of metal mounting strips. These strips are also used to install a switch box between wall studs, figure 2-20 (page 22). The use of a bracket box in such an installation may result in an off-center receptacle outlet. When an outlet box is to be mounted at a specific location between joists, as for ceiling-mounted fixtures, an offset bar hanger is used (figure 2-11).

The Code states that when a switch box or outlet box is mounted to a stud or ceiling joist by nailing through the box, the nails must be not more than 1/4 inch (6.35 mm) from the back or ends of the box, figure 2-21 (page 22). This requirement insures that when the nail passes through the box, it does not interfere with the wiring devices in the box.

BOXES FOR CONDUIT WIRING

Some cities' electrical ordinances require conduit rather than cable wiring. Conduit wiring is discussed in unit 18. Examples of conduit fill (how many wires are permitted in a given size conduit) are presented.

When conduit is installed in a residence, it is quite common to use 4-inch square boxes trimmed with suitable plaster covers. The type of box as shown in figure 2-22A (page 23) is the most popular. There are sufficient knockouts in the top, bottom, sides, and back of the box to permit a number of conduits to run to the box. Plenty of room is available for the conductors and wiring devices. Note how easily these 4-inch square outlet boxes can be mounted "back-to-back" by installing a small fitting between the boxes. This is illustrated in figures 2-22C(2) and (3) (page 24).

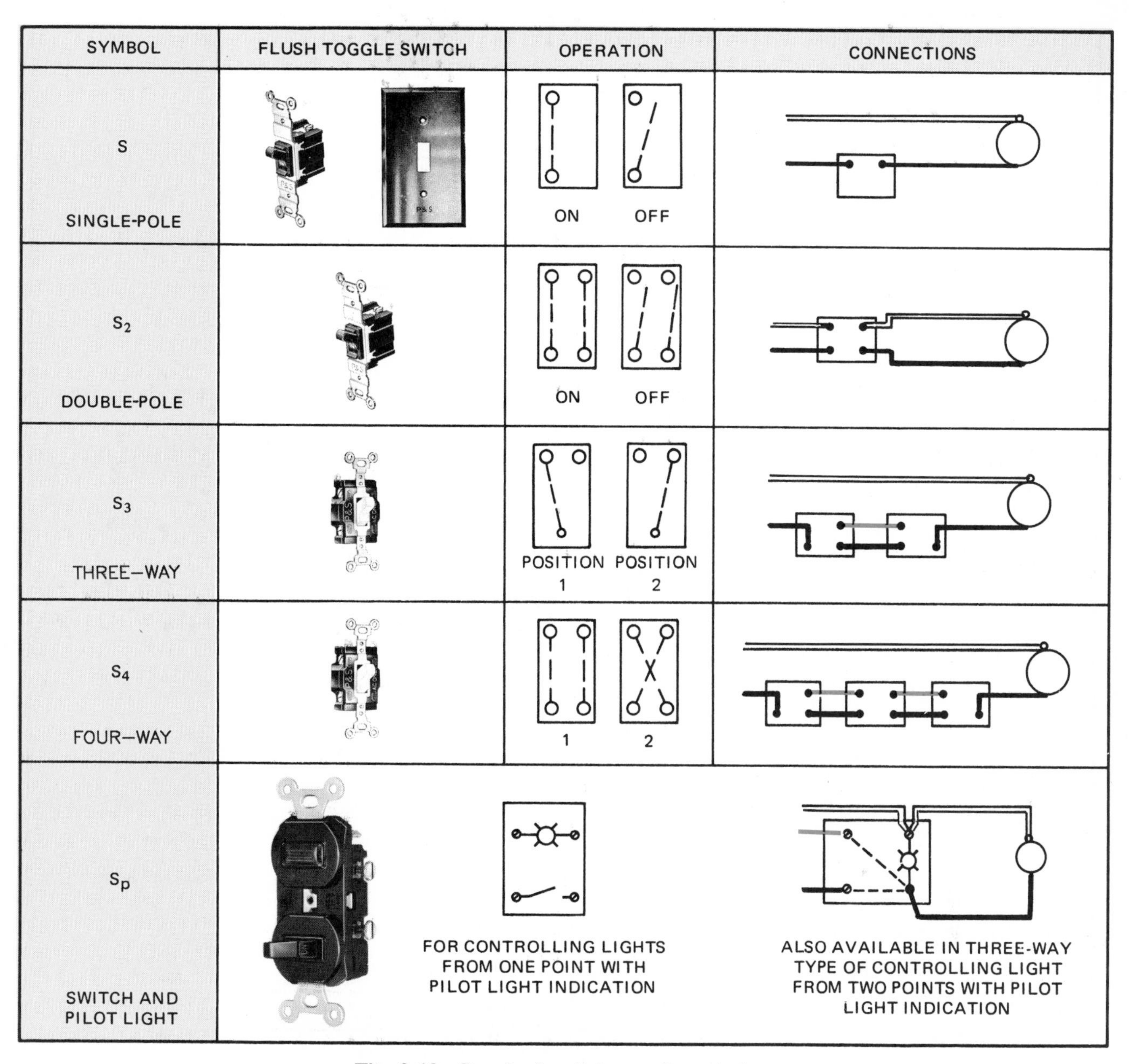

Fig. 2-13 Standard switches and symbols.

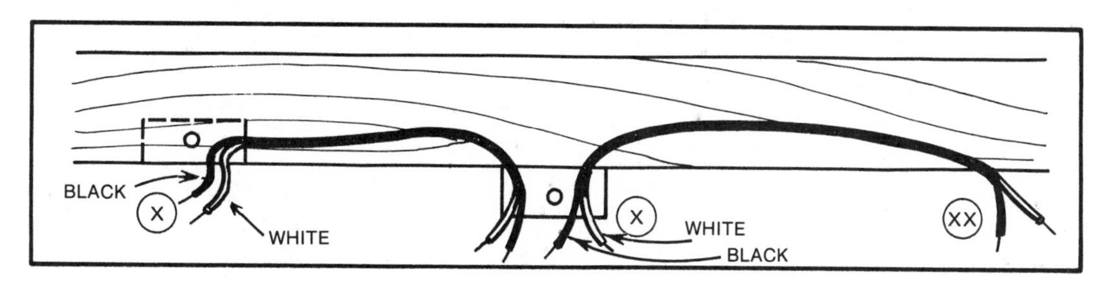

Fig. 2-14 A box (or fitting) must be installed wherever there are splices, outlets, switches, or other junction points. Refer to the points marked X. A Code violation is shown at point XX, *Section 300-15(a)(b)*.

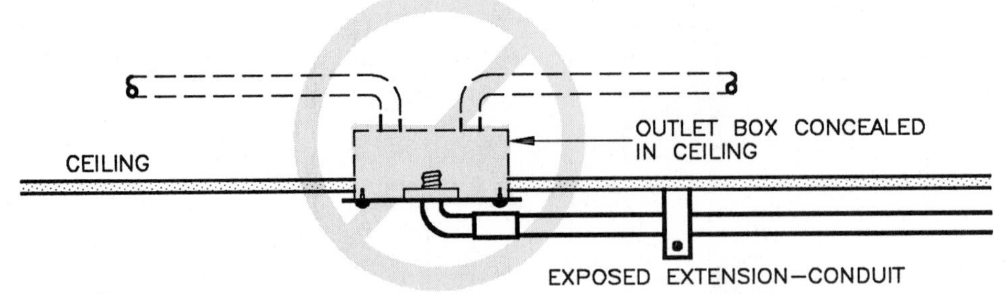

(A)

VIOLATION. IT IS AGAINST THE CODE TO MAKE A RIGID EXTENSION FROM A COVER THAT IS ATTACHED TO AN OUTLET BOX, JUNCTION BOX, OR DEVICE BOX, SECTION 370—12.

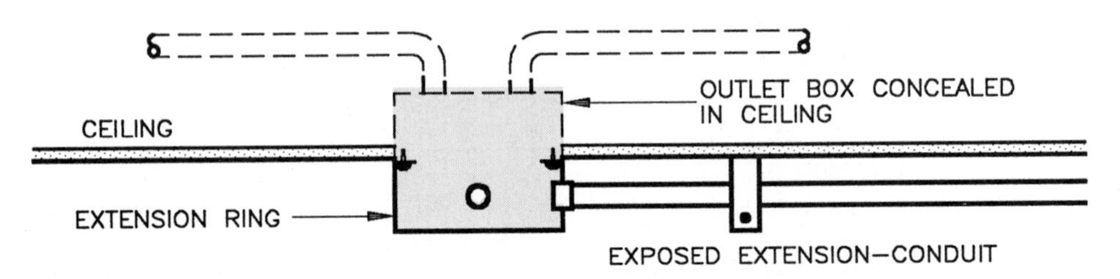

(B)

THIS MEETS CODE. A BOX OR EXTENSION RING MUST BE MOUNTED OVER AND MECHANICALLY SECURED TO THE ORIGINAL BOX, SECTION 370—12.

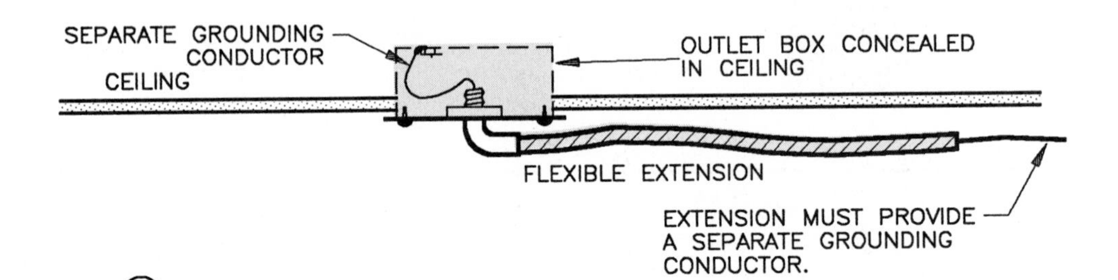

(C)

THIS MEETS CODE. IT IS PERMITTED TO MAKE A SURFACE EXTENSION FROM A COVER FASTENED TO A CONCEALED BOX WHERE THE EXTENSION WIRING METHOD IS FLEXIBLE AND WHERE A SEPARATE GROUNDING CONDUCTOR IS PROVIDED SO THAT THE GROUNDING PATH IS NOT DEPENDENT UPON THE SCREWS THAT ARE USED TO FASTEN THE COVER TO THE BOX.

IT WOULD BE A CODE VIOLATION TO MAKE THE EXTENSION WITH A RIGID WIRING METHOD SUCH AS A CONDUIT THAT WOULD MAKE IT DIFFICULT TO GAIN ACCESS TO THE CONNECTIONS INSIDE THE BOX, OR IF A FLEXIBLE WIRING METHOD WAS USED BUT HAD NO PROVISIONS FOR A SEPARATE GROUNDING CONDUCTOR.

Fig. 2-15 How to make an extension from a flush-mounted box.

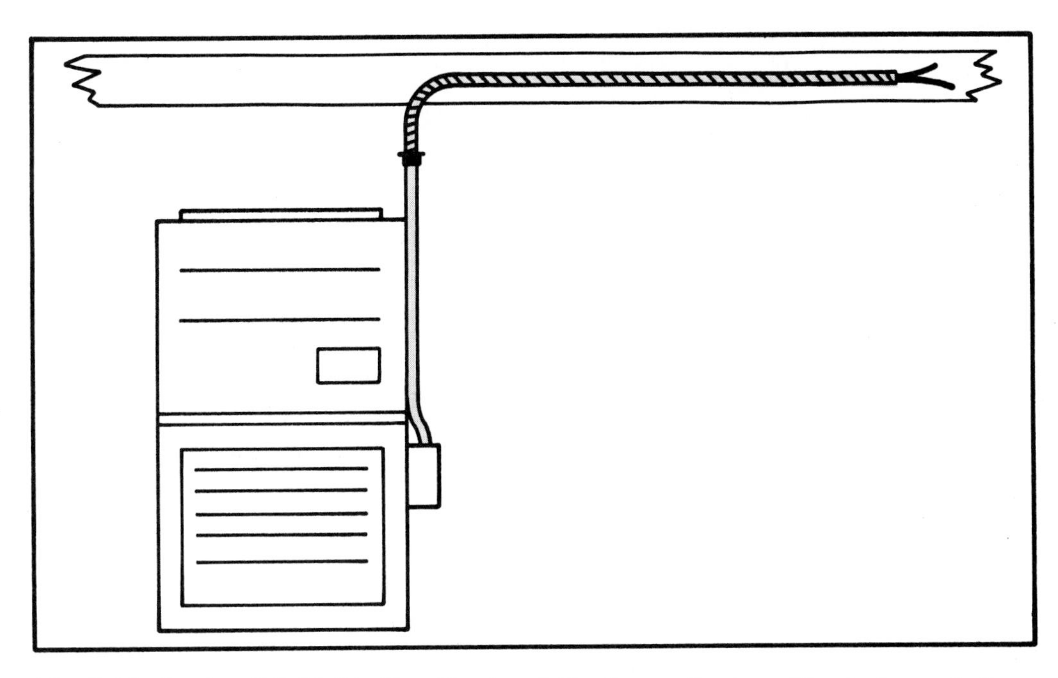

Fig. 2-16 *Section 300-15 (b), Exc. No. 1* and *No. 9* permit a transition to be made from one wiring method to another wiring method. In this case, the armor of the Type AC cable is removed, allowing sufficient length of the conductors to be run through the conduit. A proper fitting must be used at the transition point, and the fitting must be accessible after installation.

Fig. 2-17 Standard flush switches installed in ganged sectional device boxes.

Four-inch square outlet boxes can be trimmed with one-gang or two-gang plaster rings where wiring devices will be installed. Where lighting fixtures will be installed, a plaster ring having a round opening should be installed.

Any unused openings in outlet and device boxes must be closed per *Sections 110-12* and *370-8* of the Code, figure 2-23 (page 25).

The number of conductors allowed in outlet and device boxes is covered elsewhere in this text.

QUIK–CHEK BOX SELECTION GUIDE

FOR BOXES GENERALLY USED FOR RESIDENTIAL WIRING

DEVICE BOXES

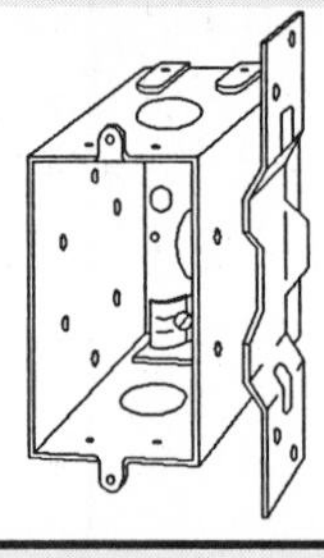

WIRE SIZE	3X2X1½ (7.5 in³)	3X2X2 (10 in³)	3X2X2¼ (10.5 in³)	3X2X2½ (12.5 in³)	3X2X2¾ (14 in³)	3X2X3 (16 in³)	3X2X3½ (18 in³)
#14	3	5	5	6	7	7	9
#12	3	4	4	5	6	7	8

SQUARE BOXES

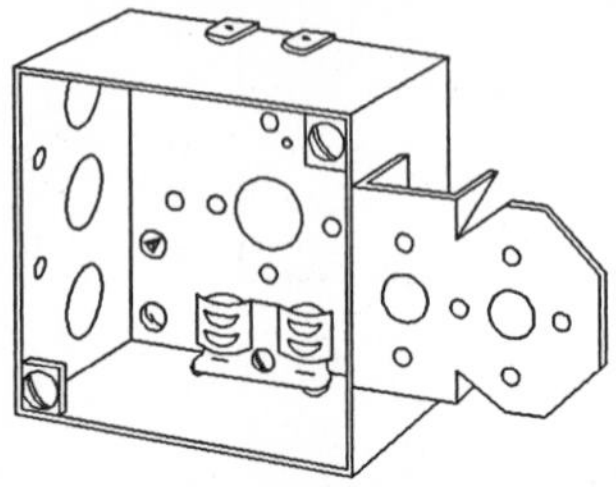

WIRE SIZE	4X4X1½ (21 in³)	4X4X2⅛ (30.3 in³)
#14	10	15
#12	9	13

OCTAGON BOXES

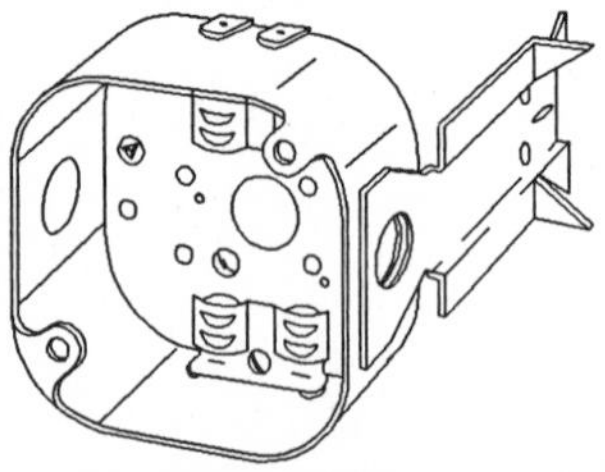

WIRE SIZE	4X1½ (15.5 in³)	4X2⅛ (21.5 in³)
#14	7	10
#12	6	9

HANDY BOXES

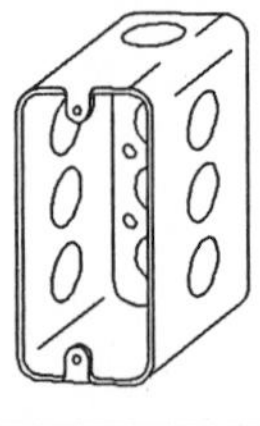

WIRE SIZE	4X2⅛X1½ (10.3 in³)	4X2⅛X1⅞ (13 in³)	4X2X2⅛ (14.5 in³)
#14	5	6	7
#12	4	5	6

RAISED COVERS

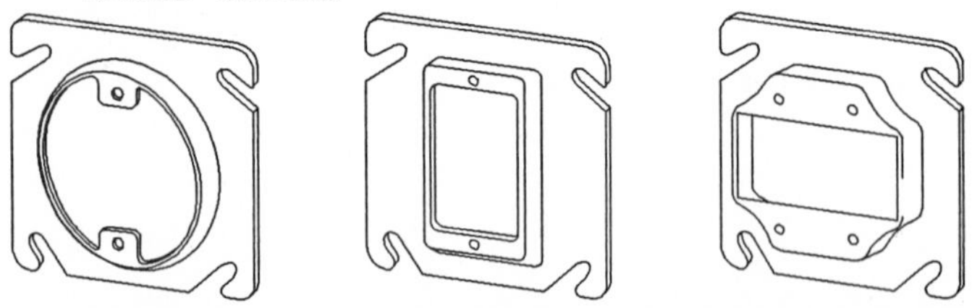

WHERE RAISED COVERS ARE MARKED WITH THEIR VOLUME IN CUBIC INCHES, THAT VOLUME MAY BE ADDED TO THE BOX VOLUME TO DETERMINE MAXIMUM NUMBER OF CONDUCTORS IN THE COMBINED BOX AND RAISED COVER.

NOTE: BE SURE TO MAKE DEDUCTIONS FROM THE ABOVE MAXIMUM NUMBER OF CONDUCTORS PERMITTED FOR WIRING DEVICES, CABLE CLAMPS, FIXTURE STUDS, AND GROUNDING CONDUCTORS. THE CUBIC INCH (IN³) VOLUME IS TAKEN DIRECTLY FROM *TABLE 370–6A* OF THE CODE.

Fig. 2-18 Quik-chek box selector guide.

INTERCHANGEABLE WIRING DEVICES

In the space of one standard wiring device, it is possible to install one, two, or three wiring devices by using an interchangeable line of devices. Up to three switches, pilot lights, receptacle outlets, or any combination of interchangeable wiring devices can be installed, one above the other, on a single strap in a standard 2″ × 3″ box opening. A total of six devices can be installed using two ganged boxes with standard openings.

Interchangeable devices are available in the same types and classifications of switches and outlets as standard wiring devices. In other words, single-pole switches, three-way switches, four-way switches, pilot lights, and other devices are available in an interchangeable style. Both standard and interchangeable devices are available in silent style that makes little or no noise when the switch is actuated.

When interchangeable devices are to be installed, standard sectional switch boxes generally

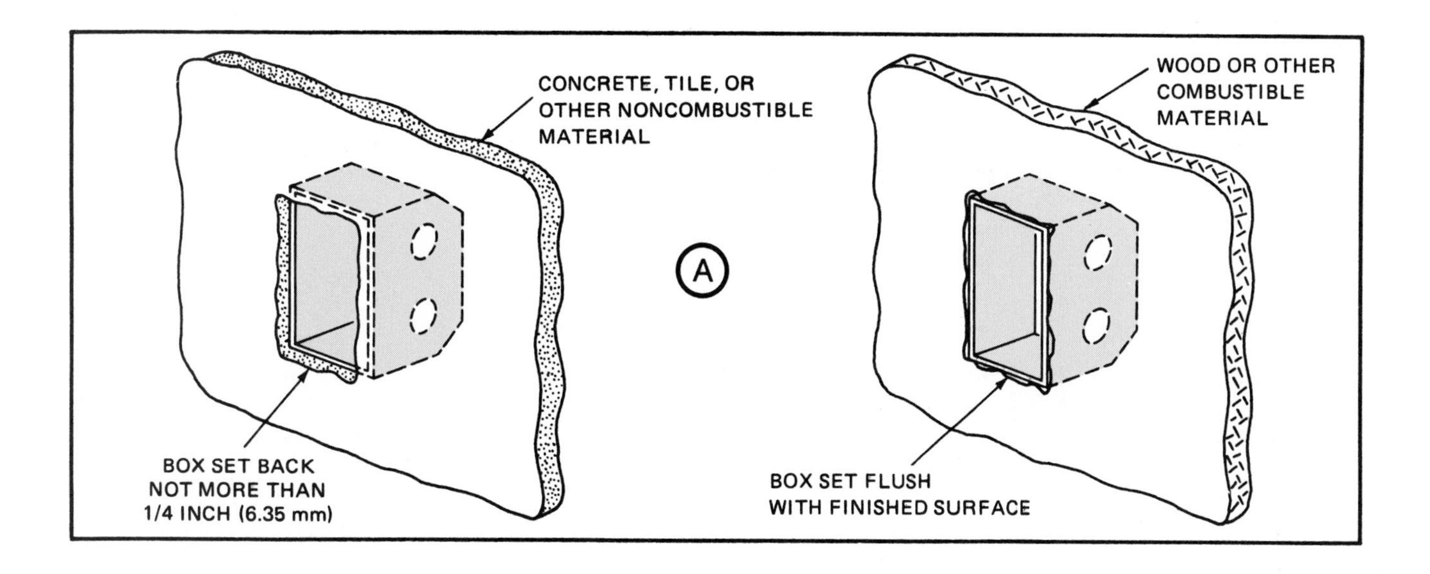

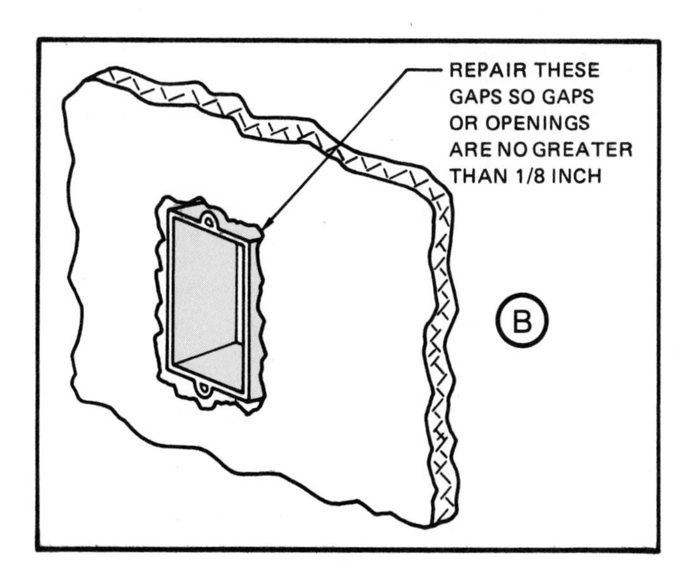

Fig. 2-19 Box position in walls and ceilings constructed of various materials, ***Sections 370-10 and 370-11.***

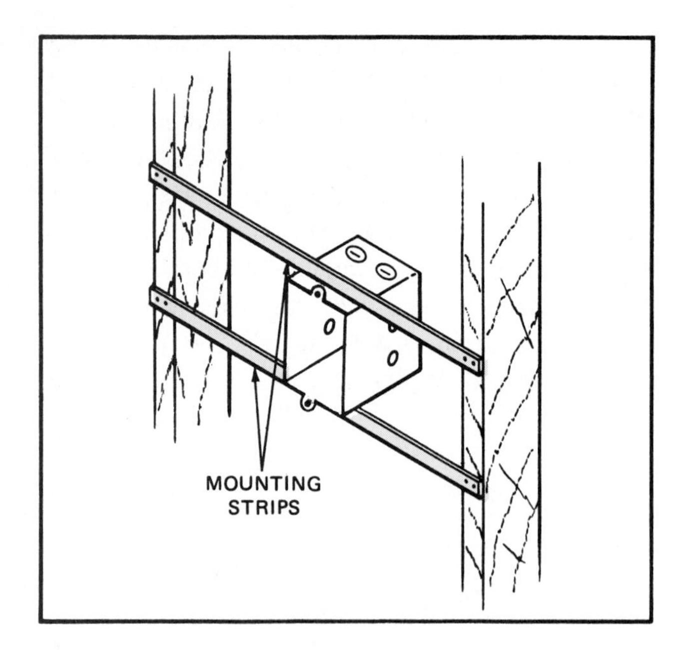

Fig. 2-20 Switch (device) boxes installed between studs using metal mounting strips.

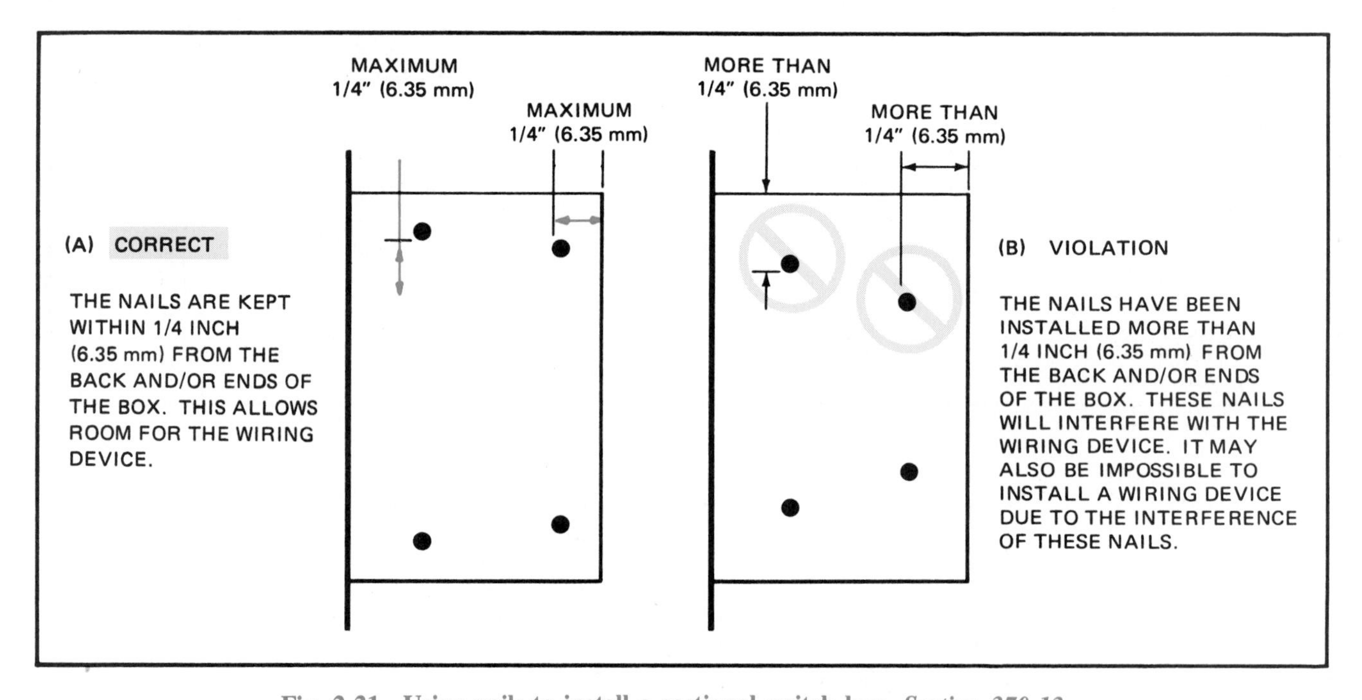

Fig. 2-21 Using nails to install a sectional switch box, *Section 370-13.*

are not used because of the lack of wiring space in the box. Instead, 4-inch square, 4 11/16-inch square, or 4″ × 6″ boxes with raised plaster covers are used. (This is not a requirement of the Code, however.) Under certain conditions, 3-inch and 3 1/2-inch deep sectional boxes are large enough for a group of interchangeable devices.

The Code requires that faceplates for switches and receptacles must completely cover the wall opening and must seat against the surface of the wall.

Figure 2-22 shows two interchangeable switches and one pilot light mounted on a single strap in a 4-inch square outlet box. The box has a plaster cover and a bracket. A single-gang, three-hole wall plate is required. This arrangement is useful where limited wall space (such as the space between a door casing and a window casing) prohibits the use of standard two-gang or three-gang wall plates.

SPECIAL-PURPOSE OUTLETS

Special-purpose outlets are usually indicated on the plans. These outlets are described by a notation and are also detailed in the specifications. The plans included in this text indicate special-purpose outlets by a triangle inside a circle with subscript letters. In some cases, a subscript number is added to the letter.

When a special-purpose outlet is indicated on the plans or in the specifications, the electrician must check for special requirements. Such a requirement may be a separate circuit, a special 240-volt circuit, a special grounding or polarized receptacle, or other preparation.

NUMBER OF CONDUCTORS IN BOX

The Code (*Section 370-6*) dictates that outlet boxes, switch boxes, and device boxes should be large enough to provide ample room for the wires in that box, without having to jam or crowd the wires into the box. The Code specifies the maximum number of conductors allowed in standard outlet boxes and switch boxes, figure 2-24 (page 25). A conductor running through the box is counted as one conductor. Each conductor originating outside of the box and terminating inside the box is counted as one conductor. Conductors which originate and terminate within the box are not counted.

When conductors are the same size, the proper box size can be selected by referring to *Table 370-6(a)*. When conductors are of different sizes, refer to *Table 370-6(b)*.

Tables 370-6(a) and *370-6(b)* do not consider fittings or devices such as fixture studs, cable clamps, hickeys, switches, or receptacles which may be in the box, figure 2-25 (page 26). When the box contains one or more fittings, such as fixture studs,

A.

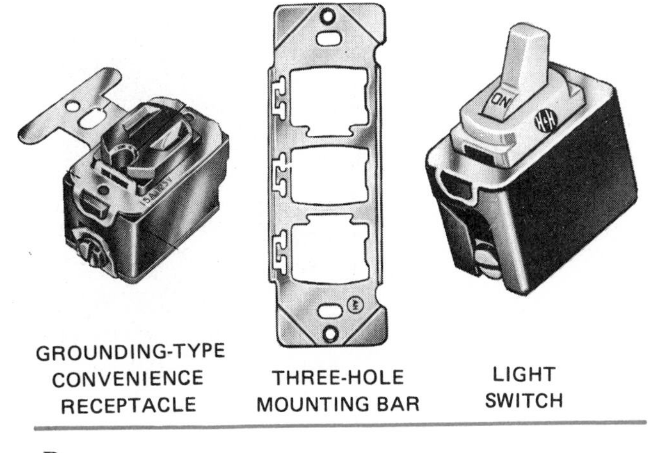

B.

Fig. 2-22 Interchangeable devices. (*continues*)

1.

2.

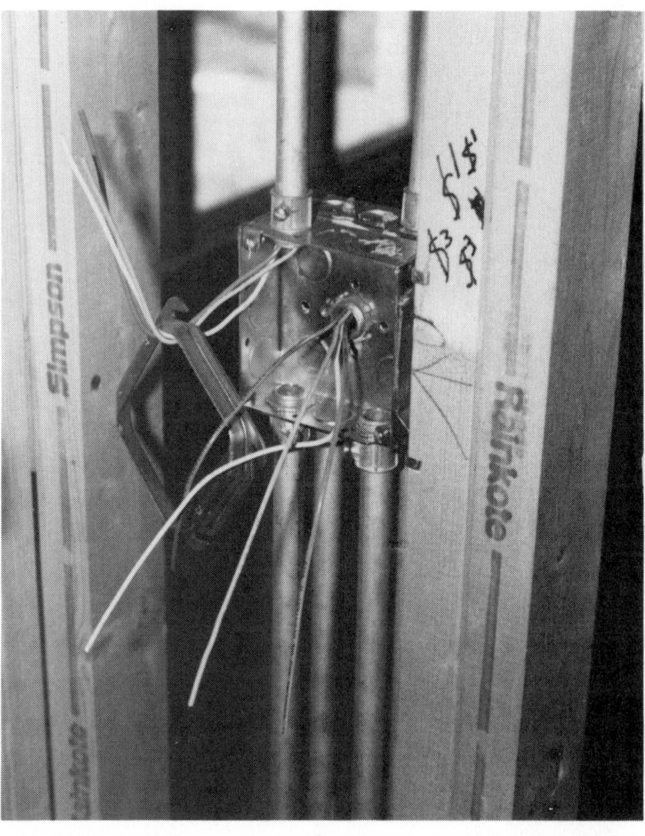

3.

C. **Examples of electrical metallic tubing showing how 4-inch square boxes are used for single outlets and switches. Note how the boxes are attached together for "back-to-back" installations.**

Fig. 2-22 Interchangeable devices.

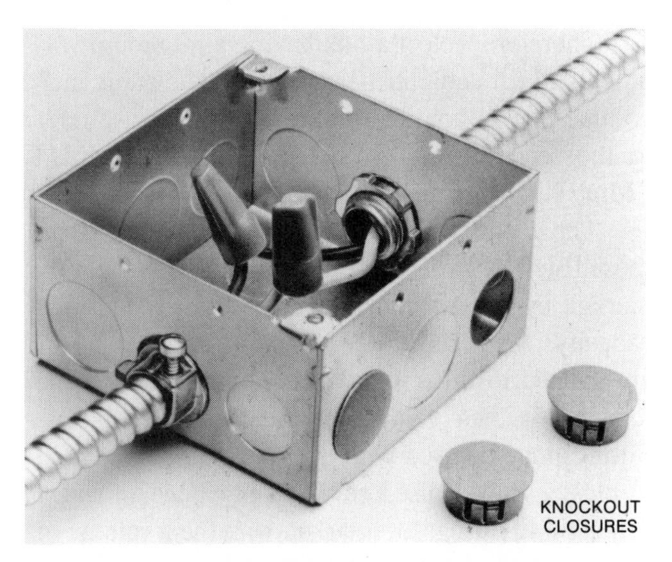

Fig. 2-23 Unused openings in boxes must be closed according to the Code, *Sections 110-12 and 370-8*. This is done to contain electrical short-circuit problems inside the box or panel and to keep rodents out.

Table 370-6(a). Metal Boxes

Box Dimension, Inches Trade Size or Type	Min. Cu. In. Cap.	Maximum Number of Conductors						
		No. 18	No. 16	No. 14	No. 12	No. 10	No. 8	No. 6
4 x 1¼ Round or Octagonal	12.5	8	7	6	5	5	4	2
4 x 1½ Round or Octagonal	15.5	10	8	7	6	6	5	3
4 x 2⅛ Round or Octagonal	21.5	14	12	10	9	8	7	4
4 x 1¼ Square	18.0	12	10	9	8	7	6	3
4 x 1½ Square	21.0	14	12	10	9	8	7	4
4 x 2⅛ Square	30.3	20	17	15	13	12	10	6
4¹¹⁄₁₆ x 1¼ Square	25.5	17	14	12	11	10	8	5
4¹¹⁄₁₆ x 1½ Square	29.5	19	16	14	13	11	9	5
4¹¹⁄₁₆ x 2⅛ Square	42.0	28	24	21	18	16	14	8
3 x 2 x 1½ Device	7.5	5	4	3	3	3	2	1
3 x 2 x 2 Device	10.0	6	5	5	4	4	3	2
3 x 2 x 2¼ Device	10.5	7	6	5	4	4	3	2
3 x 2 x 2½ Device	12.5	8	7	6	5	5	4	2
3 x 2 x 2¾ Device	14.0	9	8	7	6	5	4	2
3 x 2 x 3½ Device	18.0	12	10	9	8	7	6	3
4 x 2⅛ x 1½ Device	10.3	6	5	5	4	4	3	2
4 x 2⅛ x 1⅞ Device	13.0	8	7	6	5	5	4	2
4 x 2⅛ x 2⅛ Device	14.5	9	8	7	6	5	4	2
3¾ x 2 x 2½ Masonry Box/Gang	14.0	9	8	7	6	5	4	2
3¾ x 2 x 3½ Masonry Box/Gang	21.0	14	12	10	9	8	7	4
FS—Minimum Internal Depth 1¾ Single Cover/Gang	13.5	9	7	6	6	5	4	2
FD—Minimum Internal Depth 2⅜ Single Cover/Gang	18.0	12	10	9	8	7	6	3
FS—Minimum Internal Depth 1¾ Multiple Cover/Gang	18.0	12	10	9	8	7	6	3
FD—Minimum Internal Depth 2⅜ Multiple Cover/Gang	24.0	16	13	12	10	9	8	4

Table 370-6(b). Volume Required per Conductor

Size of Conductor	Free Space Within Box for Each Conductor
No. 18	1.5 cubic inches
No. 16	1.75 cubic inches
No. 14	2. cubic inches
No. 12	2.25 cubic inches
No. 10	2.5 cubic inches
No. 8	3. cubic inches
No. 6	5. cubic inches

Fig. 2-24 Allowable number of conductors in boxes.

cable clamps, or hickeys, the number of conductors must be one less than shown in the table for each type of device.

► Deduct two conductors for each mounting yoke or strap that has one or more wiring devices (switches, receptacles). The deduction of two conductors has become necessary because of severe crowding of conductors in a box when more than one switch or receptacle is mounted on one yoke, or when dimmer switches are installed. In most cases, dimmer switches are larger than conventional switches. ◄

Be sure to include conductor sizes No. 18, 16, 14, 12, 10, 8 and 6 AWG when determining the size of the box to be installed. See figures 2-26 and 2-27 for examples of this Code requirement, *Section 370-6(a)*.

SELECTING BOX WHEN ALL CONDUCTORS ARE THE SAME SIZE

EXAMPLE: A box contains one fixture stud and two cable clamps. The number of conductors permitted in the box shall be two less than shown in the table. (Deduct one conductor for the fixture stud; deduct one conductor for the two cable clamps.)

► A further deduction of two conductors is made for one or several wiring devices mounted on the same strap. For example, if switches or receptacles of an interchangeable line are mounted on the same strap, then two conductors are deducted. A further deduction of one conductor must be made for one or more grounding conductors entering the box. ◄

When the box contains different size wires, *Section 370-6(a)(2)*, do the following:

- Size the box based upon the total cubic inch volume required for conductors according to *Table 370-6(b)*.
- Then make the adjustments necessary due to devices, cable clamps, fixture studs, etc. This volume adjustment must be based on the cubic inch volume of the largest conductor in the box as determined in *Table 370-6(b)*.

SELECTING BOX WHEN CONDUCTORS ARE DIFFERENT SIZES

EXAMPLE: What is the minimum cubic inch volume required for a box that will contain one internal cable clamp, one switch, two No. 14 wires, and two No. 12 wires? The cable is armored cable.

SOLUTION:

2 No. 14 wires × 2 cubic inches per wire = 4 cubic inches

2 No. 12 wires × 2.25 cubic inches per wire = 4.5 cubic inches

1 cable clamp @ 2.25 cubic inches = 2.25 cubic inches

► 1 switch @ 2.25 cubic inches × 2 = 4.5 cubic inches ◄

TOTAL CUBIC INCH VOLUME = 15.25 cubic inches

• If box contains NO fittings, devices, fixture studs, cable clamps, hickeys, switches, receptacles, or grounding conductors...	• refer directly to *Table 370-6(a)*.
• If box contains ONE or MORE fittings, fixture studs, cable clamps, or hickeys...	• deduct ONE from maximum number of conductors permitted in *Table 370-6(a)* for each type.
• For each "yoke" or strap containing ONE or MORE devices such as switches, receptacles, or pilot lights...	• deduct ANOTHER TWO from *Table 370-6(a)*.
• For ONE or MORE grounding conductors in the box...	• deduct ANOTHER ONE from *Table 370-6(a)*.
• For ONE or MORE isolated (insulated) grounding conductors...	• deduct ANOTHER ONE from *Table 370-6(a)*.
• For conductors running through the box...	• count ONE conductor for each conductor running through the box.
• For conductors that originate outside of box and terminate inside of box – for example fixture wires (No. 16 AWG and No. 18 AWG)...	• count ONE conductor for each conductor originating outside and terminating in box.
• If no part of the conductor leaves the box – for example, a "jumper" wire used to connect three devices on one strap (yoke) or pigtails used as illustrated in figure 6-5...	• don't count this (these) conductor(s).

Fig. 2-25 Quick checklist for determining proper size boxes.

Therefore, select a box having a minimum volume of 15.25 cubic inches of space. The cubic inch volume may be marked on the box; otherwise refer to the second column of *Table 370-6(a)* entitled "Min. Cu. In. Cap."

The Code requires that all boxes *other* than those listed in *Table 370-6(a)* be durably and legibly marked by the manufacturer with their cubic inch capacity, *Section 370-6(b)*. When sectional boxes are ganged together, the volume to be filled is the total cubic inch volume of the assembled boxes. Fittings may be used with the sectional boxes, such as plaster rings, raised covers, and extension rings. When these fittings are marked with their volume in cubic inches, or have dimensions comparable to those boxes shown in *Table 370-6(a)*, then their volume may be considered in determining the total cubic inch volume to be filled [see *Section 370-6(a)(1)*]. Figure 2-26 shows how a 3/4-inch raised cover (plaster ring) increases the wiring

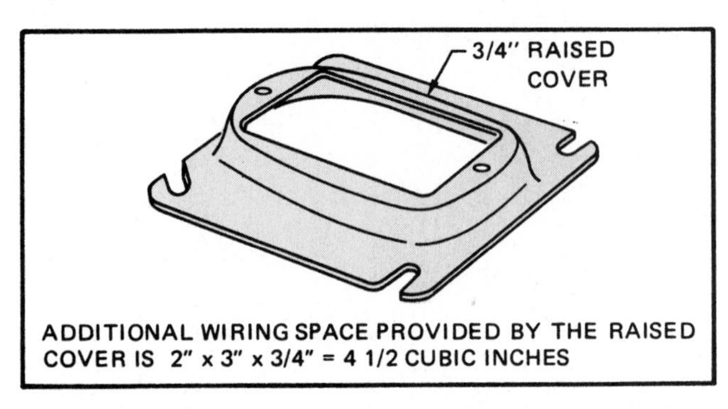

Fig. 2-26 Raised cover. Raised covers are sometimes called plaster rings.

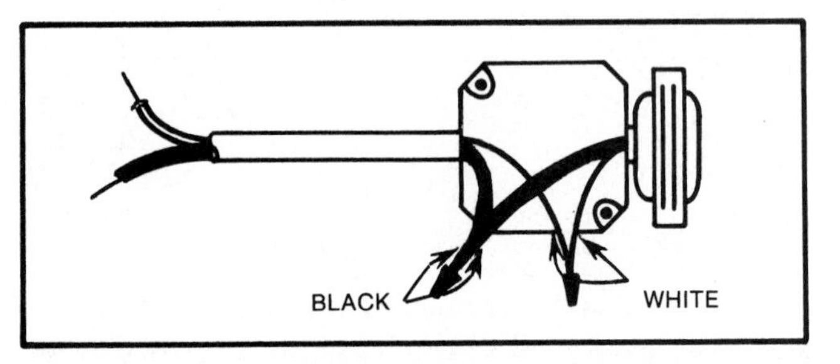

Fig. 2-27 Transformer leads No. 18 AWG or larger must be counted when selecting the proper size box, *Section 370-6(a)(1)*. In the example shown, the box is considered to contain four conductors.

space above the cubic inch capacity of the box to which it is attached.

Electrical inspectors have become very aware of the fact that GFCI receptacles, dimmers, and certain types of timers take up a lot more space than regular receptacles. Therefore, it is a good practice to install switch (device) boxes that will provide "lots" of room for the wires, instead of pushing, jamming, and crowding the wires into the box.

EXAMPLE: How many No. 12 conductors are permitted in this box and raised plaster ring?

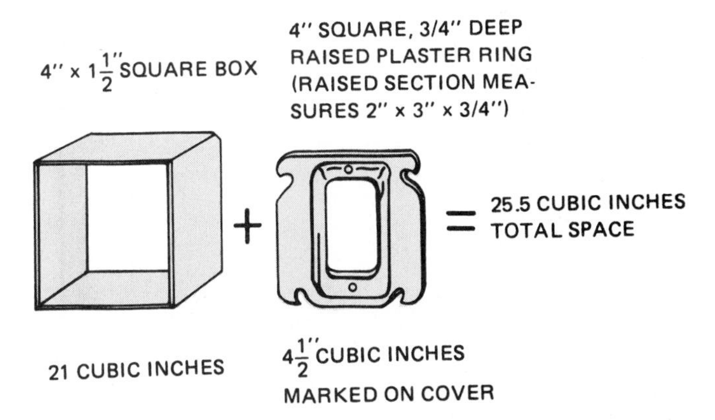

SOLUTION: See *Section 370-6* and *Tables 370-6(a)* and *370-6(b)* for volume required per conductor.

This box and cover will take:

$$\frac{\text{25.5 cubic inch space}}{\text{2.25 cubic inches per No. 12 conductor}} = \text{11 No. 12 conductors maximum, less the deductions for devices, clamps, and so forth, per } \textit{Section 370-6(a)(1)}$$

Figures 2-27 and 2-28 show that transformer leads No. 18 AWG or larger must be counted when selecting a box.

The size of equipment grounding conductors is shown in *Table 250-95*. The grounding conductors are the same size as the circuit conductors in cables having No. 14 AWG, No. 12 AWG, and No. 10 AWG circuit conductors. Thus, box sizes can be calculated using *Table 370-6(a)*. Refer to figure 2-28.

Figure 2-18 shows some of the most popular types of boxes used in residential wiring, and can be referred to as you work your way through this text. Figure 2-18 is called ***quik-chek box selector guide.***

When wiring with cable, "feeding through" a box is impossible. Also, when installing the wiring using conduit, an internal cable clamp is not used. The conduit connector does not need to be counted the same way as cable clamps are, figure 2-29.

BOX FILL

Another way that makes "box fill" easier to determine the proper size junction box or wall box to install is to look at the situation as follows:

1. Count the number of circuit wires.
2. Add one wire for fixture stud (if any).
3. Add one wire for each cable clamp (if any).
4. Add one wire for one or more grounding conductors (if any).
5. Add one wire for one or more isolated (insulated) grounding conductors (if any).
6. Add two wires for each wiring device.

The total count should then be looked up in *Table 370-6* to find a box appropriate for the intended use and one that will hold the number of conductors required, as totaled above.

EXAMPLE:		
	Six circuit conductors	6
	► One wiring device (switch)	+ 2 ◄
	One cable clamp	+ 1
	Two grounding conductors	+ 1
	Total	10

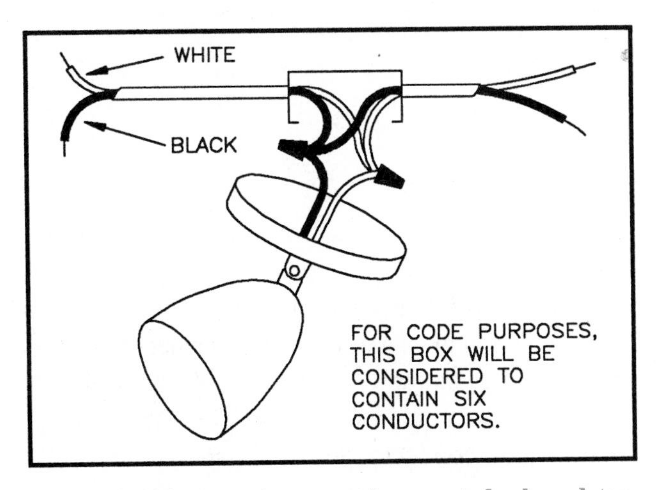

Fig. 2-28 Fixture wires *must* be counted when determining correct box size. The Code in *Section 370-6(a)(1)* states, "...each conductor originating outside of the box and terminating inside the box is counted as one conductor." *Table 370-6(a)* includes conductor sizes No. 18, 16, 14, 12, 10, 8, and 6 AWG.

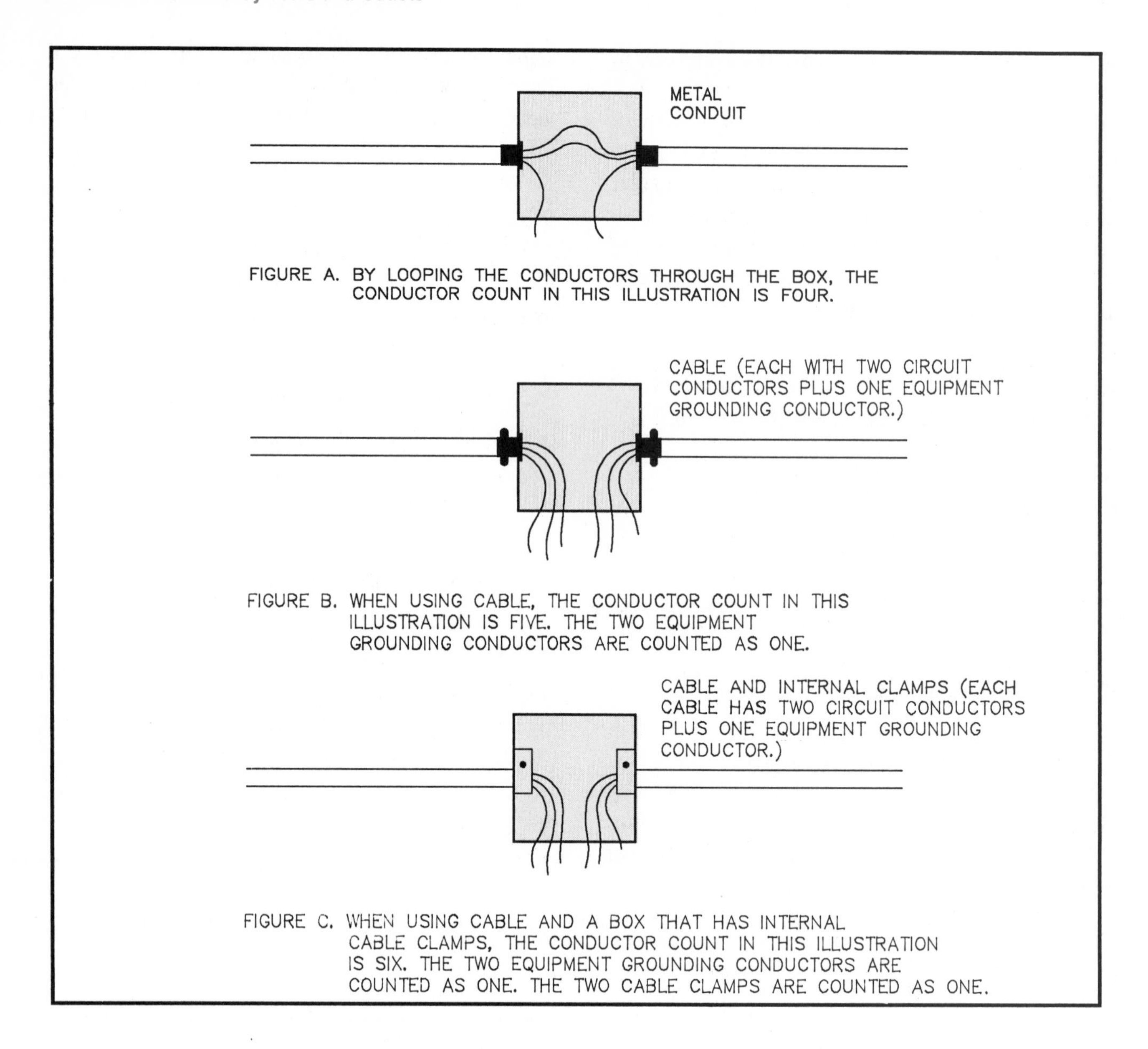

FIGURE A. BY LOOPING THE CONDUCTORS THROUGH THE BOX, THE CONDUCTOR COUNT IN THIS ILLUSTRATION IS FOUR.

FIGURE B. WHEN USING CABLE, THE CONDUCTOR COUNT IN THIS ILLUSTRATION IS FIVE. THE TWO EQUIPMENT GROUNDING CONDUCTORS ARE COUNTED AS ONE.

FIGURE C. WHEN USING CABLE AND A BOX THAT HAS INTERNAL CABLE CLAMPS, THE CONDUCTOR COUNT IN THIS ILLUSTRATION IS SIX. THE TWO EQUIPMENT GROUNDING CONDUCTORS ARE COUNTED AS ONE. THE TWO CABLE CLAMPS ARE COUNTED AS ONE.

Therefore, select a box capable of containing 10 or more conductors. See *Table 370-6(a)*.

Using *EMT* (electrical metallic tubing) results in the selection of a smaller box than when using the cable wiring method because it is possible to "loop" conductors through the box. These only count as one conductor for the purpose of box fill count.

HEIGHT OF RECEPTACLE OUTLETS

There are no hard-and-fast rules for locating most outlets. A number of conditions determine the proper height for a switch box. For example, the height of the kitchen counter backsplash determines where the switches and receptacle outlets are located between the kitchen countertop and the cabinets.

The residence featured in this text is electrically heated and is discussed in unit 23. The type of electric heat could be

- electric furnace (as in this text).
- electric resistance heating buried in ceiling

plaster or "sandwiched" between two layers of drywall material.
- electric baseboard heaters.

Let us consider the electric baseboard heaters. In most cases, the height of these electric baseboard units from the top of the unit to the finished floor seldom exceeds 6 inches (152 mm). The important issue here is that the manufacturer's receptacle accessories may have to be used to conform to the receptacle spacing requirements as covered in *Section 210-52* of the Code.

Electrical receptacle outlets are not permitted to be located above an electric baseboard heating unit. Refer to the section "Location of Electric Baseboard Heaters" in unit 23.

The location of lighting outlets is determined by the amount and type of illumination required to provide the desired lighting effects. It is not the intent of this text to describe how proper and adequate lighting is determined. Rather, the text covers the proper methods of installing the circuits for such lighting. If the student is interested, standards have been developed to guide the design of adequate lighting. The local electric utility company can supply information on these standards. The Instructor's Guide lists excellent publications relating to proper residential lighting.

It is common practice among electricians to consult the plans and specifications to determine the proper heights and clearances for the installation of electrical devices. The electrician then has these dimensions verified by the architect, electrical engineer, designer, or homeowner. This practice avoids unnecessary and costly changes in the locations of outlets and switches as the building progresses.

POSITIONING OF RECEPTACLES

Although no actual Code rules exist on positioning receptacles, there is a popular concept in the electrical industry that there is always the possibility of a metal wall plate coming loose and falling downward onto the blades of an attachment plug cap that is loosely plugged into the receptacle, thereby creating a potential shock and fire hazard. Here are the options.

Recommended

Ground*ing* hole to the top. A loose metal plate could fall onto the grounding blade of the attachment plug cap, but no sparks would fly.

Not Recommended

Ground*ing* hole to the bottom. A loose metal plate could fall onto both the grounded neutral and "hot" blades. Sparks would fly.

Recommended

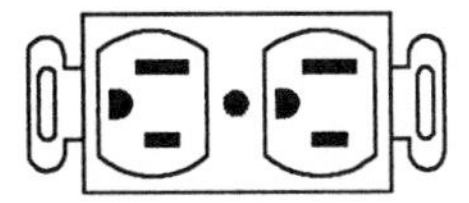

Ground*ed* neutral blades on top. A loose metal plate could fall onto these grounded neutral blades. No sparks would fly.

Not Recommended

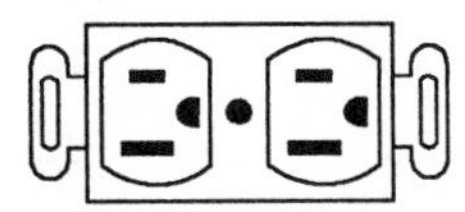

"Hot" terminal on top. A loose metal plate could fall onto these live blades. If this was a split-circuit receptacle fed by a 3-wire, 120/240 volt circuit, the short would be across the 240 volt line. Sparks would really fly!!

To insure uniform installation and safety, and in accordance with long-established custom, standard electrical outlets are located as shown in figure 2-30. These dimensions usually are satisfactory. However, the electrician must check the blueprints, specifications, and details for measurements which may affect the location of a particular outlet or switch. The cabinet spacing, available space between the countertop and the cabinet, and the tile height may influence the location of the outlet or switch. For example, if the top of wall tile is exactly 48 inches (1.22 m) from the finished floor line, a wall switch should not be mounted 48 inches (1.22 m) to center. This is considered poor workmanship. The switch should be located entirely within the tile area or entirely out of the tile area, figure 2-31. This situation requires the full cooperation of all crafts involved in the construction job.

Faceplates

Faceplates for switches shall be installed so as to completely cover the wall opening and seat against the wall surface, *Section 380-9*.

Faceplates for receptacles shall be installed so as to completely cover the opening and seat against the mounting surface. The mounting surface might be the wall, or it might be the gasket of a weatherproof box.

In either case, the intent of the Code is to prevent access to live parts that could cause electrical shock.

SWITCHES	* Inches above floor
Regular	46 (1.17 m)
Between counter and kitchen cabinets (depends on backsplash)	44-46 (1.12-1.17 m)
RECEPTACLE OUTLETS	* Inches above floor
Regular (not permitted above electric baseboard heaters)	12 (305 mm)
Between counter and kitchen cabinets (depends on backsplash)	44-46 (1.12-1.17 m)
In garage	48 (1.22 m)
WALL BRACKETS	* Inches above floor
Outside	66 (1.68 m)
Inside	60 (1.52 m)
Side of medicine cabinet	60 (1.52 m)

***Note:**
All dimensions given are from the finished floor to the center of the outlet box. Verify all dimensions before roughing in.

Fig. 2-30 Outlet locations.

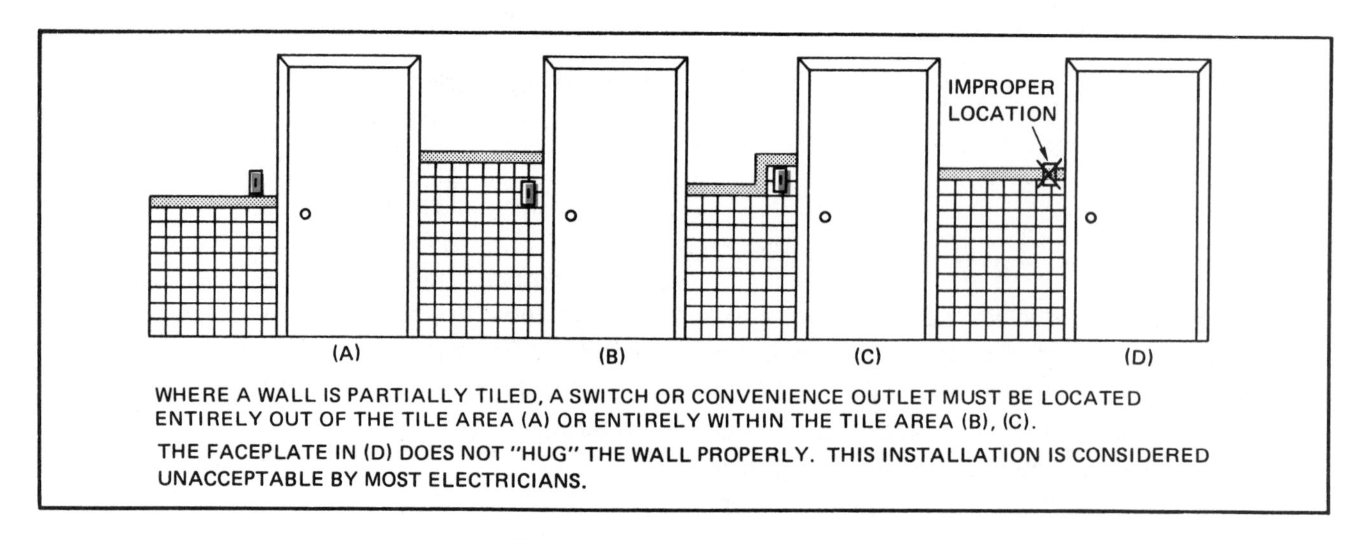

Fig. 2-31 Locating an outlet on a tiled wall.

REVIEW

Note: Refer to the *National Electrical Code®* or the plans where necessary.

PART 1 – ELECTRICAL FEATURES

1. What does a plan show about electrical outlets? ______________________

____

____

2. What is an outlet? ______________________________

____

3. Match the following switch types with the proper symbol.
   a. single-pole S_P
   b. three-way S_4
   c. four-way S
   d. single-pole with pilot light S_3

4. The plans show dash lines running between switches and various outlets. What do these dash lines indicate? ______________________________

____

5. Why are dash lines usually curved? ______________________

6. a. What are junction boxes used for? ______________________

   b. Are junction boxes normally used in wiring the first floor? Explain. ________

   ____

   ____

   c. Are junction boxes normally used to wire exposed portions of the basement? Explain. ______________________________

   ____

7. How are standard sectional switch (device) boxes mounted? ______________

____

____

____

8. a. What is an offset bar hanger? ______________________

   ____

   ____

   b. What types of boxes may be used with offset bar hangers? ______________

   ____

9. What methods may be used to mount lighting fixtures to an outlet box fastened to an offset bar hanger? ______________________________

____

____

10. What advantage does a 4-inch octagon box have over a 3 1/4-inch octagon box?

____

11. What is the size of the opening of a switch (device) box for a single device? ____

____

12. The space between a door casing and a window casing is 3 1/2 inches (88.9 mm). Two switches are to be installed at this location. What type of switches will be used?

____

____

13. Three switches are mounted in a three-gang switch (device) box. The wall plate for this assembly is called a ______________________ plate.

14. For each cable clamp inside a box (increase) (decrease) the number of conductors allowed by one. Circle correct answer.

15. a. How high above the finish floor are the switches located in the garage of this dwelling? ______________________

    b. In the living room of this dwelling? ______________________

16. How high above the finish floor are the receptacle outlets in the garage located? ______________ In the living room? ______________

17. Outdoor receptacle outlets in this dwelling are located ______________ ________ inches above grade.

18. In the spaces provided, draw the correct symbol for each of the descriptions listed in (a) through (r).

a. ________ Lighting panel	j. ________ Special-purpose outlet
b. ________ Clock outlet	k. ________ Fan outlet
c. ________ Duplex outlet	l. ________ Range outlet
d. ________ Outside telephone	m. ________ Power panel
e. ________ Single-pole switch	n. ________ Three-way switch
f. ________ Motor	o. ________ Push button
g. ________ Duplex outlet, split-circuit	p. ________ Thermostat
h. ________ Lampholder with pull switch	q. ________ Electric door opener
i. ________ Weatherproof outlet	r. ________ Multioutlet assembly

19. The front edge of a box installed in a combustible wall must be ____________ with the finished surface.

20. List the maximum number of No. 12 AWG conductors permitted in a

    a. 4″ × 1 1/2″ octagon box. ______________________

    b. 4 11/16″ × 1 1/2″ square box. ______________________

    c. 3″ × 2″ × 3 1/2″ device box. ______________________

21. When a switch (device) box is nailed to a stud, and the nail runs through the box, the nail must not interfere with the wiring space. To accomplish this, keep the nail ____________ (Select a, b, or c.)
    a. halfway between the front and rear of the box.
    b. a maximum of 1/4 inch (6.35 mm) from the front edge of the box.
    c. a maximum of 1/4 inch (6.35 mm) from the rear of the box.

22. Hanging a ceiling fixture directly from a plastic outlet box is permitted only if ____
____
____
____

23. It is necessary to count fixture wires when counting the permitted number of conductors in a box according to *Section 370-6.*
(True) (False) Underline or circle the correct answer.

24. *Table 370-6(a)* allows a maximum of ten wires in a certain box. However, the box will have two cable clamps and one fixture stud in it. What is the maximum number of wires allowed in this box? ____
____

25. When laying out a job, the electrician will usually make a layout of the circuit, taking into consideration the best way to run the cables and/or conduits and how to make up the electrical connections. Doing this ahead of time, the electrician determines exactly how many conductors will be fed into each box. With experience, the electrician will probably select two or three sizes and types of boxes that will provide adequate space that will "meet code." *Table 370-6(a)* of the Code shows the maximum number of conductors permitted in a given size box. However, the number of conductors shown in the table must be reduced by ____________ conductor(s) for each clamp, by ____________ conductor(s) for each fixture stud, ► by ____________________ conductor(s) for each wiring device contained on one yoke, and by ____________________ conductor(s) for one or more grounding conductors. ◄

26. Is it permissible to install a receptacle outlet above an electric baseboard heater?
____
____

PART 2 – STRUCTURAL FEATURES

1. To what scale is the basement plan drawn? ____________________________
2. What is the size of the footing for the steel Lally columns in the basement? ____
____
3. To what kind of material will the front porch lighting bracket fixture be attached to?
____

4. Give the size, spacing, and direction of the ceiling joists in the workshop. ______

5. What is the size of the lot on which this residence is located? ______

6. The front of the house is facing in what compass direction? ______
7. How far is the front garage wall from the curb? ______
8. How far is the side garage wall from the property lot line? ______
9. How many steel Lally columns are in the basement, and what size are they? ______

10. What is the purpose of the I-beams that rest on top of the steel Lally columns? ______

11. Is the entire garage area to have a concrete floor? ______
12. Where is access to the attic provided? ______
13. Give the thickness of the outer basement walls. ______
14. What material is indicated for the foundation walls? ______
15. Where are the smoke detectors located in the basement? ______

16. What is the ceiling height in the basement workshop from bottom of joists to floor? ______

17. Give the size and type of the front door. ______

18. What is the stud size for the partitions between the bathrooms in the bedroom area where substantial plumbing is to be installed? ______
19. Who is to furnish the range hood? ______

20. Who is to install the range hood? ______

UNIT 3

Determining the Required Number and Location of Lighting and Small Appliance Circuits

OBJECTIVES

After studying this unit, the student will be able to

- understand the fundamental Code requirements for calculating branch-circuit sizing and loading.
- understand "volt-amperes per square foot" for arriving at total calculated general lighting loading.
- calculate the occupied floor area of a residence.
- determine the minimum number of lighting branch circuits required.
- determine the minimum number of small appliance circuits required.
- know where switched and nonswitched lighting outlets must be installed in homes.
- know where receptacle outlets must be installed in homes.

It is standard practice in the design and planning of dwelling units to permit the electrician to plan the circuits. Thus, the residence plans do not include layouts for the various branch circuits. The electrician may follow the general guidelines established by the architect. However, any wiring systems designed and installed by the electrician must conform to the standards of the *National Electrical Code*®, as well as local and state code requirements.

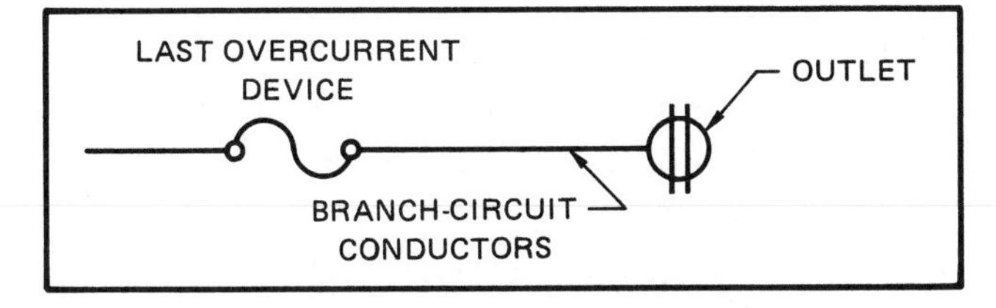

Fig. 3-1 Branch circuit.

This unit focuses on lighting branch circuits and small appliance circuits. The circuits supplying the electric range, oven, clothes dryer, and other specific circuitry not considered to be a lighting branch circuit or a small appliance circuit are covered later in this text. Refer to the index for the specific circuit being examined.

BASICS OF WIRE SIZING AND LOADING

The *National Electrical Code*® establishes some very important fundamentals that weave their way through the decision-making process for an electrical installation. They are presented here in brief form, and are covered in detail as required throughout this text.

A *branch circuit* is defined as the circuit conductors that extend beyond the final overcurrent device protecting that circuit and the outlets served by that circuit, figure 3-1.

The ampacity (current-carrying capacity) of a conductor must not be less than the rating of the overcurrent device protecting that conductor, *Section 210-19*. A common exception to this is a motor branch circuit, where it is quite common to have overcurrent devices (fuses or breakers) sized larger than the ampacity of the conductor. Motors and motor circuits are covered specifically in *Article 430* of the Code. The rating of the branch-circuit overcurrent device determines the rating of the branch circuit. For example, if a 20-ampere conductor is protected by a 15-ampere fuse, then the circuit is considered to be a 15-ampere branch circuit, *Sections 210-3, 210-19*.

The ampacity of branch-circuit conductors must not be less than the maximum load to be served. Where the circuit supplies receptacle outlets, as is typical in homes, the conductor's ampacity must not be less than the "rating" of the branch-circuit overcurrent device, *Sections 210-19* and *210-20*.

Where the ampacity of the conductor does not match up with a standard rating of a fuse or breaker, the next higher standard size overcurrent device may be used, provided the overcurrent device does not exceed 800 amperes, *Section 240-3*. This exception is *not* permitted, however, when the circuit supplies receptacles where "plug-connected" appliances, etc., could be used, because too many "plug-in" loads could result in an overload condition, *Section 210-19(a)*. You may go to the next standard size overcurrent device *only* when the circuit supplies fixed lighting outlets.

For instance, when a conductor having an original ampacity of 25 amperes (see *Table 310-16*) No. 12 Type TW is derated to 70%:

$$25 \times 0.70 = 17.5 \text{ amperes}$$

It is permitted to use a 20-ampere overcurrent device if the circuit supplies *only* fixed lighting outlets or other fixed loads.

If the above example were to supply receptacle outlets, then the rating of the overcurrent device would have to be dropped to 15 amperes; otherwise it is possible to overload the conductors by "plugging-in" more load than the conductors can safely carry.

The ampacity of conductors commonly used in residential occupancies is found in *Table 310-16* of the Code. The ampacities in *Table 310-16* are subject to *correction factors* that must be applied where high ambient temperatures are encountered — for example, in attics. See *footnote* to *Table 310-16*.

Conductor ampacities are also *derated* when four or more conductors are installed in a single raceway or cable. See *Note 8* to *Table 310-16*.

For loads that might be termed *continuous*, the load should not exceed 80% of the rating of the branch circuit. The Code defines continuous as "a load where the maximum current is expected to continue for three hours or more." The Code permits 100% loading on an overcurrent device *only* if that overcurrent device is *listed* for 100% loading. At the time of this writing, Underwriters Laboratories has *no* listing of circuit breakers of the molded-case type used in residential installations that are suitable for 100% loading.

This is another judgment call on the part of the electrician and/or the electrical inspector. Will a branch circuit supplying a heating cable in a driveway likely be on for 3 hours or more? The answer is "probably." Good design practice and experience tell us "never load a circuit conductor or overcurrent protective device to more than 80% of its rating."

The branch-circuit rating shall *not* be less than any noncontinuous load plus 125% of the continuous load. See *Sections 220-3(a), (b), (c),* and *(d)*.

The minimum lighting load per square foot of floor area for dwellings is 3 volt-amperes, *Section 220-3(b)* and *Table 220-3(b)*.

Most receptacle outlets in a residence are considered by the Code to be part of "general illumination," and additional load calculations are *not* necessary. See Footnote to *Table 220–3(b)*. Appliance receptacle outlets in the kitchen, dining room, laundry, and workshop are *not* to be considered part of general illumination. Additional loads are figured in for these receptacle outlets and will be discussed in detail later in this text.

To determine the number of lighting circuits required, divide the total volt-amperes by the ampere rating of the circuit(s) to be used. In this residence, all lighting circuits are rated 15 amperes. Therefore,

$$\frac{3 \text{ volt-amperes} \times \text{square feet}}{120 \text{ volts}} = \text{amperes}$$

$$\frac{\text{amperes}}{15} = \text{minimum number of circuits required}$$

This equates to

ONE 15-AMPERE CIRCUIT FOR EVERY 600 SQUARE FEET
ONE 20-AMPERE CIRCUIT FOR EVERY 800 SQUARE FEET

Load calculations for electric water heaters, ranges, motors, electric heat, and other specific loads are in addition to the general lighting load, and are discussed throughout this text as they appear.

VOLTAGE

All calculations throughout this text use voltages of 120 volts and 240 volts. This is in accordance with *NEC Section 220-2* and *Chapter 9, Part B*, on "Voltage."

CALCULATING FLOOR AREA

First Floor Area

To estimate the total load for a dwelling, the occupied floor area of the dwelling must be calculated. Note in the residence plans that the first floor area has an irregular shape. In this case, the simplest method of calculating the occupied floor area is to determine the total floor area using the outside dimensions of the dwelling. Then, the areas

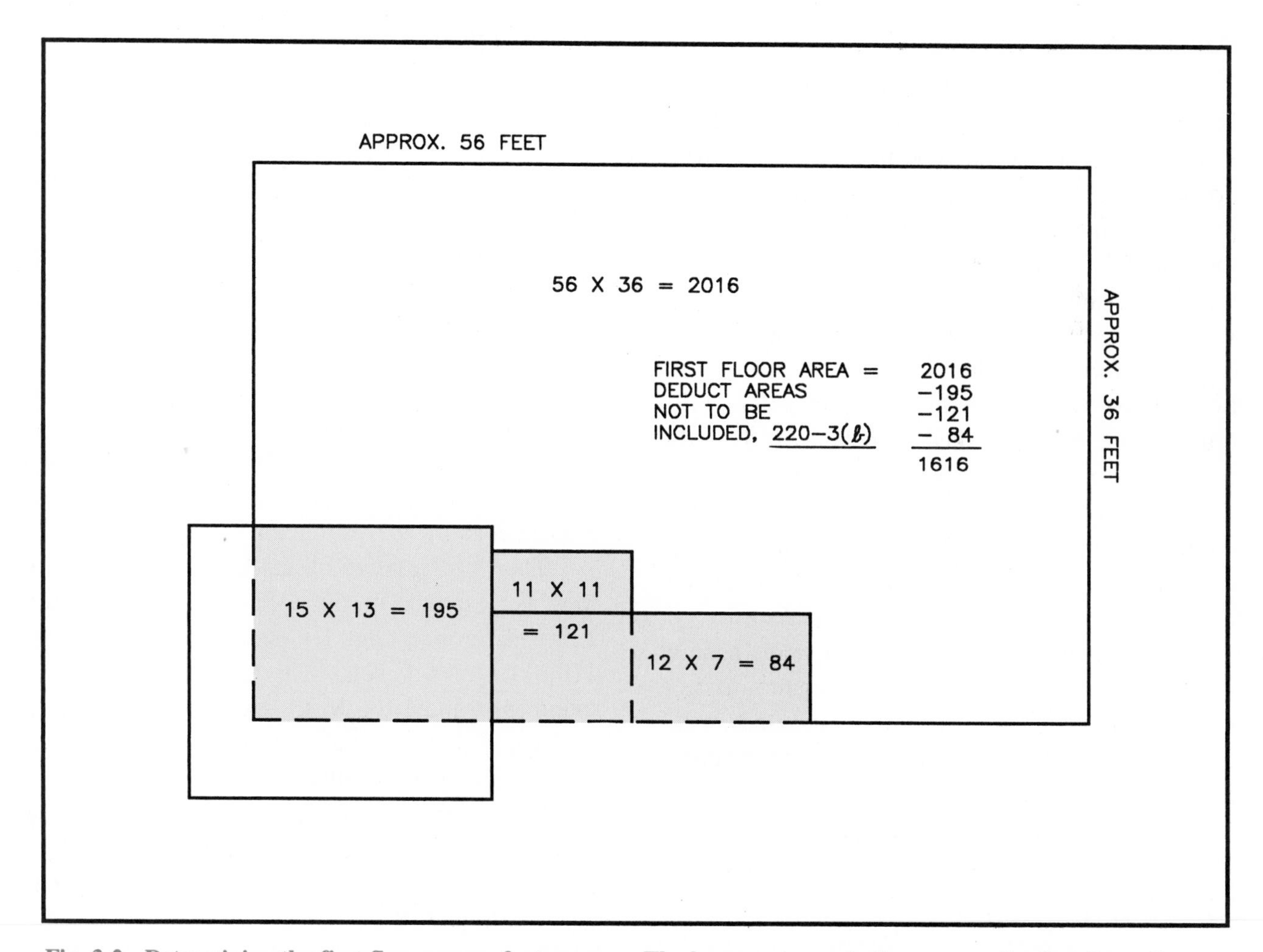

Fig. 3-2 Determining the first floor square footage area. The basement area is the same as the first floor. See text for explanation of how the dimensions of the residence were rounded off to make the calculations much simpler and much more practical for the electrician and electrical inspector to compute.

of the following spaces are subtracted from the total area: open porches, garages, or other unfinished or unused spaces if they are not adaptable for future use, *Section 220-3(b)*.

Many open porches, terraces, patios, and similar areas are commonly used as recreation and entertainment areas. Therefore, adequate lighting and receptacle outlets must be provided.

For practicality the author has chosen to round up dimensions for the determining of total square footage and to round down dimensions for those areas (garage, porch, and portions of the inset at the front of the house) not to be included in the computation of the general lighting load. This results in a more "plus" answer as opposed to being on the conservative side.

Figure 3-2 shows the procedure for calculating the volt-amperes needed for the first floor of this residence.

Basement Area

The basement area of a home is not generally included in the calculating of general lighting loads since it could be classified as an "unfinished space not adaptable for future use," *Section 220-3(b)*. Yet in this residence, more than half of the total basement area is finished off as a recreation room, which certainly is considered living area. The workshop area also is intended to be used.

Therefore, the author again has taken the position that the basement in this residence will be fully utilized and has "invoked" *Section 90-4* of the Code, as do many electrical inspectors. In part, *Section 90-4* states that "the authority having jurisdiction of enforcement of the Code will have the responsibility for making interpretation of the rules. . . ."

By making this decision based upon *Section 90-4* of the Code, the basement square footage becomes the same as the area of the first floor, making it a simple addition problem to find total square footage of the residence.

The combined occupied area of the dwelling is found by adding the first floor and basement areas together:

First Floor	1616 ft^2
Basement	1616 ft^2
Total	3232 ft^2

DETERMINING THE MINIMUM NUMBER OF LIGHTING CIRCUITS

Table 220-3(b) shows that the minimum load required for dwelling units is 3 volt-amperes (VA) per square foot (0.093 m^2) of occupied area. The calculated load for the total occupied area, 3232 ft^2, is:

$$1616 + 1616 = 3232 \text{ ft}^2$$

$$3232 \text{ (ft}^2) \times 3 \text{ (VA/ft}^2) = 9696 \text{ volt-amperes}$$

The total required amperage is determined as follows:

$$\text{Amperes} = \frac{\text{volt-amperes}}{\text{volts}}$$

$$= \frac{9696}{120} = 80.8 \text{ amperes}$$

The minimum number of 15-ampere lighting circuits required is equal to the total amperage divided by the maximum amperes of each circuit, *Section 220-4*.

$$\frac{80.8 \text{ amperes}}{15 \text{ amperes}} = 5.39 \text{ circuits}$$

Therefore, a minimum of six lighting branch-circuits is required.

The *National Electrical Code®* in *Chapter 9, Part B* illustrates a number of examples that show how to calculate circuit feeder, service-entrance, and motor circuit ampacities.

At the beginning of *Part B*, the Code states that "for uniform application of *Articles 210, 215* and *220*, a nominal voltage of 120, 120/240, 240 and 208Y/120 volts shall be used in computing the ampere load on the conductor."

The word *nominal* means **in name only, not a fact**. For example, in our calculations, we will use 120 volts, even though the actual voltage might be 110, 115, or 117 volts. The term *nominal* can be found throughout the *NEC®*. It provides us with a uniformity that, without it, would lead to misleading and confusing results in our computations.

In the same part B, the Code further states that "except where the computations result in a major fraction of an ampere (0.5 or larger) such fractions may be dropped."

This residence actually has 13 branch lighting circuits, as shown in Table 3-1.

Table 310-16 (footnote) shows that the maximum overcurrent protection is 15 amperes when No. 14 AWG copper wire is used, and 20 amperes when No. 12 AWG copper wire is used. It is significant to note that the lighting circuits are still classified as 15-ampere circuits even if No. 12 AWG copper wire is installed because the overcurrent protection for these lighting circuits is rated at 15 amperes, *Section 210-3*.

TRACK LIGHTING LOADS

Track lighting loads are considered to be part of the general lighting load for residences based upon the 3 volt-amperes per square foot as discussed previously. There is no need to add more wattage (volt-amperes) for track lighting. ► This is verified by the exception to *Section 410-102*, which exempts residential track lighting. ◄ However, for commercial installations the Code requires that every 2 feet (609.6 mm) of track lighting must be included in load calculations at 180 VA.

DETERMINING THE NUMBER OF SMALL APPLIANCE CIRCUITS

Sections 220-16(a) and *(b)* state that an additional load of not less than 1500 watts shall be included for each two-wire small appliance circuit when calculating feeders and services.

► *Section 220-4(b)* requires that two or more 20-ampere small appliance circuits must be provided for supplying receptacle outlets as specified in *Section 210-52*.

Section 210-52(b)(1) states that the 20-ampere appliance circuits mentioned in *Section 220-4(b)* shall feed only the receptacle outlets in the kitchen, pantry, dining room, breakfast room, or similar rooms of dwelling units. This section also requires that these 20-ampere small appliance circuits shall not supply other outlets.

Exceptions in *Section 210-52(b)(1)* allow outdoor receptacles and/or a clock outlet to be connected to these 20-ampere small appliance circuits, figure 3-3.

Section 210-52(b)(2) requires that not less than two 20-ampere small appliance circuits must supply the receptacle outlets serving the countertop areas in kitchens. ◄

Food waste disposers and/or dishwashers shall not be connected to these required 20-ampere small appliance circuits. They must be supplied by other circuits, as will be discussed later. No. 12 AWG wire is used in small appliance circuits to minimize the voltage drop in the circuit. The use of this larger wire, rather than No. 14 AWG wire, helps improve

LIGHTING CIRCUITS 15 AMPERE	
1	Entry — Porch
2	Workshop
3	Master Bedroom
4	Study — Bedroom
5	Bathrooms — Hall
6	Front Bedroom
7	Recreation Room
8	Living Room
9	Laundry, Service Entrance, Wash Room, Attic
10	Garage
11	Recreation Room Receptacles
12	Wet Bar
13	Kitchen

Table 3-1.

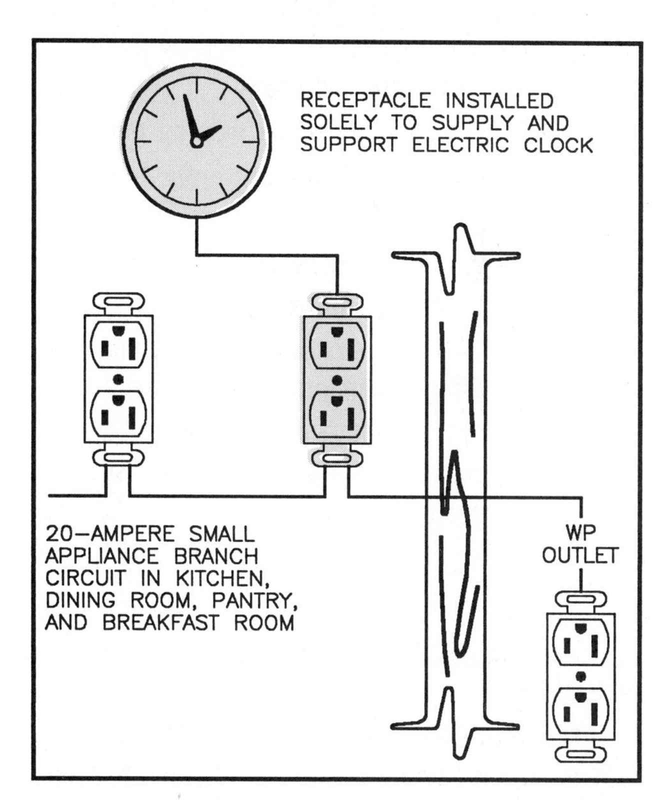

Fig. 3-3 A receptacle installed solely to supply and support an electrical clock may be connected to a 20-ampere small appliance circuit or to a general-purpose lighting circuit. It is also permissible to connect outdoor receptacles to a small appliance circuit, *Section 210-52(b), Exc. No. 1* and *No. 2*. The outdoor receptacles must be GFCI protected, *Section 210-8(a)(3)*.

appliance performance and lessens the danger of overloading the circuits.

Automatic washers draw a large amount of current during certain portions of their operating cycles. Thus, *Section 220-4(c)* requires that at least one additional 20-ampere branch circuit must be provided to supply the laundry receptacle outlet(s) as required by *Section 210-52(f)*.

A complete list of the branch circuits for this residence is found in unit 28. The circuit directories for Main Panel A and Subcenter B show clearly the number of lighting, appliance, and special-purpose circuits provided.

It is recommended that the student now study the requirements of the following sections of the Code.

- *Section 210-7* — methods of connecting grounding-type receptacles.
- *Section 210-8* — ground-fault circuit interrupter (GFCI) protection for receptacle outlets.
- *Sections 210-50* and *210-52* — number of and spacing requirements for receptacle outlets in specific rooms and hallways.
- *Section 210-70* — switched lighting outlets.
- *Section 250-57* — methods of grounding fixed equipment.
- *Section 250-74* — connecting the receptacle grounding terminal to the box.
- *Section 250-114* — continuity and attachment of grounding conductors to boxes.
- *Section 305-4* — branch circuits and receptacle requirements for temporary wiring, such as during construction.

RECEPTACLE OUTLET BRANCH-CIRCUIT RATINGS

The residence plans show that the receptacle outlets in the living room, study, dining room, recreation room, bedrooms, baths, entry, and lavatory adjacent to the laundry room are connected to the 15-ampere general lighting circuits. The receptacle outlets in the kitchen, breakfast nook, workshop, and laundry room are connected to 20-ampere small appliance circuits.

SUMMARY OF WHERE RECEPTACLE AND LIGHTING OUTLETS MUST BE INSTALLED IN RESIDENCES

The following is a brief explanation of where receptacle outlets and lighting outlets must be installed in residential occupancies. Throughout this text, each room is discussed in detail regarding outlet location. The following recaps all of the *National Electrical Code*® rules that relate to receptacle and lighting outlet requirements. In the following paragraphs, the abbreviation GFCI stands for Ground Fault Circuit Interrupter.

Receptacle Outlets (125-Volt, Single-Phase, 15 or 20 Ampere)

General Lighting

- Wall receptacle outlets must be placed so that no point along the floor line is more than 6 feet (1.83 m) from an outlet.

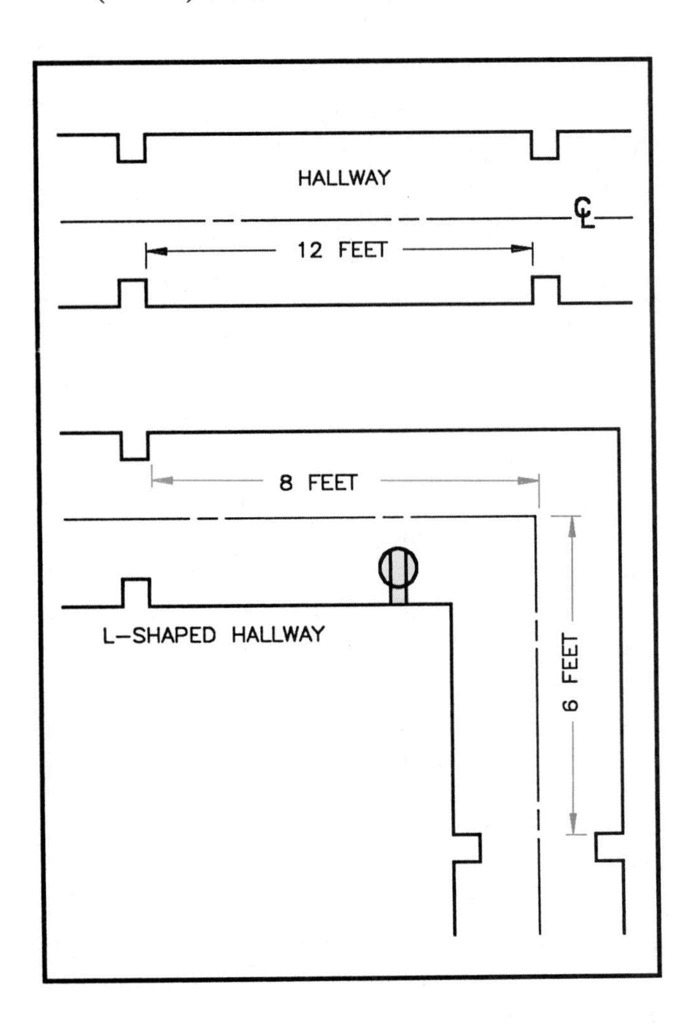

► • Nonsliding fixed glass panels on exterior walls are considered to be wall space and are to be figured in when applying the rule that no point along the floor line is more than 6 feet (1.83 m) from a receptacle. Sliding panels on exterior walls are not considered to be a wall, *Section 210-52(a)*.

• Hallways 10 feet or longer in homes must have at least one receptacle outlet per *Section 210-52(h)*. See figure 3-4. ◄

• Any wall space 2 feet (610 mm) or less must have a receptacle outlet.

• Basement areas not generally "lived in" are not included in the above spacing requirements.

• Circuits supplying wall receptacle outlets for lighting loads are generally 15-ampere circuits, but are permitted by the Code to be 20 amperes.

► • In crawl spaces, and in attics, a receptacle outlet must be installed in an accessible location within 75 feet of heating, air-conditioning, or refrigeration equipment. Do *not* connect this receptacle outlet to the load side of the equipment disconnecting means. See *Section 210-63*. ◄

• For buildings containing more than two family dwellings (for example, a four-flat building), a receptacle outlet must also be installed on the roof if there is heating, air-conditioning, or refrigeration equipment that requires servicing on the roof. This receptacle outlet must be located in an accessible location within 75 feet of the equipment on the same level as the equipment, *Section 210-63, Exception*.

• Circuits supplying electrical outlets in kitchens, dining rooms, and pantries are required to be 20 amperes where appliances will be plugged-in. Refer to *Section 210-52(b)* and *220-4(b)* of the Code.

Basements

• At least one receptacle outlet must be installed in addition to the laundry outlet.

• Connect to 15- or 20-ampere circuit.

► • In unfinished basements, ALL 125-volt, single-phase, 15- or 20-ampere receptacles must have GFCI protection, *Section 210-8(a)(4)*. The Code does not require GFCI protection:

a. for the receptacle(s) installed for laundry equipment as required by *Section 210-52(f)* and *Section 220-4(c)*.
b. for a single receptacle supplying a permanently installed sump pump.
c. for a single receptacle that is supplied by a dedicated branch-circuit which is located and identified for specific use by a cord- and plug-connected appliance; for example, a freezer or refrigerator.
d. for the receptacles in habitable finished rooms in basements. ◄
e. Porcelain lampholders that have a receptacle outlet as part of the lampholder must be GFCI protected.

• Refer to *Sections 210-8(a), 210-52(f)*, and *210-52(g)* of the Code.

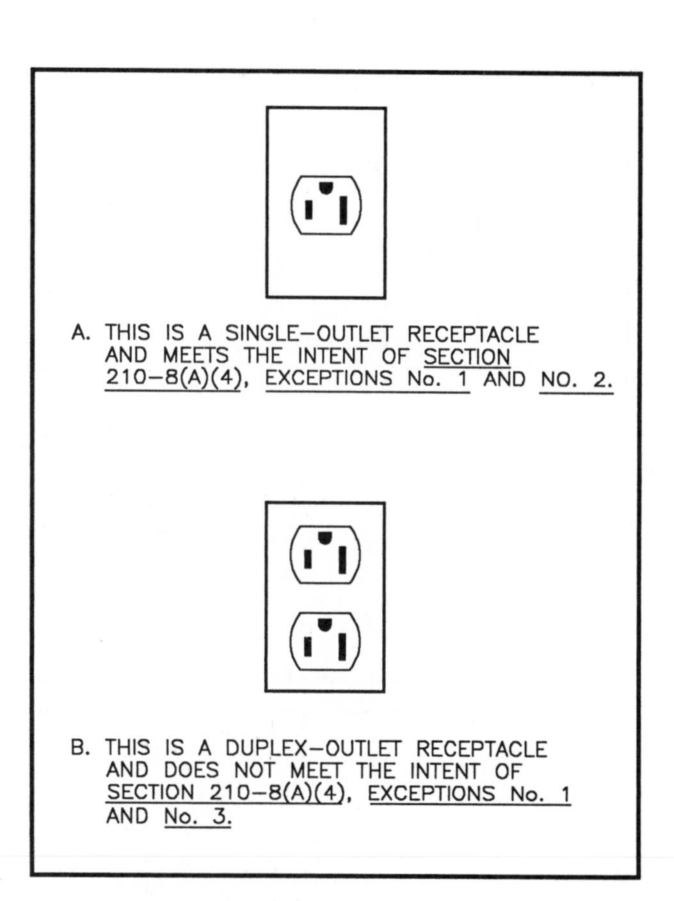

Fig. 3-5 Illustrations of single outlet receptacle (A) and duplex outlet receptacle (B).

Figure 3-5 illustrates the difference between a single-outlet receptacle and a duplex-outlet receptacle necessary to interpret the statement properly in *Section 210-8(a)(4)*.

Appliance Receptacle Outlets

- For the receptacle outlets in the kitchen, dining areas, or pantry, at least two 20-ampere small appliance circuits must supply these receptacle outlets.
- See *Sections 220-4(b)*, *220-16*, and *210-52* of the Code.

Countertops *(See figure 3-6.)*

- In kitchens and dining areas, receptacle outlets must be installed above all countertops wider than 12 inches (305 mm).
- ► Receptacle outlets that serve countertops must be supplied by at least two 20-ampere small appliance circuits. These two 20-ampere circuits are permitted to supply other receptacle outlets in the kitchen, dining room, pantry, breakfast room, or similar location. ◄
- Must be GFCI protected if located within 6 feet (1.83 m) of kitchen sink and if they serve countertop surfaces. A receptacle outlet under the sink serving the food waste disposer and/or dishwasher does not have to be GFCI protected.
- ► Peninsulas and freestanding islands must have a receptacle outlet installed for each 4 feet of countertop. These receptacle outlets must be GFCI protected if located within 6 feet (1.83 m) of the kitchen sink. ◄
- ► Receptacle outlets in kitchen and dining areas located above countertops shall be located so that no point along the countertop surface is more than 24 inches (609 mm) from a receptacle outlet. ◄

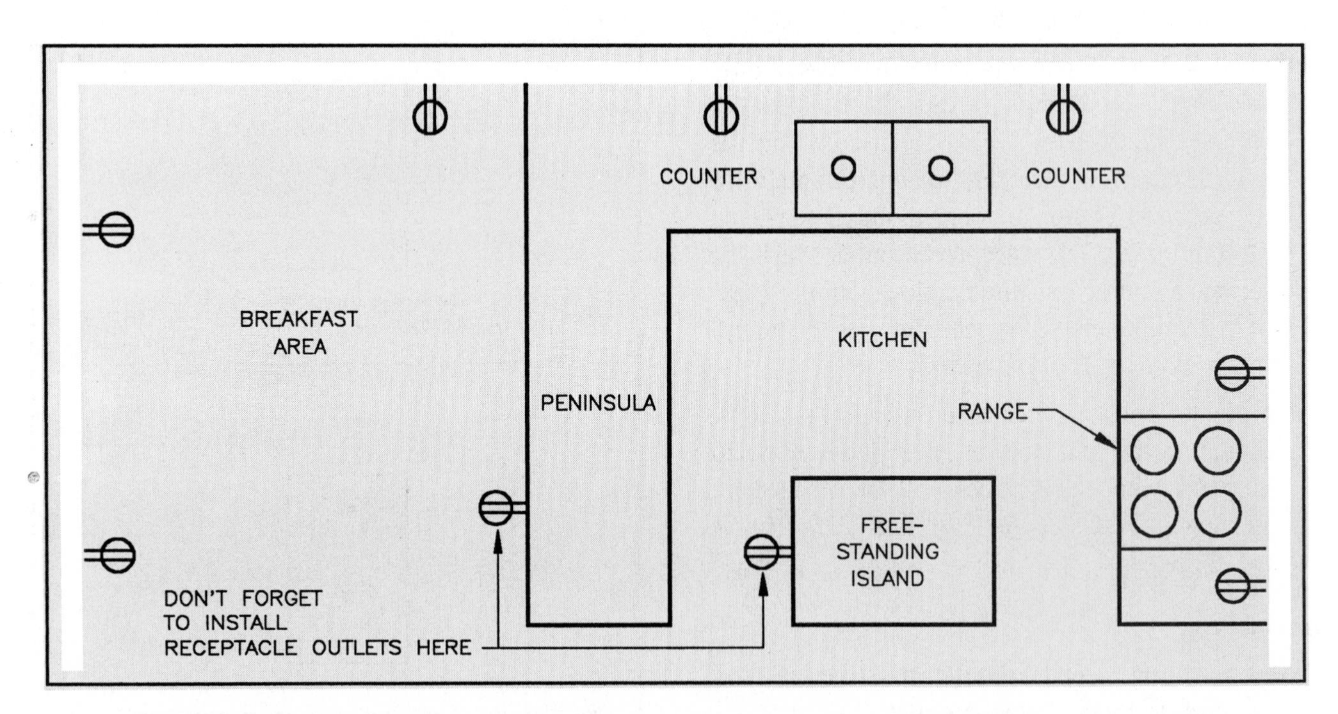

Fig. 3-6 Receptacle outlets must be installed so that no point along the floor line is more than 6 feet (1.83 m) from an outlet, *Section 210-52(a)*. This includes the "wall space" created by fixed room dividers. Remember that GFCI protection is required for all receptacle outlets that serve countertops that are installed within 6 feet of kitchen sinks. See units 6 and 12 for more information about GFCI protection. ► Position receptacle outlets that serve countertops so that no point along the countertop is more than 24 inches (609 mm) from a receptacle outlet, *Section 210-52(c)*. Provide at least one receptacle for each 4 feet (1.22 m) of countertops on islands and peninsula-type counters, *Section 210-52(c)*. ◄

- Receptacle outlets behind refrigerators or appliances fastened in place must not be thought of as meeting the above requirements for receptacle outlets above countertops, because these outlets would be impossible to use once the refrigerator or appliance was fastened in place.
- Receptacle outlets installed specifically for refrigerators or freezers should not be GFCI protected. See *Fine Print Note* to *Section 210-8(a)(5)*.
- See *Sections 210-8(a)* and *210-52* of the Code.

Laundry

- At least one receptacle outlet is required within 6 feet (1.83 m) of the intended location of the wash machine.
- The circuit must be a 20-ampere circuit.
- This receptacle *need not* be GFCI protected because acceptable levels of leakage current of some clothes washing machines, as determined by Underwriters Laboratories, can cause the GFCI to trip.
- This outlet and circuit are in addition to the basement receptacle outlet requirements already discussed, and must be in addition to the required two small appliance circuits that serve the receptacle outlets in the kitchen and dining areas. This circuit must *not* serve other outlets.
- See *Sections 220-4(c)*, *210-8(a)(4) Exception No. 2*, *210-50(c)*, *210-52(f)*, and *220-16(b)* of the Code.

Bathrooms

- At least one receptacle outlet must be installed next to the basin.
- An outlet that is an integral part of a fixture or medicine cabinet does *not* qualify for this required outlet.
- Connect to 15- or 20-ampere circuit.
- Must be GFCI protected.
- See *Sections 210-8(a)(1)* and *210-52(d)*.

Outdoors

- ► At least one receptacle outlet must be provided. figure 3-7. If there is no direct access from front yard to back yard, then at least one

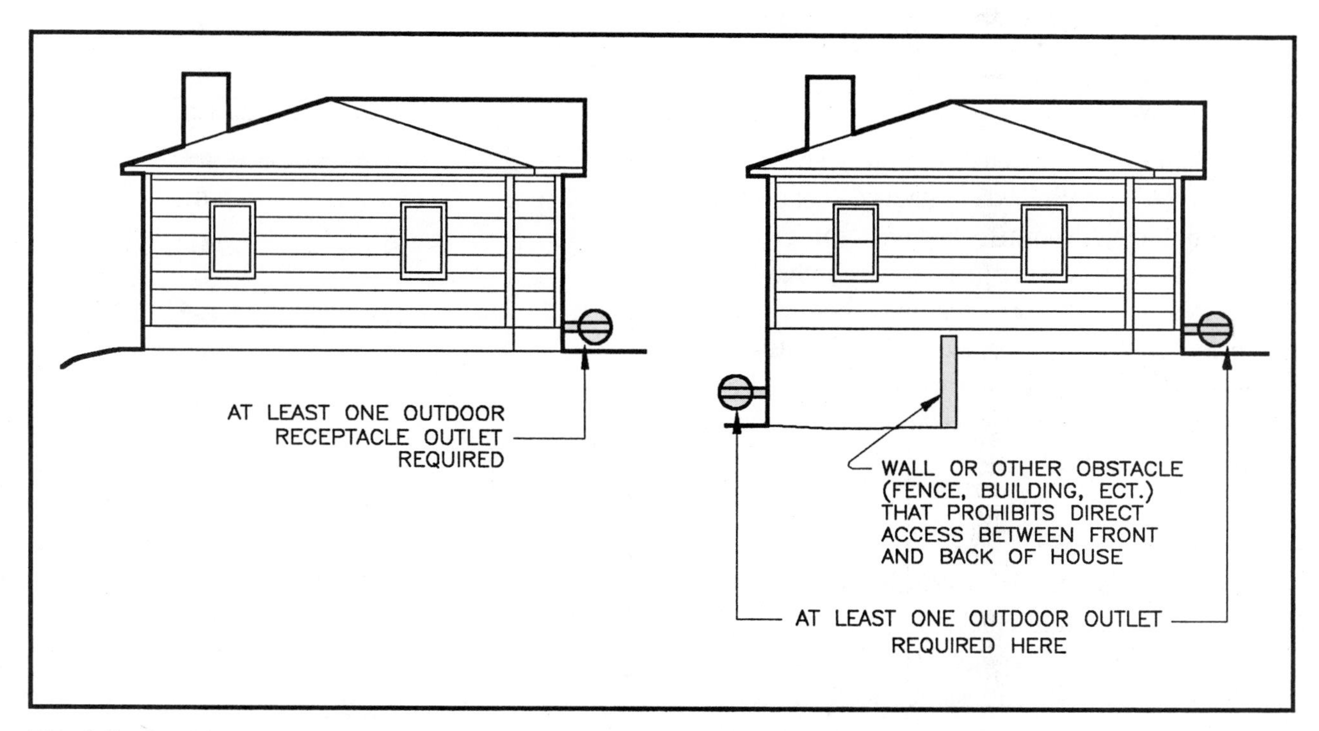

Fig. 3-7 Outdoor receptacle requirements for residences. *Section 210-52(e)* requires that for one-family homes, at least one receptacle outlet must be installed outdoors. ► If there is no direct outdoor grade level access from the front of the house to the back of the house, then at least one receptacle must be installed in the back of the house and one receptacle must be installed in the front of the house. ◄

receptacle outlet must be installed in the front yard and another in the back yard. ◄

- Connect to 15- or 20-ampere circuit.
- Must be GFCI protected.
- See *Sections 210-8(a)(3)* and *210-52(e)*.

Garages

- At least one receptacle outlet must be provided.
- Connect to 15- or 20-ampere circuit.
- Must be GFCI protected.
- If the garage is detached, the Code does not require electric power, but if electric power is provided then at least one GFCI receptacle outlet must be installed.
- See *Sections 210-8(a)(2)* and *210-52(g)*.

GFCI Exemptions

►• Where dedicated receptacle outlets are installed for such things as overhead door openers, refrigerators, or freezers, GFCI protection is not a Code requirement but is permitted if you want to provide GFCI protection. Nuisance tripping has been reported, so it might be advisable *not* to install GFCIs for these locations. See *Sections 210-8(a)(2), 210-8(a)(4), Exc. 1, 2, 3, and 210-8(a)(5), FPN.* ◄

Specific Loads

- Appliances such as electric ranges and ovens, fans, air conditioners, electric heat, heat pumps, electric water heaters, garage door openers, and similar loads are purchased as desired by the homeowner, *but* when included in the design of a residence the circuitry and installation must conform to the Code. See index for location of coverage for specific appliance or load.

Lighting Outlets (*Section 210-70*)

The *National Electrical Code*® does contain certain minimum requirements for providing lighting for dwellings. Per *Section 210-70*, lighting outlets in dwellings must be installed as follows:

Switched Lighting Outlets (See figure 3-10.)

At least one switch-controlled lighting outlet shall be installed

- in every habitable room.
- in bathrooms.
- in hallways.

Fig. 3-8 When a switch-controlled receptacle outlet is installed in rooms where the receptacle outlets normally would be 20-ampere small appliance circuits (breakfast nook, dining room, etc.), it must be in addition to the required small appliance receptacle outlets, *Section 210-52(b)(1), Exc. No. 3.*

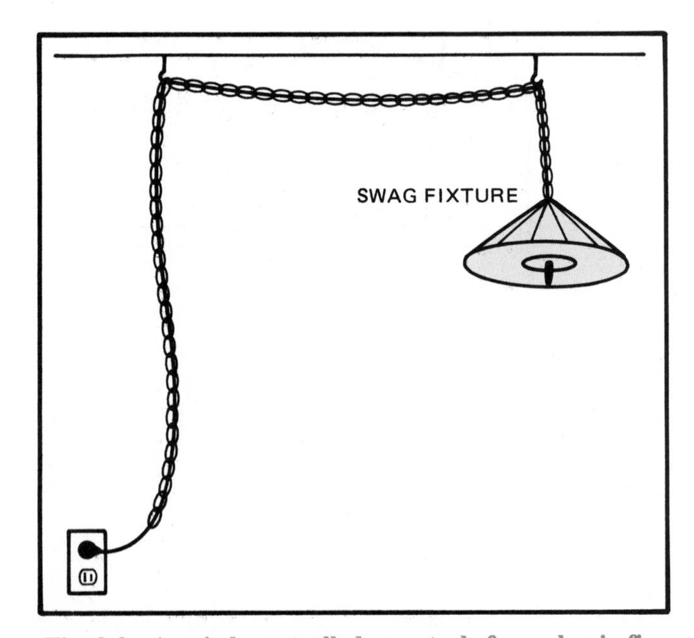

Fig. 3-9 A switch-controlled receptacle for a plug-in fixture (swag fixture) is supplied by a general lighting circuit and is *not* to be connected to the 20-ampere small appliance circuit that supplies the other three receptacles. The switch-controlled receptacle is in addition to the required small appliance outlets, *Section 210-52(b)(1), Exc. 3.*

- ► in "interior" stairways (if stairway has six steps or more, then switch control must be from both levels). ◄
- in attached garages.
- in detached garages if they have electric power.
- ► in attics, underfloor spaces, utility rooms, and basements where these areas are used for storage, or if these areas contain equipment that will require servicing (furnace, air-conditioning, heat pumps, etc.). This lighting outlet must be near the equipment and must be switch-controlled near the point of entry. ◄
- at outdoor entrances or exits (a vehicle door in an attached garage is *not* considered to be an outdoor entrance).

Important! There are two exceptions to *Section 210-70*. The first exception recognizes the fact that in most residences, lighting is accomplished by plugging table and floor lamps into wall receptacle outlets. Other than kitchens and bathrooms, switch-controlled wall receptacles are permitted in lieu of ceiling and/or wall lighting outlets.

► The Code requires that the receptacle outlets in kitchens, pantries, breakfast rooms (dinettes), and

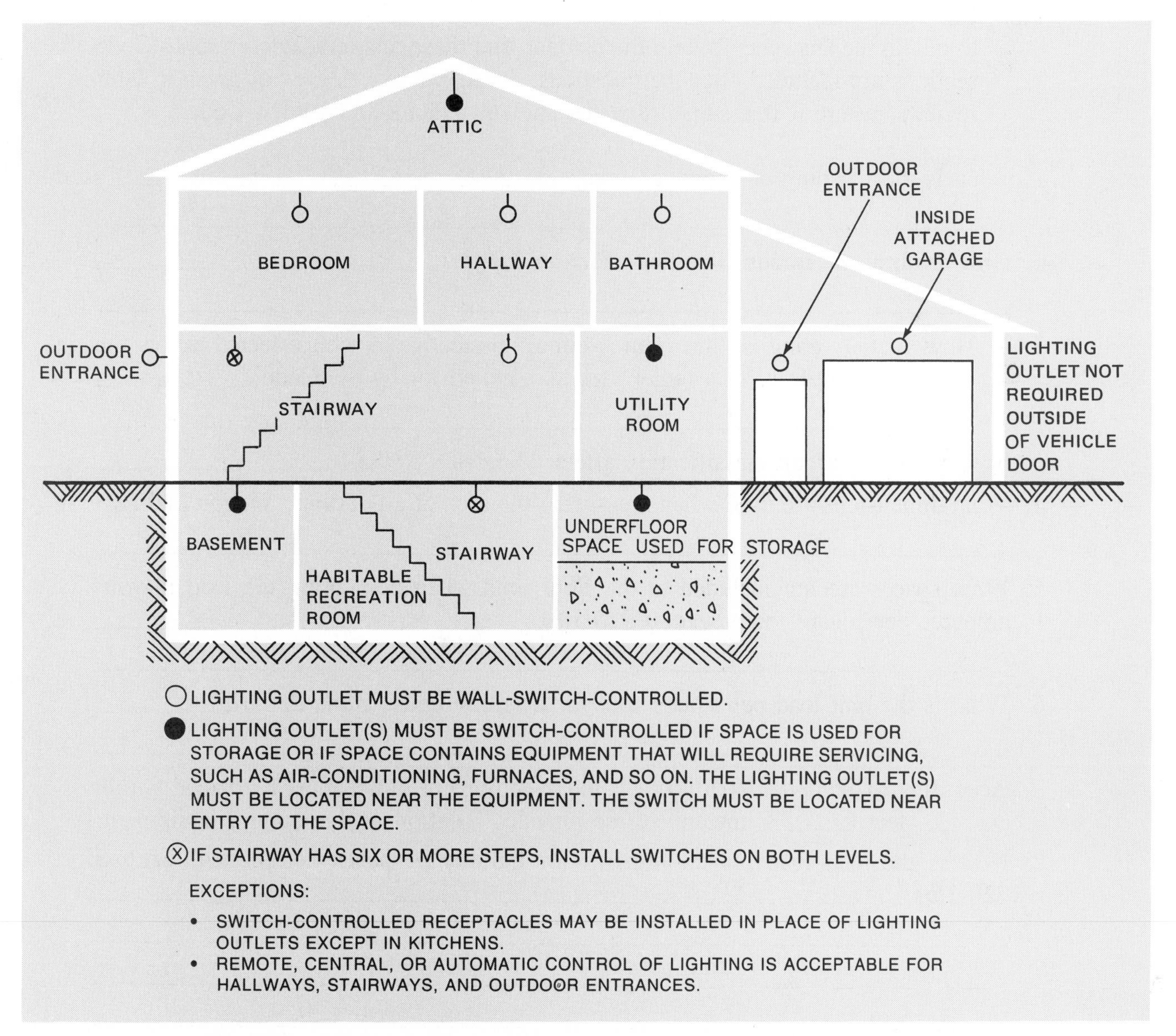

Fig. 3-10 Lighting outlets required in a typical dwelling unit, *Section 210-70*.

dining rooms be supplied by 20-ampere small appliance circuits, *Sections 210-52(b)* and *220-4(b)*. ◄

Section 210-70 lists the rooms that require switched lighting outlets. If the homeowner wishes to install swag lighting, figure 3-8, in the dining room instead of a regular ceiling fixture, a switched receptacle connected to a general lighting circuit would have to be installed. The switched receptacle must be in addition to the 20-ampere small appliance receptacle outlets, *Section 210-52(b)(1), Exc. 3*, figure 3-9.

The Code, in *Section 210-70(a), Exc. 1*, prohibits the switching of receptacles in kitchens and bathrooms for the purpose of providing lighting rather than installing actual lighting fixtures.

The second exception is that a separate switch is not required for hallways, stairways, or outdoor entrances when automatic timers, photocells, central or remote control systems have been installed for the control of these lighting outlets. For example, a post light controlled by a photocell does not require a separate switch. See *Section 210-70(a), Exc. 2*.

REVIEW

Note: In the following questions the student may refer to the *National Electrical Code®* to obtain the answers. To help the student find these answers, references to Code sections are included after the questions. These references are not given in later reviews, where it is assumed that the student is familiar with the Code.

1. What is the meaning of computed load? ______________________________

____

2. How are branch circuits classified? *Section 210-3* ______________________

____

3. a. How is the rating of the branch-circuit protective device affected when the conductors used are of a larger size than called for by the Code? __________

____

b. How is the circuit classification affected? *Section 210-3* ______________

4. What dimensions are used when measuring the area of a building? *Section 220-3(b)*

____

5. What spaces are not included in the floor area when computing the load in volt-amperes per square foot? *Section 220-3(b)* ______________________

____

6. What is the unit load per square foot for dwelling units? *Table 220-3(b)* _______

____

7. According to *Section 210-50(c)*, a laundry equipment outlet must be placed within _______ feet (_______ meters) of the intended location of the laundry equipment.

8. How is the total load in volt-amperes for lighting purposes determined? *Section 220-3(b)* ____

____

____

____

9. How is the total lighting load in amperes determined? ______________________

____

10. How is the required number of branch circuits determined? ________________

____

11. What is the minimum number of 15-ampere lighting circuits required if the dwelling has an occupied area of 4000 square feet? Show all calculations.

12. How many branch lighting circuits are provided in this dwelling? ____________

13. What is the minimum load allowance for small appliance circuits for dwellings? *Section 220-16(a)* ____

____

14. What is the smallest size wire that can be used in a branch circuit rated at 20 amperes?

____

15. How is the load determined for outlets supplying the specific appliances? *Section 220-3(c)* ____

____

16. What type of circuits must be provided for receptacle outlets in the kitchen, pantry, dining room, and breakfast room? *Section 220-4(b)* ____________________

____

17. How is the minimum number of receptacle outlets determined for most occupied rooms?

____

____

____

____

____

____

18. In a single-family dwelling, what types of overload protection for circuits are used?

____

____

19. Is a grounded circuit conductor included in this dwelling for

a. a 120-volt, 2-wire branch circuit? ________________________

b. a 240-volt, 2-wire branch circuit? ________________________

c. a 240-volt, 3-wire branch circuit? ________________________

20. May receptacle outlets in electric baseboard heaters be connected to the heater circuit?

____

21. The Code indicates the rooms in a dwelling that are required to have switched lighting outlets or switched receptacles. Write yes (switch required) or no (switch not required) for the following areas:
    a. attic ____________
    b. stairway ____________
    c. crawl space (where used for storage) ____________
    d. hallway ____________
    e. bathroom ____________
22. Is a receptacle required in a bedroom on a 3-foot (914-mm) wall space behind the door? The door is normally left open. ____________
23. The Code requires that at least one 20-ampere circuit feeding a receptacle outlet must be provided for the laundry. May this circuit supply other outlets? ____________
24. In a basement, at least one receptacle outlet must be installed in addition to the receptacle outlet installed for the laundry equipment. This additional receptacle outlet and any other receptacle outlets in unfinished areas must be ____________ protected.
25. The Code in *Section 210-8(a)(2)* requires that ALL receptacles installed in garages must have GFCI protection, but there are certain exceptions. What are these exceptions?

____

____

____

____

____

____

26. How would you define a branch circuit? ____________

____

27. Other than specific circuits, the ampacity of the conductor must generally be (not less than) (not more than) the rating of the overcurrent device. Circle correct answer.
28. The rating of a branch circuit is based upon
    a. the rating of the overcurrent device
    b. the length of the circuit
    c. the branch-circuit wire size
29. a. A 25-ampere branch circuit conductor is derated to 70%. It is important to provide proper overcurrent protection for these conductors. The derated conductor ampacity is ____________

    ____

    b. If the connected load is a "fixed" nonmotor, noncontinuous load, the branch-circuit overcurrent device may be sized at (20) (25) amperes. Circle correct answer.

c. If the above circuits supply receptacle outlets, the branch-circuit overcurrent device must be sized at not over (15) (20) (25) amperes. Circle correct answer.

30. Small appliance receptacle outlets are (included) (not included) in the 3 volt-amperes per square foot calculations. Circle correct answer.

31. If a homeowner wishes to have a switched receptacle outlet for a swag lamp in a dining room, may this switched receptacle outlet be considered to be one of the receptacle outlets required for the 20-ampere small appliance circuits? ________

32. A wall space between two doors measures 13 feet. How many receptacle outlets are required by the Code according to *Section 210-52*? Draw the outlets in on the diagram.

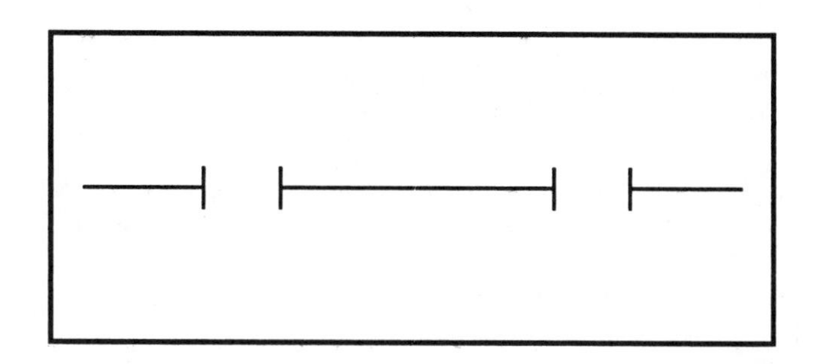

33. Is a receptacle required in a hallway in a home when the hallway is 8 feet long?

____

34. Is a receptacle required in a hallway in a home when the hallway is 13 feet long?

____

► 35. A sliding glass door is installed in a family room that leads to an outdoor deck. The sliding door has one 4-foot sliding section and one 4-foot permanently mounted glass section. Inside the recreation room, what does the Code say about wall receptacle outlets relative to the receptacle's position near the sliding glass door? ◄

____

____

____

____

____

____

36. The Code permits certain other receptacles to be connected to a 20-ampere small appliance circuit. What are they? ____

____

► 37. What is the minimum number of 20-ampere small appliance circuits required by the Code to feed the receptacle outlets above the countertops in a kitchen? In what section of the Code is this requirement found? ◄ ________________

UNIT 4

Conductor Sizes and Types, Wiring Methods, Wire Connections, Voltage Drop, Neutral Sizing for Services

OBJECTIVES

After studying this unit, the student will be able to

- define the terms used to size and rate conductors.
- discuss the subject of aluminum conductors.
- describe the types of cables used in most dwelling unit installations.
- list the installation requirements for each type of cable.
- describe the uses and installation requirements for electrical conduit systems.
- describe the requirements for service grounding conductors.
- describe the use and installation requirements for flexible metal conduit.
- make voltage drop calculations.
- understand the fundamentals of the markings found on terminals of wiring devices and wire connectors.
- understand why the ampacity of high-temperature insulated conductors may not necessarily be suitable at that same rating on a particular terminal on a wiring device, breaker, or switch.
- use a reduced neutral on residential services if certain criteria are met.

CONDUCTORS

Throughout this text, all references to conductors are for copper conductors, unless otherwise stated.

Wire Size

The copper wire used in electrical installations is graded for size according to the American Wire Gauge Standard (AWG). The wire diameter in the AWG standard is expressed as a whole number. AWG sizes vary from fine, hairlike wire used in coils and small transformers to very large diameter wire required in industrial wiring to handle heavy loads.

The wire may be a single strand (solid conductor), or it may consist of many strands. Each strand of wire acts as a separate conducting unit. Conductors which are No. 8 AWG and larger generally are stranded. The wire size used for a circuit depends upon the maximum current to be carried. The *National Electrical Code*® states that the minimum conductor size permitted in house wiring is No. 14 AWG. Exceptions to this rule are covered in *Section 210-19(c)* and *Article 725* for the wires used in lighting fixtures, bell wiring, and remote-control low-energy circuits. See conductor application chart that follows.

Ampacity

Ampacity means "the current in amperes a conductor can carry continuously under the condition of use without exceeding its temperature rating." This value depends on the area of the wire. Since the area of a circle is proportional to the square of the diameter, the ampacity of a round wire varies with the diameter squared. The diameter of a wire is given by a unit called a mil. A *mil* is defined as one-thousandth of an inch (0.001 inch). Mils squared are known as circular mils. The circular mil area (CMA) of a wire determines the current-carrying capacity of the wire. In other words, the larger the circular mil area of a wire, the greater is its current-carrying capacity. AWG sizes are also expressed in mils and range from No. 40 (10 circular mils) to No. 4/0 (211 600 circular mils). Wire sizes larger than No. 4/0 are expressed in circular mils only. ► Since the letter "k" designates 1000, large conductors, such as 500,000 circular mils, can also be expressed as 500 kcmils. However, many texts and electricians will continue to use and refer to MCM, in which "M" is a Roman Numeral representing 1000. Thus, 500 MCM also means 500,000 circular mils. For a further discussion on conductor ampacities, correction factors, and derating factors, see unit 15. ◄

CONDUCTOR APPLICATIONS CHART

CONDUCTOR SIZE	APPLICATIONS
No. 16 and No. 18 AWG	cords, low-voltage control circuits, bell and chime wiring
No. 14 and No. 12 AWG	normal lighting circuits, and circuits supplying receptacle outlets
No. 10, No. 8, No. 6, and No. 4 AWG	clothes dryers, ovens, ranges, cooktops, water heaters, heat pumps, central air conditioners, furnaces, feeders to sub-panels
No. 3, No. 2, No. 1 (and larger) AWG	main service entrances, feeders to sub-panels

Conductors must have an ampacity not less than the maximum load that they are supplying, figure 4-1. All conductors of a specific branch circuit must have an ampacity of the branch circuit's rating, figure 4-2. There are some exceptions to this rule, such as taps for electric ranges (unit 20).

Circuit Rating

The rating of the overcurrent device (OCD) determines the rating of a circuit, figure 4-3.

Aluminum Conductors

The conductivity of aluminum wire is not as great as that of copper wire for a given size. For example, checking *Table 310-16*, it is found that a No. 8 THHN copper wire has an ampacity of 55 amperes, whereas a No. 8 THHN aluminum or copper-clad aluminum wire has an ampacity of only 45 amperes. These ampacities are further reduced if the conductors are installed in high-ambient-temperature locations. (See ampacity correction

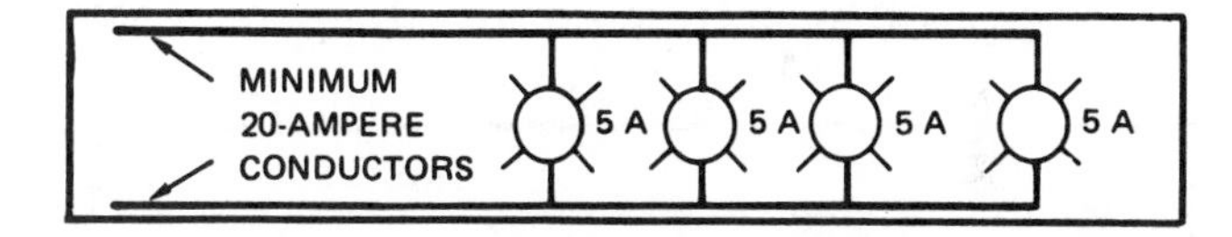

Fig. 4-1 Branch-circuit conductors shall have an ampacity not less than the maximum load to be served, *Section 210-19(a)*. See *Section 210-22* for maximum loads. See *Section 210-23* for permissible loads.

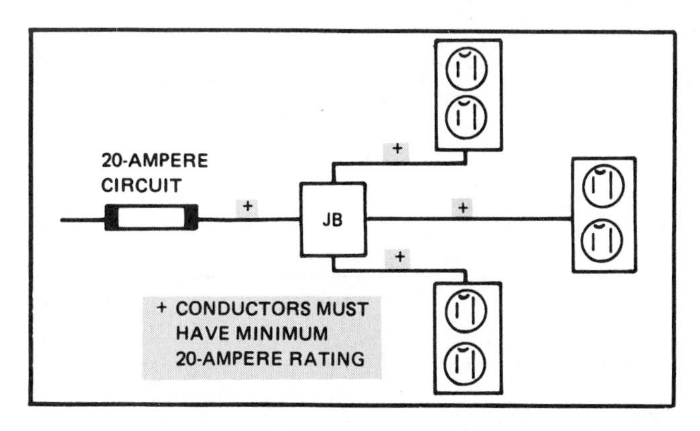

Fig. 4-2 All conductors in this circuit supplying readily accessible receptacles shall have an ampacity of not less than the rating of the branch circuit. In this example, 20-ampere conductors must be used, *Section 210-19(a)*.

factors below the table.) Ampacities are also reduced when there are more than three current-carrying conductors in a raceway or cable. See *Note 8* to *Table 310-16*.

Another example is found in the *footnotes* to *Table 310-16* where the maximum overcurrent protection for No. 12 AWG copper wire is 20 amperes, but only 15 amperes for No. 12 AWG aluminum or copper-clad aluminum wire.

Resistance is an important consideration when installing aluminum conductors. An aluminum conductor has a higher resistance compared to a copper conductor for a given wire size which, therefore, causes a greater voltage drop.

$$\text{Voltage Drop}(E_d) = \text{Amperes}(I) \times \text{Resistance}(R)$$

Common Connection Problems

Some common problems associated with aluminum conductors when not properly connected may be summarized as follows:

- a corrosive action is set up when dissimilar wires come in contact with one another when moisture is present.
- the surface of aluminum oxidizes as soon as it is exposed to air. If this oxidized surface is not broken through, a poor connection results. When installing aluminum conductors, particularly in large sizes, an inhibitor is brushed onto the aluminum conductor, then the conductor is scraped with a stiff brush where the connection is to be made. The process of scraping the conductor breaks through the oxidation, and the inhibitor keeps the air from coming into contact with the conductor. Thus, further oxidation is prevented. Aluminum connectors of the compression type usually have an inhibitor paste already factory installed inside of the connector.
- aluminum wire expands and contracts to a greater degree than does copper wire for an equal load. This factor is another possible cause of a poor connection. Crimp connectors for aluminum conductors are usually longer than those for comparable copper conductors, thus resulting in greater contact surface of the conductor in the connector.

Proper Installation Procedures

Proper, trouble-free connections for aluminum conductors require terminals, lugs, and/or connectors which are suitable for the type of conductor being installed.

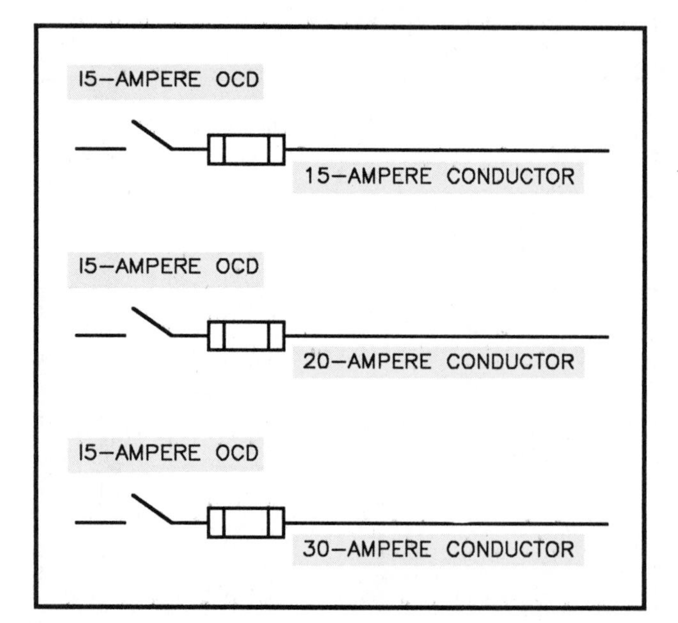

Fig. 4-3 All three of the above circuits are classified as 15-ampere circuits, even though larger conductors were used for some other reason, such as solving a voltage drop problem. The rating of the overcurrent device (OCD) determines the rating of a circuit, *Section 210-3*.

TYPE OF DEVICE	MARKING ON TERMINAL OR CONNECTOR	CONDUCTOR PERMITTED
15- or 20-ampere receptacles and switches	CO/ALR	aluminum, copper, copper-clad aluminum
15- and 20- ampere receptacles and switches	NONE	copper, copper-clad aluminum
30-ampere and greater receptacles and switches	AL/CU	aluminum, copper, copper-clad aluminum
30-ampere and greater receptacles and switches	NONE	copper only
Screwless pressure terminal connectors of the push-in type	NONE	copper or copper-clad aluminum
Wire connectors	AL/CU	aluminum, copper, copper-clad aluminum
Wire connectors	NONE	copper only
Wire connectors	AL	aluminum only
Any of the above devices	COPPER OR CU ONLY	copper only

Table 4-1.

CRIMP CONNECTORS USED TO SPLICE AND TERMINATE NO. 20 AWG TO 500 KCMILS ALUMINUM-TO-ALUMINUM, ALUMINUM-TO-COPPER, OR COPPER-TO-COPPER.	PROPERLY CRIMP, THEN TAPE
CONNECTORS USED TO CONNECT WIRES TOGETHER IN COMBINATIONS OF NO. 18 AWG THROUGH NO. 6 AWG. THEY ARE TWIST-ON, SOLDERLESS, AND TAPELESS.	WIRE CONNECTORS VARIOUSLY KNOWN AS WING NUT, WIRE NUT, AND SCOTCHLOK.
CONNECTORS USED TO CONNECT WIRES TOGETHER IN COMBINATIONS OF NO. 16, NO. 14, AND NO. 12 AWG. THEY ARE CRIMPED ON WITH A SPECIAL TOOL, THEN COVERED WITH A SNAP-ON INSULATING CAP OR WRAPPED WITH INSULATING TAPE.	CRIMP-TYPE WIRE CONNECTOR AND INSULATING CAP.
SOLDERLESS CONNECTORS ARE AVAILABLE IN SIZES NO. 14 AWG THROUGH 500 MCM. THEY ARE USED FOR ONE SOLID OR ONE STRANDED CONDUCTOR ONLY, UNLESS OTHERWISE NOTED ON THE CONNECTOR OR ON ITS SHIPPING CARTON. THE SCREW MAY BE OF THE STANDARD SCREWDRIVER SLOT TYPE, OR IT MAY BE FOR USE WITH AN ALLEN WRENCH.	SOLDERLESS CONNECTORS
COMPRESSION CONNECTORS ARE USED FOR NO. 8 AWG THROUGH 1000 MCM. THE WIRE IS INSERTED INTO THE END OF THE CONNECTOR, THEN CRIMPED ON WITH A SPECIAL COMPRESSION TOOL.	COMPRESSION CONNECTOR
SPLIT-BOLT CONNECTORS ARE USED FOR CONNECTING TWO CONDUCTORS TOGETHER, OR FOR TAPPING ONE CONDUCTOR TO ANOTHER. THEY ARE AVAILABLE IN SIZES NO. 10 AWG THROUGH 1000 MCM. THEY ARE USED FOR TWO SOLID AND/OR TWO STRANDED CONDUCTORS ONLY, UNLESS OTHERWISE NOTED ON THE CONNECTOR OR ON ITS SHIPPING CARTON.	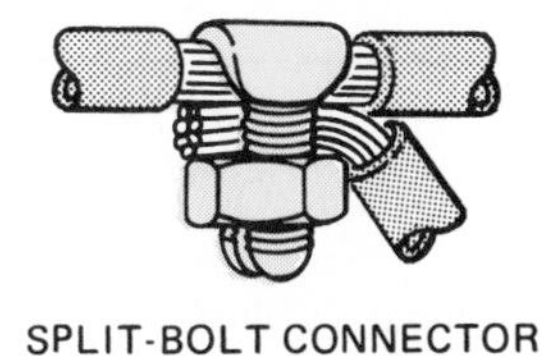 SPLIT-BOLT CONNECTOR

Fig. 4-4 Types of wire connectors.

Terminals on receptacles and switches must be suitable for the conductors being attached. Table 4-1 shows how the electrician can identify these terminals. Listed connectors provide proper connection when properly installed.

See also *Sections 110-14, 380-14(c),* and *410-56(b).*

Wire Connections

When splicing wires or connecting a wire to a switch, fixture, circuit breaker, panelboard, meter socket, or other electrical equipment, the wires may be twisted together; soldered; then taped. Usually, however, some type of wire connector is required.

Wire connectors are known in the trade by such

names as *screw terminal*, *pressure terminal connector*, *wire connector*, *wing nut*, *wire nut*, *Scotchlok*, *split-bolt connector*, *pressure cable connector*, *solderless lug*, *soldering lug*, *solder lug*, and others. Solder-type lugs are not often used today. In fact, connections that depend upon solder are *not* permitted for connecting service-entrance conductors to service equipment. The labor costs and time spent make the use of solder-type connections prohibitive.

Solderless connectors, designed to establish connections by means of mechanical pressure, are quite common. Examples of some types of wire connectors, and their uses, are shown in figure 4-4.

As with the terminals on wiring devices (switches and receptacles) wire connectors must be marked *AL* when they are to be used with aluminum conductors. This marking is found on the connector itself, or appears on or in the shipping carton.

Connectors marked *AL/CU* are suitable for use with aluminum, copper, or copper-clad aluminum conductors. This marking is found on the connector itself, or it appears on or in the shipping carton.

Connectors not marked *AL* or *AL/CU* are for use with copper conductors only.

Unless specially stated on or in the shipping carton, or on the connector itself, conductors made of copper, aluminum, or copper-clad aluminum may not be used in combination in the same connector. Combinations, when permitted, are usually limited to dry locations only.

The preceding data is found in the *National Electrical Code®* and in the UL Standards.

Insulation of Wires

The Code requires that all wires used in electrical installations be insulated, *Section 310-2*. Exceptions to this rule are clearly indicated in the Code.

The most common type of insulation used on wires is a thermoplastic material, although rubber compound conductor insulations are still recognized by the Code. *Table 310-13* lists the various conductor insulations and applications. The insulation completely surrounds the metal conductor, has a uniform thickness, and runs the entire length of the wire or cable.

The allowable ampacities of copper conductors are given in *Table 310-16* for various types of insulation. A deviation from the ampacity ratings given in *Table 310-16* is permitted for single-phase, three-wire residential services only. For conductors having RH, RHH, RHW, THW, THWN, THHN, and XHHW insulation, the ratings shown in Table 4-2 are permitted. See *Note 3* to *Table 310-16*.

The insulation covering wires and cables used in house wiring is usually rated at 600 volts or less. Exceptions to this statement are low-voltage bell wiring and fixture wiring.

SPECIAL AMPACITY RATINGS FOR SINGLE-PHASE, THREE-WIRE RESIDENTIAL SERVICES AND FEEDERS ONLY.
(Note 3, Table 310-16)

Copper Conductor AWG for Insulation of RH-RHH-RHW-THW-THWN-THHW-THHN-XHHW-USE	Aluminum or Copper-Clad Aluminum Conductors (AWG)	Service Entrance Ampacity Rating
4	2	100
3	1	110
2	1/0	125
1	2/0	150
1/0	3/0	175
2/0	4/0	200
3/0	250 kcmil	225
4/0	300 kcmil	250
250 kcmil	350 kcmil	300
350 kcmil	500 kcmil	350
400 kcmil	600 kcmil	400

Table 4-2.

Temperature Considerations

Conductors are also rated as to the temperature they can withstand. For example:

	Celsius	Fahrenheit
Type TW	60°	140°
Type THW	75°	167°
Type THHN	90°	194°

Without question, conductors with Type THHN insulation are the most popular and most commonly used, particularly in the smaller sizes, because of their small diameter (easy to handle; more conductors of a given size permitted in a given size raceway) and their ability to be installed where high temperatures are encountered, such as attics, buried in insulation, and supplying recessed fixtures.

The temperature rating of the conductors found in nonmetallic-sheathed cable, most commonly used for house wiring, is 90°C. However, the permitted ampacity of the conductors in nonmetallic, sheathed cable is that of the 60°C column of *Table 310-16*. See *Section 336-26*.

Conductor Temperature Ratings

Table 310-16 shows the many types of insulations available on conductors for building wire of the types used in most electrical installations. It is this table that electricians refer to most often when selecting wire sizes for specific loads. These are the types of conductors found in nonmetallic-sheathed cable (Romex), armored cable (BX), and the type commonly installed in conduit.

Note that in the common building wire categories, we find 60°C, 75°C, and 90°C temperature ratings.

UL standards state that, unless otherwise noted, terminals and switches, breakers, motor controllers, and so forth, are based upon the use of 60°C conductors for wire sizes No. 14 through No. 1 AWG, and upon the use of 75°C conductors for wire sizes No. 1/0 and larger.

UL standards also state in general that for switches, breakers, contactors, controllers, and so on, rated 125 amperes or less are for use with 60°C conductors with ampacities based upon the 60°C column of *Table 310-16*, and when rated over 125 amperes they are for use with 75°C conductors with ampacities based upon the 75°C column of *Table 310-16*.

Therefore, we cannot arbitrarily check *Table 310-16* to find the ampacity of a given conductor merely by selecting a conductor that has a high temperature rating, such as 90°C column. We can use the high temperature ratings when high temperatures are encountered, such as in attics, or when we must derate because of the number of conductors in a conduit, but we must always check the temperature ratings of terminals to make sure that we are not creating a "hot spot."

Terminals and/or the label in the equipment might be marked "75°C only" or "60°/75°C."

It is recommended that the reader study *NEC® Article 310, Conductors for General Wiring*.

Neutral Size

► The grounded neutral conductor Ⓝ is permitted to be *two* AWG *sizes smaller* than the "hot" ungrounded phase conductors Ⓓ in figure 4-5 (*Note 3, Table 310-16*) *only* if it can be proved that ALL three service-entrance conductors are properly and adequately sized to carry the loads computed by the Code rule as stated in *Sections 215-2, 220-22*, and *230-42*. ◄

Section 215-2 relates to feeders and refers us to *Article 220* for computation requirements.

Note that *Section 230-42* makes direct reference to *Article 220*, where branch-circuit and feeder calculation requirements are found.

Focusing on the neutral conductor, *Section 220-22* states that "the feeder neutral load shall be the maximum unbalance of the load determined by this Article."

Section 220-22 further states that "the maximum unbalanced load shall be the maximum net computed load between the neutral and any one ungrounded conductor. . . ."

In this residence we find a number of loads that carry little or no neutral currents, such as the electric water heater, electric clothes dryer, electric oven and range, electric furnace, and air conditioner. These appliances for the most part are drawing their biggest current through the "hot" conductors because they are fed with 240-volt Ⓒ or 120/240-volt circuits Ⓔ. Loads Ⓑ are connected line to neutral only.

Thus we find logic in the Code, *Note 3, Table 310-16*, which permits reducing the neutral size on services, feeders, and branch circuits where the computations prove that the neutral will be carrying less current than the "hot" phase conductors.

VOLTAGE DROP

Low voltage can cause lights to dim, television pictures to "shrink," motors to run hot, electric heaters to not produce their rated heat output, and appliances to not operate properly.

Low voltage in a home can be caused by:

- wire that is too small for the load being served,
- a circuit that is too long,
- poor connections at the terminals,
- conductors operating at high temperatures having higher resistance than when operating at lower temperatures.

A simple formula for calculating voltage drop on single-phase systems considers only the dc resistance of the conductors and the temperature of the conductor. The more accurate formulas consider ac resistance, reactance, temperature, spacing, in metal conduit, and in nonmetallic conduit, *NEC® Table 9, Chapter 9.* Voltage drop is covered in great detail in *Electrical Wiring — Commercial.* The simple voltage drop formula is more accurate with smaller conductors, and gets less and less accurate as conductor size increases. It is sufficiently accurate for voltage drop calculations necessary for residential wiring.

To find voltage drop:

$$E_d = \frac{K \times I \times L \times 2}{CSA}$$

where: E_d = allowable voltage drop in volts.

K = resistance in ohms per circular-mil foot at approximately 75°C.
- For copper, approximately 12 ohms.
- For aluminum, approximately 19 ohms.

I = the current in amperes flowing through the conductors.

L = the length from the beginning of the circuit to the load.

CSA = cross-sectional area of the conductor in circular mils. Refer to Table 4-3.

To find conductor size:

$$CSA = \frac{K \times I \times L \times 2}{E_d}$$

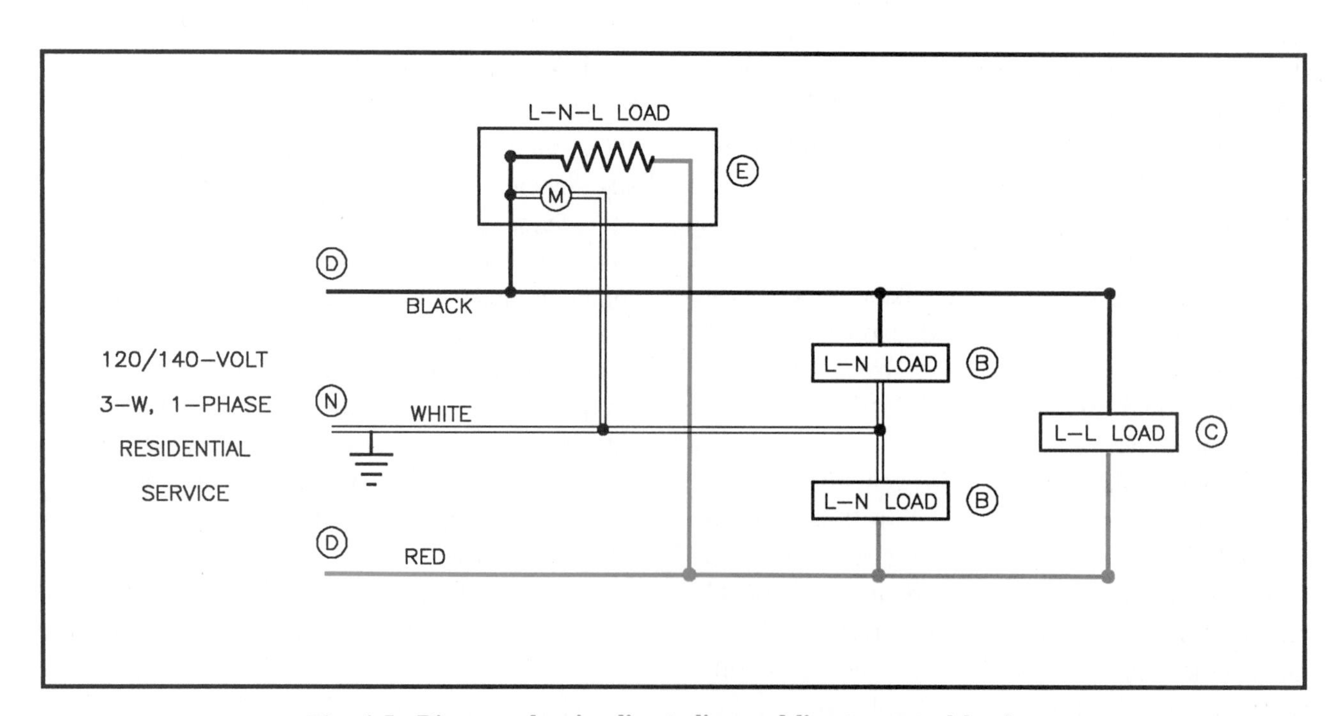

Fig. 4-5 Diagram showing line-to-line and line-to-neutral loads.

CONDUCTOR SIZE	CROSS-SECTIONAL AREA IN CIRCULAR MILS
18	1,620
16	2,580
14	4,110
12	6,530
10	10,380
8	16,510
6	26,240
4	41,740
3	52,620
2	66,360
1	83,690
0	105,600
00	133,100
000	167,800
0000	211,600

Table 4-3.

Code Reference to Voltage Drop

The Code states that the voltage drop on a branch circuit should not exceed 3%. Refer to figure 4-6.

When the voltage drop issue involves both a feeder and a branch circuit, figure 4-7 explains the Code requirements.

Caution Using High-Temperature Wire

When trying to use smaller size conduits, be careful about selecting conductors on the ability of their insulation to withstand high temperatures. A quick check of *Table 310-16* of the Code will confirm that the ampacity of high-temperature conductors is greater than the ampacity of lower temperature conductors. For instance:

- No. 8 THHN copper has an ampacity of 55 amperes.
- No. 8 TW copper has an ampacity of 40 amperes.

Therefore, always seriously consider doing a voltage-drop calculation, particularly when the decision to use the higher-temperature-rated conductors was based on the fact that the use of smaller size conduit would be possible. Select the proper size conductor to use for the job according to the load requirements, then run a voltage-drop calculation to see that the maximum permitted voltage drop is not exceeded.

EXAMPLE 1: What is the approximate voltage drop on a 120-volt, single-phase circuit consisting of No. 14 AWG copper conductors where the load is 11 amperes and the distance of the circuit from the panel to actual load is 85 feet?

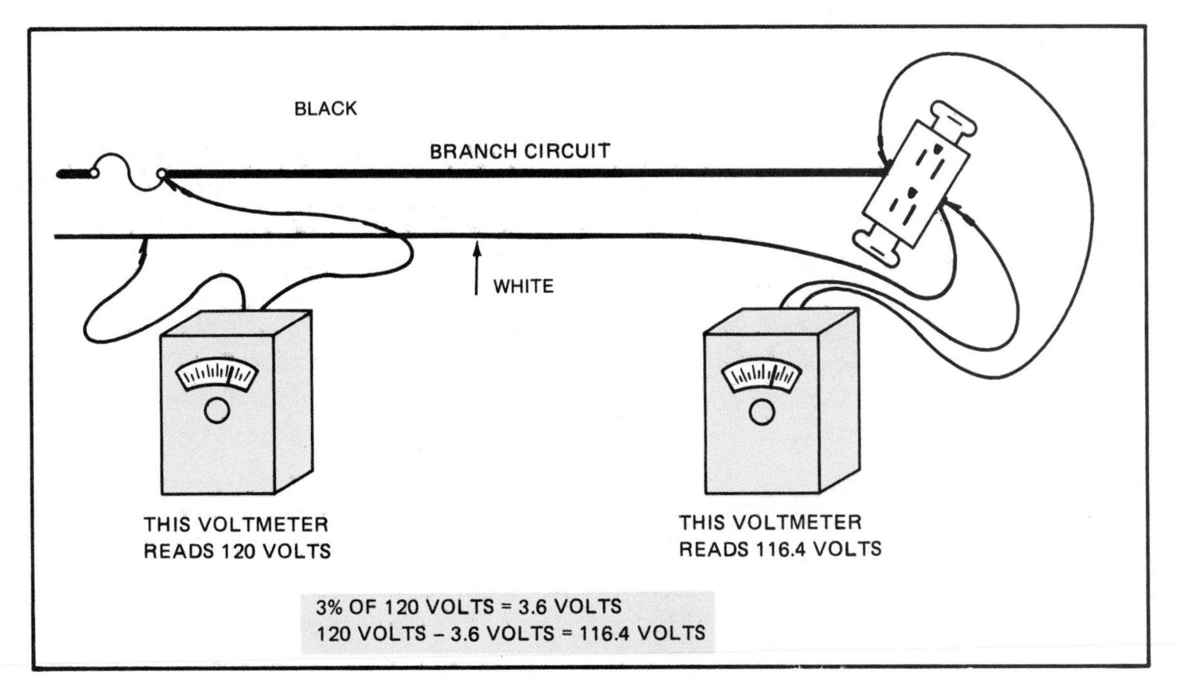

Fig. 4-6 Maximum allowable voltage drop on a branch circuit is 3%, *Section 210-19, Fine Print Notes (FPN)*.

SOLUTION:

$$E_d = \frac{K \times I \times L \times 2}{CSA}$$

$$E_d = \frac{12 \times 11 \times 85 \times 2}{4110}$$

$$E_d = 5.46 \text{ volts drop}$$

Note: Refer to Table 4-3 for the CSA value.

This exceeds the voltage drop permitted by the Code, which is:

$$3\% \text{ of } 120 \text{ volts} = 3.6 \text{ volts}$$

Let's try it again using No. 12 AWG:

$$E_d = \frac{12 \times 11 \times 85 \times 2}{6530}$$

$$E_d = 3.44 \text{ volts drop}$$

This "Meets Code."

EXAMPLE 2: Find the wire size needed to keep the voltage drop to no more than 3% on a single-phase, 240-volt circuit that feeds a 44-ampere air conditioner. The circuit originates at the main panel which is located approximately 65 feet from the air conditioner. No neutral is required for this equipment.

SOLUTION: *First*, size the conductors at 125% of the air-conditioner load.

$44 \times 1.25 = 55$-ampere conductors are required.

Checking *Table 310-16*, the conductors could be No. 6 Type TW copper or No. 8 THHN copper.

The permitted voltage drop is:

$$E_d = 240 \times 0.03$$

$$= 7.2 \text{ volts}$$

$$CSA = \frac{K \times I \times L \times 2}{E_d}$$

$$= \frac{12 \times 44 \times 65 \times 2}{7.2}$$

$$= 9533 \text{ circular mils minimum}$$

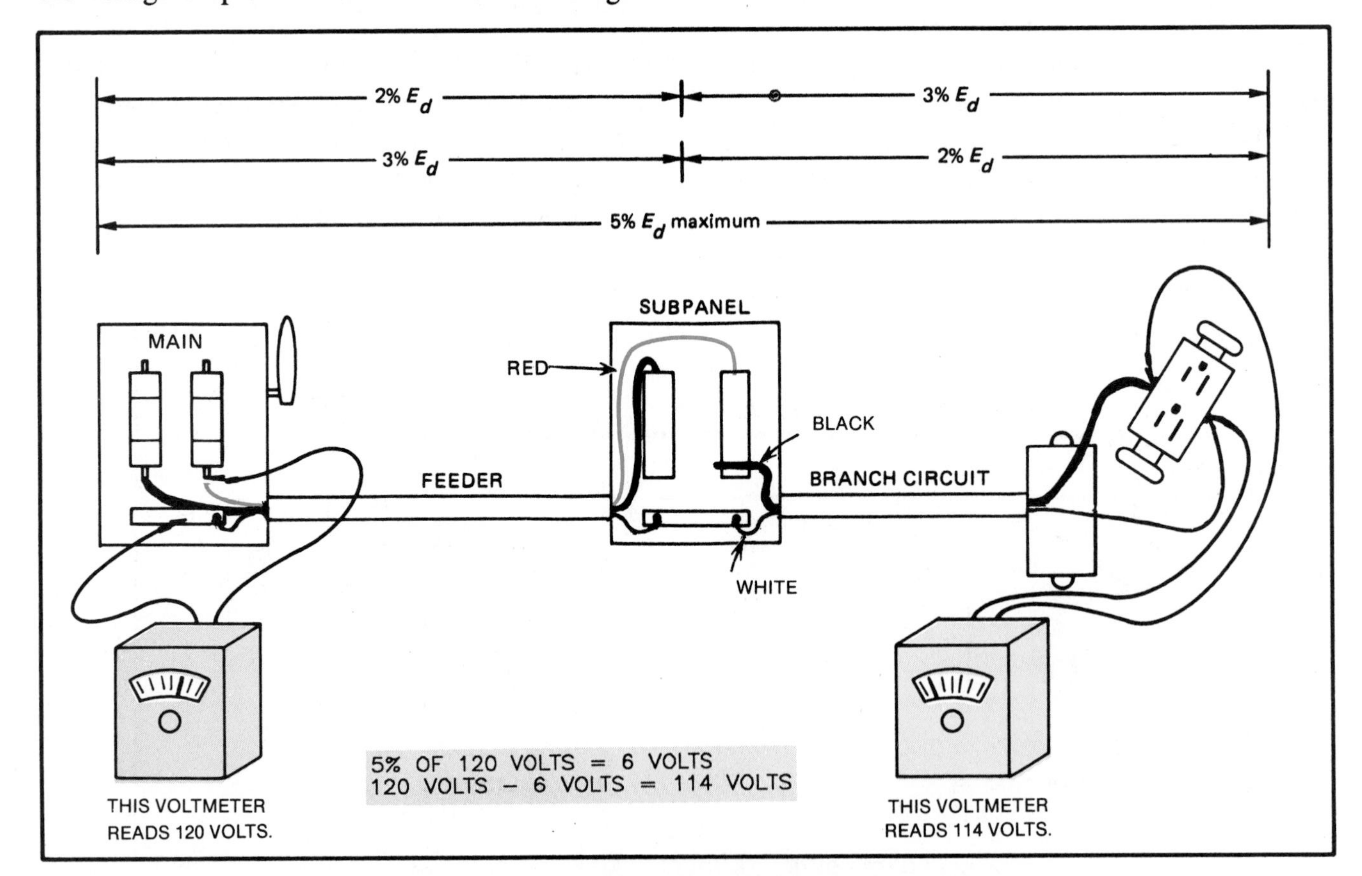

Fig. 4-7 *Sections 210-19 FPN No. 4* and *215-2 FPN No. 2* of the Code state that the total voltage drop from the beginning of a feeder to the farthest outlet on a branch circuit should not exceed 5%. In this figure, if the voltage drop in the feeder is 3%, then do not exceed 2% voltage drop in the branch circuit. If the voltage drop in the feeder is 2%, then do not exceed 3% voltage drop in the branch circuit.

According to Table 4-3 this would be a No. 10 AWG conductor, but this size conductor would be too small for the load. So, we are back to No. 6 TW or No. 8 THHN conductors.

Let's see what the voltage drop is when we install No. 8 THHN:

$$E_d = \frac{12 \times 44 \times 65 \times 2}{16500}$$

$$= 4.16 \text{ volts}$$

This is well below the permissible 7.2 volts drop, so the No. 8 THHN would be a good choice. There would also be a savings in conduit size since two No. 8 THHN fit in a 1/2″ conduit, whereas two No. 6 TW require a 3/4″ conduit. See *Table 3A*, Chapter 9 of the Code.

APPROXIMATE CONDUCTOR SIZE RELATIONSHIP

Rule One: For wire sizes up through 0000, every third size doubles or halves in circular mil area.

Thus, a No. 1 AWG conductor is 2X larger than a No. 4 AWG conductor (83 690 versus 41 470). Thus, a "0" wire is one-half the size of "0000" wire (105 600 versus 211 600).

Rule Two: For wire sizes up through 0000, every consecutive wire size is approximately 1.26 larger or smaller than the preceding wire size.

Thus, a No. 3 AWG conductor is approximately 1.26 larger than a No. 4 AWG conductor (41 740 × 1.26 = 52 592). Thus, a No. 2 AWG wire is approximately 1.26 smaller than a No. 1 AWG wire (66 360 ÷ 1.26 = 52 667).

Try to fix in your mind that a No. 10 AWG conductor has a cross-sectional area of 10 380 CM, and that it has a resistance of 1.2 ohms per 1000 feet. The resistance of aluminum wire is approximately 2 ohms per 1000 feet. By remembering these numbers, you will be able to perform voltage drop calculations without having the wire tables readily available.

EXAMPLE: What is the approximate CMA and resistance of a No. 6 AWG copper conductor?

SOLUTION:

WIRE SIZE	CMA	OHMS PER 1000 FEET
10	10 380	1.2
9		
8		
7	20 760	0.6
6	26 158	0.0476

Note in this chart that when the CMA of a wire is doubled, then its resistance is cut in half. Inversely, when a given wire size is reduced to one-half, its resistance doubles.

NONMETALLIC-SHEATHED CABLE (*ARTICLE 336*)

Description

Nonmetallic-sheathed cable is defined as a factory assembly of two or more insulated conductors having an outer sheath of moisture-resistant, flame-retardant, nonmetallic material. This cable is available with two or three current-carrying conductors. The conductors range in size from No. 14 through No. 2 for copper conductors, and from No. 12 through No. 2 for aluminum or copper-clad aluminum conductors. Two-wire cable contains one black conductor, one white conductor, and one bare grounding conductor, or green insulated conductor. Three-wire cable contains one black, one white, one red, and one bare grounding conductor, or green insulated conductor.

Underwriters Laboratories, Inc. lists two classifications of nonmetallic-sheathed cable:

- Type NM has a flame-retardant, moisture-resistant, nonmetallic covering over the conductors.

Fig. 4-8 Nonmetallic-sheathed cable with an uninsulated copper conductor. Note: ① "grounded" conductor; ② bare "wrapped" equipment "grounding" conductor; ③ black "ungrounded" (hot) conductor.

- Type NMC has a flame-retardant, moisture-resistant, fungus-resistant, and corrosion-resistant nonmetallic covering over the conductors.

Most electricians refer to nonmetallic-sheathed cable as *Romex*.

The Code, in *Section 336-26*, requires that the conductors found in types NM and NMC cable must be rated at 90 °C (194 °F). The cable would have the suffix "B" on the jacket, i.e., NMC-B. This ruling was put into effect because of the extreme high temperatures found in attics and when cables are "buried" in insulation and/or near recessed lighting fixtures.

Even though the conductor insulation is rated at 90 °C, the actual ampacity of the conductors must be based on the ampacity of 60 °C conductors. Refer to *Section 336-26.*

Nonmetallic-sheathed cable is also available with an uninsulated copper conductor which is used for grounding purposes only, figure 4-8. This grounding conductor is not intended for use as a current-carrying circuit wire.

Equipment grounding requirements are specified in *Sections 250-42*, *250-43*, *250-44*, and *250-45*. It is for the reasons stated in these sections that all boxes and fixtures in the residence are to be grounded.

Table 250-95 of the Code lists the sizes of the grounding conductors used in cable assemblies. Note that the copper grounding conductor shown in the portion of *Table 250-95* is the same size as the circuit conductors for 15-, 20-, and 30-ampere ratings.

Table 4-4 shows the uses permitted for type NM and type NMC cable.

Installation

Nonmetallic-sheathed cable is the least expensive of the various wiring methods. It is relatively light in weight and easy to install. It is widely used for dwelling unit installations on circuits of 600 volts or less. The installation of both types of nonmetallic-sheathed cable must conform to the requirements of *NEC Article 336*. Refer to figures 4-10, 4-11, 4-11A, 4-12, and 4-13 (pages 62–63).

- The cable must be strapped or stapled not more than 12 inches (305 mm) from a box or fitting.

Table 310-16. Ampacities of Insulated Conductors Rated 0-2000 Volts, 60° to 90°C (140° to 194°F) Not More Than Three Conductors in Raceway or Cable or Earth (Directly Buried), Based on Ambient Temperature of 30°C (86°F)

Size	Temperature Rating of Conductor. See Table 310-13.								Size
	60°C (140°F)	75°C (167°F)	85°C (185°F)	90°C (194°F)	60°C (140°F)	75°C (167°F)	85°C (185°F)	90°C (194°F)	
AWG kcmil	TYPES †TW, †UF	TYPES †FEPW, †RH, †RHW, †THHW, †THW, †THWN, †XHHW †USE, †ZW	TYPE V	TYPES TA, TBS, SA SIS, †FEP, †FEPB, †RHH, †THHN, †THHW, †XHHW	TYPES †TW, †UF	TYPES †RH, †RHW, †THHW, †THW, †THWN, †XHHW †USE	TYPE V	TYPES TA, TBS, SA, SIS, †RHH, †THHW, †THHN, †XHHW	AWG kcmil
	COPPER				ALUMINUM OR COPPER-CLAD ALUMINUM				
18	....	....	....	14	....	....	....	....	....
16	....	....	18	18	....	....	....	....	....
14	20†	20†	25	25†	....	....	....	....	....
12	25†	25†	30	30†	20†	20†	25	25†	12
10	30	35†	40	40†	25	30†	30	35†	10
8	40	50	55	55	30	40	40	45	8
6	55	65	70	75	40	50	55	60	6
4	70	85	95	95	55	65	75	75	4
3	85	100	110	110	65	75	85	85	3
2	95	115	125	130	75	90	100	100	2
1	110	130	145	150	85	100	110	115	1
1/0	125	150	165	170	100	120	130	135	1/0
2/0	145	175	190	195	115	135	145	150	2/0
3/0	165	200	215	225	130	155	170	175	3/0
4/0	195	230	250	260	150	180	195	205	4/0
250	215	255	275	290	170	205	220	230	250
300	240	285	310	320	190	230	250	255	300
350	260	310	340	350	210	250	270	280	350
400	280	335	365	380	225	270	295	305	400
500	320	380	415	430	260	310	335	350	500
600	355	420	460	475	285	340	370	385	600
700	385	460	500	520	310	375	405	420	700
750	400	475	515	535	320	385	420	435	750
800	410	490	535	555	330	395	430	450	800
900	435	520	565	585	355	425	465	480	900
1000	455	545	590	615	375	445	485	500	1000
1250	495	590	640	665	405	485	525	545	1250
1500	520	625	680	705	435	520	565	585	1500
1750	545	650	705	735	455	545	595	615	1750
2000	560	665	725	750	470	560	610	630	2000

AMPACITY CORRECTION FACTORS

Ambient Temp. °C	For ambient temperatures other than 30°C (86°F), multiply the ampacities shown above by the appropriate factor shown below.								Ambient Temp. °F
21-25	1.08	1.05	1.04	1.04	1.08	1.05	1.04	1.04	70-77
26-30	1.00	1.00	1.00	1.00	1.00	1.00	1.00	1.00	79-86
31-35	.91	.94	.95	.96	.91	.94	.95	.96	88-95
36-40	.82	.88	.90	.91	.82	.88	.90	.91	97-104
41-45	.71	.82	.85	.87	.71	.82	.85	.87	106-113
46-50	.58	.75	.80	.82	.58	.75	.80	.82	115-122
51-55	.41	.67	.74	.76	.41	.67	.74	.76	124-131
56-60	....	.58	.67	.71	....	.58	.67	.71	133-140
61-70	....	.33	.52	.58	....	.33	.52	.58	142-158
71-80	....	....	.30	.41	....	....	.30	.41	160-176

† Unless otherwise specifically permitted elsewhere in this Code, the overcurrent protection for conductor types marked with an obelisk (†) shall not exceed 15 amperes for 14 AWG, 20 amperes for 12 AWG, and 30 amperes for 10 AWG copper; or 15 amperes for 12 AWG and 25 amperes for 10 AWG aluminum and copper-clad aluminum after any correction factors for ambient temperature and number of conductors have been applied.

Table 250-95. Minimum Size Equipment Grounding Conductors for Grounding Raceway and Equipment

Rating or Setting of Automatic Overcurrent Device in Circuit Ahead of Equipment, Conduit, etc., Not Exceeding (Amperes)	Size	
	Copper Wire No.	Aluminum or Copper-Clad Aluminum Wire No.*
15	14	12
20	12	10
30	10	8
40	10	8
60	10	8
100	8	6
200	6	4
300	4	2
400	3	1
500	2	1/0
600	1	2/0
800	1/0	3/0
1000	2/0	4/0
1200	3/0	250 kcmil
1600	4/0	350 "
2000	250 kcmil	400 "
2500	350 "	600 "
3000	400 "	600 "
4000	500 "	800 "
5000	700 "	1200 "
6000	800 "	1200 "

* See installation restrictions in Section 250-92(a).

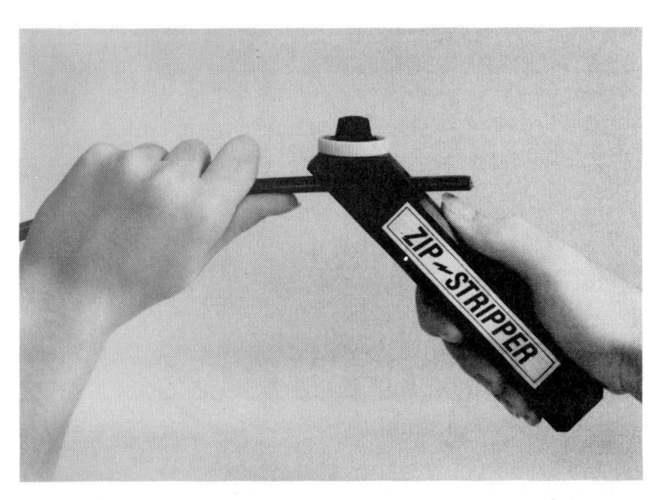

Fig. 4-9 This is an example of a stripper from one manufacturer. It is used to strip the outer jacket from nonmetallic-sheathed cable. Courtesy of Seatek Co., Inc.

- The intervals between straps or staples must not exceed 4 1/2 feet (1.37 m).
- The cable must be protected against physical damage where necessary.
- The cable must not be bent to a radius less than five times its diameter.
- The cable must not be used in circuits of more than 600 volts.
- Nonmetallic-sheathed cable must be protected where passing through a floor by at least 6 inches (152 mm) of rigid metal conduit, intermediate metal conduit, electrical metallic tubing, or other metal pipe, *Section 336-6(b)*. A fitting (bushing or connector) must be used at both ends of the conduit to protect the cable from abrasion, *Section 300-15(b), Exception 4.*
- When cables are "bundled" together for distances longer than 24 inches (610 mm), figure 4-12, the heat generated by the conductors cannot easily dissipate. These conductors *must be* derated according to *Note 8* to *Tables 310-16* through *310-19*. (See unit 15.)

  Figure 4-14 shows the installation of nonmetallic-sheathed cable in unfinished basements, *Section 336-12*.

Figures 4-14A and 4-14B (page 64) show the Code requirements for securing nonmetallic sheathed cable when using nonmetallic boxes.

Additional requirements for the installation of nonmetallic-sheathed cable are given in *Articles 200, 210, 220, 240, 250, 300*, and *310*.

The Code, in *Section 370-20(d)*, requires that all metal boxes for use with nonmetallic cable must have provisions for the attachment of the grounding conductor. Figure 4-15 (page 65) shows a gang-type switch (device) box that is tapped for a screw by which the grounding conductor may be connected underneath. Figure 4-16 (page 65) shows an outlet box, also with provisions for attaching grounding conductors. Figure 4-17 (page 65) illustrates the use of a small grounding clip.

Section 410-20 of the Code requires that there be a provision whereby the equipment grounding conductor can be attached to the exposed metal parts of lighting fixtures.

FOR TYPICAL RESIDENTIAL WIRING TYPE NM AND NMC CABLE	TYPE NM	TYPE NMC
• May be used on circuits of 600 volts or less	Yes	Yes
• Has flame-retardant and moisture-resistant outer covering	Yes	Yes
• Has fungus-resistant and corrosion-resistant outer covering	No	Yes
• May be used to wire one- and two-family dwellings, or multifamily dwellings that do not exceed three floors above grade	Yes	Yes
• May be installed exposed or concealed in damp location	No	Yes
• May be embedded in masonry, concrete, plaster, adobe, fill	No	No
• May be exposed to corrosive fumes	No	Yes
• May be installed in dry, hollow voids in masonry blocks and similar locations	Yes	Yes
• May be installed in moist, damp, hollow voids in masonry blocks and similar locations	No	Yes
• May be used as service-entrance cable	No	No
• Must be protected against damage	Yes	Yes
• May be run in shallow chase of masonry, concrete, or adobe if protected by a steel plate at least 1/16 inch thick, then covered with plaster, adobe, or similar finish	No	Yes

Table 4-4. Uses permitted in typical residential wiring for type NM and type NMC nonmetallic-sheathed cable.

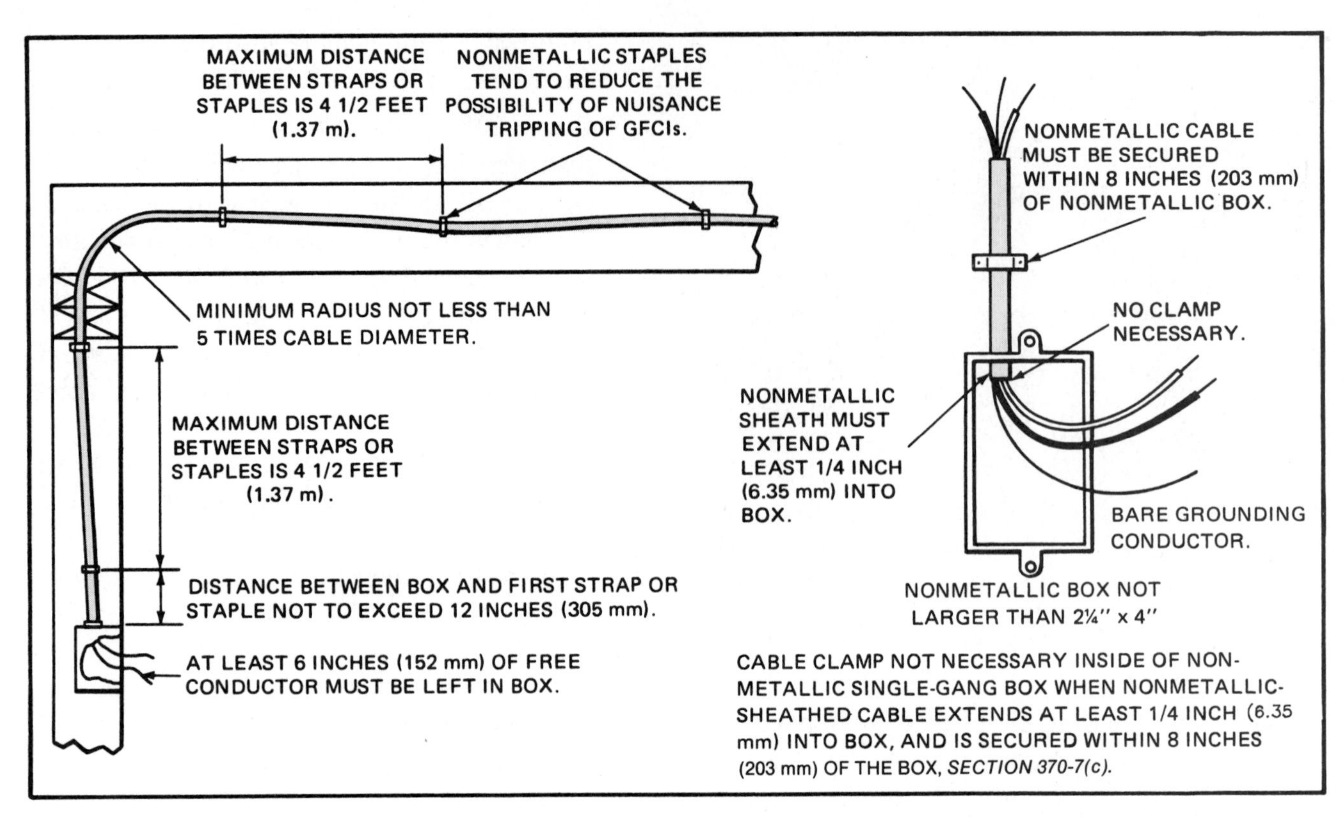

Fig. 4-10 Installation of nonmetallic-sheathed cable and armored cable.

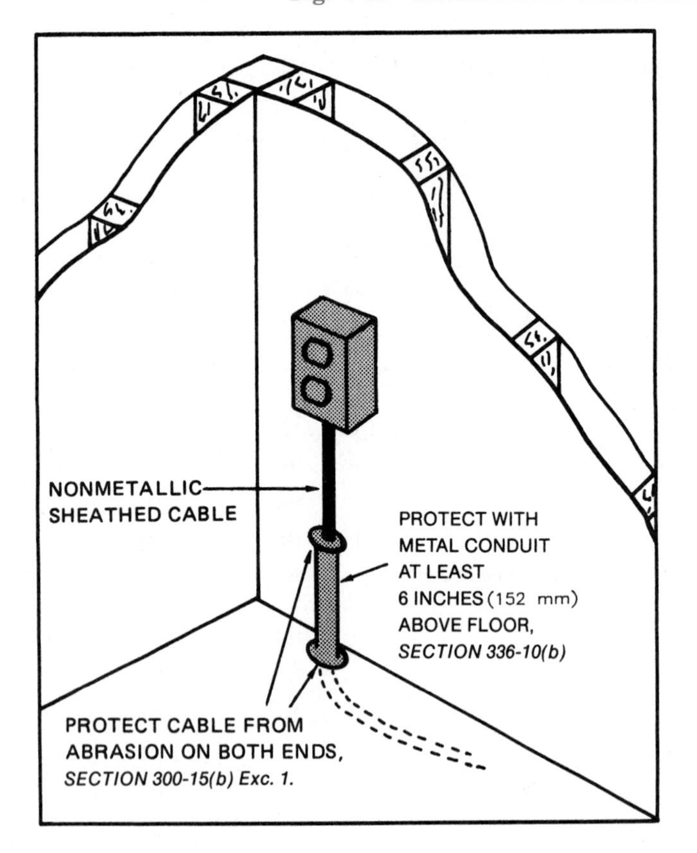

Fig. 4-11 Installation of nonmetallic-sheathed cable where passing through a floor.

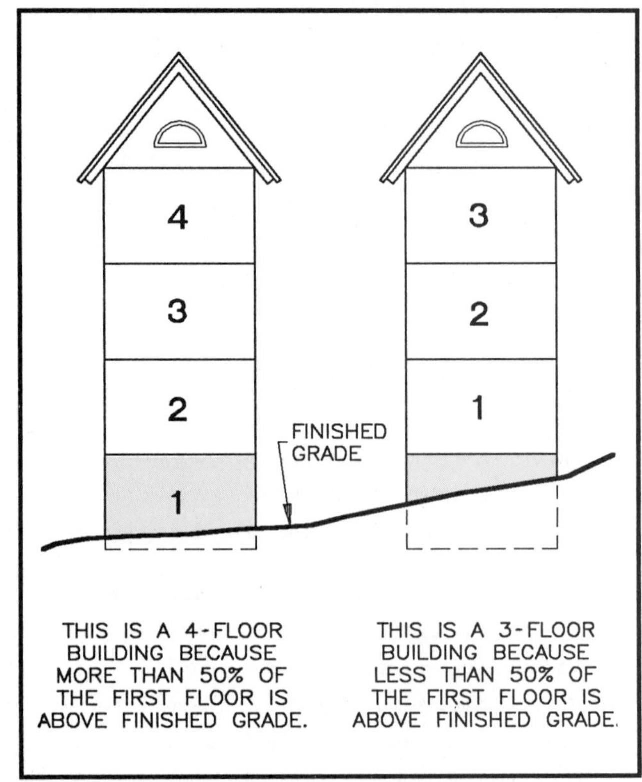

► **Fig. 4-11A Illustration of how *Section 336-4(a)* defines the first floor of a building.** ◄

Fig. 4-12 "Bundled" cables. When cables are bundled for distances over 24 inches, their ampacity must be derated. See Unit 18.

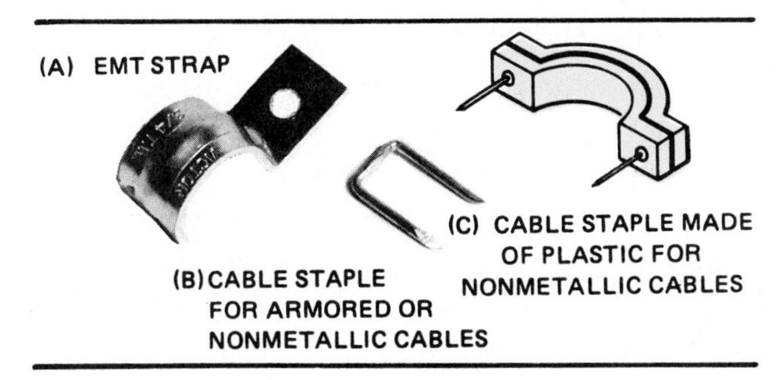

Fig. 4-13 Devices used for attaching conduit and cable to masonry surfaces (A) and wood surfaces (B) or (C).

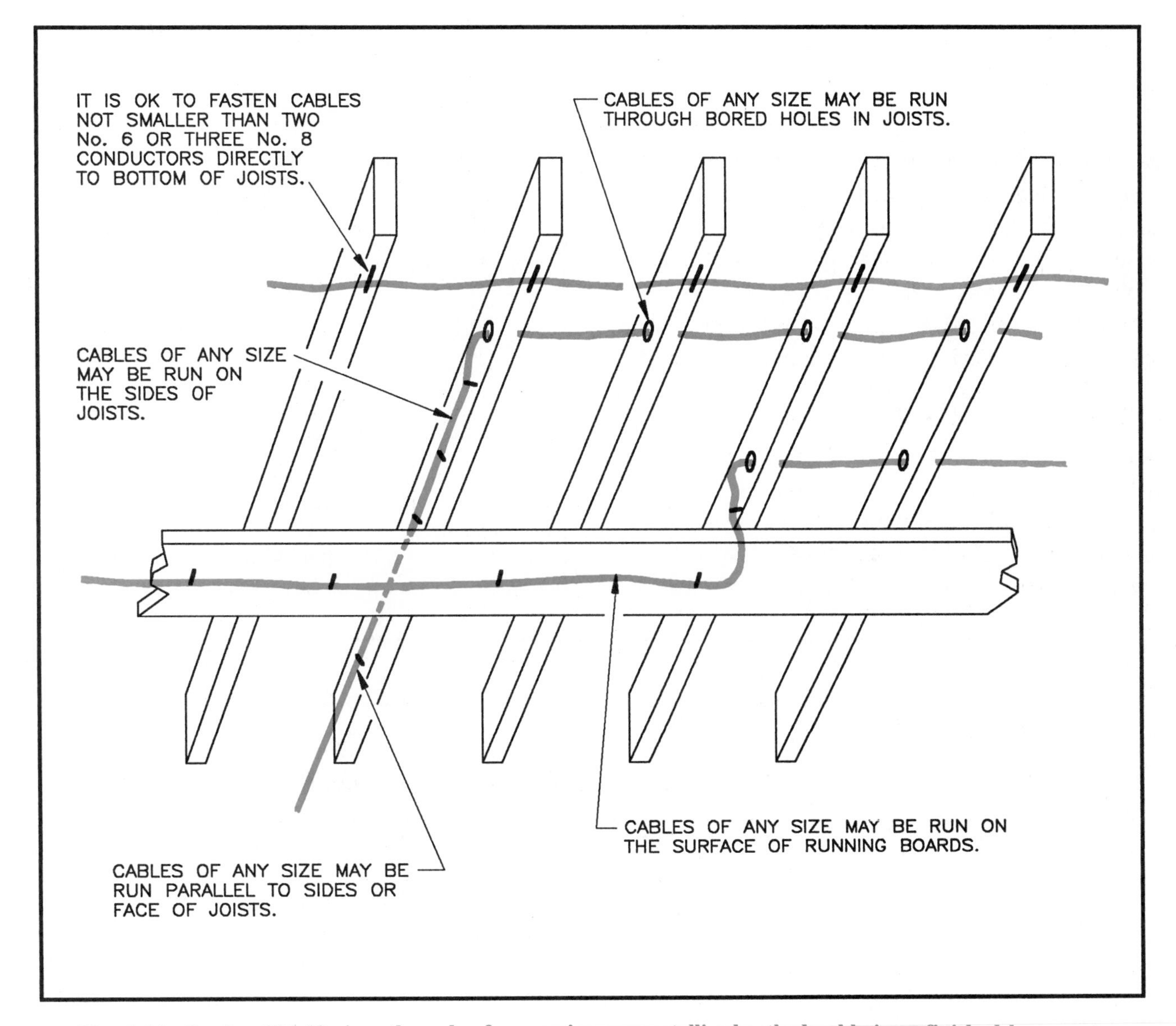

Fig. 4-14 *Section 336-12* gives the rules for running nonmetallic-sheathed cable in unfinished basements.

Multistory Buildings

► Nonmetallic-sheathed cable is not permitted to be used where the building exceeds three floors above grade, as might be the case in multifamily dwellings such as condominiums and apartments. *Section 336-4(a)* defines what constitutes the first floor of a building. The first floor of a building is that floor that has 50% or more of the exterior wall surface area level with or above finished grade. The Code allows one additional level for parking, storage, or similar use when such level is not used for human habitation. Walk around the entire building to determine the 50% criteria. Refer to Figure 4-11A. ◄

ARMORED CABLE (*ARTICLE 333*)

Description

The *National Electrical Code®* describes type AC armored cable. Underwriters Laboratories lists types AC and ACT armored cable (which are the same as the cable described by the Code). The terms are used interchangeably by electricians. Many electricians call this type of cable BX.

Although not authenticated, the term "BX" supposedly was derived from an abbreviation of the Bronx in New York City, where armored cable was once manufactured. At one time BX was a trademark owned by the General Electric Company. It has now become a generic term.

Section 333-1 describes armored cable as an assembly of insulated conductors in a flexible metallic enclosure, figure 4-18. Armored cable is designated as type ACT if the conductors have thermoplastic insulation, or as type AC if the conductors have thermosetting insulation.

The conductor insulation is rated at 60°C, unless a suffix "H," indicating a 75°C rating (ACTH), or a suffix "HH," indicating a 90°C rating (ACTHH), is added to the type AC or type ACT label.

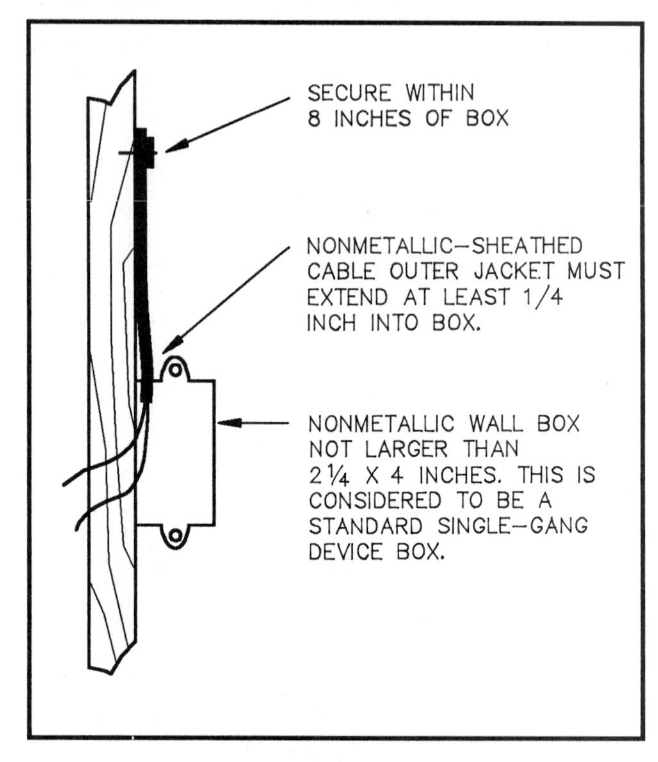

Fig. 4-14A *Section 370-7(c), Exception* states that nonmetallic-sheathed cable *need not* be clamped into a single-gang nonmetallic wall box not larger than 2 1/4 × 4 inches if secured within 8 inches of the box. This exception is applicable to wall boxes only.

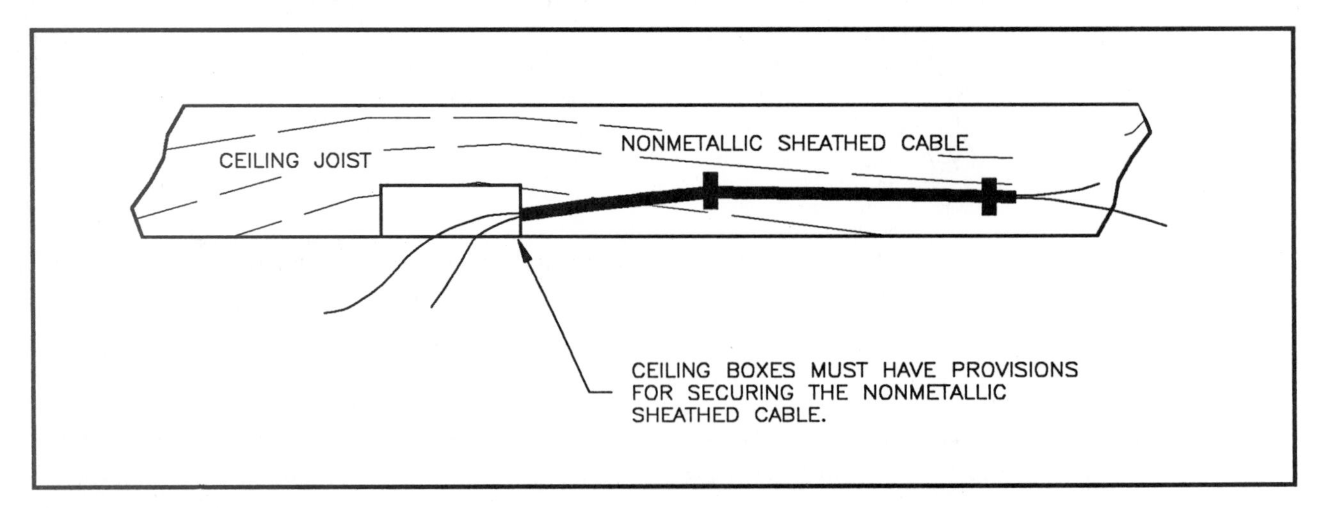

Fig. 4-14B *Section 370-7(c), Exception* requires nonmetallic-sheathed cable to be secured to nonmetallic ceiling boxes. The exception that permits "no securing" is for wall boxes only.

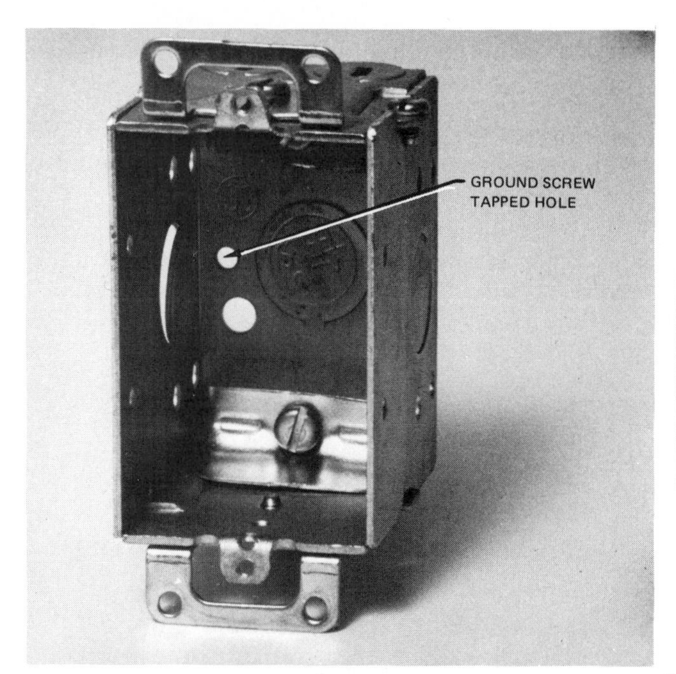

Fig. 4-15 Gang-type switch (device) box.

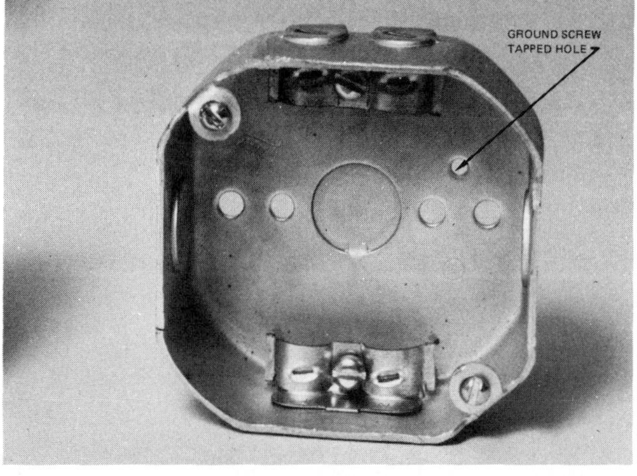

Fig. 4-16 Outlet box.

Armored cable is generally available with two or three conductors in sizes from No. 14 AWG to No. 1 AWG inclusive.

- Two-wire armored cable contains one black and one white conductor.
- Three-wire armored cable contains one each of black, white, and red conductors.

Armored cable (except ACL) must have an internal bonding strip of copper or aluminum in close contact with the armor for its entire length. This type of cable is used in a grounded system because of the metal bonding strip and the flexible steel armor. The armor also provides mechanical protection to the conductors.

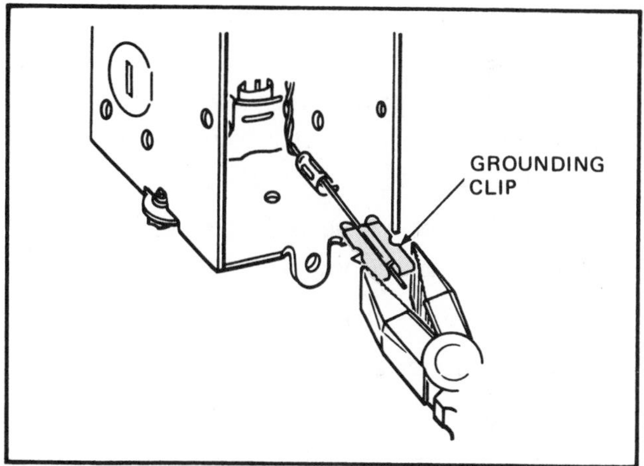

Fig. 4-17 Method of attaching ground clip to switch (device) box. See *Section 250-114*.

Use and Installation

Armored cable can be used in more applications than nonmetallic-sheathed cable. Certain conditions govern the use of types AC and ACT armored cable in dwelling unit installations. AC and ACT armored cable:

- may be used on circuits and feeders for applications of 600 volts or less.
- may be used for open and concealed work in dry locations.
- may be run through walls and partitions.

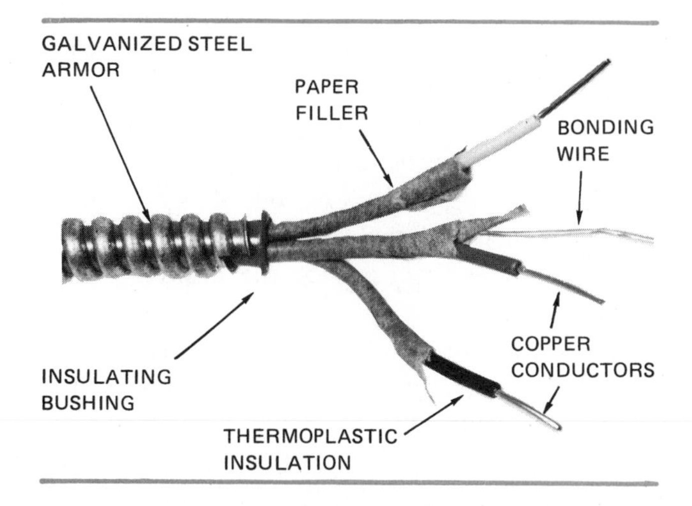

Fig. 4-18 Flexible armored cable.

- may be embedded in the plaster finish on masonry walls or run through the hollow spaces of such walls if these locations are not considered damp or wet (see *Locations, Article 100*).
- must be secured within 12 inches (305 mm) from every outlet box or fitting and at intervals not exceeding 4 1/2 feet (1.37 m).
- must not be bent to a radius of less than five times the diameter of the cable (see figure 4-10).
- ► may be used in lengths of not over 6 feet (1.83 m) to run between an outlet box and a lighting fixture (or other equipment) above an accessible ceiling. The cable must be secured (stapled) within 12 inches (305 mm) of the outlet box. See *Section 333-11, Exception No. 3.* ◄
- must have an approved fiber insulating bushing (antishort) at the cable ends to protect the conductor insulation, figure 4-19.

Armored cable (Type AC and ACT) is *not* approved for use in the following situations:

- underground installations.
- burying in masonry, concrete, or fill of building during construction.
- installation in any location which is exposed to weather.
- installation in any location exposed to oil, gasoline, or other materials which have a destructive effect on the insulation.

To remove the outer metal cable armor, it is recommended that a tool similar to the one shown in figure 4-20 be used. Otherwise, a hacksaw can be used, figure 4-21, to cut through one of the raised convolutions of the cable armor. Be very careful not to cut too deep or you will cut into the conductor insulation. Then bend the cable armor at the cut. It will snap off easily, exposing the desired 8–10 inches of conductor.

To prevent cutting of the conductor insulation by the sharp metal armor, an antishort bushing is inserted at the cable ends, as shown in figure 4-19. Neither the Code nor UL specifies exactly how to insert the antishort bushing. The figure shows four ways to insert the bushings. Photos (A) and (B) show that the bonding wire holds the antishort bushing in place.

Nonmetallic-sheathed cable and armored cable each has advantages which make it suitable for par-

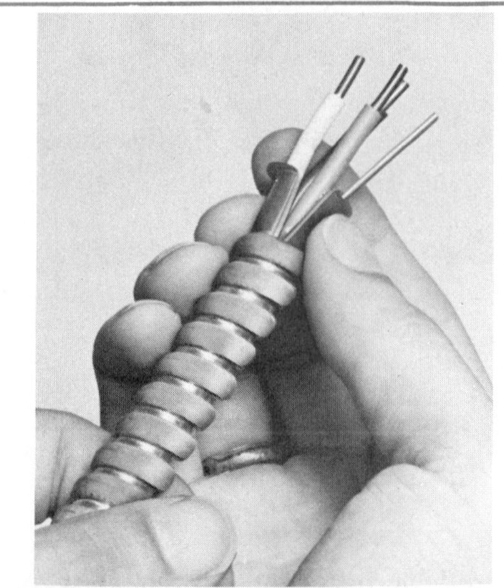

(A)

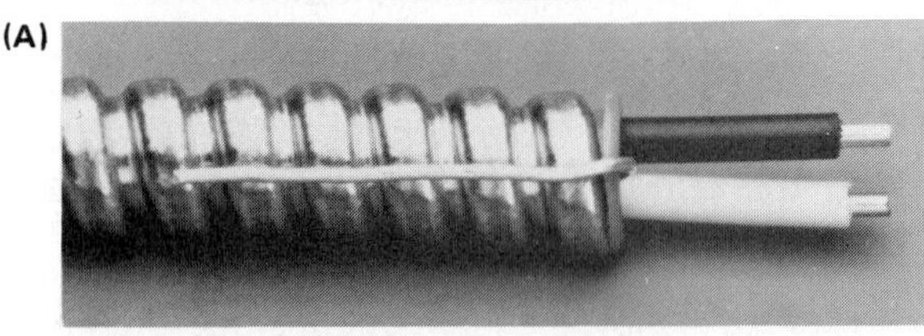

(B)

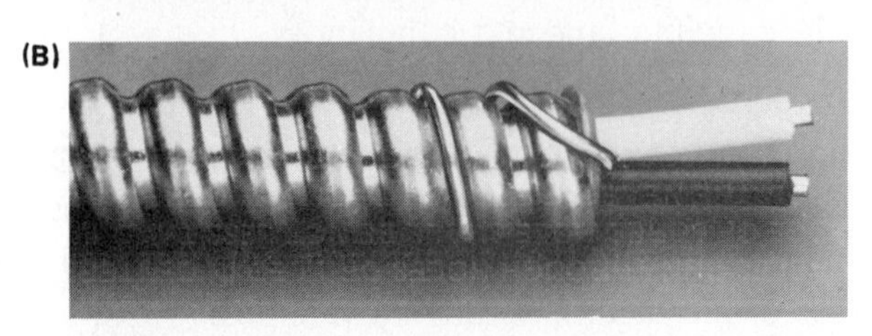

(C)

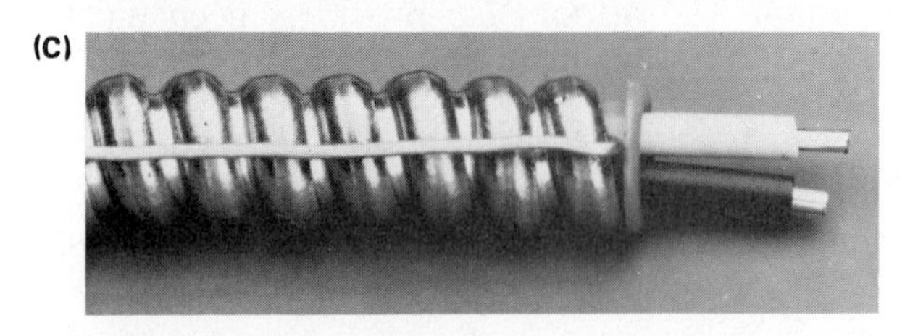

(D)

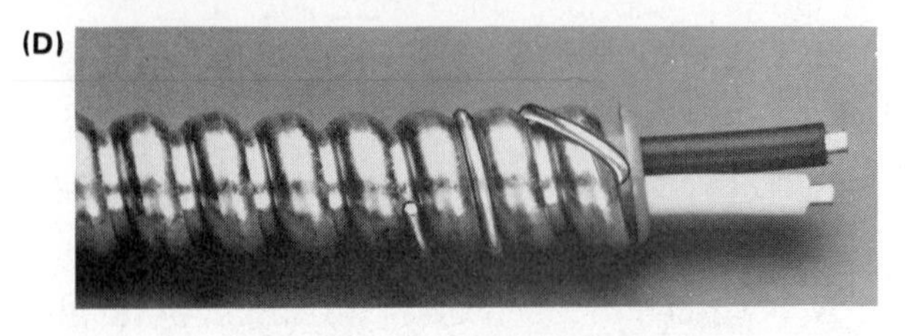

Fig. 4-19 Antishort bushing prevents cutting of the conductor insulation by the sharp metal armor.

ticular types of installations. However, the type of cable to be used in a specific situation depends largely on the wiring method permitted or required by the local building code.

INSTALLING CABLES THROUGH WOOD AND METAL FRAMING MEMBERS (*SECTION 300-4*)

To complete the wiring of a residence, cables must be run through studs, joists, and rafters. One method of installing cables in these locations is to run the cables through holes drilled at the approximate centers of wood building members, or at least 1 1/4 inches (31.8 mm) from the nearest edge. Holes bored through the center of standard 2 × 4s meet the requirements of *Section 300-4*. Refer to figure 4-22.

If the 1 1/4-inch distance cannot be maintained, or if the cables are to be laid in a notch, a metal plate at least 1/16 inch (1.59 mm) thick must cover the notch to protect the cable from nails, *Section 300-4*. Refer to figure 4-22.

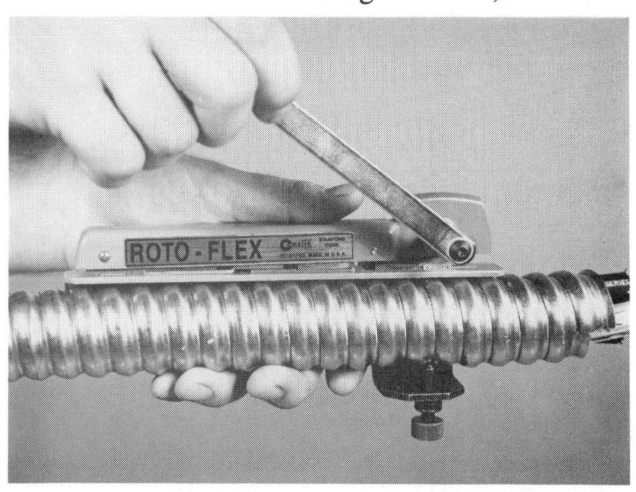

Fig. 4-20 This is an example of a tool from one manufacturer. It precisely cuts the outer armor of armored cable, making it easy to remove the armor with a few turns of the handle. Courtesy of Seatek Co., Inc.

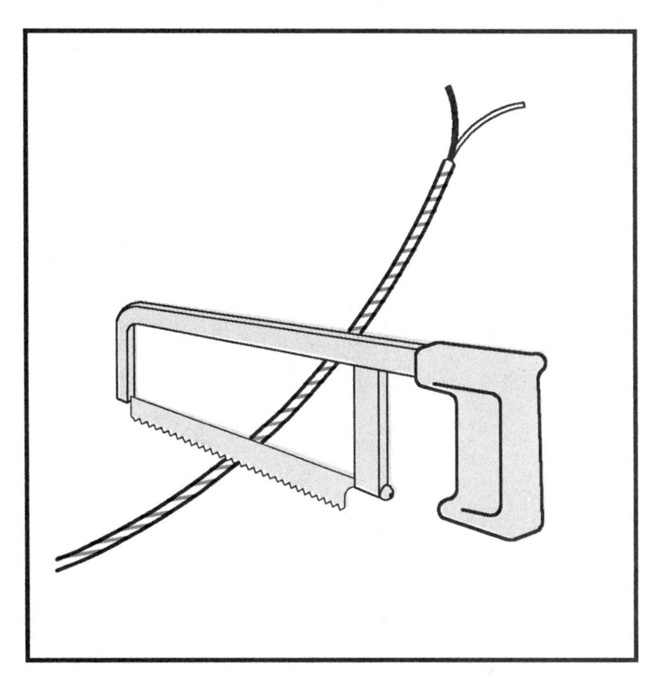

Fig. 4-21 Using a hacksaw to cut through a raised convolution of the cable armor.

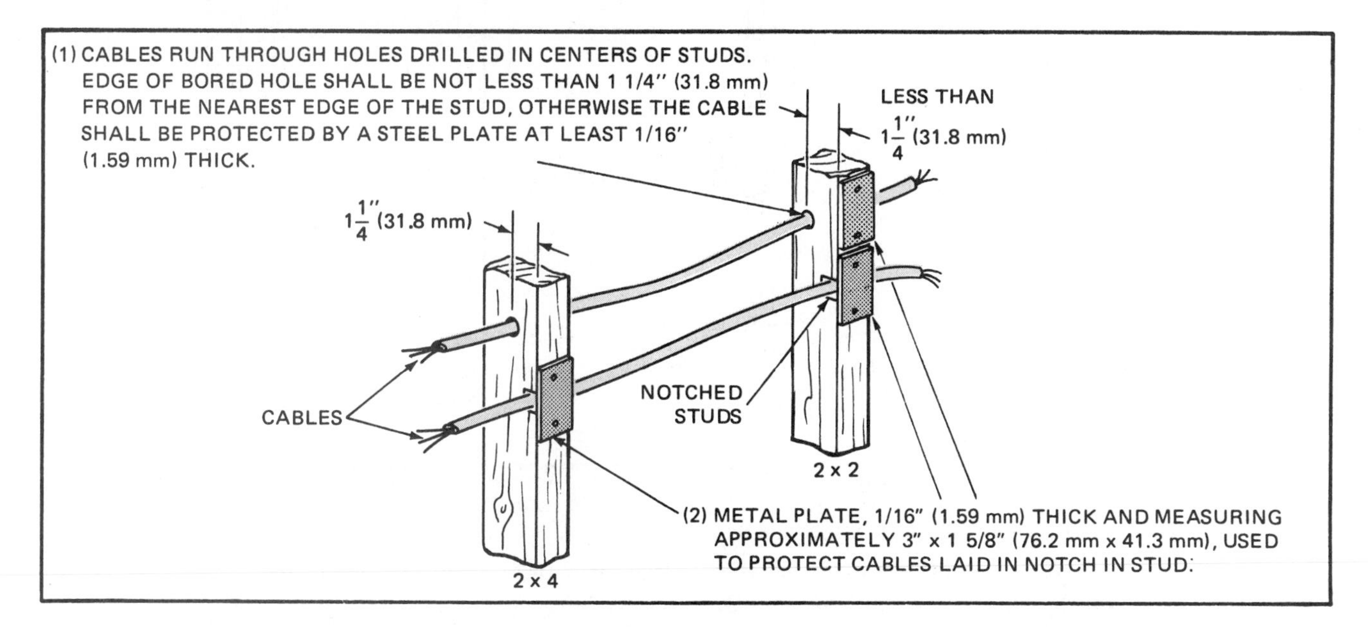

Fig. 4-22 Methods of protecting cables, *Section 300-4*. Because of their strength, intermediate metal conduit, rigid metal conduit, and electrical metallic tubing are exempt from this rule per the exceptions found in *Section 300-4*.

When running nonmetallic-sheathed cable through holes in metal studs or joists, the cable must be protected by bushings or grommets securely fastened in place prior to installing the cable, figure 4-23. These bushings, up to 2 inches in diameter, are available from various manufacturers. They easily snap into place in holes having the same size as box knockouts — for example, 1/2-inch, 3/4-inch, or 1-inch knockouts.

When running nonmetallic-sheathed cable, armored cable, or electrical nonmetallic tubing through metal framing members where there is a likelihood that nails or screws could be driven into the cables or tubing, a steel plate, steel sleeve, or steel clip at least 1/16 inch thick must be installed to protect the cables and/or tubing. This is spelled out in *Section 300-4(b)(2)* of the Code.

► Where cables and certain types of raceways are

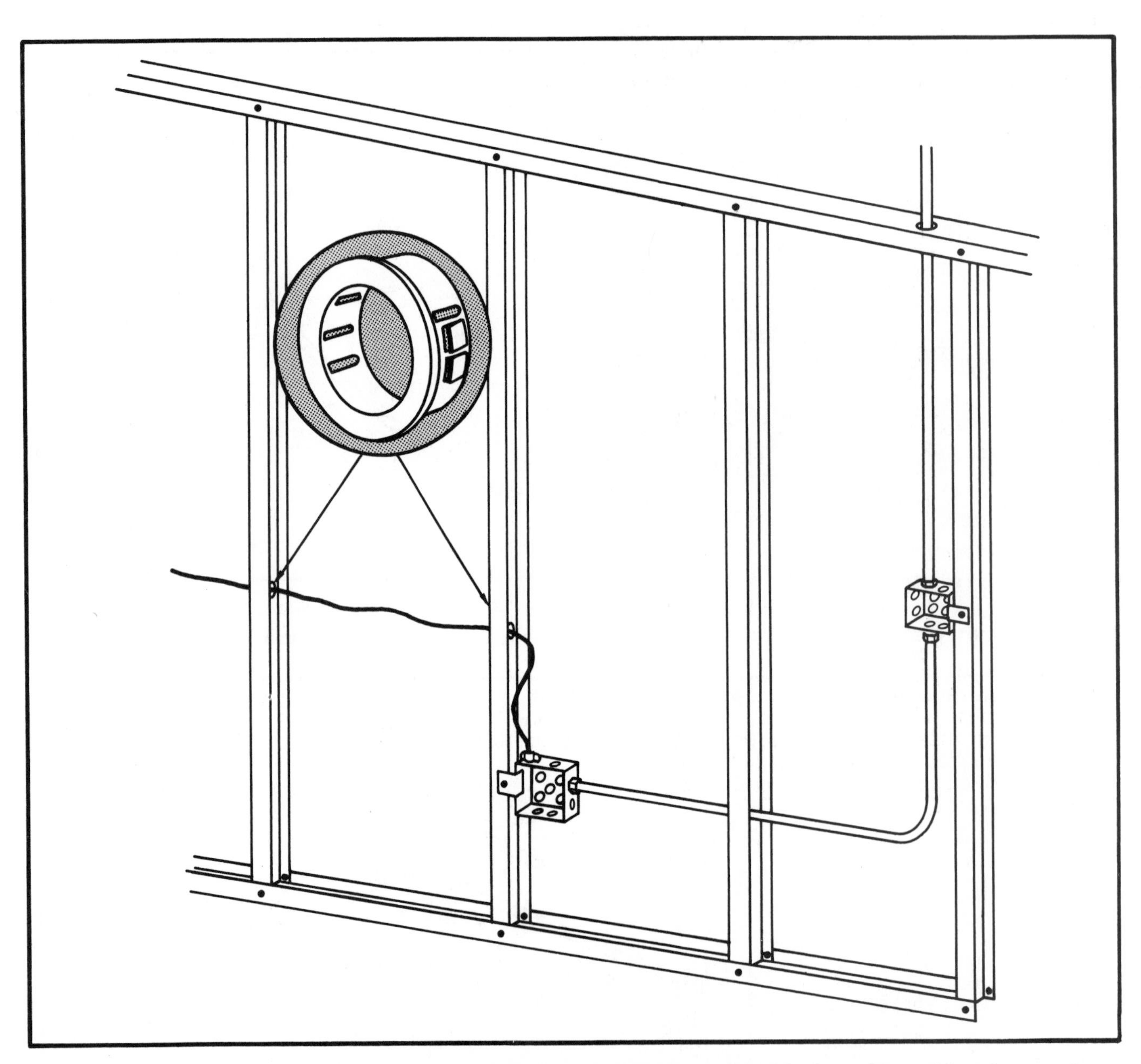

Fig. 4-23 Bushings must be installed prior to installing nonmetallic-sheathed cable through metal framing members. This protects the cable from abrasion. As a general rule, most electricians would avoid using nonmetallic-sheathed cable through metal studs and joists. They would use electrical metallic tubing, flexible metal conduit, armored cable, or electrical nonmetallic tubing. Where there is a likelihood that nails or screws might penetrate nonmetallic-sheathed cable or electrical nonmetallic tubing, 1/16″ thick steel plates, sleeves, or clips must be installed to protect the cable or tubing.

run concealed or exposed, parallel to a stud, joist or rafter (any building framing member), the cables and raceways must be supported and installed in such a manner that there is not less than 1 1/4 inches (31.8 mm) from the edge of the stud or joist (framing member) to the cable or raceway. If the 1 1/4-inch clearance cannot be maintained, then a steel plate, sleeve, or equivalent that is at least 1/16 inch thick must be provided to protect the cable or raceway from damage that can occur when nails or screws are driven into the wall or ceiling, *Section 300-4(d)*, figure 4-24.

This additional protection is not required when the raceway is intermediate metal conduit (*Article 345*), rigid metal conduit (*Article 346*), rigid nonmetallic conduit (*Article 347*), or electrical metallic tubing (*Article 348*).

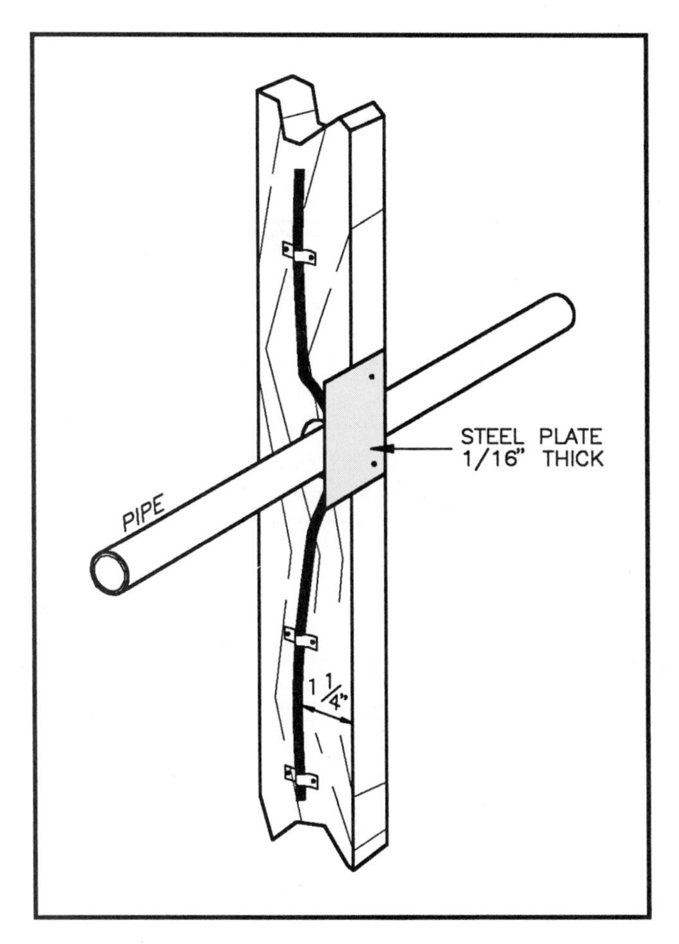

► **Fig. 4-24 Where cables or certain types of raceways are run parallel to a joist or stud (any framing member), stay 1 1/4 inches from the edge of the framing member with the cable. Where 1 1/4-inch clearance is impossible to maintain, then install a steel plate that is at least 1/16 inch thick, *Section 300-4(d)*.** ◄

Additional protection obviously cannot be installed inside the walls or ceilings of an existing building where the walls and ceilings are already closed in.

Exception No. 2 to *Section 300-4(d)* permits cables to be "fished" between boxes or other access points without additional protection. The logic is similar to the exception that permits nonmetallic-sheathed cable to be "fished" between outlets without requiring supports as stated in *Section 333-7.* ◄

In the recreation room of the residence discussed in this text, precaution must be taken to assure that adequate protection against physical damage is provided for the wiring. The walls in the recreation room are to be paneled. Therefore, if the carpenter uses 1- × 2-inch or 2- × 2-inch furring strips, figure 4-24A, nonmetallic-sheathed cable would require the additional 1/16-inch steel plate protection up the walls from each receptacle outlet and switch. This could be quite costly and time consuming and would make a good case for doing the installation in electrical metallic tubing.

Some building contractors attach 1-inch insulation to the walls, then construct a 2- × 4-inch wall in front of the basement structural foundation wall, leaving 1-inch spacing between the insulation and the back side of the 2- × 4-inch studs. This makes it easy to run cables or conduits behind the 2- × 4-inch studs, and eliminates the need for the additional mechanical protection. See figure 4-24B.

► Watch out for any wall partitions where 2- × 4-inch studs are installed "flat," see figure 4-24C. In this case, the choice is to provide the required mechanical protection as required by *Section 300-4*, or to install the wiring in electrical metallic tubing. ◄

See unit 15 for the methods of cable installation and protection in attic areas. *Sections 333-10* and *336-7* of the Code refer to cables run through studs, joists, and rafters. These sections, in turn, refer to *Section 300-4* for more detailed information.

INSTALLATION OF CABLES THROUGH DUCTS

Section 300-22 of the Code is extremely strict as to what types of wiring methods are permitted for installation of cables through ducts or plenum chambers. These stringent rules are for fire safety.

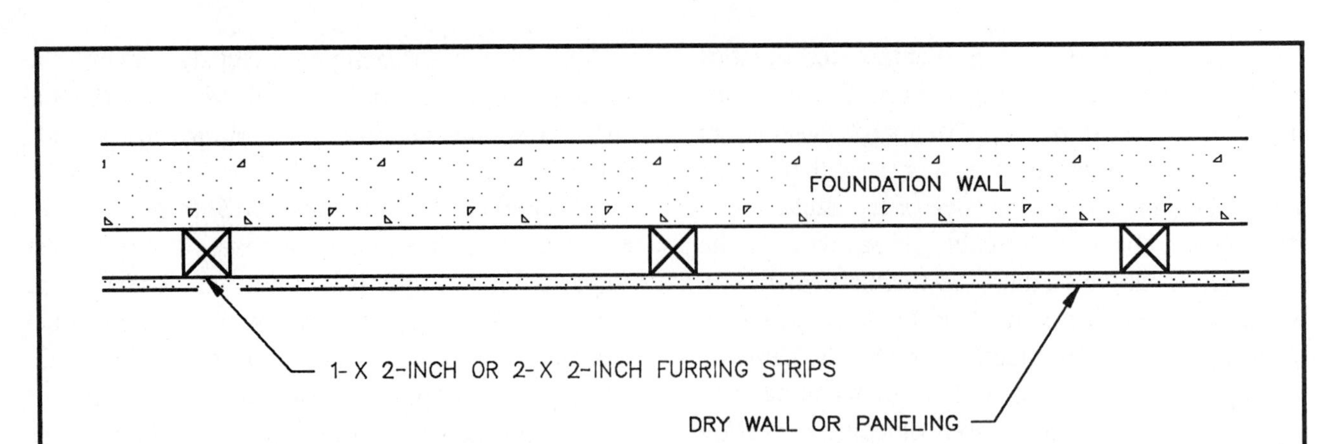

► **Fig. 4-24A** When 1- × 2-inch or 2- × 2-inch furring strips are installed on the surface of a basement wall, additional protection as discussed in figure 4–24C must be provided. ◄

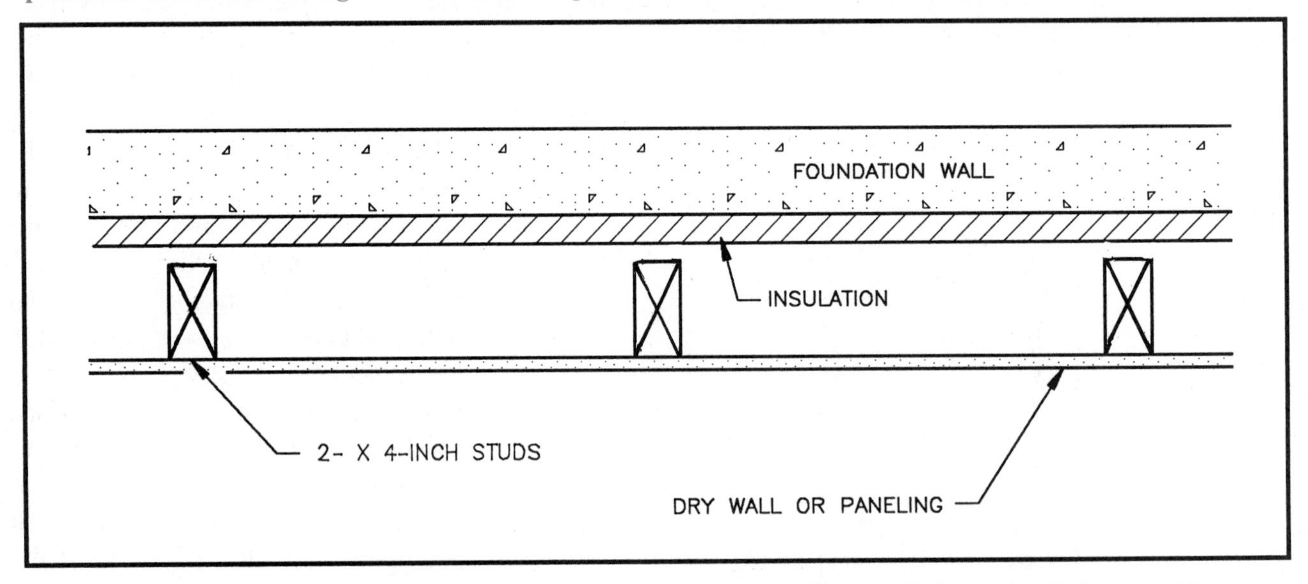

► **Fig. 4-24B** This sketch shows how to insulate a foundation wall and how to construct a wall that provides space to run cables and/or raceways that will not require the additional protection as discussed in figure 4-24A and 4-24C. ◄

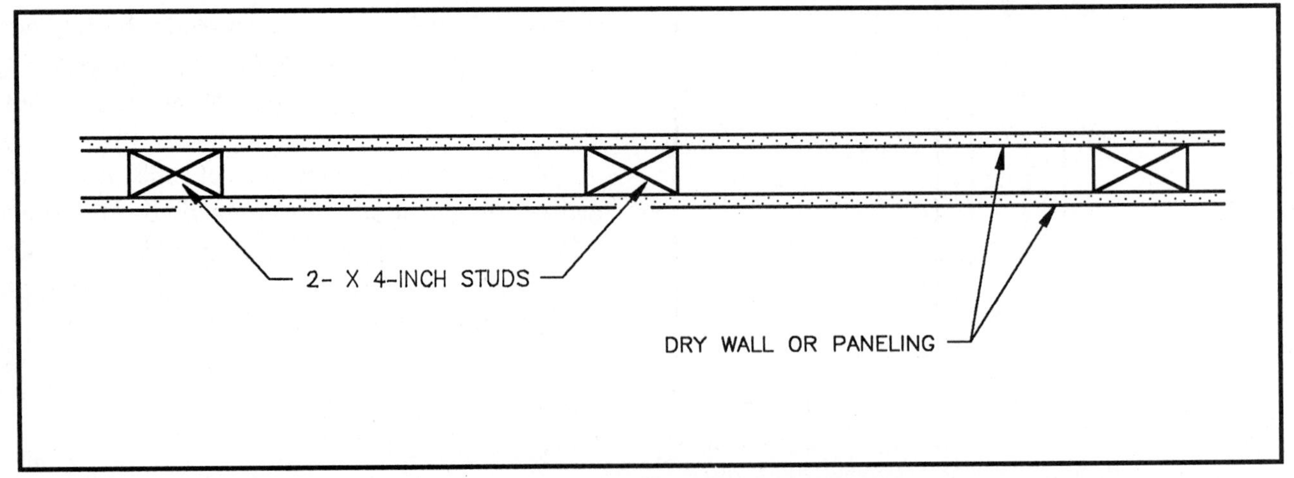

► **Fig. 4-24C** When 2- × 4-inch studs are installed "flat," such as might be found in nonbearing partitions, cables running parallel to or through the studs must be protected against the possibility of having nails or screws driven through the cables. This means protecting the cable with 1/16-inch steel plates or equivalent for the entire length of the cable, or installing the wiring using intermediate metal conduit, rigid metal conduit, rigid nonmetallic conduit, or electrical metallic tubing. ◄

The Code requirements have been relaxed slightly, however, to permit types NM and NMC cable to be installed in joist and stud spaces (i.e., cold air returns) in dwellings, *Section 300-22, Exception No. 5.* This exception permits types NM and NMC to pass through such spaces only if they are run perpendicular to the long dimensions of such spaces, as illustrated in figure 4-25.

CONNECTORS FOR INSTALLING NONMETALLIC-SHEATHED AND ARMORED CABLE

The connectors shown in figure 4-26 are used to fasten nonmetallic-sheathed cable and armored cable to the boxes and panels in which they terminate. These connectors clamp the cable securely to each outlet box. Many boxes have built-in clamps and do not require separate connectors.

The question continues to be asked: "may more than one cable be inserted in one connector?" Unless the listing by Underwriters Laboratories of a specific cable connector indicates that the connector has been tested for use with more than one cable, the rule is *one cable, one connector.*

ELECTRICAL METALLIC TUBING (*ARTICLE 348*), INTERMEDIATE METAL CONDUIT (*ARTICLE 345*), RIGID METAL CONDUIT (*ARTICLE 346*), AND RIGID NONMETALLIC CONDUIT (*ARTICLE 347*)

Some communities do not permit the installation of cable of any type in residential buildings. These communities require the installation of a

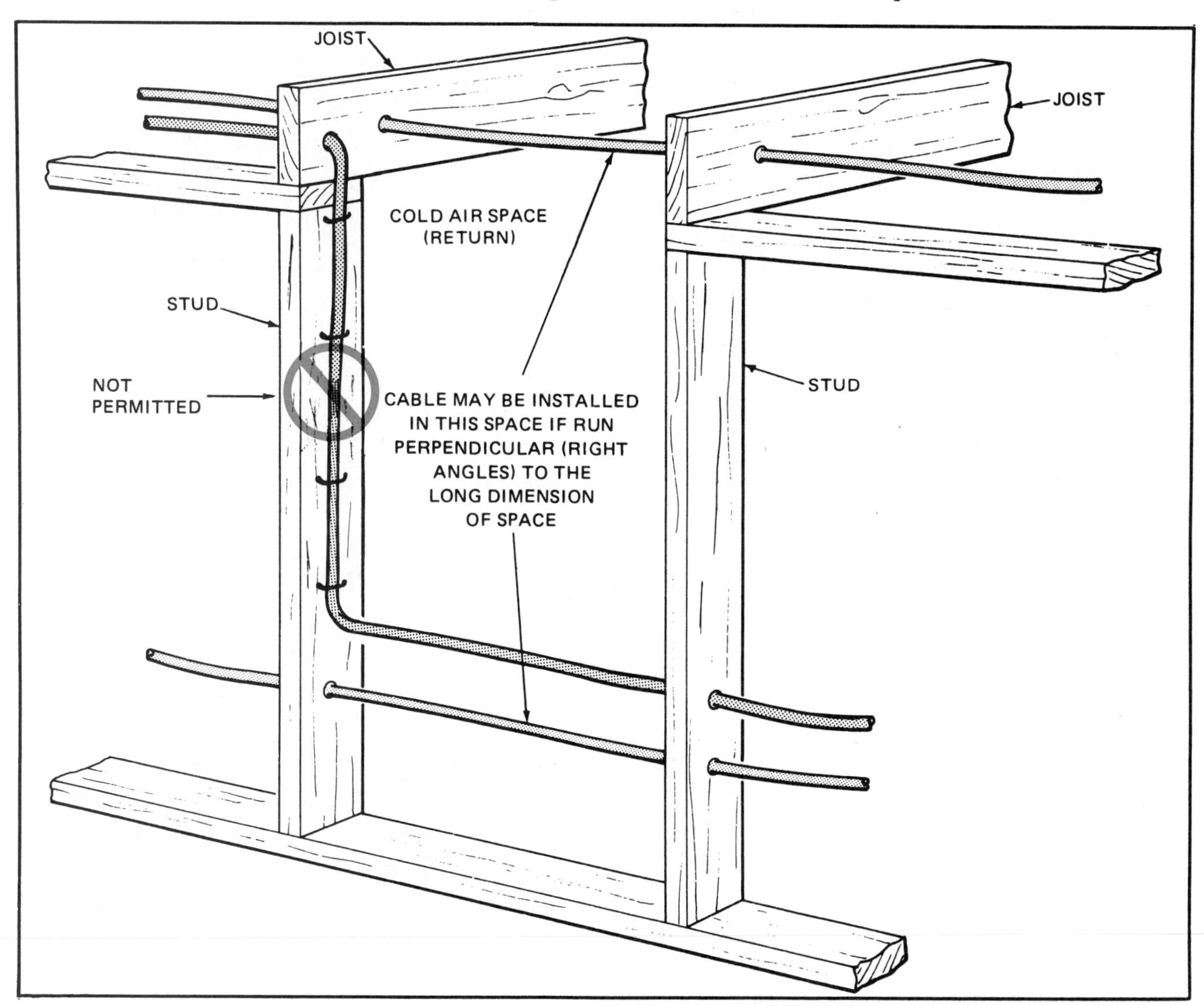

Fig. 4-25 Nonmetallic-sheathed cable may pass through cold air return, joist, or stud spaces in dwellings, but only if it is run at right angles to the long dimension of the space, *Section 300-22, Exc. No. 5.*

raceway system of wiring, such as electrical metallic tubing (thinwall), intermediate metal conduit, rigid metal conduit, or rigid nonmetallic conduit.

Where the building construction is cement block, cinder block, or poured concrete, it will be necessary to make the electrical installation in conduit.

Sufficient data is given in this text for both cable and conduit wiring methods. Thus, the student will be able to complete the type of installation permitted or required by local codes.

According to installation requirements for a raceway system of wiring, electrical metallic tubing (EMT), intermediate metal conduit, and rigid metal conduit:

- may be buried in concrete or masonry and may be used for open or concealed work.
- may *not* be installed in cinder, concrete, or fill unless (1) protected on all sides by a layer of noncinder concrete at least 2 inches (50.8 mm) thick, or (2) the conduit is at least 18 inches (457 mm) below the fill.

In general, conduit must be supported within 3 feet (914 mm) of each box or fitting and at 10-foot (3.05 m) intervals along runs, figure 4-27.

The number of conductors permitted in EMT, intermediate metal conduit, and rigid conduit is given in *Table 1, Chapter 9* of the Code. Heavy-wall rigid conduit provides greater protection against mechanical injury to conductors than does either EMT or intermediate metal conduit, which have much thinner walls (*Articles 345*, *346*, and *348*).

The residence specifications indicate that a meter pedestal is to be furnished by the utility and

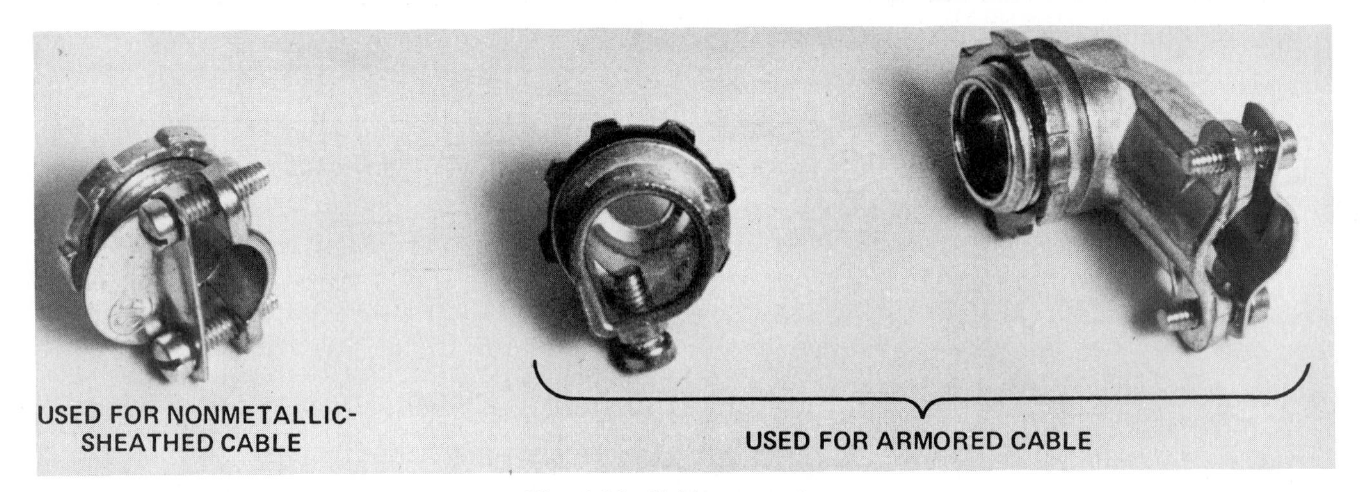

Fig. 4-26 Cable connectors.

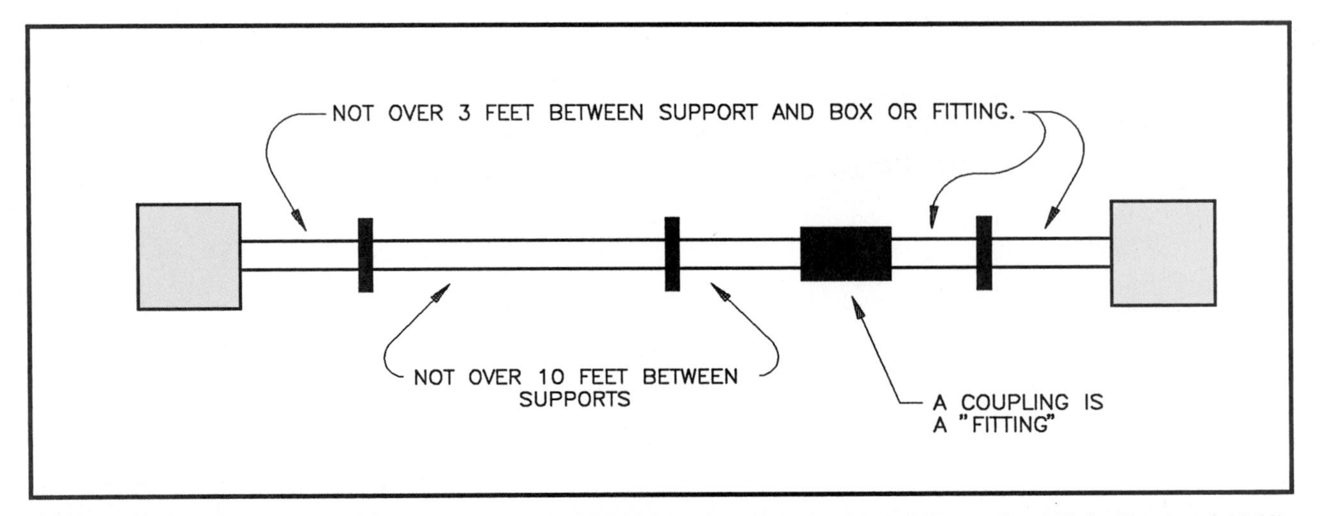

Fig. 4-27 Code requirements for support of metal conduit (*Section 345-12*), rigid metal conduit (*Section 346-12*), electrical metallic tubing (*Section 348-12*). For supporting rigid nonmetallic conduit, refer to *Section 347-8*.

installed by the electrical contractor. Main Panel A is located in the workshop. The meter pedestal is located on the outside of the house near the air-conditioning unit. The electrical contractor must furnish and install a 1 1/2-inch electrical metallic tubing between the meter pedestal and Main Panel A. *Section 373-6(c)* requires that insulating bushings or equivalent be used where No. 4 or larger conductors enter a raceway. In addition to the requirements of *Section 373-6(c)*, a bonding-type bushing must be used where the service-entrance conduit enters Main Panel A. (See *Sections 250-72* and *250-73* for the method of bonding service equipment.)

The plans also indicate that electrical metallic tubing is to be used in the workshop. Since the recreation room and basement stairwell have finished ceilings and walls, they will be wired with cable.

Many of the *National Electrical Code*® rules are applicable to both cable and conduit installations. Some Code requirements are applicable only to cable installations. For example, there are special requirements for the attachment of the grounding conductors and when protection from physical damage must be provided. Many of the Code requirements governing the installation of cables do not apply to a raceway system. Properly installed electrical metallic tubing provides good continuity for equipment grounding, a means of withdrawing or pulling in additional conductors, and excellent protection against physical damage to the conductors.

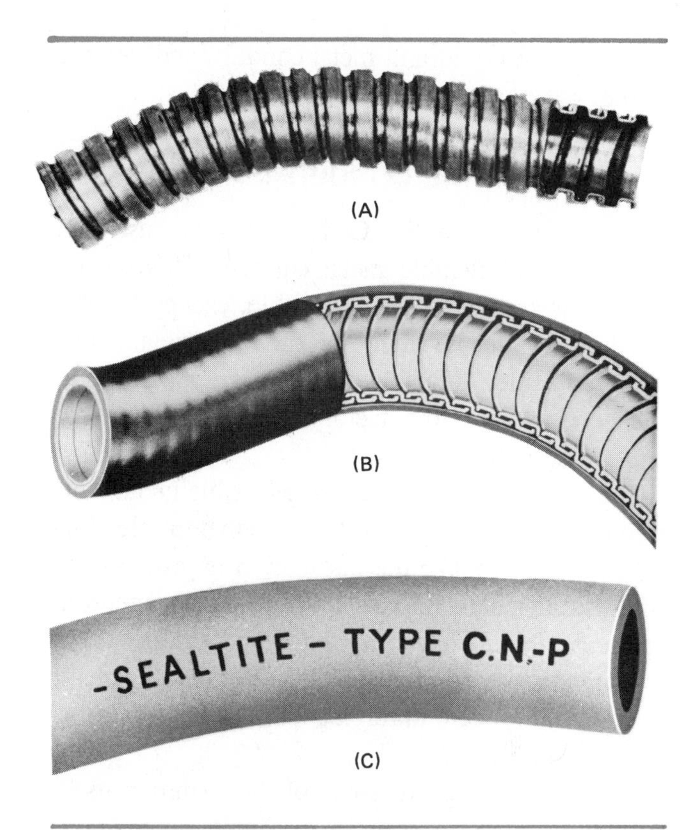

Fig. 4-28 (A) Flexible metal conduit. (B) Flexible liquidtight metal conduit. (C) Flexible liquidtight nonmetallic conduit.

FLEXIBLE CONNECTIONS (*ARTICLES 350* AND *351*)

The installation of certain equipment requires flexible connections, both to simplify the installation and to stop the transfer of vibrations. In residential wiring, flexible connections are used to wire attic fans, food waste disposers, dishwashers, air conditioners, heat pumps, recessed fixtures, and similar equipment.

The three types of flexible conduit used for these connections are: flexible metal conduit, figure 4-28(A); flexible liquidtight metal conduit, figure 4-28(B); and flexible liquidtight nonmetallic conduit, figure 4-28(C). Figure 4-29 shows many of the types of connectors used with flexible metal con-

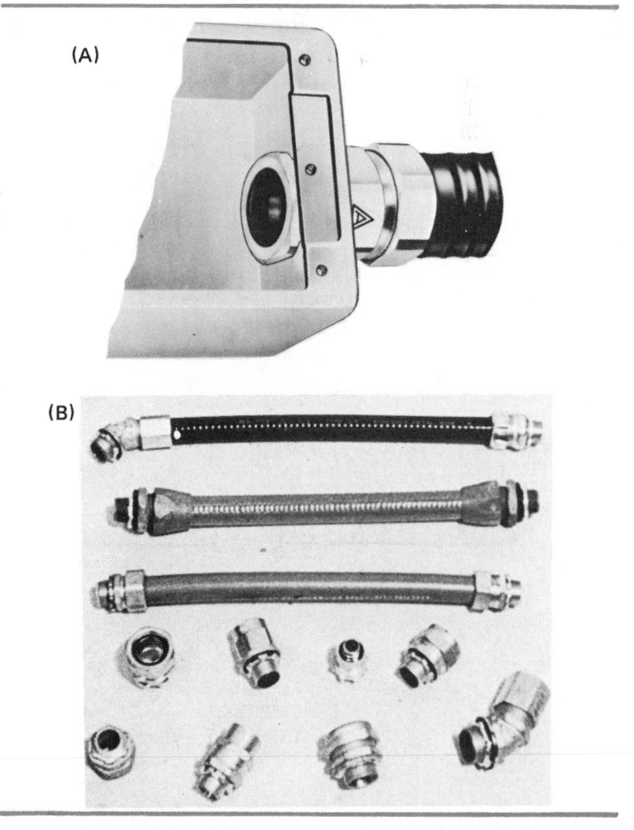

Fig. 4-29 (A) Liquidtight fitting. (B) Various types of connectors.

duit, flexible liquidtight metal conduit, and flexible liquidtight nonmetallic conduit.

FLEXIBLE METAL CONDUIT (*ARTICLE 350*)

Article 350 of the Code covers the use and installation of flexible metal conduit. This wiring method is similar to armored cable, except that the conductors are installed by the electrician. For armored cable, the cable armor is wrapped around the conductors at the factory to form a complete cable assembly.

Common installations using flexible metal conduit are shown in figure 4-30. Note that the flexibility required to make the installation is provided by the flexible metal conduit. The figure calls attention to the *National Electrical Code®* and Underwriters Laboratories restrictions on the use of flexible metal conduit with regard to relying on the metal armor as a grounding means.

The following are some of the common uses, limitations, and applications for flexible metal conduit, combining the material found in *Article 350* of the Code, and in the UL White Book.

- do not install in wet locations unless the conductors are suitable for use in wet locations (TW, THW, THWN), and the installation is made so water will not enter the enclosure or other raceways to which the flex is connected.
- do not bury in concrete.
- do not bury underground.
- do not use in locations subject to corrosive conditions.
- is acceptable as a grounding means if not over 6 feet (1.83 m), has overcurrent protection not over 20 amperes, is not over 3/4-inch trade size, and is connected with fittings listed for the purpose.
- if longer than 6 feet (1.83 m), regardless of trade size, is not acceptable as grounding means. You must install a separate grounding conductor.
- fittings larger than 3/4 inch that have been listed as OK for grounding means will be marked "GRND" or some equivalent marking.
- 3/8-inch trade size not permitted longer than 6 feet (1.83 m). Fixture "whips" are a good example. See recreation room unit for a discussion of fixture whips.
- must be supported every 4 1/2 feet (1.37 m).
- must be supported within 12 inches (305 mm) of box, cabinet, or fitting.
- if flexibility is needed, support within 36 inches (914 mm) of termination.
- if the flexible metal conduit is not acceptable as a grounding means, then install a bonding jumper inside or outside the flexible metal conduit.
- a bonding jumper must be installed inside the flexible metal conduit if the conduit is over 6 feet (1.83 m) long.
- do not have more than four quarter bends (360°) between boxes.
- do not conceal angle-type fittings because of the difficulty that would present itself when pulling wires in.
- if flexibility is needed, always install a separate grounding conductor regardless of trade size.
- may be used for exposed or concealed installations.

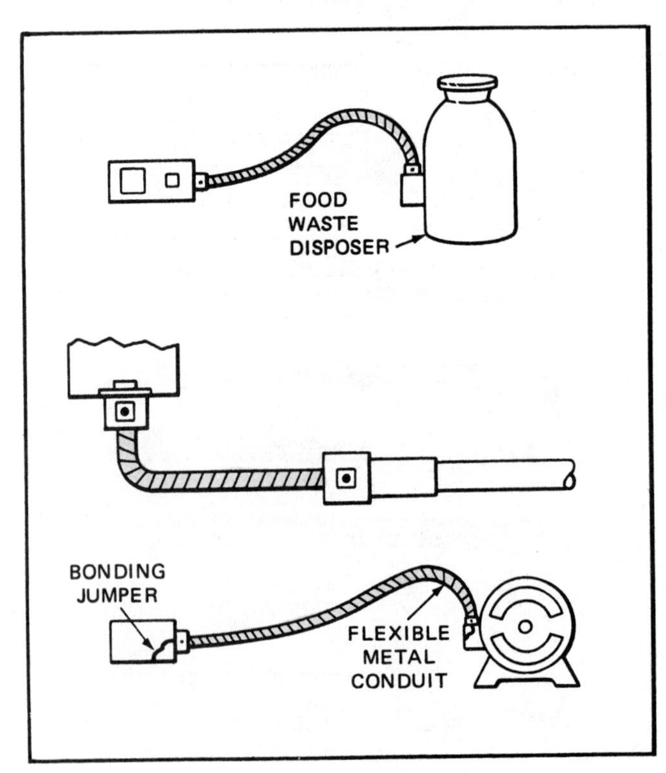

Fig. 4-30 Some of the more common places where flexible metal conduit may be used.

- conductor fill same as regular conduit. See *Table 1, Chapter 9, NEC®*.
- conductor fill for 3/8-inch size. See *Table 350-3, NEC®*.
- fittings used with flexible metal conduit must be listed.

Section 250-91(b), Exception 1 also discusses flexible conduit installations and repeats the requirements mentioned above.

LIQUIDTIGHT FLEXIBLE METAL CONDUIT (*ARTICLE 351*)

The use and installation of liquidtight flexible metal conduit are described in *Article 351* of the Code. Liquidtight flexible metal conduit has a "tighter" fit of its spiral turns as compared to standard flexible metal conduit. Liquidtight conduit also has a thermoplastic outer jacket that is liquidtight. Liquidtight flexible metal conduit is commonly used as the flexible connection to a central air-conditioning unit located outdoors, figure 4-31.

Figure 4-32 shows the Code rules for the use of liquidtight flexible metal conduit as a grounding means. These limitations are given in the Underwriters Laboratories Standards.

Listed here are some of the common uses, limitations, and applications for liquidtight flexible metal conduit, combining *National Electrical Code®* rules and UL White Book requirements.

- may be used for exposed and concealed installations.
- may be buried directly in the ground if so listed and marked.
- may not be used where subject to physical abuse.
- may not be used if ambient temperature and heat from conductors will exceed the temperature limitation of the nonmetallic outer jacket. UL listing will indicate maximum temperature.
- 3/8-inch trade size not permitted longer than 6 feet (1.83 m).
- conductor fill same as regular conduit. See *Table 1, Chapter 9, NEC®*.
- conductor fill for 3/8-inch size. See *Table 350-3, NEC®*.

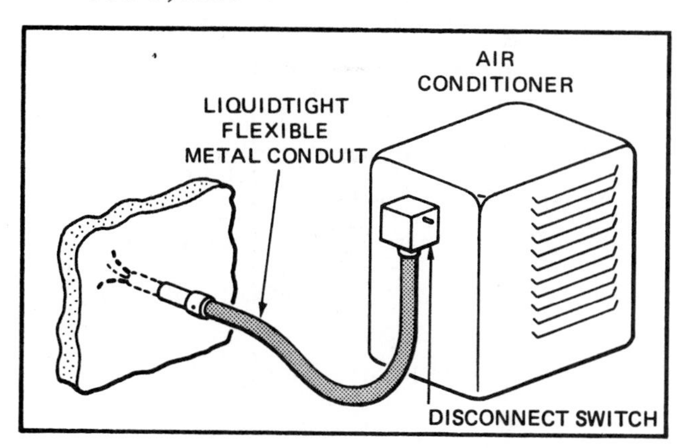

Fig. 4-31 Use of liquidtight flexible metal conduit or flexible liquidtight nonmetallic conduit.

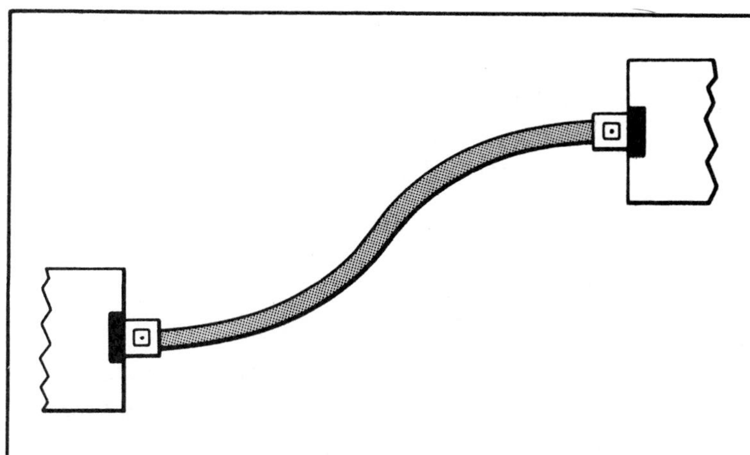

A. LIQUIDTIGHT FLEXIBLE METAL CONDUIT MAY BE USED AS A GROUNDING MEANS IF IT IS NOT OVER 1 1/4 INCH TRADE SIZE, IS NOT OVER 6 FEET (1.83 m) LONG, AND IS CONNECTED BY FITTINGS LISTED FOR GROUNDING PURPOSES.

B. MINIMUM SIZE 1/2 INCH, EXCEPT 3/8 INCH WHEN NOT LONGER THAN 6 FEET (1.83 m) PERMITTED ONLY AS PART OF AN APPROVED ASSEMBLY OR TO CONNECT LIGHTING FIXTURES. WHEN USED AS THE GROUNDING MEANS, THE MAXIMUM RATED OVERCURRENT DEVICE IS 20 AMPERES FOR 3/8- AND 1/2-INCH SIZES, AND 60 AMPERES FOR 3/4 INCH, 1 INCH, AND 1 1/4 INCHES.

C. LIQUIDTIGHT FLEXIBLE METAL CONDUIT MAY NOT BE USED AS A GROUNDING MEANS IN ANY SIZE WHEN LONGER THAN 6 FEET (1.83 m), OR IN SIZES 1 1/2 INCH AND LARGER OF ANY LENGTH. IN SUCH CASES, A BONDING JUMPER IS REQUIRED. SEE FIGURE 4—30.

Fig. 4-32 Liquidtight flexible metal conduit Code rules, *Article 351*.

- must be used only with approved fittings.
- must be supported every 4 1/2 feet (1.37 m).
- must be supported within 12 inches (305 mm) of box, cabinet, or fitting.
- if flexibility is needed, support within 3 feet (914 mm) of termination.
- support not required when connecting up a lighting fixture when the flex is not over 6 feet (1.83 m).
- 3/8-inch and 1/2-inch trade sizes OK for grounding purposes if fittings are listed for grounding, the flex is not over 6 feet (1.83 m), and the overcurrent device is not over 20 amperes.
- 3/4-inch, 1-inch, and 1 1/4-inch trade sizes OK for grounding purposes if fittings are listed for grounding, the flex is not over 6 feet (1.83 m), and the overcurrent device is not over 60 amperes.
- liquidtight flexible metal conduit is *not* suitable as a grounding means:

  if larger than 1 1/2-inch trade size.

  if trade size is 3/8 inch or 1/2 inch, the length is longer than 6 feet (1.83 m), and the overcurrent protection exceeds 20 amperes.

  if trade size is 3/4 inch, 1 inch, or 1 1/4 inch, the run is longer than 6 feet (1.83 m), and the overcurrent protection exceeds 60 amperes.
- when the flex is not acceptable as a grounding means as mentioned above, then a separate grounding conductor must be installed.
- if flexibility is needed, always install a separate grounding conductor regardless of trade size of the flex.
- do not conceal angle-type fittings.
- do not have more than four quarter bends (360°) between boxes.

Section 250-91(b), Exception 2 also discusses liquidtight flexible metal conduit installations, and repeats the requirements mentioned above.

LIQUIDTIGHT FLEXIBLE NONMETALLIC CONDUIT *(ARTICLE 351)*

The following are some of the important Code and UL rules pertaining to liquidtight flexible nonmetallic conduit.

- do not use in direct sunlight unless specifically marked for use in direct sunlight.
- may be used in exposed or concealed installations. ◄
- can become brittle in extreme cold applications.
- may be used outdoors when listed and marked as suitable for this application.
- ►► may be buried directly in the earth when listed and market as suitable for this application. ◄
- may not be used where subject to physical abuse.
- may not be used if ambient temperature and heat from the conductors will exceed the temperature limitation of the nonmetallic material. UL listing will indicate this.
- may not be used in lengths over 6 feet (1.83 m).
- may not be used in sizes smaller than 1/2 inch except that 3/8-inch trade size OK for enclosing the leads to a motor per *Section 430-145(c)*.
- conductor fill same as for regular conduit. See *Table 1, Chapter 9, NEC®*
- must be used with listed fittings.
- if an equipment grounding conductor is needed, the equipment grounding conductor may be run inside or outside of the flex. The equipment grounding conductor shall not be longer than 6 feet (1.83 m) if run on the outside of the flex.

GROUNDING ELECTRODE CONDUCTOR

The sizing requirements for the grounding electrode conductor, which is the conductor between the main service-entrance equipment and the grounding electrode (street side of water meter, ground rods, concrete encased ground, etc., as mentioned in *Section 250-91*) is covered in unit 28. The grounding electrode conductor in this residence is a No. 4 AWG copper conductor contained in a flexible steel sheath, figure 4-33.

The Code specifies, in *Section 250-92(a)*, that a No. 4 or larger grounding conductor may be attached to the surface on which it is carried without the use of knobs, tubes, or insulators. Protection is not required unless the ground wire is expected to be exposed to severe physical damage.

If it is to be free from exposure to physical damage, a No. 6 grounding conductor may be run along the surface of the building construction without additional protection. If subject to physical damage, the conductor must be installed in rigid metal conduit, intermediate metal conduit, rigid nonmetallic conduit, electrical metallic tubing, or cable armor. Grounding conductors smaller than No. 6 must be enclosed in rigid conduit, electrical metallic tubing, rigid nonmetallic conduit, or cable armor, *Section 250-92*.

In this residence, a No. 4 AWG armored ground wire (service grounding electrode conductor) is run out of the top of the main service panel, across the ceiling of the workshop, over to the water pipe that comes through the basement wall in the front outside corner of the workshop area. Here the grounding electrode conductor is connected to the water pipe that leads to the well with a ground clamp. Good workmanship dictates that this armored ground wire closely follow the structure of the building. If possible, staple this armored ground wire up and between the joists, out of sight and not subject to physical damage.

Fig. 4-33 Armored ground wire.

The grounding electrode conductor must be connected to a metal underground water system when the water pipe is in direct contact with the earth for at least 10 feet (3.05 m). The well casing must also be bonded to the water pipe. A supplemental (additional) grounding electrode must be installed. In this residence, a driven ground rod has been installed.

Service-entrance grounding and bonding are discussed in detail in unit 28.

See unit 19 for additional reference material relating to submersible pump grounding and bonding.

See unit 18 for additional discussion on grounding.

REVIEW

Note: Refer to the Code or the plans where necessary.

1. What is the largest size of solid building wire cable that is generally used rather than the stranded variety? ______________________________

2. What is the minimum branch circuit wire size that may be installed in a dwelling?
____

3. What exceptions, if any, are there to the answer for question 2? ______________
____
____

4. What determines the ampacity of a wire? ______________________________
____
____

5. What unit of measurement is used for the diameter of wires? ______________

6. What unit of measurement is used for the cross-sectional area of wires? ________
____

7. What is the maximum voltage rating of all building wire and cable? __________
8. Indicate the ampacity of the following type THHN (copper) conductors. Refer to *Table 310-16*.
   a. 14 AWG ________ amperes
   b. 12 AWG ________ amperes
   c. 10 AWG ________ amperes
   d. 8 AWG ________ amperes
   e. 6 AWG ________ amperes
   f. 4 AWG ________ amperes
9. What is the maximum operating temperature of the following conductors? Give the answer in degrees Fahrenheit and degrees Celsius.
   a. Type XHHW ________________
   b. Type RH ________________
   c. Type THHN ________________
   d. Type TW ________________
10. What are the colors of the conductors in nonmetallic-sheathed cable for
    a. two-wire cable? ____________, ____________
    b. three-wire cable? ____________, ____________, ____________
11. For nonmetallic-sheathed (type NM) cable, can the uninsulated conductor be used for purposes other than grounding? ____________________
12. What size ground wire is used with the following sizes of nonmetallic-sheathed cable?
    a. 14 AWG ________________
    b. 12 AWG ________________
    c. 10 AWG ________________
    d. 8 AWG ________________
13. Under what condition may nonmetallic-sheathed cable (type NM) be fished in the hollow voids of masonry block walls? ____________________

    ____
14. a. What is the maximum distance permitted between straps on a cable installation?

    ____

    b. What is the maximum distance permitted between a box and the first strap in a cable installation? ____________________
15. What is the difference between type AC and type ACT cable? ____________

    ____

    ____

    ____

    ____
16. Type ACT cable may be bent to a radius of not less than ____________ times the diameter of the cable.
17. When armored cable is used, what protection is provided at the cable ends?

    ____

    ____
18. What protection must be provided when installing a cable in a notched stud or joist, or when a cable is run through bored holes in a stud or joist where the distance is less than 1¼ inches from the edge of the framing member to the cable, or where the cable is run parallel to a stud or joist and the distance is less than 1¼ inches from the edge of the framing member to the cable? ____________________

    ____

    ____

    ____

19. For installing directly in a concrete slab, (armored cable, nonmetallic-sheathed cable, conduit) may be used. Circle the correct method of installation.
20. Describe the Code requirements for the mechanical protection of service grounding electrode conductors. ____

____

____

____

____

____

____

21. The edge of a bored hole in a stud shall not be less than __________ inches from the edge of the stud.
22. Where is the main service entrance panel located in this residence? __________

____

23. a. Is nonmetallic-sheathed cable permitted in your area for residential occupancies?

____

b. From what source is this information obtained? ______________________
24. Is it permitted to use flexible metal conduit over 6 feet (1.83 m) in length as a grounding means? (Yes) (No) Circle one.
25. Liquidtight flexible metal conduit may serve as a grounding means in sizes up to and including __________ inches where used with approved fittings.
26. The allowable current-carrying capacity (ampacity) of aluminum wire, or the maximum overcurrent protection in the case of No. 14, No. 12, and No. 10 AWG conductors, is less than that of copper wire for a given size, insulation, and temperature of 86 °F. Refer to *Table 310-16* and footnotes and complete the following table:

	COPPER		ALUMINUM	
WIRE	Ampacity	Overcurrent Protection	Ampacity	Overcurrent Protection
No. 12 THHN				
No. 10 THHN				
No. 3 THW				
0000 THWN				
500 KCMIL THWN				

* ENTER BOTH AMPACITY AND MAXIMUM OVERCURRENT PROTECTION VALUES.

27. It is permissible for an electrician to connect aluminum, copper, or copper-clad aluminum conductors together in the same connector. True or False. Circle one.
28. Terminals of switches and receptacles marked CO/ALR are suitable for use with ______________, ______________, and ______________ conductors.
29. Wire connectors marked AL/CU are suitable for use with ________________, ________________, and ________________ conductors.

30. A wire connector bearing no marking or reference to AL, CU, or ALR is suitable for use with (copper) (aluminum) conductors only. Underline the correct answer.

31. When type NM or NMC cable is run through a floor, it must be protected by at least ________ inches (________ mm) of ________________________________

____

____.

32. When nonmetallic-sheathed cables are "bunched" or "bundled" together for distances longer than 24 inches (610 mm), what happens to their current-carrying ability?

____

33. In diagrams A and B, nonmetallic-sheathed cable is run through the cold air return. Which diagram "Meets Code"? A ______ B ______ Check one.

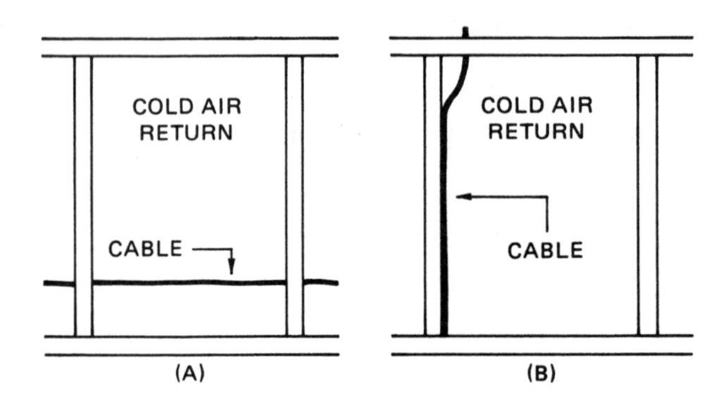

34. The marking on the outer jacket of a nonmetallic-sheathed cable indicates the letters NMC-B. What does the letter B signify? ________________________________

____

____

35. A 120-volt branch circuit supplies a resistive heating load of 10 amperes. The distance from the panel to the heater is approximately 140 feet. Calculate the voltage drop using (a) No. 14, (b) No. 12, (c) No. 10, (d) No. 8 AWG copper conductors.

36. In question 35, it is desired to keep the voltage drop to 3% maximum. What minimum size wire would be installed to accomplish this 3% maximum voltage drop?

37. *Section 310-16, Note 3* states that the neutral conductor of a residential service or feeder can be reduced to two sizes smaller than the phase conductors, but only

38. The ampacity of a No. 4 THHN from *Table 310-16* is 95 amperes. What is this conductor's ampacity if connected to a terminal listed for use with 60°C wire?

39. If, because of some obstruction in a wall space, it is impossible to keep an NMC cable at least 1/4 inch from the edge of the stud, then it shall be protected by a metal plate at least __________ inch thick.

40. The recessed fluorescent fixtures installed in the ceiling of the recreation room of this residence are connected with 3/8-inch flexible metal conduit. These flexible connections are commonly referred to as "fixture whips." Does the flexible metal conduit provide adequate grounding for the fixtures, or must a separate grounding conductor be installed?

41. Flexible liquidtight metal conduit will be used to connect the air-conditioner unit. It will be 3/4-inch trade size. Must a separate equipment grounding conductor be installed in this flex to ground the air conditioner properly?

42. What size overcurrent device protects the air-conditioning unit?

UNIT 5

Switch Control of Lighting Circuits, Receptacle Bonding, and Induction Heating Resulting from Unusual Switch Connections

OBJECTIVES

After studying this unit, the student will be able to

- identify the grounded and ungrounded conductors in cable or conduit (color coding).
- identify the various types of toggle switches for lighting circuit control.
- select a switch with the proper rating for the specific installation conditions.
- describe the operation that each type of toggle switch performs in typical lighting circuit installations.
- demonstrate the correct wiring connections for each type of switch per Code requirements.
- understand the various ways to bond wiring devices to the outlet box.
- understand how to design circuits to avoid heating by induction.

The electrician installs and connects various types of lighting switches. To do this, both the operation and method of connection of each type of switch must be known. In addition, the electrician must understand the meanings of the current and voltage ratings marked on lighting switches, as well as the *National Electrical Code*® requirements for the installation of these switches.

CONDUCTOR IDENTIFICATION (*ARTICLES 200* AND *210*)

Before making any wiring connections to devices, the electrician must be familiar with the ways in which conductors are identified. For alternating-current circuits, the Code requires that the grounded (identified) circuit conductor have an outer covering that is either white or a natural gray. In multiwire circuits, the grounded circuit conductor is also called a *neutral* conductor, *Sections 200-6(a)* and *210-5*.

The ungrounded (unidentified) conductor of a circuit must be marked in a color other than green, white, or gray. This conductor generally is called the *hot* conductor. A shock is felt if this conductor and the grounded conductor are held at the same time, or if this conductor and a grounded surface such as a water pipe are touched at the same time.

Neutral Conductor

In all residential, commercial, and industrial wiring, the grounded neutral conductor's insulation is white or natural gray. As mentioned in the previous section on "Conductor Identification," the grounded circuit conductor is also called a neutral conductor when it is part of a multiwire branch circuit.

Technically speaking, when a two-wire circuit is used, the white grounded circuit conductor is not truly a neutral conductor. To be called a neutral conductor correctly:

- it should be the conductor that carries only the unbalanced current from the other conductors, as in the case of multiwire circuits of *three or more* conductors, *Note 8* to *Table 310-16.*
- it should be the conductor where the voltage from every other conductor to it is equal under normal operating conditions.

Therefore, we can see that the white grounded circuit conductor of a two-wire circuit is not really a "neutral," even though many electricians refer to it as such. See figure 5-1.

Color Coding (Cable Wiring)

Nonmetallic-sheathed cable (Romex) and armored cable (BX) are color-coded as follows:

Two-Wire: one black ("hot" phase conductor)

one white (grounde*d* "identified" conductor)

one bare (equipment groundi*ng* conductor)

Three-Wire: one black ("hot" phase conductor)

one white (grounde*d* "identified" conductor)

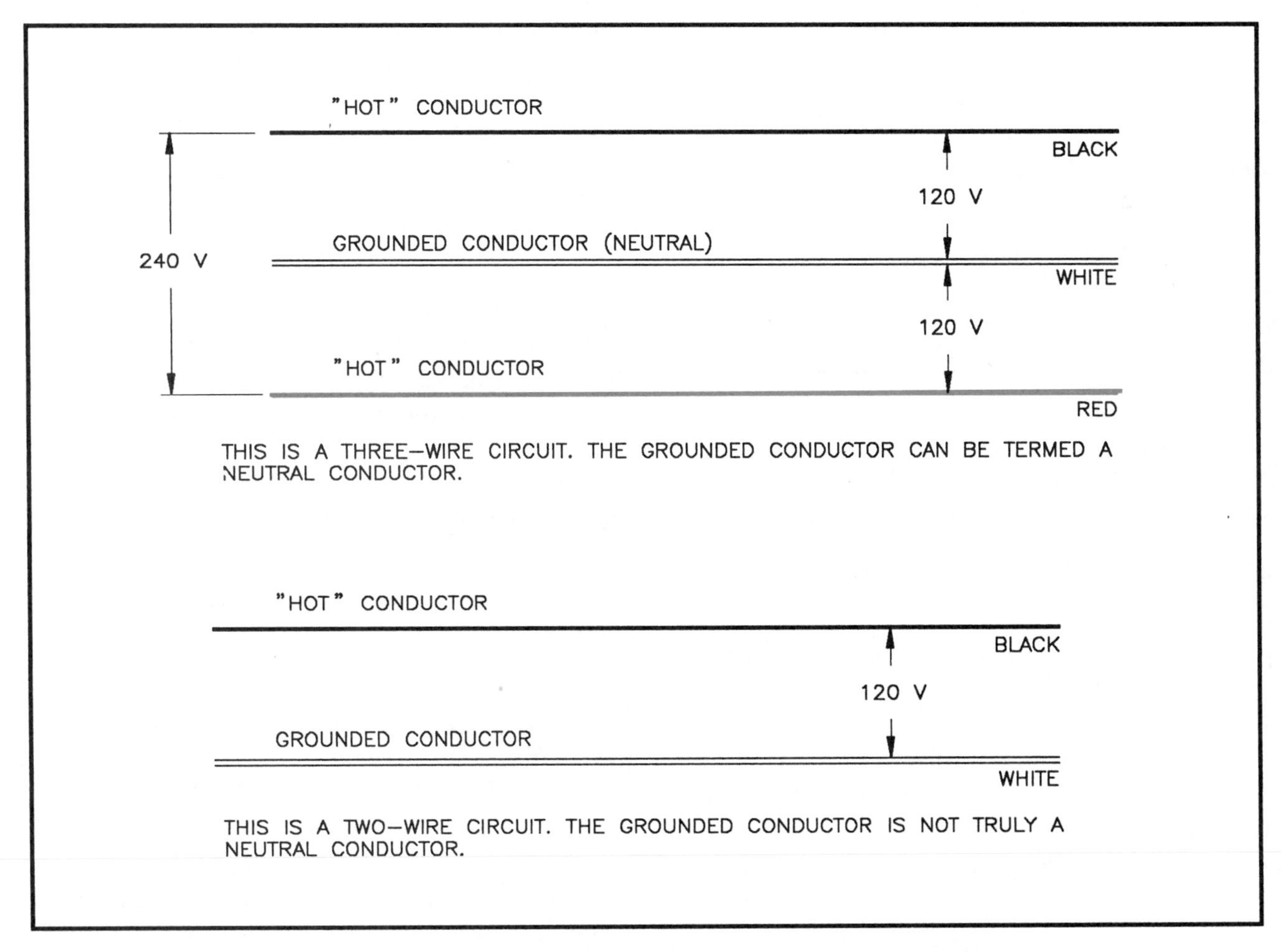

Fig. 5-1 Illustrations showing definition of a true neutral conductor.

one red ("hot" phase conductor)
one bare (equipment grounding conductor)

Four-Wire:* one black ("hot" phase conductor)
one white (grounded "identified" conductor)
one red ("hot" phase conductor)
one blue ("hot" phase conductor)
one bare (equipment grounding conductor)

Color Coding (Conduit Wiring)

When the installation is conduit, the electrician is permitted (*Section 210-5*) to use any color for the hot phase conductor except:

Green	reserved for use as a ground*ing* conductor only
White or Gray	reserved for use as the ground*ed* identified circuit conductor

Changing Colors

Should it become necessary to change the actual color of the conductor to meet Code requirements, the electrician may change the colors as follows:

FROM	TO	DO
red, black, blue, etc.	ground*ing* conductor	for conductors larger than No. 6 AWG, strip off insulation to make it bare — or paint it green where exposed in the box — or mark exposed insulation with green tape *(Section 250-57)*
red, black, blue, etc.	ground*ed* identified conductor	remark with colored tape or paint white or gray
white, gray, or green	red, black, blue, etc.	reidentify with colored tape or paint, figure 5-2

For cable installations only, *Section 200-7* of the Code, *Exception No. 2*, permits the white conductor to be used as a switch "loop" for single-pole, three-way, or four-way switches, but further states that the white conductor must feed the switch. Therefore, the connection must be made up so that the return conductor from the switch is a colored conductor. When the connections are done in this manner, the Code does not require reidentification, but instead, the Code accepts the actual splices and attachment of the conductors to the terminals of the switch as adequate identification.

A good example of where this reidentification is applicable is in figure 5-6, where the white wire in the octagon box and the white wire in the switch box are in reality hot phase conductors.

Although *Section 200-7, Exception No. 2* specifically states that reidentification is *not* required, some electrical inspectors demand that these white wires be reidentified with black plastic tape to avoid any confusion. The electrical inspector has the right to require this according to the authority given him by *Section 90-4* of the Code, which states in part:

"The authority having jurisdiction of enforcement of the Code will have responsibility for making interpretations of the rules, for deciding upon the approval of equipment and materials, and for granting the special permission contemplated in a number of the rules."

TOGGLE SWITCHES *(ARTICLE 380)*

The most frequently used switch in lighting circuits is the toggle flush switch, figure 5-3. When mounted in a flush switch box, the switch is concealed in the wall with only the insulated handle or toggle protruding.

Figure 5-4 shows how switches and receptacles are weatherproofed.

Toggle Switch Ratings

Underwriters Laboratories classifies toggle switches used for lighting circuits as *general-use*

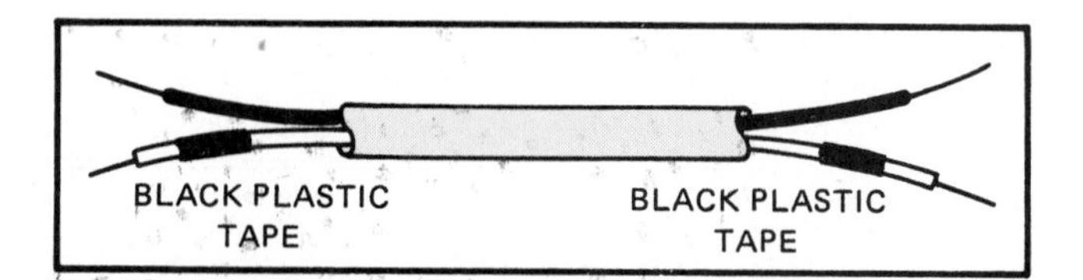

Fig. 5-2 A white wire may be wrapped with a piece of black plastic tape so that everyone will know that this wire is *not* a grounded conductor. Use red tape if it is desired to mark the conductor red, *Section 200-7, Exc. 1* and *2.*

*Almost impossible to find at supply houses.

snap switches. These switches are divided into two categories.

Category 1 contains those ac/dc general-use snap switches which are used to control

- alternating-current (ac) or direct-current (dc) circuits.
- resistive loads not to exceed the ampere rating of the switch at rated voltage.
- inductive loads not to exceed one-half the ampere rating of the switch at rated voltage unless otherwise marked.
- tungsten filament lamp loads not to exceed the ampere rating of the switch at 125 volts when marked with the letter "T."

A tungsten filament lamp draws a very high momentary inrush current at the instant the circuit is energized. This is because the *cold resistance* of tungsten is very low. For instance, the cold resistance of a typical 100-watt lamp is approximately 9.5 ohms. This same lamp has a *hot resistance* of 144 ohms when operating at 100% of its rated voltage.

Normal operating current would be

$$I = \frac{E}{R} = \frac{120}{144} = 0.83 \text{ ampere}$$

But, *maximum* instantaneous inrush current could be as high as

$$I = \frac{E}{R} = \frac{170 \text{ (peak voltage)}}{9.5} = 17.9 \text{ amperes}$$

This instantaneous inrush current drops off to normal operating current in about 6 cycles (0.10 second). The contacts of T-rated switches are designed to handle these momentary high inrush currents. See unit 13 for more information pertaining to inrush currents.

The ac/dc general-use snap switch normally is not marked ac/dc. However, it is always marked with the current and voltage rating, such as 10A-125V or 5A-250V-T.

Category 2 contains those ac general-use snap switches which are used to control

- alternating-currents only.
- resistive, inductive, and tungsten filament lamp loads not to exceed the ampere rating of the switch at 120 volts.
- motor loads not to exceed 80% of the ampere rating of the switch at rated voltage, but not exceeding 2 horsepower.

Ac general-use snap switches are marked ac only, in addition to identifying their current and voltage ratings. A typical switch marking is 15A, 120-277V ac. The 277-volt rating is required on 277/480-volt systems.

Terminals of switches rated at 15 or 20 amperes, when marked *CO/ALR*, are suitable for use with aluminum, copper, and copper-clad aluminum conductors. Switches not marked *CO/ALR* are suitable for use with copper and copper-clad conductors only.

Screwless pressure terminals of the conductor push-in type may be used with copper and copper-clad aluminum conductors only. These push-in type terminals are not suitable for use with ordinary aluminum conductors.

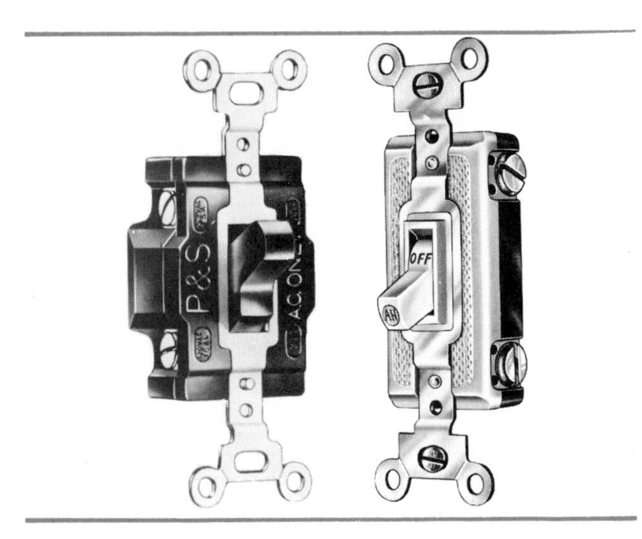

Fig. 5-3 Toggle flush switches.

Fig. 5-4 Receptacle outlet and toggle switch are protected by a weatherproof cover.

Further information on switch ratings is given in *NEC® Section 380-14* and in the Underwriters Laboratories *Electrical Construction Materials List*.

Refer to Table 4-1 for more information regarding the markings found on wiring devices indicating the types of conductors the devices are listed for.

Toggle Switch Types

Toggle switches are available in four types: single-pole switch, three-way switch, four-way switch, and double-pole switch.

Single-pole Switch: A single-pole switch is used when a light or group of lights or other load is to

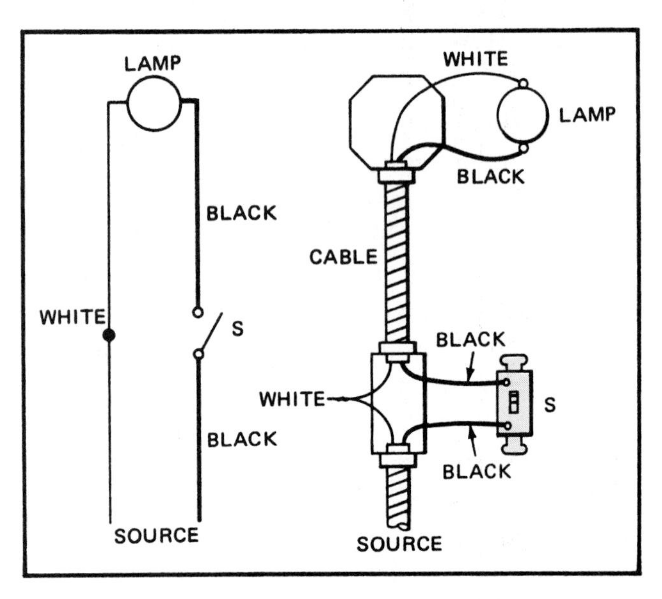

Fig. 5-5 Single-pole switch in circuit with feed at switch.

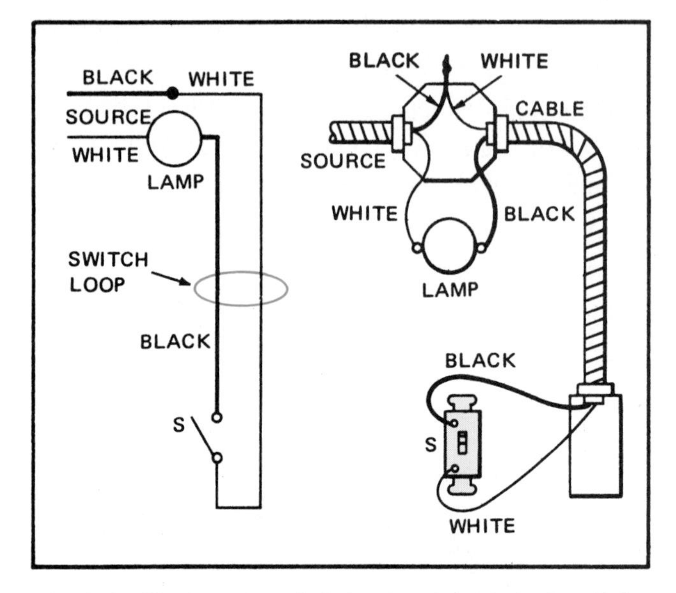

Fig. 5-6 Single-pole switch in circuit with feed at light.

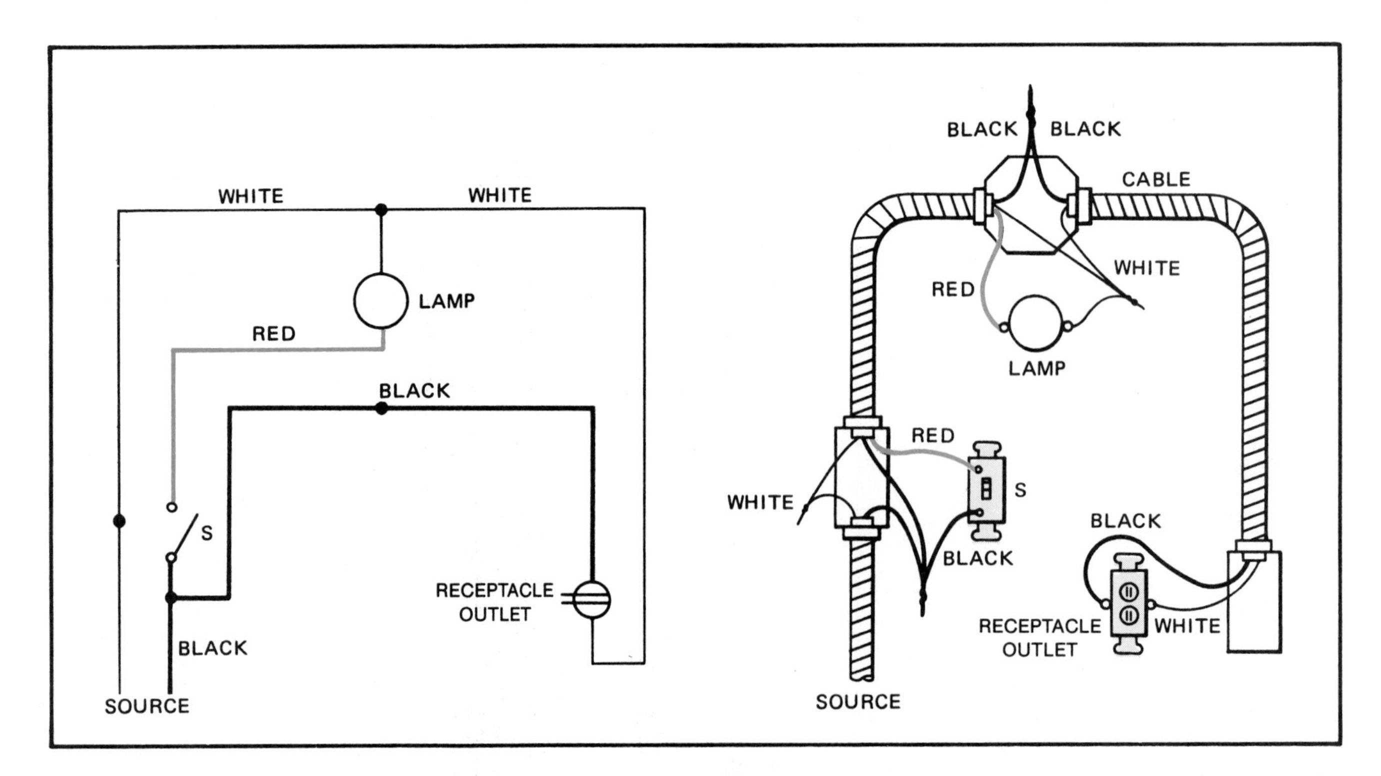

Fig. 5-7 Ceiling outlet controlled by single-pole switch with live receptacle outlet and feed at switch.

be controlled from one switching point, figure 5-5. The switch is identified by its two terminals and the toggle which is marked ON/OFF. The single-pole switch is connected in series with the ungrounded or hot wire feeding the load.

Figure 5-5 shows a single-pole switch controlling a light from one switching point. Note that the 120-volt source feeds directly through the switch location. In addition, the identified white wire goes directly to the load and the unidentified black wire is broken at the single-pole switch.

In figure 5-6, the 120-volt source feeds the light outlet directly so that a two-wire cable with black and white wires is used as a switch loop between the light outlet and the single-pole switch. The use of a white wire in a single-pole switch loop is permitted in *Section 200-7, Exc. No. 2*. The unidentified or black conductor must connect between the switch and the load as in figure 5-6.

Figure 5-7 shows another application of a single-pole switch. The feed is at the switch which controls the light outlet. The receptacle outlet is independent of the switch.

Three-way Switch: The name "three-way" is somewhat misleading since a three-way switch is used to control a light from two locations. A three-way switch has one terminal to which the internal switch mechanism is always connected. This terminal is called the *common* terminal. The two other terminals are called *traveler wire terminals*. The switching mechanism alternately switches between the common terminal and either one of the traveler terminals. Figure 5-8 shows the two positions of the three-way switch. Note that a three-way switch is actually a single-pole, double-throw switch.

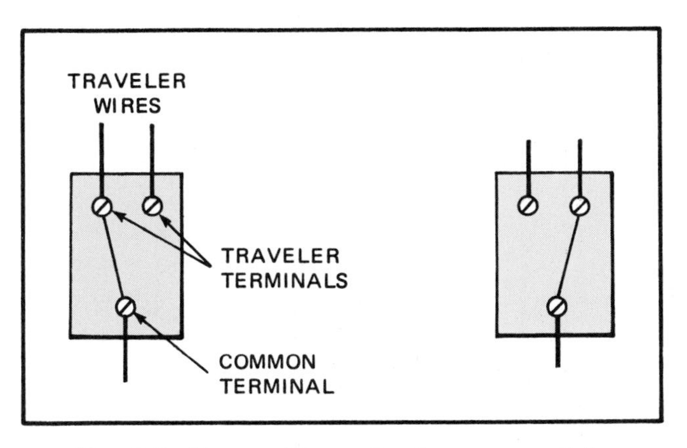

Fig. 5-8 Two positions of a three-way switch.

A three-way switch differs from a single-pole switch in that the three-way switch does not have an ON or OFF position. Thus, the switch handle does not have ON/OFF markings, figure 5-9. The three-way switch can be identified further by its three terminals. The common terminal is darker in color than the two traveler wire terminals which are usually natural brass in color.

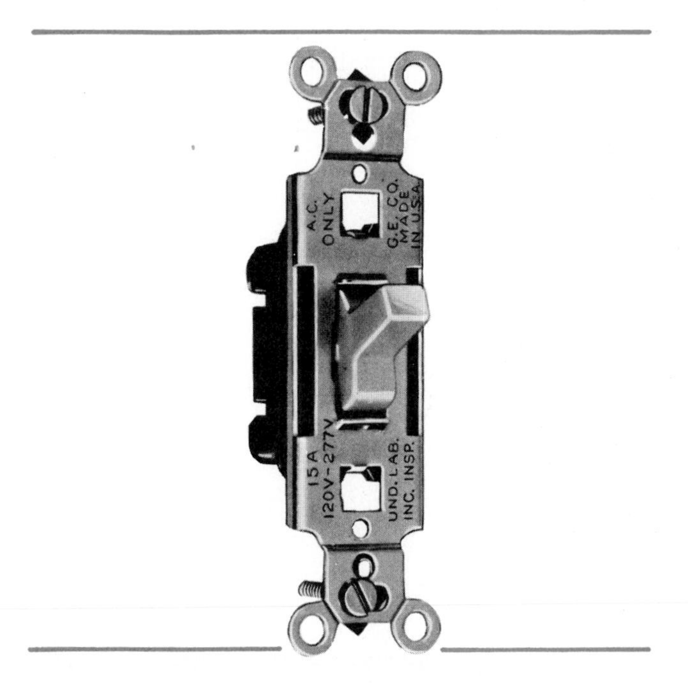

Fig. 5-9 Toggle switch: three-way flush switch.

Three-way Switch Control with Feed at Switch. Three-way switches are used when a load (or loads) is to be controlled from two different switching points. As shown in figure 5-10, two three-way switches are used. Note that the feed is at the first switch control point.

Three-way Switch Control with Feed at Light. The circuit in figure 5-11 uses three-way switch control with the feed at the light. For this circuit, the white wire in the cable must be used as part of the three-way switch loop. The unidentified or black wire is used as the return wire to the light outlet, in compliance with *NEC Section 200-7, Exception 2*. This exception makes it unnecessary to paint or identify with tape the terminal of the identified conductor at the switch outlet when wiring single-pole, three-way, or four-way switch loops.

Alternate Configuration for Three-way Switch Control with Feed at Light. Figure 5-12 shows another arrangement of components in a three-way switch circuit. The feed is at the light with cable runs from the ceiling outlet to each of the three-way switch control points located on either side of the light outlet.

The need for three-way switches occurs in rooms that have entry from more than one location. In this residence the garage lighting, entry hall lighting, living room receptacle outlets, rear outdoor lighting located between the living room slid-

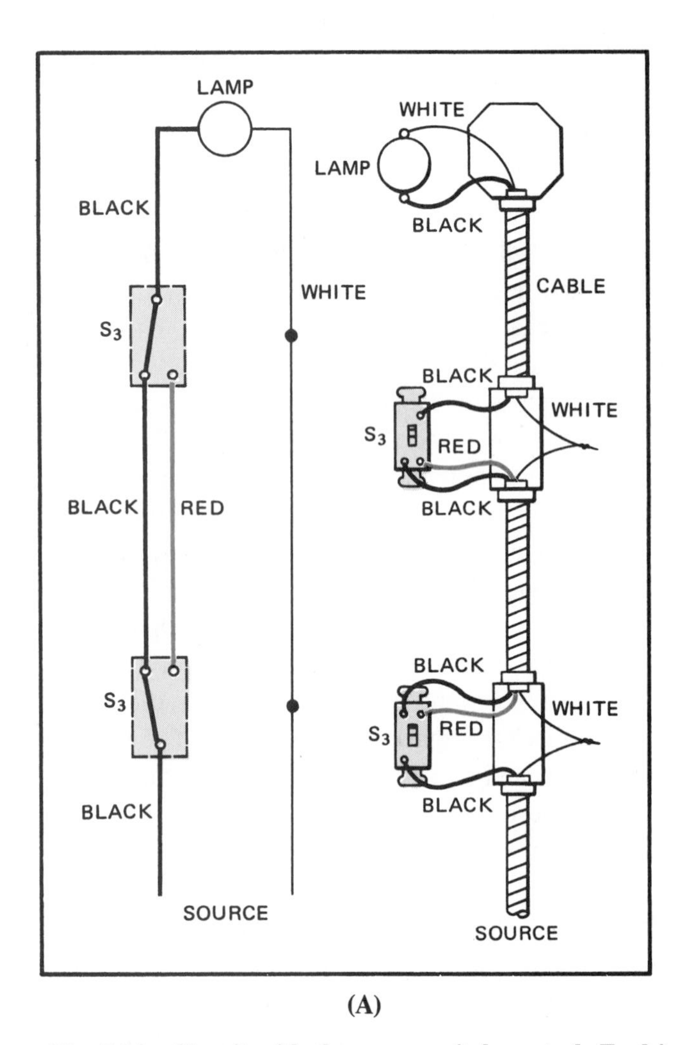

(A)

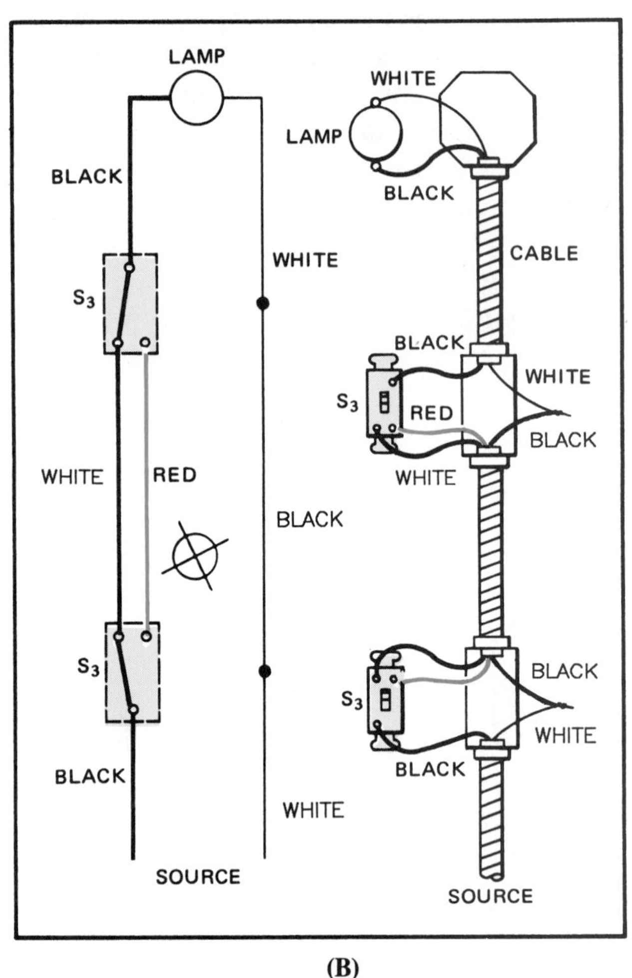

(B)

Fig. 5-10 Circuit with three-way switch control. Feed is at first switch control point. The load is downstream from the second switch control point. Diagram "A" uses black and red for the travelers. Diagram "B" uses red and white for the travelers. Either way is permitted by the Code. Be sure that the white conductor at the lampholder is connected to the lampholder's white (silver) terminal. See *Section 200-10, NEC®*

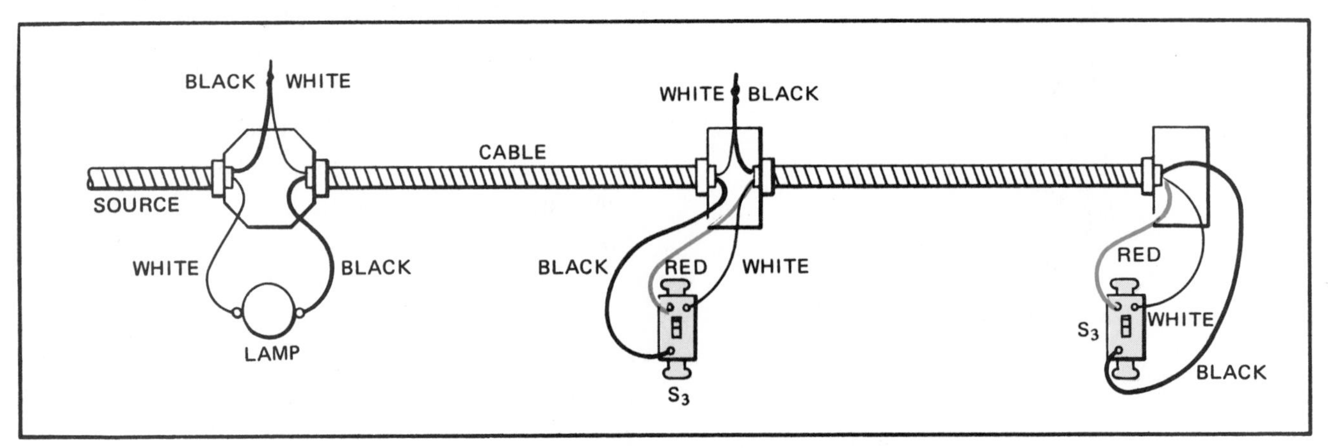

Fig. 5-11 Circuit with three-way switch control and feed at light.

ing doors and the kitchen sliding doors, and the receptacle outlets in the study are excellent examples of the installation of three-way switches.

Conductor Color Coding for Switch Connections

The color coding for the travelers (dummies) for three-way and four-way switch connections when using cable can vary. The Code, in *Article 200*, requires that the white (identified) conductor always be connected to the white (silver) terminal of a lampholder or receptacle. Because of the many ways that 3-way and 4-way switches can be connected, the choice of colors to use for the travelers is left to the electrician. Most electrical contractors and electricians will establish some sort of color coding that works well for them.

One way is to establish white and red for the travelers, then make up the electrical splices accordingly. Refer to figure 5-11. This is not a Code requirement, but rather, a matter of choice.

Another way is illustrated in figure 5-10(A). The white conductor is spliced straight through to the lampholder, while using black and red for the travelers. Figure 5-10(B) illustrates still another way to connect the switches, where red and white is used for the travelers.

This issue does not arise when the wiring method is conduit, because there is an unlimited choice of conductor insulation colors to use for the travelers, switch returns, and so forth.

Four-way Switch. The four-way switch is constructed so that the switching contacts can alternate their positions as shown in figure 5-13. The four-way switch has two positions, but neither position is ON or OFF. The four-way switch can be identified readily by its four terminals and by the fact that the toggle does not have ON or OFF markings.

Four-way switches are used when a load must be controlled from more than two switching points. To accomplish this, three-way switches are connected to the source and to the load. The switches at all other control points, however, must be four-way switches.

Figure 5-14 shows how a lamp can be controlled from any one of three switching points. Care must be used in connecting the traveler wires to the proper terminals of the four-way switch: the two traveler wires from one three-way switch are con-

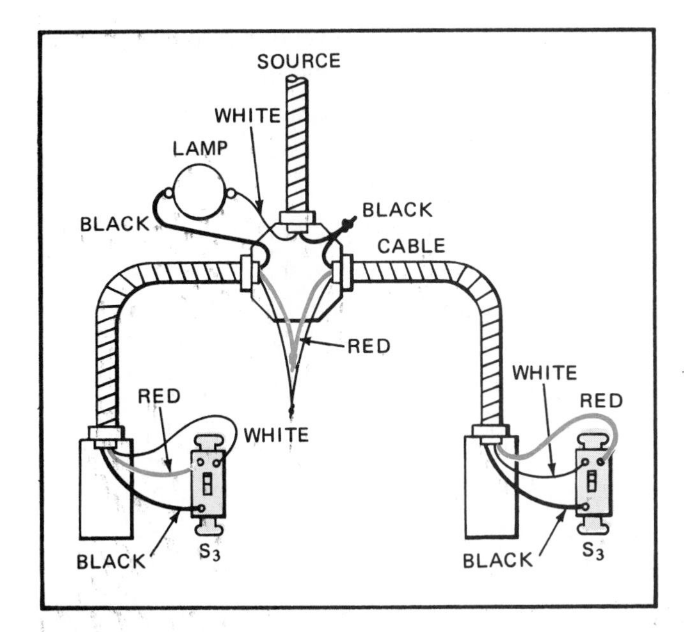

Fig. 5-12 Alternative circuit with three-way switch control and feed at light.

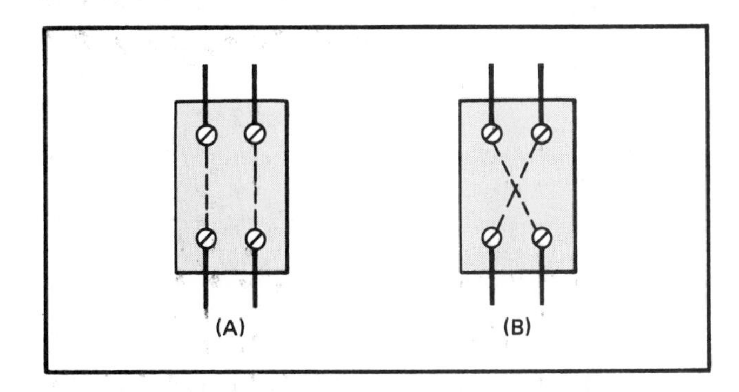

Fig. 5-13 Two positions of a four-way switch.

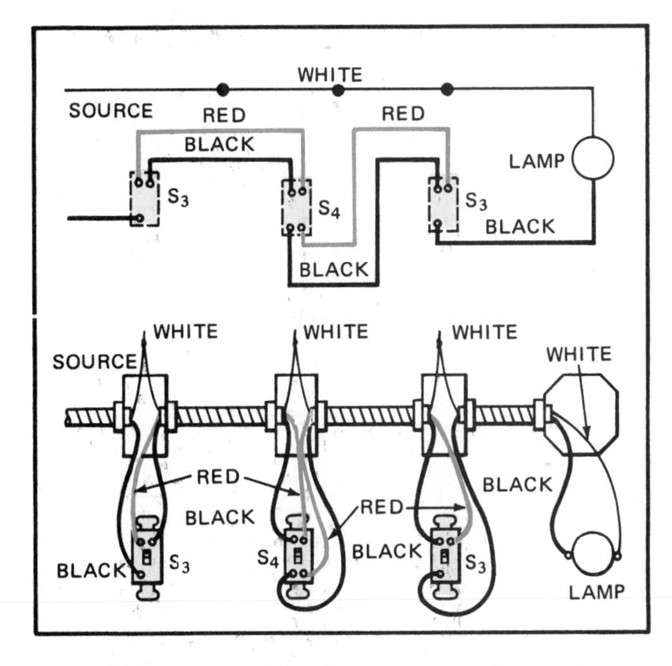

Fig. 5-14 Circuit with switch control at three different locations.

nected to the two terminals on one side of the four-way switch, and the other two traveler wires from the second three-way switch are connected to the two terminals on the other side of the four-way switch.

Double-pole Switch. A double-pole (2-pole) switch may be used when two separate circuits must be controlled with one switch, figure 5-15. A 2-pole switch may also be used to provide 2-pole disconnecting means for a 240-volt load, figure 5-16.

Double-pole toggle switches are not commonly used in residential work. Double-pole disconnect switches, however, are used quite often in residences for the furnace, water pump motors, and other 240-volt feeders.

Switches with Pilot Lights. There are instances when a pilot light is desired at the switch location, as is the case for the attic lighting in this residence.

Pilot light switches are available in several styles. See figures 5-17, 5-18, and 5-19.

BONDING AT RECEPTACLES

Ground-fault protection is covered in detail in unit 6. However, the subject of how to provide bonding of the equipment grounding of conductor to a metal box and to the receptacles ground*ing* terminal is important.

To wire grounding-type receptacles with armored cable, the metal armor and the integral bonding strip under the armor are adequate to ground the metal outlet box and the grounding terminal of the receptacle. If nonmetallic-sheathed cable is used, the cable must contain a separate grounding conductor, *Sections 210-7, 250-45,* and *250-57.* Figure 5-20 shows the groundng conductor

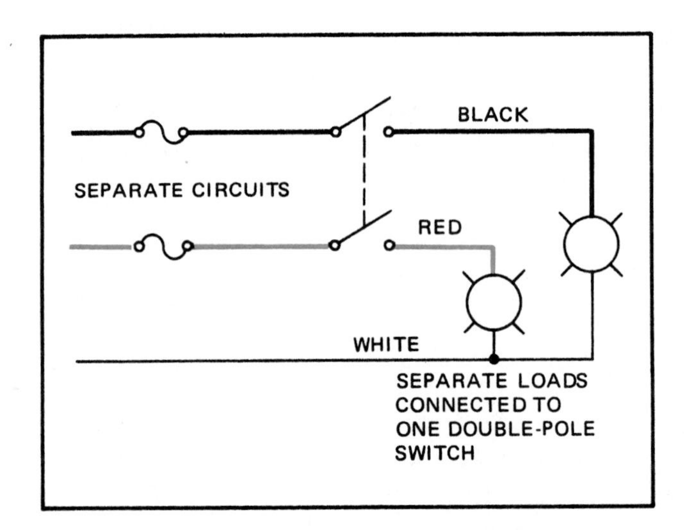

Fig. 5-15 Application of a double-pole switch.

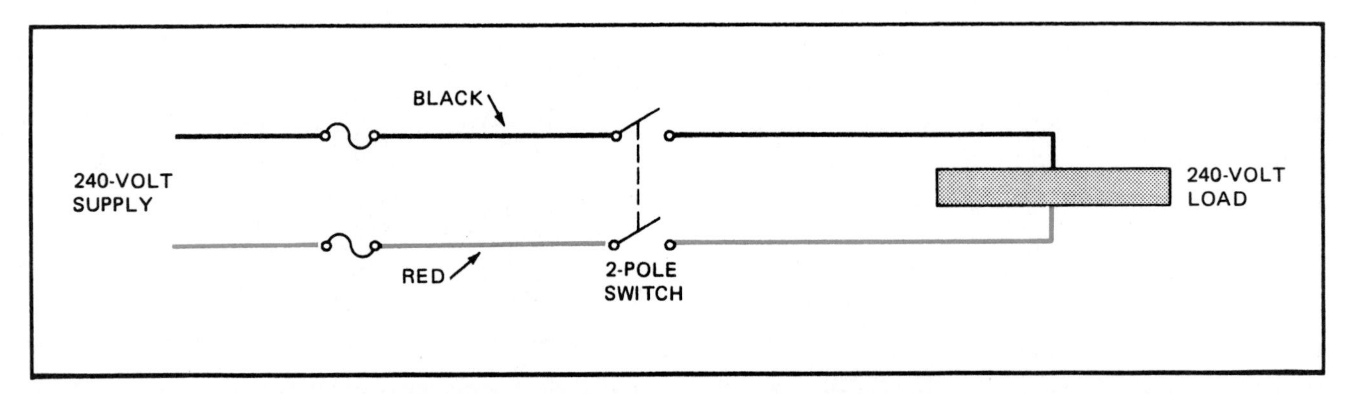

Fig. 5-16 Double-pole (2-pole) disconnect switch.

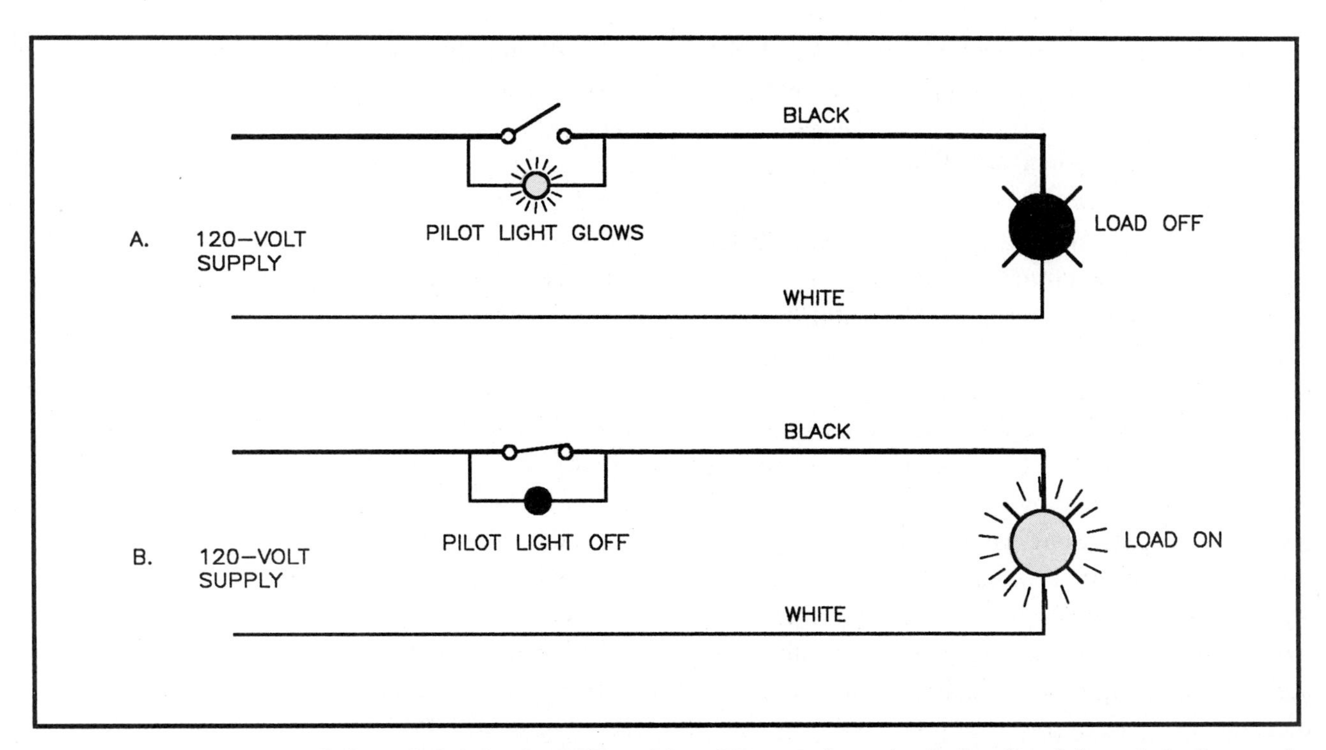

Fig. 5-17 In diagram "A," the switch is in the OFF position. The neon lamp in the handle of the switch glows and the load is off. This is a series circuit. The neon lamp has extremely high resistance. In a series circuit voltage divides proportionately to the resistance. Therefore, the neon lamp "sees" line voltage (120 volts) for all practical purposes, and the load "sees" zero voltage. The neon lamp glows.

In diagram "B," the switch is in the ON position. The neon lamp is shunted out and therefore has zero voltage across it. Thus, the neon lamp does not glow. Full voltage is supplied to the load. This type of switch might be referred to as a *locator* because it glows when the load is *off* and does not glow when the load is *on*.

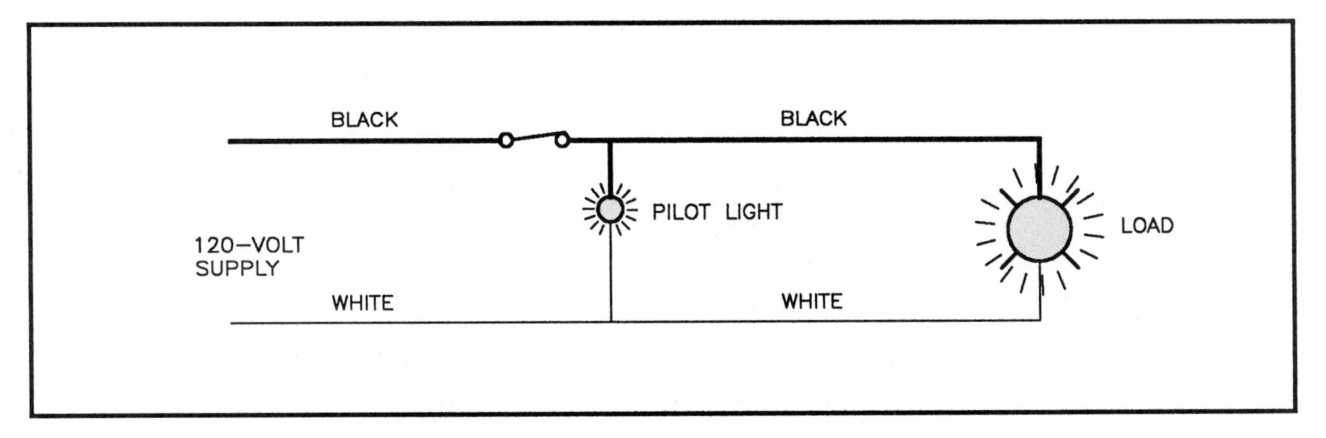

Fig. 5-18 This illustrates a true pilot light. The pilot lamp is an integral part of the switch. When the load is turned on, the pilot light is also on. When the load is turned off, the pilot light is also off. This switch has three terminals because it requires a grounded neutral circuit conductor.

of nonmetallic-sheathed cable attached to the receptacle per *Section 250-74*.

Almost all of the grounding terminals of the type shown in figure 5-20 (page 93) are bonded to the metal yoke of the receptacle. There is a question, however, of whether this grounding terminal provides grounding continuity between a grounded outlet box and the grounding circuit of the receptacle.

An outlet box is said to be grounded when it is connected to grounded cable armor or metal raceway, *Section 250-57*. When a grounding wire is not connected to the receptacle grounding terminal and the receptacle is fastened to a grounded device box, then the grounding terminal is automatically grounded through the threads of the No. 6-32 mounting screws. For a 4-inch square outlet box, grounding to the raised cover takes place through the threads of the No. 8-32 screws. Grounding to the outlet box occurs through the threads of the No. 8-32 cover mounting screws. However, it is doubtful that the continuity of the grounding circuit is reliable under these conditions.

Section 250-74 requires the connection of a separate grounding conductor to the grounding screw of the grounding-type receptacle. This requirement holds regardless of the wiring method used—metal conduit, armored cable, or nonmetallic-sheathed cable with a ground conductor. In other words, the metal outlet box must be grounded in addition to using the grounding terminal of the receptacle, figure 5-20.

To make sure that the continuity of the equipment grounding conductor path is insured, *Section 250-114* of the Code requires that when more than one grounding conductor enters a box, they shall be spliced with devices "suitable for the use." Splices shall not depend on solder. The splicing must be done so that if a receptacle or other wiring device is removed the continuity of the equipment grounding path shall not be interrupted. This is clearly shown in figures 5-21 and 5-22. When wiring with nonmetallic-sheathed cable, the equipment grounding conductor is a noninsulated, bare conductor. When wiring with conduit, an equipment grounding conductor, when required, must be a green insulated conductor, sized per *Table 250-95* of the Code.

Splices must be made in accordance with *Section 110-14(b)*.

A bonding jumper, figure 5-20, is not necessary when the receptacle is fastened to a metal surface-mounted box such as the one in figure 18-7. In this case, there is direct metal-to-metal contact between the receptacle yoke and the box (*Section 250-74, Exception No. 1*). The No. 6-32 mounting screws of most receptacles and switches are held captive in the yoke by a small fiber or cardboard washer. Removing this washer gives metal-to-metal contact.

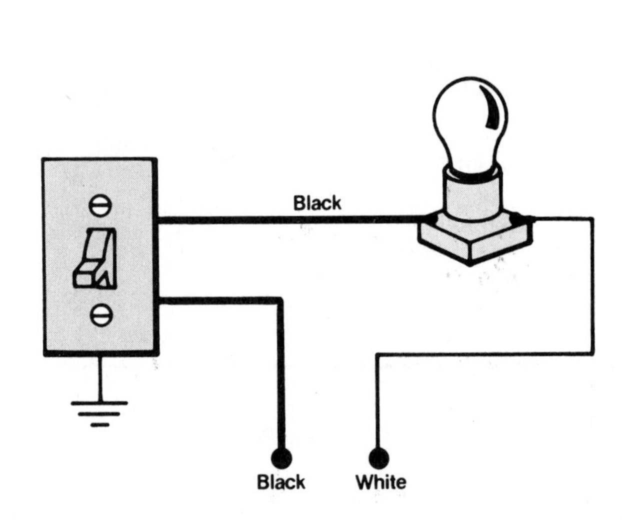

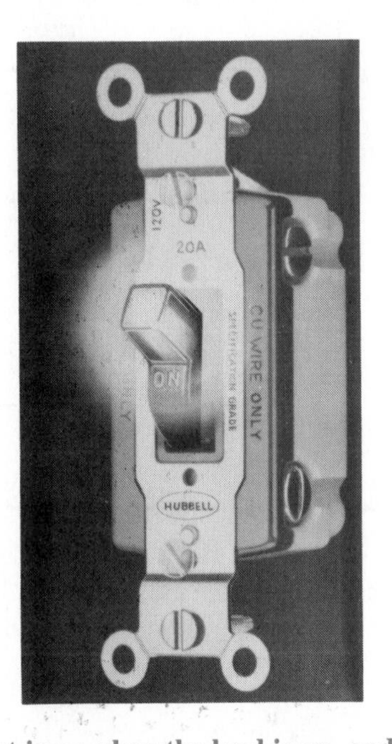

Fig. 5-19 This pilot light switch is a true pilot light. The pilot light is on when the load is on, and is off when the load is off. This is possible through its internal electronic circuit. This type of switch can be used on any 120-volt, ac grounded electrical circuit. It makes use of the system's equipment ground as its reference to permit the pilot light to glow. It does *not* require a grounded neutral conductor. The pilot light "pulsates" at approximately 60 times per minute. Courtesy of Hubbell Incorporated.

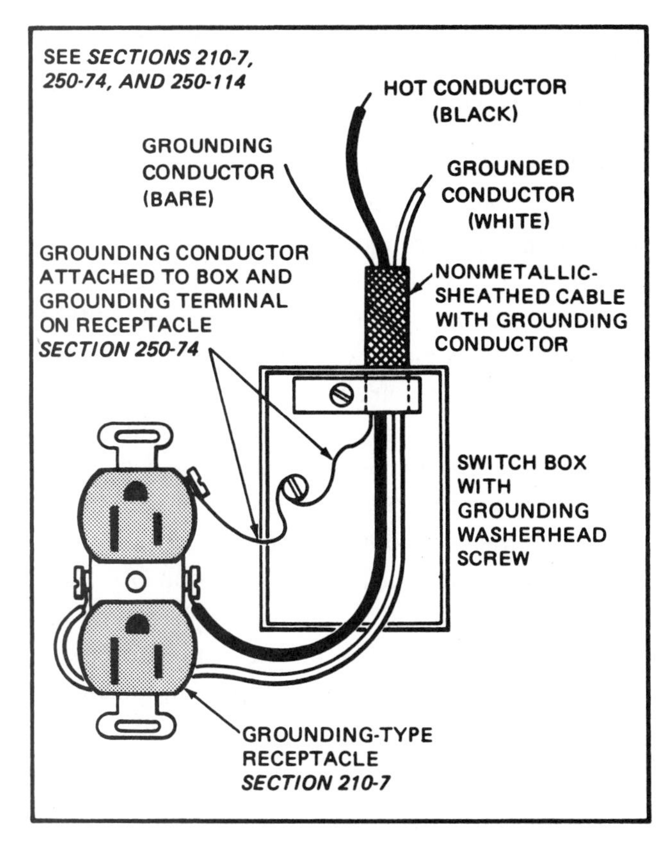

Fig. 5-20 Detail showing connections for grounding-type receptacles. See *Sections 210-7, 250-74*, and *250-114*.

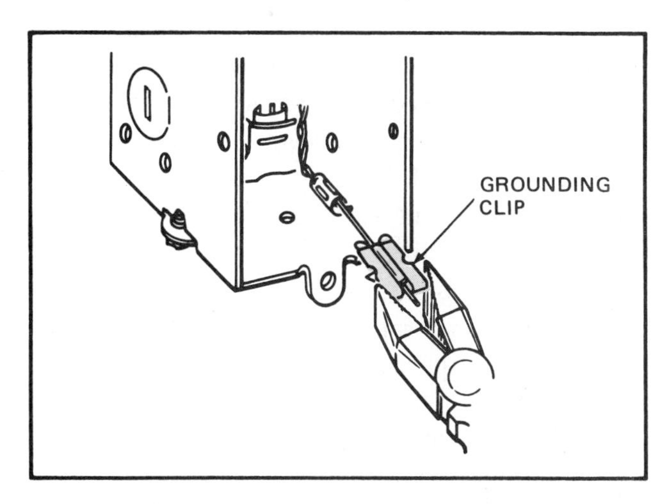

Fig. 5-22 Method of attaching grounding clip to switch (device) box. See *Sections 250-114* and *250-74.*

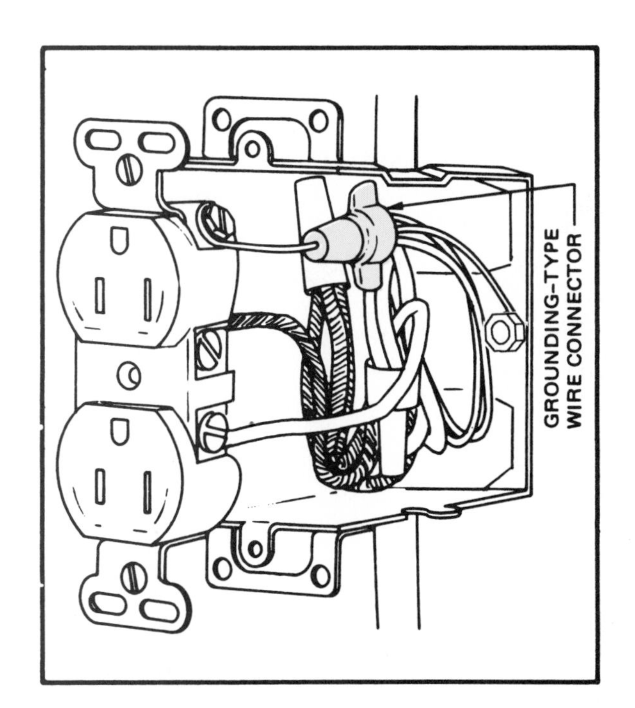

Fig. 5-21 An approved method of connecting the grounding conductor using a special grounding-type wire connector. See *Sections 250-114 and 250-74.*

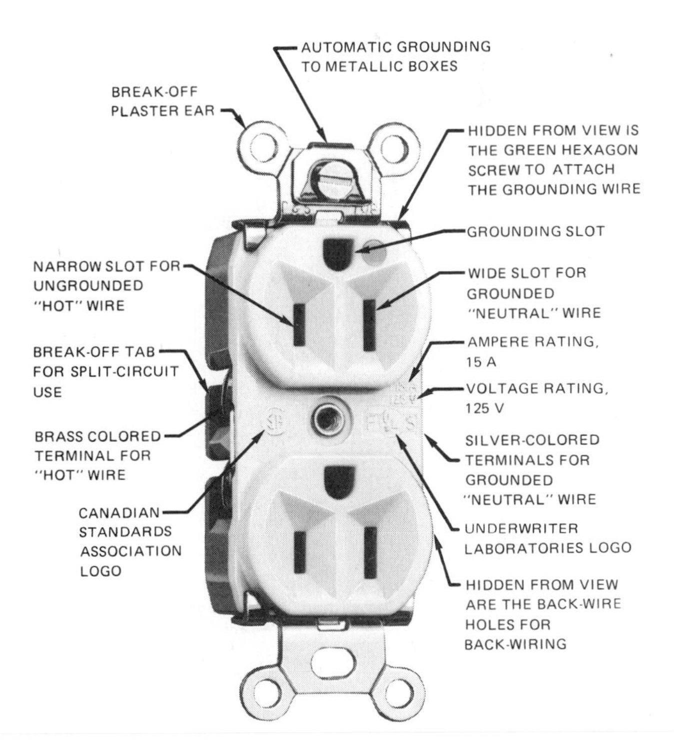

Fig. 5-23 Grounding-type receptacle detailing various parts of the receptacle. Courtesy of Pass & Seymour, Inc.

Some outlet and switch box manufacturers use washer head screws so that a grounding conductor can be placed under them. This method is useful when installing armored cable or nonmetallic-sheathed cable. That is, a means is provided to terminate the grounding conductor in the metal outlet or switch box. Grounding clips listed by Underwriters Laboratories can be used to clamp the small equipment grounding conductor to the edge of the outlet box or switch box, figure 5-21. Grounding clips are convenient since special tools are not required to fasten them.

Exception No. 2 to *Section 250-74* states that the bonding jumper need not be used when the receptacle has a special self-grounding strap, figures 5-23 and 5-24. When such a receptacle is to be installed, a bonding jumper is not wired from the receptacle to the grounded outlet box.

NONGROUNDING-TYPE RECEPTACLES

Modern appliances and portable hand-held tools often have a three-wire cord and a three-wire grounding-type attachment plug cap. However, these tools and appliances commonly must be plugged into nongrounding-type receptacles. *Section 410-58* permits the use of movable self-restoring grounding contacts on grounding-type attachment plugs of power supply cords. Such a movable grounding contact can be pushed out of the way when the attachment plug is placed in a nongrounding receptacle outlet. The appliance is *not* grounded when the contact (prong) is pushed out of the way. If the grounding prong does not fold back to its proper position, the user will probably cut it off. As a result, it will be impossible to ground the appliance even when it is plugged into a properly connected three-wire grounding-type receptacle.

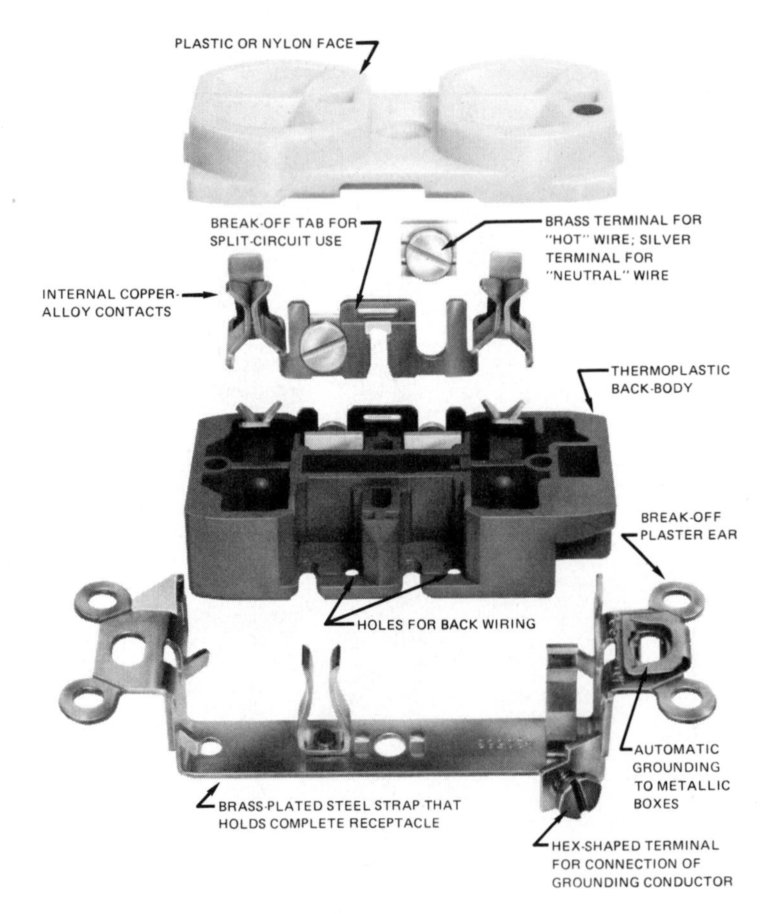

Fig. 5-24 Exploded view of a grounding-type receptacle, showing all internal parts. Courtesy of Pass & Seymour, Inc.

INDUCTION HEATING

When alternating-current circuits are run in metal raceways, trenches, and through openings in metal boxes, the circuiting must be arranged to prevent *induction heating* of the metal, *Section 300-20*. This means that the conductors of a given circuit must be arranged so the flux surrounding each conductor will cancel each other, figure 5-25. This also means that when individual conductors of the same circuit are run in trenches, they should be kept together, not spaced far apart. See figures 5-26, 5-27, and 5-28.

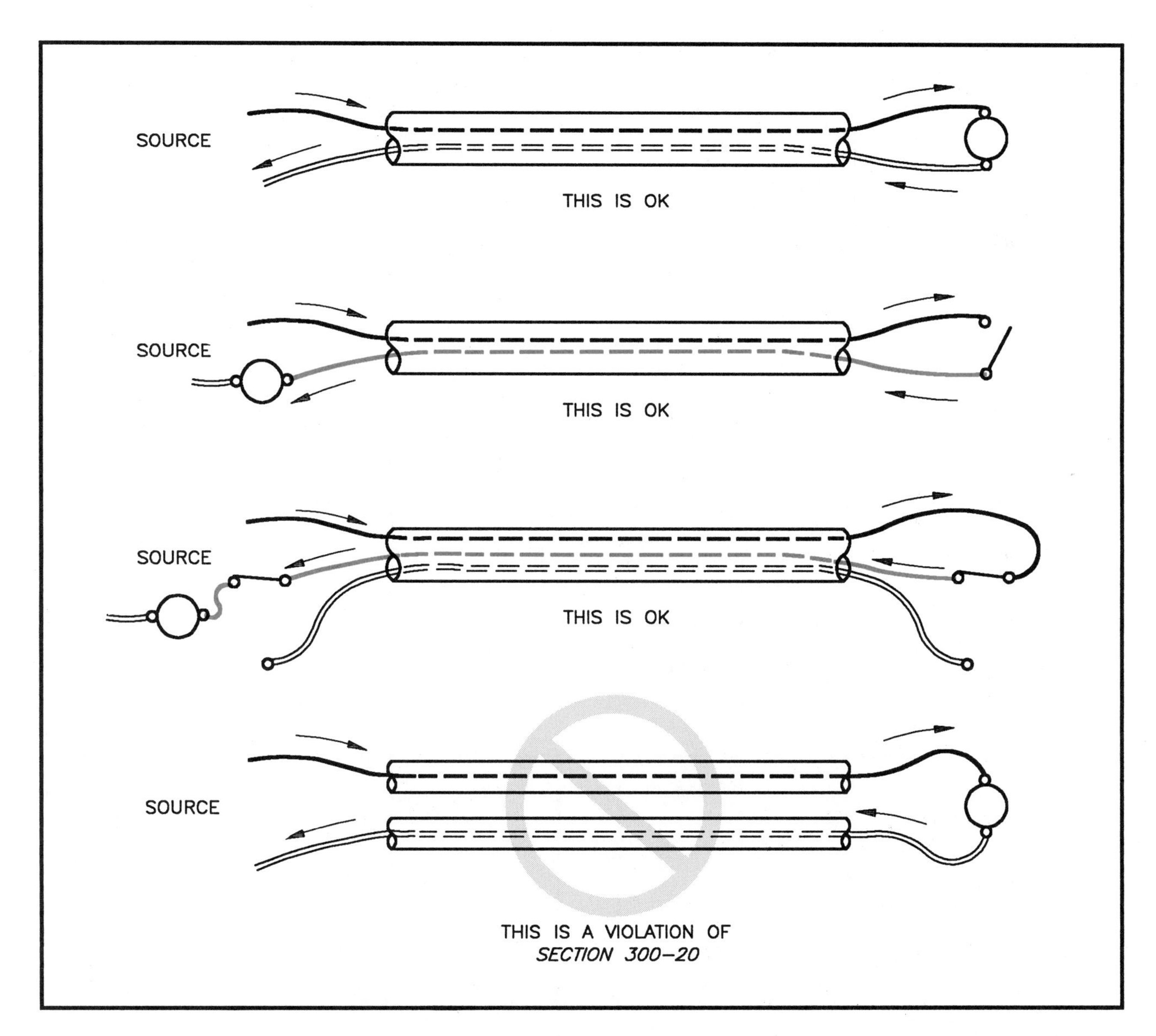

Fig. 5-25 A few examples of arranging circuitry to avoid induction heating according to *Section 300-20* when metal raceway or metal cables (BX) are used. Always ask yourself, "Is the same amount of current flowing in both directions in the metal raceway?" If the answer is "no," there will be induction heating that can damage the insulation on the conductors.

Obviously, the above situation is rarely if ever encountered in house wiring, *but* some electricians have learned that they can cut corners and save substantial lengths of three-wire cable in hooking up three-way and four-way switches in unusual ways. They use two-wire cable and make connections as shown in figure 5-29. This unusual hook-up, however, is permitted when nonmetallic boxes are used, but it violates *Section 300-20* when metal boxes are installed. The question is, is this unorthodox connection worth the confusion during installation and later on when troubleshooting the circuit might become necessary? Check with the electrical inspection authority before attempting to use this type of circuitry.

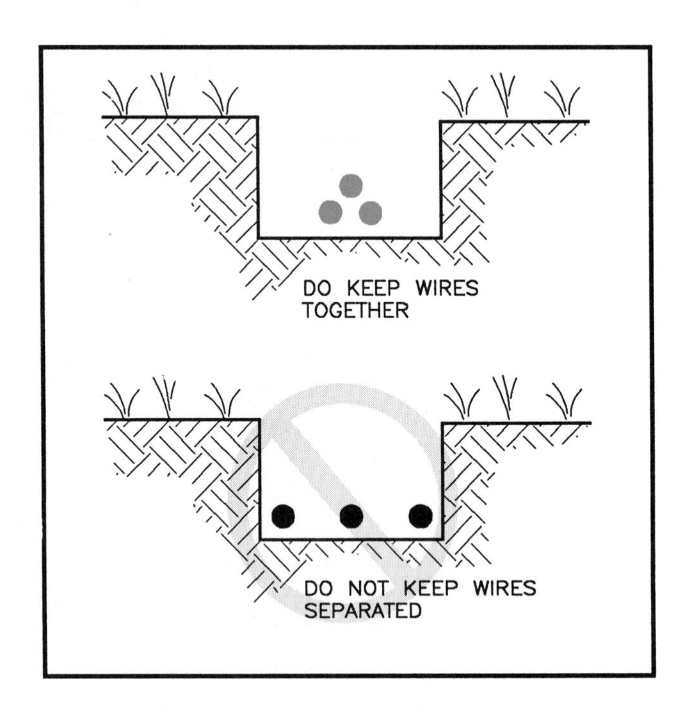

Fig. 5-26 Arrangement of conductors in trenches.

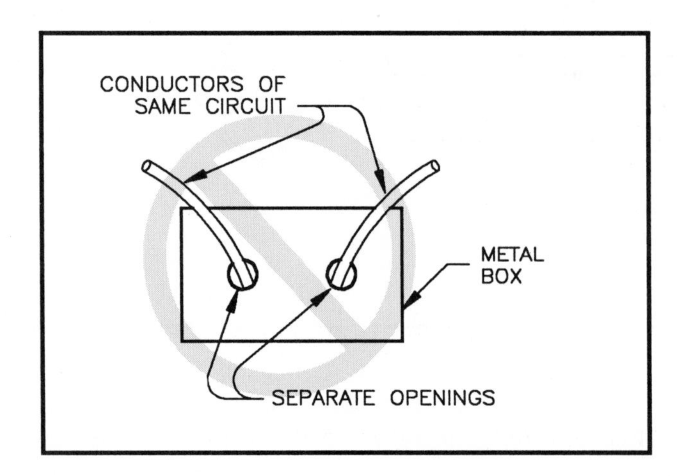

Fig. 5-27 Another typical example of how induction heating can occur is to run conductors of the same circuit through different openings of a metal box.

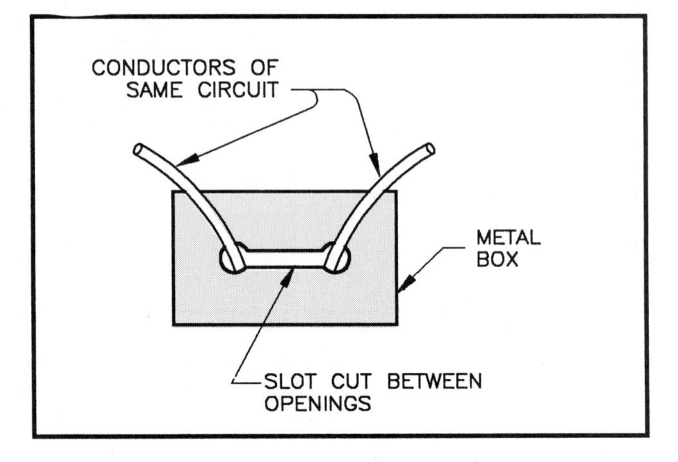

Fig. 5-28 To reduce the inductive effects the foregoing example can be minimized by cutting slots between the openings.

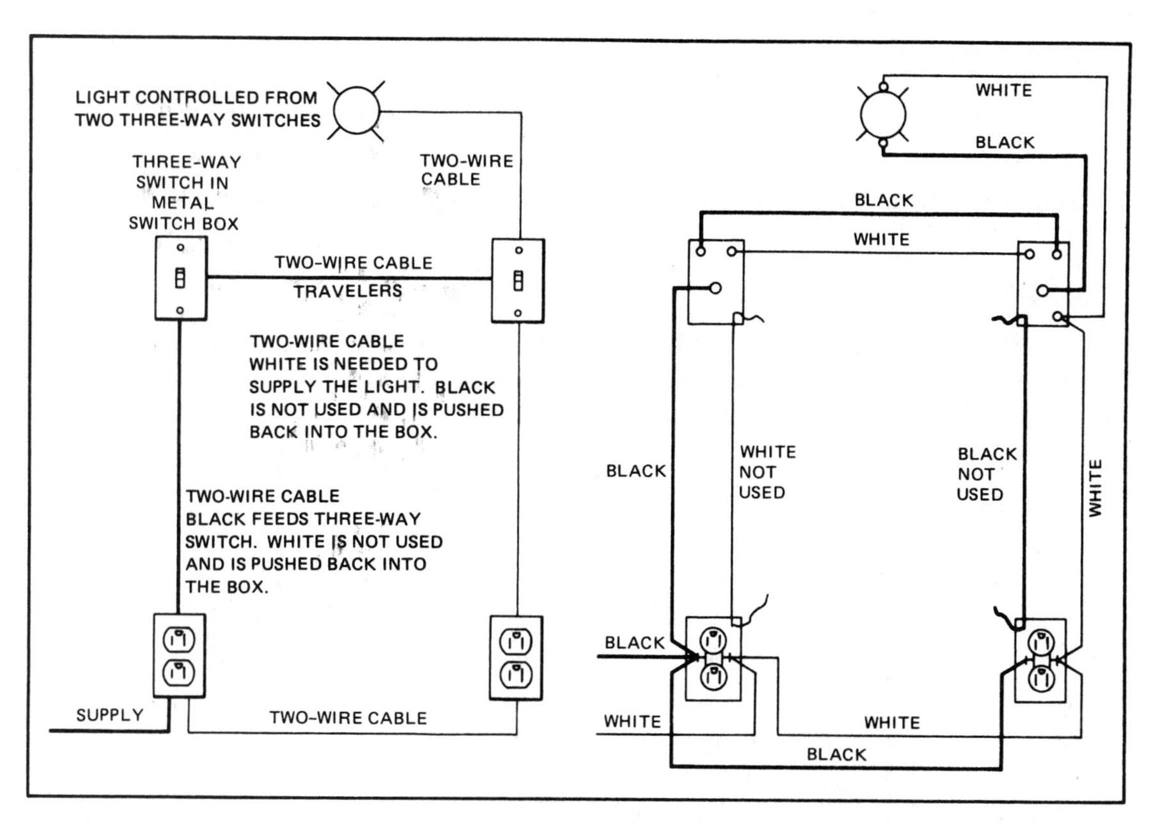

Fig. 5-29 Example of violation of *Section 300-20* when metal switch boxes are used. When non-metallic boxes are used, there is no Code violation.

REVIEW

Note: Refer to the Code or the plans where necessary.

1. The identified grounded circuit conductor must be ____________________ or ____________________ in color.

2. Explain how lighting switches are rated. ____________________

____

____

____

____

____

3. A T-rated switch may be used to its ____________________ current capacity when controlling an incandescent lighting load.

4. What switch type and rating is required to control five 300-watt tungsten filament lamps on a 120-volt circuit? Show calculations. ____________________

____

____

____

____

5. List four types of lighting switches.

   a. ______________________ c. ______________________

   b. ______________________ d. ______________________

6. To control a lighting load from one control point, what type of switch would be used?

   ____

7. Single-pole switches are always connected to the ______________________ wire.

8. Complete the connections in the following arrangement so that both ceiling light outlets are controlled from the single-pole switch. Assume the installation is in cable.

WHITE
120–VOLT
SOURCE
BLACK
LAMP
WHITE
RED
BLACK
LAMP
WHITE
BLACK
S

9.

WHITE
120–VOLT
SOURCE
BLACK
LAMP
SWITCH

   a. Complete the connections for the diagram. Installation is cable.

   b. Which conductor of the cable feeds the switch? ______________________

   c. Which conductor is used as the return wire? ______________________

   d. From which wire does the switch feed tap? ______________________

   e. What are the colors of the conductors connected to the fixture? ______________________

   ____

10. A three-way switch may be compared to a ______________________ switch.

11. What type of switch is installed to control a lighting fixture from two different control points? How many switches are needed and what type are they? ______________________

    ____

12. Complete the connections in the following arrangement so that the lamp may be controlled from either three-way switch.

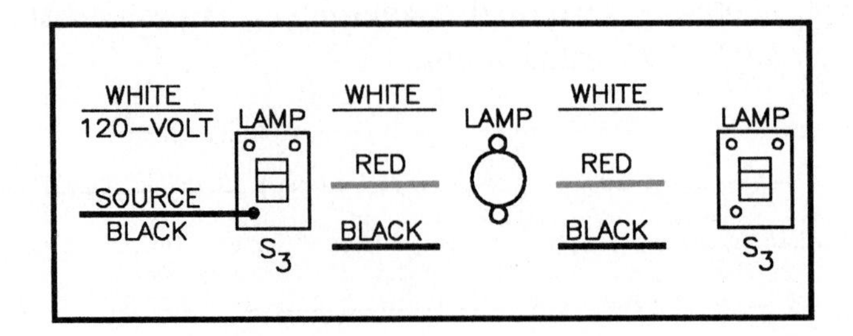

13. When connecting four-way switches, care must be taken to connect the travelers to the ______________________________ terminals.

14. Show the connections for a ceiling outlet which is to be controlled from any one of three switch locations. The 120-volt feed is at the light, *Section 200-7.* Label the conductor colors. Assume the installation is in cable.

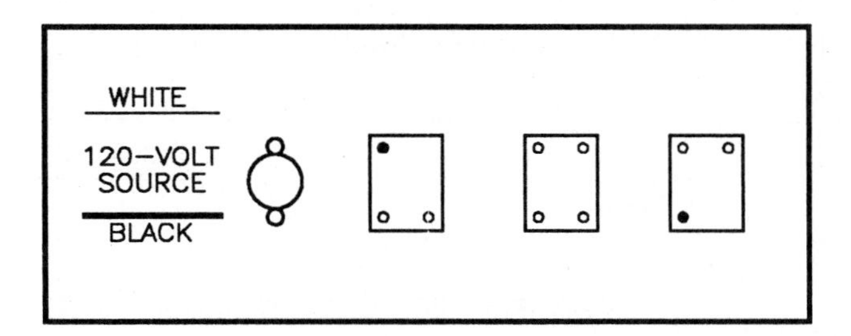

15. Match the following switch types with the correct number of terminals for each.

Three-way switch	Two terminals
Single-pole switch	Four terminals
Four-way switch	Three terminals

16. When connecting single-pole, three-way, and four-way switches, they must be wired so that all switching is done in the ______________________ circuit conductor.

17. What section of the Code emphasizes the fact that all circuiting must be done so as to avoid the damaging effects of induction heating? ______________________

18. If you had to install an underground three-wire feeder to a remote building using three individual conductors, which of the following installations "meet Code"? Circle correct installation.

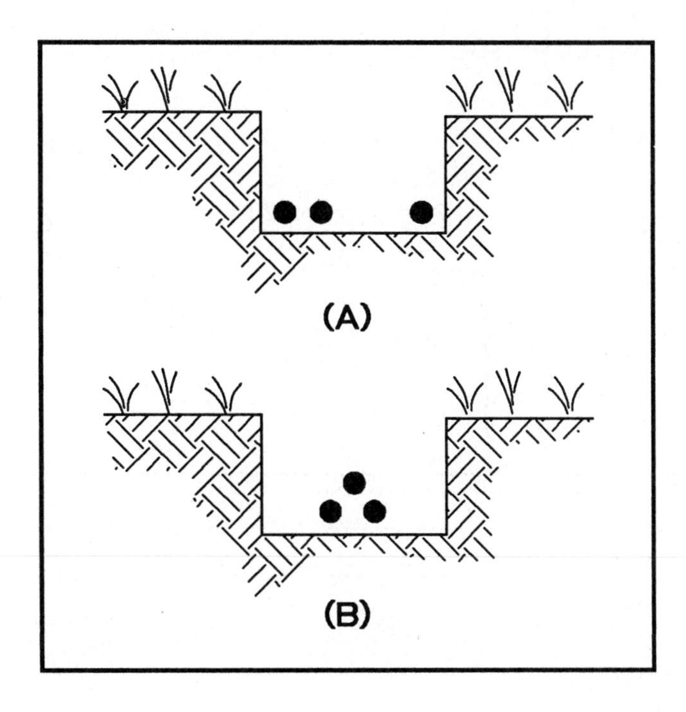

19. Is it always necessary to attach the bare equipment grounding conductor of a nonmetallic-sheathed cable to the green hexagon-shaped grounding screw on a receptacle? Explain.

____

20. List the methods by which a grounding conductor is connected to a device box.

____

21. When two nonmetallic-sheathed cables (Romex) enter a box, is it permitted to bring both of the bare grounding conductors directly to the grounding terminal of a receptacle, using the terminal as a splice point? _______________

22. This installation is in electrical metallic tubing. Since receptacle outlet "B" is only a few feet from switch "C," and switch "D" is only a few feet from receptacle outlet "A," the electrician saved a considerable amount of wire by picking up the feed for switch "C" from receptacle "B." The switch leg return is at switch "D." Please comment on this installation.

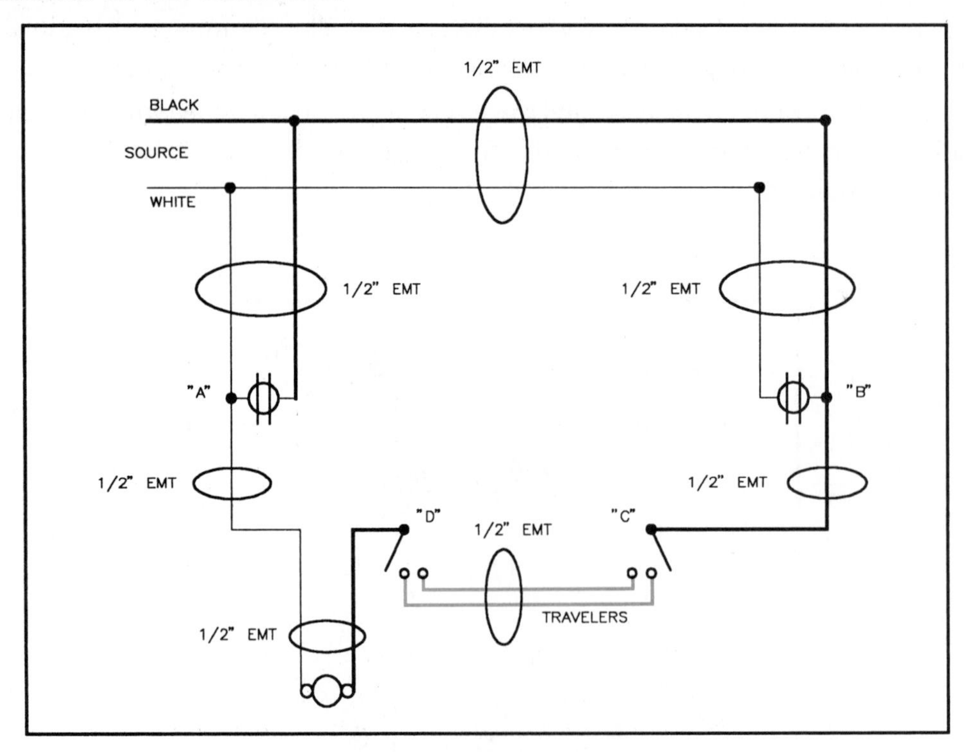

____

UNIT 6

Ground-Fault Circuit Interrupters, Transient Voltage Surge Suppressors, Isolated Ground Receptacles, Immersion Detection Circuit Interrupters, and Appliance Leakage Circuit Interrupters

OBJECTIVES

After studying this unit, the student will be able to

- understand the theory of operation of ground-fault circuit-interrupting (GFCI) devices.
- explain the operation and connection of GFCIs.
- discuss the do's and don'ts relative to the use of GFCI devices.
- explain why GFCIs are required.
- discuss locations where GFCIs must be installed in homes.
- understand the proper way to install and utilize feedthrough GFCI receptacles.
- understand testing, recording, and identification of GFCI receptacles.
- discuss the Code rules relating to the replacement of existing receptacles with GFCI receptacles.
- understand the Code requirements for GFCI protection on construction sites.
- understand the logic of the exemptions to GFCI mandatory requirements for receptacles in certain locations in kitchens, garages, and basements.
- discuss and understand the basics of transient voltage surge suppressors, and isolated ground receptacles.
- ► discuss and understand the principles of *immersion detection circuit interrupters*. ◄
- ► discuss and understand the principles of *appliance leakage circuit interrupters*. ◄

ELECTRICAL HAZARDS

Many lives have been lost because of electrical shock from an appliance or a piece of equipment that is *hot*. This means that the hot circuit conductor in the appliance is contacting the metal frame of the appliance. This condition may be due to the breakdown of the insulation because of wear and tear, defective construction, or accidental misuse of the equipment.

The shock hazard exists whenever the user can touch both the defective equipment and grounded surfaces, such as water pipes, metal sink rims, grounded metal lighting fixtures, earth, concrete in contact with the earth, water, or any other grounded surface.

Figure 6-1 shows a time-current curve that indicates *the amount of current* that a normal healthy adult can stand *for a certain time*. Just as in overcurrent protection for branch circuits, motors, appliances, and so on, it is always a matter of ***how much?*** for ***how long?***

CODE REQUIREMENTS FOR GROUND-FAULT CIRCUIT INTERRUPTERS *(SECTION 210-8)*

To protect against shocks, the *National Electrical Code®* requires that ground-fault circuit interrupters (GFCIs) be provided for the receptacle outlets of dwellings as follows:

- for *all* 125-volt, single-phase, 15- and 20-ampere receptacles in bathrooms and outdoors where there is direct grade-level access to the dwelling. Thus, it is not required to provide a GFCI-protected receptacle on a second floor balcony, such as in condominium and apartment wiring. See unit 10, figure 10-5 for the definition of a bathroom.

- for *all* 125-volt, single-phase 15- and 20-ampere receptacles in garages, attached or detached. GFCI protection, however, is not required if the receptacle is not readily accessible, such as one which is installed on the ceil-

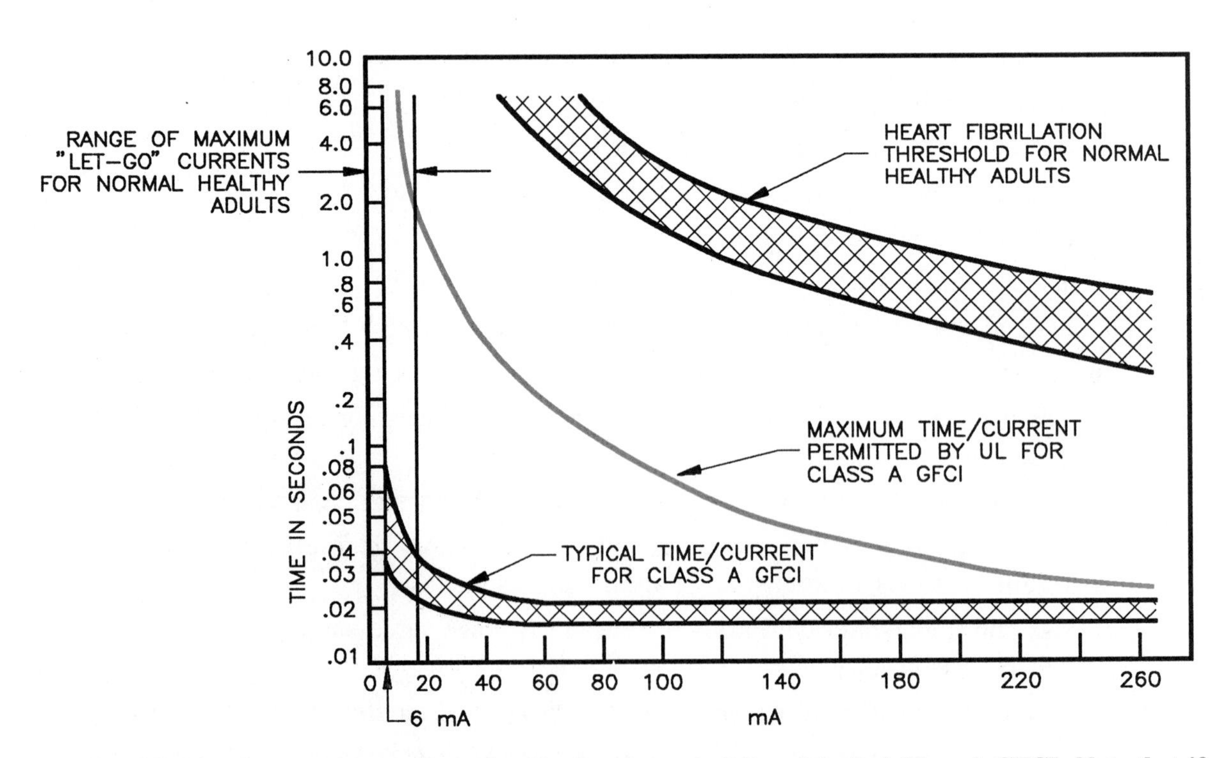

Fig. 6-1 The time/current curve shows the tripping characteristics of typical Class A GFCI. Note that if you follow the 6-mA line vertically to the crosshatched typical time/current curve, you will find that the GFCI will open in from approximately 0.035 second to just less than 0.1 second. One electrical cycle is 1/60 of a second (0.0167 second).

ing for the overhead garage door opener. GFCI protection is not required if the receptacle is installed for a cord-connected appliance occupying a dedicated space, such as a freezer.

- in unfinished basements, and in crawl spaces that are at or below grade level, all 125-volt, single-phase, 15- or 20-ampere receptacles must have GFCI protection. The Code does *not* require GFCI receptacles for the following:
  1. the receptacle for laundry equipment.
  2. a *single-outlet* receptacle that supplies a permanently installed sump pump.
  3. a *single-outlet* receptacle that is supplied by a *dedicated* branch circuit which is *located* and *identified* for specific use by a cord-and-plug-connected appliance, for example, a refrigerator or freezer.
  4. habitable, finished rooms in basements.
  5. Porcelain lampholders that have a receptacle outlet as part of the lampholder must be GFCI protected.
- in kitchens, all 125-volt, single-phase, 15- and 20-ampere receptacles that serve countertops and that are located within 6 feet (1.83 m) of the sink must be GFCI protected. This includes receptacles located on an island or peninsula, figure 6-2.
- a receptacle installed below the sink for plug-in connection of a food waste disposer is *not* required to be GFCI protected. This receptacle shall *not* be connected to the 20-ampere small appliance circuits provided in the kitchen.
- a receptacle installed solely for a clock is not required to be GFCI protected.
- in boathouses, all 125-volt, single-phase 15- or 20-ampere receptacles must have GFCI protection.
- residential underground circuits rated 120 volts or less, 20 amperes or less, and buried no less than 12 inches below grade level must be GFCI protected. Refer to *Table 300-5*. An example of this would be the Type UF cable supplying the post light in front of the residence. Type UF cable would have to be buried no less than 24 inches if the circuit is not GFCI protected. If the circuit is installed in rigid or intermediate metal conduit, then a depth of not less than 6 inches is required. *Table 300-5* shows all possible types of underground installations. The Notes to *Table 300-5* are very important for a clear understanding of the table.

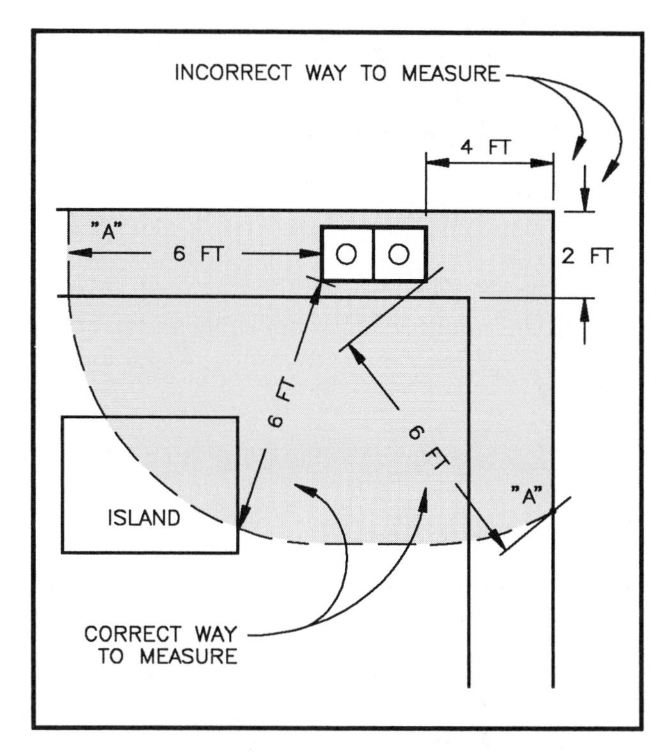

Fig. 6-2 Watch out for the 6-foot rule for GFCI receptacles in relationship to the sink. The 6-foot distance "is the shortest distance between the sink and the receptacle." *Do not* measure the distance following the wall line. All receptacle outlets located within the 6-foot shaded area that serve countertop surfaces must be GFCI protected.

NEC® *Article 680* gives the electrical requirements for swimming pools and the use of GFCI devices. (See unit 30.)

The Code requirements for ground-fault circuit protection can be met in many ways. Figure 6-3 illustrates a GFCI circuit breaker installed on a specified circuit on Panel B. A fault or current in excess of 6 milliamperes shuts off the entire circuit. For example, a ground fault at any point on circuit B14 will shut off the garage lighting, garage receptacles, rear garage door outside bracket fixture, post light, and overhead door opener.

When a GFCI receptacle is installed, then only that receptacle is shut off when a ground fault greater than 6 milliamperes occurs, figure 6-4.

Figure 6-5 shows the effect of a GFCI installed as part of a main feeder supplying 15- and 20-ampere receptacle branch circuits. This is rarely used in residential wiring.

Regardless of the location of the GFCI in the circuit, it must open the circuit when a current to ground exceeds 6 milliamperes (0.006 ampere).

- Class A GFCI — designed to trip when a ground fault of 6 milliamperes or more occurs.
- Class B GFCI — designed to trip when a ground fault of 20 milliamperes or more occurs.

According to the National Electrical Manufacturers Standard No. 280, the Class B GFCI would only be used for swimming pool underwater lighting that was installed before May 1965. This corresponds to the adoption date of the *National Electrical Code*.®

A GFCI does *not* limit the magnitude of ground-fault current. It limits the time that a ground-fault current will flow.

Figure 6-6 is a pictorial view of how a ground-fault circuit interrupter operates. Figure 6-7 shows the actual internal wiring of a GFCI.

The GFCI monitors the current balance between the hot conductor and the neutral conductor. As

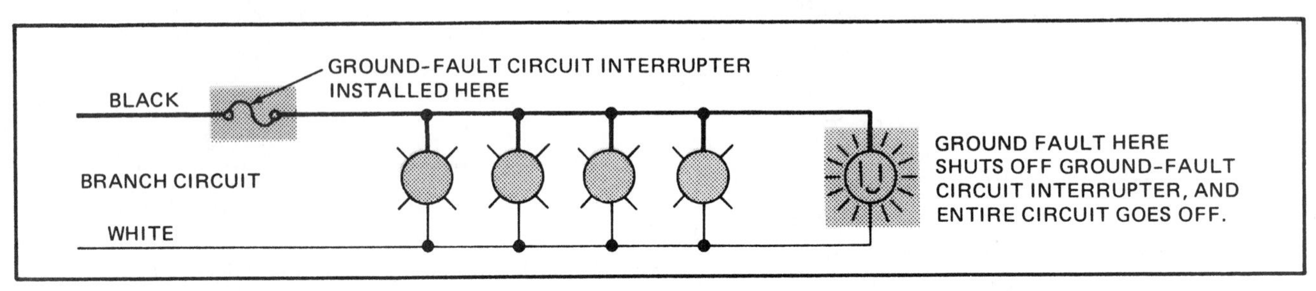

Fig. 6-3 Ground-fault circuit interrupter as a part of the branch-circuit overcurrent device.

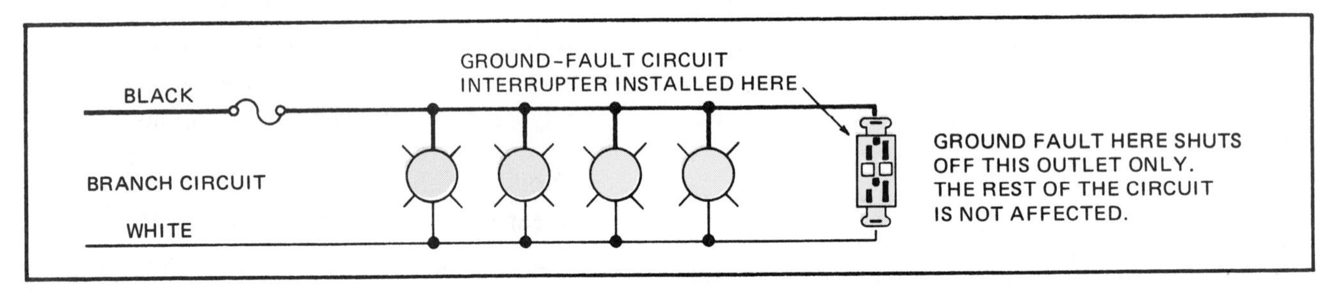

Fig. 6-4 Ground-fault circuit interrupter as an integral part of a convenience receptacle outlet.

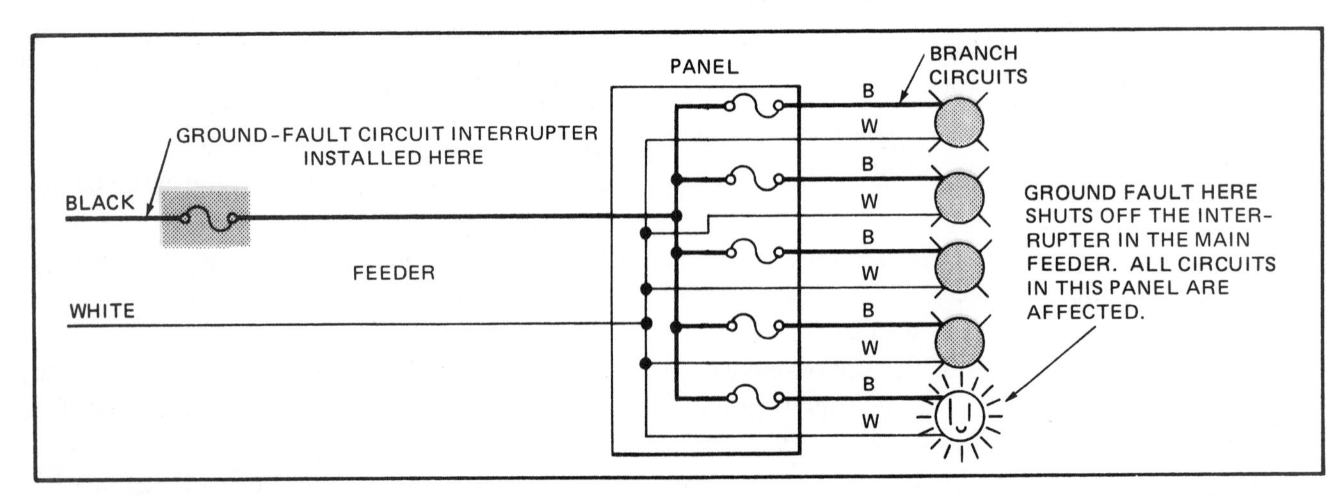

Fig. 6-5 Ground-fault circuit interrupter as part of the main feeder supplying 15- and 20-ampere receptacle circuits, *Section 215-9*. This is permitted by the Code, but rarely if ever used in residential wiring.

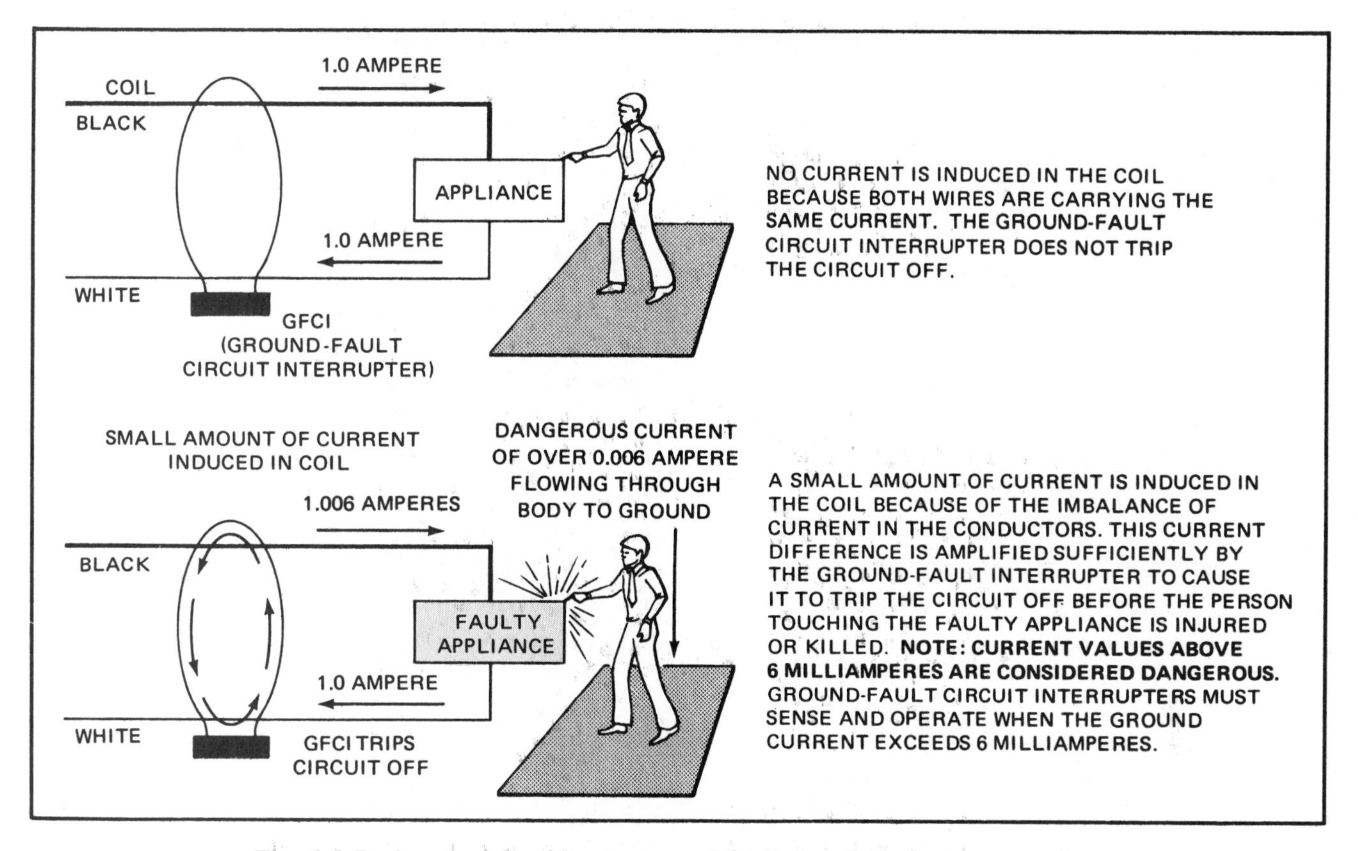

Fig. 6-6 Basic principle of how a ground-fault circuit interrupter operates.

soon as the current in the neutral conductor is less than the current in the hot conductor, the GFCI senses this imbalance and opens the circuit. The imbalance indicates that part of the current in the circuit is being diverted to some path other than the normal return path along the neutral conductor. Thus, if the GFCI trips off, it is an indication of a possible shock hazard from a line-to-ground fault.

Nuisance tripping of GFCIs is known to occur. Sometimes this can be attributed to extremely long runs of cable for the protected circuit. Consult the manufacturer's instructions to determine if maximum lengths for a protected circuit are recommended. Some electricians advise the use of nonmetallic staples (see figure 4-10C) or nonmetallic straps instead of metallic as a means of preventing nuisance tripping of GFCIs.

WORD OF CAUTION: The subject of *interrupting ratings* for overcurrent devices is covered in unit 27. For life safety reasons, electricians and electrical inspectors always concern themselves with the *interrupting ratings* of fuses and circuit breakers, particularly at the main service-entrance equipment of a residence, condo, or apartment building. A GFCI receptacle *also* has the ability to interrupt current when a fault occurs. According to the UL standard for GFCI receptacles, the current-withstand rating is 5000 amperes, RMS symmetrical.

So, unless the overcurrent protective device protecting the circuit that feeds the GFCI receptacle can "limit" the line-to-ground fault current to a maximum of 5000 amperes RMS symmetrical, installation of any GFCI receptacles close to the main panel should only be done when a "short-circuit" study is made to insure that the available line-to-ground fault current is 5000 amperes or less.

Without this precaution, the GFCI sensing and tripping mechanisms could be rendered inoperable should a line-to-ground fault occur. When the GFCI is called upon to prevent injury or electrocution of a person, it might not operate safely. This is a mighty strong reason for periodically testing all GFCI devices. Instructions furnished with GFCIs emphasize "operate upon installation and at least monthly" and record the date and results of the test on the form provided by the manufacturer. *Never* test a GFCI receptacle by shorting out line-to-neutral. The mechanism will be damaged so that it will be inoperable.

Remember, on single-phase systems, the line-to-ground fault current can exceed the line-to-line fault current. See unit 27.

Check fuse and circuit-breaker manufacturers' current-limiting charts to determine their current-limiting ability.

A GFCI does *not* protect against shock when a person touches both circuit conductors at the same time (two hot wires, or one hot wire and one grounded neutral wire).

Do not reverse the line and load connections on a feedthrough GFCI. This would result in the receptacle still being "live" even though the GFCI mechanism has tripped.

A GFCI receptacle *does not* provide overload protection for the circuit conductor. It provides *ground-fault protection only.*

GROUND-FAULT CIRCUIT INTERRUPTER IN RESIDENCE CIRCUITS

In this residence receptacle outlets installed outdoors, in the bathrooms, in the garage, in the workshop, specific receptacles in the basement, and receptacles within 6 feet of the kitchen sink serving the countertop are protected by ground-fault circuit interrupters, figure 6-8. These receptacle outlets are connected to the circuits listed in Table 6-1.

Swimming pools also have special requirements for GFCI protection. These requirements are covered in unit 30.

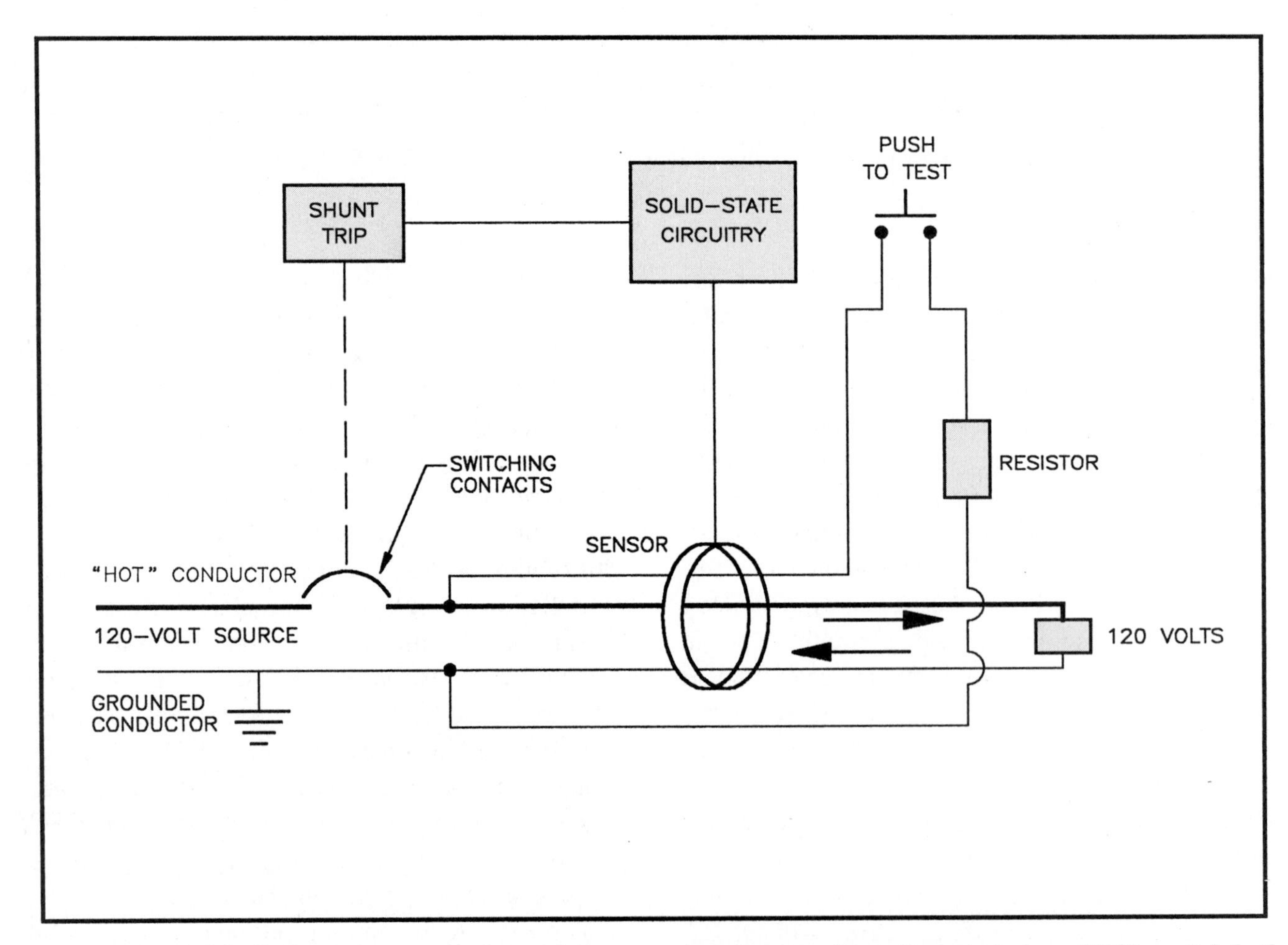

Fig. 6-7 Ground-fault circuit interrupter internal components and connections. Receptacle-type GFCIs switch both the phase (hot) and grounded conductors. Note that when the test button is pushed, the test current passes through the test button, the sensor, then back around (bypasses, outside of) the sensor, then back to the opposite circuit conductor. This is how the "unbalance" is created, then monitored by the solid-state circuitry to signal the GFCI's contacts to open. Note that since both "load" currents pass through the sensor, no unbalance is present.

GFCIs operate properly only on grounded electrical systems, as is the case in all residential, condo, apartment, commercial, and industrial wiring. The GFCI will operate on a two-wire circuit even though an equipment grounding conductor is not included with the circuit conductor. In this residence, equipment grounding conductors are in the cables. In the case of metal conduit, the equipment grounding conductor is the metal conduit.

CIRCUITS	RECEPTACLE LOCATION
A19	Outdoor receptacle on rear of residence outside of Master Bedroom
A14	Bathrooms receptacles, Bedroom area
A15	Front porch receptacle
A16	Outdoor receptacle on front of residence outside of Front Bedroom
A18	Workbench receptacle
A20	Workshop receptacles on window wall
B9	Wet-bar receptacles
B10	Washroom receptacle near rear entry hall
B13	Kitchen receptacles
B14	Garage receptacles
B15	Kitchen receptacles
B16	Kitchen receptacles
B17	Outdoor receptacle outside of Living Room next to sliding door
B20	Outdoor receptacle outside of laundry

Table 6-1. Circuit number and general location of GFCI receptacles in this residence.

Never ground a system neutral conductor except at the service equipment. A GFCI would be inoperative. An exception would be the frames of electric ranges and clothes dryers, *Section 250-60.*

Long branch-circuit runs can cause nuisance tripping of the GFCI due to leakage currents in the circuit wiring. GFCI receptacles at the point of utilization tend to minimize this problem. A circuit supplied by a GFCI circuit breaker in the main panel could cause nuisance tripping if the branch circuit is long — some electricians say 50 feet or more can cause nuisance tripping of the GFCI breaker.

Never connect the neutral of one circuit to the neutral of another circuit.

When a GFCI feeds an isolation transformer (separate primary winding and separate secondary winding), as might be used for swimming pool underwater fixtures, the GFCI *will not* detect any ground faults on the secondary of the transformer.

Ground-fault circuit interrupters may be installed on other circuits, in other locations, and even when rewiring existing installations where the Code does not specifically call for GFCI protection. See figure 6-9.

FEEDTHROUGH GROUND-FAULT CIRCUIT INTERRUPTER

The decision to use more GFCIs rather than trying to protect many receptacles through one GFCI becomes one of economy and practicality. GFCI receptacles are more expensive than regular

(A) Ground-fault circuit interrupter as an integral part of a duplex grounding-type convenience receptacle.

(B) Ground-fault circuit interrupter as an integral part of a duplex grounding-type convenience receptacle mounted in a weatherproof cover for outdoor installation.

Fig. 6-8 Ground-fault circuit interrupters (GFCIs).

receptacles. Here is where knowledge of material and labor costs comes into play. The decision must be made separately for each installation, keeping in mind that GFCI protection is a safety issue recognized and clearly stated in the Code. However, the actual circuit layout is left up to the electrician.

Figure 6-10 illustrates how a feedthrough GFCI receptacle supplies many other receptacles. Should a ground fault occur anywhere on this circuit, all 11 receptacles lose power — not a good circuit layout. Attempting to locate the ground-fault problem, unless obvious, can be very time-consuming. The use of more GFCI receptacles is generally the more practical approach.

Figure 6-11 shows how the feedthrough GFCI receptacle is connected into the circuit. When connected in this manner, the feedthrough GFCI receptacle and all downstream outlets on the same circuit will have ground-fault protection, figure 6-12.

The most important factors to be considered are the continuity of electrical power and the economy of the installation. The decision must be made separately for each installation.

► *CAUTION*: DO NOT connect the freezer receptacle in the workshop or the refrigerator receptacle in the kitchen of this residence to a GFCI-protected circuit, as exempted by *Section 210-8(a)(4), Exception 1.* This exemption calls for a single receptacle (not duplex or triplex), and it requires that this receptacle must be identified for specific use by a cord-and-plug connected appliance, for example, a refrigerator or freezer. The leakage current tolerances allowed by Underwriters Laboratories approaches the minimum trip settings of Class A GFCIs (4 to 6 milliamperes). Nuisance tripping can happen, thus shutting off the power to the refrigerator or freezer. Coming home to spoiled food is not a pleasant experience. That is why the *National Electrical Code®* exempts certain receptacles from the mandatory requirements as found in *Section 210-8.* ◄

See *Section 210-8(a)(2), Exceptions 1* and *2.* This section exempts receptacles in *garages* that are not readily accessible, such as a receptacle for a garage door operator on the ceiling and receptacles for appliances that are cord-and-plug connected, and that occupy "dedicated" spaces, such as a specific location in the garage for a freezer.

Section 210-8(a)(5) in the *Fine Print Note (FPN)* also exempts the GFCI requirement in the kitchen where the refrigerator/freezer will be plugged in.

► Sump pump receptacle outlets are also exempt from requiring ground-fault circuit interrupter protection per *Section 210-8(a)(4), Exception 3.* The laundry circuit is also exempted from requiring ground-fault protection per *Section 210-8(a)(4), Exception 2.* ◄

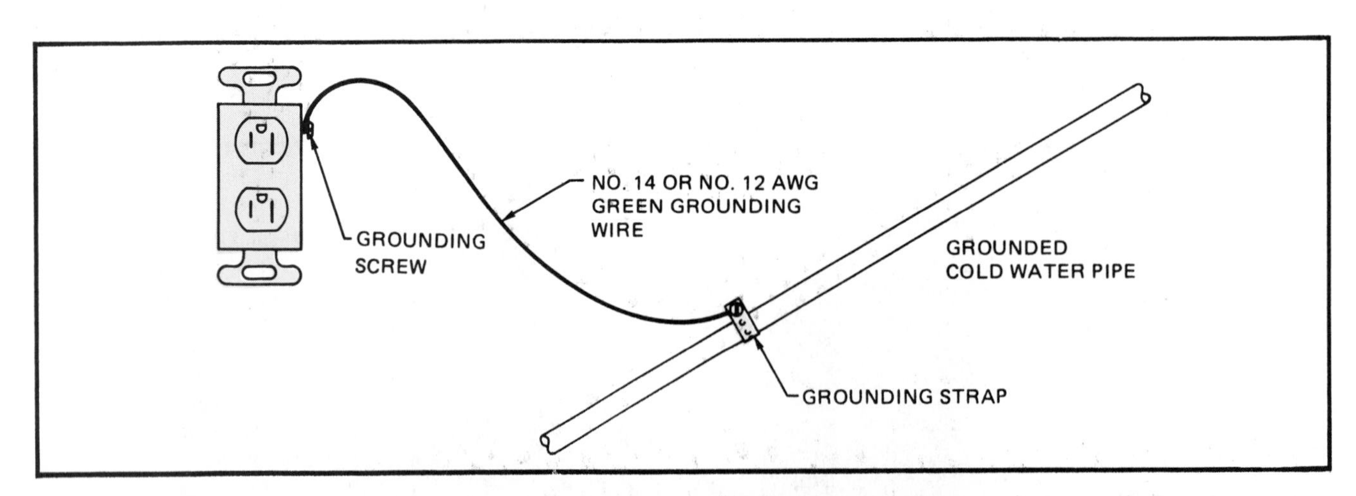

Fig. 6-9 In existing installations that do not have an equipment grounding wire as part of the branch-circuit wiring, the Code permits replacing an old-style nongrounding-type receptacle with a grounding-type receptacle, *Section 210-7(d).* The grounding terminal on the receptacle must be properly grounded. One of the acceptable ways to ground the grounding terminal is to run a conductor to an effectively grounded water pipe, *Section 250-50(a)* and *(b).* See *Section 250-91(b)* for other acceptable means of providing proper grounding of the receptacle's grounding terminal.

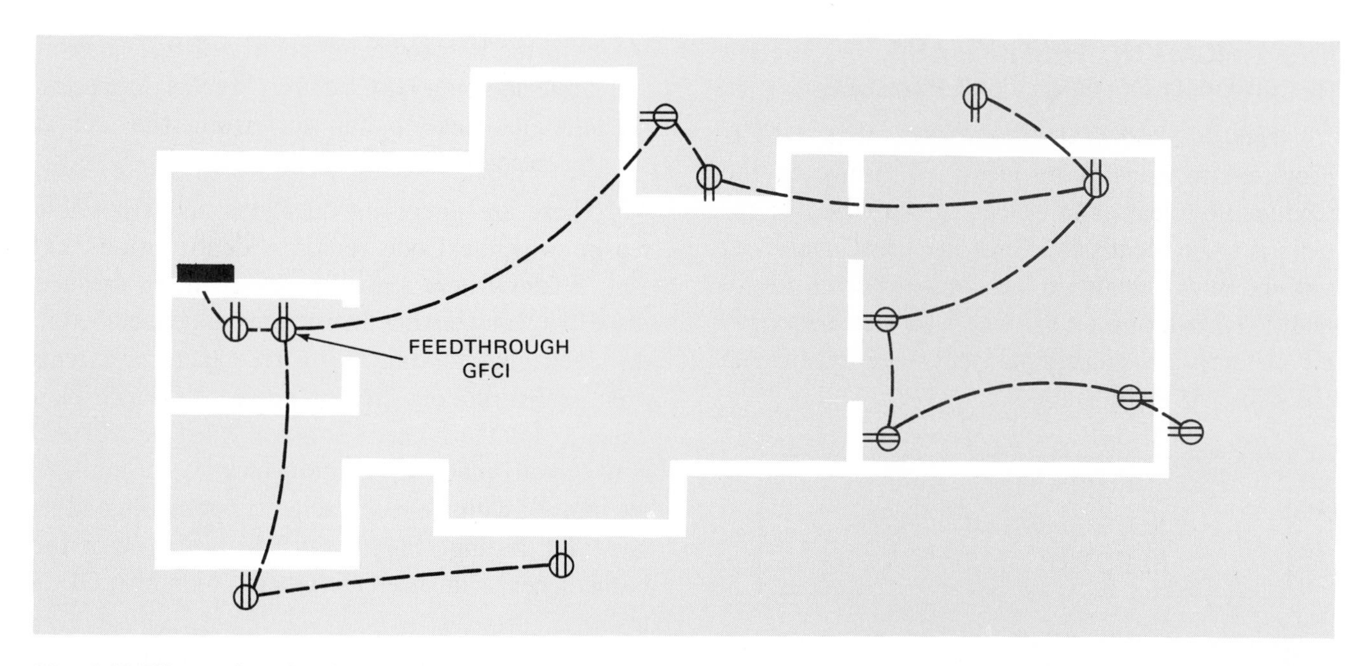

Fig. 6-10 Illustration showing one GFCI feedthrough receptacle protecting 10 other receptacles. Although this might be an economical way to protect many receptacle outlets with only one GFCI feedthrough receptacle, there is great likelihood that the GFCI will nuisance trip because of leakage currents. Use common sense and manufacturer's recommendations as to how many devices should be protected from one GFCI.

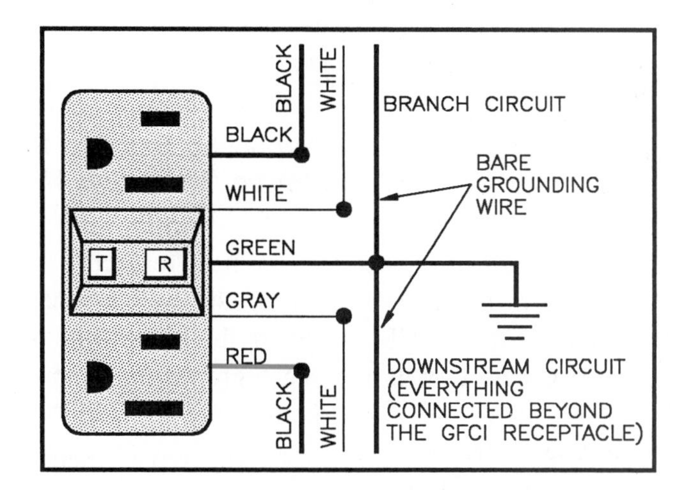

Fig. 6-11 Connecting feedthrough ground-fault circuit interrupter into circuit.

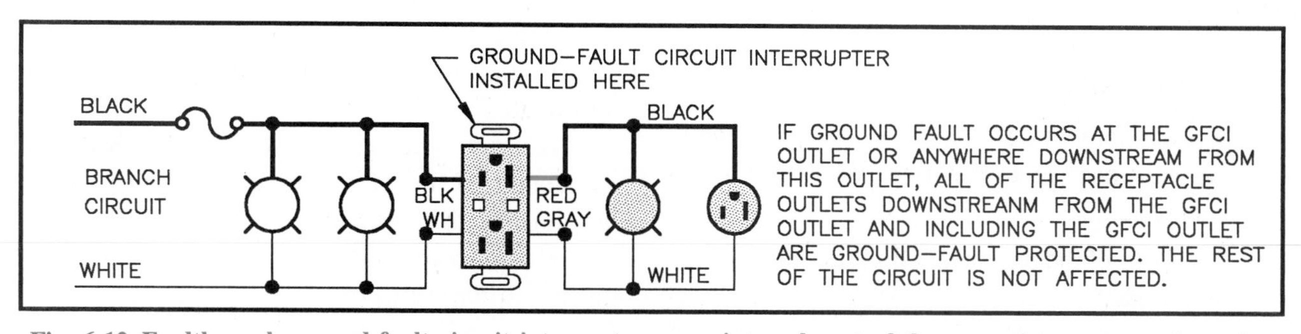

Fig. 6-12 Feedthrough ground-fault circuit interrupter as an integral part of the convenience receptacle outlet.

IDENTIFICATION, TESTING, AND RECORDING OF GFCI RECEPTACLES

Section 210-8(a)(4) requires that GFCI-protected receptacles be identified. However, the Code is not clear as to exactly how a GFCI receptacle is to be identified. Some electrical inspectors will interpret "identified" as requiring some sort of additional marking on the GFCI-protected receptacle plate or immediately adjacent to the receptacle. Examples of markings are:

THIS RECEPTACLE PROVIDES PROTECTION AGAINST ELECTRICAL SHOCK

THIS RECEPTACLE IS GROUND FAULT PROTECTED

GROUND FAULT PROTECTED Instructions at power panel

Refer to figure 6-8 (page 107) and note that these GFCI receptacles have "T" (Test) and "R" (Reset) buttons. Pushing the test button places a small ground fault on the circuit. If operating properly, the GFCI receptacle should trip to the OFF position. The operation is the same for the GFCI circuit breakers. Pushing the reset button will restore power.

Detailed information and testing instructions are furnished with GFCI receptacles and GFCI circuit breakers. Monthly testing is recommended to insure that the GFCI mechanism will operate properly if a human being is subjected to an electrical shock.

- Will the homeowner remember to do the monthly testing?
- Will the homeowner recognize and understand the purpose and function of the GFCI receptacle?
- Should all of the basement GFCI receptacles be identified?
- Will the homeowner recognize a receptacle if protected by a GFCI circuit breaker located in an electrical panel far from the actual receptacle?

These are important questions and a possible reason why the Code requires identification, and why Underwriters Laboratories requires detailed installation and testing instructions to be included in the packaging for the GFCI receptacle or circuit breaker. Instructions are not only for the electrician but also for the homeowner, and must be left in a conspicuous place so the homeowner can familiarize himself with the receptacle, its operation, and its need for testing. Figure 6-13 shows a chart the homeowner can use to record monthly GFCI testing.

REPLACING EXISTING RECEPTACLES

The *National Electrical Code®* is very specific on the type of receptacle permitted to be used as a replacement for an existing receptacle. These Code rules are found in *Sections 210-7, 250-50, 250-74,* and *250-91(b)*.

Replacing Existing Receptacles Where Grounding Means Does Exist
(Refer to figure 6-14)

When replacing an existing receptacle (A) where the wall box is properly grounded (E) or where the branch-circuit wiring contains an equipment grounding conductor (D), at least two easy choices are possible for the replacement receptacle:

1. The replacement receptacle must be of the grounding type (B) unless.....
2. The replacement receptacle is of the GFCI type (C).

In either case, be sure that the equipment grounding conductor of the circuit is connected to the receptacle's green hexagon-shaped grounding terminal.

Do not connect the white grounded circuit conductor to the green hexagon-shaped grounding terminal of the receptacle.

Do not connect the white grounded circuit conductor to the wall box.

Replacing Existing Receptacles Where Grounding Means Does Not Exist

(Refer to figures 6-15 and 6-15A)

When replacing an existing nongrounding type receptacle (A), where the box is not grounded (E), or where an equipment grounding conductor has not been run with the circuit conductors (D), four choices are possible for selecting the replacement receptacle:

1. The replacement receptacle may be a nongrounding type (A).
2. The replacement receptacle may be a GFCI type (C).
   - The green hexagon grounding terminal of the GFCI replacement receptacle does not have to be connected to any grounding means. It can be left "unconnected." The GFCI's trip mechanism will operate properly when ground-faults occur anywhere on the load side of the GFCI replacement receptacle. Ground-fault protection is still there. Refer to figure 6-7 (page 106).

OCCUPANT'S TEST RECORD

TO TEST, depress the "TEST" button, the "RESET" button should extend. Should the "RESET" button not extend, the GFCI will not protect against electrical shock. Call qualified electrician.

TO RESET, depress the "RESET" button firmly into the GFCI unit until an audible click is heard. If reset properly, the RESET button will be flush with the surface of the test button.

This label should be retained and placed in a conspicuous location to remind the occupants that for maximum protection against electrical shock, each GFCI should be tested monthly.

YEAR	JAN	FEB	MAR	APR	MAY	JUN	JUL	AUG	SEP	OCT	NOV	DEC

Fig. 6-13 Homeowner's testing chart for recording GFCI testing dates.

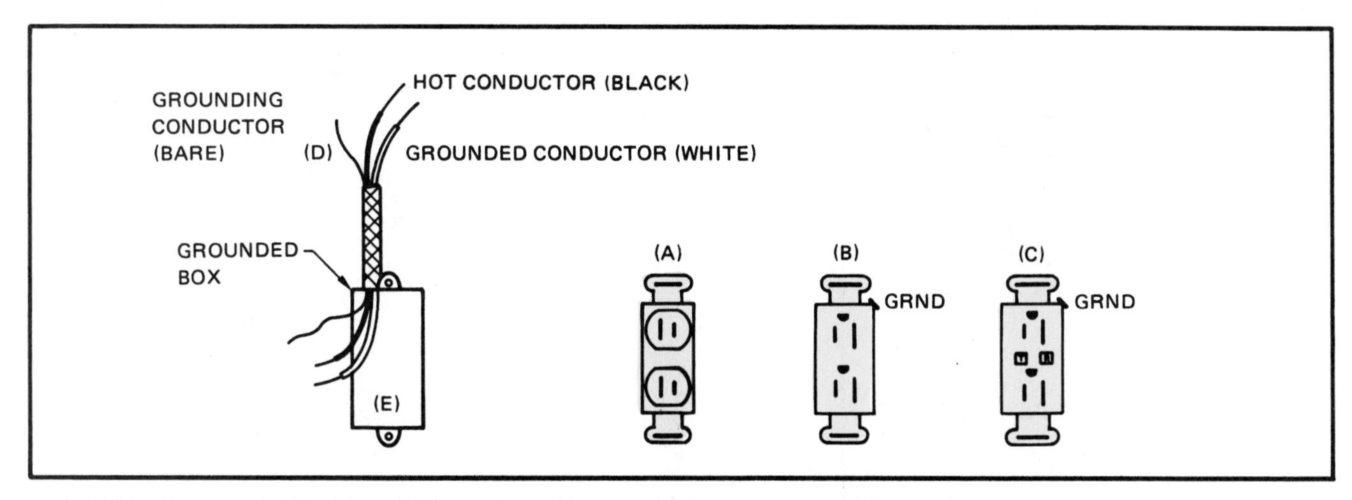

Fig. 6-14 Where box (E) is properly grounded, an existing receptacle of the type shown in (A) *must* be replaced with a grounding-type receptacle (B). It is also OK to replace (A) with GFCI receptacle (C). See text for further discussion.

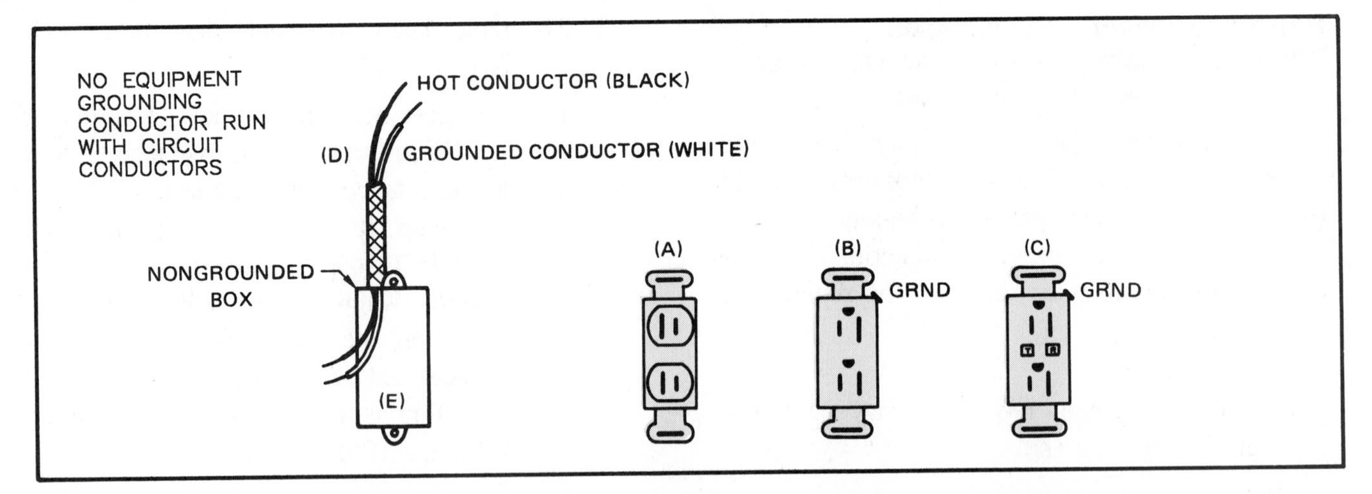

Fig. 6-15 Where box (E) is *not* grounded, or where an equipment grounding conductor has *not* been run with the circuit conductors (D), an existing nongrounding-type receptacle (A) may be replaced with a nongrounding-type receptacle (A), a GFCI-type receptacle (C), ► a grounding-type receptacle (B) if supplied through a GFCI receptacle (C) ◄, or a grounding-type receptacle (B) if a separate equipment grounding conductor is run from the receptacle to a cold-water pipe or other effective grounding means.

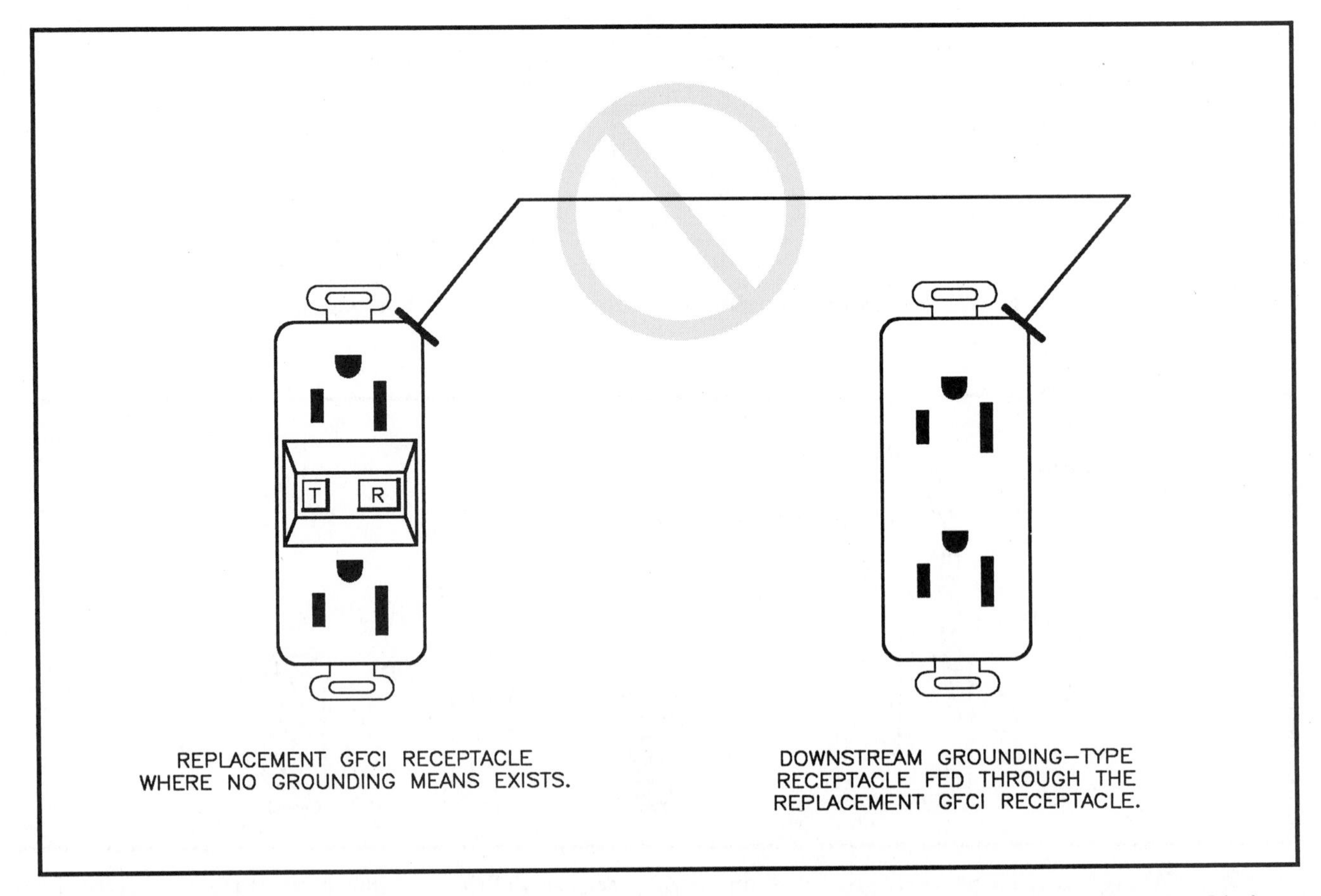

Fig. 6-15A *Do not* connect a grounding conductor between the above receptacles. The text explains why this is not permitted.

- Do not connect an equipment grounding conductor from the green hexagon grounding terminal of a replacement GFCI receptacle (C) to any other downstream receptacles that are fed through the replacement GFCI receptacle.

  The reason this is not permitted is that if at a later date someone saw the conductor connected to the green hexagon grounding terminal of the downstream receptacle, there would be an immediate assumption that the other end of that conductor had been properly connected to an acceptable grounding point of the electrical system. The fact is that the so-called equipment grounding conductor had been connected to the replacement GFCI receptacle's green hexagon grounding terminal that was not grounded in the first place. We now have a false sense of security, a real shock hazard.

►3. The replacement receptacle may be a grounding type (B) if it is supplied through a GFCI-type receptacle (C).

- The green hexagon equipment grounding terminal of the replacement grounding type receptacle (B) need not be connected to any grounding means. It may be left "unconnected." The upstream feedthrough GFCI receptacle (C) trip mechanism will work properly when ground-faults occur anywhere on its load-side. Ground-fault protection is still there. Refer to figure 6-7. ◄

4. The replacement receptacle may be a grounding type (B) if:

- an equipment grounding conductor, sized per *Table 250-95*, is run between the replacement receptacles green hexagon grounding terminal and a grounded water pipe. See *Article 250, Part K* for the acceptable means of connecting the equipment grounding conductor to the water pipe.

GROUND-FAULT PROTECTION FOR CONSTRUCTION SITES

Because of the nature of construction sites, there is a continual presence of shock hazard that can lead to serious personal injury or death through electrocution. Workers are standing in water, standing on damp or wet ground, or in contact with steel framing members. Electric cords and cables are lying on the ground, subject to severe mechanical abuse. All of these conditions spell "DANGER."

Therefore, *Section 305-6(a)* of the Code requires that all 125-volt, single-phase, 15- and 20-ampere receptacle outlets that are not part of the permanent wiring of the building and that will be used by the workers on the construction site must be GFCI protected. Receptacle outlets that are part of the actual permanent wiring of the building or structure are exempt from this requirement.

There is an alternative to requiring GFCI protection on all construction sites. *Section 305-6(b)* does permit a written procedure that must be continuously enforced at the construction site. A designated person must keep a written log, insuring that all electrical equipment is properly installed and maintained according to *Sections 210-7(c), 250-45, 250-59,* and *305-4(d)*. This option has proved to be difficult to enforce. Thus, most "authorities having jurisdiction" require GFCI protection as stated in *Section 305-6(a)*.

IMMERSION DETECTION CIRCUIT INTERRUPTERS

A new type of circuit interrupter is beginning to find its way into the marketplace. This device is called an *immersion detection circuit interrupter (IDCI)*. It can be an integral part of the attachment plug cap found on personal grooming appliances that are commonly used near water and basins, such as hair dryers, heated stylers, heated air combs, heated air curlers, curling iron–hair dryer combinations, and so on.

These electrical appliances are most commonly used at a vanity having a wash basin with water in it, or even worse, next to a water filled bathtub with a person (child or adult) in it. Accidentally dropping or pulling the appliance into the tub when reaching for a towel can be lethal.

Electricity and water do not mix! The shock hazard protection of these appliances is permitted to be in the form of a GFCI device or in the form of an immersion detection device. If any of these appliances fall into a basin or tub that is filled with water, the IDCI will sense the leakage current and will trip to the OFF position. Both GFCI and IDCI devices disconnect both conductors of the branch circuit. These devices are also required to open the circuit regardless of whether the appliance switch is in the ON or OFF position.

The operation of a GFCI device has been discussed previously. The operation of an IDCI depends upon a "third" wire in the cord, referred to as a *probe*, which is connected to a sensor within the appliance. It is this probe that will detect leakage current flow when conductive liquid enters the appliance and makes contact with any live part inside the appliance and the internal "sensor." When this happens, the IDCI is instructed to trip to the OFF position.

It is not permitted to replace the special GFCI or IDCI attachment plug cap on these appliances because that would make the appliance absolutely unsafe. The UL requirement is that the plug cap shall be marked: "WARNING: TO REDUCE THE RISK OF ELECTRIC SHOCK, DO NOT REMOVE, MODIFY, OR IMMERSE THIS PLUG."

An IDCI must open the circuit within 25 milliseconds from the time that the sensor detects a leakage current that exceeds 5 milliamperes. ► *Section 422-24* of the *National Electrical Code®* requires protection against electrocution for freestanding hydromassage appliances and hand-held hair dryers. This is intended to include heated stylers, heated air combs, heated air curlers, and curling iron–hair dryer combinations. ◄

APPLIANCE LEAKAGE CIRCUIT INTERRUPTER (ALCI)

Also covered by the requirements of *Section 422-23* of the Code is a device called an *Appliance Leakage Circuit Interrupter.* This device will de-energize the supply to an appliance when a leakage current value of some predetermined value for some specified time period is detected.

An Overview

Underwriters Laboratories lists GFCIs, IDCIs, and ALCIs.

Hydromassage tubs that are cord-and-plug connected and that are of the freestanding portable type must also provide "people" protection against electrocution regardless if the unit is in the "on" or "off" position. This is covered in *Section 422-23* of the *National Electrical Code®.*

Because all of the appliances mentioned here can and will be used in new and older homes, the protection against electrocution must be provided by the manufacturer of the appliances. Most older homes do not have GFCI protection, thus, the logic of requiring the manufacturer to provide the protection device.

TRANSIENT VOLTAGE SURGE SUPPRESSION (TVSS)

In today's homes, we find many electronic appliances (television, stereo, personal computers, facsimile equipment, word processors, video cassette recorders, digital stereo equipment, microwave ovens), all of which contain many sensitive, delicate electronic components (printed circuit boards, chips, microprocessors, transistors, etc.).

Voltage transients, called *surges* or *spikes*, can stress, degrade, and/or destroy these components; they can cause loss of memory in the equipment or "lock up" the microprocessor.

The increased complexity of electronic integrated circuits makes this equipment an easy target for "dirty" power that can and will affect the performance of the equipment.

Voltage transients cause abnormal current to flow through the sensitive electronic components. This energy is measured in joules. A *joule* is the unit of energy when one ampere passes through a one-ohm resistance for one second (like wattage, only on a much smaller scale).

Line surges can be line-to-neutral, line-to-ground, and line-to-line.

Transients

Transients are generally grouped into two categories:

- Ring wave — These transients originate within the building and are caused by welders, elevators, copiers, motors, air conditioners, or other inductive loads.
- Impulse — These transients originate outside of the building and are caused by utility company switching, lightning, and so on.

To minimize the damaging results of these transient line surges, service equipment, panels, load centers, feeders, branch circuits, and individual receptacle outlets can be protected with a device called a *transient voltage surge suppressor* (TVSS), figure 6-16.

A TVSS contains one or more metal-oxide varistors (M.O.V.) that clamp the transients by absorbing the major portion of the energy (joules) created by the surge, allowing only a small, safe amount of energy to enter the actual connected load.

The M.O.V. clamps the transient in times of less than 1 nanosecond, which is one billionth of a second, and keeps the voltage spike passed through to the connected load to a maximum range of from 400 to 500 peak volts.

Typical TVSS devices for homes are available as an integral part of receptacle outlets that mount in the same wall boxes as do regular receptacles. They look the same as a normal receptacle, and may have an audible alarm that sounds when a M.O.V. has failed. The TVSS device may also have a visual indication, such as an LED (light-emitting diode) that glows continuously until an M.O.V. fails. Then the LED starts to flash on and off. See figure 6-17.

TVSS devices are also available in plug-in strips, figure 6-18.

One TVSS on a branch circuit will provide surge suppression for other receptacles nearby that are on the same circuit.

Noise

Low-level nondamaging transients can also be present. These can be caused by fluorescent lamps and ballasts, other electronic equipment in the area, x-ray equipment, motors running or being switched on and off, improper grounding, and so forth. Although not physically damaging to the electronic equipment, computers can lose memory or perform

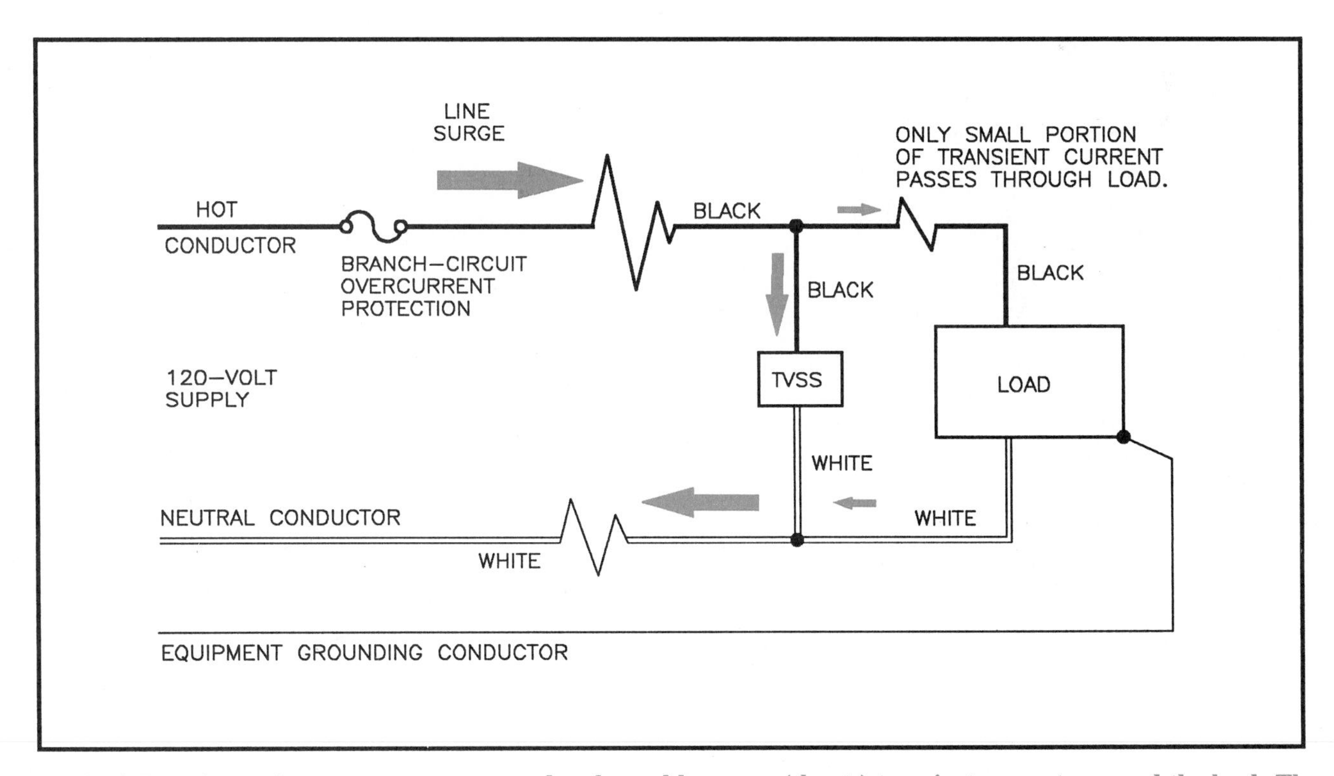

Fig. 6-16 Transient voltage surge suppressor absorbs and bypasses (shunts) transient currents around the load. The M.O.V. dissipates the surge in the form of heat.

wrong calculations. Their intended programming can malfunction.

"Noise" comes from *electromagnetic interference* (EMI) and *radio frequency interference* (RFI). EMI is usually caused by ground currents of very low values from motors, utility switching loads, lightning, and so forth, and are transmitted through metal conduits. RFI is noise, like the buzzing heard on a car radio when driving under a high-voltage transmission line. This interference "radiates" through the air from the source and is picked up by the grounding system of the building.

This undesirable noise can be reduced by reducing the number of ground reference points on a system. This can be accomplished by installing an *isolated ground receptacle.*

ISOLATED GROUND RECEPTACLE

In a standard conventional receptacle outlet, figure 6-19, the green grounding hexagon screw, the grounding contacts, the yoke (strap), and the metal wall box are all "tied" together to the building's equipment grounding system. Picture then the many receptacles in a building, all having and creating the multiple-ground situation.

In an isolated ground receptacle, the green grounding hexagon screw and the grounding contacts of the receptacle are isolated from the metal yoke (strap) of the receptacle and also from the building's equipment grounding system. A separate green insulated grounding conductor is then installed from the green hexagon screw on the receptacle all the way back to the main service disconnect. This separate green grounding conductor does *not* connect to any panels, load centers, or other ground reference points in between. The system grounding is now "clean," and the result is

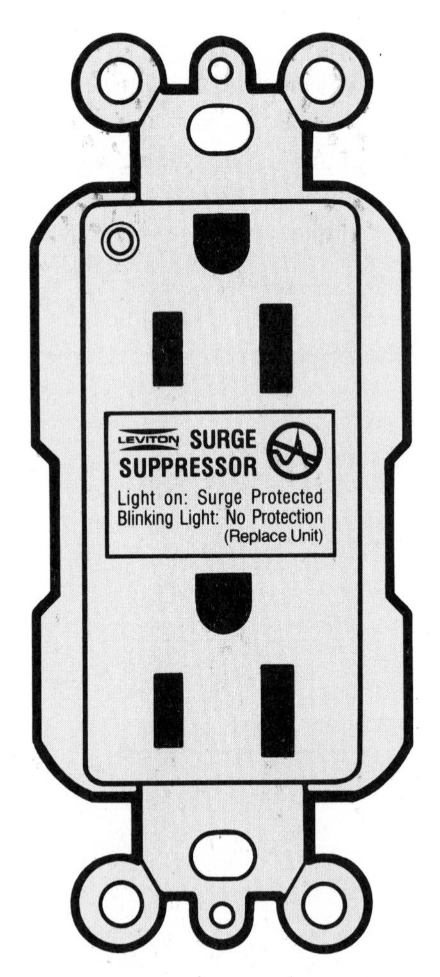

Fig. 6-17 Surge protector. Courtesy of Leviton.

Fig. 6-18 Surge suppressor. Courtesy of Leviton.

less transient noise (disturbances) transmitted to the connected load.

Isolated ground receptacles are identified by an orange triangle imprinted on the face of the receptacle; or they may be made entirely of orange material, in which case the triangle is of some other color.

In this residence, a TVSS and/or isolated grounding-type receptacle could be installed in the Study/Bedroom at the request of the owner. This might be desirable should the owner have a personal computer (PC), a word processor, or other electronic equipment subject to the problems that arise when transients appear on the wiring.

As more and more electronic equipment is brought onto the marketplace, the subject of transient voltage surge suppression and isolated grounding receptacles will be given even more attention than it is today.

Underwriters Laboratories Standards 1449 and 498 cover transient voltage surge suppressors and isolated ground receptacles.

The *National Electrical Code®* in *Section 210-7(c), Fine Print Note* refers us to *Section 250-74, Exception 4.* Here we find the permission to install isolated ground receptacles to reduce noise. ► Permission is granted in *Section 384-20, Exception* to pass the separate grounding conductor through one or more panels without connecting it to that panel's equipment grounding terminal bar. ◄ The FPN of this Code reference reminds us, however, that the metal raceway and metal outlet both must still be grounded.

This separate grounding conductor is carried all the way back to the main service, where it is connected to the point where the neutral of the system is bonded to the grounding electrode conductor and the equipment grounding bus. In a residence, this would be the terminal bus provided for the connection of equipment grounding conductors as shown in figure 26-11.

When a transformer is installed between the equipment and the main service, the isolated grounding conductor is then run to the panelboard connected to the secondary of the transformer where the conductor is connected to the terminal. At the terminal, the grounding electrode conductor, equipment grounding conductor, and neutral conductors are bonded together.

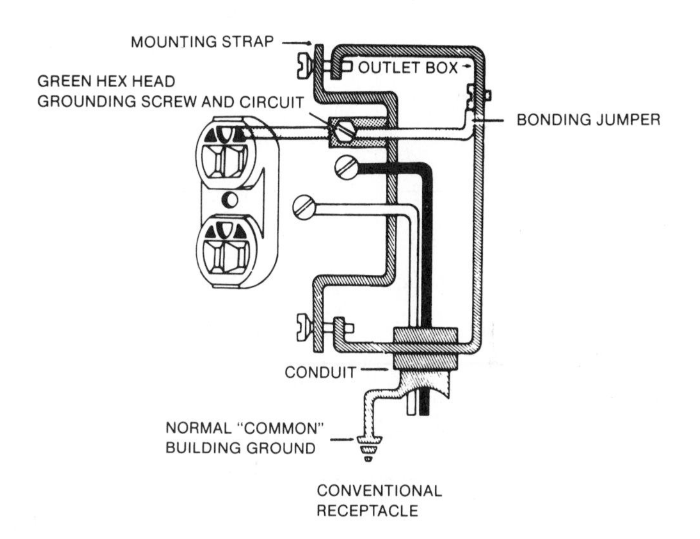

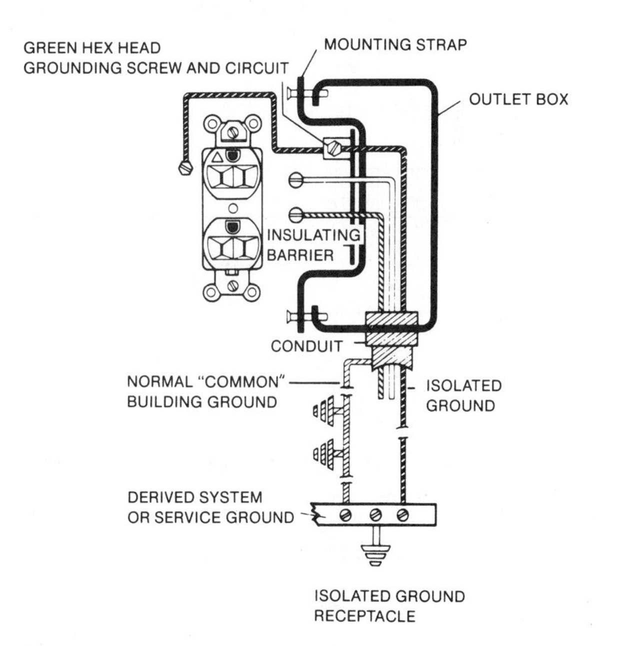

Fig. 6-19 Standard conventional receptacle and isolated ground receptacle. Courtesy of Hubbell Corporation.

REVIEW

1. Explain the operation of a ground-fault circuit interrupter. Why are GFCI devices used? Where are GFCI receptacles required? ______________________

2. Residential GFCI devices are set to trip a ground-fault current above ________ milliamperes.

3. Where must GFCI receptacles be installed in residential garages? ____________

4. The Code requires GFCI protection for certain receptacles in the kitchen. Explain where these are required. ______________________

5. Is it a Code requirement to install GFCI receptacles in a fully carpeted, finished recreation room in the basement? ______________________

6. A homeowner calls in an electrical contractor to install a separate circuit in the basement (unfinished) for a freezer. Is a GFCI receptacle required? ____________

7. GFCI protection is available as (a) branch-circuit breaker GFCI, (b) feeder circuit breaker GFCI, (c) individual GFCI receptacles, (d) feedthrough GFCI receptacles. In your opinion, for residential use, what type would you install? ___________

8. Extremely long circuit runs connected to a GFCI branch-circuit breaker might result in
   a. nuisance tripping of the GFCI
   b. loss of protection
   c. the need to reduce the load on the circuit
   Circle the correct answer.

9. If a person comes in contact with the hot and grounded conductors of a two-wire branch circuit that is protected by a GFCI, will the GFCI trip "off"? Why?

   ____

   ____

10. What might happen if the line and load connections of a feedthrough GFCI receptacle were reversed? ____

    ____

    ____

    ____

    ____

    ____

11. May a GFCI receptacle be installed on an old installation where the two-wire circuit has no equipment grounding conductor? ___________________________

12. What two types of receptacles may be used to replace a defective receptacle in an older home that is wired with knob-and-tube wiring where no grounding means exists in the box? ____

    ____

13. You are asked to replace a receptacle. Upon checking the wiring, you find that the wiring method is conduit and that the wall box is properly grounded. The receptacle is of the older style two-wire type that does not have a grounding terminal. You remove the old receptacle and replace it with (Circle the letter of the correct answer.)
    a. the same type of receptacle as the type being removed.
    b. a receptacle that is of the grounding type.
    c. a GFCI receptacle.

14. What color are the terminals of a standard grounding-type receptacle? __________

    ____

    ____

15. What special shape are the grounding terminals of receptacle outlets and other devices? ____

    ____

16. Construction sites can be dangerous because of the manner in which extension cords, portable electrical tools, and other electrical equipment are used and abused. *Article 305* of the Code offers two options, either of which may be followed to reduce the hazards of electrical shock. In your own words, what are these options?

____

____

____

____

____

____

17. In your own words, explain why the Code does not require certain receptacle outlets in kitchens, garages, and basements to be GFCI protected. ______________

____

____

____

____

____

____

18. Circuit A1 supplies GFCI receptacles ① and ② . Circuit A2 supplies GFCI receptacles ③ and ④ . Receptacles ① and ③ are feedthrough type. Using colored pencils or marking pens, complete all connections.

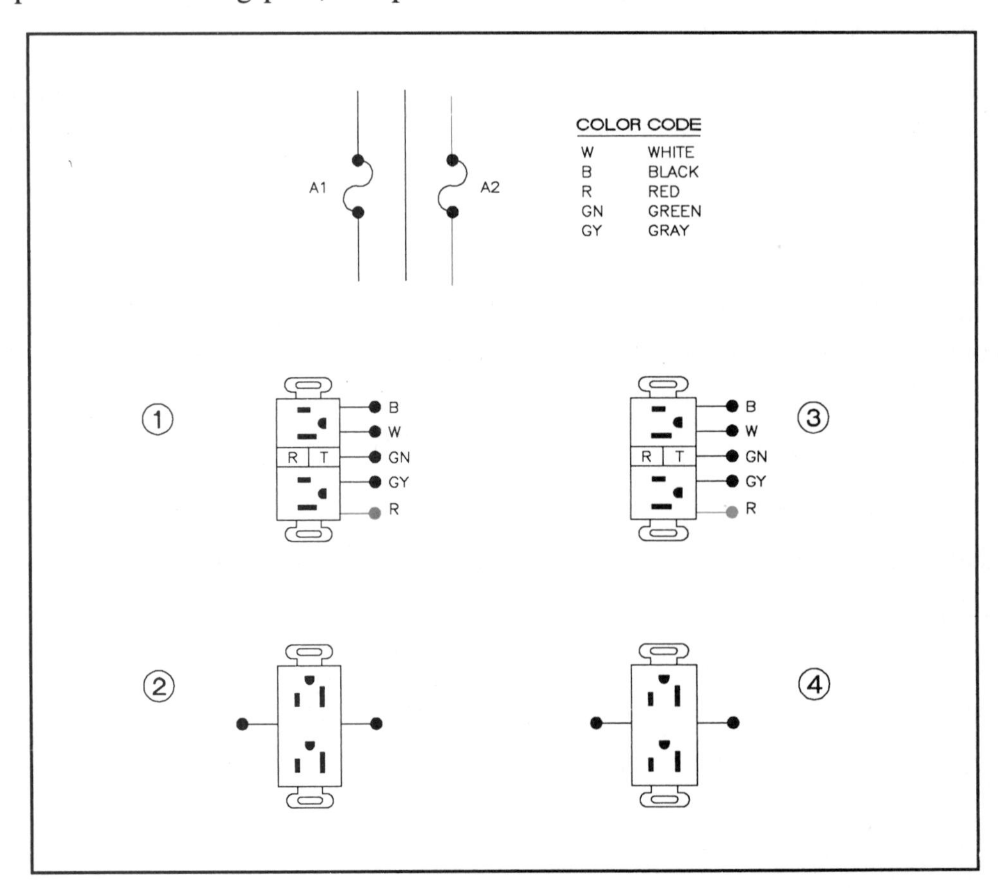

19. The term TVSS is becoming quite common. What do the letters stand for? ______

______

______

20. Transients (surges) on a line can cause spikes or surges of energy that can damage delicate electronic components. A TVSS device contains one or more ______ ______ ______ that bypass and absorb the energy of the transient.

21. Undesirable noise on a circuit can cause computers to lock up, lose their memory, and/or cause erratic performance of the computer. This noise does not damage the equipment. The two types of this noise are EMI and RFI. What do these letters mean?

______

______

______

______

22. Can TVSS receptacles be installed in standard device boxes? ______

23. Some line transients are not damaging to electronic equipment, but can cause the equipment to operate improperly. The effects of these transients can be minimized by installing ______ ______ ______.

24. When an isolated ground receptacle is installed, the Code permits us to carry the separate equipment grounding conductor back through the raceway, through one or more panels, back to the point in the panel where the grounding electrode conductor, the equipment grounding conductors, and the neutral circuit conductors are connected together. What sections of the Code reference this topic? ______

______

______

25. Briefly explain the operation of an immersion detection circuit interrupter (IDCI).

______

______

______

______

______

______

26. What is the time/current characteristic of an IDCI? ______

______

______

UNIT 7

Lighting Fixtures and Ballasts

OBJECTIVES

After studying this unit, the student will be able to

- understand fixture terminology, such as Type IC, Type NON IC, suspended ceiling fixtures, recessed fixtures, and surface-mounted fixtures.
- connect recessed fixtures, both prewired and nonprewired types, according to Code requirements.
- specify insulation clearance requirements.
- discuss the importance of temperature effects while planning recessed fixture installations.
- describe thermal protection for recessed fixtures.
- understand class P ballasts.

TYPES OF LIGHTING FIXTURES

This unit begins an in-depth coverage of the various types of recessed fixtures commonly installed in homes. It also considers other types of available lighting fixtures.

The Code in *Article 410* sets forth the requirements for the installation of lighting fixtures. These requirements are discussed throughout this text as necessary. Although not involved in the actual manufacture of lighting fixtures, the electrician must "meet Code" when installing fixtures, including mounting, supporting, grounding, live-parts exposure, insulation clearances, supply conductor types, maximum lamp wattages, and so forth.

The homeowner, interior designer, architect, or electrician has literally thousands of different types of fixtures from which to choose to satisfy certain needs, space requirements, and price considerations, among other factors.

It is absolutely essential for the electrician to know early in the roughing-in stage of wiring a house what types of fixtures are to be installed. This is particularly true for recessed-type fixtures. The electrician must work closely with the general building contractor, carpenter, plumber, heating contractor, and the other building trades people. This is to insure that such factors as location, proper and adequate framing, clearances from pipings and ducts, clearances from combustible material, and insulation restrictions as concern the fixture installations are being complied with during construction.

Underwriters Laboratories provides the safety standards to which a fixture manufacturer must conform. Common fixture types are:

FLUORESCENT	INCANDESCENT
–surface	–surface
–recessed	–recessed
–suspended ceiling	–suspended ceiling

IMPORTANT: Always carefully read the label on the fixture. The information found on the label, together with conformance to *Article 410* of the Code, should result in a safe installation. The label will provide such information as:

- For wall mount only
- Ceiling mount only
- Maximum lamp wattage
- Type of lamp
- Access above ceiling required
- Suitable for air handling use
- For chain or hook suspension only
- Suitable for operation in ambient temperatures not exceeding ________°F (°C)
- Suitable for installation in poured concrete
- For installation in poured concrete only
- For line volt-amperes, multiply lamp wattage by 1.25
- Suitable for use in suspended ceilings
- Suitable for use in noninsulated ceilings
- Suitable for use in insulated ceilings
- Suitable for damp locations (such as bathrooms and under eaves)
- Suitable for wet locations
- Suitable for use as a raceway
- Suitable for mounting on low-density cellulose fiberboard
- For supply connections, use wire rated for at least ________C°
- Not for use in dwellings
- Thermally protected fixture

Fig. 7-1 Typical recessed fixture.

- Type IC
- Type NON IC
- Inherently protected

The Underwriters Laboratories *Electrical Construction Materials Directory* (Green Book), and *General Information* (White Book), and the fixture manufacturers' catalogs and literature are excellent sources of information.

Underwriters Laboratories list, test, and label fixtures for conformance to their standards and to the *National Electrical Code*®.

Recessed fixtures, figure 7-1, in particular, have an inherent heat problem. Therefore, they must be suitable for the application and must be properly installed.

To protect against overheating, Underwriters Laboratories requires that recessed fixtures be equipped with an integral thermal protector, figure 7-2. These devices will cycle ON and OFF repeatedly until the heat problem is removed.

In this residence, recessed fixtures are installed in

LOCATION	TYPE
Closets	Incandescent
Recreation room	Fluorescent
Wet bar	Incandescent
Above kitchen sink	Incandescent
Bathrooms (heat/vent/light)	Incandescent
Hall leading to kitchen	Incandescent

Figure 7-3 shows a fluorescent fixture mounted on low-density cellulose fiberboard. Because of the potential fire hazard when mounting lighting fixtures on combustible materials, *Section 410-76(b)* states that these fixtures must be listed for such

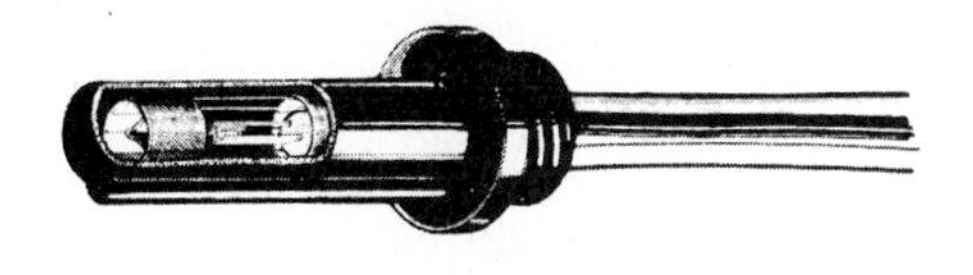

Fig. 7-2 Recessed fixture thermal cutout.

mounting, or must be spaced at least 1 1/2 inches (38 mm) from the fiberboard surface. It is important to refer to the *Fine Print Note (FPN)* below *Section 410-76(b)*. This FPN explains combustible low-density cellulose fiberboard as sheets, panels, and tiles that have a density of 20 pounds per cubic foot or less that are formed of bonded plant fiber material. Solid or laminated wood does not come under the definition of "combustible low-density cellulose fiberboard." It does not include fiberboard that has a density of over 20 pounds per cubic foot or material that has been integrally treated with fire-retarding chemicals to meet specific standards.

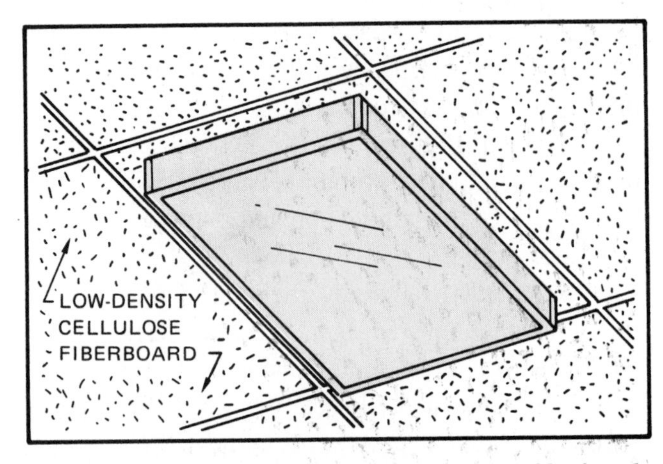

Fig. 7-3 When mounted on *low-density ceiling fiberboard*, surface-mounted fluorescent fixtures must be marked "Suitable for Surface Mounting on Low-Density Cellulose Fiberboard."

CODE REQUIREMENTS FOR INSTALLING RECESSED FIXTURES

The Code requirements for the installation and construction of recessed fixtures are given in *Sections 410-64* through *410-72*. Of particular importance are the Code restrictions on conductor temperature ratings, fixture clearances from combustible materials, and maximum lamp wattages. Recessed fixtures generate a great deal of heat within the enclosure. **Thus, these fixtures are a fire hazard if they are not wired and installed properly,** figure 7-4. Figure 7-5 shows the roughing-in box of a recessed fixture with mounting brackets and junction box.

The branch-circuit conductors are run to the junction box for the recessed fixture. Here they are connected to conductors whose insulation can

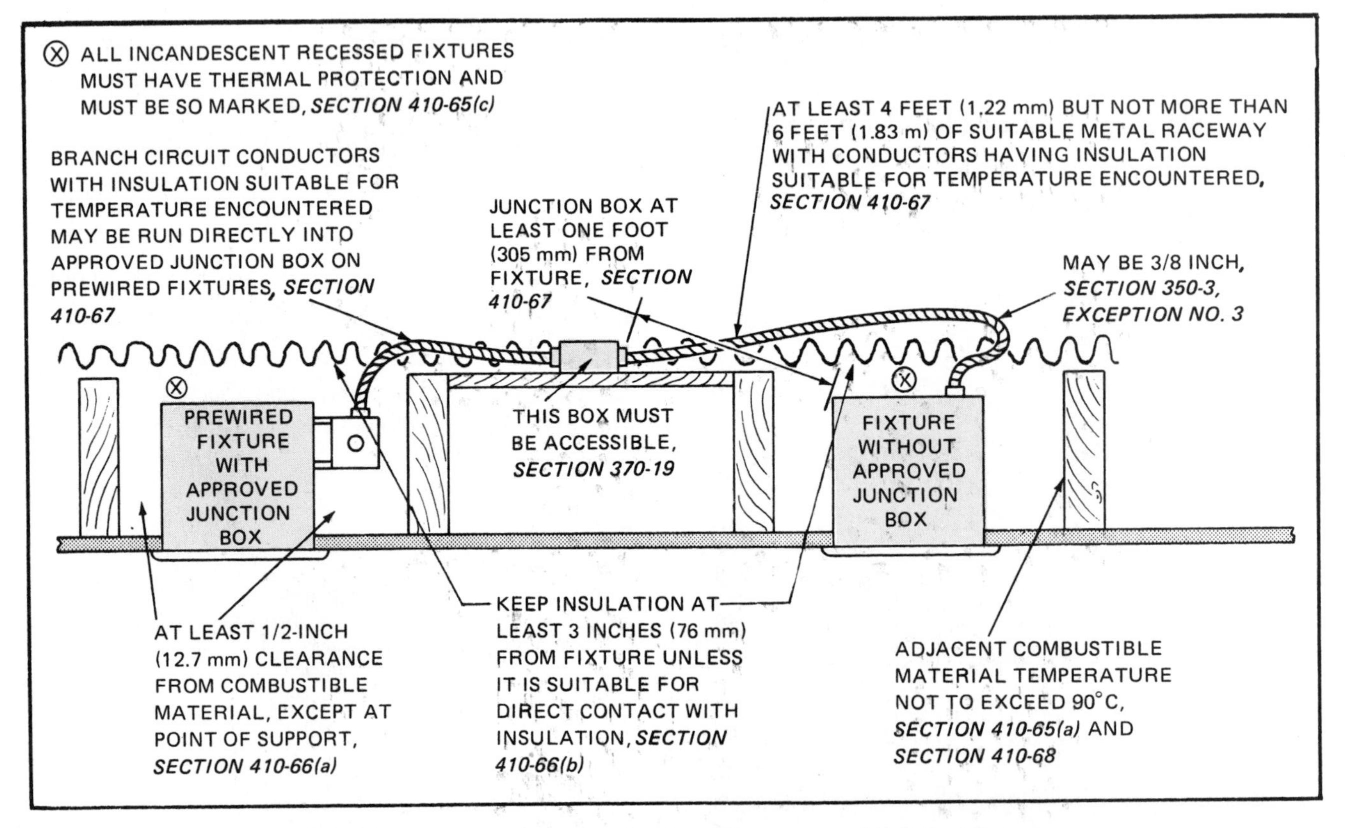

Fig. 7-4 Clearance requirements for installing recessed lighting fixtures.

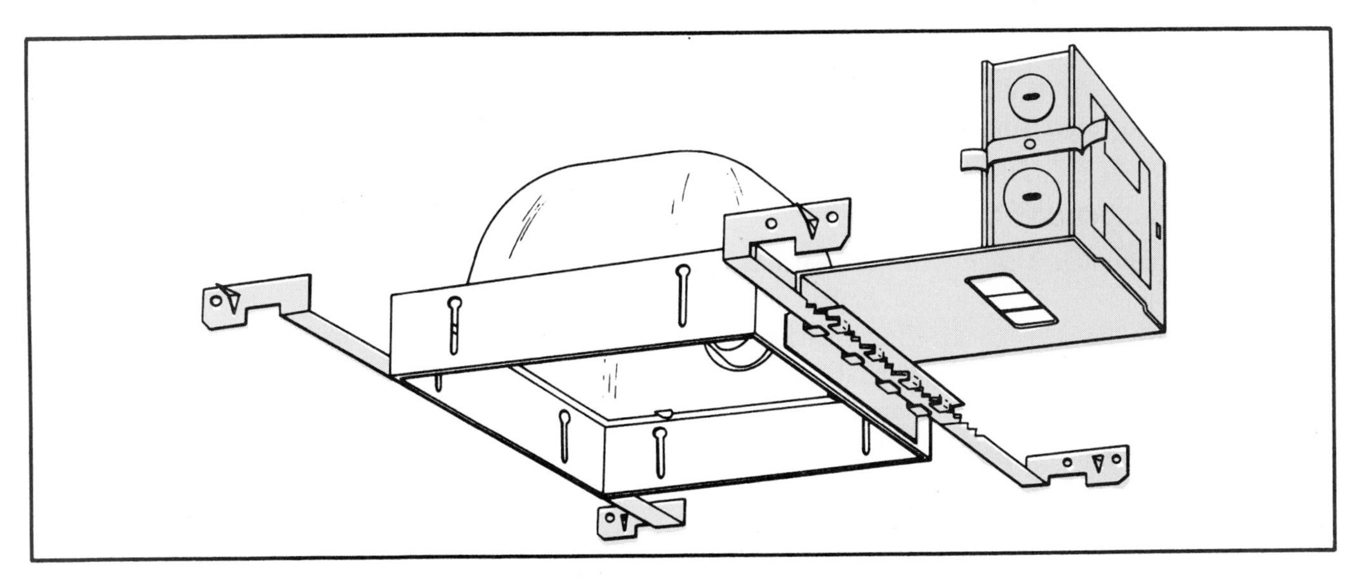

Fig. 7-5 Roughing-in box of a recessed fixture with mounting brackets and junction box.

handle the temperature at the fixture lampholder. The junction box is placed at least one foot (305 mm) from the fixture. Thus, heat radiated from the fixture cannot overheat the wires in the junction box. Conductors rated for higher temperatures must run through at least four feet (1.22 m) of metal raceway, but not to exceed six feet (1.83 m) between the fixture and the junction box. As a result, any heat conducted from the fixture along the metal raceway will be reduced before reaching the junction box. Many recessed fixtures are factory equipped with a flexible metal raceway containing high-temperature (150 °C) wires that meet the requirements of *Section 410-67.*

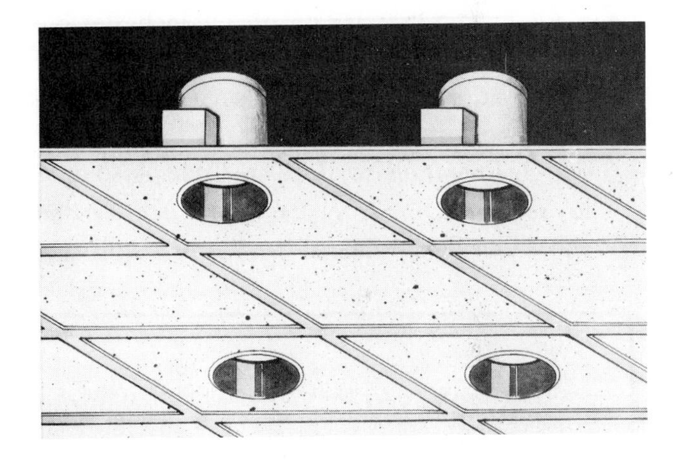

Fig. 7-6 Suspended ceiling fixtures.

Suspended Ceiling Lay-in Fixtures (Figure 7-6)

The recreation room of the residence discussed in this text has "dropped" suspended acoustical paneled ceiling. The fixtures installed in this ceiling bear a label stating "SUSPENDED CEILING FIXTURE." Note that these fixtures are *not* classified as recessed fixtures because, in most cases, there is a great deal of open space above and around these fixtures.

Because of this, it is not necessary to follow the Code rules relating to recessed fixtures as stated in *Part N* of *Article 410* of the Code.

Support of Suspended Ceiling Fixtures

Section 410-16(c) requires that when framing members of a suspended ceiling grid are used to support lighting fixtures, then all members must be securely fastened together and to the building structure. It is up to the installer of the ceiling grid to do this.

All lay-in suspended ceiling fixtures must be securely fastened to the ceiling grid members by bolts, screws, or rivets. Most fixture manufacturers supply special clips that are suitable for fastening the fixture to the grid.

The logic behind all of the "securely fastening"

requirements is that in the event of a major problem (i.e., earthquake or fire), the lighting fixtures will not fall down and injure someone.

Connecting Suspended Ceiling Fixtures

The most common way to connect suspended ceiling lay-in lighting fixtures is to complete all of the wiring above the suspended ceiling using normal wiring methods, such as electrical metallic tubing, nonmetallic-sheathed cable, armored cable, or whatever type of wiring method is acceptable by the governing electrical code. Then from outlet boxes strategically placed above the ceiling near the intended location of the lay-in lighting fixtures, a flexible connection is made between the outlet box and the fixture.

The flexible connection is usually made by installing

- a 6-foot (1.83 m) length of 3/8-inch flexible metal conduit using conductors suitable for the temperature requirement as stated on the fixture label, usually at least 90 °C. See *Section 350-3*, *Exception No. 3*, of the Code. Also refer to *Section 250-91(b)*, *Exception No. 1*, that states that flexible metal conduit in lengths not over 6 feet (1.83 m) and protected by overcurrent devices not over 20 amperes, and using fittings listed for grounding is acceptable as a means of grounding the fixture. See figure 7-7A.

or

- ► a 6-foot (1.83 m) length of armored cable containing conductors suitable for the temperature requirements as stated on the fixture label, usually 90 °C. Be sure to staple the armored cable within 12 inches of the outlet box. See *Section 333-7*, *Exception No. 3*. See figure 7-7B. ◄

In the electrical trade, these flexible connections between an outlet box and a fixture are called *fixture whips*.

A factor to be considered by the electrician when installing recessed fixtures is the necessity of working with the installer of the insulation to be sure that the clearances, as required by the Code, are maintained.

A boxlike device, figure 7-8(A), is available which snaps together around the recessed fixture, thus preventing the insulation from coming into contact with the fixture, as required by Code. The

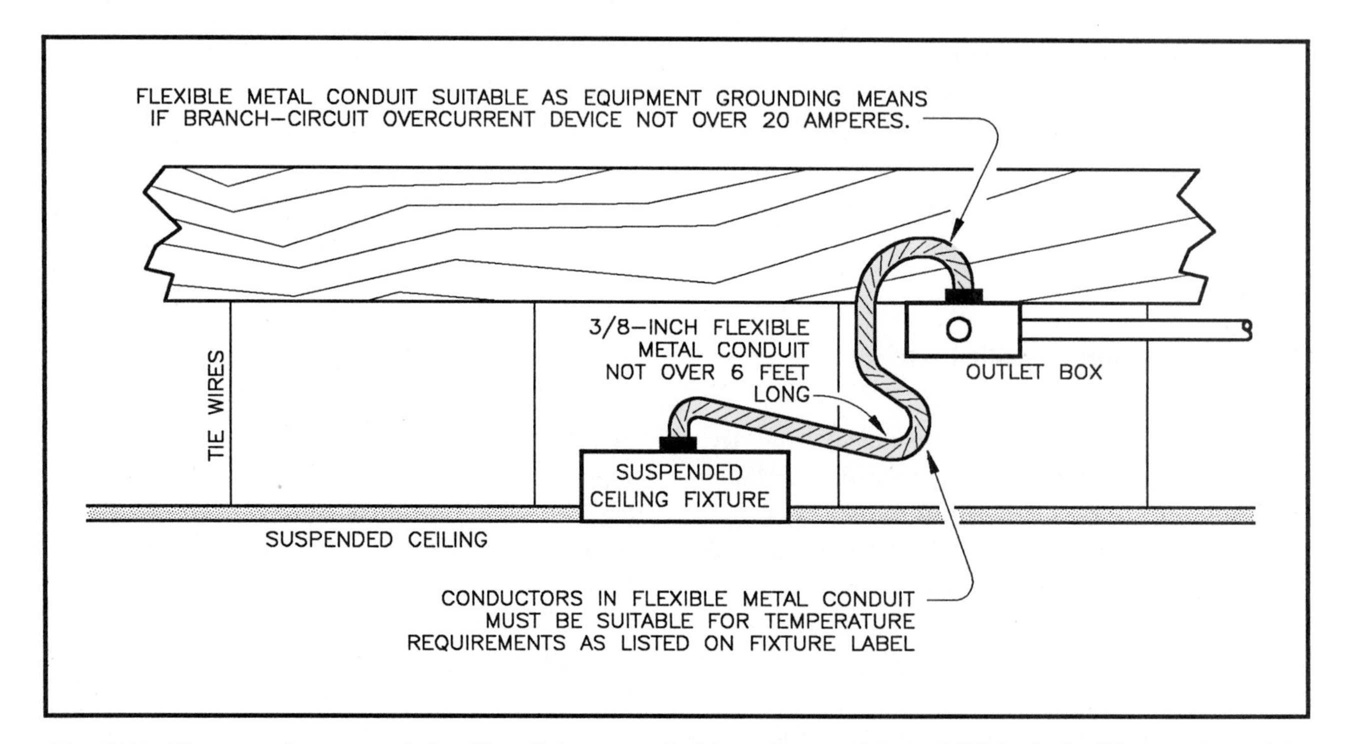

Fig. 7-7A Diagram of a suspended ceiling fixture supplied by not over 6 feet of 3/8-inch flexible metal conduit. Although not required to be secured within 12 inches of the outlet box, *Section 350-4, Exception 3*, some electricians do secure (strap) the flex within 12 inches of the outlet box.

material used in these boxes is fireproof. Figure 7-8(B) indicates the clearances for a recessed lighting fixture installed near thermal insulation (*Section 410-66*).

Recessed fixtures of the type installed in ceilings usually have a box mounted on the side of the fixture. See figure 7-5. The branch-circuit conductors can be run directly into this box where they are connected to the conductors entering the fixture.

Prewired fixtures do not require additional wiring, figure 7-9. *Section 410-11* states that branch-circuit wiring shall not be passed through an

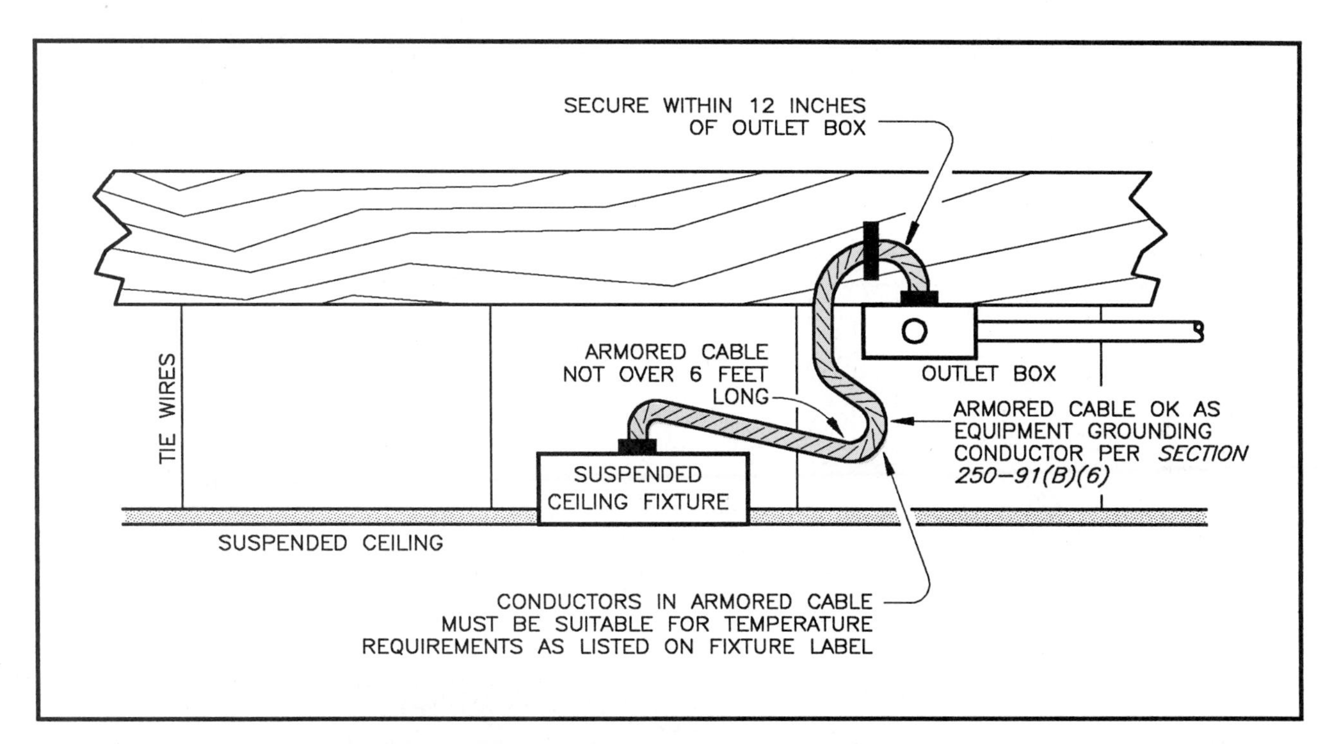

Fig. 7-7B Diagram of a suspended ceiling fixture supplied by armored cable not over 6 feet in length. Cable must be secured (stapled) within 12 inches of the outlet box, *Section 333-7*.

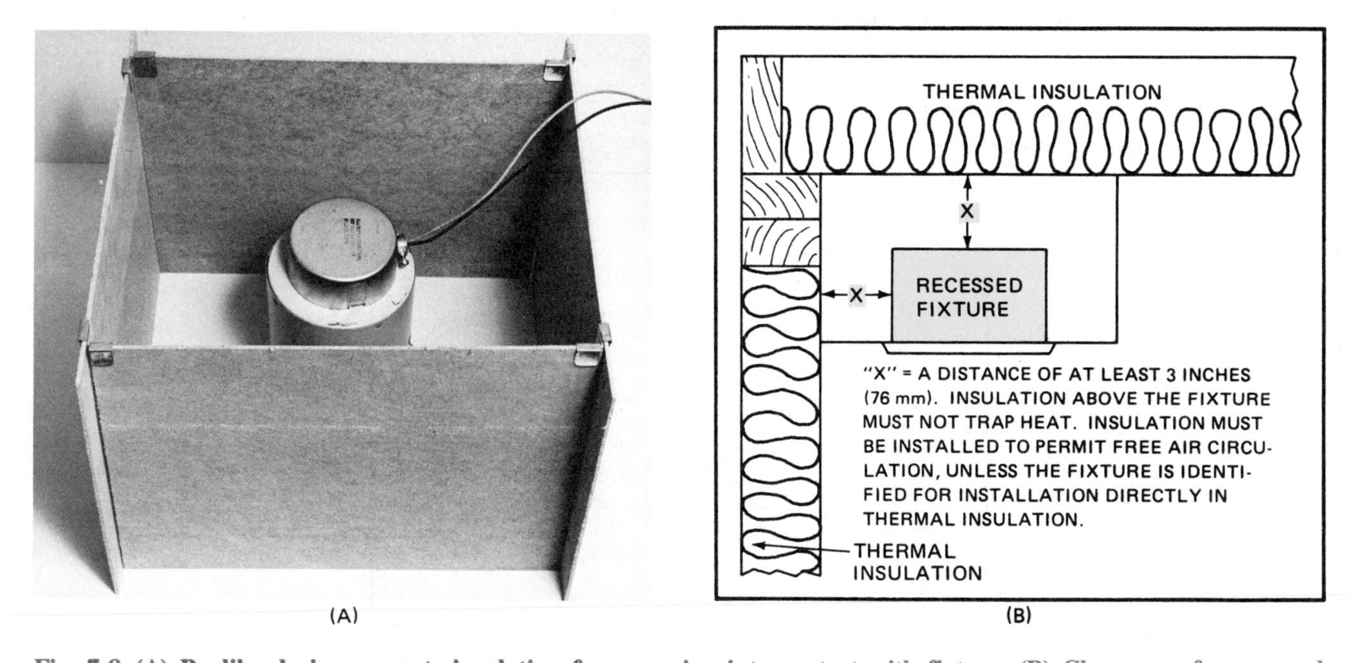

Fig. 7-8 (A) Boxlike device prevents insulation from coming into contact with fixture. (B) Clearances for recessed lighting fixture installed near thermal insulation, *Section 410-66*.

outlet box that is an integral part of an incandescent fixture unless the fixture is identified for through wiring.

For a recessed fixture that is not prewired, the electrician must check the fixture for a label indicating what insulation temperature rating is required.

The cables to be installed in the residence have 90°C (194°F) insulation. If the temperature will exceed this value, conductors with other types of insulation must be installed, *Table 310-13*.

BALLAST PROTECTION

Section 410-73(e) states that all fluorescent ballasts installed indoors (except simple reactance-type ballasts) both for new and replacement installations, must have thermal protection built into the ballast by the manufacturer of the ballast, figure 7-10. Ballasts provided with built-in thermal protection are listed by the Underwriters Laboratories as Class P ballasts. Under normal conditions, the Class P ballast has a case temperature not exceeding

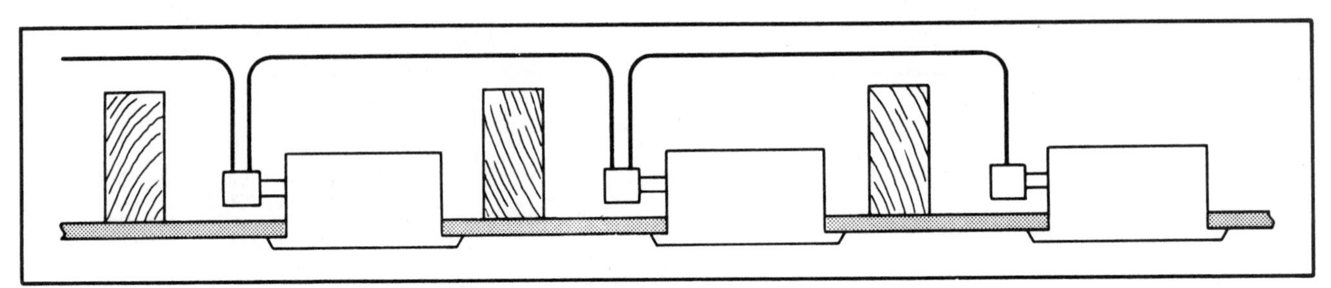

Fig. 7-9 Installation permissible only with prewired recessed fixtures with approved junction box, *Section 410-11.*

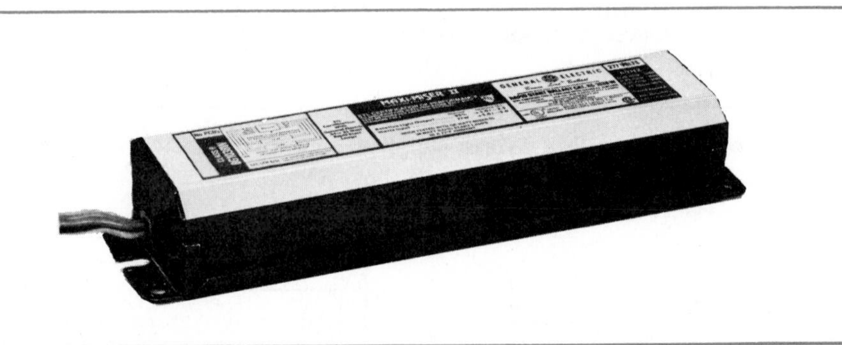

Fig. 7-10 Fluorescent ballasts installed indoors on new or replacement installations are required to have built-in thermal protection. These ballasts are called Class P ballasts, *Section 410-73(e).*

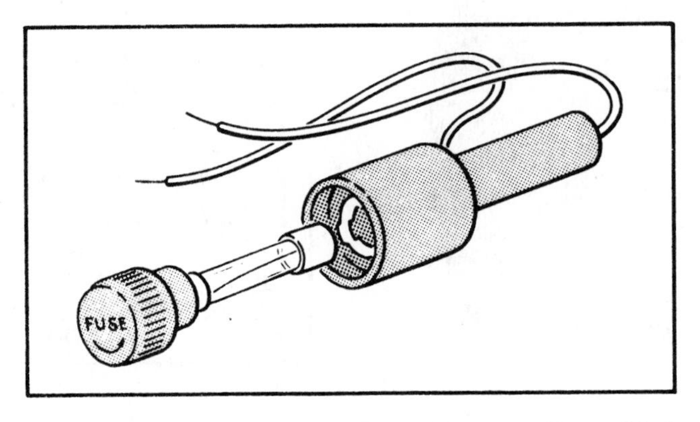

Fig. 7-11 In-line fuseholder for ballast provides added protection.

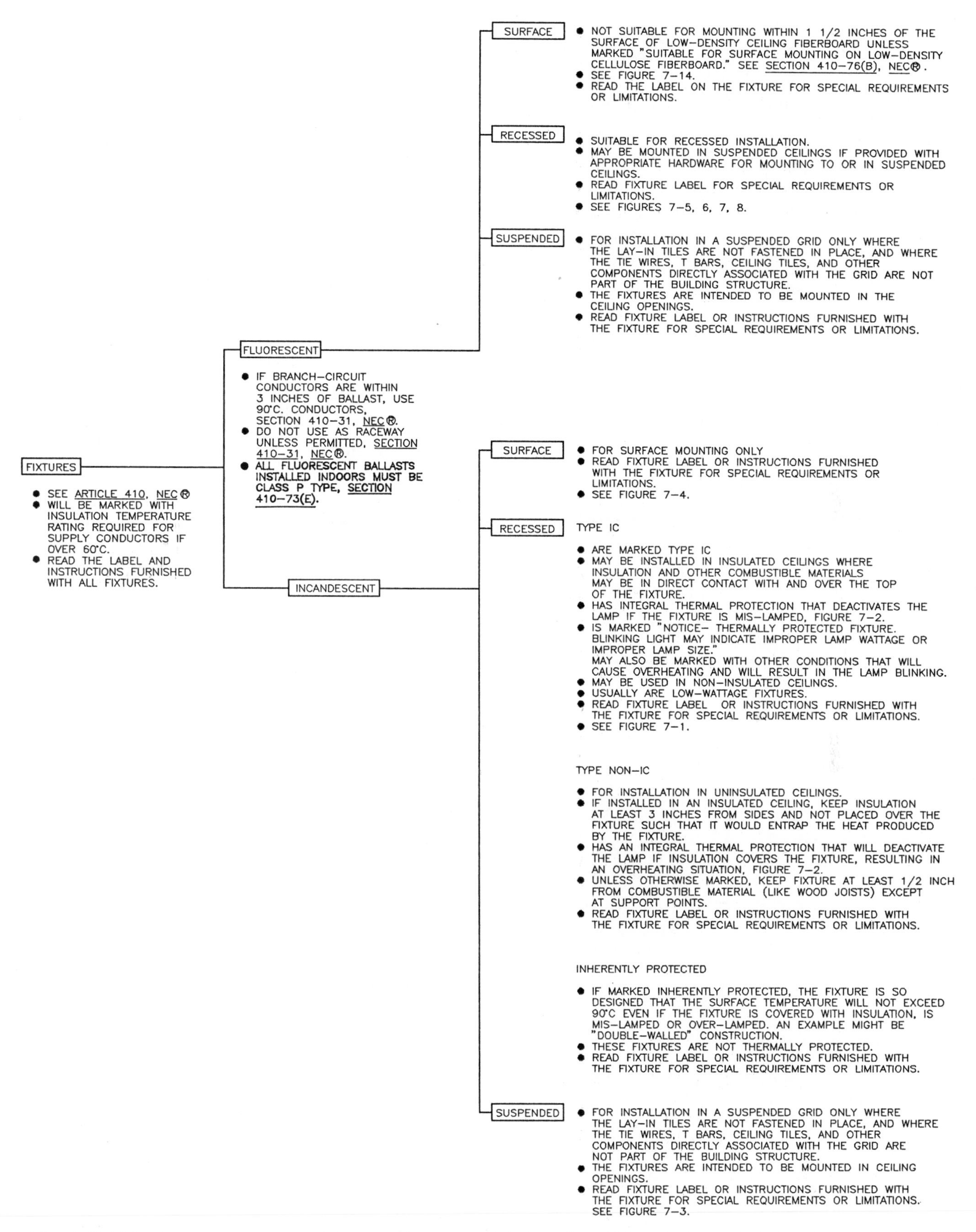

Fig. 7-12 This figure shows some of the most important Underwriters Laboratories and *National Electrical Code®* requirements for recessed fixtures. Always refer to the UL Standards, the *NEC®* and the label and/or instructions furnished with the fixture.

90 °C. The thermal protector must open within two hours when the case temperature reaches 110 °C.

Some Class P ballasts also have a nonresetting fuse integral with the capacitor to protect against capacitor leakage and violent rupture. The Class P ballast's internal thermal protector will disconnect the ballast from the circuit in the event of overtemperature. Excessive temperatures can be caused by abnormal voltage and improper installation, such as being covered with insulation.

The reason for thermal protection is to reduce the hazards of possible fire due to an overheated ballast when the ballast becomes shorted, grounded, covered with insulation, lacking in air circulation, and so on. Ballast failure has been a common cause of electrical fires.

Additional backup protection can be provided by an in-line fuseholder connected in series with the black line lead and the black ballast lead, figure 7-11 (page 128). The in-line fuseholder acts as a disconnect for the ballast. Should a ballast fail (short out), the individual fuse opens, thus isolating the faulty ballast but not affecting the rest of the circuit. When the ballast is to be replaced, the fuse is removed to disconnect the ballast, but the entire circuit need not be turned off. In general, the ballast manufacturer will recommend the correct type and ampere rating of the fuse to be used for the particular ballast. Refer to figure 7-12 (page 129).

LIGHTING FIXTURE VOLTAGE LIMITATIONS

The maximum voltage allowed for residential lighting fixtures is 120 volts between conductors per *Section 210-6* of the Code.

► *Section 410-80(b)* makes a further restriction stating that for dwelling occupancies, no lighting equipment shall be used if it operates with an open-circuit voltage over 1000 volts, figure 7-13. This section more or less focuses on the desire to use neon lighting for more decorative lighting purposes. ◄

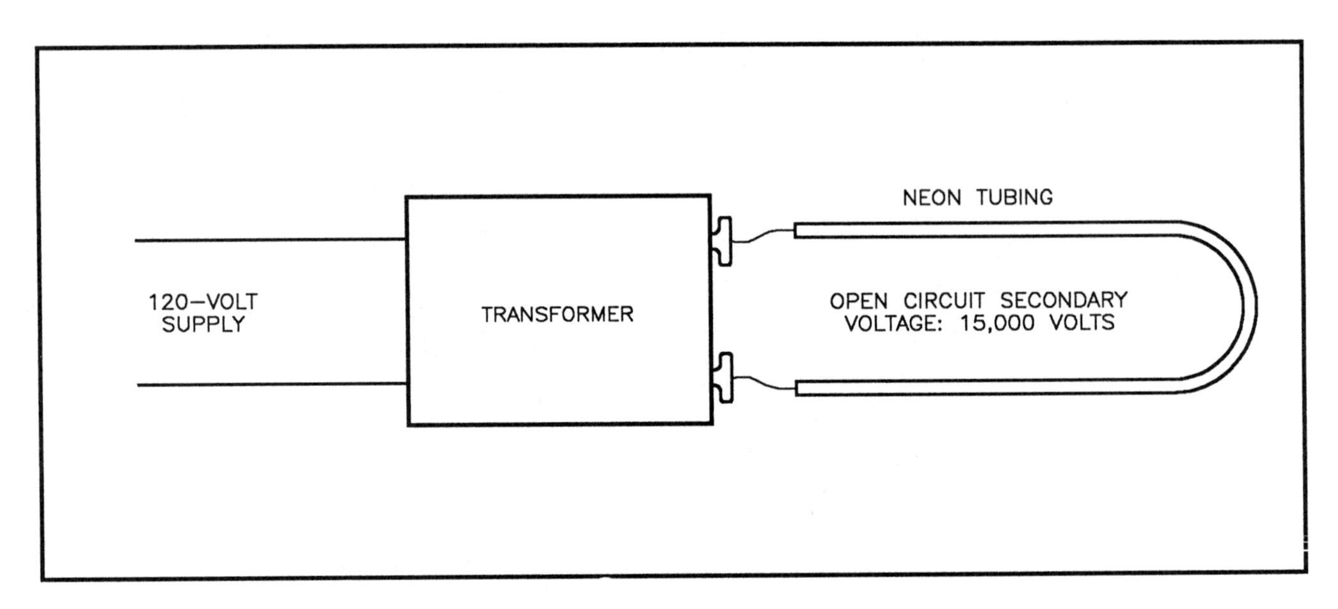

Fig. 7-13 It is NOT permitted to use neon lighting fixtures in residences where the open-circuit voltage is over 1,000 volts.

REVIEW

1. Is it permissible to install a recessed fixture directly against wood ceiling joists?

   ____

2. If a recessed fixture without an approved junction box is installed, what extra wiring must be provided? ______________________________

   ____

   ____

   ____

   ____

   ____

   ____

3. Thermal insulation shall not be installed within ______________ inches (millimeters) of the top or ______________ inches (millimeters) of the side of a recessed fixture unless the fixture is identified for use in direct contact with thermal insulation.

4. Recessed fixtures are available for installation in direct contact with thermal insulation. These fixtures bear the UL mark "Type ______________."

5. Unless specifically designed, all recessed incandescent fixtures must be provided with factory-installed ______________________________.

6. Plans require the installation of a surface-mounted fluorescent fixture on the ceiling of a recreation room that is finished with low-density ceiling fiberboard. What sort of mark would you look for on the label of the fixture? ______________

   ____

   ____

   ____

7. If a recessed fixture bears no marking that it is listed for branch circuit feedthrough wiring, is it permitted to run the circuit conductors from fixture to fixture? What section of the Code covers this? ______________________________

8. Fluorescent ballasts for all indoor applications must be ______________ type. These ballasts contain internal ______________ protection to protect against overheating.

9. Additional backup protection for ballasts can be provided by connecting a(an) ______________________________ with the proper size fuse as recommended by the ballast manufacturer.

10. You are called upon to install a number of lighting fixtures in a suspended ceiling. The ceiling will be dropped approximately 8 inches from the ceiling joists. Briefly explain how you might go about wiring these fixtures. ______________________

► 11. The Code places a maximum open-circuit voltage on lighting equipment in homes. This maximum voltage is (600) (750) (1000). (Circle the correct answer.) Where in the *National Electrical Code®* is this voltage maximum referenced? ◄

UNIT 8

Lighting Branch Circuit for Front Bedroom

OBJECTIVES

After studying this unit, the student will be able to

- explain the factors which influence the grouping of outlets into circuits.
- estimate loads for the outlets of a circuit.
- draw a cable layout and a wiring diagram based on information given in the residence plans, the specifications, and Code requirements.
- select the proper wall box for a particular installation.
- explain how wall boxes can be grounded.
- list the requirements for the installation of fixtures in clothes closets.

This unit is longer than the other units because it is the first exposure to real-life installations. Most issues confronting the electrician for a typical residential installation are covered in this unit. Repetition in later units is kept to a minimum.

GROUPING OUTLETS

The grouping of outlets into circuits must conform to *National Electrical Code®* standards and good wiring practices. There are many possible combinations or groupings of outlets. In most residential installations, circuit planning is usually done by the electrician, who must insure that the circuits conform to Code requirements. In larger, more costly residences, the circuit layout may be completed by the architect and included on the plans.

Because many circuit arrangements are possible, there are few guidelines for selecting outlets for a particular circuit. An electrician plans circuits that are economical without sacrificing the quality of the installation.

For example, some electricians prefer to have more than one circuit feed a room. Should one circuit have a problem, the second circuit would still continue to supply power to the other outlets in that room. See figure 8-1.

Another excellent way to provide good wiring and at the same time economize on wiring materials is to connect receptacle outlets back-to-back, as illustrated in figure 8-2.

Some electricians consider it poor practice to include outlets on different floors on the same circuit. Here, too, the decision can be a matter of personal choice. Some local building codes limit this type of installation to lights at the head and foot of a stairway.

Residential Lighting

Residential lighting is a personal thing. The homeowner, builder, and electrical contractor must meet to decide on what types of lighting fixtures are to be installed in the residence. Many variables (cost, personal preference, construction obstacles, etc.) must be taken into consideration.

Residential lighting can be segmented into four groups.

General Lighting: General lighting is the lighting required to provide overall illumination for a given area. A few examples in this residence would be the front entry hall, the bedroom hall, and the garage lighting.

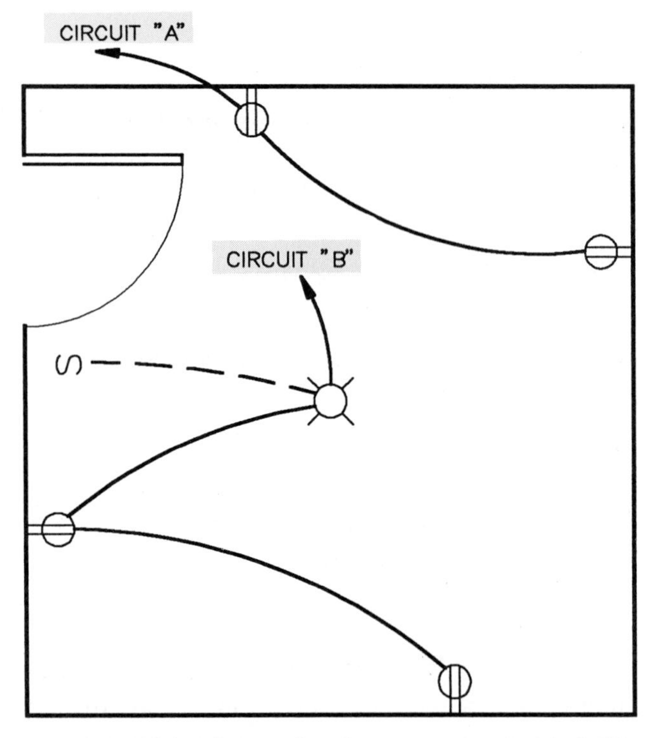

Fig. 8-1 Wiring layout showing one room that is fed by two different circuits. If one of the circuits goes out, the other circuit will still provide electricity to the room.

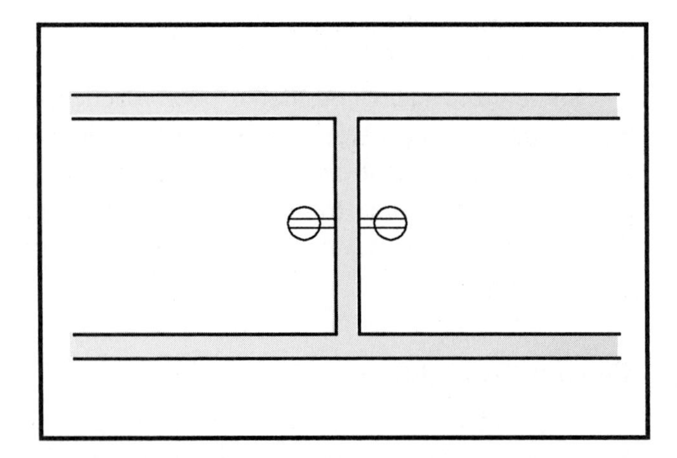

Fig. 8-2 Receptacle outlets connected back-to-back. This can reduce the cost of the installation because of the short distance between the outlets.

Accent Lighting: Accent lighting is that lighting which provides a "focus" or "attention" or "accent" on an object or area in the home. Examples are the recessed "spots" over the fireplace and the track lighting on the ceiling of the living room. Track lighting accents a picture, photo, painting, or sculpture that might be hung or displayed on that wall of the living room.

Task Lighting: Task lighting provides proper lighting where "tasks" are performed. In this residence the fluorescent fixtures over the workbench, the lighting inside the range hood in the kitchen, and the recessed fixture over the kitchen sink are all examples of "task" lighting.

Security Lighting: Security lighting generally includes outdoor lighting, such as post lamps, wall fixtures, walkway lighting, and all other lighting that serves the purpose of providing lighting for security and safety reasons. Security lighting in many instances is provided by the normal types of lighting fixtures found in the typical residence. In the residence discussed in this text, the outdoor bracket fixtures in front of the garage and next to the entry doors, as well as the post light, might be considered to be security lighting even though they add to the beauty of the residence.

An organization called the American Lighting Association offers a number of excellent reading materials with residential lighting suggestions and other materials relating to residential lighting. This organization is made up of the many manufacturers of lighting fixtures, as well as electrical distributors that maintain extensive lighting showrooms across the country. A visit to one of these showrooms offers the prospective buyer of lighting fixtures an opportunity to select from hundreds (in some cases thousands) of lighting fixtures. Most of the finer lighting showrooms are staffed with individuals highly qualified to make recommendations to homeowners so that they will get the most value for their money.

The lighting provided throughout the residence conforms to the residential lighting standards as recommended by the American Lighting Association. Certainly many other variations are possible. For the purposes of studying an entire wiring installation for the residence, all of the load calculations, wiring diagrams, and so on, have been accomplished using the fixture selection as indicated on the plans and in the specifications.

CABLE RUNS

A student studying the great number of wiring diagrams throughout this text will find many different ways to run the cables and make up the circuit connections. Sizing of boxes that will conform to the Code for the number of wires and devices is covered in great detail in this text. When a circuit is run "through" a recessed fixture like the type shown in figure 7-1, that recessed fixture must be identified for "through wiring." This subject is discussed in detail in unit 7. When in doubt as to whether or not the fixture is suitable for running wires to and beyond it to another fixture or part of the circuit, then it is recommended that the circuitry be designed so as to end up at the recessed fixture with only the two wires that will connect to that fixture.

To summarize, the grouping of outlets into circuits must satisfy the requirements of the *National Electrical Code*®, local building codes, good wiring practices, and common sense. A good wiring practice is to divide all loads as evenly as possible among the circuits, as required by *Section 220-4(d)*.

ESTIMATING LOADS FOR OUTLETS

Although this text is not intended to be a basic electrical theory text, it should be mentioned that, when calculating loads,

VOLTS × AMPERES = VOLT-AMPERES

Yet, many times we say that

VOLTS × AMPERES = WATTS

What we really mean to say is that

VOLTS × AMPERES × POWER FACTOR = WATTS

In a pure resistive load such as a simple light bulb, a toaster, a flat iron, or a resistance electric heating element, the power factor is 100%. Then

VOLTS × AMPERES × 1 = WATTS

When we are involved with transformers, motors, ballasts, and other "inductive" loads, wattage is not necessarily the same as volt-amperes.

EXAMPLE: Calculate the wattage and volt-amperes of a 120-volt, 10-ampere resistive load.

SOLUTIONS:

a. $120 \times 10 \times 1 = 1200$ watts
b. $120 \times 10 = 1200$ volt-amperes

EXAMPLE: Calculate the wattage and volt-amperes of a 120-volt, 10-ampere motor load at 50% power factor.

SOLUTIONS:

a. $120 \times 10 \times 0.5 = 600$ watts
b. $120 \times 10 = 1200$ watts

Therefore, to be sure that adequate ampacity is provided for in branch circuit wiring, feeder sizing, and service-entrance calculations, the Code requires that we use the term **VOLT-AMPERES**. This allows us to ignore power factor, and address the *true* current draw that will enable us to determine the correct ratings of electrical equipment.

However, in some instances the terms **WATTS** and **VOLT-AMPERES** can be used interchangeably without creating any problems. For instance, the Code, in *Section 220-19*, recognizes that for electric ranges and other cooking equipment, the kVA and kW ratings shall be considered to be equivalent for the purpose of branch-circuit and feeder calculations.

Examples No. 2(a) and (b) in *Chapter 9* of the Code indicate that for wall-mounted ovens, counter-mounted cooking units, water heaters, dishwashers, and combination clothes dryers, their kW ratings are equivalent to kVA values. Therefore, throughout this text the terms **WATTAGE** and **VOLT-AMPERES** are used when calculating and/or estimating loads.

Building plans typically do not specify the ratings in watts for the outlets shown. When planning circuits, the electrician must consider the types of fixtures which may be used at the various outlets. To do this, the electrician must know the general uses of the receptacle outlets in the typical dwelling.

Unit 3 shows that the general lighting load of a residence is determined by allowing a load of 3 volt-amperes for each square foot (0.093 m^2) of floor area, *Section 220-3(b)*. For the residence in the plans, it is shown that six lighting circuits meet the minimum standards set by the Code. However, to provide sufficient capacity, 13 lighting circuits are to be installed in this residence.

The Code does not specify the maximum number of receptacle outlets or lighting outlets and fixtures that may be connected to one circuit in a residence. It may seem wrong, ridiculous, and illogical that 10, 20, 30, or more receptacle outlets can be connected to one circuit and not be in violation of the Code. Consider the fact, however, that what we really have is many "convenience" receptacle outlets that could be deemed "safer" since many receptacle outlets will virtually eliminate the use of extension cords, one of the highest reported causes of electrical fires. Rarely, if ever, would all of the outlets be fully loaded at the same time.

As discussed in unit 3, the total calculated load is based on the volt-amperes per square foot method.

Circuit Loading Guidelines

A good rule to follow in residential wiring (in fact *any* wiring) is *never load the circuit to more than 80% of its capacity.* A 15-ampere, 120-volt branch circuit would be calculated as

$$15 \times 0.80 = 12 \text{ amperes}$$

OR

$$12 \text{ amperes} \times 120 \text{ volts} = 1440 \text{ volt-amperes}$$

Although 20-ampere lighting circuits are generally not installed in residences, the maximum allowable load in volt-amperes for such a circuit is

$$20 \times 0.80 = 16 \text{ amperes}$$

OR

$$16 \text{ amperes} \times 120 \text{ volts} = 1920 \text{ volt-amperes}$$

This is one method that can be used to *estimate* residential lighting loads.

Certain fixtures, such as recessed lights and fluorescent lights, are marked with their maximum lamp wattage and ballast current (for fluorescent fixtures only). Other fixtures, however, are not marked. Furthermore, the electrician does not know the exact load that will be connected to the receptacle outlets. Also unknown is the size of the lamps that will be installed in the lighting fixtures (other than the recessed and fluorescent types). In other words, it is difficult to anticipate what the homeowner may do after the installation is complete. The electrician should remember that the room in which the outlets are located does give some indications as to their possible uses. The circuits should be planned accordingly.

Load Estimation

The lamp loads in the lighting fixtures can be estimated by assuming the lamp wattages that will probably be needed in each fixture to provide adequate lighting for the area involved. Recommendations are found in American Lighting Association publications and in various manufacturers' publications. It is recommended that the student write to these organizations and request copies of the latest lighting publications. The Instructors Guide for Residential Wiring lists the names, addresses, and titles of available publications.

Estimating Number of Outlets by Assigning an Amperage Value to Each

One method of determining the number of lighting and receptacle outlets to be included on one circuit is to assign a value of 1 to 1 1/2 amperes to each outlet to a total of 15 amperes. Thus, a total of 10 to 15 outlets can be included in a 15-ampere circuit.

As stated previously, the *National Electrical Code®* does not limit the number of outlets on one circuit. However, many local building codes do specify the maximum number of outlets per circuit. Before planning any circuits, the electrician must check the local building code requirements.

All outlets will not be required to deliver 1 1/2 amperes. For example, closet lights, night lights, and clocks will use only a small portion of the allowable current. A 60-watt closet light will draw less than 1 ampere.

$$I = \frac{W}{E} = \frac{60}{120} = 0.5 \text{ ampere}$$

For example, if low-wattage fixtures are connected to a circuit, it is quite possible that 15 or more lighting outlets (many times referred to as "openings") could be connected to the circuit without a problem. On the other hand, if the load consists of high-wattage lamps, the number of outlets would be less. In this residence, estimated loads were figured at 120 volt-amperes (1 ampere) for those receptacle outlets obviously intended for general use. Receptacle outlets in the garage and workshop were figured at 180 volt-amperes (1 1/2 amperes) because of the use these outlets will be supplying (i.e., tools, drills, table saws, etc.).

Receptacle outlets in the kitchen and laundry have been included in the small appliance circuit calculations.

If an electrician were to *estimate* the number of 15-ampere lighting branch circuits desired for a new residence where the total count of lighting and receptacle outlets is, for example, 80, then

$$\frac{80}{10} = 8 \quad \text{(eight 15-ampere lighting circuits)}$$

If the circuit consists of low-wattage loads, then

$$\frac{80}{15} = 5.3 \quad \text{(six 15-ampere lighting circuits)}$$

In residential occupancies there is great diversity in the loading of lighting branch circuits.

CAUTION: The foregoing procedure for estimating the number of lighting and receptacle outlets is for branch circuits only. Obviously, where the small appliance circuits in the kitchen, laundry area, and other areas supply heavy concentrations of plug-in appliances, the 1 1/2-amperes per receptacle outlet would not be applicable. These small appliance circuits are discussed later on in this text.

Review of How to Determine the Minimum Number of Lighting Circuits

1. Use the volt-amperes per square foot method as required by the Code and discussed in unit 3.
2. Estimate the probable loading for each lighting outlet and receptacle outlet (do not include the small appliance receptacle outlets). Try to have a total VA not over 1440 for a 15-ampere lighting circuit, and not over 1920 for a 20-ampere lighting circuit.
3. Connect approximately 10 to 15 outlets per circuit — more than 10 if the probable connected load is small, less than 10 if the probable connected load is large. For instance, the wattage of a 120-volt smoke detector is approximately 1/2 watt.

Whereas item 1 is the method required by the Code, items 2 and 3 are guidelines and are not to be considered "cast-in-concrete" rules, because general lighting loads in homes are so diversified. As the electrician lays out the circuits of a residence, these guidelines will generally provide adequate circuitry. This would include both lighting outlets and receptacle outlets that are intended for general lighting, *not* appliance circuit receptacle outlets. Be practical. Consider the wide diversity factor that is present.

Divide Loads Evenly

Section 220-4(d) makes it mandatory to divide loads evenly between the various circuits. The obvious reason is not to experience overload conditions on some circuits, whereas other circuits might be lightly loaded. This is common sense. At the Main Service, don't connect the branch circuits and feeders to result in, for instance, 120 amperes on phase A and 40 amperes on phase B. Look at the probable and/or calculated loads to attain as close a balance as possible, like 80 amperes on each A and B phase.

SYMBOLS

The symbols used on the cable layouts in this text are the same as those found on the actual electrical plans for the residence. Refer to figure 2-4.

Pictorial illustrations are used on all wiring diagrams in this text to make it easy for the reader to complete the wiring diagrams, figure 8-3.

DRAWING THE WIRING DIAGRAM OF A LIGHTING CIRCUIT

The electrician must take information from the building plans and convert it to the forms that will be most useful in planning an installation which is economical, conforms to all Code regulations, and follows good wiring practice. To do this, the electrician first makes a cable layout. Then a wiring diagram is prepared to show clearly each wire and all connections in the circuit.

A skilled electrician, after many years of experience, will not prepare wiring diagrams for most residential circuitry because he has the ability to "think" the connections through mentally. An unskilled individual should make detailed wiring diagrams.

The following steps will guide the student in the preparation of wiring diagrams. (The student is required to draw wiring diagrams in later units of this text.) As the student works on the wiring diagrams of the many circuits provided in this text, note that the receptacles are positioned exactly as though you were standing in the center of the room. Then, as you turn to face each wall, you will find the receptacles positioned in that wall with the grounded conductor slot on the top, as shown in figure 8-4.

1. Refer to plans and make a cable layout of all lighting and receptacle outlets, figure 8-5.
2. Draw a wiring diagram showing the traveler conductors for all three-way switches, if any, figure 8-6 (page 140).
3. Draw a line between each switch and the outlet or outlets it controls. For three-way switches, do this for one switch only. This is the "switch leg."
4. Draw a line from the grounded terminal on the lighting panel to each current-consuming outlet. This line may pass through switch boxes, but must not be connected to any switches. This is the white grounded circuit conductor, often called the "neutral" conductor. It is not truly a neutral conductor unless it is part of a multiwire (3-wire) circuit.

   NOTE: An exception to step 4 may be made for double-pole switches. For these switches, all conductors of the circuit are opened simultaneously. They are rarely used in residential wiring.
5. Draw a line from the ungrounded (hot) terminal on the panel to connect to each switch and unswitched outlet. Connect to one three-way switch only.
6. Show splices as small dots where the various wires are to be connected together. In the wiring diagram, the terminal of a switch or outlet may be used for the junction point of wires. In actual wiring practice, however, the Code does not permit more than one wire to be connected to a terminal unless the terminal is a type approved for more than one conductor. The standard screw-type terminal is *not* approved for more than one wire, *Section 110-14.*
7. The final step in preparing the wiring diagram is to mark the color of the conductors, figure 8-7 (page 140). Note that the colors selected

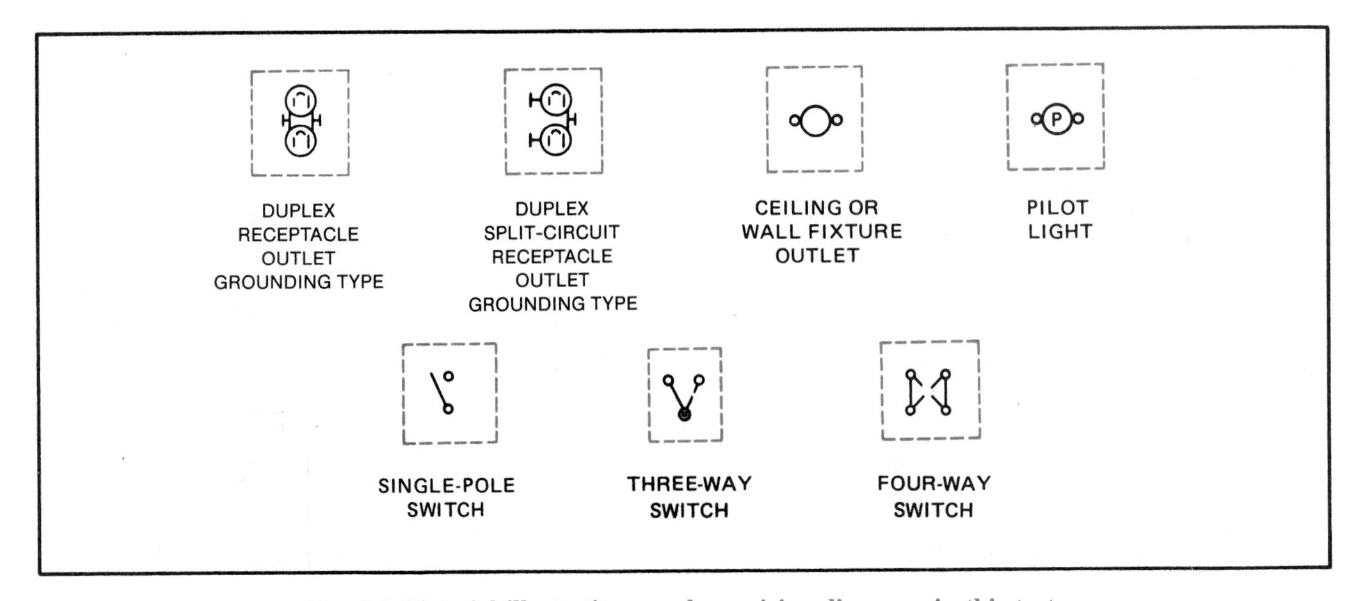

Fig. 8-3 Pictorial illustrations used on wiring diagrams in this text.

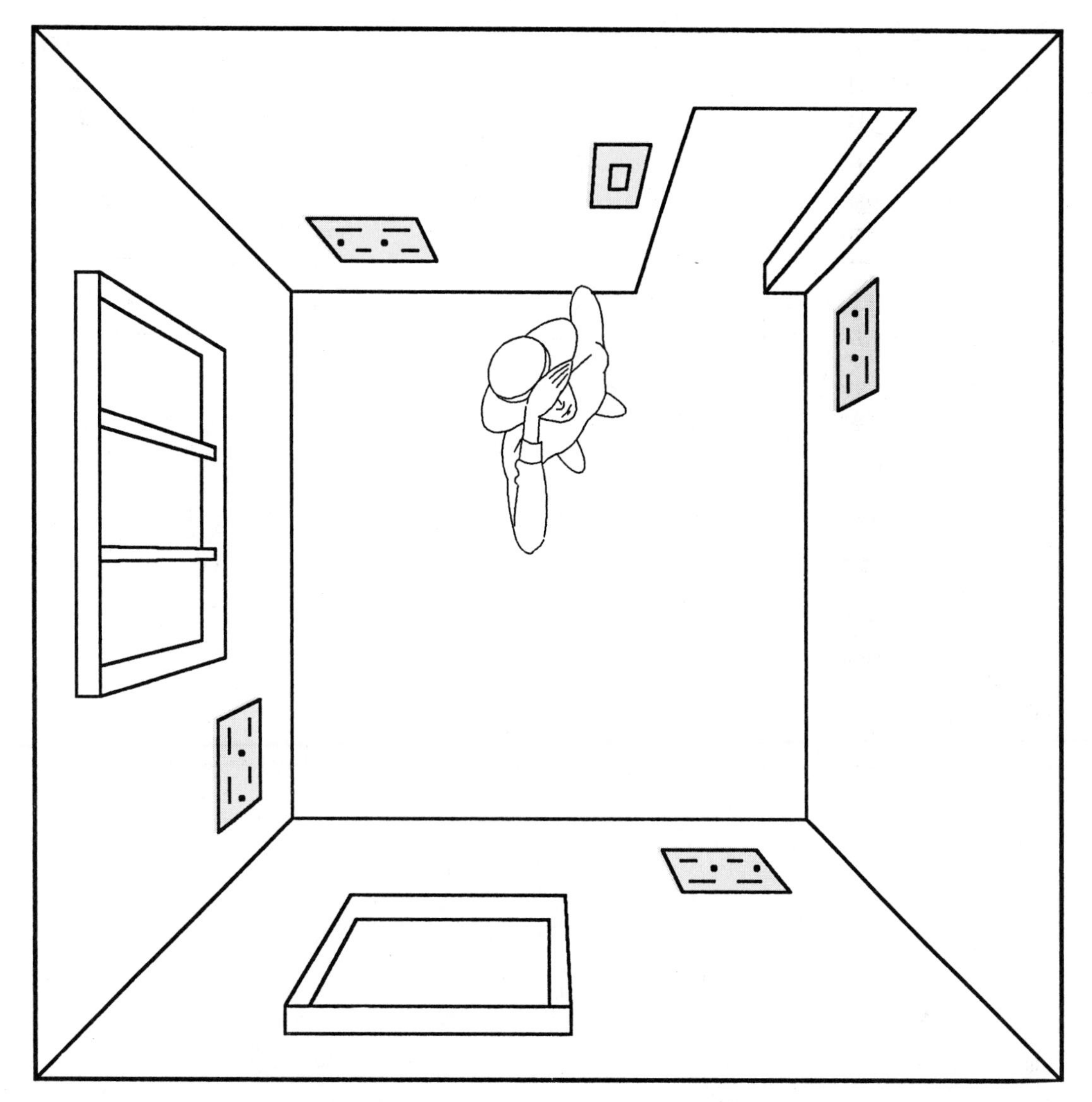

Fig. 8-4 This top view of a room shows the relative positioning of receptacle outlets so that the wide grounded terminal slot is at the top. This provides for added safety if a metal face plate loosens or if a metal clip or other object falls on top of the blades of an attachment plug cap that is not plugged in all the way. All of the wiring diagrams throughout this text will illustrate the receptacles in this manner.

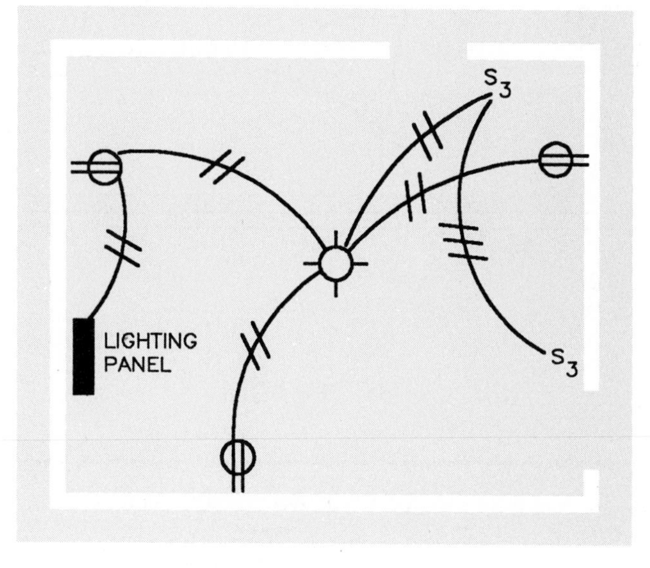

Fig. 8-5 Typical cable layout.

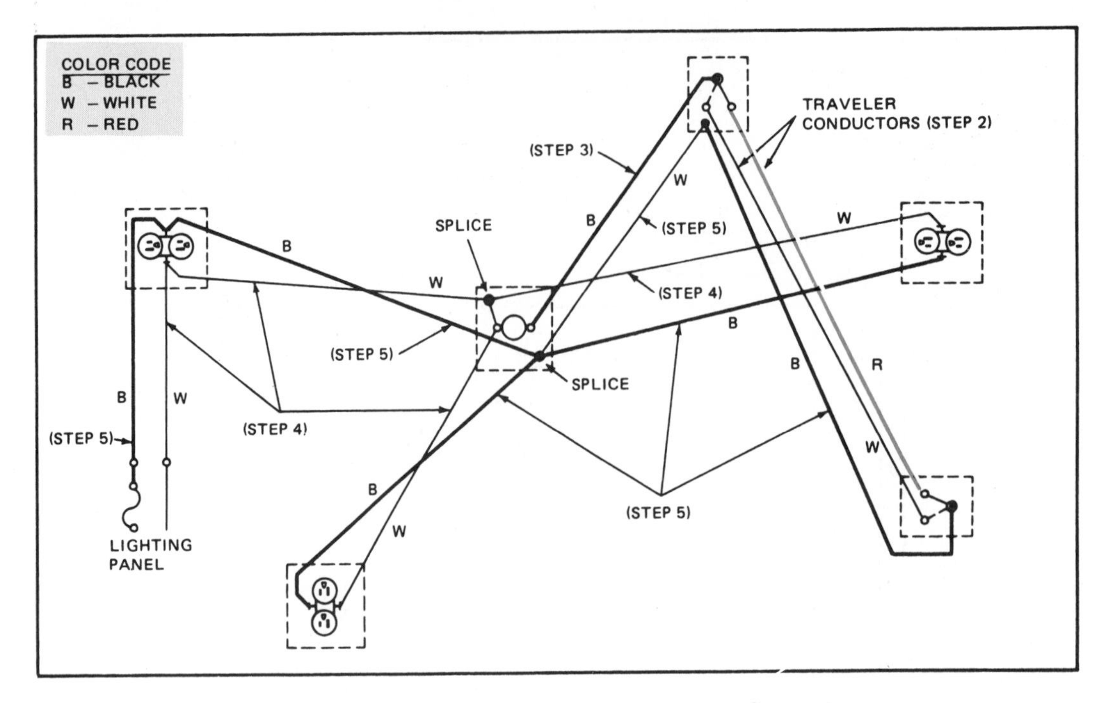

Fig. 8-6 Wiring diagram of circuit shown in figure 8-5.

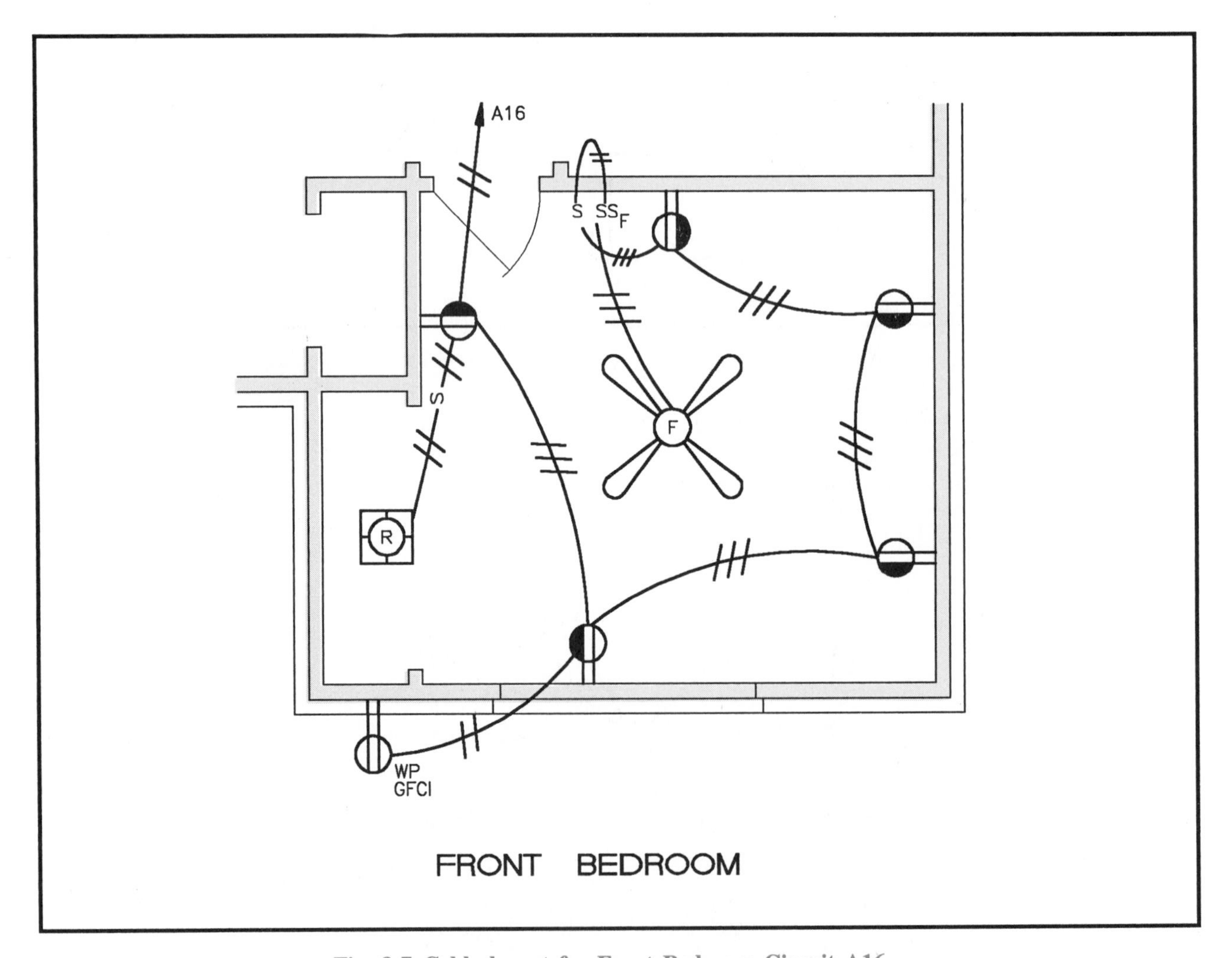

Fig. 8-7 Cable layout for Front Bedroom Circuit A16.

— black (B), white (W), and red (R) — are the colors of two- and three-conductor cables (refer to unit 5). It is suggested that the student use colored pencils or markers for different conductors when drawing the wiring diagram. Yellow can be used to indicate white conductors to provide better contrast with white paper. A note to this effect should be placed on the diagram, *Section 200-7.*

As students work on the wiring diagrams of the many circuits throughout this text, note that the receptacles are positioned exactly as though the students were standing in the center of the room. Then, as they turn to face each wall, they will find the receptacles positioned in that wall with the grounded conductor slot on the top, as shown in figure 8-4.

LIGHTING BRANCH CIRCUIT A16 FOR FRONT BEDROOM

This is a 15-ampere branch circuit. As we look at this circuit, we find five split-circuit receptacle outlets. These will provide the general lighting through the use of table or swag lamps that will be plugged into the switched receptacles.

Radio, television, and other electrical items not intended to be controlled by the wall switch will be plugged into the "hot continuously" receptacle.

Next to the wall switch we find the control for the ceiling fan, which also has a lighting fixture as an integral part of the fan, figure 8-8 (page 142). A single-pole switch controls the ON/OFF and speed of the fan motor.

The recessed closet light is controlled by a single-pole switch outside and to the right of the closet door.

Note that one outside weatherproof receptacle is also connected to circuit A16.

Checking the actual electrical plans, we find one television outlet and one telephone outlet are to be installed in the Front Bedroom. Television and telephones are discussed in unit 25. Table 8-1 summarizes the outlets in the front bedroom and the estimated load.

DETERMINING THE WALL BOX SIZE

Several factors must be considered when the electrician determines the size of the wall box. These factors include:

- the number of conductors entering the box.
- the types of boxes available.
- the space allowed for the installation of the box.

Box Size According to the Number of Conductors in a Box *(Section 370-6)*

To determine the proper box size for any location, the total number of conductors entering the box must be determined. The following example shows how the proper wall box is determined for a particular installation, Table 8-2 (page 143).

1. Add the No. 14 AWG circuit conductors:
   2 + 3 + 3 = 8
2. Add grounding wires
   (count one only) 1
3. Add two conductors for
   the receptacle 2

Total 11 Conductors

Note that the pigtails connected to the receptacle, figure 8-9 (page 142), need *not* be counted when determining the correct box size. The Code in *Section 370-6(a)(1)* states, "Conductors, no part of which leave the box, shall not be counted."

Once the total number of conductors is known, refer to *Table 370-6(a)* (figure 2-14) to find the box that can hold the conductors. For example, a 4″ × 2 1/8″ square box with a suitable plaster ring can be used.

DESCRIPTION	QUANTITY	WATTS	VOLT-AMPERES
Receptacles @ 120 watts each	5	600	600
Weatherproof receptacle	1	120	120
Closet recessed fixture	1	75	75
Paddle fan/light	1		
3 - 50-W lamps		150	150
Fan motor (0.75 A @ 120 V)		80	90
TOTALS	8	1025	1035

Table 8-1. Front bedroom: outlet count and estimated load. Circuit A16.

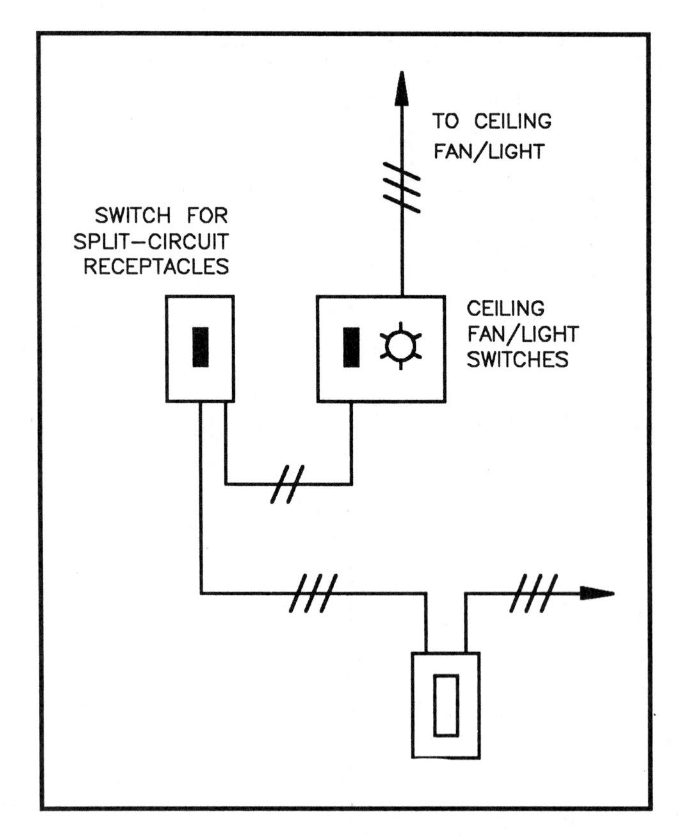

Fig. 8-8 Conceptual view of how the Front Bedroom switch arrangement is to be accomplished.

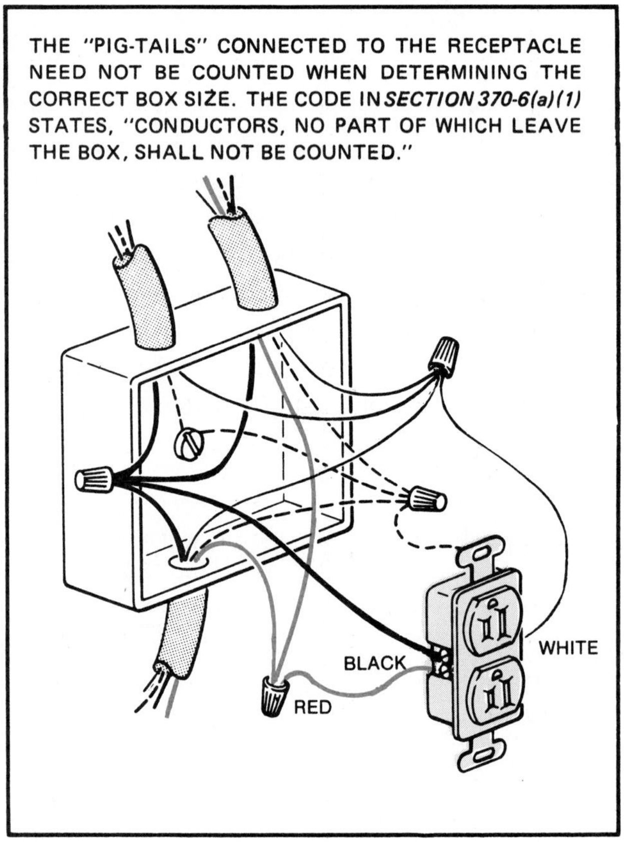

THE "PIG-TAILS" CONNECTED TO THE RECEPTACLE NEED NOT BE COUNTED WHEN DETERMINING THE CORRECT BOX SIZE. THE CODE IN *SECTION 370-6(a)(1)* STATES, "CONDUCTORS, NO PART OF WHICH LEAVE THE BOX, SHALL NOT BE COUNTED."

BLACK
WHITE
RED

Fig. 8-9 Determining size of box according to number of conductors.

The volume of the box plus the space provided by plaster rings, extension rings, and raised covers may be used to determine the total available volume. In addition, it is desirable to install boxes with external cable clamps. Remember that if the box contains one or more devices, such as cable clamps, fixture studs, or hickeys, the number of conductors permitted in the box shall be one less than shown in *Table 370-6(a)* for *each type* of device contained in the box. (See the example under "Number of Conductors" in unit 2.)

GROUNDING OF WALL BOXES

The specifications for the residence state that *all* metal grounding boxes are to be grounded. The means of grounding is armored cable or nonmetallic-sheathed cable containing an extra grounding conductor. This conductor is used only to ground the metal box. This bare grounding conductor must *not* be used as a current-carrying conductor since severe shocks can result.

According to *Section 210-7*, grounding-type receptacles must be installed on 15-ampere and 20-ampere branch circuits. The methods of attaching the grounding conductor to the proper terminal on the convenience receptacle are covered in unit 5.

POSITIONING OF SPLIT-CIRCUIT RECEPTACLES

The receptacle outlets shown in the Front Bedroom are called two-circuit or split-circuit receptacles. The top portion of such a receptacle is hot at all times, and the bottom portion is controlled by the wall switch, figure 8-10. It is recommended that the electrician wire the bottom section of the

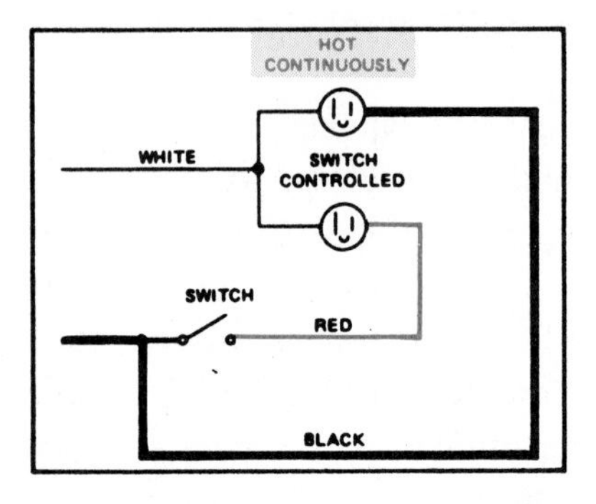

Fig. 8-10 Split-circuit wiring for receptacles in Bedroom.

QUIK-CHEK BOX SELECTION GUIDE

FOR BOXES GENERALLY USED FOR RESIDENTIAL WIRING

DEVICE BOXES

WIRE SIZE	3X2X1½ (7.5 in^3)	3X2X2 (10 in^3)	3X2X2¼ (10.5 in^3)	3X2X2½ (12.5 in^3)	3X2X2¾ (14 in^3)	3X2X3 (16 in^3)	3X2X3½ (18 in^3)
#14	3	5	5	6	7	7	9
#12	3	4	4	5	6	7	8

SQUARE BOXES

WIRE SIZE	4X4X1½ (2.1 in^3)	4X4X2⅛ (30.3 in^3)
#14	10	15
#12	9	13

OCTAGON BOXES

WIRE SIZE	4X1½ (15.5 in^3)	4X2⅛ (21.5 in^3)
#14	7	10
#12	6	9

HANDY BOXES

WIRE SIZE	4X2⅛X1½ (10.3 in^3)	4X2⅛X1⅞ (13 in^3)	4X2X2⅛ (14.5 in^3)
#14	5	6	7
#12	4	5	6

RAISED COVERS

WHERE RAISED COVERS ARE MARKED WITH THEIR VOLUME IN CUBIC INCHES, THAT VOLUME MAY BE ADDED TO THE BOX VOLUME TO DETERMINE MAXIMUM NUMBER OF CONDUCTORS IN THE COMBINED BOX AND RAISED COVER.

NOTE: BE SURE TO MAKE DEDUCTIONS FROM THE ABOVE MAXIMUM NUMBER OF CONDUCTORS PERMITTED FOR WIRING DEVICES, CABLE CLAMPS, FIXTURE STUDS, AND GROUNDING CONDUCTORS. THE CUBIC INCH (IN^3) VOLUME IS TAKEN DIRECTLY FROM *TABLE 370–6A* OF THE CODE.

Table 8-2 Quik-chek box selection guide.

receptacle as the switched section. As a result, when the attachment plug cap of a lamp is inserted into the bottom switched portion of the receptacle, the cord does hang in front of the unswitched section. This unswitched section can be used as a receptacle for clock, vacuum cleaner, radio, stereo, compact disc player, personal computer, television, or other appliances where a switch control is not necessary or desirable.

When split-circuit receptacles are horizontally mounted, which is common when 4″ square boxes are used because the plaster ring is easily fastened to the box in either a vertical or horizontal position, locate the switched portion to the right.

POSITIONING OF RECEPTACLES NEAR ELECTRIC BASEBOARD HEATING

This unit is the first unit in this text where we begin to discuss actual circuit layout of receptacle outlets, lighting outlets, and switches.

Later we will discuss electric heat. Electric heating for a home can be accomplished with an

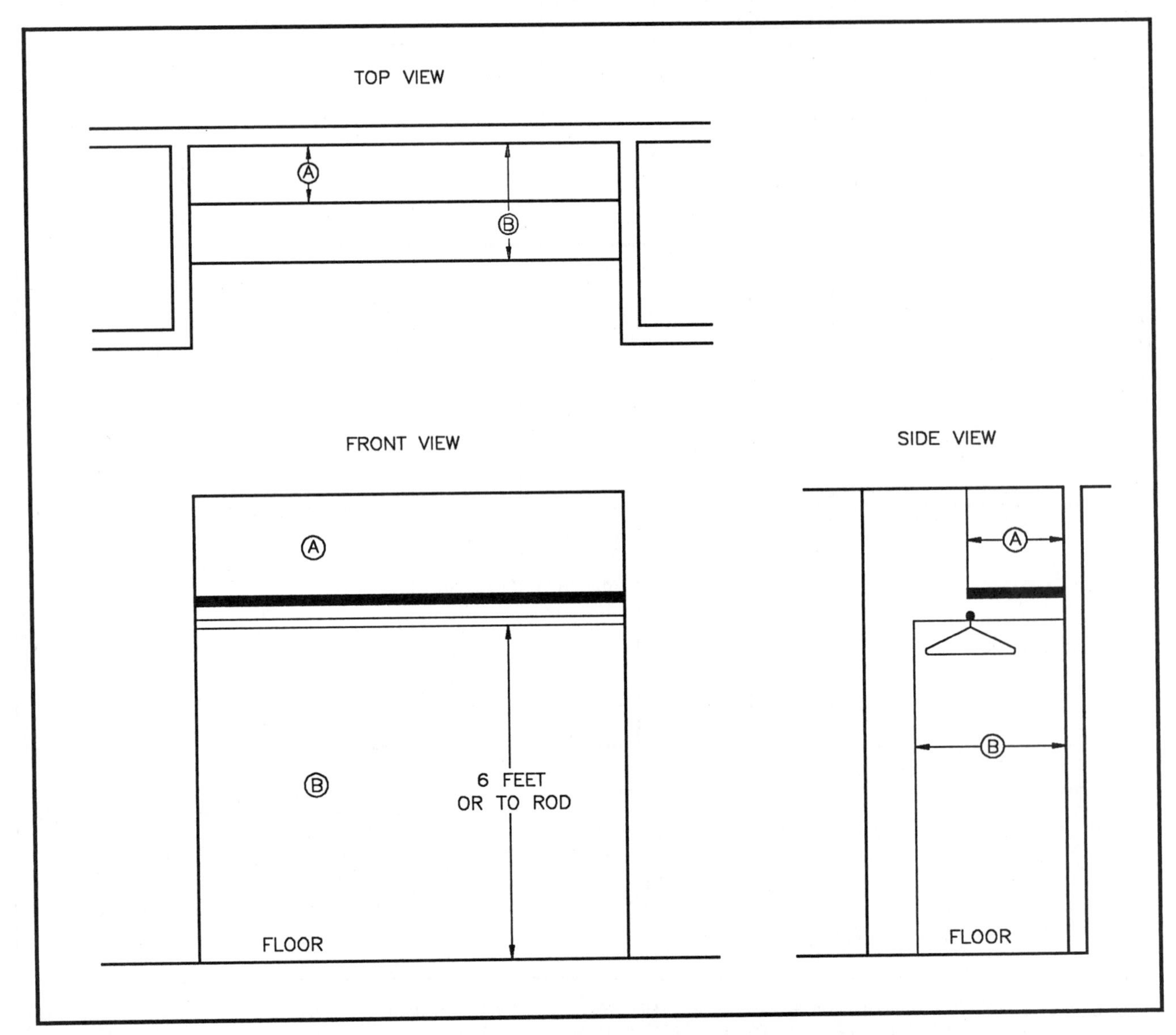

Fig. 8-11 Typical closet with one shelf and one rod. The shaded area defines "storage area." Dimension A is width of shelf or 12 inches from wall, whichever is greater. Dimension B is below rod, 24 inches from wall. See *Section 410-8(a)*.

electric furnace, electric baseboard heating units, heat pump, or resistance heating cables embedded in plastered ceilings or sandwiched between two layers of drywall sheets on the ceiling.

The important thing to remember at this point is that there are stringent *National Electrical Code*® rules governing the positioning of wall electrical baseboards in relationship to the location of wall receptacle outlets. See *Section 210-52(a), Exception* and *FPN.*

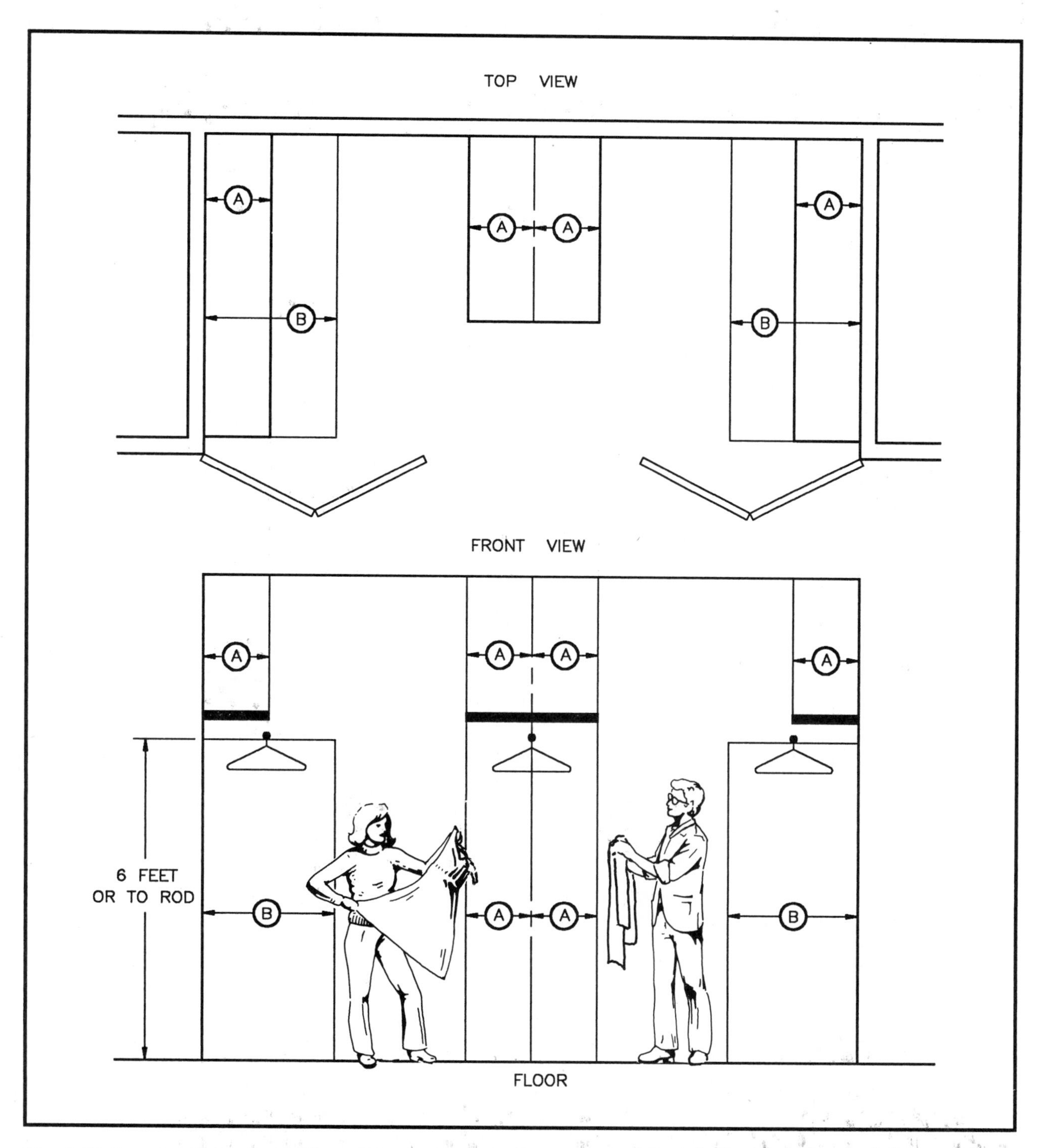

Fig. 8-12 Large walk-in closet where there is access to the center rod from both sides. The shaded area defines "storage area." Dimension A is width of shelf or 12 inches, whichever is greater. Dimension B is 24 inches from wall. See *Section 410-8(a).*

See unit 23 for a detailed discussion of installing electric baseboard heating units and their relative positions below receptacle outlets.

The Front Bedroom does have a paddle fan/light on the ceiling. Paddle fans are discussed in unit 9.

► FIXTURES IN CLOTHES CLOSETS

Clothing, boxes, and other material normally stored in clothes closets are a potential fire hazard. These items may ignite on contact with the hot surface of exposed lamps. Incandescent lamps have a hotter surface temperature than fluorescent lamps. *Section 410-8* of the Code gives very specific rules relative to the location and types of lighting fixtures permitted to be installed in clothes closets. Figures 8-11 through 8-15 (pages 144–147) illustrate the requirements of *Section 410-8*. ◄

Figure 8-16 (page 148) shows typical bedroom-type lighting fixtures.

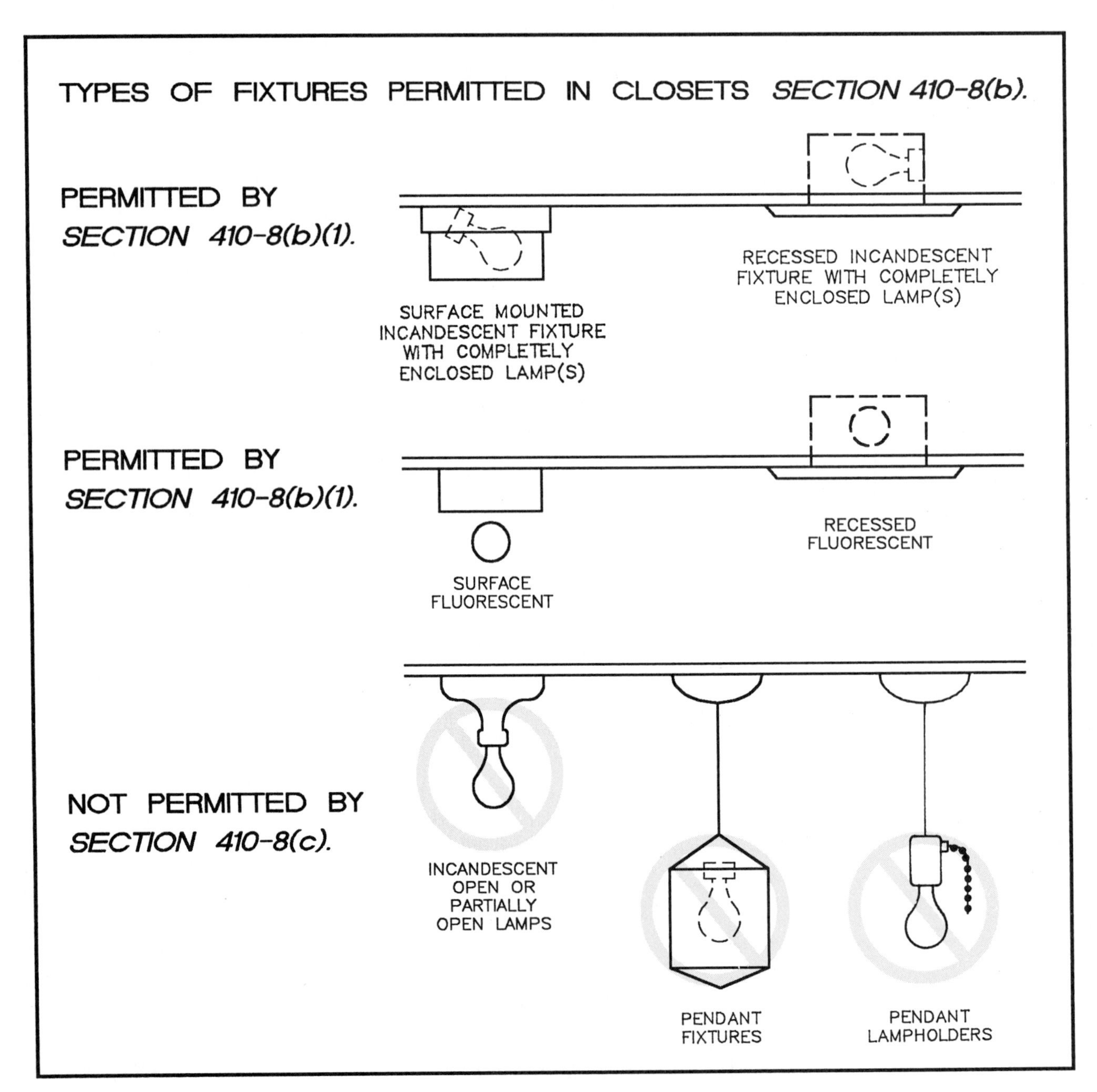

► Fig. 8-13 Illustrations of the types of lighting fixtures permitted in clothes closets. The Code does not permit bare incandescent lamps, pendant fixtures, or pendant lampholders to be installed in clothes closets. ◄

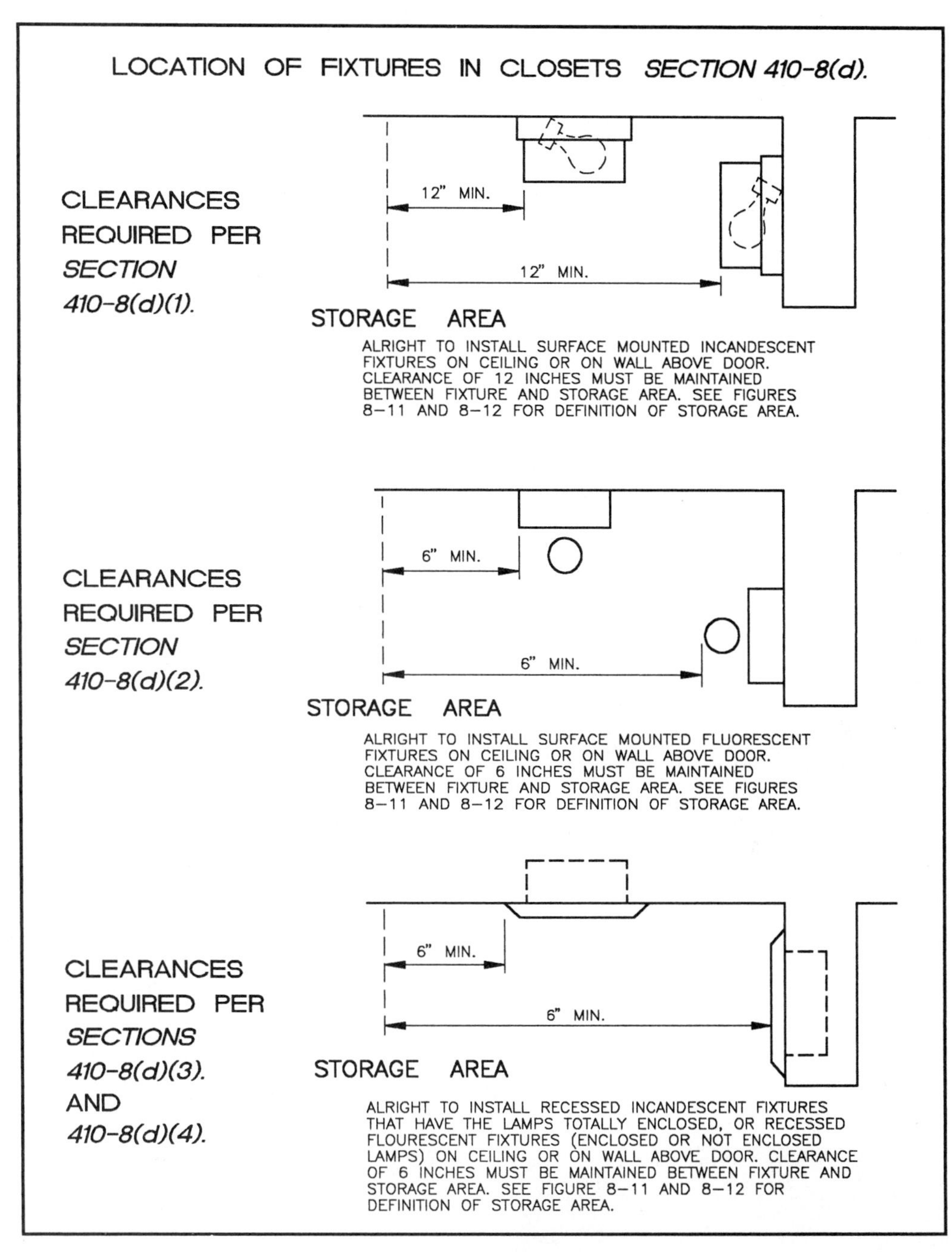

► Fig. 8-14 The above illustrations show the minimum required clearances between fixtures and the storage area. ◄

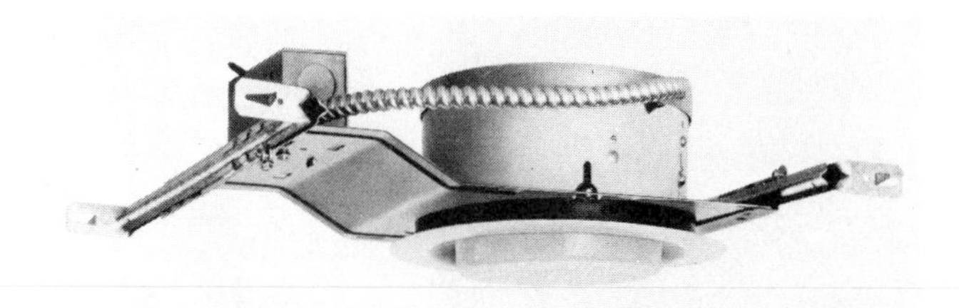

Fig. 8-15 Recessed incandescent closet lighting fixture with pull-chain switch that may be used where a separate wall switch is not installed.

Fig. 8-16 Typical bedroom-type lighting fixtures.

REVIEW

Note: Refer to the Code or the plans where necessary.

1. Can the outlets in a circuit be arranged in different groupings to obtain the same result? Why? ____________________

2. Is it good practice to have outlets on different floors on the same circuit? Why? ____________________

3. What usually determines the grouping of outlets into a circuit? ____________________

4. The continuous load on a lighting branch circuit must not exceed __________ percent of the branch-circuit rating.
5. To determine the maximum number of outlets in a circuit, __________ amperes per outlet are allowed. For a 15-ampere circuit, this results in a maximum of __________ outlets.
6. For this residence, what are the estimated wattages used in determining the loading of branch circuit A16?

   Receptacles __________ watts (volt-amperes)

   Closet recessed fixture __________ watts (volt-amperes)
7. What is the ampere rating of circuit A16? __________
8. What size wire is used for the lighting circuit in the Front Bedroom? __________
9. How many receptacles are connected to this circuit? __________

10. What main factor influences the choice of wall boxes? __________

11. How is a wall box grounded? __________

12. What is a split-circuit receptacle? ______________________

13. Is the switched portion of an outlet mounted toward the top or the bottom? Why?

14. The following questions pertain to lighting fixtures in clothes closets.
    a. Does the Code allow bare incandescent lamp fixtures such as porcelain keyless or porcelain pull-chain lampholders to be installed? ______________
    b. Does the Code allow bare fluorescent lamp fixtures to be installed? ________
    c. Does the Code permit pendant fixtures or pendant lampholders to be installed?
    d. What is the minimum clearance from the storage area to surface-mounted incandescent fixtures? ______________
    e. What is the minimum clearance from the storage area to surface-mounted fluorescent fixtures? ______________
    f. What is the minimum distance between recessed incandescent or recessed fluorescent fixtures and the storage area? ______________
    g. Define the "storage area."
    h. If a clothes hanging rod is installed where there is access from both sides, such as might be found in a large walk-in closet, define the storage area under that rod.

15. How many switches are in the bedroom circuit and of what type are they? ______

16. The following is a layout of the lighting circuit for the Front Bedroom. Using the cable layout shown in figure 8-8, make a complete wiring diagram of this circuit. Indicate the color of each conductor.

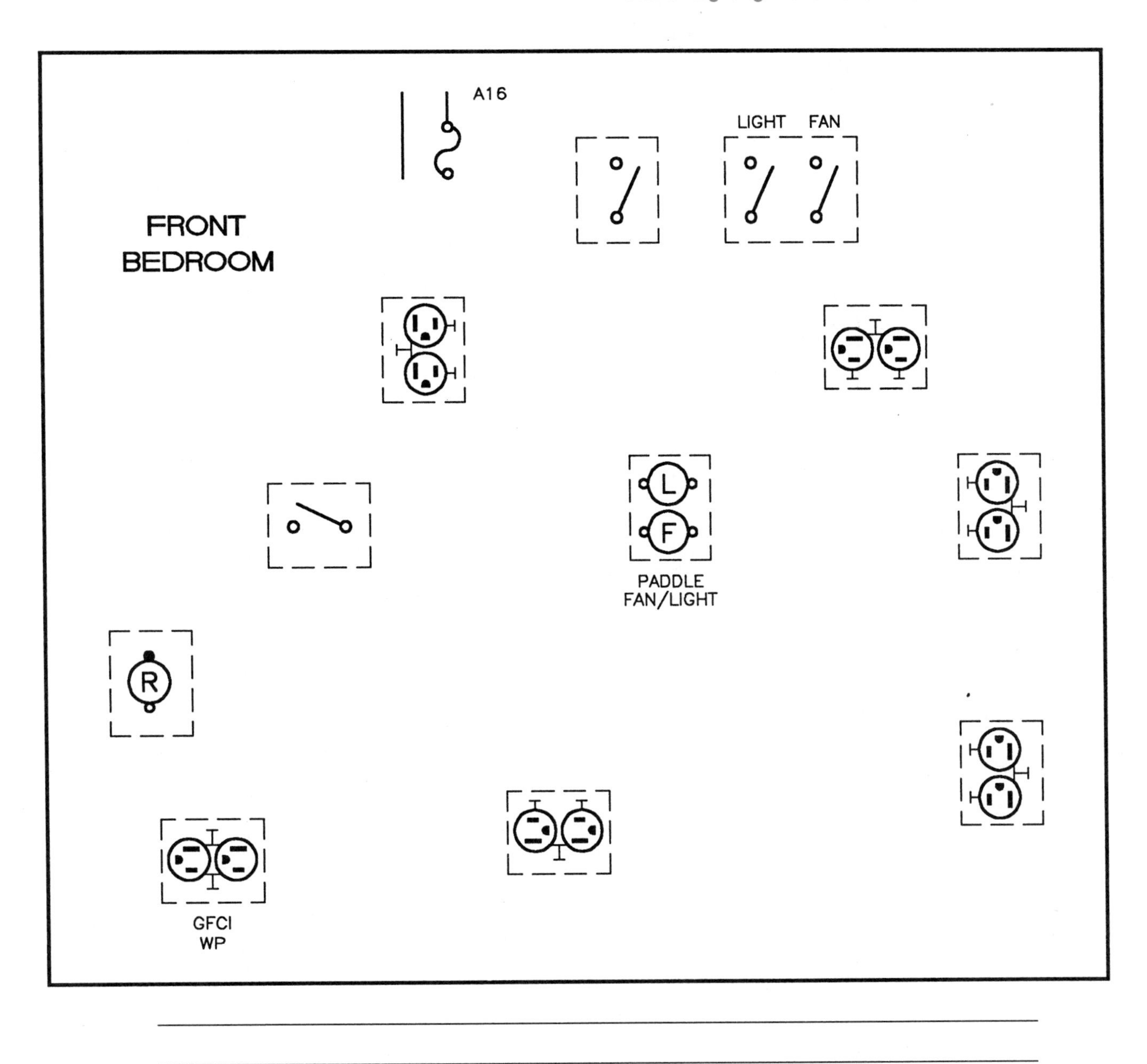

17. When planning circuits, what common practice is followed regarding the division of loads?

18. The Code uses the terms *watts*, *volt-amperes*, *kW*, and *kVA*. Explain their significance in calculating loads.

19. How many No. 14 AWG conductors are permitted in a device box that measures 3″ × 2″ × 2 3/4″? ______________

20. A 4″ × 1 1/2″ octagon box has one cable clamp and one fixture stud. How many No. 14 AWG conductors are permitted? ______________

____

____

____

____

____

21. Is it OK to install a surface-mount porcelain pull fixture in the closet in the Front Bedroom? ______________

UNIT 9

Lighting Branch Circuit for Master Bedroom

OBJECTIVES

After studying this unit, the student will be able to

- draw the wiring diagram of the cable layout for the Master Bedroom.
- study Code requirements for the installation of paddle fans.
- estimate the probable connected load for a room based on the number of fixtures and outlets included in the circuit supplying the room.
- gain more practice in determining box sizing based upon the number of conductors, devices, and clamps in the box.
- make the connections for three-way switches.

The discussion in unit 3 of grouping of outlets, estimating loads, selecting wall box sizes, and drawing wiring diagrams can also be applied to the circuit for the Master Bedroom.

The residence panel schedules show that the Master Bedroom is supplied by Circuit A19. Because Panel A is located in the basement below the wall switches next to the sliding doors, the home run for Circuit A19 is brought into the outdoor weatherproof receptacle. This *results* in six conductors in the outdoor receptacle box. The home run could have been brought into the corner receptacle outlet in the bedroom. Again, it is a matter of studying the circuit to determine the best choice for conservation of cable or conduit in these runs and to economically select the correct size of wall boxes.

LIGHTING BRANCH CIRCUIT A19 FOR MASTER BEDROOM

Figure 9-1 and Table 9-1 and the electrical plans show that Circuit A19 has four split-circuit receptacle outlets in this bedroom, one outdoor weatherproof GFCI receptacle outlet, two recessed closet fixtures, each on a separate switch, plus a paddle fan/light fixture, one telephone outlet, and one television outlet.

In addition, an outdoor bracket fixture is located adjacent to the sliding door and is controlled by a single-pole switch just inside the sliding door.

DESCRIPTION	QUANTITY	WATTS	VOLT-AMPERES
Receptacles @ 120 watts each	4	480	480
Weatherproof receptacle	1	120	120
Outdoor bracket fixture	1	150	150
Closet recessed fixtures One 75-W lamp each	2	150	150
Paddle fan/light	1		
Three 50-W lamps		150	150
Fan motor (0.75 @ 120 V)		80	90
TOTALS	9	1025	1035

Table 9-1. Master bedroom outlet count and estimated load. Circuit A19.

Because of the sliding-door arrangement, the split-circuit receptacle outlets are controlled by two three-way switches. One is located next to the sliding door. As in the Front Bedroom, Living Room, and Study/Bedroom, the use of split-circuit receptacles offers the advantage of having switch control of one of the receptacles at a given outlet, while the other receptacle remains "live" at all times. See figure 12-13 (page 187) for definition of a receptacle.

Next to the bedroom door we find one three-way switch, plus the paddle fan/light controls, which are installed in a separate two-gang box. See figure 9-2.

► SLIDING GLASS DOORS

See unit 3 for a discussion on how to space receptacle outlets on walls where sliding glass doors are installed. ◄

SELECTION OF BOXES

As discussed in unit 7, the selection of outlet boxes and switch boxes is made by the electrician. These decisions are based on Code requirements, space allowances, good wiring practices, and common sense.

For example, in the Master Bedroom, the electrician may decide to install a 4-inch square box with a two-gang raised plaster cover at the box location next to the sliding door. Or two sectional switch boxes ganged together may be installed at that location.

The type of box to be installed depends upon the number of conductors entering the box. The suggested cable layout, figure 9-1, shows that four cables enter this box for a total of 14 conductors (10 No. 14 AWG circuit conductors and 4 grounding conductors). Since the Code stipulates in *Section 370-6(a)(1)* that only one of the grounding conduc-

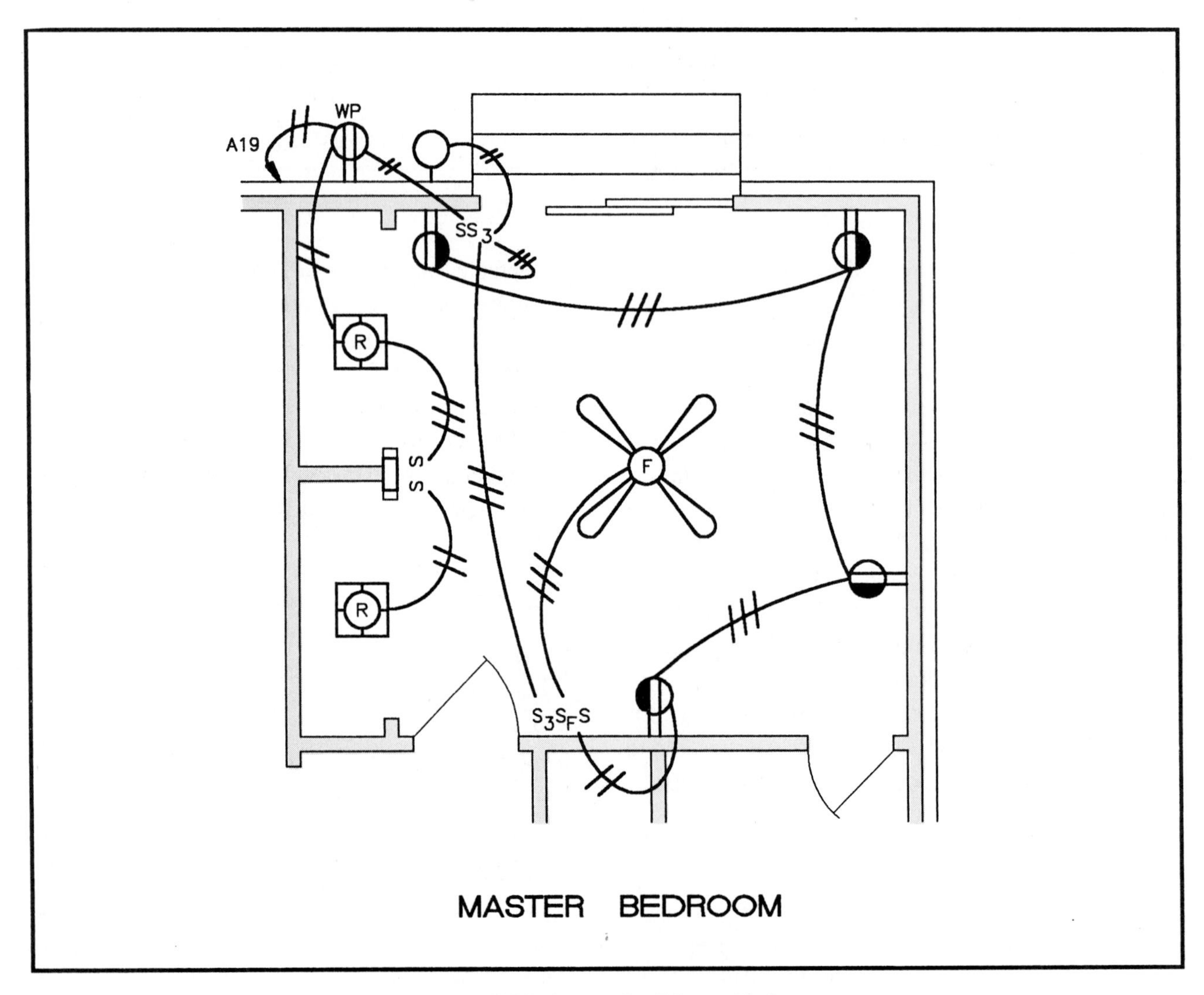

Fig. 9-1 Cable layout for Master Bedroom.

tors in the cable need be counted for the purpose of box fill, the number of conductors to be included in the determination of proper box size is 10 circuit conductors plus one grounding conductor, for a total of 11 conductors.

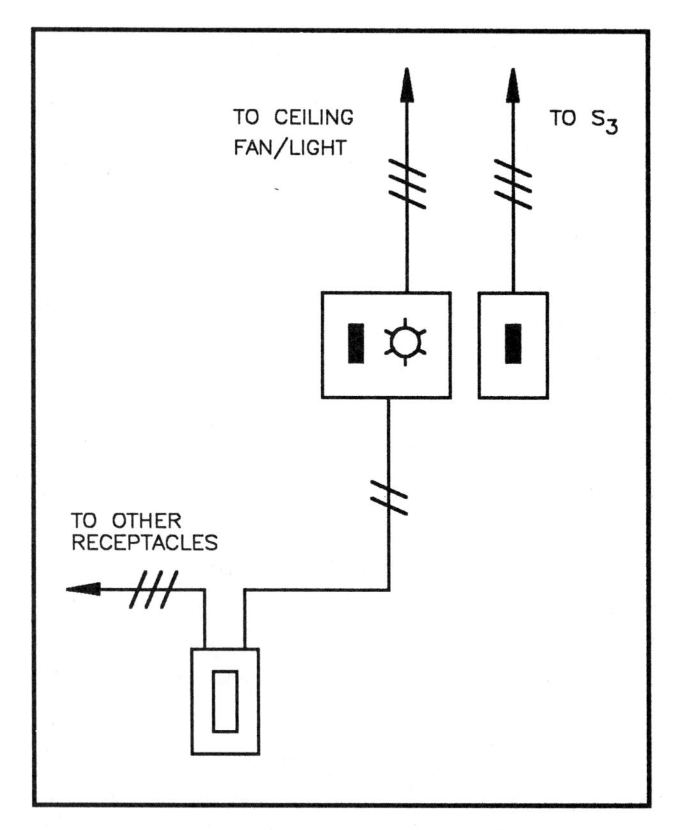

Fig. 9-2 Conceptual view of how the switching arrangement is to be accomplished in the Master Bedroom Circuit A19.

1. Add the circuit conductors
   2 + 2 + 3 + 3 = 10
2. Add grounding wire (count 1 only) 1
3. Add two for each switch (2 + 2) 4
4. Add two for the cable clamps 2

Total 17

Now look at the Quik-Check Box Selector, Table 8-2, and select a box or a combination of gangable device boxes that are permitted to hold 17 conductors.

Possibilities are:

1. Gang two 3″ × 2″ × 3 1/2″ device boxes together (9 × 2 = 18 No. 14 AWG conductors allowed).
2. Install a 4″ × 4″ × 2 1/8″ square box (15 No. 14 AWG conductors allowed). Note that the raised plaster cover, if marked with its cubic-inch volume, can increase the maximum number of conductors permitted for the combined box and raised cover. See figures 9-3 and 9-4.

ESTIMATING CABLE LENGTHS

The length of cable needed to complete an installation can be estimated roughly. It may be possible to run the cable in a straight line directly from one wall outlet to another. In some cases,

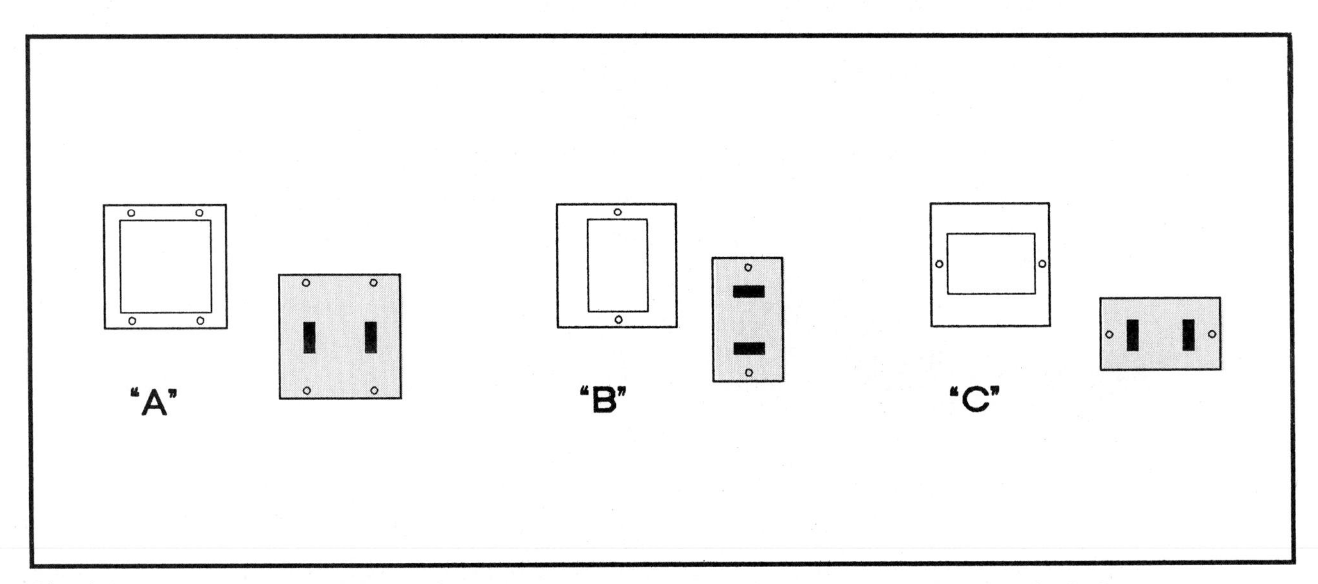

Fig. 9-3 "A" shows two switches and wall plate that attach to a two-gang, 4″ square raised plaster cover or to two device boxes that have been ganged together. "B" shows two interchangeable type switches and wall plate that attach to a one-gang, 4″ square raised plaster cover or to a single device (switch) box. These types of wiring devices may be mounted vertically or horizontally.

obstacles such as steel columns, sheet-metal ductwork, and plumbing may require the cable routing to follow a longer path.

To insure that the estimate is not short, all measurements are to be made "square." For example, measure from the ceiling outlet straight to the wall, and finally measure straight over to the wall outlet. Add one foot (305 mm) of cable at each cable termination. This is an allowance to permit the outer jacket of the cable to be stripped at all junction and outlet boxes. Check the plans carefully as an aid in determining where the cables can be routed.

No. of Gangs	Height	Width
1	4 1/2" (114.3 mm)	2 3/4" (69.85 mm)
2	4 1/2" (114.3 mm)	4 9/16" (115.9 mm)
3	4 1/2" (114.3 mm)	6 3/8" (161.9 mm)
4	4 1/2" (114.3 mm)	8 3/16" (207.9 mm)
5	4 1/2" (114.3 mm)	10" (254 mm)
6	4 1/2" (114.3 mm)	11 13/16" (300 mm)

Fig. 9-4 Height and width of standard wall plates.

PADDLE FANS

For appearance and added comfort, paddle fans have become extremely popular in recent years, figure 9-5. These fans rotate slowly (60 r/min to 250 r/min for home-type fans). The air currents they create can save energy during the heating season because they destratify the warm air at the ceiling, bringing it down to living levels. In the summer, even with air conditioning, a slight circulation of air creates a wind-chill factor and causes a person to feel cooler.

Fig. 9-5 Typical home-type paddle fan.

Residential paddle fans usually extend approximately 12 inches (305 mm) below the ceiling; so, for 8-foot (2.44 m) ceilings, the fan blades are about 9 to 10 inches (229–254 mm) below the ceiling, figure 9-6.

The safe supporting of a paddle fan involves three issues.

- the actual weight of the fan.
- the twisting and turning motion when started.
- vibration.

To address these issues, the Code in *Section 370-17(c)* states that outlet boxes shall *not* be used as the sole support of ceiling (paddle) fans unless the boxes are UL listed as capable of being used for the sole support of the fan. Complete requirements relative to electric fans mounted to outlet boxes are found in UL Standard 507 and in UL Standard 514A for metallic outlet boxes and Standard 514C for nonmetallic outlet boxes, flush device boxes, and covers.

Section 422-18 states that listed ceiling fans not over 35 pounds (15.8 kg) may be supported by the outlet box, but again, only if the box is so listed by UL.

Always check the instructions furnished with the ceiling fan to be sure that the mounting and supporting methods will be safe. Boxes that are permitted to support a fan are marked ACCEPTABLE FOR FAN SUPPORT.

Section 410-16 of the Code requires that a lighting fixture must be supported independently of the outlet box if the fixture weighs more than 50 pounds (22.7 kg).

Figure 9-7 shows the wiring for a fan/light combination with the supply at the fan/light unit. Figure 9-8 shows the fan/light combination with the supply at the switch.

The electrician must install a junction box where the fan or fan/light is to be located. Since the fan is quite heavy, the box must be solidly secured and it must be approved for the purpose. A mounting kit furnished by one manufacturer of paddle fans is shown in figure 9-9.

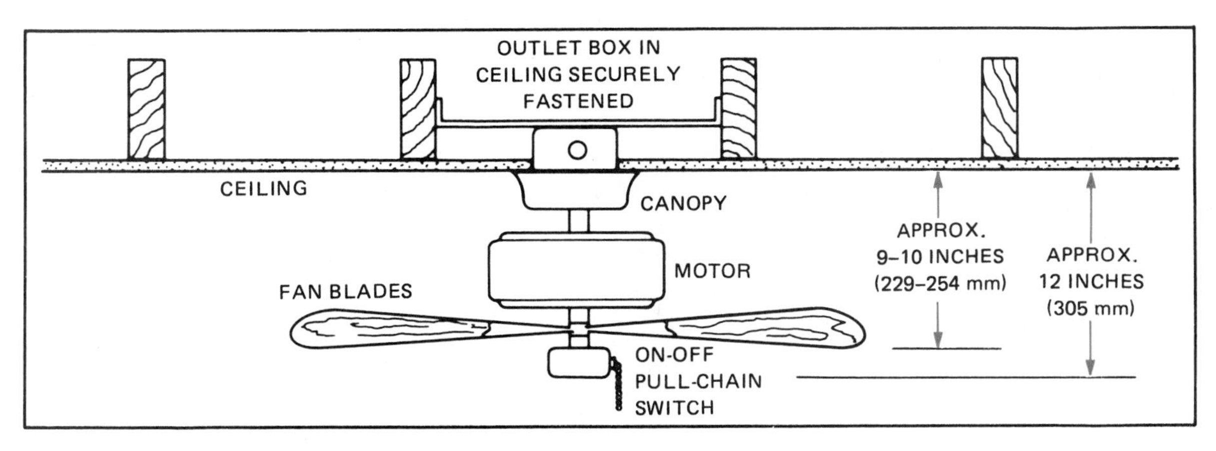

Fig. 9-6 Typical mounting of a ceiling paddle fan.

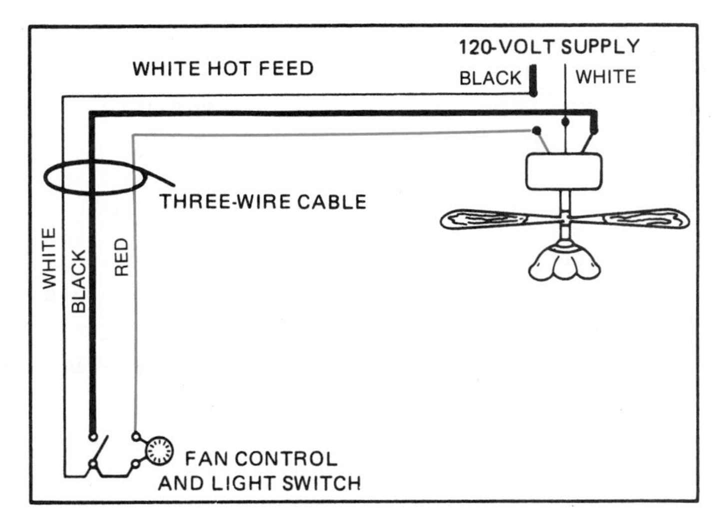

Fig. 9-7 Fan/light combination with the supply at the fan/light unit.

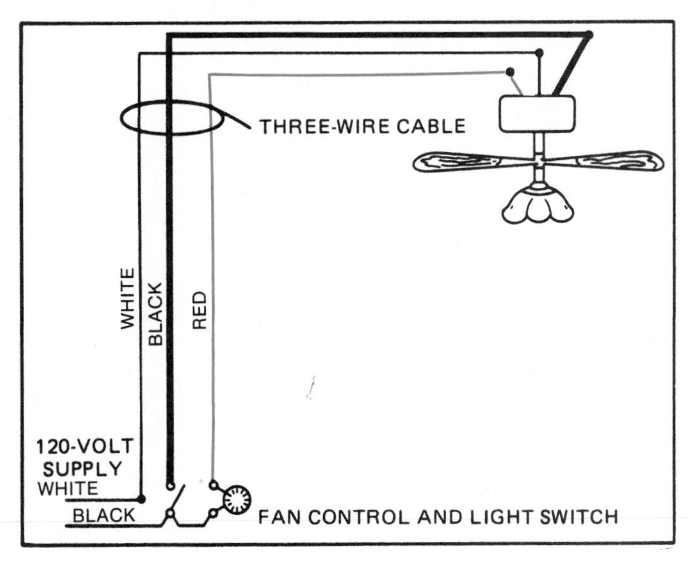

Fig. 9-8 Fan/light combination with the supply at the switch.

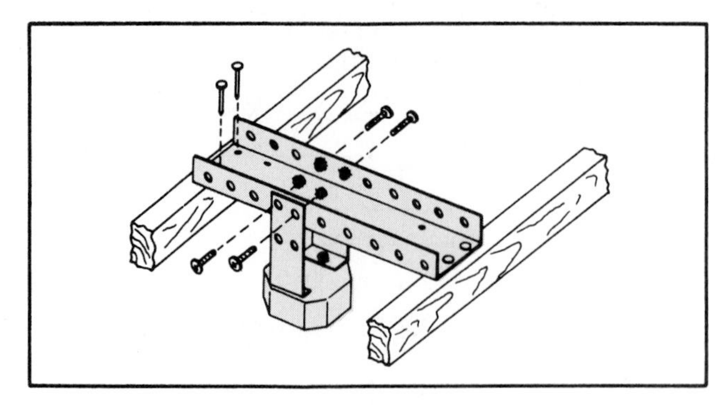

Fig. 9-9 A mounting kit furnished by one manufacturer of paddle fans.

Figure 9-10 shows an adjustable-type ceiling fan hanger installed.

Figure 9-11 illustrates a typical combination light switch and three-speed fan control. The electrician is required to install a deep 4-inch box with a two-gang raised plaster ring for this light/fan control.

A typical home-type paddle fan motor draws 50 to 100 watts (50 to 100 volt-amperes). This is approximately 0.4 to slightly over 0.8 ampere.

Paddle fan/light combinations increase the load requirements somewhat. Some paddle fan/light units have one lamp socket, whereas others have four or five lamp sockets.

For the paddle fan/light unit in the Master Bedroom, 240 volt-amperes were included in the load calculations. The current draw is

$$I = \frac{VA}{V} = \frac{240}{120} = 2 \text{ amperes}$$

For most fans, switching on and off may be accomplished by the pull-switch, an integral part of the fan (OFF-HIGH-MED-LOW), or with a solid-state switch having an infinite number of speeds. Some fan motors are of the reversible type.

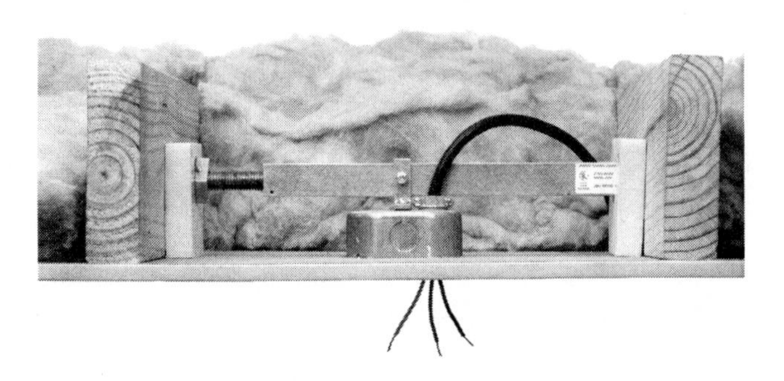

Fig. 9-10 An adjustable-type ceiling fan hanger that is UL listed to support 100 pounds. The hanger adjusts between 16-inch and 24-inch spaces. It can be installed in new construction. It can also be installed from below an existing ceiling through a hole cut in the ceiling.

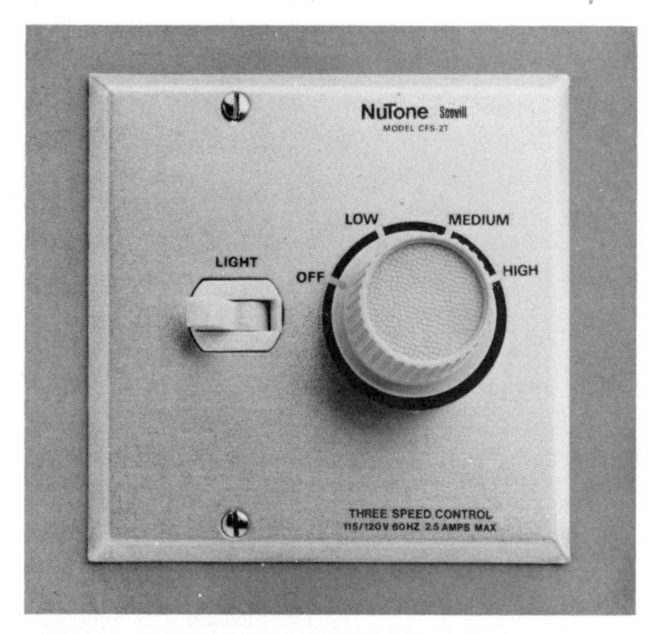

Fig. 9-11 Combination light switch and three-speed fan switch.

REVIEW

Note: Refer to the Code or the plans where necessary.

1. What circuit supplies the Master Bedroom? ______________________
2. What lighting besides the bedroom is supplied by this circuit? ______________________

   ____
3. What type of receptacles are provided in this bedroom? How many receptacles are there? ______________________
4. How many ceiling outlets are included in this circuit? ______________________

   ____
5. What wattage for the recessed closet fixtures was used for calculating their contribution to the circuit? ______________________
6. What is the current draw for the paddle fan/light ceiling fixture? ______________________
7. What is the estimated load in volt-amperes for the circuit supplying the Master Bedroom? ______________________
8. Which is installed first, the switch and outlet boxes or the cable runs? ______________________

   ____
9. How many conductors enter the paddle fan/light wall box? ______________________
10. What type and size of box may be used for the paddle fan/light wall box? ______

    ____

    ____

    ____
11. What type of covers are used with 4-inch square outlet boxes? ______________________

    ____

    ____
12. a. Does the circuit for the Master Bedroom have a grounded conductor? ______
    b. Does it have a grounding conductor? ______________________
    c. Explain the answers to (a) and (b). ______________________

    ____

    ____

    ____

    ____

    ____
13. Approximately how many feet (meters) of two-wire cable and three-wire cable are needed to complete the circuit supplying the Master Bedroom? Two-wire cable ________ feet (________ meters). Three-wire cable ________ feet (________ meters).

14. If the cable is laid in notches in the corner studs, what protection for the cable must be provided? ____

____

____

15. How high are the receptacles mounted above the finish floor in this Bedroom?

____

16. Approximately how far from the bedroom door is the first receptacle mounted? (See the plans.) ____

17. What is the distance from the finish floor to the center of the wall switches in this bedroom? ____

18. The Master Bedroom features a sliding glass door. For the purpose of providing the proper receptacle outlets, answer the following statements true or false.
    - Sliding glass panels are considered to be wall space. ______
    - Sliding glass panels are not considered to be wall space. ______
    - Fixed panels of glass doors are considered to be wall space. ______
    - Fixed panels of glass doors are not considered to be wall space. ______

19. What type of receptacle will be installed outdoors, just outside of the Master Bedroom? ____

____

20. Show your calculations of how to select a proper wall box for the closet fixture switch. Keep in mind that the available space between the wood casings is small. See figure 9-1. ____

____

____

____

____

____

____

____

____

21. When an outlet box is to be installed to support a paddle fan, how must it be marked?

____

____

22. The following is a layout of the lighting circuit for the Master Bedroom. Using the cable layout shown in figure 9-1, make a complete wiring diagram of this circuit. Indicate the color of each conductor.

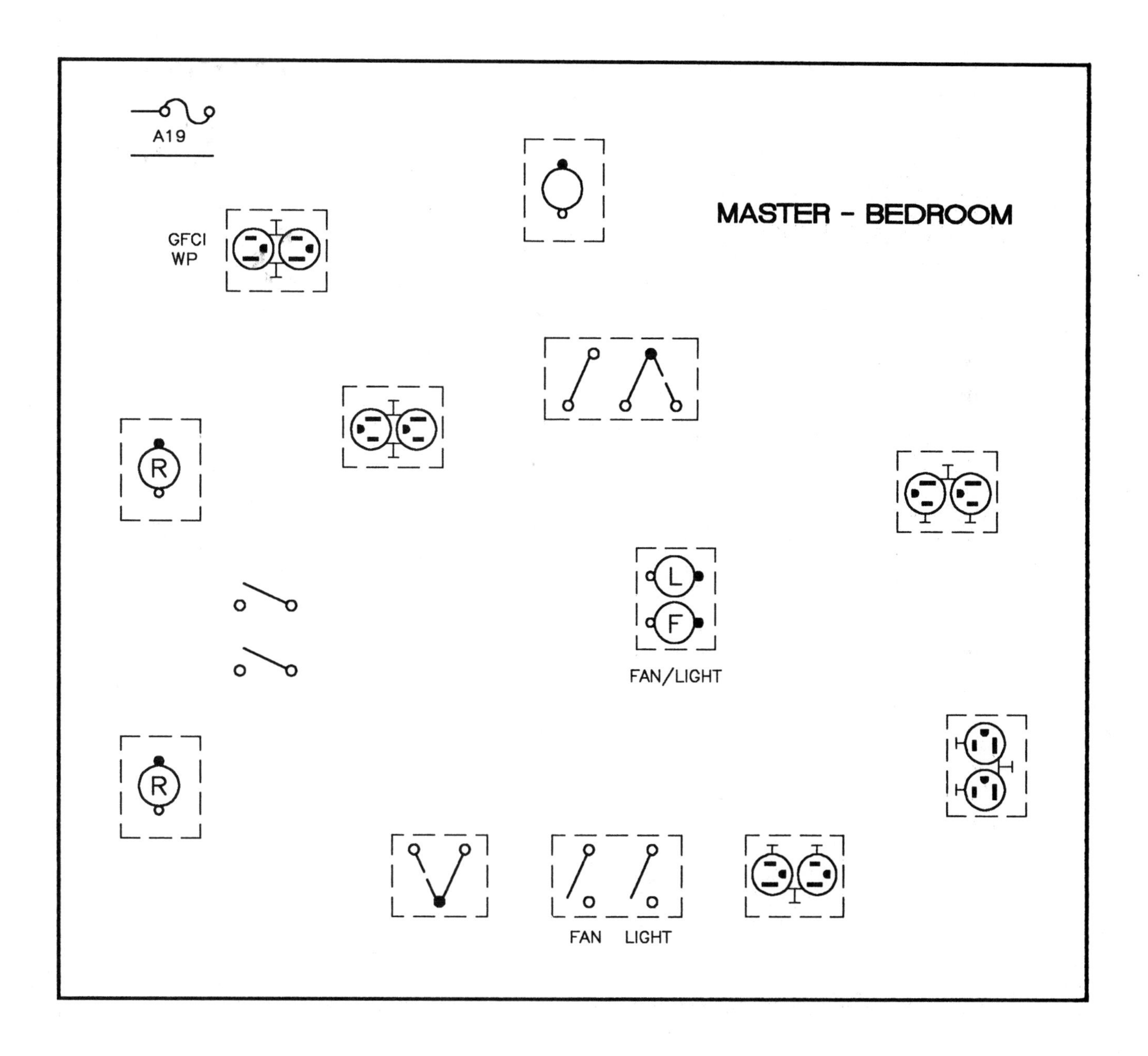

23. Connect the circuit shown so that the receptacle is on CONTINUOUS.

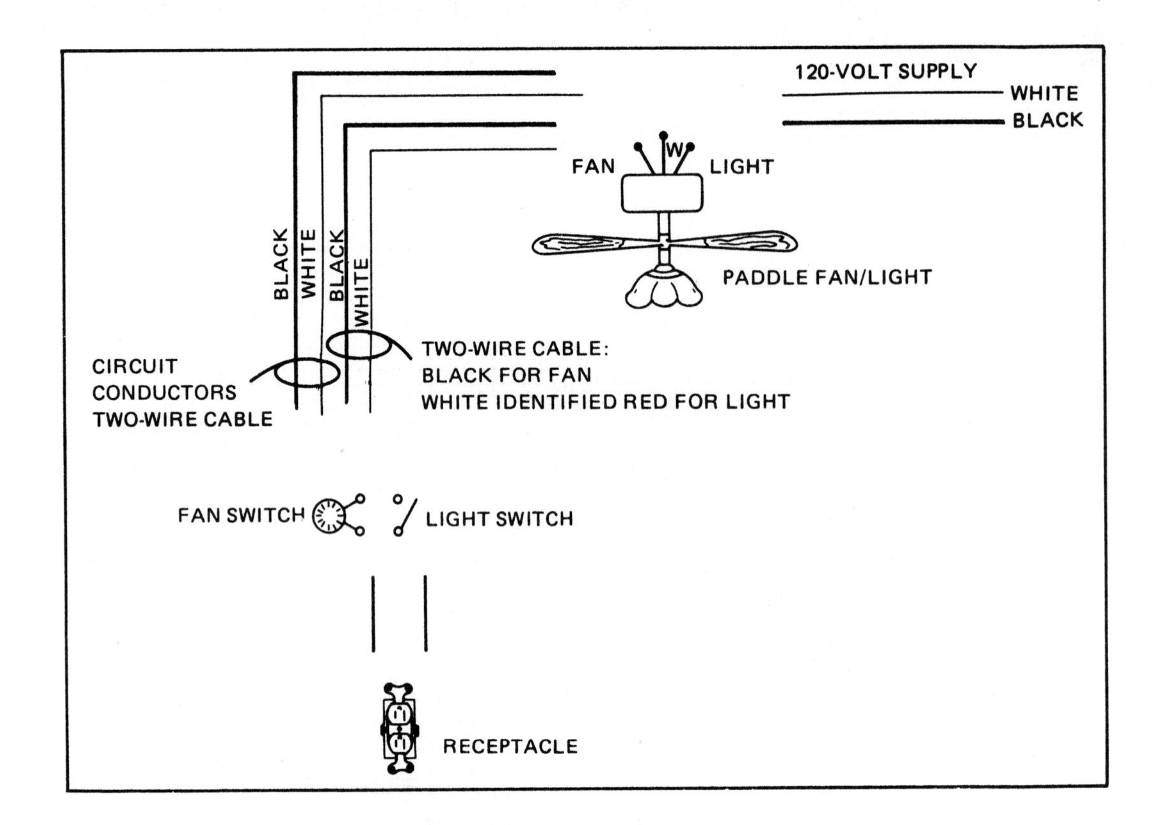

24. What is the width of a two-gang wall plate? ______________________

UNIT 10

Lighting Branch Circuit — Bathrooms, Hallway

OBJECTIVES

After studying this unit, the student will be able to

- list equipment grounding requirements for bathroom installations.
- draw a wiring diagram for the bathroom and hallway.
- understand Code requirements relating to receptacle outlets in hallways.
- discuss fundamentals of proper lighting for bathrooms.

Circuit A14 supplies the lighting outlets and receptacle outlets in the bedroom hall and both bathrooms.

Table 10-1 summarizes outlets and estimated load for the bathrooms and bedroom hall.

Note that each bathroom shows a ceiling heater/light/fan that is connected to separate circuit ▲J and ▲K. These are discussed in detail in unit 22.

A hydromassage tub is located in the bathroom serving the master bedroom. It is connected to a separate circuit ▲A and is also discussed in unit 22.

The attic exhaust fan in the hall is supplied by a separate circuit ▲L, also covered in unit 22.

LIGHTING BRANCH CIRCUIT A14 FOR HALLWAY AND BATHROOMS

Figure 10-1 and the electrical plans for this area of the home show that each bathroom has a fixture above the vanity mirror. Some typical fixtures for this purpose are shown in figure 10-2. Of course, the homeowner might decide to purchase a medicine cabinet complete with a self-contained lighting fixture. See figure 10-3(A) and (B), which illustrate how to rough-in the wiring for each type. These fixtures are controlled by single-pole switches at the doors.

You must remember that a bathroom (powder room) should have proper lighting for shaving, combing hair, and so on. Mirror lighting can accomplish this, since a mirror will reflect what it "sees." If the face is poorly lit, with shadows on the face, that is precisely what will be reflected in the mirror.

A lighting fixture directly overhead will light the top of one's head, but will cause shadows on the

DESCRIPTION	QUANTITY	WATTS	VOLT-AMPERES
Receptacles @ 120 W	3	360	360
Vanity fixtures @ 200 W each	2	400	400
Hall fixture	1	100	100
TOTALS	6	860	860

Table 10-1. Bathrooms and bedroom hall: outlet count and estimated load. Circuit A14.

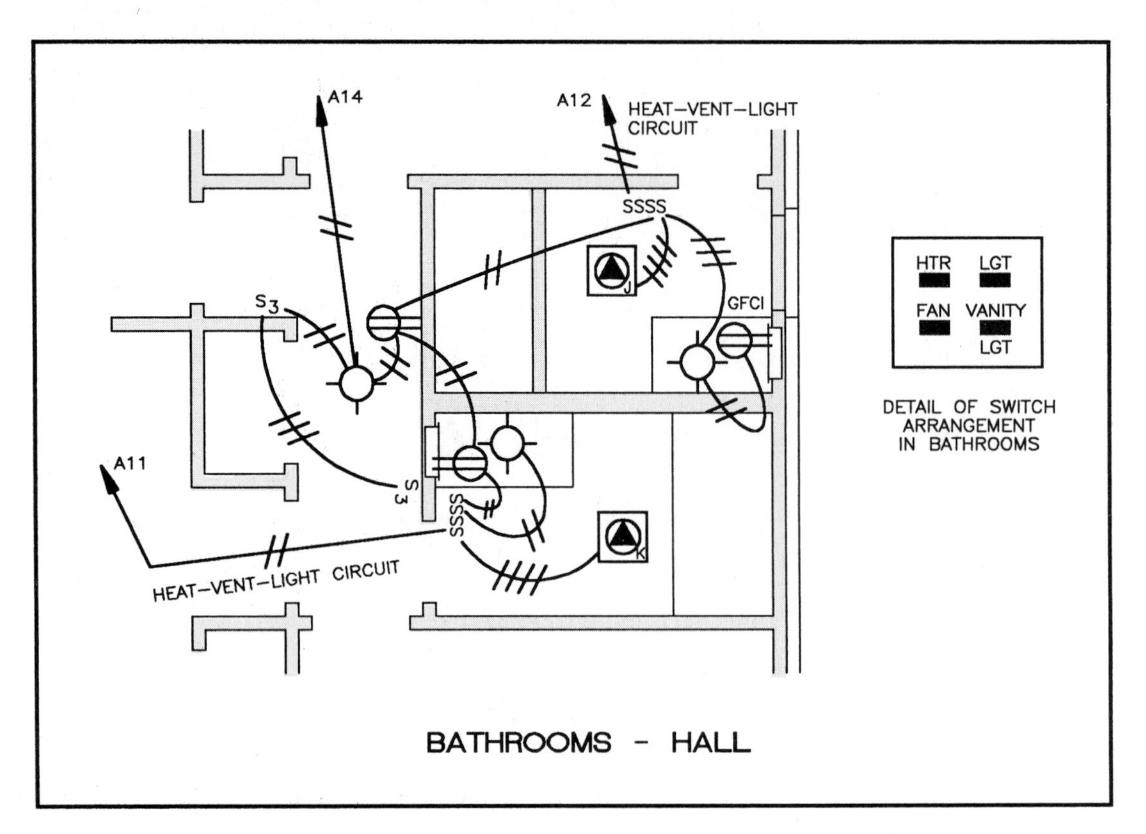

Fig. 10-1 Cable layout for bathrooms and hall. Layout includes general lighting circuit plus the two separate circuits for the heat/vent/lights. Not shown here are the special purpose outlets for the hydromassage tub (A) or the attic exhaust fan (L), nor is the smoke detector indicated. These are all connected to circuits other than Circuit A14 and are covered in other units.

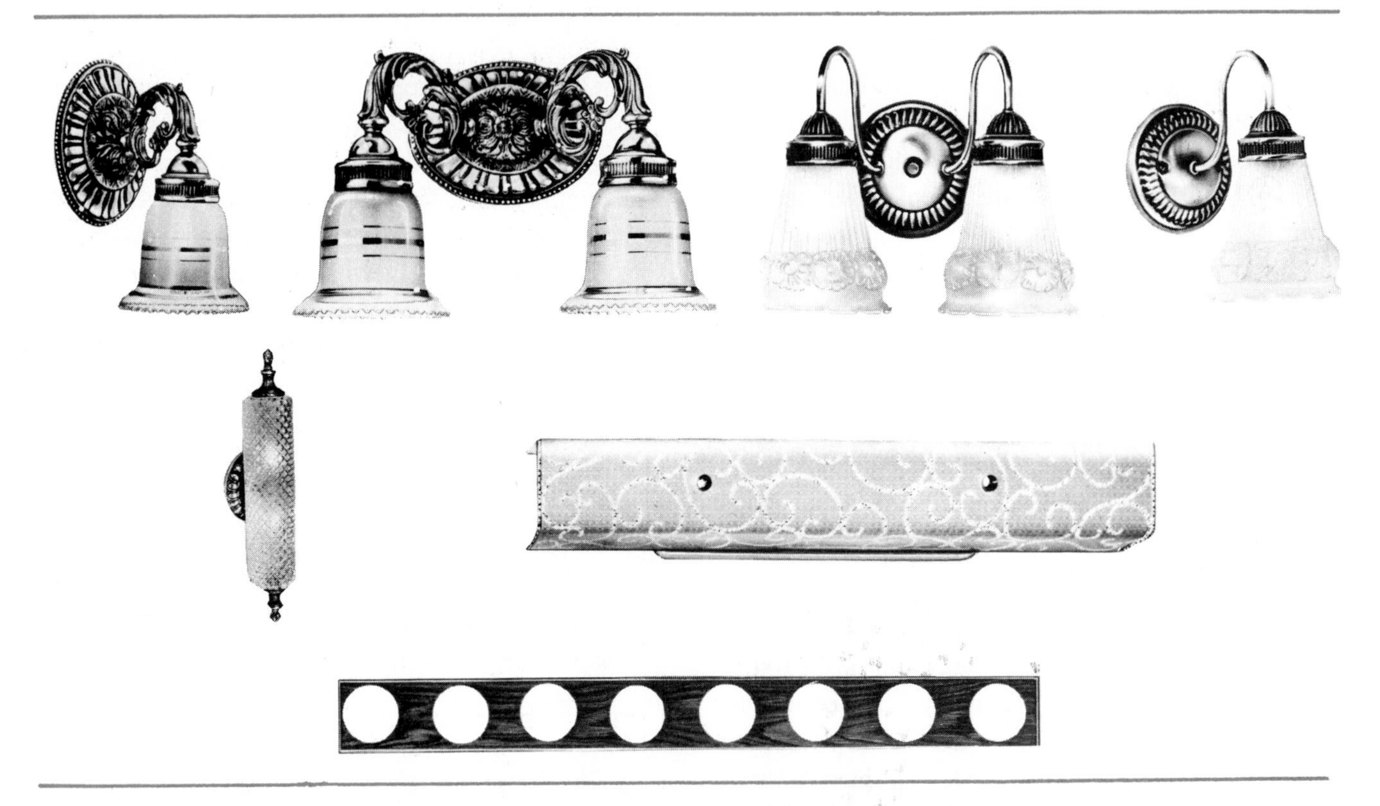

Fig. 10-2 Typical vanity (bathroom) lighting fixtures of the side bracket and strip types.

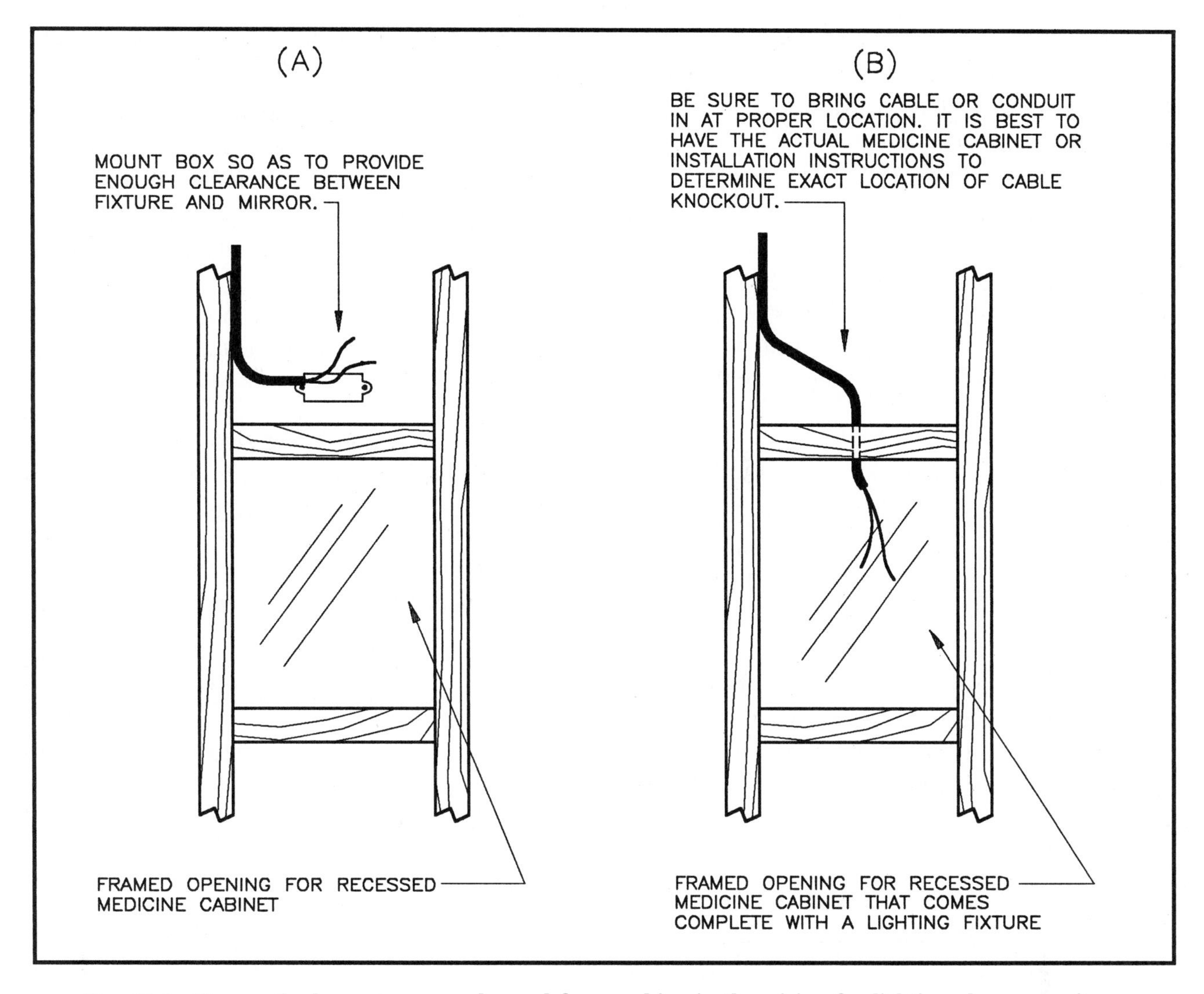

Fig. 10-3 Two methods most commonly used for roughing-in the wiring for lighting above a vanity.

face. Mirror lighting and/or adequate lighting above and forward of the standing position at the vanity can provide excellent lighting in the bathroom. Figures 10-4 to 10-6 show pictorial as well as section views of typical soffit lighting above a bathroom vanity.

The receptacles in each bathroom are located adjacent to the vanity. As required by *Section 210-8* and discussed in unit 6, these receptacles are GFCI type. The Code defines a bathroom as "an area including a basin, plus one or more of the following: a toilet, a tub or a shower." See figure 10-7 (page 168).

GENERAL COMMENTS ON LAMPS AND COLOR

Incandescent lamps (light bulbs) provide pleasant color tones, bringing out the warm red flesh tones similar to those of natural light.

When installing fluorescent lighting, excellent color rendition is achieved by using fluorescent lamps that bring out the warm red tones also. These would be lamps such as warm white (WW) and deluxe warm white (WWX), which simulate incandescent lamps, or deluxe cool white (CWX), which simulates outdoor daylight on a cloudy day.

White (W) or cool white (CW) fluorescents downplay red tones and tend to give a person a "cool" feeling and give the skin a pale appearance.

HANGING FIXTURES IN BATHROOMS

Section 410-4(d) of the Code states that no parts of hanging fixtures or pendants shall be located within a zone measured 3 feet (914 mm) horizontally and 8 feet (2.44 m) vertically from the top of the bathtub rim. Figure 10-8 projects the top view of the 3-foot restriction. Figure 10-9 shows both the allowable installation and Code violations of the given dimensions.

HALLWAY LIGHTING

The hallway lighting is provided by one ceiling fixture that is controlled with two three-way switches located at either end of the hall. The home run to Main Panel A has been brought into this ceiling outlet box.

► RECEPTACLE OUTLETS IN HALLWAYS

One receptacle outlet has been provided in the hallway as required in *Section 210-52(h)*, which states that "for hallways of 10 feet (3.05 m) or more in length, at least one receptacle outlet shall be required."

For the purpose of determining the length of a hallway, the measurement is taken down the centerline of the hall, turning corners if necessary, but not passing through a doorway, figure 10-10 (page 169). ◄

EQUIPMENT GROUNDING REQUIREMENTS FOR A BATHROOM CIRCUIT

All exposed metal equipment, including fixtures, electric heaters, faceplates, and similar items must be grounded. The equipment grounding requirements for bathroom circuits and any other equipment in a residential building are contained in *Section 250-42*.

In general, all exposed noncurrent-carrying metal parts of electrical equipment must be grounded:

- if they are within 8 feet (2.44 m) vertically or 5 feet (1.52 m) horizontally of the ground or other grounded metal objects.
- if the ground and the electrical equipment can be touched at the same time.
- if they are located in wet or damp locations, such as in bathrooms, showers, and outdoors.
- if they are in electrical contact with metal. This requirement includes metal lath and aluminum foil insulation.

Fig. 10-4 Positioning of bathroom lighting fixtures. Note the wrong way and the right way to achieve proper lighting.

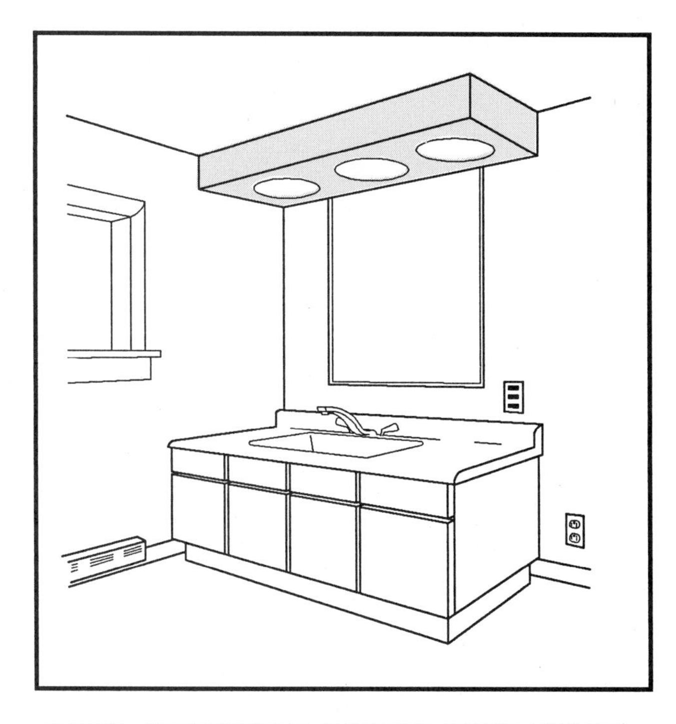

TYPICAL INCANDESCENT RECESSED SOFFIT LIGHTING OVER BATHROOM VANITY

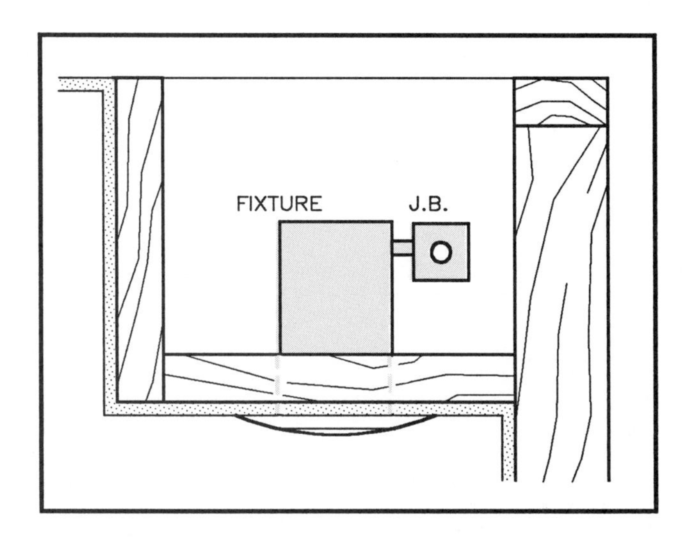

END CUTAWAY OF SOFFIT SHOWING RECESSED INCANDESCENT FIXTURES IN TYPICAL SOFFIT ABOVE BATHROOM VANITY. TWO OR THREE FIXTURES GENERALLY INSTALLED TO PROVIDE PROPER LIGHTING.

Fig. 10-5 Incandescent soffit lighting.

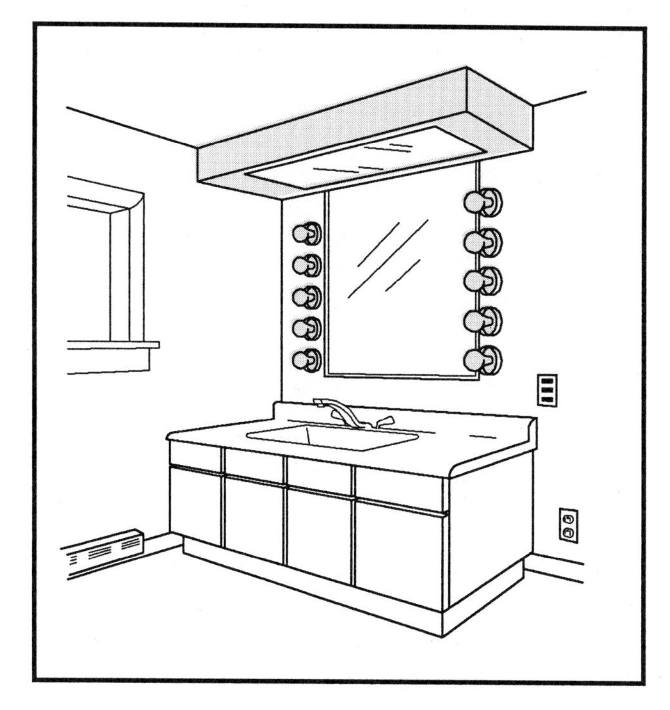

TYPICAL FLUORESCENT RECESSED SOFFIT LIGHTING OVER BATHROOM VANITY. NOTE ADDITIONAL INCANDESCENT "SIDE-OF-MIRROR" LIGHTING.

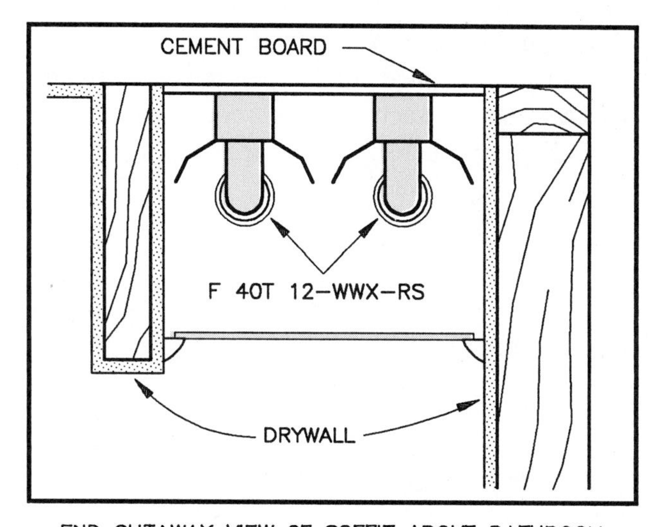

END CUTAWAY VIEW OF SOFFIT ABOVE BATHROOM VANITY SHOWING RECESSED FLUORESCENT FIXTURES CONCEALED ABOVE TRANSLUCENT ACRYLIC LENS

Fig. 10-6 Combination fluorescent and incandescent bathroom lighting.

Equipment is considered grounded when it is properly and permanently connected to metal raceway, the armor of armored cable, the grounding conductor in nonmetallic-sheathed cable, or a separate grounding conductor. Of course, the means of grounding (the metal raceway, the armored cable, or the grounding conductor in the nonmetallic-sheathed cable) must itself be properly grounded.

According to *Section 250-45(c)*, all residential electric cord-connected appliances must be grounded. There are two exceptions to this basic grounding requirement. *Exception No. 1* accepts

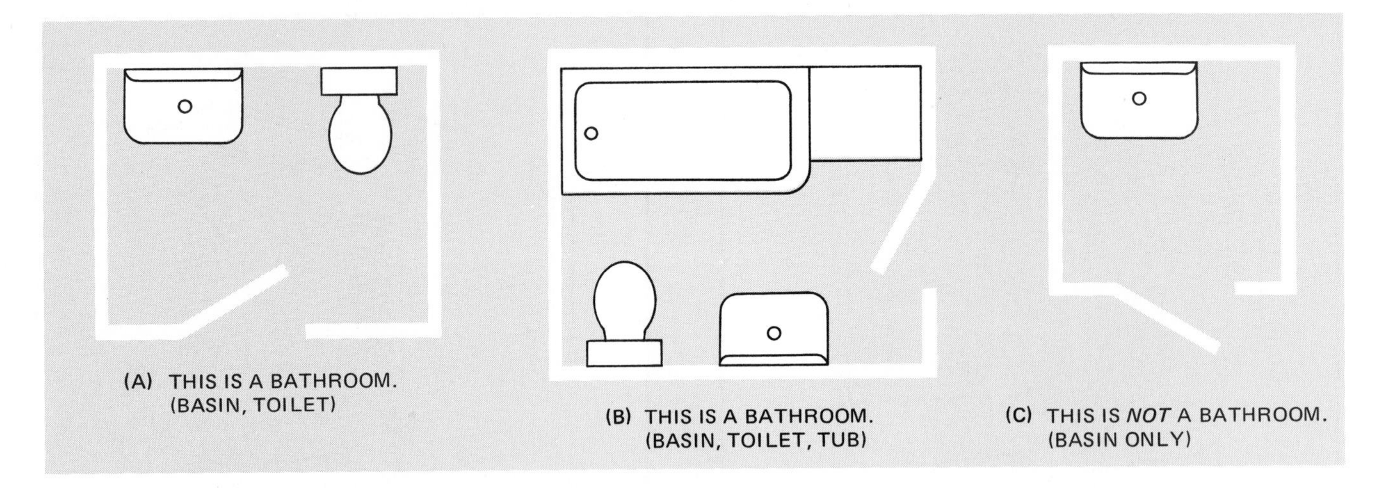

Fig. 10-7 Definition of a bathroom, *Section 210-8.*

"double insulating" or equivalent in lieu of grounding. ► *Exception No. 2* permits utilization equipment to be connected to a GFCI protected outlet, such as would be the case in an existing installation where non-grounding type receptacle outlets may already be installed. It is possible to protect these existing receptacle outlets by installing a feed-thru GFCI receptacle outlet somewhere on that circuit. This is discussed in *Section 210-7(d).* ◄

Double-insulated appliances must be clearly marked and as such do not require a separate grounding conductor. These appliances are furnished with a two-wire cord.

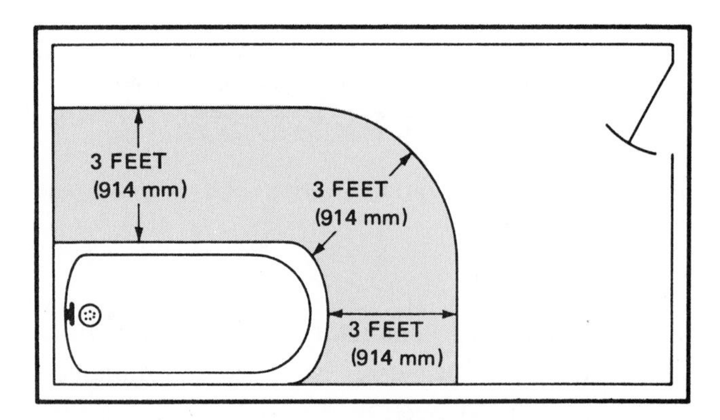

Fig. 10-8 No parts of hanging fixtures or pendants shall be located within 3 feet (914 mm) measured horizontally from the top of the bathtub rim (top view), *Section 410-4(d).*

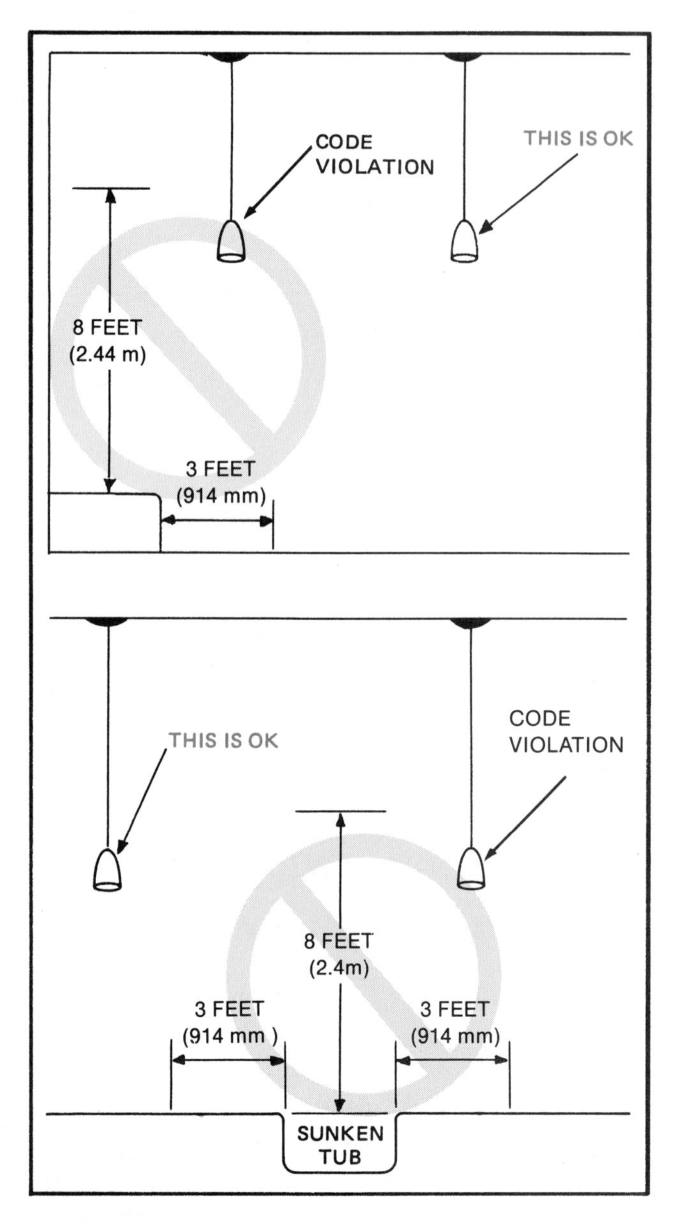

Fig. 10-9 No parts of hanging fixtures or pendants allowed in shaded X areas, *Section 410-4(d).*

Those residential electrical appliances that must be grounded according to *Section 250-45(c)* are

Toothbrushes	Hair dryers
Curling irons	Hot rollers
Facial saunas	Refrigerators
Freezers	Air conditioners
Clothes washers	Clothes dryers
Dishwashers	Sump pumps
Wet scrubbers	Aquarium equipment

Hand-held motor-operated tools
Electric motor-operated hedge trimmers
Electric motor-operated lawn mowers
Electric motor-operated snow blowers
Portable hand lamps

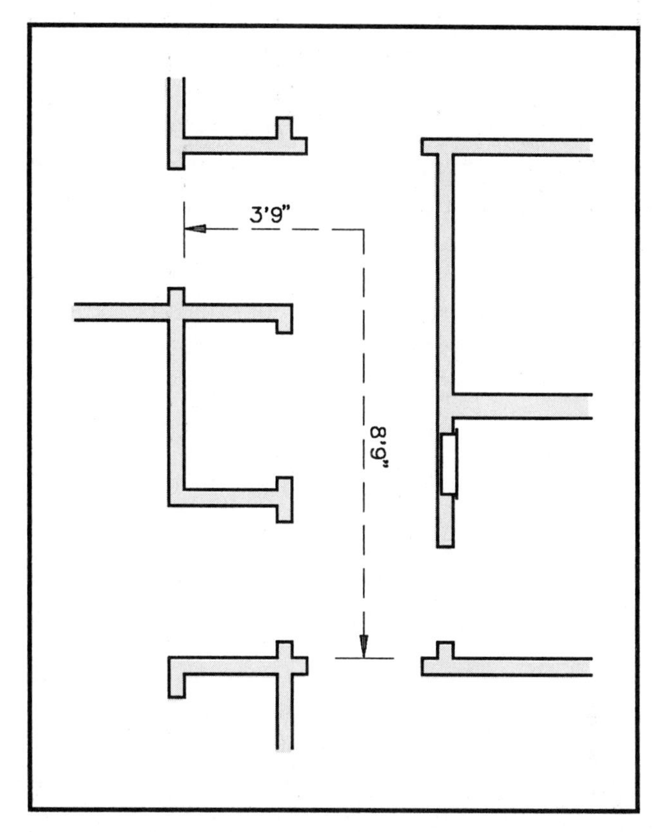

► **Fig. 10-10 The centerline measurement of the bedroom hallway in this residence is 12′6″, which requires at least one wall receptacle outlet, *Section 210-52(h)*.** ◄

REVIEW

1. List the number and types of switches and receptacles used in Circuit A14. ____

2. There is a three-way switch in the bedroom hallway leading into the living room. Show your calculation of how to determine the box size for this switch. Refer to the Quik-Check Box Selection Guide.

3. What wattage was used for each vanity fixture to calculate the estimated load on Circuit A14? ____

4. What is the current draw for the answer given in question 3? ____

5. Exposed noncurrent-carrying metallic parts of electrical equipment must be grounded if installed within ________ feet (________ meters) vertically or ________ feet (________ meters) horizontally of bathtubs, plumbing fixtures, pipes, or other grounded metal work or grounded surfaces.
6. The faceplates to be installed in the bathrooms are made of what material?

   ____

7. Most appliances of the type commonly used in bathrooms, such as hair dryers, electric shavers, and curling irons, have two-wire cords. These appliances are ____________________ insulated.
8. The Code requires that all receptacles in bathrooms be ______________ protected.
9. Hanging lighting fixtures must be kept at least ________ feet (________ meters) from the edge of the tub as measured horizontally. In bathrooms with high ceilings, where the hanging fixture is installed directly over the tub, it must be kept at least ________ feet (________ meters) above the edge of the tub.
10. The following is a layout of a lighting circuit for the bathroom and hallway. Using the cable layout shown in figure 10-1, make a complete wiring diagram of this circuit. Use colored pencils to indicate the conductors.

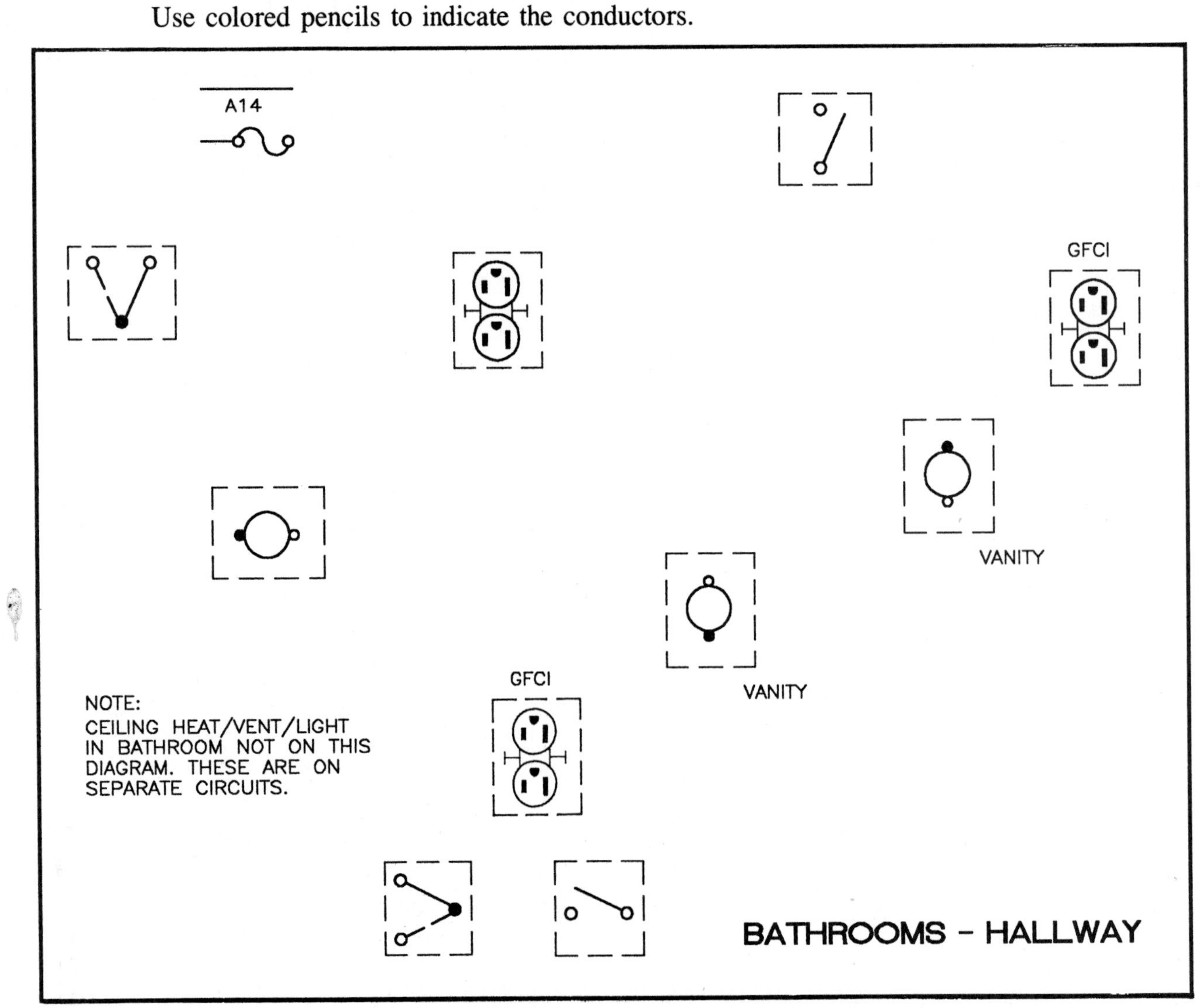

► 11. Circle the correct answer as to whether or not a receptacle outlet is required in the following hallways. ◄

RECEPTACLES

(A) SCALE: 1/4" = 1 FT. REQUIRED NOT REQUIRED

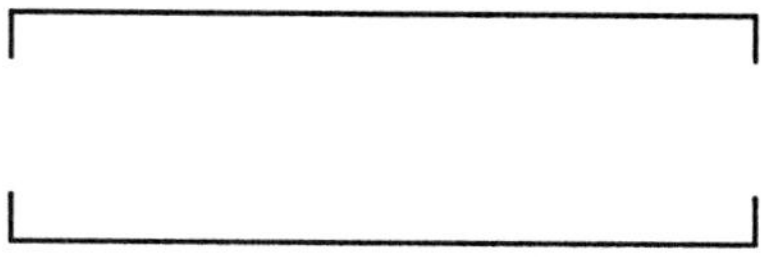

(B) SCALE: 1/4" = 1 FT. REQUIRED NOT REQUIRED

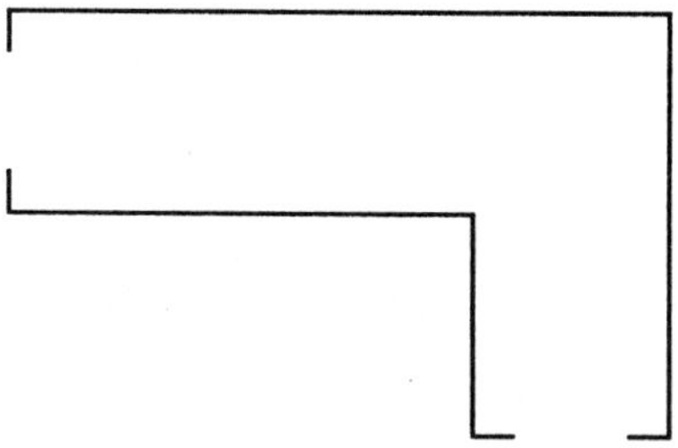

(C) SCALE: 3/8" = 1 FT. REQUIRED NOT REQUIRED

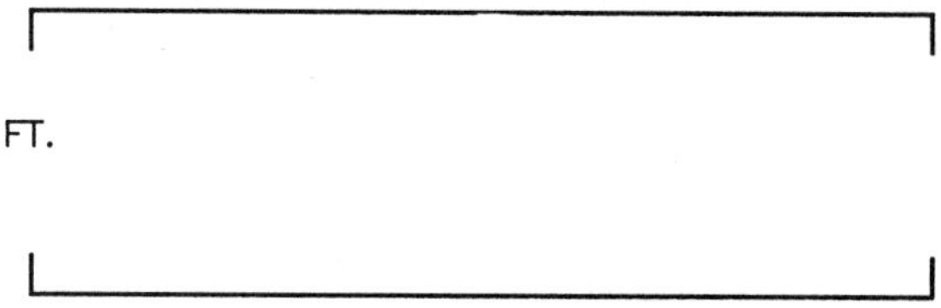

(D) SCALE: 1/8" = 1 FT. REQUIRED NOT REQUIRED

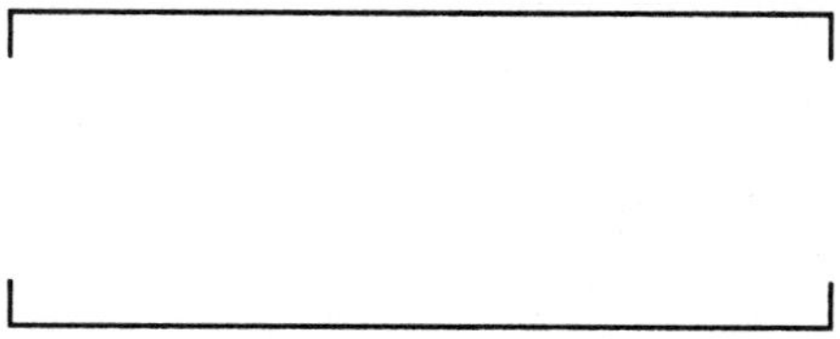

(E) SCALE: 1/2" = 1 FT. REQUIRED NOT REQUIRED

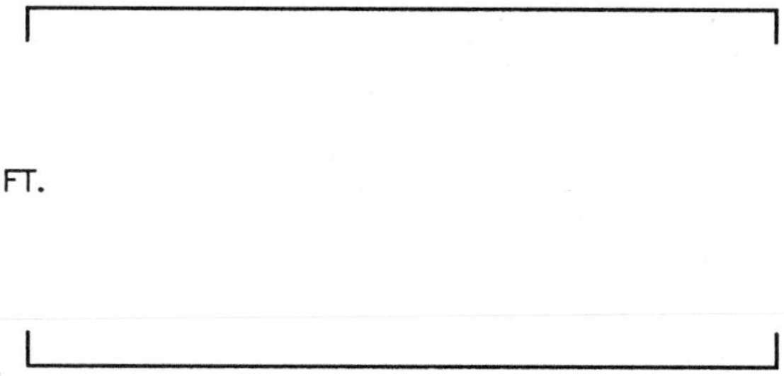

UNIT 11

Lighting Branch Circuit — Front Entry, Porch

OBJECTIVES

After studying this unit, the student will be able to

- understand how to install a switch in a doorjamb for automatic ON/OFF when the door is opened or closed.
- discuss type of fixtures recommended for porches and entries.
- complete the wiring diagram for the Entry-Porch circuit.
- discuss the advantages of switching outdoor receptacles from indoors.
- define wet and damp locations.
- understand box fill for sectional ganged device boxes.

The Front Entry-Porch is connected to Circuit A15. The home run enters the ceiling box in the front entry. From this box, the circuit spreads out, feeding the recessed closet light, the porch bracket fixture, the two bracket fixtures on the front of the garage, one receptacle outlet in the entry, and one outdoor weatherproof GFCI receptacle on the porch. See figure 11-1. Table 11-1 summarizes the outlets and estimated load for the entry and porch.

Note that the receptacle on the porch is controlled by a single-pole switch just inside the front door. This allows the homeowner to plug in outdoor lighting, such as strings of ornamental Christmas lights, or decorative lighting, as illustrated in figure 11-2 and figure 11-3, and have convenient switch control of that receptacle from inside the house, a nice feature.

Typical ceiling fixtures commonly installed in front entryways, where it is desirable to make a good first impression on guests, are shown in figure 11-4.

Typical outdoor ceiling-mounted and wall-bracket style porch and entrance lighting fixtures are shown in figure 11-5.

DESCRIPTION	QUANTITY	WATTS	VOLT-AMPERES
Receptacles @ 120 W	1	120	120
Weatherproof receptacles @ 120 W	1	120	120
Outdoor porch bracket fixture	1	100	100
Outdoor garage bracket fixtures @ 100 W each	2	200	200
Ceiling fiixture	1	150	150
Closet recessed fixture	1	75	75
TOTALS	7	765	765

Table 11-1. Entry and porch: outlet count and estimated load. Circuit A15.

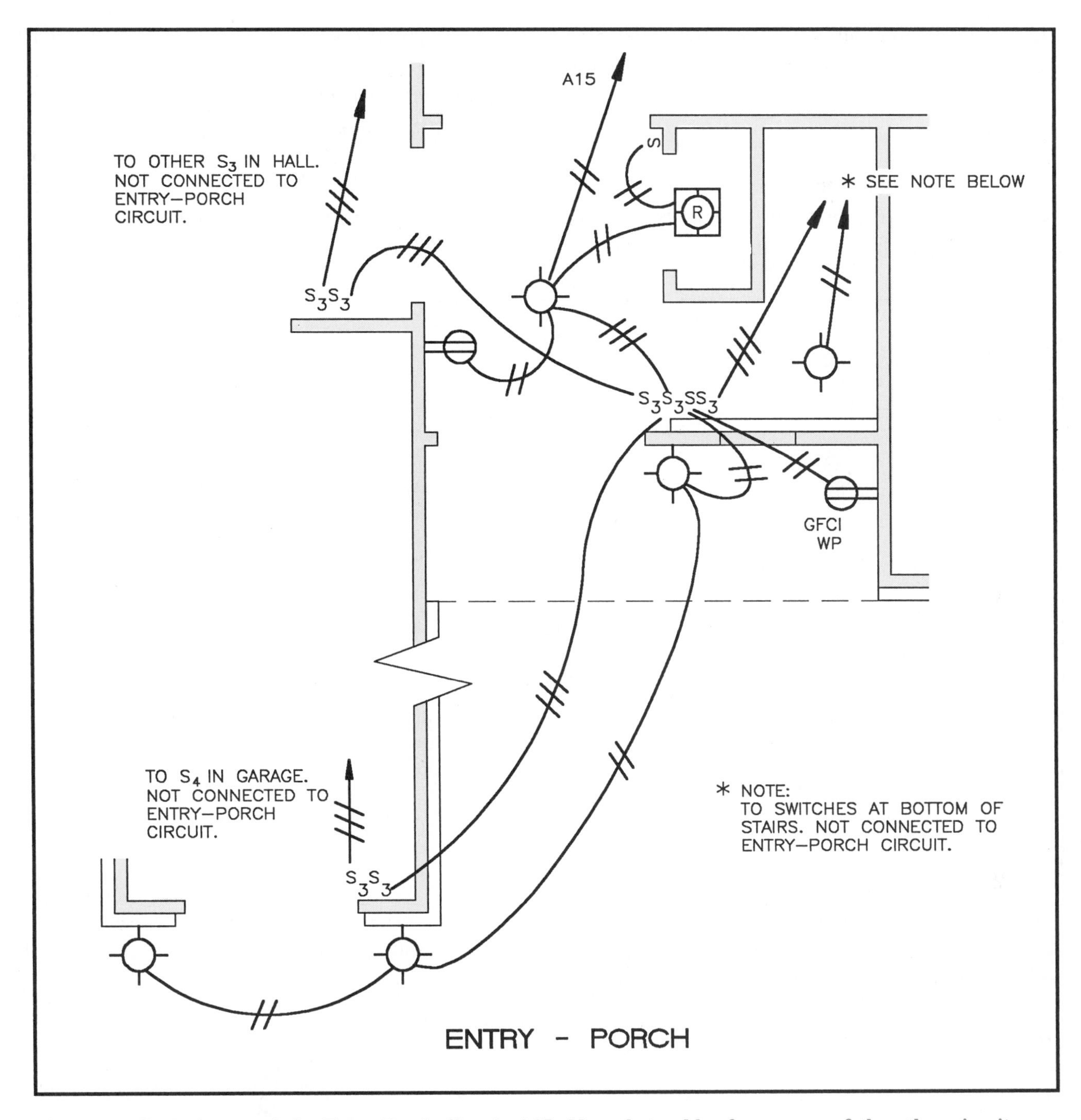

Fig. 11-1 Cable layout of the Entry-Porch Circuit A15. Note that cables from some of the other circuits are shown so that you can get a better idea of exactly how many wires will be found at the various locations.

According to the Code, any location exposed to the weather is considered to be a wet location. *Section 410-4(a)* states that outdoor fixtures must be constructed so that water cannot enter or accumulate in lampholders, wiring compartments, or other electrical parts. These fixtures must be marked "Suitable for Wet Locations."

Partially protected areas under roofs, open porches, or areas under canopies are defined by the Code as damp locations. Fixtures to be used in these locations must be marked "Suitable for Damp Locations." Many types of fixtures are available. Therefore, it is recommended that the electrician check the UL label on the fixture or refer to the UL

Electrical Construction Materials List to determine the suitability of the fixture for a wet or damp location.

CIRCUIT A15

This circuit is a rather simple circuit that has two sets of three-way switches: one set controlling the front entry ceiling fixture, the other set controlling the porch bracket fixture and the two bracket fixtures on the front of the garage. Also one single-pole switch controls the weatherproof porch receptacle outlet, and another single-pole switch controls the recessed closet light. In the Review questions you will be asked to follow the suggested cable layout for making up all of the circuit connections.

Probably the most difficult part of this circuit

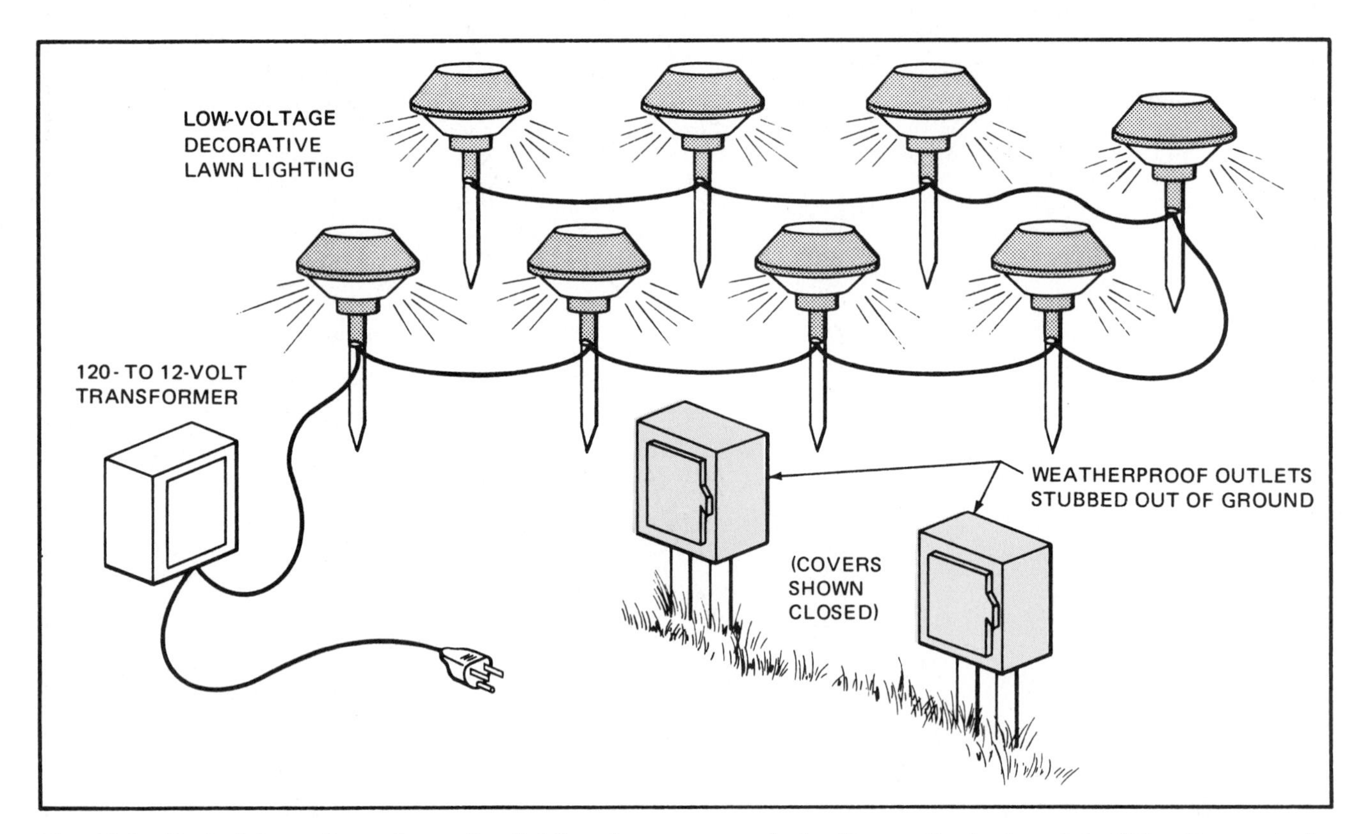

Fig. 11-2 Typical low-voltage decorative lighting for use around shrubbery. Cord plugs into 120-volt receptacle, feeds transformer when voltage is reduced to 120 volts.

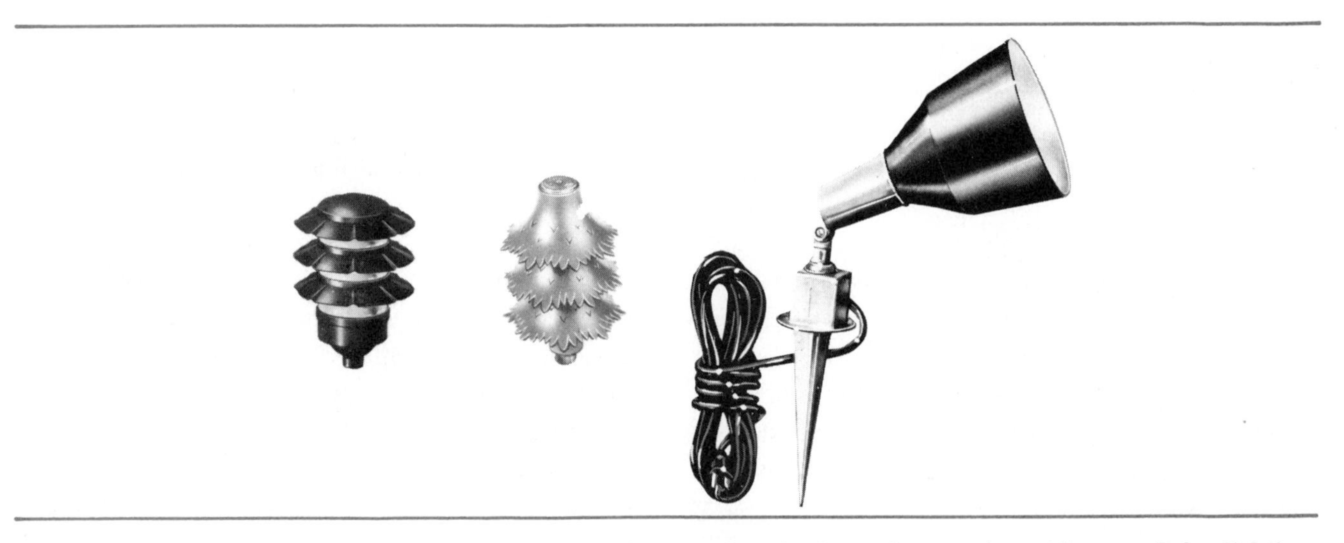

Fig. 11-3 Fixtures used for decorative purposes outdoors under shrubs and trees, in gardens, and for lighting paths and driveways.

Fig. 11-4 Hall lighting fixtures: ceiling mount and chain mount.

Fig. 11-5 Outdoor porch and entrance lighting fixtures: ceiling mount and wall bracket styles.

is planning and making up the connections at the switch location just inside the front door. Here we find four toggle switches: (1) a three-way switch for ceiling fixture; (2) a three-way switch for porch and garage fixtures; (3) a single-pole switch for weatherproof receptacle on porch; and (4) a three-way switch for the fixture over the stairway leading to the basement (connected to the Recreation Room lighting circuit B12, *not* the entry-porch circuit A15).

This location is another challenge for determining the proper size wall box. Let's give it a try.

1. Add the circuit conductors $2 + 2 + 3 + 3 + 3 + 3 =$	16
2. Add for grounding wire (count 1 only)	1
► 3. Add eight for the switches	8 ◄
4. Add four for the cable clamps	4
Total	29

Checking figure 2-18, the Quik-Check Box Selector Guide, and figures 2-24 and 2-25, we find many possibilities. For example, four $3 \times 2 \times 3\ 1/2$ device boxes could be ganged together ($4 \times 9 = 36$ conductors).

An interesting possibility presents itself for the front entry closet. Although the plans show that the recessed closet fixture is turned on and off by a standard single-pole switch to the left as you face the closet, a doorjamb switch could have been installed. A doorjamb switch is usually mounted about 6 feet (1.83 m) above the floor, on the inside of the 2×4 framing for the closet door. The electrician will run the cable to this point, and let it hang out until the carpenter can cut the proper size opening into the doorjamb for the box. After the finished woodwork is completed, the electrician then finishes installing the switch. Figure 11-6 illustrates a door switch. Note that the plunger on the switch is pushed inward when the edge of the door pushes on it as the door is closed, shutting off the light. The plunger can be adjusted in or out to make the switch work properly. These doorjamb switches come complete with their own special wall box.

Fig. 11-6 Door switch.

REVIEW

1. How many wires enter the entry ceiling box? ______________________

2. Assuming that an outlet box for the entry ceiling has a fixture stud but has external cable clamps, what size box could be installed?

____

____

____

____

____

____

3. How many receptacle outlets and lighting outlets are supplied by Circuit A15?

____

4. Outdoor fixtures directly exposed to the weather must be marked as:
   a. suitable for damp locations
   b. suitable for dry locations only
   c. suitable for wet locations

   Circle correct answer.

5. Make a material list of all types of switches and receptacles connected to Circuit A15.

   ____
   ____
   ____
   ____

6. From left to right, facing the switches, what do the switches next to the front door control?

   ____
   ____
   ____
   ____
   ____

7. Who is to select the entry ceiling fixture? ______________________

8. The following layout is for Lighting Circuit A15, the entry, porch, and front garage lights. Using the cable layout shown in figure 11-1, make a complete wiring diagram of this circuit. Use colored pencils to indicate the color of the conductors' insulation.

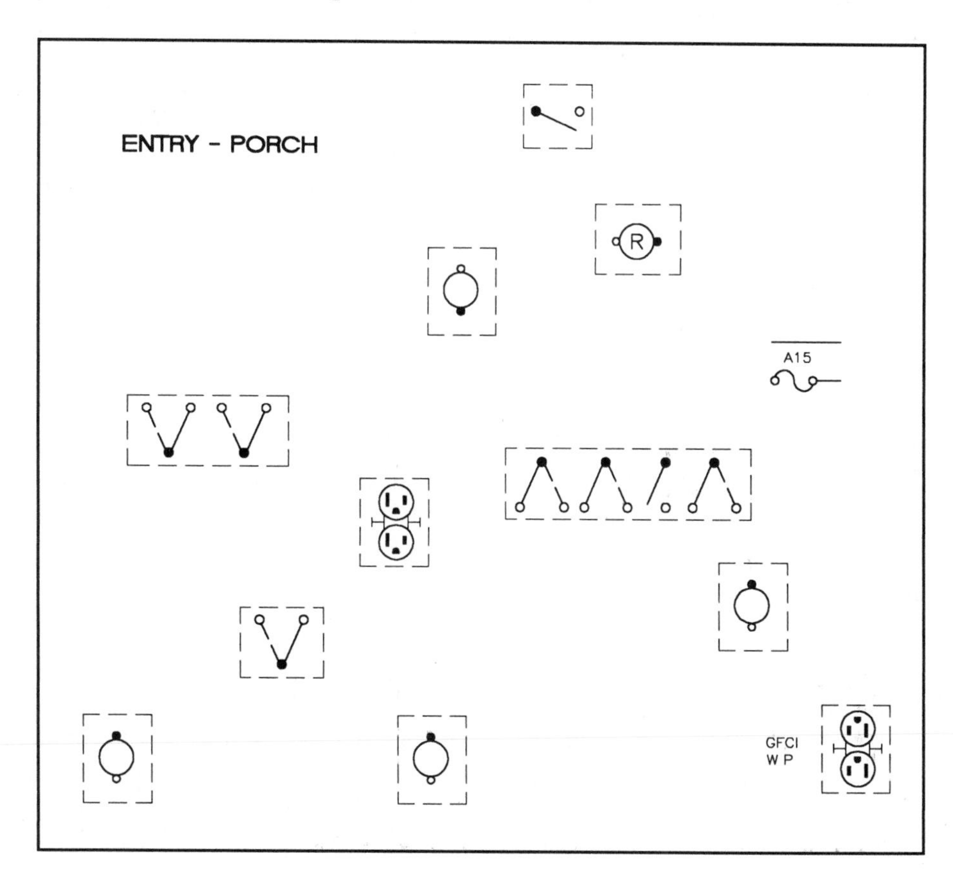

UNIT 12

Lighting Branch Circuit and Small Appliance Circuits for Kitchen

OBJECTIVES

After studying this unit, the student will be able to

- discuss the features of an exhaust fan for the removal of kitchen cooking odors and humidity.
- explain the Code requirements for small appliance circuits in kitchens.
- understand the Code requirements for ground-fault protection for receptacles that serve countertops.
- discuss split-circuit receptacles.
- discuss multiwire circuits.
- discuss exhaust fan noise ratings.
- understand the overall general concept of grounding electrical equipment.
- know the difference between a receptacle outlet and a lighting outlet.
- discuss typical kitchen, dining room, undercabinet, and swag fixtures.

LIGHTING CIRCUIT B7

The lighting circuit supplying the kitchen and nook originates at Panel B. The circuit number is B7. The home run is very short, leading from Panel B in the corner of the recreation room to the switch box to the right of the kitchen sink. Here, the circuit continues upward through the junction box on the recessed fixture above the sink which is approved for feedthrough wiring (see unit 7 for complete explanation of Code requirements for recessed fixtures) through the clock outlet box, and then across the kitchen ceiling connecting the ceiling light fixtures, the track light, the outdoor bracket fixture, and range hood exhaust fan/light unit, figure 12-1. Table 12-1 shows the outlets and estimated loads for the kitchen/nook area.

KITCHEN LIGHTING

The general lighting for the kitchen is provided by two four-lamp fluorescent surface-mounted fixtures of the type shown in figure 12-2. These fixtures are controlled by three-way switches: one located adjacent to the sliding door, the other at the doorway leading to the living room.

An adjustable chain swag fixture, figure 12-3A, is mounted on the track to provide lighting above the breakfast nook table. Many fixtures of this type are

DESCRIPTION	QUANTITY	WATTS	VOLT-AMPERES
Ceiling fixtures (eight 20-W FL lamps)	2	160	180
Recessed fixture over sink	1	100	100
Track light (nook)	1	100	100
Outdoor rear bracket fixture	1	100	100
Clock outlet	1	15	15
Range hood fan/light	1		
Two 60-W lamps		120	120
Fan motor (1 1/3 A @ 120 V)		160	180
TOTALS	7	755	795

Table 12-1. Kitchen/Nook: outlet count and estimated load. Circuit B7.

provided with a three-position switch that allows one to select low, medium, or high light levels. For instance, a three-way bulb might provide these levels of light, such as 50-100-150 watts. This switch is built into the fixture itself. Other types of ceiling-mount lighting fixtures are shown in figure 12-3B.

One goal of good lighting is to reduce shadows in work areas. This is particularly important in the kitchen. The two ceiling fixtures, the fixture above the sink, plus the fixture within the range hood will provide excellent lighting in the kitchen area. Of course, the track lighting in the nook area offers good lighting for that area.

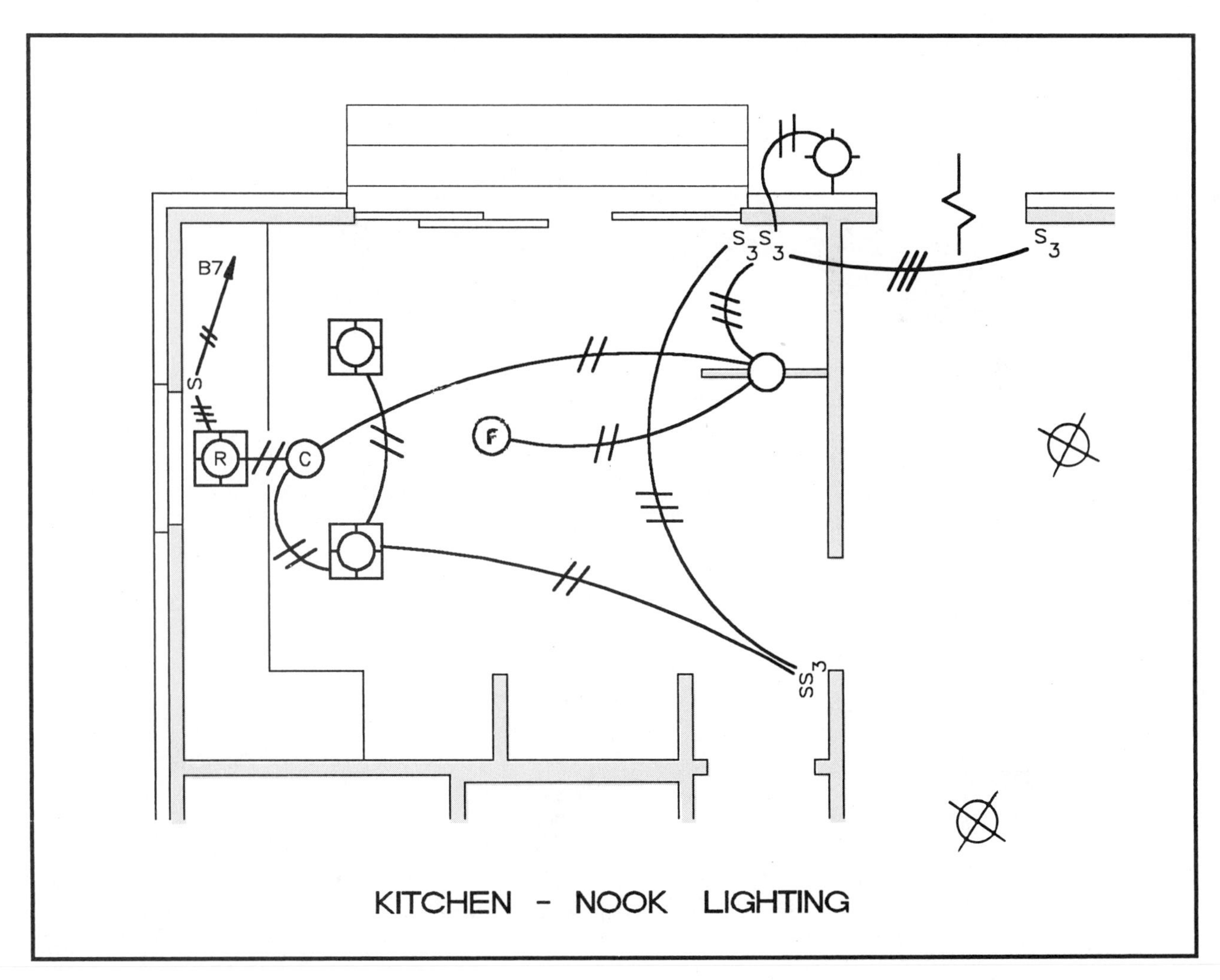

Fig. 12-1 Cable layout for the kitchen lighting. This is Circuit B7. The appliance circuit receptacles are connected to 20-ampere circuits B13, B15, and B16 and are not shown on this lighting circuit layout.

Undercabinet Lighting

In some instances, particularly in homes with little outdoor exposure, the architect might specify strip fluorescent fixtures under the kitchen cabinets. Several methods may be used to install undercabinet lighting fixtures. They can be fastened to the wall just under the upper cabinets, figure 12-4(A), installed in a recess that is part of the upper cabinets so that they are hidden from view, figure 12-4(B), or fastened under and to the front of the upper cabinets, figure 12-4(C). All three possibilities require close coordination with the cabinet installer to be sure that the wiring is brought out of the wall at the proper location to connect to the undercabinet fixtures.

Lamp Type

Good color rendition in the kitchen area is achieved by using incandescent lamps or warm white deluxe (WWX) or warm white (WW) or deluxe cool white (CWX) 3000-K lamps wherever fluorescent lamps are used.

Fan Outlet

The fan outlet is connected to branch lighting circuit B7. Fans can be installed in a kitchen either to exhaust the air to the outside of the building or to filter and recirculate the air. Both types of fans can be used in a residence. The ductless hood fan does not exhaust air to the outside and does not remove humidity from the air.

Homes heated electrically with resistance-type units usually have excess humidity. The heating units neither add nor remove moisture from the air unless humidity control is provided. The proper use of weather stripping, storm doors, and storm windows, and a vapor barrier of polyethylene plastic film tends to retain humidity.

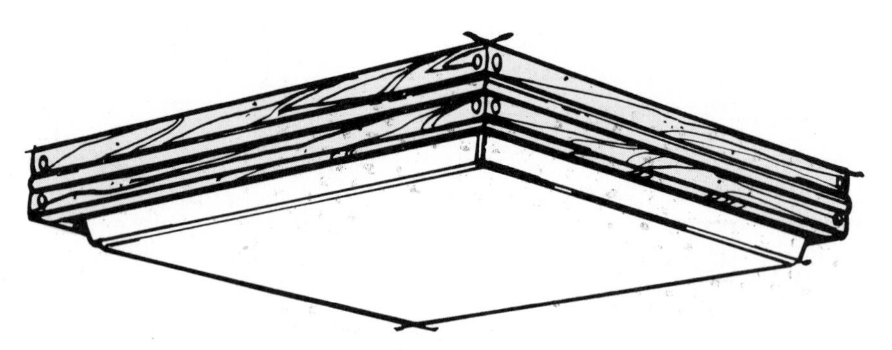

Fig. 12-2 Typical ceiling fixture of the type installed in the kitchen of this residence. This type of fixture might have four 20-watt fluorescent lamps, or it might have two 40-watt U-shaped lamps.

Fig. 12-3A Fixtures that can be hung either singly or on a track over dinette tables, such as in the nook area of this kitchen.

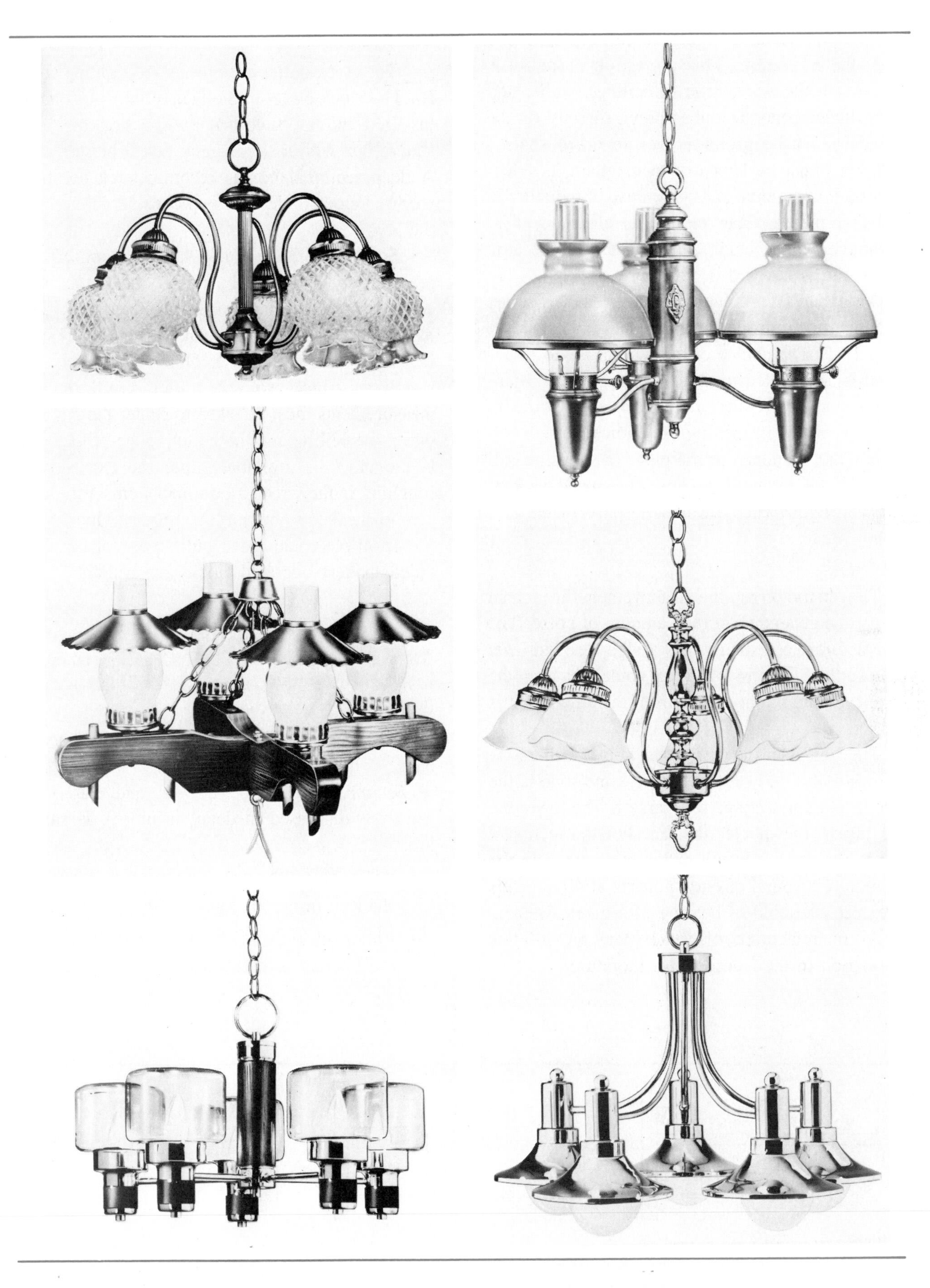

Fig. 12-3B Types of fixtures used over dining room tables.

Vapor barriers are installed to keep moisture out of the insulation. This protection normally is installed on the warm side of ceilings, walls, and floors, under concrete slabs poured directly on the ground, and as a ground cover in crawl spaces. Insulation must be kept dry to maintain its effectiveness. For example, a 1% increase of moisture in insulating material can reduce its efficiency 5%. (Humidity control using a humidistat on the attic exhaust fan is covered in unit 22.)

An exhaust fan simultaneously removes odors and moisture from the air and exhausts heated air, figure 12-5. The choice of the type of fan to be installed is usually left to the owner. The electrician, builder, and/or architect may make suggestions to guide the owner in making the choice.

For the residence in the plans, a separate wall switch is not required because the speed switch, light, and light switch are integral parts of the fan.

Fan Noise

The fan motor and the air movement through an exhaust fan generate a certain amount of noise. The manufacturers of exhaust fans assign a sound-level rating so that it is possible to get some idea as to the noise level one might expect from the fan after it is installed.

The unit used to define fan noise is the *sone* (rhymes with tone). For simplicity, one sone is the noise level of an average refrigerator. The lower the sone rating, the quieter the fan. For the technical person, the sone is a unit of loudness equal to the loudness of a sound of one kilohertz at 40 decibels above the threshold of hearing of a given listener.

All manufacturers of exhaust fans provide this information in their descriptive literature.

Clock Outlets

The clock outlet is connected to lighting circuit B7. The clock outlet in the face of the soffit may be installed in any sectional switch box or 4-inch square box with a single-gang raised plaster cover. A deep sectional box is recommended because a recessed clock outlet takes up considerable room in the box.

Some decorative clocks have the entire motor recessed so that only the hands and numbers are exposed on the surface of the soffit or wall. Some of these clocks require special outlet boxes (usually furnished with the clock). Other clocks require a standard 4-inch square outlet box. Accurate measurements must be taken to center the *clock* between the ceiling and the bottom edge of the soffit. If the clock is available while the electrician is roughing in the wiring, the dimensions of the clock can be checked to help in locating the clock outlet.

If an electrical clock outlet has not been provided, battery-operated clocks are available.

SMALL APPLIANCE BRANCH CIRCUITS FOR CONVENIENCE RECEPTACLES IN KITCHEN

The Code requirements for small appliance circuits in the kitchens of dwellings are covered in ► *Sections 210-52(b), 220-4,* and *220-16.* ◄ This was discussed in detail in unit 3. Highlights are:

- At least two 20-ampere small appliance circuits shall be installed, figure 12-6.
- Either (or both) of the two circuits required in the kitchen is permitted to supply receptacle

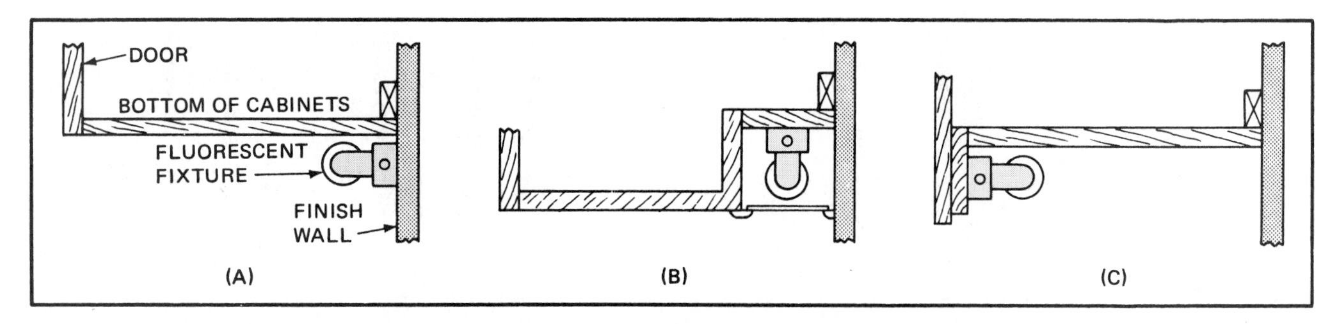

Fig. 12-4 Methods of installing undercabinet lighting fixtures.

Fig. 12-5 Typical range hood exhaust fan. Speed control and light are integral part of the unit.

outlets in other rooms, such as the dining room, breakfast room, or pantry. Figure 12-6 illustrates this portion of the Code.

- These small appliance circuits shall be assigned a load of 1500 volt-amperes each when calculating feeders and service-entrance requirements.
- The clock outlet may be connected to an appliance circuit when this outlet is to supply and support the clock only (see figure 3-3). In the residence in the plans, the clock outlet is connected to the lighting circuit.
- Countertop receptacles must be supplied by at least two 20-ampere small appliance circuits, figure 12-6.
- All 125-volt, single-phase, 15- and 20-ampere receptacles installed within 6 feet (1.83 m) of the kitchen sink that serve countertops must have GFCI protection, figure 12-7.
- Outdoor receptacles may be connected to the small appliance circuits. The outdoor receptacles must be GFCI protected.
- The small appliance circuits are provided to serve plug-in portable appliances. They are not permitted by the Code to serve lighting fixtures or appliances such as dishwashers, garbage disposals, or exhaust fans. Some local electrical codes require that the receptacle outlet for the refrigerator be on a separate circuit.

The small appliance circuits prevent circuit overloading in the kitchen as a result of the heavy concentration and use of electrical appliances. For example, one type of cord-connected microwave oven is rated 1500 watts at 120 volts. The current required by this appliance alone is:

$$I = \frac{W}{E} = \frac{1500}{120} = 12.5 \text{ amperes}$$

Section 90-1(b) states that the Code requirements are actually the minimum provisions for safety. According to the Code, compliance with

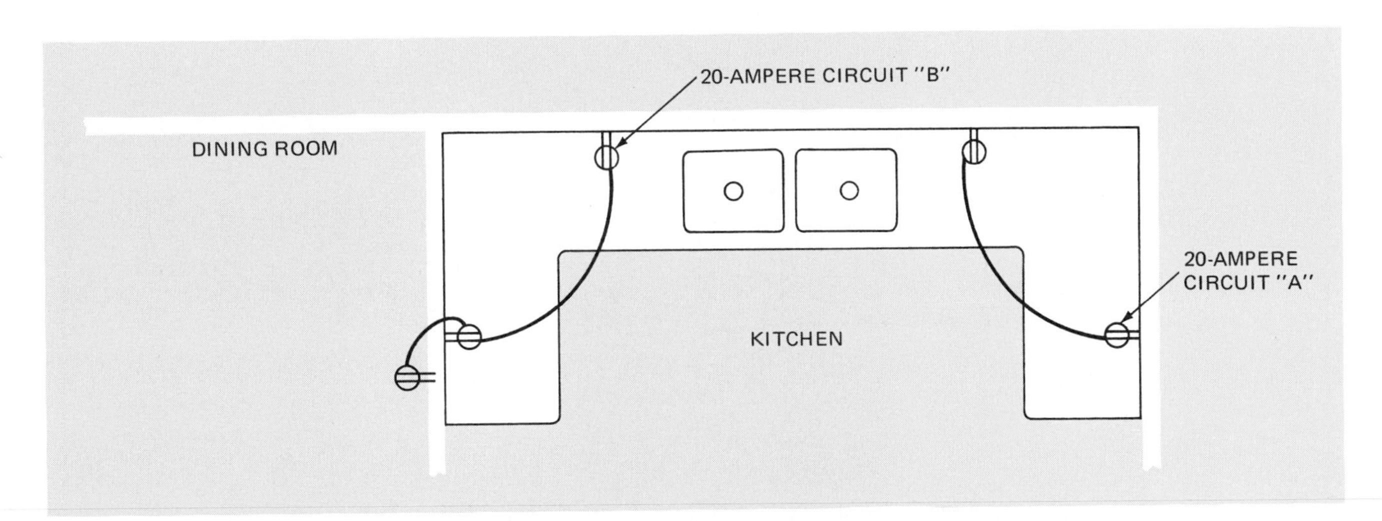

► Fig. 12-6 A receptacle outlet in a dining room may be connected to one of the two required small appliance circuits in the kitchen, *Section 210-52(b)(1)*. Countertop receptacles must be supplied by at least two 20-ampere small appliance circuits, *Section 210-52(b)(2)*. ◄

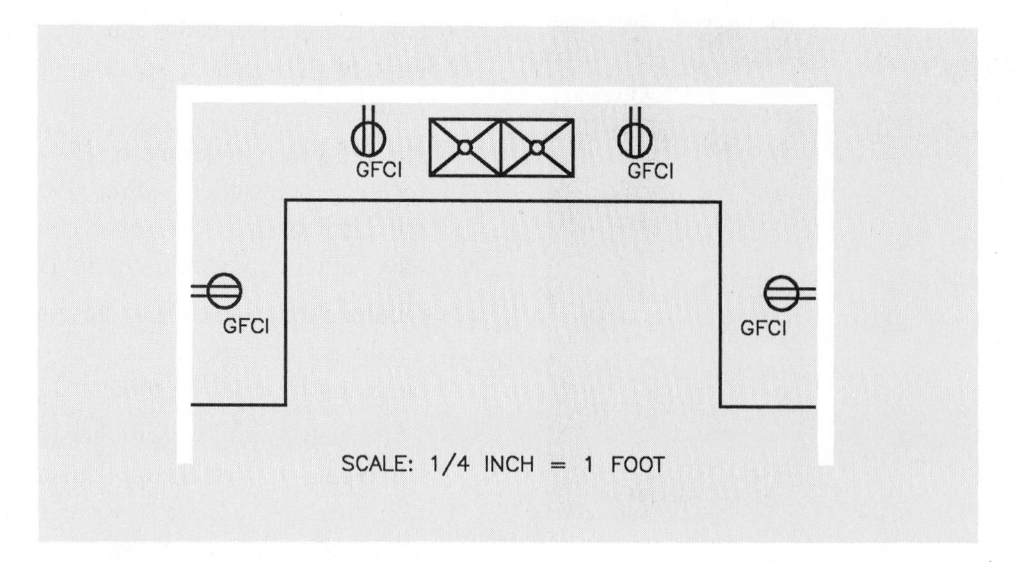

► Fig. 12-7 In kitchens, all 125-volt, single-phase, 15- and 20-ampere receptacles must be GFCI protected if they are within 6 feet (1.83 m) of the sink, if they serve countertops, and are located above or below the surface of the countertop, island, or peninsula, *Section 210-8(a)(5)*. ◄

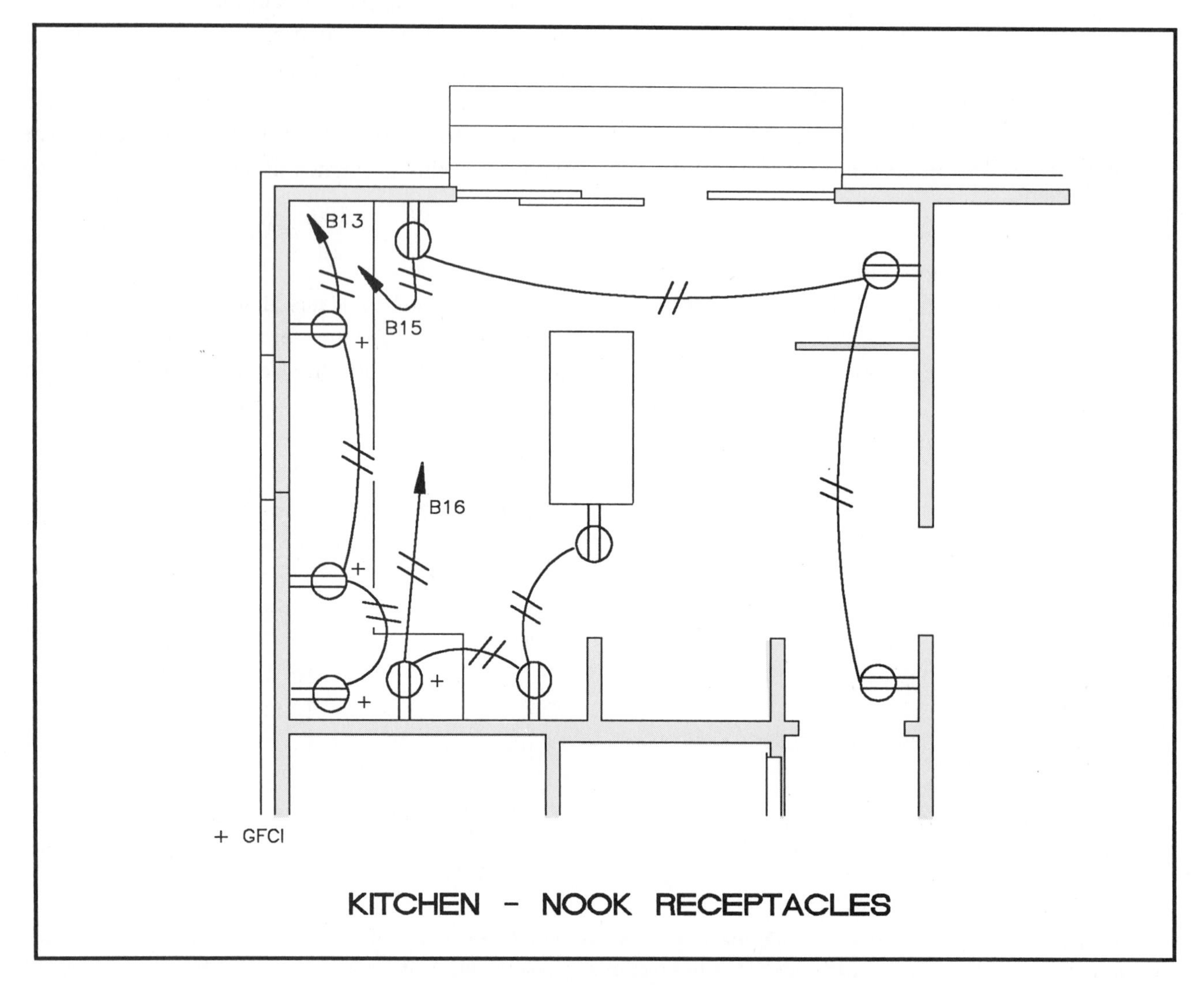

Fig. 12-8 Cable layout for kitchen/nook receptacles.

these rules will not necessarily result in an efficient, convenient, or adequate installation for good service or future expansion of electrical use.

Figure 12-8 shows three 20-ampere small appliance circuits feeding the kitchen/nook receptacles, of which there are nine. These circuits are B13, B15, B16.

Obviously, there are many ways that receptacles can be connected to the small appliance circuits. The intent is to arrange the circuitry so as to provide the best availability of electrical power for appliances that will be used in the heavily concentrated work areas in the kitchen. Keep in mind that GFCI protection is required for all receptacle outlets that serve countertops located within 6 feet (1.83 m) of the kitchen sink. See unit 6 for a discussion of GFCI receptacles.

Figure 12-8 shows one way to arrange the kitchen small appliance circuits. Note that there is a receptacle outlet on the side of the kitchen cooktop island. This is required by the Code. The receptacle is connected to Circuit B16. It is *not* connected to the cooktop circuit.

When a receptacle outlet is installed below the sink for easy plug-in connection of a food waste disposer, it is *not* to be included in the Code ► requirements for small appliance circuits, *Sections 210-52(c)* and *220-4(b)*, and it is *not* required to be GFCI protected, *Section 210-8(a)(5)*. ◄

Sliding Glass Doors

► See unit 3 for a discussion on how to space receptacle outlets on walls where sliding glass doors are installed. ◄

Small Appliance Branch-Circuit Load Calculations

The receptacle load for the three small appliance branch circuits in the kitchen is calculated using a load of 1500 volt-amperes per circuit. Thus, the computed load is

$$1500 \text{ volt-amperes} \times 3 = 4500 \text{ volt-amperes}$$

Section 220-16(a) requires that *feeder* loads be calculated on the basis of 1500 volt-amperes for each two-wire small appliance circuit. Using the same basis for estimating the branch-circuit loading will result in ample circuit capacity.

SPLIT-CIRCUIT RECEPTACLES AND MULTIWIRE CIRCUITS

Split-circuit receptacles may be installed where a heavy concentration of plug-in load is anticipated, in which case each receptacle is connected to a separate circuit as in figures 12-9 and 12-10.

A popular use of split-circuit receptacles is to control one receptacle with a switch and leave one receptacle hot at all times, as is done in the bedrooms and living room of this residence.

Split-circuit GFCI receptacles are not available in the marketplace. Refer to unit 6 where GFCIs are discussed in great detail.

A feedthrough GFCI receptacle could be installed to the right of the sink, with regular grounding-type receptacles installed for the other two receptacles on Circuit B13.

Another feedthrough GFCI receptacle could be installed for the first receptacle fed by Circuit B16.

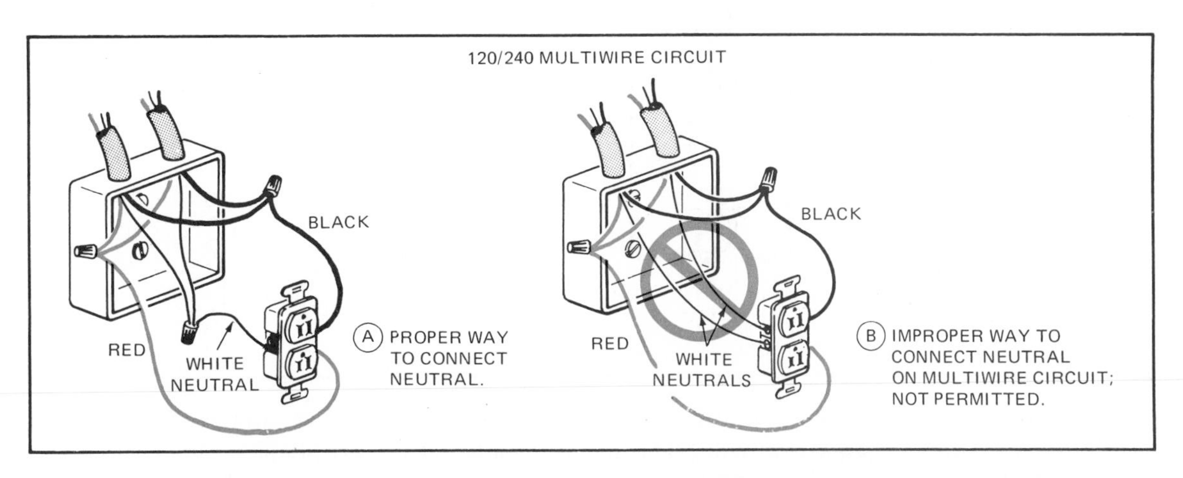

Fig. 12-9 Connecting the neutral in a three-wire (multiwire) circuit, *Section 300-13(b)*.

Another way to connect Circuit B16 would be to use regular grounding-type receptacles for the first two receptacles fed by Circuit B16, then install a GFCI receptacle on the side of the range cooktop island where the plans indicate a receptacle outlet.

Circuit B15 does not require any GFCI receptacles.

Therefore, the use of split-circuit receptacles is rather difficult to incorporate where GFCI receptacles are required. A multiwire branch circuit could be run to a box where proper connections are made, with one circuit feeding one feedthrough GFCI receptacle and its downstream receptacles.

This would be used when it becomes more economical to run one three-wire circuit rather than installing two two-wire circuits. Again, it is a matter of knowing what to do, and when it makes sense to do it. See figure 12-11.

When multiwire circuits are connected, the neutral conductor must *not* be broken at the recep-

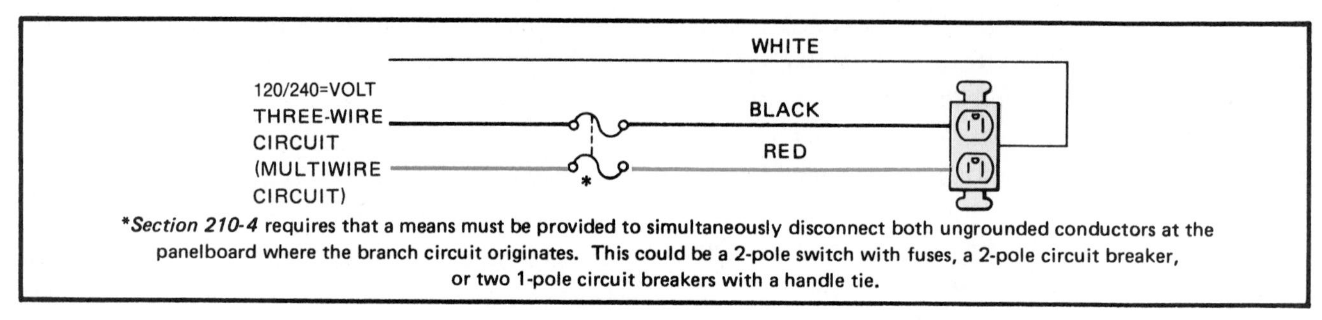

Fig. 12-10 Split-circuit receptacle connected to multiwire circuit.

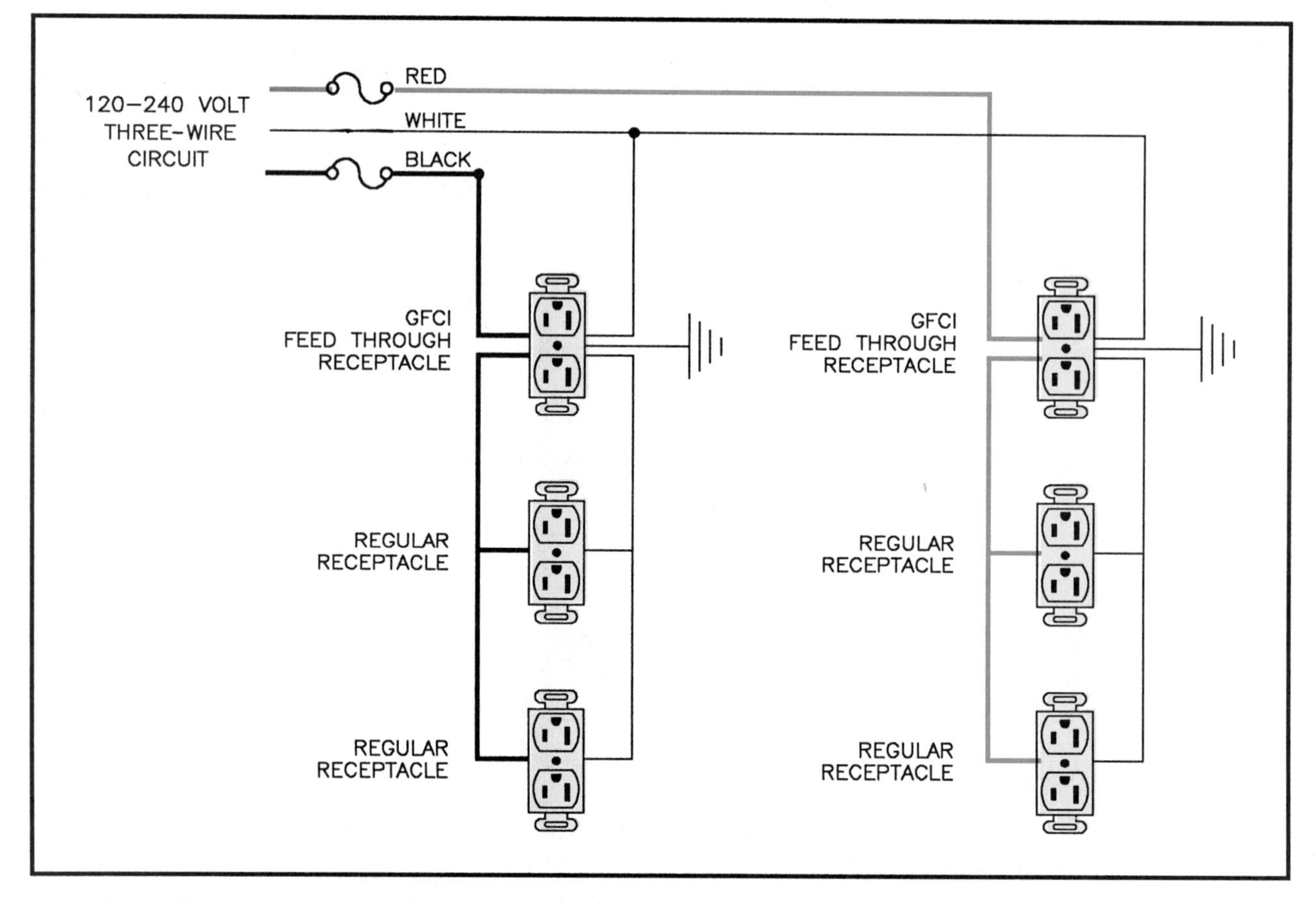

Fig. 12-11 This diagram shows how a multiwire circuit can be used to carry two circuits to a box, then split the three-wire circuit into two two-wire circuits. This circuitry does not require simultaneous disconnect of the ungrounded conductors as shown in figure 12-10.

tacles. *NEC® Section 300-13(b)* states that where more than one conductor is to be spliced, all connections for the grounded conductors in multiwire circuits must be made up independently of the receptacles, lampholders, and so on. This requirement is illustrated in figure 12-9. Note that the screw terminals of the receptacle *must not* be used to splice the neutral conductors. The hazards of an open neutral are discussed in unit 17. Figure 12-12 shows two types of grounding receptacles commonly used for installations of this type.

When multiwire branch circuits supply more than one receptacle on the same yoke, a means must be provided to disconnect simultaneously all of the hot conductors at the panelboard where the branch circuit originates. Figure 12-10 shows that, although 240 volts is present on the wiring device, only 120 volts is connected to each receptacle on the wiring device. ► See *Sections 210-6(a)(2)* and *210-6(b)(3)*. ◄

The "simultaneous disconnect" rule applies to any receptacles, lampholders, or switches mounted on one yoke. See *Section 210-4(b)*.

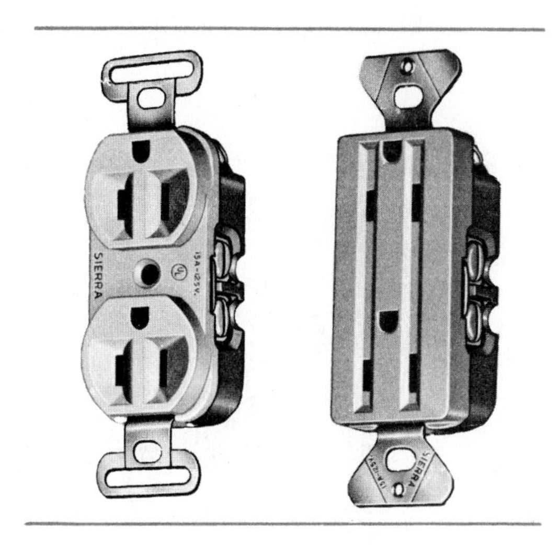

Fig. 12-12 Grounding-type duplex receptacles, 15A-125V rating.

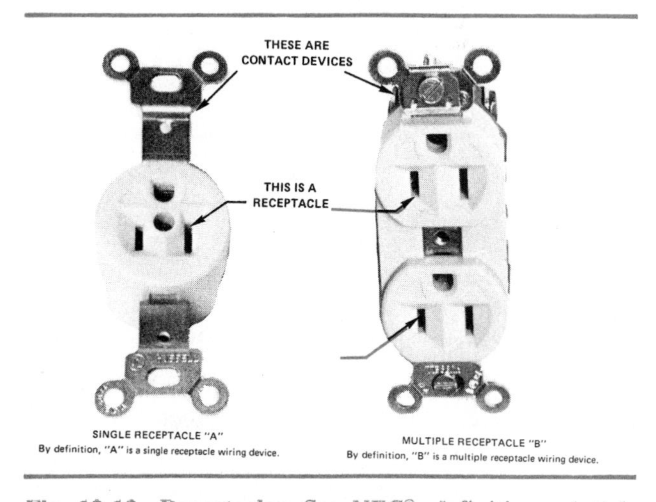

Fig. 12-13 Receptacles: See *NEC®* definitions, *Article 100*.

RECEPTACLES AND OUTLETS

Article 100 of the *NEC®* gives the definition of a receptacle and an outlet, as illustrated in figures 12-13 and 12-14.

Receptacles are selected for circuits according to the following guidelines. A single receptacle connected to a circuit must have a rating not less than the rating of the circuit, *Section 210-21(b)*. In a residence, typical examples for this requirement for single receptacles are the clothes dryer outlet (30 amperes), or the range outlet (50 amperes), or the freezer outlet (15 amperes).

Circuits rated at 15 amperes supplying two or more receptacles shall not contain receptacles rated at over 15 amperes. For circuits rated at 20 amperes supplying two or more receptacles, the receptacles connected to the circuit may be rated at 15 or 20 amperes. See *Table 210-2(b)(3)*.

GENERAL GROUNDING CONSIDERATIONS

The specifications for the residence require that all outlet boxes, switch boxes, and fixtures be grounded. In other words, the entire wiring installation must be a grounded system. The Code also requires that the outside fixtures be grounded, *Section 410-17*. Every electrical component located within reach of a grounded surface must be grounded. Thus, it can be seen that there are few locations where ungrounded boxes or fixtures are permitted. Post lights, weatherproof receptacles, basement wiring, boxes and fixtures within reach of sinks, ranges, and range hood fans must all be grounded. Hot water heating, hot air heating, and steam heating registers are all grounded surfaces. This means that any electrical equipment, boxes, or fixtures within reach of these surfaces must be grounded. *Section 250-42* requires that any boxes installed

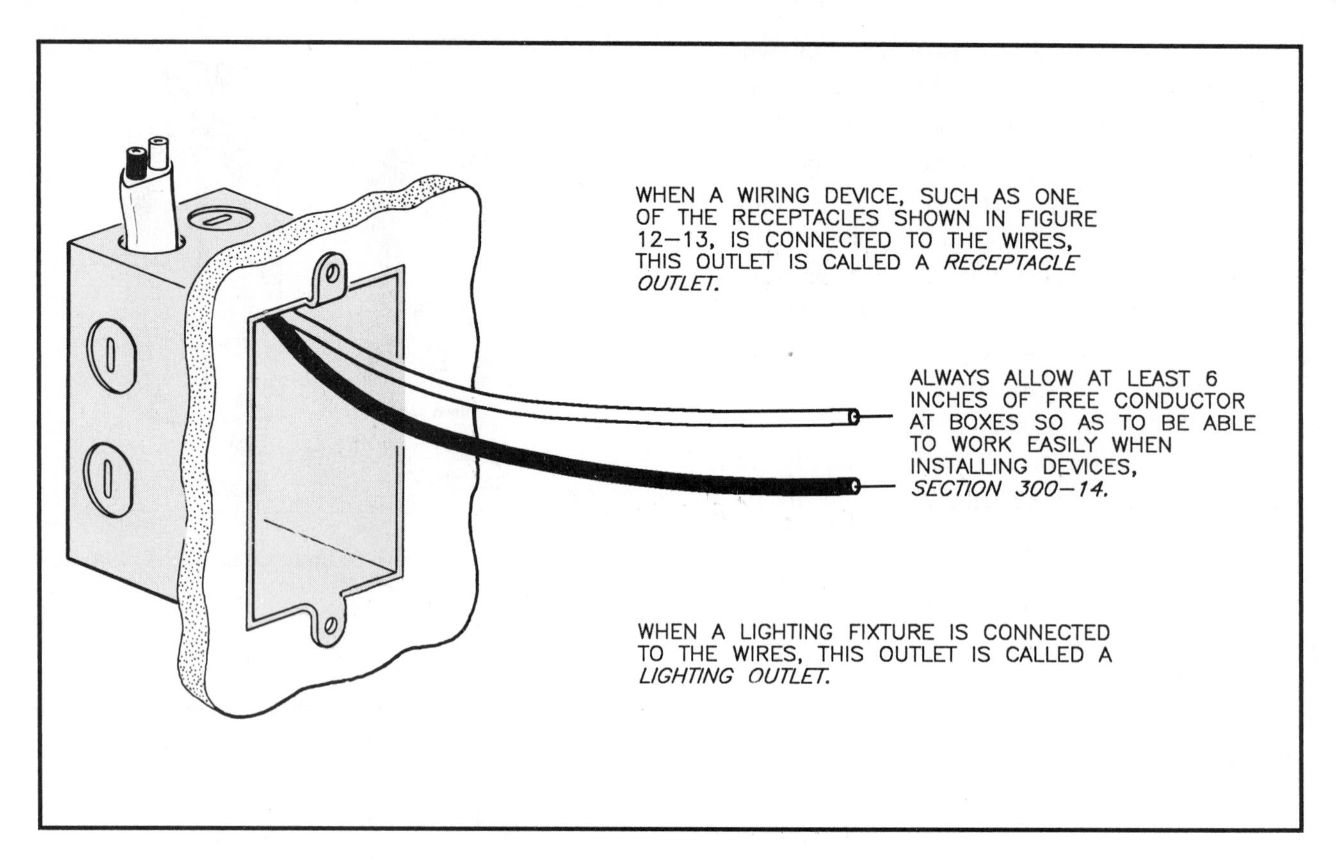

Fig. 12-14 Outlets: See *NEC*® definitions, *Article 100.*

near metal or metal lath, tinfoil, or aluminum insulation must also be grounded.

As previously discussed, grounding is accomplished through the proper use of nonmetallic-sheathed cable, armored cable, and metal conduit.

In the case of nonmetallic-sheathed cable, do not confuse the bare *grounding* conductor with the *grounded* circuit conductor. The *grounding* conductor is used to ground equipment, whereas the white *grounded* conductor is one of the branch-circuit conductors. They are *not* to be connected together except at the main service-entrance equipment.

REVIEW

Note: Refer to the Code or plans where necessary.

1. If everything on Circuit B7 were turned on, what would be the total current draw?

____

____

2. From what panel does the kitchen lighting circuit originate? What size conductors are used? ____

____

3. How many lighting fixtures are connected to the kitchen lighting circuit? ______

______

4. What color fluorescent lamps are recommended for residential installations?

______

______

5. a. What is the minimum number of 20-ampere small appliance circuits required for a kitchen according to the Code? ______

   b. How many are there in this kitchen? ______

6. How many receptacle outlets are provided in the kitchen? ______

7. What is meant by the term two-circuit (split-circuit) receptacles? ______

______

______

______

______

______

8. A single receptacle connected to a circuit must have a rating ______ than the ampere rating of the circuit.

9. In kitchens, a receptacle must be installed at each counter space wider than ______ inches.

10. A fundamental rule regarding the grounding of metal boxes, fixtures, and so on, is that they must be grounded when "in reach of ______."

11. How many circuit conductors enter the box

    a. where the clock will be installed? ______

    b. in the ceiling box over which the track will be installed? ______

    c. at the switch location to the right of the sliding door? ______

12. How much space is there between the countertop and upper cabinets? ______

13. Where is the speed control for the fan located? ______

14. Who is to furnish the range hood? ______

15. List the appliances in the kitchen that must be connected. ______

______

______

______

______

16. Complete the wiring diagram, connecting feedthrough GFCI 2 to also protect receptacle 1, both to be supplied by Circuit A1. Connect feedthrough GFCI 3 to also protect receptacle 4, both to be supplied by Circuit A2. Use colored pencils or colored markers to show proper color. Assume that the wiring method is electrical metallic tubing, where more freedom in the choice of insulation colors is possible.

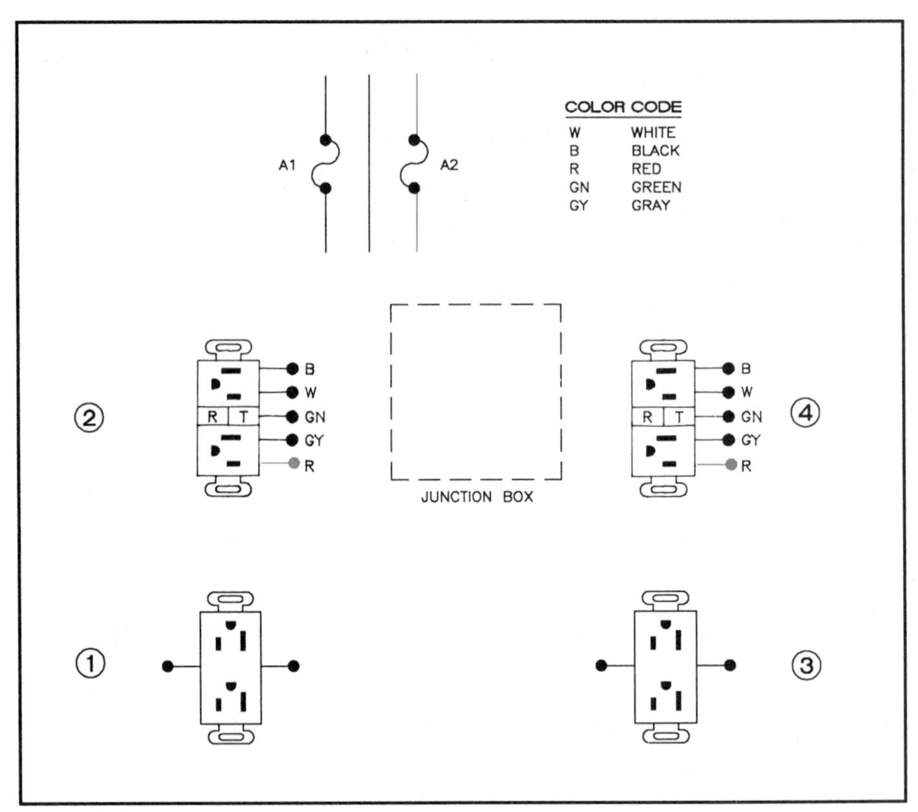

17. Each 20-ampere appliance circuit load demand shall be determined at ________.
Choose one.

a. 2400 volt-amperes b. 1500 volt-amperes c. 1920 volt-amperes

18. Is it permitted to connect an outlet supplying a clock receptacle to a 20-ampere small appliance circuit? ________________________________

____

____

19. a. The Code requires a minimum of two small appliance circuits in a kitchen. Is it permitted to connect a receptacle in a dining room to one of the kitchen small appliance circuits? ________________________________

What Code section applies to this situation? ________________

b. May outdoor weatherproof receptacles be connected to a 20-ampere small appliance circuit? ________________________________

c. Receptacles located above the countertops in the kitchen must be supplied by at least ________________ 20-ampere small appliance circuits.

20. According to *Section 210-52*, no point along the floor line shall be more than ________ feet (________ meters) from a receptacle outlet. A receptacle must be installed in any wall space ________ feet (________ meters) wide or greater.

21. The Code states that in multiwire circuits, the screw terminals of a receptacle must *not* be used to splice the neutral conductors. Why?

____

____

____

22. Electric fans produce a certain amount of noise. It is possible to compare the noise levels of different fans prior to installation by comparing their ________ ratings.

23. Is it permitted to connect the white grounded circuit conductor to the grounding terminal of a receptacle? ____________________

24. A load of ________ volt-amperes must be included for each 120-volt, 20-ampere small appliance circuit.

25. GFCI protection is required for all receptacles that serve countertops within ____ feet of a kitchen sink.

26. The following is a layout for the lighting circuit for the Kitchen/Nook. Complete the wiring diagram using colored pens or pencils to show the conductors' insulation color.

►27. The Kitchen/Nook features a sliding glass door. For the purpose of providing the proper receptacle outlets, answer the following statements *true* or *false*.

a. Sliding glass panels are considered wall space. ____________

b. Sliding glass panels are not considered wall space. ____________

c. Fixed panels of glass doors are considered wall space. ____________

d. Fixed panels of glass doors are not considered wall space. ____________ ◄

B7

DISPOSER

C

F

RANGE HOOD
FAN–LIGHT

DW

KITCHEN - NOOK

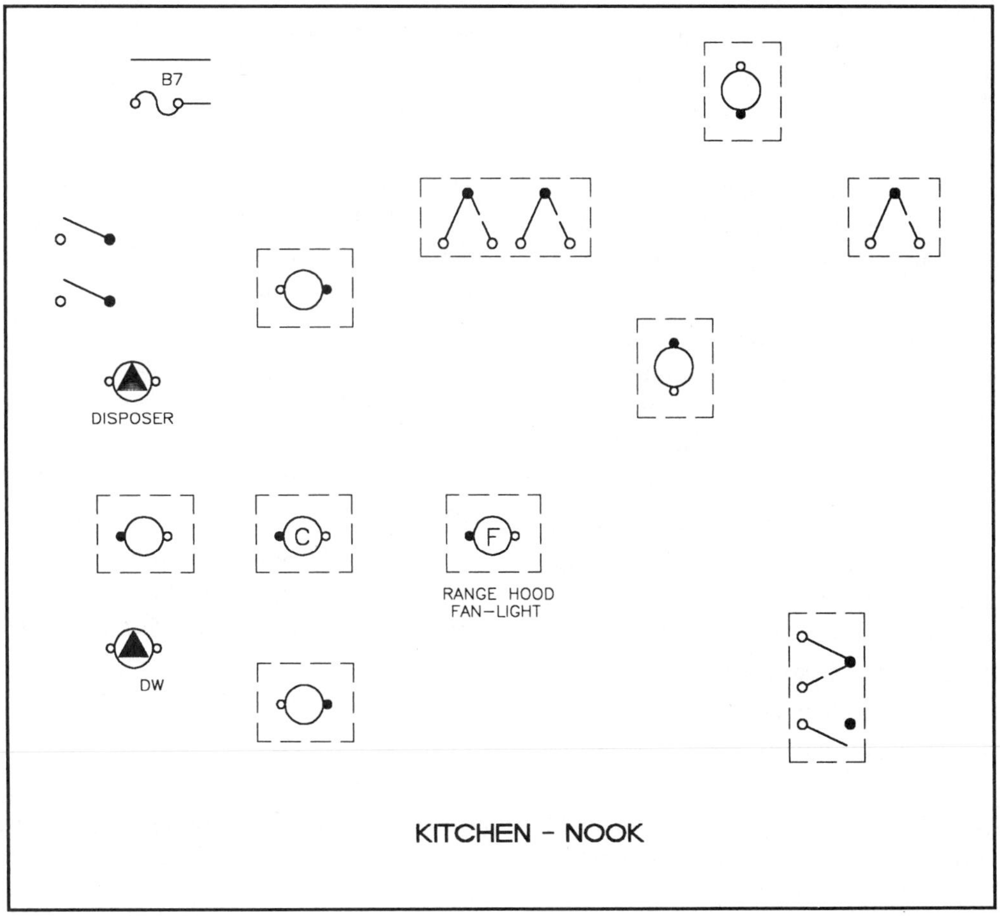

UNIT 13

Lighting Branch Circuit for the Living Room

OBJECTIVES

After studying this unit, the student will be able to

- understand various types of dimmer controls.
- connect dimmers to incandescent lamp loads and fluorescent lamp loads.
- understand the phenomenon of incandescent lamp inrush current.
- discuss Class P ballasts and ballast overcurrent protection.
- discuss the basics of track lighting.

LIGHTING CIRCUIT OVERVIEW

The feed for the living room lighting branch circuit is connected to Circuit B17, a 15-ampere circuit. The home run is brought from Panel B to the weatherproof receptacle outside of the living room next to the sliding door. The circuit is then run to the split-circuit receptacle just inside the sliding door, almost back-to-back with the outdoor receptacle, figure 13-1.

A three-wire cable is then carried around the living room feeding in and out of the eight receptacles. This three-wire cable carries the black and white circuit conductors plus the red wire, which is the switched conductor. Remember, since these receptacles will provide most of the general lighting for the living room through the use of floor lamps and table lamps, it is advantageous to have some of these lamps controlled by the three switches — two three-way switches and one four-way switch.

Note that the receptacle below the four-way switch is required according to the Code ruling which states that a receptacle outlet shall be installed in any wall space 2 feet (611 mm) or more in width, *Section 210-52(a)*. It is highly unlikely that a split-circuit receptacle connected for switch control, as are the other receptacles in the living room, will ever be needed, plus the fact that the distance to the Study/Bedroom receptacle is so short it makes economic sense to make up the connections as shown on the cable layout.

Accent lighting above the fireplace is in the form of two recessed fixtures controlled by a single-pole dimmer switch. See unit 7 for installation data and Code requirements for recessed fixtures.

Table 13-1 summarizes the outlets and estimated load for the living room circuit.

The spacing requirements for receptacle outlets and location requirements for lighting outlets are covered in Unit 3, as is the discussion on the spacing of receptacle outlets on exterior walls where sliding glass doors are involved. This is pointed out in *Section 210-52(a)* of the *National Electrical Code®*

TRACK LIGHTING (*Article 410, Part R*)

The plans show that track lighting is mounted on the ceiling of the living room on the wall opposite the fireplace.

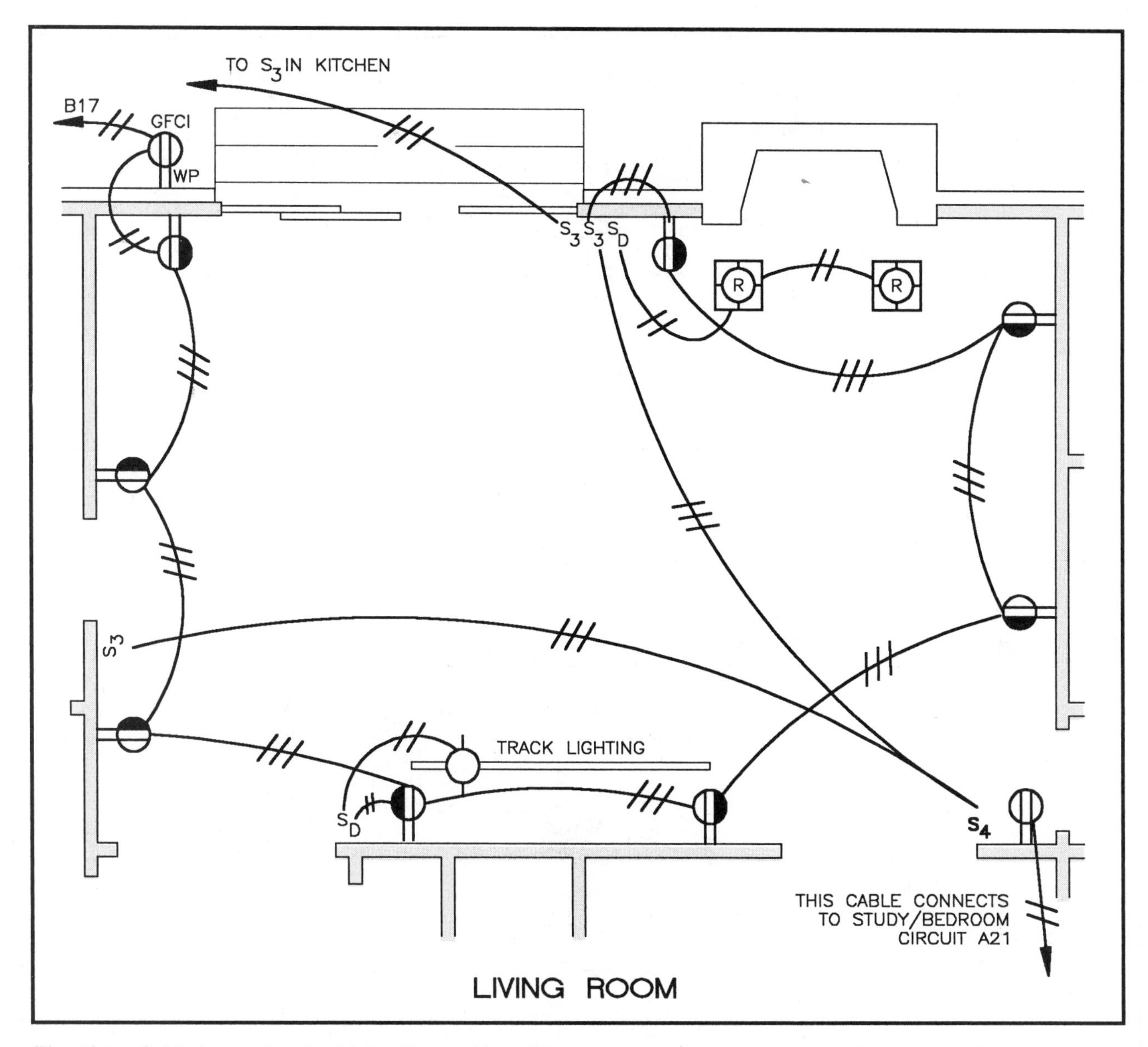

Fig. 13-1 Cable layout for the Living Room. Note that the receptacle on the short wall where the four-way switch is located is fed from the Study/Bedroom Circuit A21.

DESCRIPTION	QUANTITY	WATTS	VOLT-AMPERES
Receptacles @ 120 W Note: Receptacle under S_4 is connected to Study/Bedroom circuit.	8	960	960
Weatherproof receptacle	1	120	120
Track light five lamps @ 40 W each	1	200	200
Fireplace recessed fixtures @ 75 W each	2	150	150
TOTALS	12	1430	1430

Table 13-1. Living Room: outlet count and estimated load. Circuit B17.

Track lighting provides accent lighting for a fireplace, a painting on the wall, some item (sculpture or collection) that the homeowner wishes to focus attention on, or it may be used to light work areas such as counters, game tables, or tables in such areas as the kitchen nook.

Differing from recessed fixtures and individual ceiling fixtures that occupy a very definite space, track lighting offers flexibility because the actual lampholders can be moved and relocated on the track as desired.

Lampholders from track lighting selected from hundreds of styles are inserted into the extruded

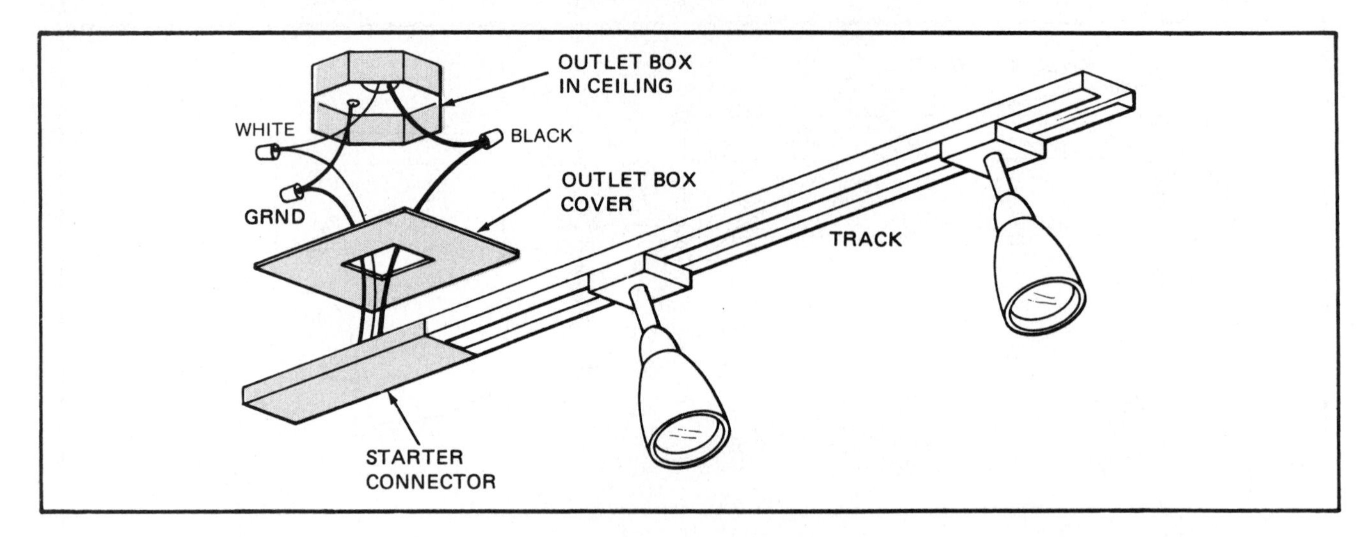

Fig. 13-2 End-feed track light.

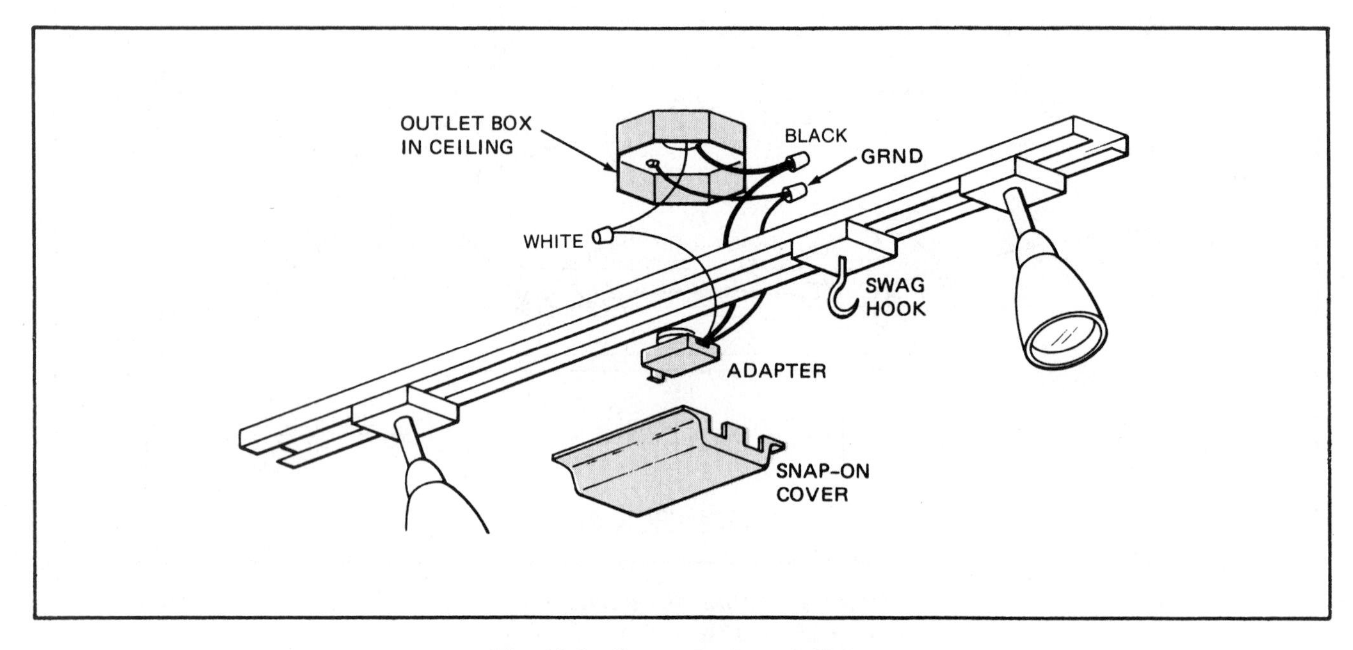

Fig. 13-3 Center-feed track light.

aluminum track at any point (the circuit conductors are in the track), and the plug-in connector on the lampholder completes the connection. In addition to being fastened to the outlet box, the track is generally fastened to the ceiling with toggle bolts or screws. Various track light installations are shown in figures 13-2, 13-3, and 13-4.

Note on the plans for the residence that the living room ceiling has wood beams. This presents a challenge when installing track lighting that is longer than the space between the wood beams, as is the case in this residence.

There are a number of things the electrician can do:

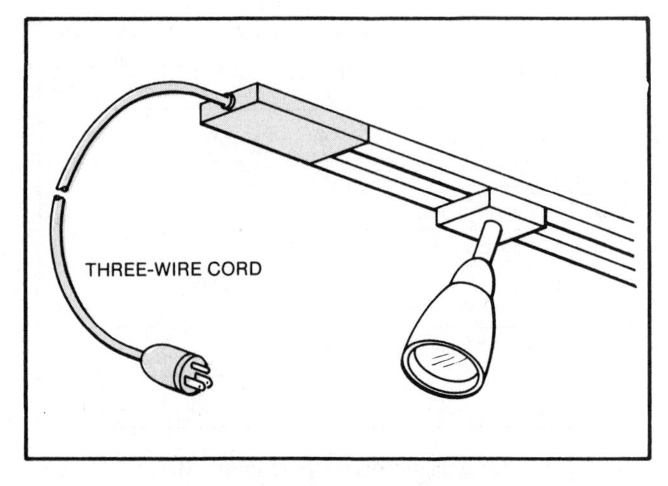

Fig. 13-4 Plug-in track light.

1. Install pendant kit assemblies that will allow the track to hang below the beam, figure 13-5.
2. Install conduit conductor fittings on the ends of sections of the track, then drill a hole in the beam through which 1/2-inch electrical metallic tubing can be installed and connected to the conduit connector fittings, figure 13-6. Running track lighting through walls or partitions is prohibited by the *National Electrical Code®, Section 410-101(c)(7).*
3. Run the branch-circuit wiring concealed above the ceiling to outlet boxes located at the points where the track is to be fed by the branch-circuit wiring, figure 13-7.

Where to Mount Track Lighting

The ceiling height, the aiming angle, the type of lampholder, the type of lamp, and the type of lighting to be achieved are all important factors that must be considered in order to install the lighting track at the proper distance from the wall.

For example, if we are designing an installation that will illuminate an oil painting that is to be centered 5′5″ (average eye level) from the floor, where the ceiling height is 9′0″, and the lampholder and lamp information tell us that the aiming angle should be 60°, we would find in the manufacturer's installation data that the track should be mounted 2 feet from the wall.

If, in this example, the ceiling height was 8′0″, then the distance to the track from the wall would be 18″.

Since all of the factors are variables that affect the positioning of the track, it is absolutely necessary to consult the track lighting manufacturer's catalog for recommendations that will achieve the desired results.

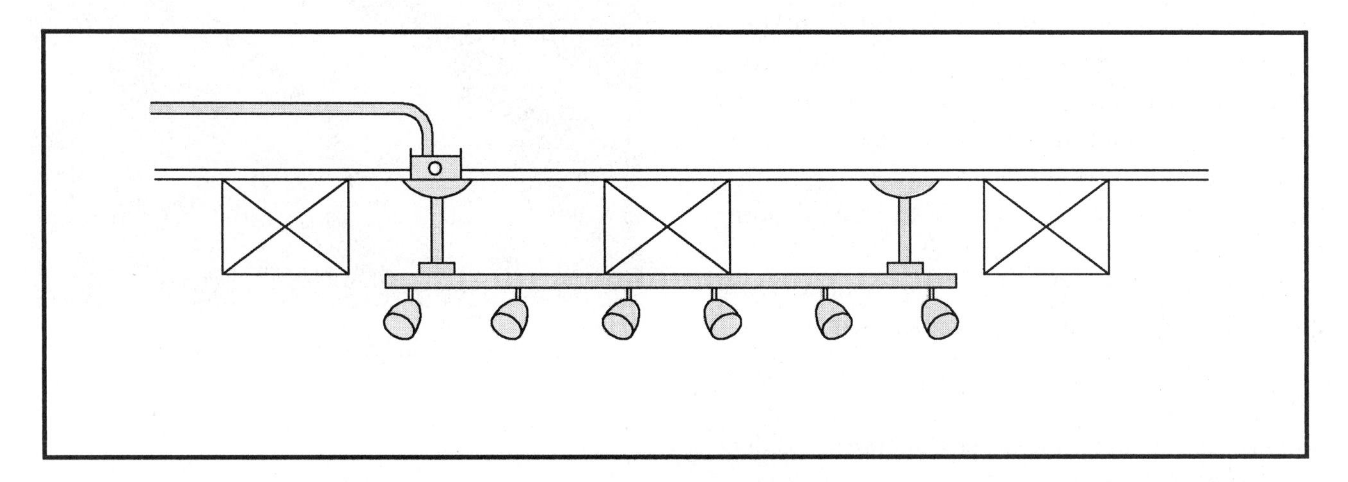

Fig. 13-5

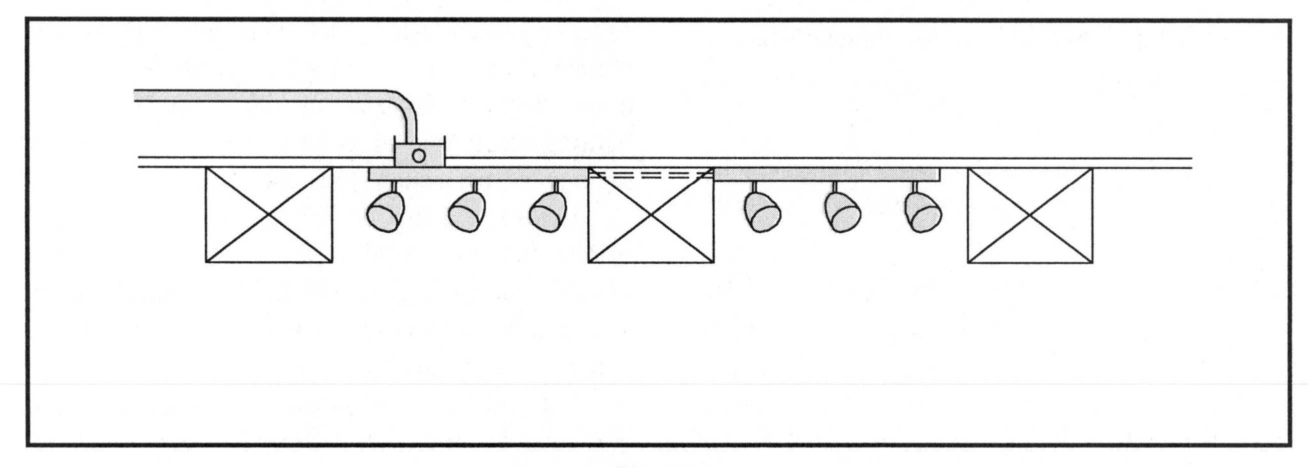

Fig. 13-6

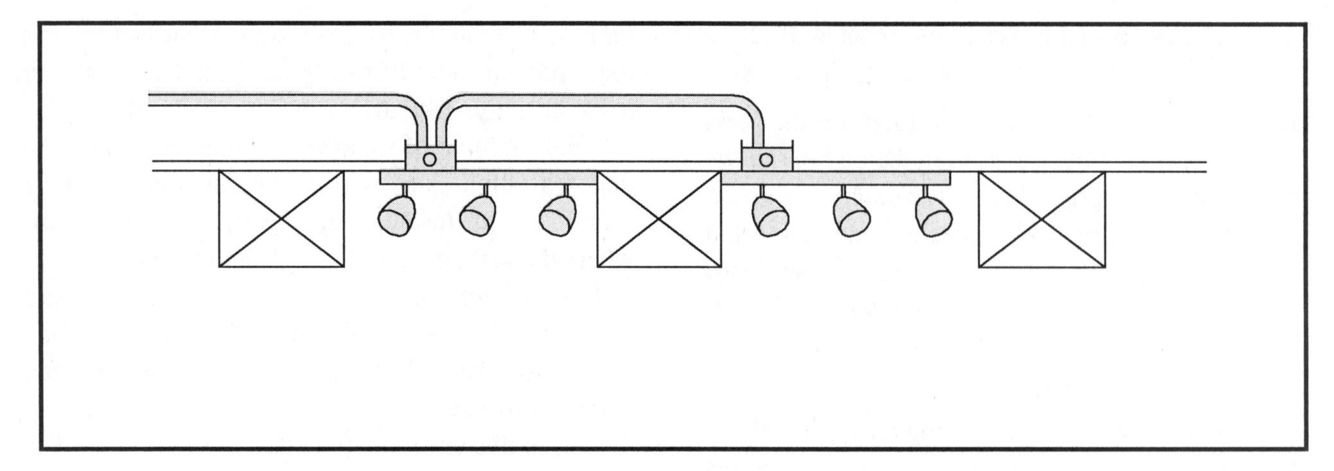

Fig. 13-7

Track lighting Code requirements are found in *Part R* of *Article 410* in the *National Electrical Code*®.

The track itself must be grounded properly to the outlet box. The plug-in connector also has a ground contact to insure that the track lighting fixture is safely grounded. Figure 13-8 shows tracks in cross-sectional views.

Typical fixtures that attach to track lighting systems are shown in figure 13-9.

The track lighting in the living room is turned on and off by a single-pole dimmer switch.

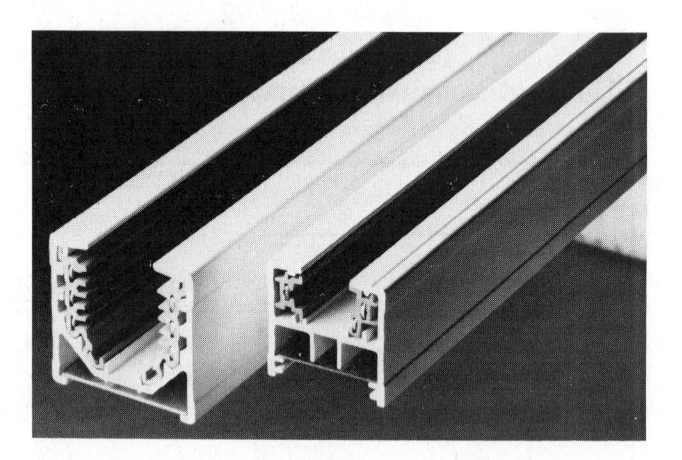

Fig. 13-8 Tracks in cross-sectional views.

Portable Lighting Tracks

Portable lighting tracks per Underwriters Laboratories' standards must:

- not be longer than 4 feet.
- have a cord no longer than 9 feet.
- have integral switching in the lampholders.
- have integral overcurrent protection in the lampholders.

The length of portable lighting tracks must not be altered in the field, per *Section 410-100*.

Supporting Track Lighting (per *Section 410-104*)

Lighting tracks that are 4 feet or less in length must be fastened at two places minimum. Then, if more track is needed, install at least one additional support for each individual extension that is 4 feet long or less.

Load Calculations for Lighting Track

► For track lighting installed in homes, it is not necessary to add additional loads for the purpose of branch-circuit, feeder, or service-entrance calculations. *Section 410-102* of the Code states that lighting track load is to be considered part of the regular VA-per-square-foot computations. ◄

However, if the lighting track is to be installed in a commercial building, then a loading factor of 180 VA for each 2 feet of lighting track must be added to the calculations for branch-circuit, feeder, and service-entrance computations.

For further information regarding track lighting, refer to *Part R* of *Article 410* of the *National Electrical Code*®.

DIMMER CONTROLS

Autotransformer Dimmer Control

One type of dimmer control has a continuously adjustable autotransformer which changes the lighting intensity by varying the voltage to the lamps, figure 13-10. When the knob of the transformer is rotated, a brush contact moves over a bared portion of the transformer winding to change the lighting intensity from complete darkness to full brightness. This type of control uses only the current required to produce the lighting desired.

Dimmer controls are used on 120-volt, 60-hertz, single-phase, alternating-current systems. The operation of such controls depends on transformer action. Thus, they will not operate on direct-current lines.

The electrician can prevent overloading of the dimmer control transformer by checking the total load in watts to be controlled by the dimmer against the rated capacity of the dimmer. The maximum wattage that can be connected safely is shown on the nameplate of the dimmer. In general, overcurrent protection is built into dimmer controls to prevent burnout due to an overload.

Instructions furnished with most dimmers state that they are for control of lighting only, and are not to be used to control receptacle outlets. The reasons for this are:

- Serious overloading of the dimmer will result, since it may not be possible to determine or limit what other loads might be plugged into the receptacle outlets.
- Reduced voltage supplying a television or radio, stereo components, home computer, or such motors as vacuum cleaners and food processors can result in costly damage to the appliance.

Because they are available in quite large wattage

Fig. 13-9 Typical fixtures that attach to track lighting systems.

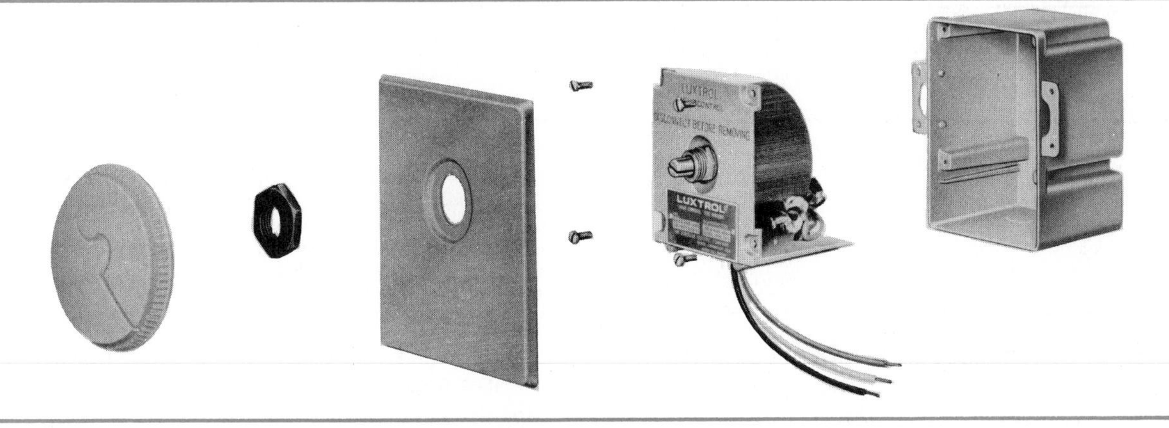

Fig. 13-10 Dimmer control, autotransformer type.

ratings, autotransformer dimmers are rather large physically when compared to solid-state dimmers. They require a special box that is furnished with the dimmer. Autotransformers are common in commercial establishments (offices, restaurants), whereas the smaller solid-state dimmer is most commonly used in residential installations.

Electronic Solid-State Dimmer Controls

According to their UL listing, solid-state electronic dimmers of the type sold for home use "are intended only for the control of permanently installed incandescent fixtures," figure 13-11. These dimmers are more compact than the autotransformer dimmers. For residential installations, the dimmer commonly used is a 600-watt unit which fits into a standard single-gang wall box. Both single-pole and three-way dimmers are available. The control knob is pushed to turn the lights on or off and is rotated to dim the lights. Compact, 1000-watt dimmer units are available for use in two-gang wall boxes. Figure 13-12 shows how single-pole and three-way solid-state dimmer controls are used in circuits.

Fig. 13-11 Solid-state dimmer control.

Dimmer Controls for Incandescent Lighting

The wiring diagram in figure 13-13 shows the circuit used to control incandescent lamps. Note that the entire load is cut off when the internal switch is in the OFF position.

Dimmer Controls for Fluorescent Fixtures

With slight changes in the wiring, dimmer controls can be used to control rapid-start, fluorescent lamps (type F40T12RS). Only special ballasts designed for dimmers can be used. These special ballasts energize the cathodes of the fluorescent lamp to a voltage that can maintain the proper operating temperature. The dimmer unit varies the current in the arc.

For fluorescent lamp installations, the electrician must install three conductors between the fluorescent dimming ballast and the dimmer control. The connections for a fluorescent lamp dimmer control are shown in figure 13-14.

The lamp reflector and the ballast case must be solidly grounded to insure proper operation of the lamps. If a nonmetallic reflector is used, a grounded metal strip must be mounted parallel to the lamp and within 1 inch (25.4 mm) of the lamp. This strip

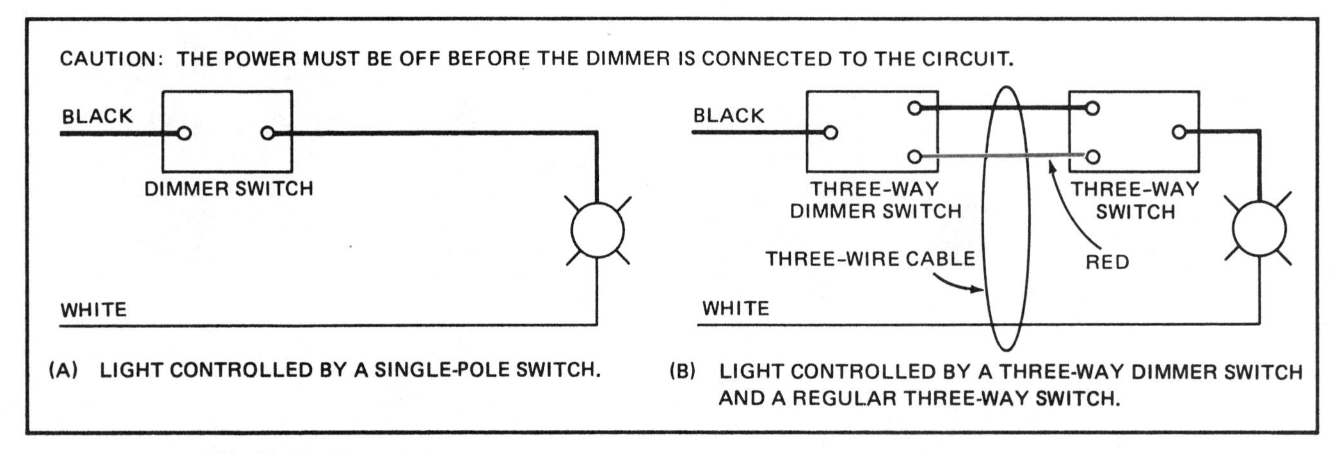

Fig. 13-12 Use of single-pole and three-way solid-state dimmer controls in circuits.

is about 1 inch (25.4 mm) wide and 45 inches (1.14 m) long. The dimming lead of the ballast must run to the controlled lead of the dimmer. Because incandescent lamps and fluorescent lamps have different characteristics, these lamps cannot be controlled simultaneously with a single dimmer control.

INCANDESCENT LAMP LOAD INRUSH CURRENTS

An unusual action occurs when a circuit is energized to supply a tungsten filament lamp load. A tungsten filament lamp has a very low resistance when it is cold. When the lamp is connected to the proper voltage, the resistance of the filament increases very quickly. This increase occurs within 1/240 second or one-quarter cycle after the circuit is energized. During this period, there is an inrush current that is 10 to 20 times greater than the normal operating current of the lamp.

In unit 5 it is stated that T-rated snap switches are available. If a T-rated switch is used in a circuit supplying tungsten filament lamps, then the momentary high surge of current has little effect on

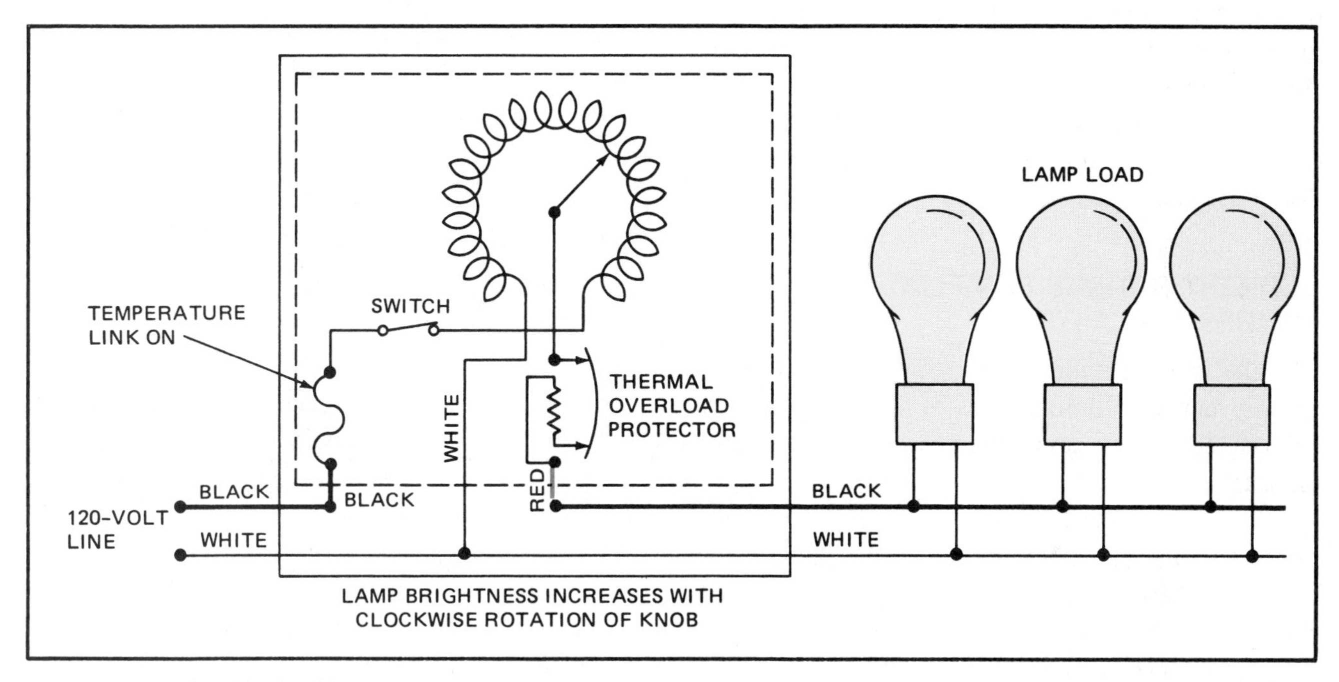

Fig. 13-13 Dimmer control (autotransformer) wiring diagram for incandescent lamps.

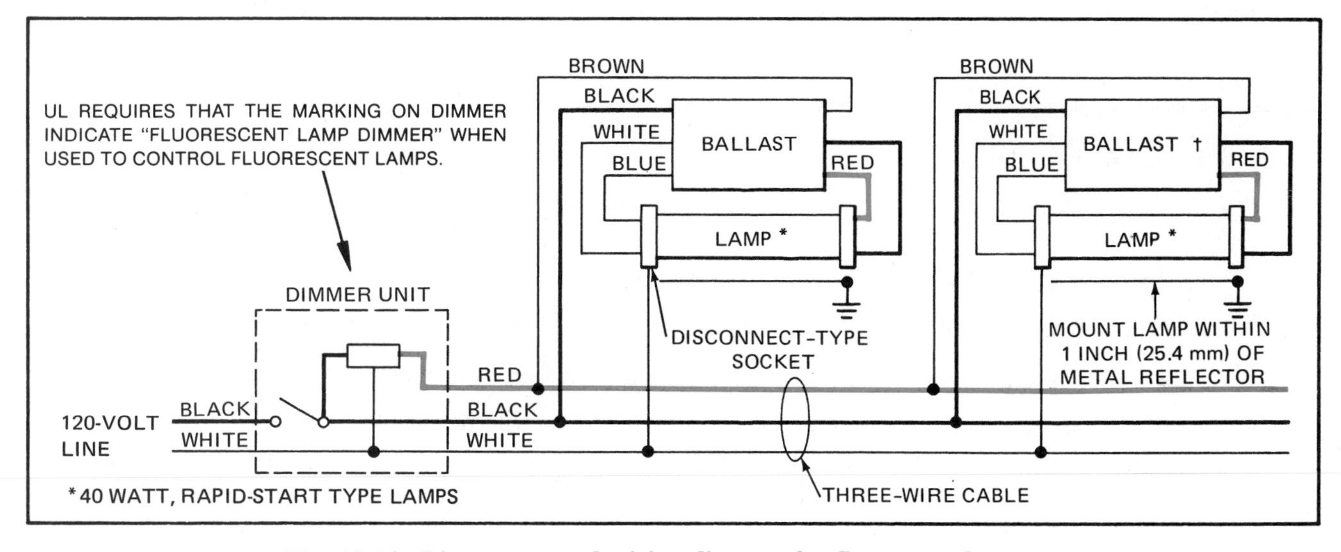

Fig. 13-14 Dimmer control wiring diagram for fluorescent lamps.

the branch circuit under ordinary conditions. However, tungsten filament lamps may be connected through a dimmer control. Assume that the dimmer is turned to a preset low (dim) setting and the circuit is turned on with a separate switch. The resulting inrush current lasts slightly longer due to the slower heating of the lamp filament.

This prolonged surge of current may cause certain types of circuit breakers to trip. These breakers have a highly sensitive magnetic tripping mechanism. When this problem occurs, the sensitive breaker must be replaced with one that has high magnetic tripping characteristics. Most manufacturers whose breakers react to the inrush current surge of tungsten filament lamp loads can supply the high magnetic tripping circuit breaker. Branch circuit fuses are not affected by the inrush current as discussed here.

CURRENT RATINGS FOR FLUORESCENT FIXTURE BALLASTS

The table in figure 13-15 shows the change in line current for different types of 40-watt dimming ballasts depending upon the power factor of the ballast.

Note: The designations A, B, C, and D in figure 13-15 are for reference only. These letters do not represent a particular model or type of ballast. The ratings are typical for dimming ballasts used with type F40T12 fluorescent lamps.

It is possible to overload a branch circuit if too many "low-power-factor" ballasts are connected to the circuit. This is not really a problem in residential installations, since many fluorescent lamps are not generally supplied by one circuit, but it is worth mentioning. Fluorescent loads are a good example of why volt-amperes, rather than watts, must be considered in order to find the current draw of the ballast.

Ballast	Line Current	Wattage	Line Power Factor
A	0.850	40	0.39 (39%)
B	0.450	40	0.74 (74%)
C	0.500	40	0.67 (67%)
D	0.550	40	0.61 (61%)

Fig. 13-15 Variation of line current for dimming ballasts, 120-volt, 40-watt lamps.

Let us make a comparison:

20 type A ballasts × 0.850 ampere each = 17 amperes
20 fluorescent lamps × 40 watts per lamp = 800 watts

$$\frac{W}{E} = \frac{800}{120} = 6.67 \text{ amperes}$$

Obviously, the correct current draw of these ballasts is 17 amperes, yet we could come to an incorrect solution (6.67 amperes) if we consider lamp wattage (40 in the example) only.

Low-power-factor ballasts are much cheaper (lower cost) than high-power-factor ballasts.

Thus, it is very important that the electrician check the current rating of all fluorescent ballasts (both dimming and regular) before the installation is started. The conductors and overcurrent devices must be rated to carry the connected load.

REVIEW

Note: Refer to the Code or the plans where necessary.

1. To what circuit is the living room connected? ____________
2. How many convenience receptacles are connected to the living room circuit? ______
3. a. How many wires enter the switch box at the four-way switch location? ______
   b. What type and size of box may be installed at this location? ____________

4. How many wires must be run between an incandescent lamp and its dimmer control? ______________________________

5. Complete the wiring diagram for the dimmer and lamp.

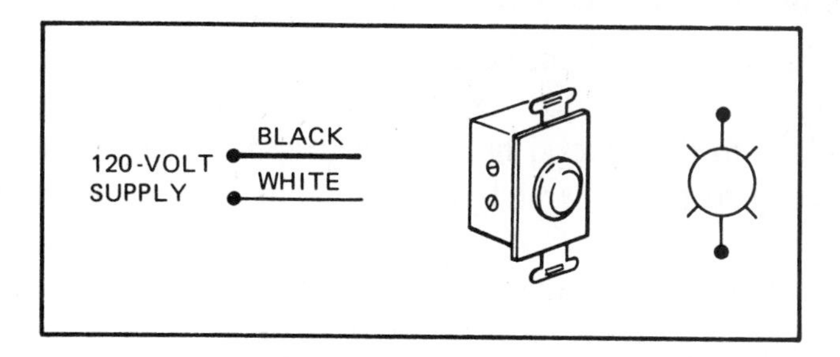

6. What is meant by incandescent lamp load inrush current? What type of switch is required? ______________________________

____

____

____

____

____

7. Is it possible to dim standard fluorescent ballasts? ______________

8. a. How many wires must be run between a dimming-type fluorescent ballast and the dimmer control? ______________________________

   b. Is a switch needed in addition to the dimmer control? ______________

____

____

____

____

9. Explain why fluorescent lamps having the same wattage draw different current values.

____

____

____

____

____

10. What is the total current consumption of the track lighting and recessed fixtures above the fireplace? Show your calculations.

11. How many television outlets are provided in the living room? ______________

12. Where is the telephone outlet located in the living room? ______________________

____

13. Solid-state electronic dimmers of the type sold for residential use (may) (may not) be used to control fluorescent fixtures. These dimmers (are) (are not) intended for speed control of small motors. Circle the correct answers.

14. When calculating a dwelling lighting load on the "volt-ampere per square foot" basis, the receptacle load used for floor lamps, table lamps, and so forth,
    a. must be added to the general lighting load at 1 1/2 amperes per receptacle.
    b. must be added to the general lighting load at 1500 volt-amperes for every 10 receptacles.
    c. is already included in the calculations as part of the volt-amperes per square foot.

15. A layout of the outlets, switches, dimmers, track lighting, and recessed fixtures is shown in the following diagram. Using the cable layout of figure 13-1, make a complete wiring diagram of this circuit. Use colored pencils or felt tip marking pens to indicate conductors.

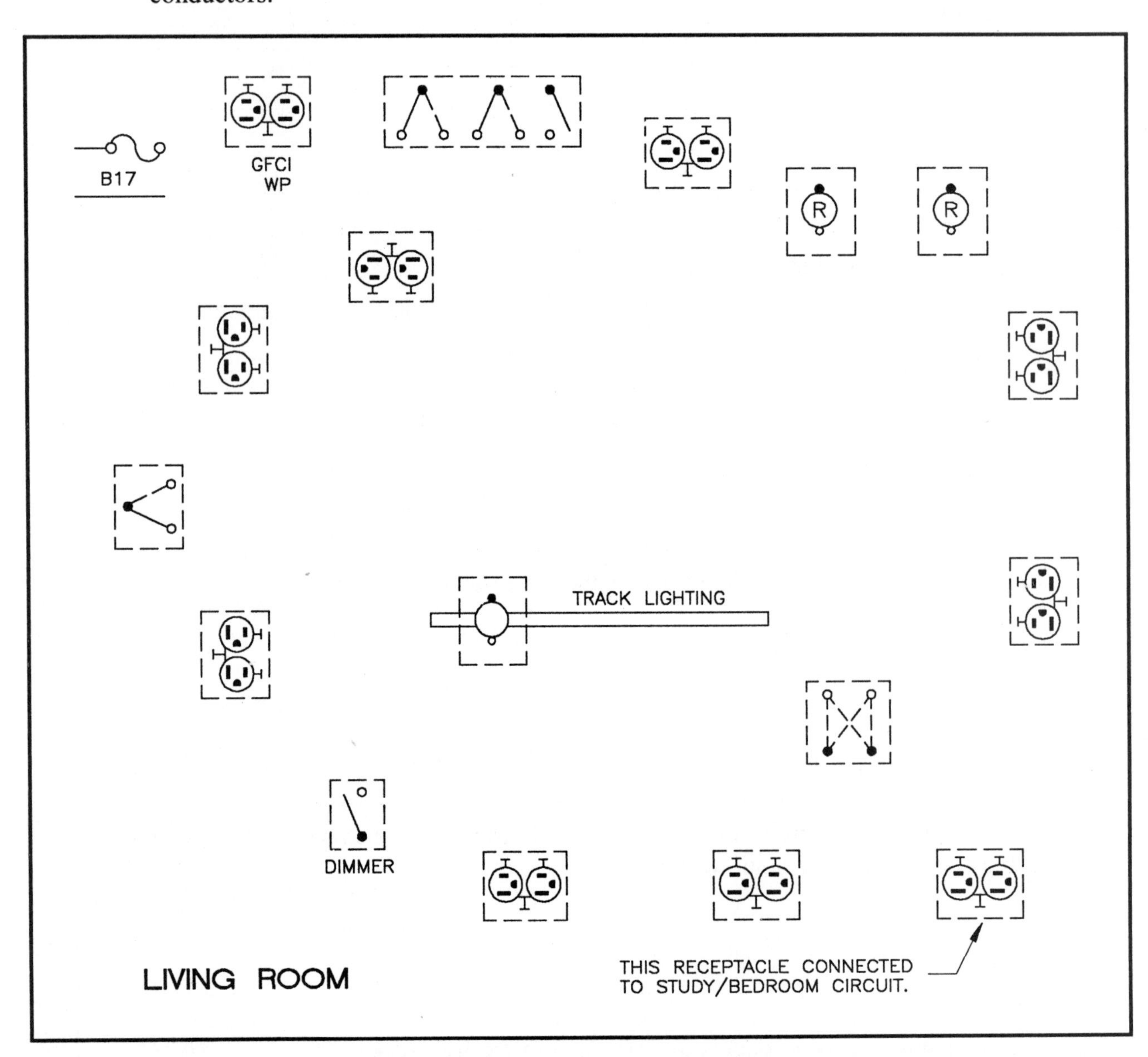

16. Prepare a list of television outlets, telephone outlets, and wiring devices shown on the plans and cable layout for the living room area. Include the number of each type present.

____
____
____
____
____
____
____
____

17. The living room features a sliding glass door. For the purpose of providing the proper receptacle outlets, answer the following statements *true* or *false*.

   a. Sliding glass panels are considered wall space. ____________

   b. Sliding glass panels are not considered wall space. ____________

   c. Fixed panels of glass doors are considered wall space. ____________

   d. Fixed panels of glass doors are not considered wall space. ____________

18. a. May lengths of track lighting be added when the track is permanently connected?

   ____

   b. May lengths of track lighting be added when the track is cord-connected?

   ____

19. Must track lighting always be fed (connected) at one end of the track? ________

____

UNIT 14

Lighting Branch Circuit for the Study/Bedroom

OBJECTIVES

After studying this unit, the student will be able to

- discuss valance lighting.
- make all connections in the Study/Bedroom for the receptacles, switches, fan, and lighting.

CIRCUIT OVERVIEW

The Study/Bedroom circuit is so named because it can be used as a study for the present, providing excellent space for home office, personal computer, or den. Should it become necessary to have a third bedroom, the changeover is easily done.

Circuit A21 originates at Main Panel A. This circuit feeds the Study/Bedroom at the receptacle in the living room just outside of the study. From this point, the circuit feeds the split-circuit receptacle just inside the door of the study. A three-wire cable runs around the room to feed the other four split-circuit receptacles. Figure 14-1 shows the cable layout for Circuit A21.

The black and white conductors carry the "live" circuit, and the red conductor is the "switch return" for the control of one portion of the split-circuit receptacle.

There are four switched split-circuit receptacles in the Study/Bedroom controlled by two three-way switches, plus one surge suppressor receptacle.

A paddle fan/light fixture is mounted on the ceiling. This installation is covered in unit 9. Table 14-1 summarizes the outlets and estimated load for the Study/Bedroom circuit.

VALANCE LIGHTING

Above the windows, indirect fluorescent lighting is provided. See figure 14-2, which illustrates one method of installing fluorescent valance lighting.

DESCRIPTION	QUANTITY	WATTS	VOLT-AMPERES
Receptacles @ 120 W each	6	720	720
Closet recessed fixture	1	75	75
Paddle fan/light	1		
Three 50-W lamps		150	150
Fan motor (0.75 A @ 120 V)		80	90
Valance lighting	1		
Two 40-W fluorescent lamps		80	90
TOTALS	9	1105	1125

Table 14-1. Study/Bedroom: outlet count and estimated load. Circuit A21.

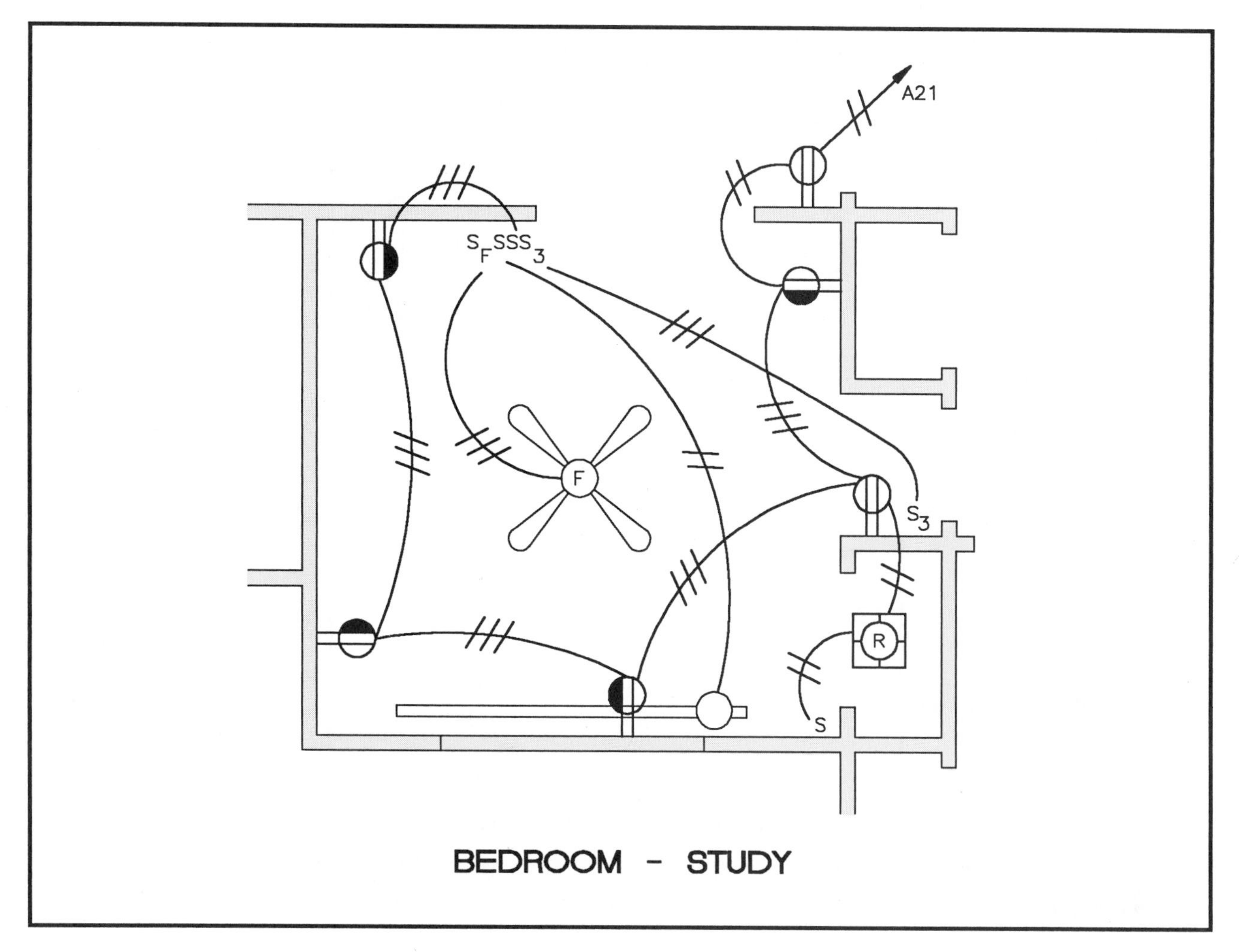

Fig. 14-1 Cable layout for the Study/Bedroom.

If the fluorescent valance lighting is to have a dimmer control, then special dimming ballasts would be required. See discussion on fluorescent lamp dimming in unit 13.

The Study/Bedroom wiring is rather simple, most of which has been covered in previous units in this text and need not be repeated here.

Figure 14-3 suggests one way to arrange the wiring for the paddle fan/light control, the valance lighting control, and the receptacle outlet control.

SURGE SUPPRESSORS

Since this room is intended initially to be utilized as a study (home office), it is very likely that personal computers will be located here. Transient voltage surge suppressors are covered in unit 6.

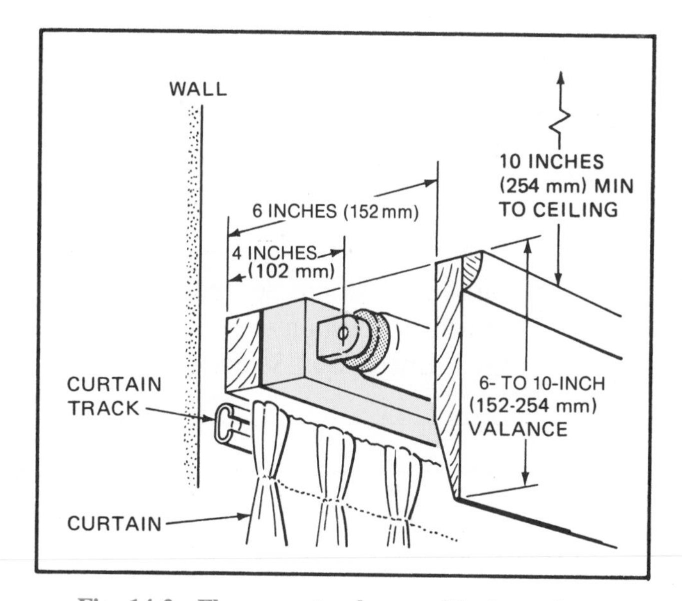

Fig. 14-2 Fluorescent valance with draperies.

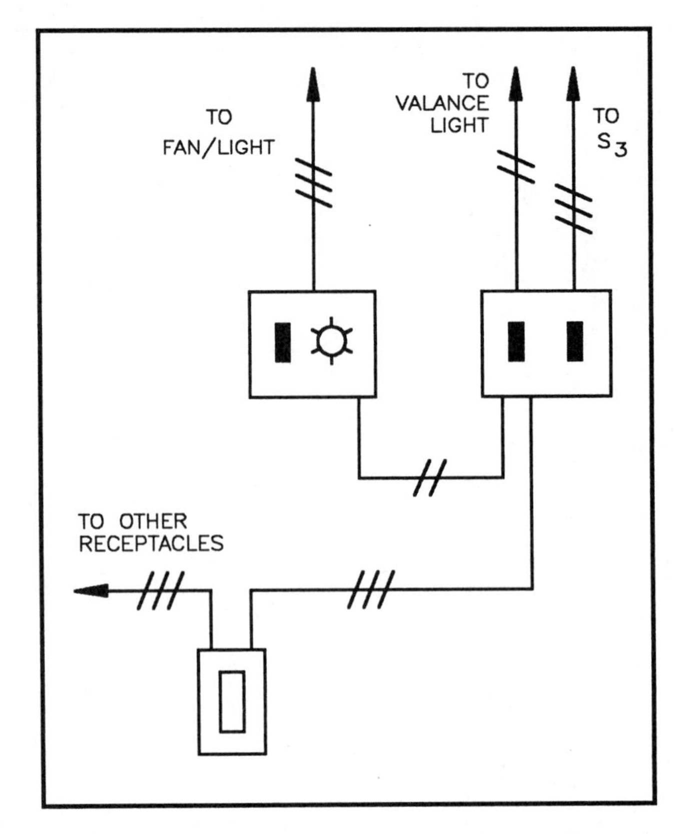

Fig. 14-3 Conceptual view of how the Study/Bedroom switch arrangement is to be accomplished, Circuit A21.

REVIEW

Note: Refer to the Code or plans where necessary.

1. Based upon the total estimated load calculations, what is the current draw on the Study/Bedroom circuit? ____________________

2. The Study/Bedroom is connected to circuit ____________________

3. The conductor size for this circuit is ____________________

4. Why is it necessary to install a receptacle in the wall space leading to the bedroom hallway? ____________________

5. Show calculations needed to select a properly sized box for the receptacle outlet mentioned in question 4. Nonmetallic-sheathed cable is the wiring method. Refer to the Quik-Check Box Selector Guide, unit 2.

6. Show calculations necessary to select a properly sized box for the receptacle outlet mentioned in question 4. The wiring method is electrical metallic tubing (EMT), usually referred to as "thin-wall conduit."

7. Prepare a list of *all* wiring devices used in Circuit A21.

8. Using the cable layout shown in figure 14-1, make a complete wiring diagram of Circuit A21.

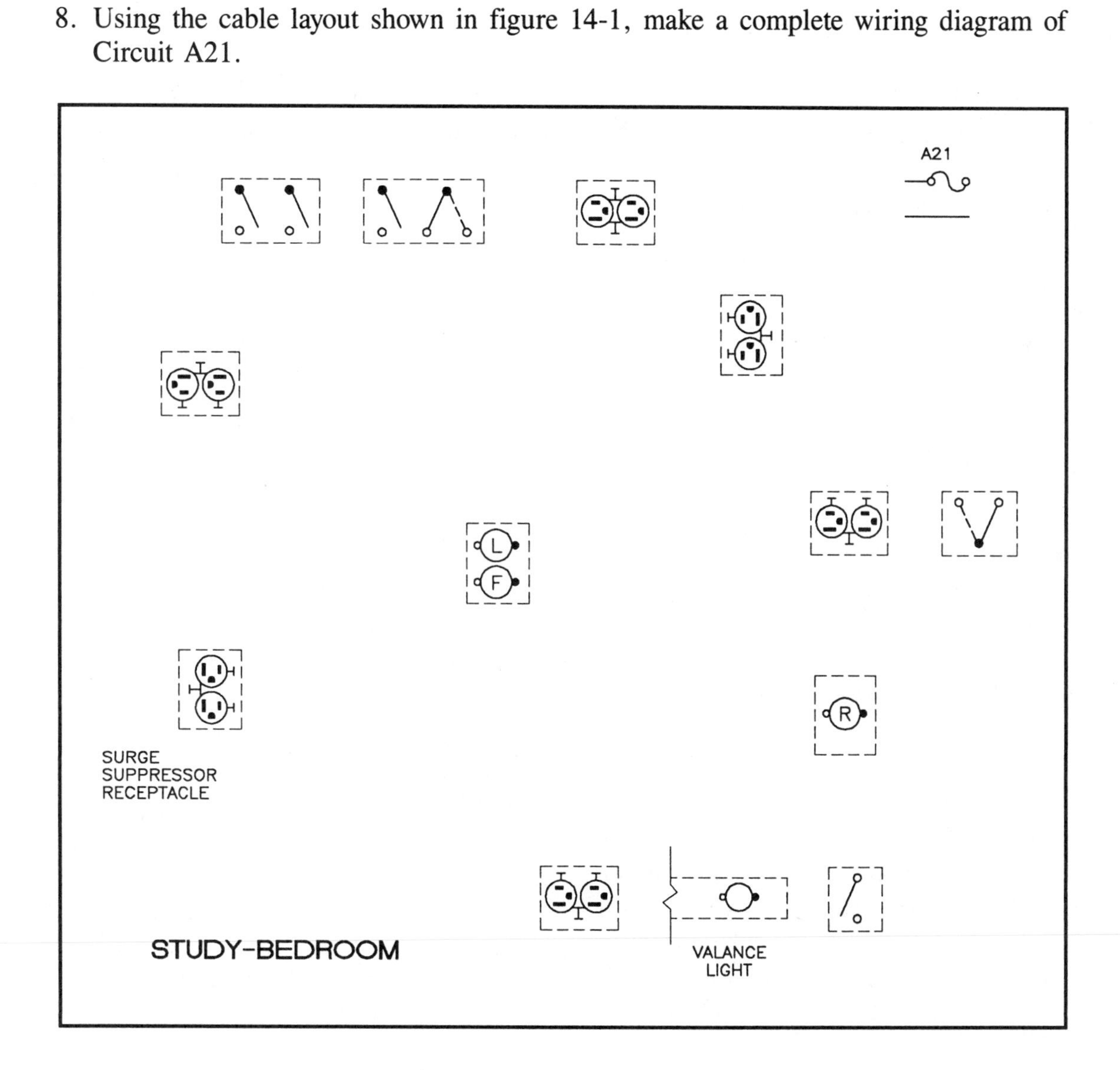

UNIT 15

Dryer Outlet and Lighting Circuit for the Laundry, Washroom, Rear Entry Hall, and Attic

OBJECTIVES

After studying this unit, the student will be able to

- make the proper wiring and grounding connections for large appliances, based on the type of wiring method used.
- list the Code regulations for using service-entrance cable to connect large appliances.
- understand the Code requirements governing the receptacle outlet(s) for laundry areas.
- calculate proper conduit sizing for conductors of the same size and for conductors of different sizes in one conduit.
- discuss the subject of "reduced size neutrals."
- discuss the Code rules pertaining to wiring methods in attics.
- demonstrate the proper way to connect pilot lights and pilot light switches.

DRYER CIRCUIT ⓐD

A separate circuit is provided in the laundry room for the electric clothes dryer. This appliance demands a large amount of power. The special circuit provided for the dryer is indicated on the plans by the symbol ⓐD. The dryer is connected to Circuit B(1–3).

The Clothes Dryer

Clothes dryer manufacturers make many dryer models with different wiring arrangements and connection provisions. All electric dryers have electric heating units and a motor-operated drum which tumbles the clothes as the heat evaporates the

DESCRIPTION	QUANTITY	WATTS	VOLT-AMPERES
Washroom receptacle (GFCI)	1	120	120
Rear entry hall receptacle	1	120	120
Washroom vanity fixture	1	200	200
Rear entrance hall recessed ceiling fixtures @ 100 W each	2	200	200
Laundry room fluorescent fixture Four 40-W lamps	1	160	180
Attic lights @ 75 W each	4	300	300
Laundry exhaust fan	1	80	90
Washroom exhaust fan	1	80	90
TOTALS	12	1260	1300

Table 15-1. Laundry–Rear Entrance–Washroom–Attic. Circuit B10.

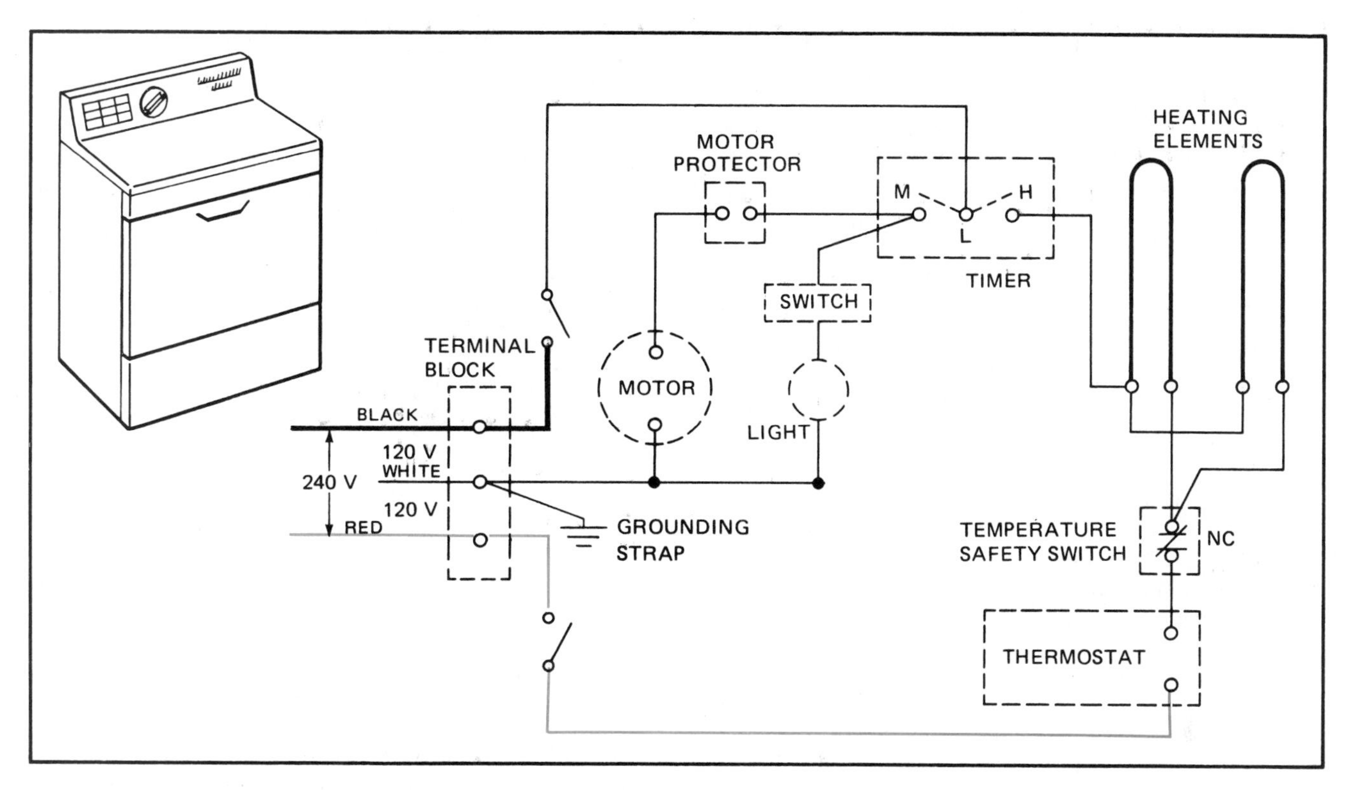

Fig. 15-1 Laundry dryer: wiring and components.

moisture. Dryers also have thermostats to regulate the temperature of the air inside the dryer. Timers are used to regulate the lengths of the various drying cycles. Drying time is determined by the type of fabric and can be set to a maximum of about 85 minutes. Dampness controls are often provided to stop the drying process at a preselected stage of dampness so that it is easier to iron the clothes.

Clothes dryers create large amounts of humid air. Therefore, be sure that proper ventilation is provided. This residence has a through-the-wall exhaust fan located in the laundry room.

Table 15-1 shows the estimated loads for the various receptacles and fixtures in the laundry, rear entry, washroom, and attic.

Electric clothes dryers operate from a 120/240-volt, single-phase, three-wire circuit. Be sure that the circuit has adequate capacity to handle the load.

Dryer Connection Methods

Electric clothes dryers can be connected in several ways. The method used depends largely on local codes. The local electrical inspector can provide information about the proper wiring procedure. Figure 15-1 shows the internal components and wiring of a typical dryer circuit.

One method of connecting the dryer is to run a three-wire armored cable to a junction box on the dryer provided by the manufacturer, figure 15-2. The cable conductors are connected to their corresponding dryer terminals in this junction box.

Another method of connecting a clothes dryer is to run conduit (EMT) to a point just behind the dryer. Flexible conduit from 24 to 36 inches (610 mm to 914 mm) in length is installed between the dryer junction box and the EMT, figure 15-3. The transition from the EMT to the flexible conduit

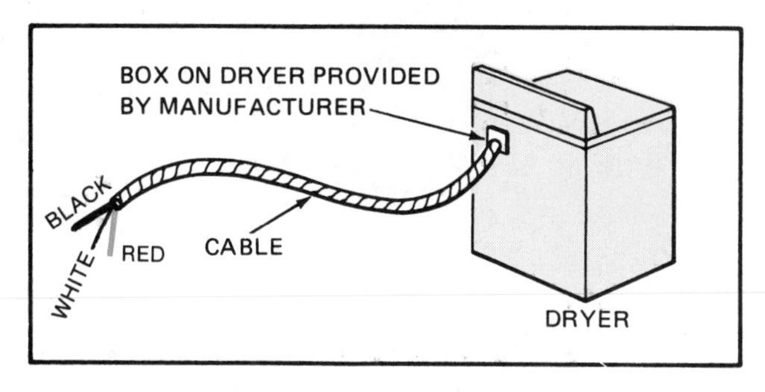

Fig. 15-2 Dryer connected by armored cable.

is made using a fitting such as the combination coupling shown in figure 15-3. This flexible connection allows the dryer to be moved for cleaning and servicing without disconnecting the wiring.

Still another method of connecting a clothes dryer is to install a receptacle on the wall at the rear of the dryer. A cord set is connected to the dryer, figure 15-4, and is plugged into the receptacle. This method satisfies the requirements of *Section 422-22(a)* for a disconnecting means.

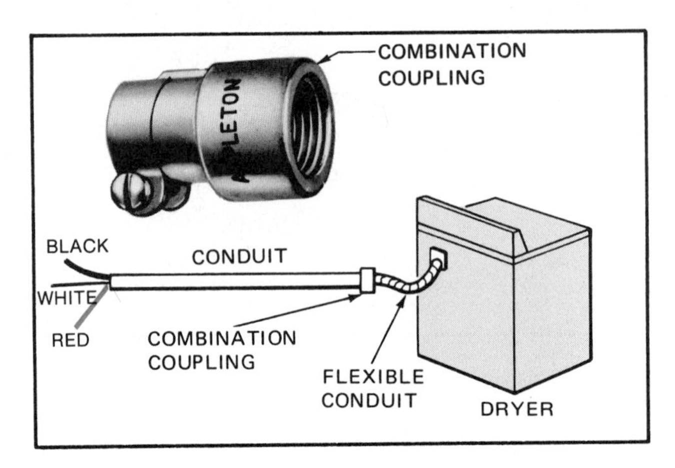

Fig. 15-3 Dryer connected by EMT and flexible metal conduit.

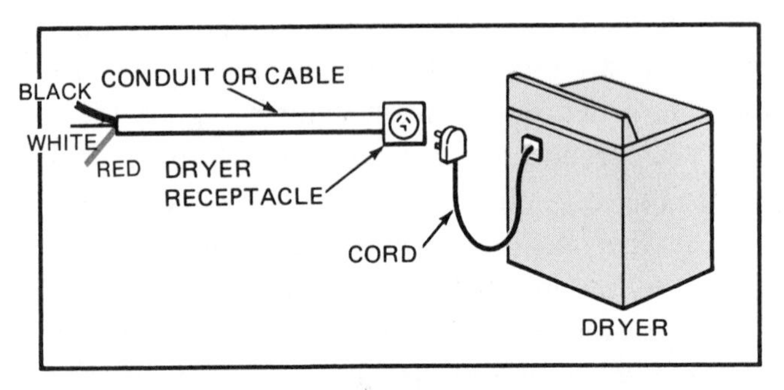

Fig. 15-4 Dryer connection using a cord set.

The receptacle can be wired with cable or conduit if concealed wiring is desired. EMT should be used if the wiring is to be exposed. In the residence, the wiring method for the dryer is nonmetallic-sheathed cable which runs from Panel B to a dryer receptacle on the wall behind the dryer. A cord set is attached to the dryer and is plugged into this receptacle. The receptacle and cord must have a current-carrying capacity not less than that of the attached appliance.

A standard dryer receptacle and cord set is shown in figure 15-5. Such a set is usually rated at 30 amperes, 250 volts. Cords and receptacles rated for higher currents are available. The L-shaped grounding and neutral slot of the outlet does not accept a 50-ampere cord set.

The surface-mounting dryer outlet, figure 15-5(A), can be wired directly with cable or conduit. The flush-mounting outlet, figure 15-5(B), will fit single-gang or two-gang sectional switch boxes. Figure 15-5(C) shows the cord set used with this receptacle. The outlet will also fit a 4-inch square or 4 11/16-inch square box if the proper single-gang or two-gang covers are used.

Code Requirements for Connecting a Dryer (*Article 422*)

The information necessary to install a dryer or any appliance is contained in *Article 422* of the *National Electrical Code®*.

In this residence, the dryer is installed in the laundry room. The schedule of special-purpose outlets in the specifications shows that the dryer is rated at 5700 watts and 120/240 volts. The schematic wiring diagram, figure 15-1, shows that the heating elements are connected to the 240-volt ter-

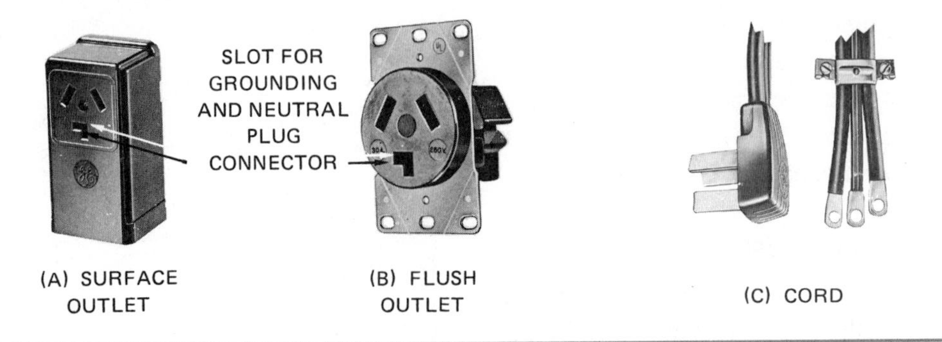

Fig. 15-5 Typical dryer receptacle and cord.

minals of the terminal block. The dryer motor and light are connected between the hot wire and the neutral terminal, 120 volts.

The motor of the dryer has integral thermal protection. This thermal protector prevents the motor from reaching dangerous temperatures as the result of an overload or failure to start. *Section 422-27* lists overcurrent protection requirements. This section also refers to parts of *Article 430.*

To determine the feeder or branch-circuit rating, *Section 220-18* requires that the minimum load included for a dryer be 5000 watts (volt-amperes) or the nameplate rating of the dryer, whichever is larger.

The electric clothes dryer in this residence has a nameplate rating of 5700 watts. Thus

$$I = \frac{W}{E} = \frac{5700}{240} = 23.75 \text{ amperes}$$

Conductor Size

Section 422-4(a) of the Code requires that individual circuit conductors supplying an appliance must not be less than the "marked" rating of the appliance. When the appliance has load in addition to the motor, as in the case of a clothes dryer (motor and heater), then *Section 422-4(a)* makes reference to *Section 422-32*, which says that the nameplate marking on the appliance must "specify the minimum circuit size and the maximum rating of the circuit overcurrent protective device."

Because the manufacturer of the appliance provides the circuit requirements on the nameplate and in the installation manual, it is a simple matter to determine the proper wire size based upon this rating.

Section 210-22(a) further states that for circuits supplying motor operated equipment plus "other loads," the computation is based on:

> 125% of the motor's full-load ampere rating plus the other loads.

On a clothes dryer it is very difficult for the installer to check the motor rating and the heating element rating. It is much easier to follow the rating requirements found on the nameplate of the appliance. If the branch circuit rating is *not* marked on the appliance, *Section 422-5(a)*, Exception No. 1, refers the reader to *Part B* of *Article 430* (Motors). Here again we find the fundamental 125% rule. That is, take the current rating times 125% to find the required circuit ampacity. Therefore,

$$23.75 \text{ amperes} \times 1.25 = 29.7 \text{ (30) amperes}$$

According to *Table 310-16,* the ampacity for No. 10 THHN conductors is 35 amperes, and the overcurrent device is sized at 30 amperes.

The dryer Circuit B(1–3) is a three-wire, 30-ampere, 240-volt circuit. The conductors in the nonmetallic-sheathed cable are No. 10 AWG.

Where conduit is used as the wiring method, *Table 3A, Chapter 9* of the Code shows that three No. 10 THHN conductors require 1/2-inch conduit.

The neutral conductor may be smaller than the phase wires because it carries the current of the motor, lamp, and timer only. In most cases, these are connected "line-to-neutral."

Section 210-19(a), first sentence, states that "branch-circuit conductors shall have an ampacity not less than the maximum load to be served."

Section 220-22 permits the neutral load for "feeders" to household electric ranges, wall-mounted ovens, counter-mounted cooking units, and electric dryers to be based upon 70% of the load on the ungrounded conductors. This 70% figure has been used in computing the service-entrance conductors for this residence as well as the feeder to Panel B.

By Code definition, a branch circuit refers to the circuit conductors between the final overcurrent device and the outlet. Therefore, the conductors between the overcurrent device in Panel A and the dryer outlet constitute the dryer's branch circuit. Knowing that the neutral conductor of a branch circuit supplying an electric clothes dryer carries only the motor, lamp, and timer loads, most electricians apply the 70% rule as stated in *Section 220-22* for sizing the neutral conductor.

In this example,

$$23.75 \times 0.70 = 16.63 \text{ amperes}$$

Therefore, the neutral conductor could be a No. 12 THHN wire. See *Table 310-16* and the *Footnotes* to the table.

Accordingly, the dryer in this residence could be connected as follows:

Phase (hot) wires — two No. 10 THHN
Neutral wire — one No. 12 THHN
} 1/2-inch conduit

Overcurrent Protection

Section 422-5 states that the branch-circuit overcurrent device rating is not to exceed the overcurrent device rating marked on the appliance. Here, again, the manufacturer of the appliance will so mark the circuit overcurrent device rating. We again apply the 125% rule:

23.75 amperes × 1.25 = 29.7 (30) amperes

Grounding Frames of Ranges and Dryers

Section 250-60 states that a dryer may be grounded directly to the grounded neutral conductor if the neutral is no smaller than No. 10 AWG copper, figure 15-6. If a three-wire cord set is used to install a dryer, the cord and conductors supplying the receptacle must have a neutral conductor no smaller than No. 10 AWG.

When the dryer is connected with flexible metal conduit, the metal conduit provides the necessary equipment ground.

Two-conductor, No. 10 AWG nonmetallic-sheathed cable containing a bare grounding conductor may be used for a dryer connection if the grounding conductor is No. 10 AWG.

The branch-circuit conductors for the dryer are connected to Panel B, Circuit B(1–3). This is a double-pole circuit which breaks only the two ungrounded conductors. The neutral conductor remains solid and unbroken.

Alternate Method of Connecting Dryers and Ranges: Service-Entrance Cable

In previous units, when cable was used as the means of wiring the general lighting and small appliance branch circuits, it was assumed that the cable was either armored or nonmetallic-sheathed cable. The residence specifications require that conduit be installed in certain areas. *Section 338-3* of the *National Electrical Code®* recognizes another wiring method that may be used under certain conditions.

To install service-entrance cable with an *insulated* neutral, the cable may start from the service-entrance equipment or a subpanelboard. If the other requirements of *Section 250-60* are met, then the metal frame of the range or clothes dryer may be grounded to this neutral.

When service-entrance cable with an *uninsulated* neutral is used to connect a range or clothes dryer, the cable must originate at the service-entrance equipment, *Section 250-60(c)* and *Section 338-3(b)*. The metal frame of a range or clothes dryer may be grounded to this neutral if the other requirements of *Section 250-60* are met, figure 15-7.

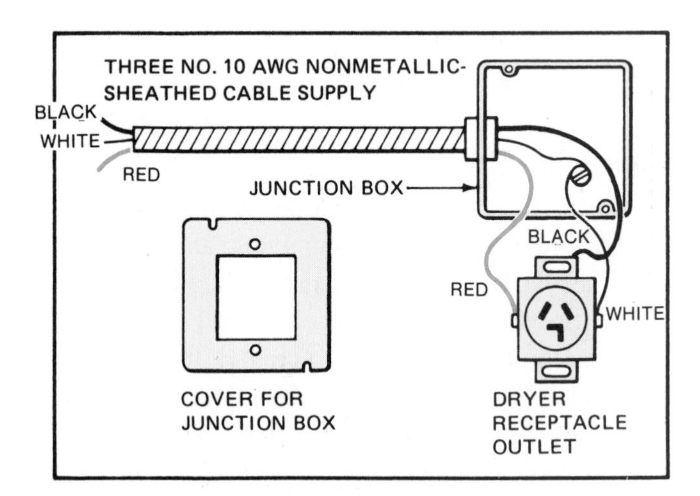

Fig. 15-6 It is permitted to ground the junction box to the neutral only when the box is part of the circuit for an electrical dryer or a range, *Section 250-60.*

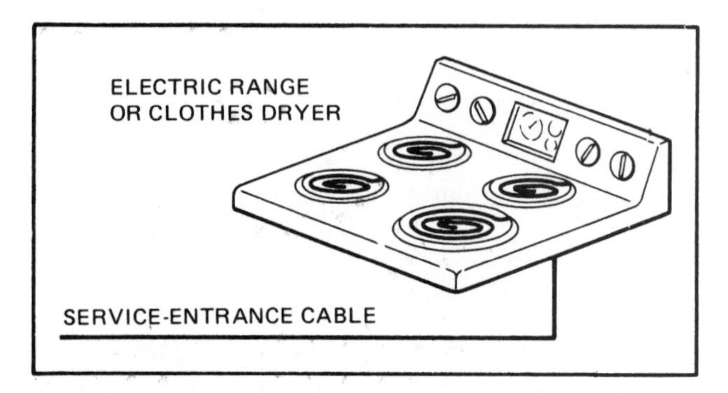

Fig. 15-7 An electric range or a clothes dryer may be connected to service-entrance cable. If the cable has an uninsulated neutral, the cable must originate at the service-entrance equipment, *Sections 250-60(c)* and *338-3(b)*.

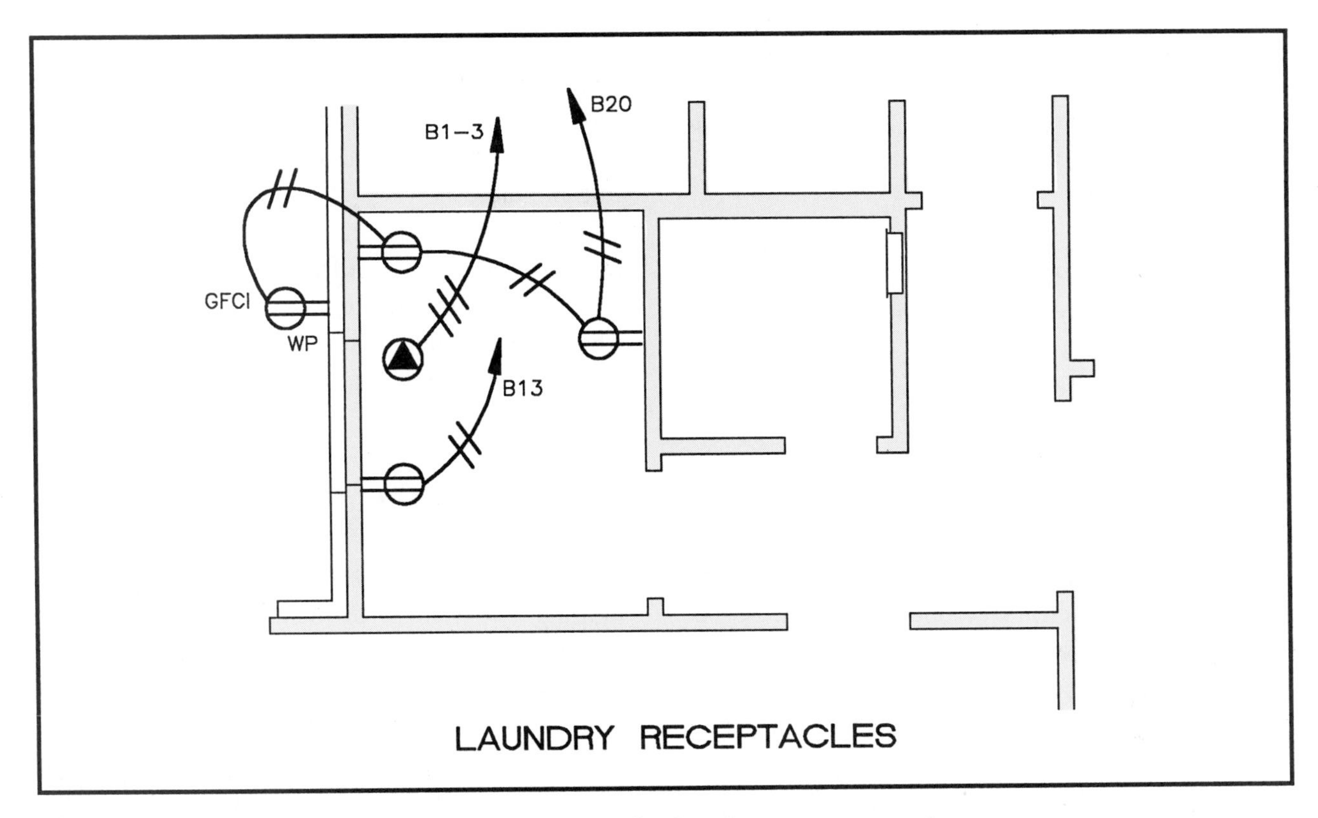

Fig. 15-8 Cable layout for laundry room receptacles.

RECEPTACLE OUTLETS — LAUNDRY

The Code in *Section 220-4(c)* requires that a separate 20-ampere branch circuit be provided for the laundry equipment and this circuit "shall have no other outlets."

Section 210-50(c) requires that a receptacle outlet must be installed within 6 feet of the intended location of the washer.

Section 210-52(f) states that *at least* one receptacle outlet must be installed for the laundry equipment.

Section 220-16(b) requires that 1500 volt-amperes must be included for each two-wire branch circuit serving laundry equipment. This is for the purpose of computing the size of service-entrance and/or feeders. See the complete calculations in unit 28.

In the laundry room in this residence, the 20-ampere Circuit B18 supplies the receptacle for the washer. Circuit B18 serves no other outlets. The receptacle is not required to be GFCI protected. Figure 15-8 shows the cable layout for the laundry room.

Circuit B20 connects to the two receptacles in the laundry, which will serve an iron, sewing machine, or other small portable appliances. Circuit B20 also serves the weatherproof receptacle on the outside wall of the laundry area.

Circuits B18 and B20 are included in the calculations for service-entrance and feeder conductor sizing at a computed load of 1500 volt-amperes per circuit (*Section 220-16(b)*). This load is included with the general lighting load for the purpose of computing the service-entrance and feeder conductor sizing and is subject to the demand factors that are applicable to these calculations. See the complete calculations in unit 28.

LIGHTING CIRCUIT

The general lighting circuit for the laundry, washroom, and hall area is supplied by Circuit B10. Note on the cable layout, figure 15-9, that in order to help balance loads evenly, the attic lights are also connected to Circuit B10.

The types of vanity lights, ceiling fixtures, and

standard wall receptacle outlets, as well as the circuitry and switching arrangements, are similar to types already discussed in other units of this text and will not be repeated.

Exhaust Fans

Ceiling exhaust fans are installed in the laundry and in the washroom. These exhaust fans are connected to Circuit B10.

The exhaust fan in the laundry will remove excess moisture resulting from the use of the clothes washer.

Exhaust fans may be installed in walls or ceilings, figure 15-10. The wall-mounted fan can be adjusted to fit the thickness of the wall. If a ceiling-mounted fan is used, sheet-metal duct must be installed between the fan unit and the outside of the house. The fan unit terminates in a metal hood or grille on the exterior of the house. The fan has a shutter which opens as the fan starts up and closes as the fan stops. The fan may have an integral pull-chain switch for starting and stopping, or it may be used with a separate wall switch. In either case, single-speed or multispeed control is available. The fan in use has a very small power demand, 90 volt-amperes.

To provide better humidity control, both ceiling-mounted and wall-mounted fans may be controlled with a humidistat. This device starts the fan when the humidity reaches a certain value. When the humidity drops to a preset level, the humidistat turns the fan off.

ATTIC LIGHTING AND PILOT LIGHT SWITCHES

► *Section 210-70(a)* requires that at least one lighting outlet must be installed in an attic, and that this lighting outlet must be controlled by a switch located near the entry to the attic. This section of the Code addresses attics that are used for storage or attics that contain equipment that might need servicing. An example of such equip-

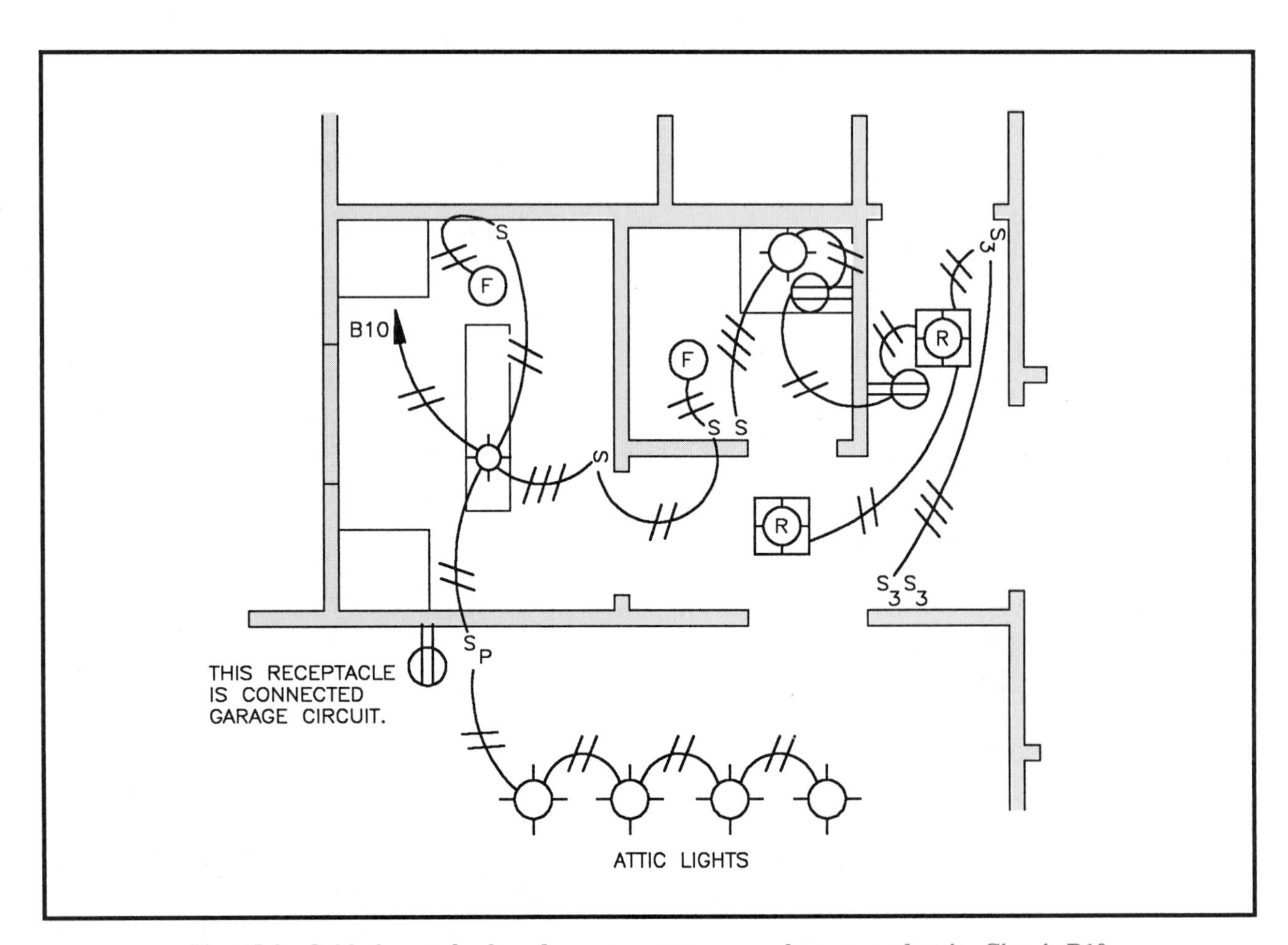

Fig. 15-9 Cable layout for laundry, rear entrance, washroom, and attic, Circuit B10.

ment would be air-conditioning equipment. *Section 210-70(a)* requires that the lighting outlet(s) be installed near this equipment.

Section 210-63 requires that a 125-volt, single-phase, 15- or 20-ampere receptacle outlet be installed in an accessible location within 75 feet (22.82 m) of air-conditioning or heating equipment in attics, in crawl space, or on the roof. Connecting this receptacle outlet to the load side of the equipment's disconnecting means is not permitted because this would mean that if you turned off the equipment, the power to the receptacle would also be off and thus would be useless for servicing the equipment. The exception to *Section 210-63* exempts requiring a receptacle outlet on the roof of one- and two-family dwellings, although installing a receptacle outlet would certainly be a good idea.

Reading *Section 210-70(a)* closely tells us that a switch controlled lighting outlet is required *only* if the attic, underfloor space, utility room, and basement are used for storage or contain equipment that might require servicing.

Reading *Section 210-63* closely tells us that a 125-volt, single-phase, 15- or 20-ampere receptacle outlet is required on rooftops, in attics, and in crawl spaces *only* if these areas contain heating, air-conditioning, and/or refrigeration equipment.

The residence discussed in this text does not have air-conditioning, heating, or refrigeration equipment in the attic or on the roof. ◄

The porcelain lighting fixtures installed in the attic of the residence are available with a receptacle outlet for convenience in plugging in an extension cord. However, most electrical inspectors (authority having jurisdiction) would not accept the porcelain lampholders receptacle in lieu of the required receptacle outlet as stated in *Section 210-63*.

The four porcelain lampholders in the attic are turned on and off by a single-pole switch on the garage wall close to the attic storable stairway. Associated with this single-pole switch is a pilot light. The pilot light may be located in the handle of the switch or it may be separately mounted. Figure 15-11 shows how pilot lamps are connected in circuits containing either single-pole or three-way switches.

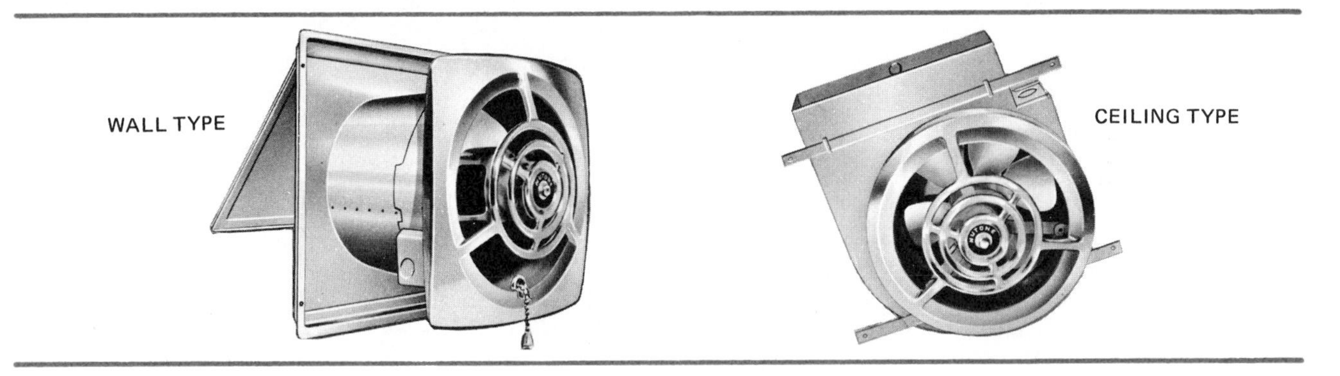

Fig. 15-10 Exhaust fans.

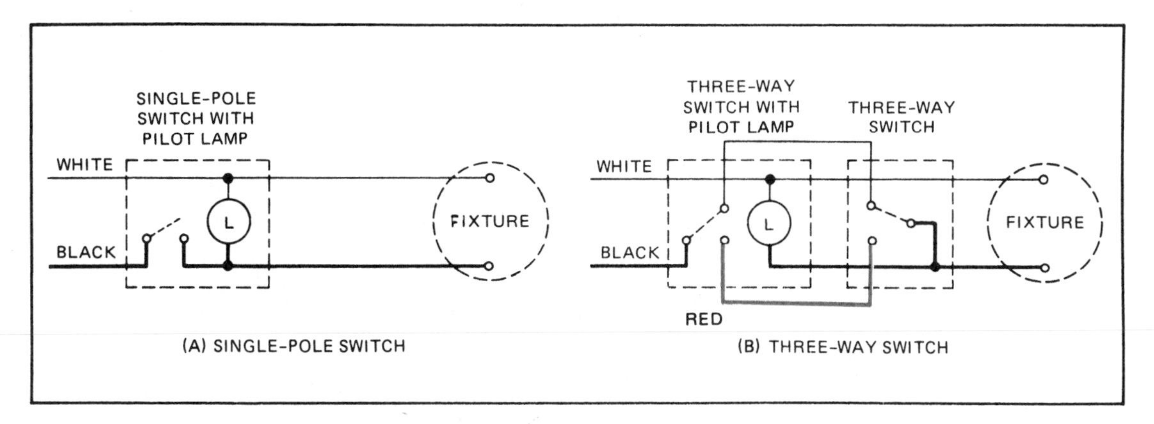

Fig. 15-11 Pilot lamp connections.

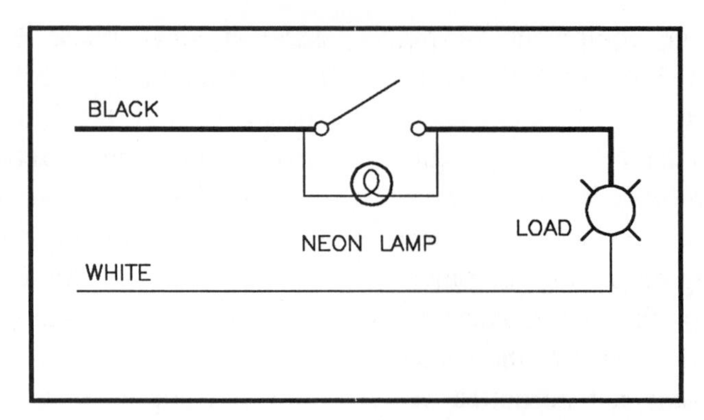

Fig. 15-12 Example of neon pilot lamp in handle of toggle switch.

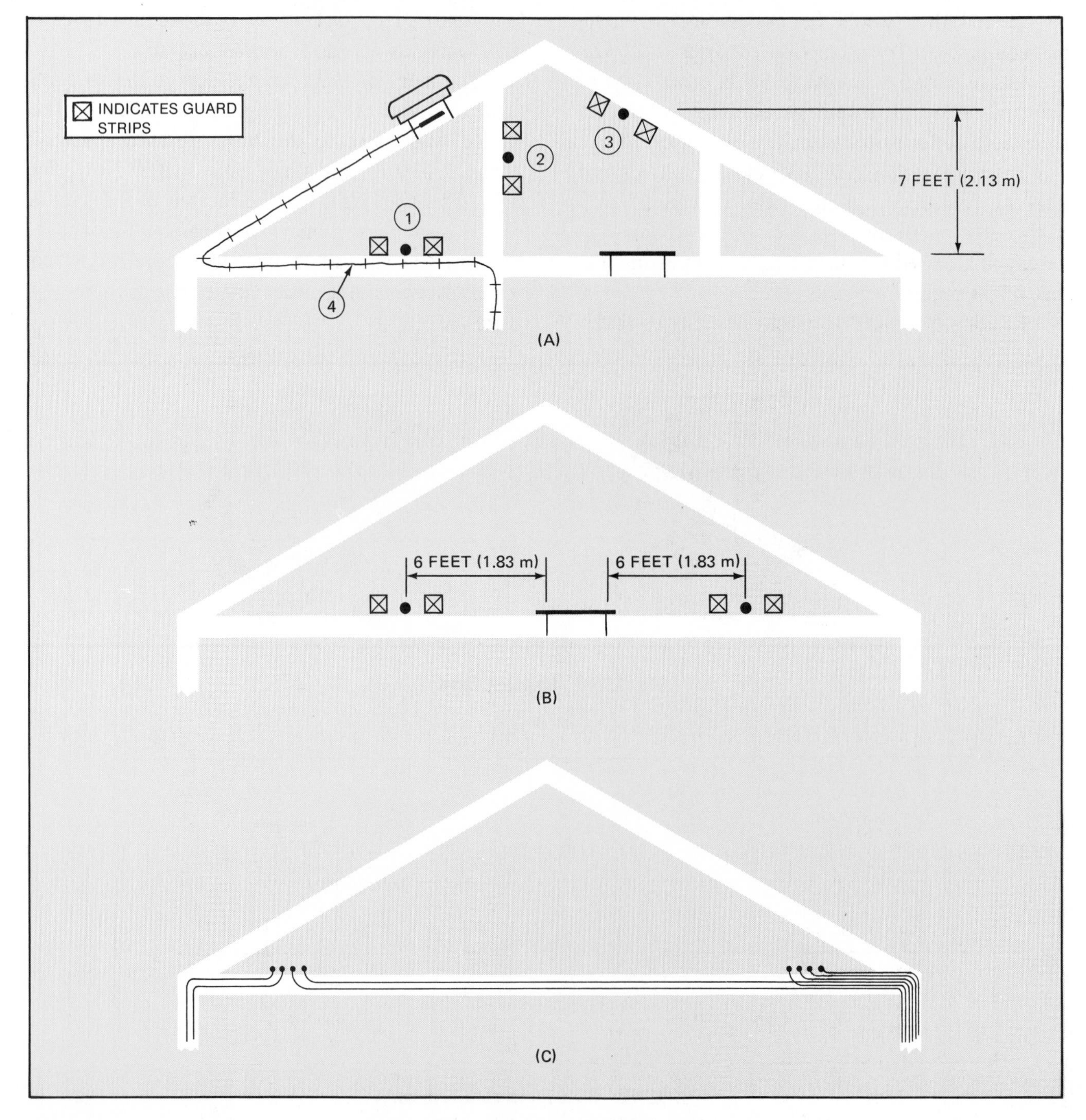

Fig. 15-13 Protection of cable in attic.

If a neon pilot lamp in the handle (toggle) of a switch does not have a separate grounded conductor connection, then it will glow only when the switch is in the OFF position, as the neon lamp will then be in series with the lamp load, figure 15-12.

The voltage across the load lamp is virtually zero, so it does not burn, and the voltage across the neon lamp is 120 volts, allowing it to glow. When the switch is turned on, the neon lamp is bypassed (shunted), causing it to turn off and the lamp load to turn on.

Use this type of switch when it is desirable to have a switch glow in the dark to make it easy to locate.

Installation of Cable in Attics

The wiring in the attic is to be done in cable and must meet the requirements of *Section 336-13* for nonmetallic-sheathed cable. This section refers the reader directly to *Section 333-12*, which describes how the cable is to be protected, figure 15-13.

In accessible attics, figure 15-13(A), cables must be protected by guard strips when

- they are run across the top of floor joists ①.
- they are run across the face of studs ② or rafters ③ within 7 feet (2.13 m) of the floor or floor joists.

Guard strips are *not* required if the cable is run along the sides of rafters, studs, or floor joists ④.

In attics not accessible by permanent stairs or ladders, figure 15-13(B), guard strips are required only within 6 feet (1.83 m) of the nearest edge of the scuttle hole or entrance.

Figure 15-13(C) illustrates a cable installation that most electrical inspectors consider to be safe. Because the cables are installed close to the point where the ceiling joists and the roof rafters meet, they are protected from physical damage. It would be very difficult for a person to crawl into this space, or store cartons in an area with such a low clearance. Although the plans for this residence show a 2-foot-wide catwalk in the attic, the owner may decide to install additional flooring in the attic to obtain more storage space. Because of the large number of cables required to complete the circuits, it would interfere with flooring to install guard strips wherever the cables run across the tops of the joists, figure 15-14. However, the cables can be run through holes bored in the joists and along the sides of the joists and rafters. In this way, the cables do not interfere with the flooring.

► When running cables parallel to framing members, be careful to maintain at least 1 - 1/4 inch (31.8 mm) between the cable and the edge of the framing member. This is a Code requirement referenced in *Section 300-4(d)* to minimize the possibility of driving nails into the cable. The entire *Section 300-4* of the Code is devoted to the subject of "protection against physical damage." ◄

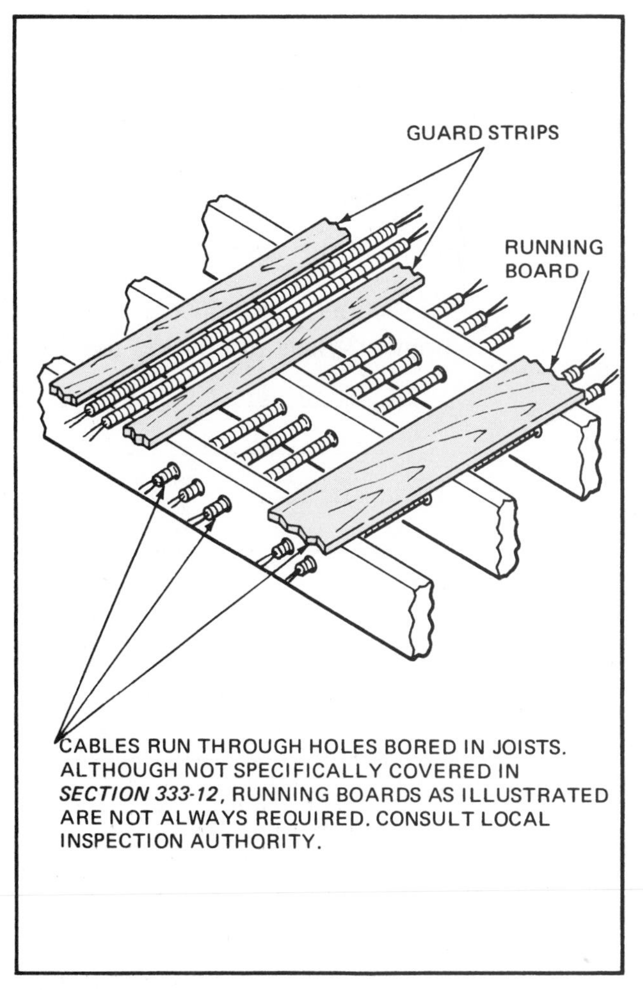

Fig. 15-14 Methods of protecting cable installations in attics.

REVIEW

Note: Refer to the Code or plans where necessary.

1. List the switches, receptacles, and other wiring devices that are connected to Circuit B10. ____________________

2. What section of the Code states that the receptacle in the washroom is to be GFCI protected? ____________________

3. What special type of switch is controlling the attic lights? ____________________

4. When installing cables in an attic along the top of the floor joists __________ or ____________________ must be installed to protect the cables.

5. The total estimated volt-amperes for Circuit B10 has been calculated to be 1300 volt-amperes. How many amperes is this at 120 volts? ____________________

6. If an attic is accessible through a scuttle hole, guard strips are installed to protect cables run across the top of the joists only within 6 feet (1.83 m), 7 feet (2.13 m), 12 feet (3.66 m) of the scuttle hole. Circle correct answer.

7. What section(s) of the Code refers (refer) to the receptacle required for the laundry equipment? State briefly the requirements.

8. What is the current draw of the exhaust fan in the laundry? ____________________

9. The following is a layout of the lighting circuit for the laundry, washroom, rear entry hall, and attic. Complete the wiring diagram using colored pens or pencils to indicate conductor insulation color.

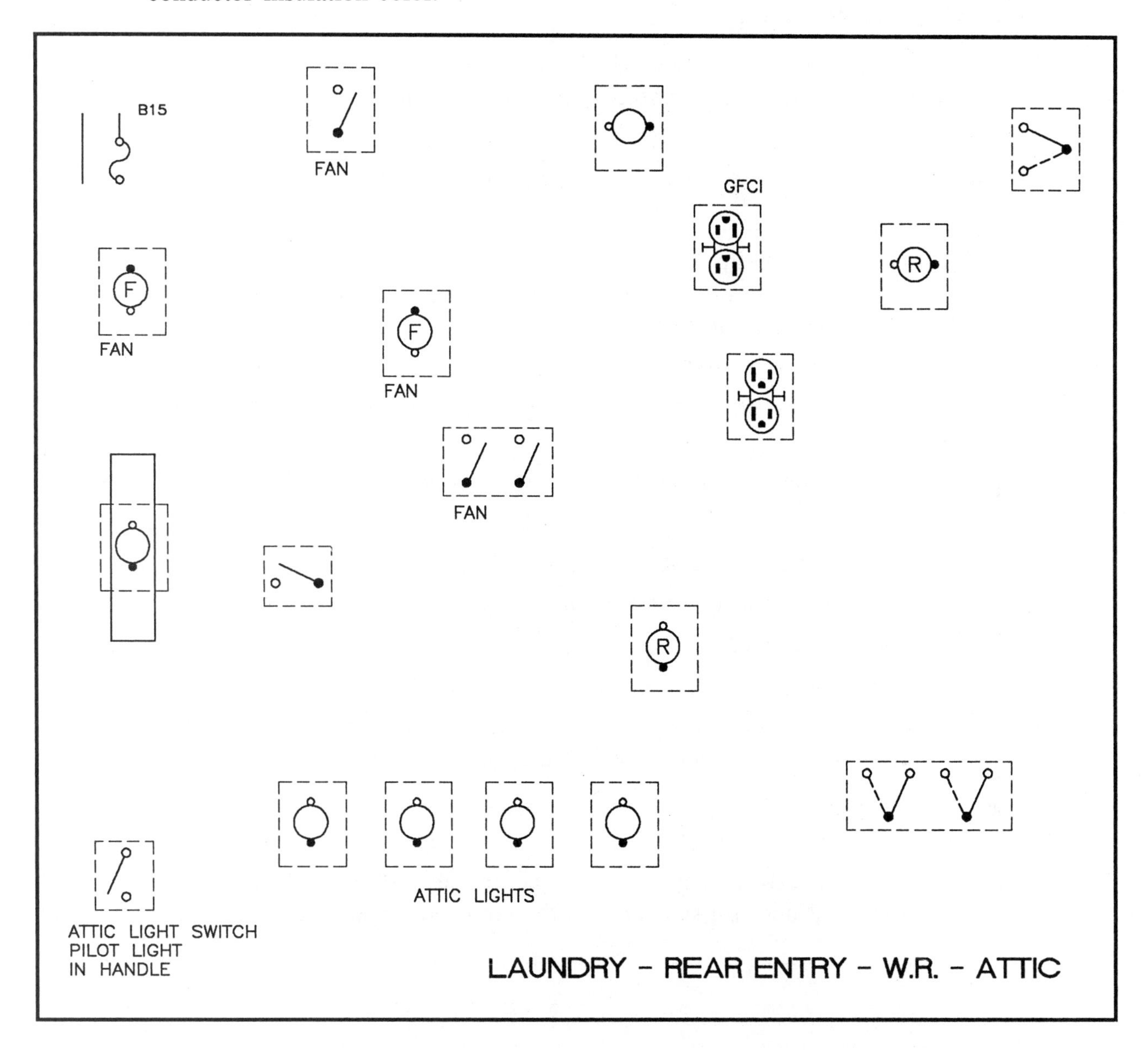

10. The receptacle outlets are connected to small appliance circuits that have been included in the residential load calculations at a value of ____________ volt-amperes per circuit. Circle one: 1000, 1500, 2000 volt-amperes.

11. a. What regulates the temperature in the dryer? ______________________

    b. What regulates drying time? ______________________

12. List the various methods of connecting an electric clothes dryer. ______________________

13. What is the unique shape of the neutral blade of a 30-ampere dryer cord set? ______________________

14. a. What is the minimum power demand allowed by the Code for an electric dryer if no actual rating is available for the purpose of calculating feeder and service-entrance conductor sizing? ______________________

    b. How much current is used? ______________________

15. What is the maximum permitted current rating of a portable appliance on a 30-ampere branch circuit? ______________________

16. What provides motor running overcurrent protection for the dryer? ______________________

17. a. Must an electric clothes dryer be grounded? ______________________

    b. May a dryer be grounded to a neutral conductor? ______________________

    c. Under what condition? ______________________

18. An electric dryer is rated at 7.5 kW and 120/240 volts, three-wire, single-phase. The terminals on the dryer and panelboard are marked 75°C.

    a. What is the wattage rating? ______________________

    b. What is the current rating? ______________________

    c. What size type THHN copper conductors are required? For ease in calculating, size the neutral one size smaller than the hot conductors. ______________________

    d. What size conduit is required? ______________________

19. When a metal junction box is installed as part of the cable wiring to a clothes dryer or electric range, may this box be grounded to the circuit neutral? ____________

►20. An attic is the location of a residential air-conditioning unit.

a. Is a lighting outlet required? ____________

b. What Code section? ____________

c. If a lighting outlet is required, how shall it be controlled? ____________

____

d. What Code section? ____________

e. Is a receptacle outlet required? ____________

f. What Code section? ____________ ◄

UNIT 16

Lighting Branch Circuit for the Garage

OBJECTIVES

After studying this unit, the student will be able to

- understand the fundamentals of providing proper lighting in residential garages.
- understand the application of GFCI protection for receptacles in residential garages.
- understand the Code requirements for underground wiring, both conduit and underground cable.
- complete the garage circuit wiring diagram.
- discuss typical outdoor lighting and Code requirements for same.
- describe how conduits and cables are brought through cement foundations to serve loads outside of the building structure.
- understand the application of GFCI protection for loads fed by underground wiring.
- make a proper installation for a residential overhead garage door opener.
- select the proper overload protective devices based on the ampere rating of the connected motor load.

LIGHTING BRANCH CIRCUIT

Circuit B14 originates at Panel B in the recreation room, and is brought into the wall receptacle box on the front wall of the garage. Figure 16-1 shows the cable layout for Circuit B14. From this point, the circuit is then carried to the switch box adjacent to the side door of the garage. From there the circuit conductors plus the switch leg are carried upward to the ceiling fixtures, and then to the wall receptacle outlets on the right-hand wall of the garage.

The garage ceiling's porcelain lampholders are controlled from switches located at all three entrances to the garage.

The post light in the front of the house is fed from the receptacle located just inside the overhead door. The post light turns on and off by a photocell, an integral part of the post light.

Note that the fixtures on the outside next to the overhead garage door are controlled by two three-way switches, one just inside the overhead door and the other in the front entry. These fixtures are not connected to the garage lighting circuit.

All of the wiring in the garage will be concealed because the walls and ceiling are to be covered with 3/4-hour fire-rated drywall.

Table 16-1 summarizes the outlets and the estimated load for the garage circuit.

LIGHTING A TYPICAL RESIDENTIAL GARAGE

The recommended approach to provide adequate lighting in a residential garage is to install one 100-watt lamp on the ceiling above each side of an automobile, figure 16-2. Figure 16-3 shows typical lampholders commonly mounted in garages.

- For a one-car garage, a minimum of two fixtures is recommended.

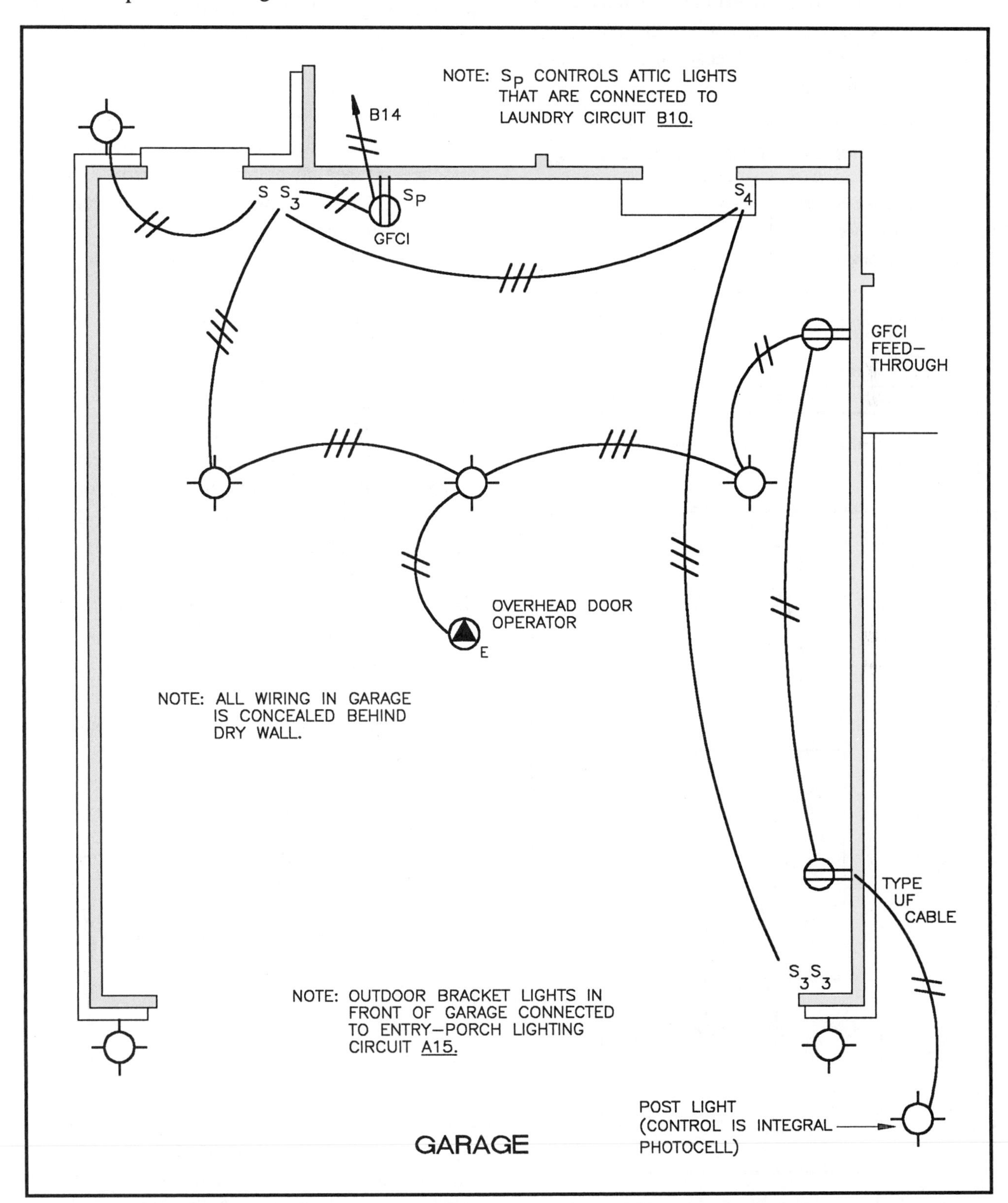

Fig. 16-1 Cable layout for the Garage Circuit B14.

- For a two-car garage, a minimum of three fixtures is recommended.
- For a three-car garage, a minimum of four fixtures is recommended.

Lighting fixtures arranged in this manner eliminate shadows between automobiles. Shadows are a hazard because they hide objects which may cause a person to trip or fall.

DESCRIPTION	QUANTITY	WATTS	VOLT-AMPERES
Receptacles @ 180 W each	3	540	540
Ceiling fixtures @ 100 W each	3	300	300
Outdoor garage bracket fixture (rear)	1	100	100
Post light	1	100	100
Overhead door opener			
Two 40-W lamps	1	80	80
Motor (1/2 A @ 120 V)		180	200
TOTALS	9	1300	1320

Table 16-1. Garage: outlet count and estimated load. Circuit B14.

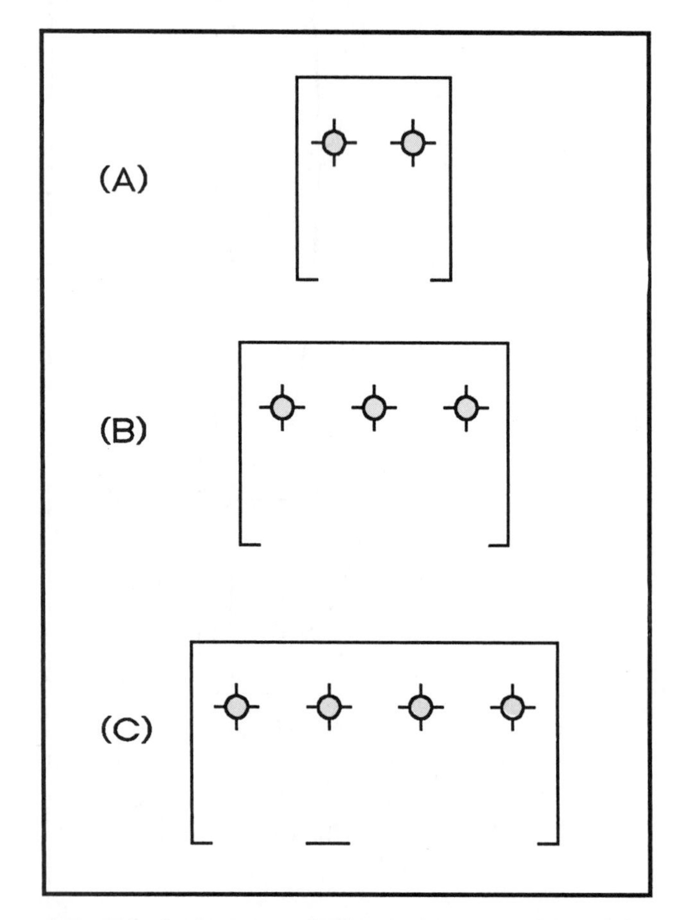

Fig. 16-2 Positioning of lights in (A) a one-car garage, (B) a two-car garage, and (C) a three-car garage.

It is highly recommended that these ceiling fixtures be mounted toward the front end of the automobile as it is normally parked in the garage. This arrangement will provide better lighting where it is most needed when the owner is working under the hood. Do not place lights where they will be covered by the open overhead door.

RECEPTACLE OUTLETS

At least one receptacle must be installed in an attached residential garage per *Section 210-52(g)*. The garage in this residence has three receptacles. If a detached garage has no electricity provided, then obviously no rules are applicable, but just as soon as a detached garage is wired, then all pertinent Code rules become effective and must be followed. These would include rules relating to grounding, lighting outlets, receptacle outlets, GFCI protection, and so on.

The Code in *Section 210-8(a)(2)* requires that "all 125-volt, single-phase, 15- or 20-ampere receptacles installed in garages shall have ground-fault circuit-interrupter protection." GFCIs are discussed in detail in unit 6.

There are a few exceptions to *Section 210-8(a)(2)* requiring GFCI protection of receptacles in residential garages. They are:

1. GFCI protection is *not* required if the receptacle is not readily accessible, such as the receptacle installed for the overhead door opener.

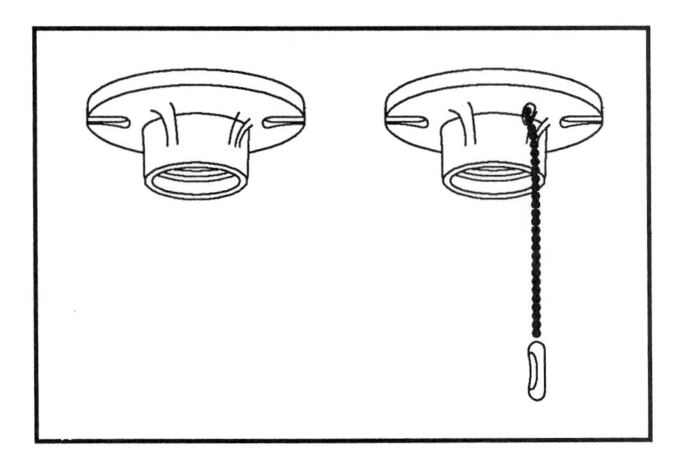

Fig. 16-3 These are typical porcelain lampholders of the type that would be installed in garages, attics, basements, crawl spaces, and similar locations. Note that one has a pull-chain and one does not.

2. GFCI protection is *not* required if the receptacle is installed for an appliance that occupies a dedicated space and which is cord-connected. This exception is difficult to understand, because how can the electrician actually know during rough-in that a certain receptacle is for the sole purpose of plugging in a specific appliance, say a large freezer. Such large appliances are not at the construction site then. Consult the local authority having jurisdiction (electrical inspector) for his interpretation of *Section 210-8(a)(2)* of the Code.

Note that one GFCI receptacle is installed on the front wall of the garage, in the same box as the single-pole switch for the attic lighting. A second feedthrough GFCI receptacle is installed on the right-hand garage wall near the entry to the house. This GFCI receptacle also provides the required GFCI protection for the other receptacle on the same wall and for the protection of the underground wiring to the post light.

If we were to install one GFCI feedthrough receptacle at the point where Circuit B14 feeds the garage, then the entire circuit would be GFCI protected as required. But nuisance tripping of the GFCI might occur because of:

- the overall length of the circuit wiring.
- possible problems in the overhead door opener.
- possible problems in the underground run to the post light.
- problems in the post lamp itself.

All of these would result in a complete outage on Circuit B14. The electrician must consider the cost of GFCI receptacles and compare it with the cost of additional cable or conduit and wire needed. He must also consider the possibility of a power outage when trying to get by with fewer GFCI receptacles.

OUTDOOR WIRING

If requested by the homeowner that 120-volt receptacles be installed away from the building structure, the electrician can provide weatherproof receptacle outlets as illustrated in figure 16-4.

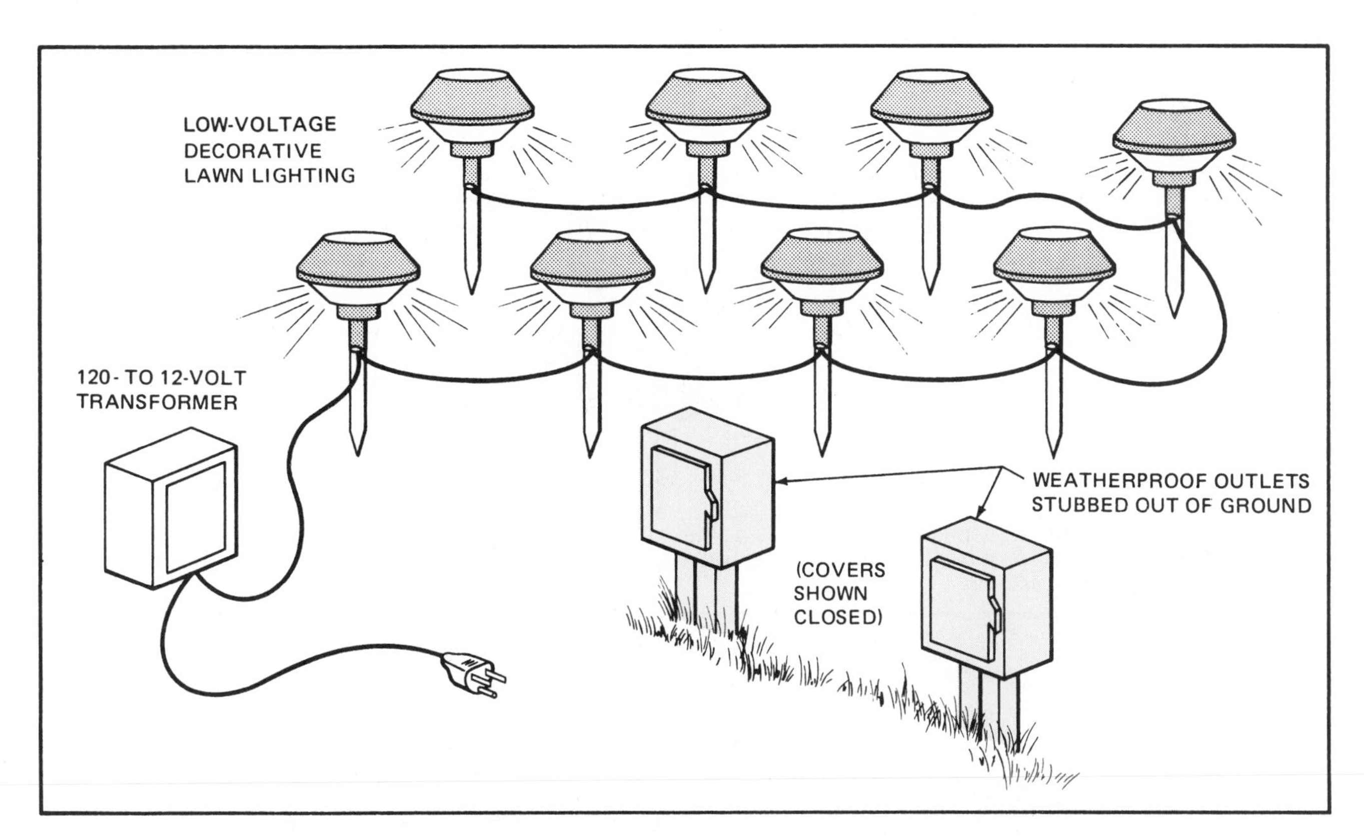

Fig. 16-4 Weatherproof receptacle outlets stubbed out of the ground; either low-voltage decorative lighting or 120-volt PAR lighting fixtures may be plugged into these outlets.

These types of boxes have 1/2-inch female threaded openings in which conduit fittings secure the conduit "stub-ups" to the box. Any unused openings are closed with 1/2-inch plugs that are screwed tightly into the threads. Refer to figure 16-5.

Wiring with Type UF Cable

The plans show that Type UF underground cable, figure 16-6, is used to connect the post light. The current-carrying capacity (ampacity) of Type UF cable is found in *Table 310-16*, using the 60°C column. According to *Article 339*, Type UF cable

- is marked underground feeder cable.
- is available in sizes from No. 14 AWG through No. 4/0 (for copper conductors) and from No. 12 AWG through No. 4/0 (for aluminum conductors). See *Table 310-16* for the ampacity ratings.

(A) WHEN TWO OR MORE CONDUITS ARE TIGHTLY THREADED INTO THE HUBS OF A CONDUIT BODY, THE CONDUIT BODY IS CONSIDERED TO BE ADEQUATELY SUPPORTED, ACCORDING TO THE LAST PARAGRAPH OF *SECTION 370-13.*

(B) THIS CONDUIT BODY IS NOT ADEQUATELY SUPPORTED BY THE ONE CONDUIT THREADED INTO THE HUB. THIS CONDUIT BODY COULD TWIST VERY EASILY, RESULTING IN DAMAGED INSULATION ON THE CONDUCTORS AND A POOR GROUND CONNECTION BETWEEN THE CONDUIT AND THE CONDUIT BODY.

(C) CONDUCTORS MAY BE SPLICED IN THESE CONDUIT BODIES ONLY IF THE CONDUIT BODY IS MARKED WITH ITS CUBIC-INCH CAPACITY SO THAT THE PERMISSIBLE CONDUCTOR FILL MAY BE DETERMINED USING THE CONDUCTOR VOLUME FOUND IN *TABLE 370-6(b).*

Fig. 16-5 Supporting threaded conduit bodies.

Fig. 16-6 Type UF underground cable.

- may be used in direct exposure to the sun if the cable is marked "sunlight resistant."
- may be used with nonmetallic-sheathed cable fittings.
- is flame-retardant.
- is moisture-, fungus-, and corrosion-resistant.
- may be buried directly in the earth.
- may be used for branch-circuit and feeder wiring.
- may be used in interior wiring for wet, dry, or corrosive installations.
- is installed by the same methods as nonmetallic-sheathed cable (*Article 336*).
- must not be used as service-entrance cable.
- must not be embedded in concrete, cement, or aggregate.
- must be buried in the same trench where single-conductor cables are installed.
- shall be installed according to *NEC® Section 300-5.*

The grounding of equipment fed by Type UF cable is accomplished by properly connecting the bare grounding conductor found in the UF cable to the equipment to be grounded, figure 16-7.

Regarding the subject of underground wiring, it is significant to note that while the Code does permit running up the side of a tree with conduit or cable (when protected from physical damage), figure 16-8, the Code does *not* permit supporting conductors between trees, figure 16-9. Figure 16-10 shows one type of lighting fixture that is permitted for the installation shown in figure 16-8.

UNDERGROUND WIRING

Underground wiring is common in residential applications. Examples would be decorative landscape lighting and wiring to post lamps and detached buildings such as garages or tool sheds.

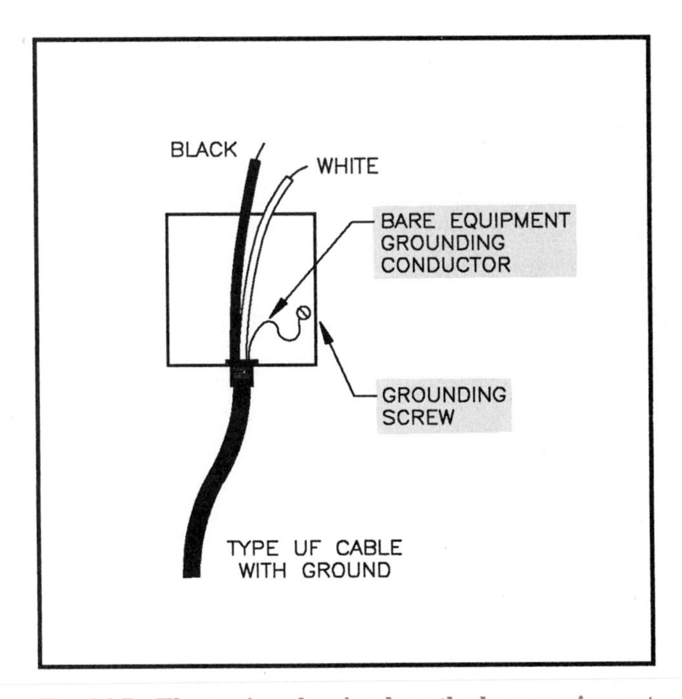

Fig. 16-7 Illustration showing how the bare equipment grounding conductor of a Type UF cable is used to ground the metal box.

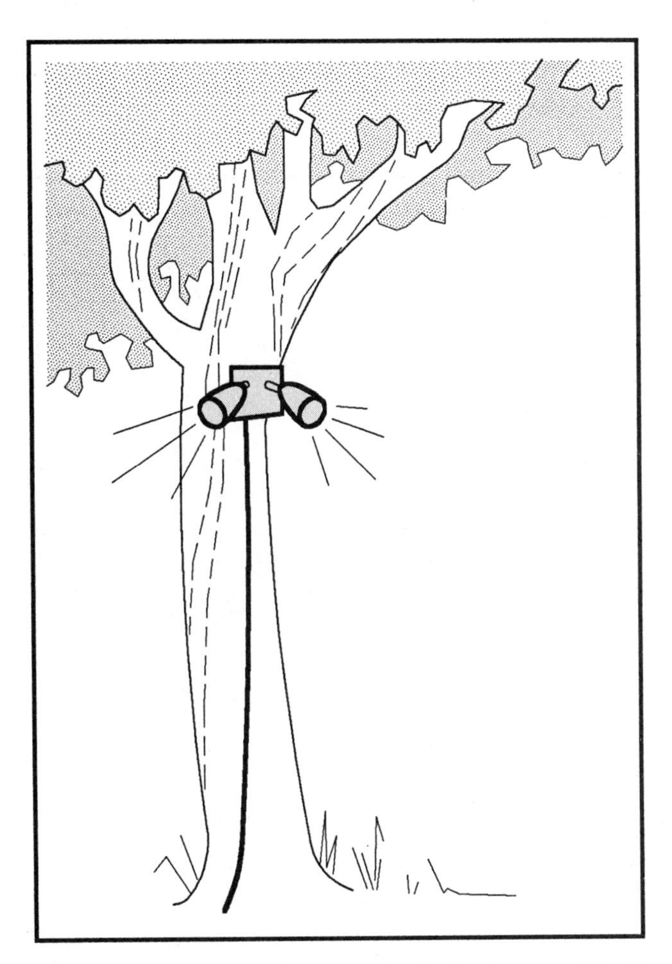

Fig. 16-8 ► This conforms to the Code. *Section 225-26* does not permit spans of conductors to be run between live trees, but it does not prohibit installing decorative lighting as shown in this figure. *Section 410-16(h)* permits outdoor lighting fixtures and associated equipment to be supported by trees. ◄ Approved wiring methods to carry the conductors up the tree must be used.

► *Table 300-5* of the Code shows the minimum depths required for the various types of wiring methods. Refer to figures 16-11B (page 231) and 16-11C (page 232).

Across the top of the table are listed wiring methods such as direct burial cables or conductors, rigid metal conduit, intermediate metal conduit, rigid nonmetallic conduit that is approved for direct burial, special considerations for residential branch circuits rated 120 volts or less having GFCI protection and overcurrent protection not over 20 amperes, and low-voltage landscape lighting supplied with Type UF cable.

Down the left-hand side of the table we find the location of the wiring methods that are listed across the top of the table. Listed are all locations, including installations in trenches covered with 2 inches of concrete, under buildings, under 4 inches of concrete, under streets, highways, roads, alleys, driveways, and parking lots, and under residential driveways and parking areas, as well as airport installations and solid rock encounters.

The notes to *Table 300-5* are very important. Note 4 allows us to select the shallower of two depths when the wiring methods in Columns 1, 2, or 3 (across the top of the table) are combined with Columns 4 or 5 (across the top of the table). For example, the basic rule for direct buried cable is that it be covered by a minimum of 24 inches. However, this minimum depth is reduced to 12 inches for residential installations when the branch circuit is not over 120-volts, is GFCI protected, and is not over 20 amperes. This is for residential installations only.

The measurement for the depth requirement is from the top of the raceway or cable to the top of the finished grade, concrete, or other similar cover. ◄

Fig. 16-9 ► VIOLATION. *Section 225-26* does not permit live vegetation such as trees to support overhead conductor spans. ◄

Installation of Conduit Underground

Some local electrical codes require the installation of conduit for all underground wiring, using conductors approved for wet locations, such as Type TW, THW, or THWN.

All conduit installed underground must be protected against corrosion. The manufacturer of the conduit and referring to Underwriters' Laboratories Standards will furnish the information as to whether or not the conduit is suitable for direct burial. Additional supplemental protection of metal conduit can be accomplished by "painting" the conduit with a nonmetallic coating. See Table 16-2.

When metal conduit is used as the wiring method, the metal boxes, post lights, and so forth that are properly connected to the metal raceways are considered to be grounded because the conduit

Fig. 16-10 Outdoor fixture with lamps.

Table 300-5. Minimum Cover Requirements, 0 to 600 Volts, Nominal, Burial in Inches
(Cover is defined as the shortest distance measured between a point on the top surface of any direct buried conductor, cable, conduit or other raceway and the top surface of finished grade, concrete, or similar cover.)

	Type of Wiring Method or Circuit				
Location of Wiring Method or Circuit	Direct Burial Cables or Conductors	Rigid Metal Conduit or Intermediate Metal Conduit	Rigid Nonmetallic Conduit Approved for Direct Burial Without Concrete Encasement or Other Approved Raceways	Residential Branch Circuits Rated 120 Volts or less with GFCI Protection and Maximum Overcurrent Protection of 20 Amperes	Circuits for Control of Irrigation and Landscape Lighting Limited to Not More than 30 Volts and Installed with Type UF or in Other Identified Cable or Raceway
All Locations Not Specified Below	24	6	18	12	6
In Trench Below 2-Inch Thick Concrete or Equivalent	18	6	12	6	6
Under a Building	0 (In Raceway Only)	0	0	0 (In Raceway Only)	0 (In Raceway Only)
Under Minimum of 4-Inch Thick Concrete Exterior Slab with no vehicular traffic and the slab extending not less than 6 inches beyond the underground installation	18	4	4	6 (Direct Burial) 4 (In Raceway)	6 (Direct Burial) 4 (In Raceway)
Under Streets, Highways, Roads, Alleys, Driveways, and Parking Lots	24	24	24	24	24
One- and Two-Family Dwelling Driveways and Parking areas, and Used for No Other Purpose	18	18	18	12	18
In or Under Airport Runways Including Adjacent Areas Where Trespassing Prohibited	18	18	18	18	18
In Solid Rock Where Covered by Minimum of 2 Inches Concrete Extending Down to Rock	2 (In Raceway Only)	2	2	2 (In Raceway Only)	2 (In Raceway Only)

Note 1. For SI Units: one inch = 25.4 millimeters
Note 2. Raceways approved for burial only where concrete encased shall require concrete envelope not less than 2 inches thick.
Note 3. Lesser depths shall be permitted where cables and conductors rise for terminations or splices or where access is otherwise required.
Note 4. Where one of the conduit types listed in columns 1-3 is combined with one of the circuit types in columns 4 and 5, the shallower depth of burial shall be permitted.

serves as the equipment ground, *Section 250-91(b)*, figure 16-12 (page 233). See *Article 410, Part E* for details on grounding requirements for fixtures.

When nonmetallic raceways are used, then a separate equipment grounding conductor, either green or bare, must be installed to accomplish the adequate grounding of the equipment served. This conductor must be sized according to *Table 250-95* of the Code. See figure 16-13 (page 233).

Figure 16-11A shows two methods of bringing the conduit into the basement: (1) it can be run below ground level and then can be brought through

	IS SUPPLEMENTAL CORROSION PROTECTION REQUIRED?		
	IN CONCRETE ABOVE GRADE?	IN CONCRETE ABOVE GRADE?	IN DIRECT CONTACT WITH SOIL?
Rigid Conduit[1]	No	No	No[2]
Intermediate Conduit[1]	No	No	No[2]
Electrical Metallic Tubing[1]	No	Yes	Yes[3]

[1] Severe corrosion can be expected where ferrous metal conduits come out of concrete and enter the soil. Here again, some electrical inspectors and consulting engineers might specify the application of some sort of supplemental nonmetallic corrosion protection.

[2] Unless subject to severe corrosive effects, different soils have different corrosive characteristics.

[3] In most instances, electrical metallic tubing is not permitted to be installed underground in direct contact with the soil because of corrosion problems.

Table 16-2.

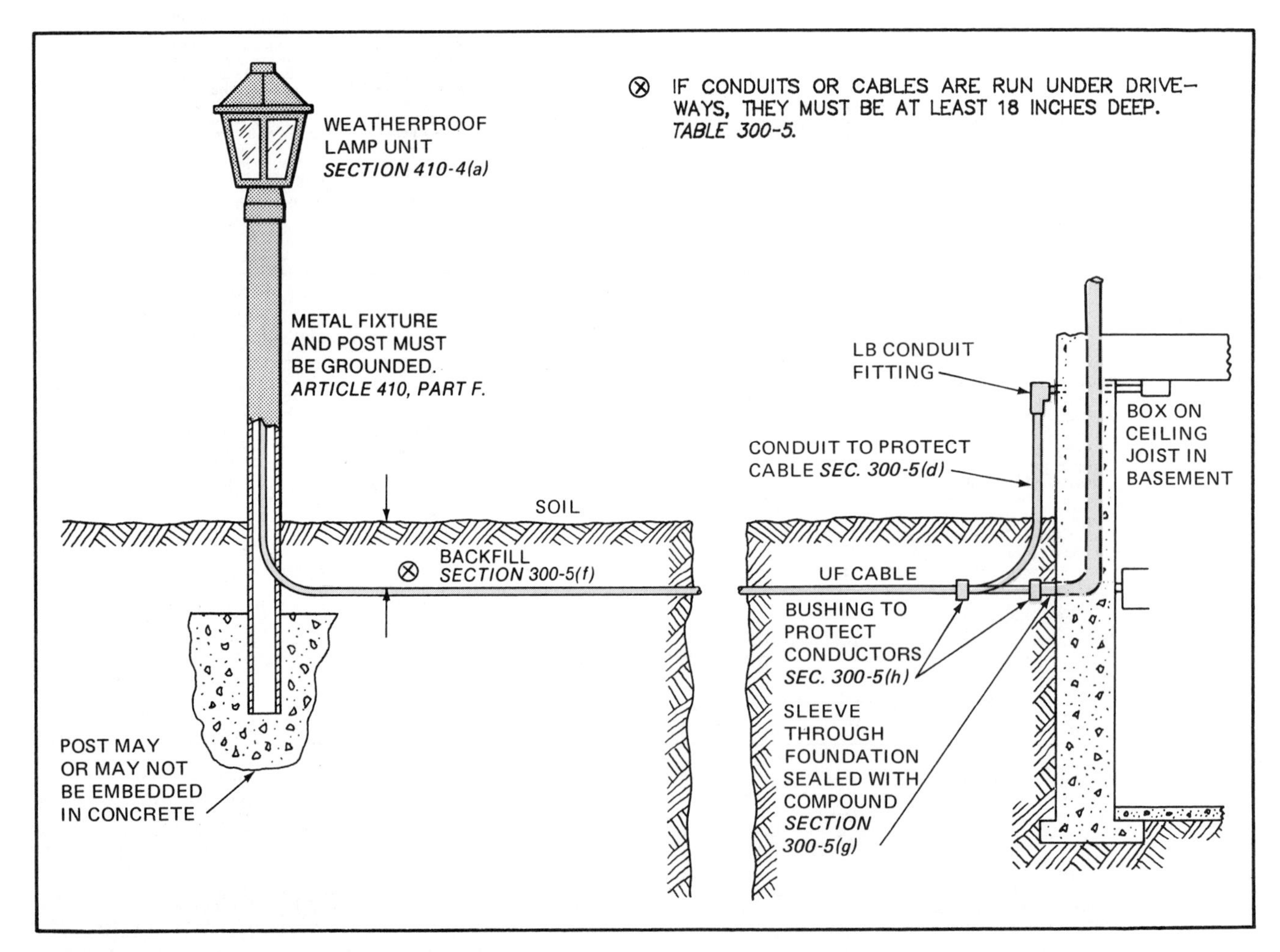

Fig. 16-11A Methods of bringing cable and/or conduit through a concrete wall and/or upward from a concrete wall into the hollow space within a framed wall.

the basement wall, or (2) it can be run up the side of the building and through the basement wall at the ceiling joist level. When the conduit is run through the basement wall, the opening must be sealed to prevent moisture from seeping into the basement. The electrician must decide which of the two methods is more suitable for each installation.

According to the Code, any location exposed to the weather is considered a wet location.

Section 410-4(a) states that outdoor fixtures must be constructed so that water cannot enter or accumulate in lampholders, wiring compartments, or other electrical parts. These fixtures must be marked "Suitable for Wet Locations."

Partially protected areas such as roofed open porches or areas under canopies are defined by the Code as damp locations. Fixtures to be used in these locations must be marked "Suitable for Damp Locations." Many types of fixtures are available. Therefore, it is recommended that the electrician check the UL label on the fixture or refer to the UL Electrical Construction Materials List to determine the suitability of the fixture for a wet or damp location.

The post in figure 16-11A may or may not be embedded in concrete, depending on the consistency of the soil, the height of the post, and the size of the lighting fixture. Most electricians prefer to embed the base of the post in concrete to prevent rotting of wood posts and rusting of metal posts.

Figure 16-14 (page 233) illustrates some typical post lights.

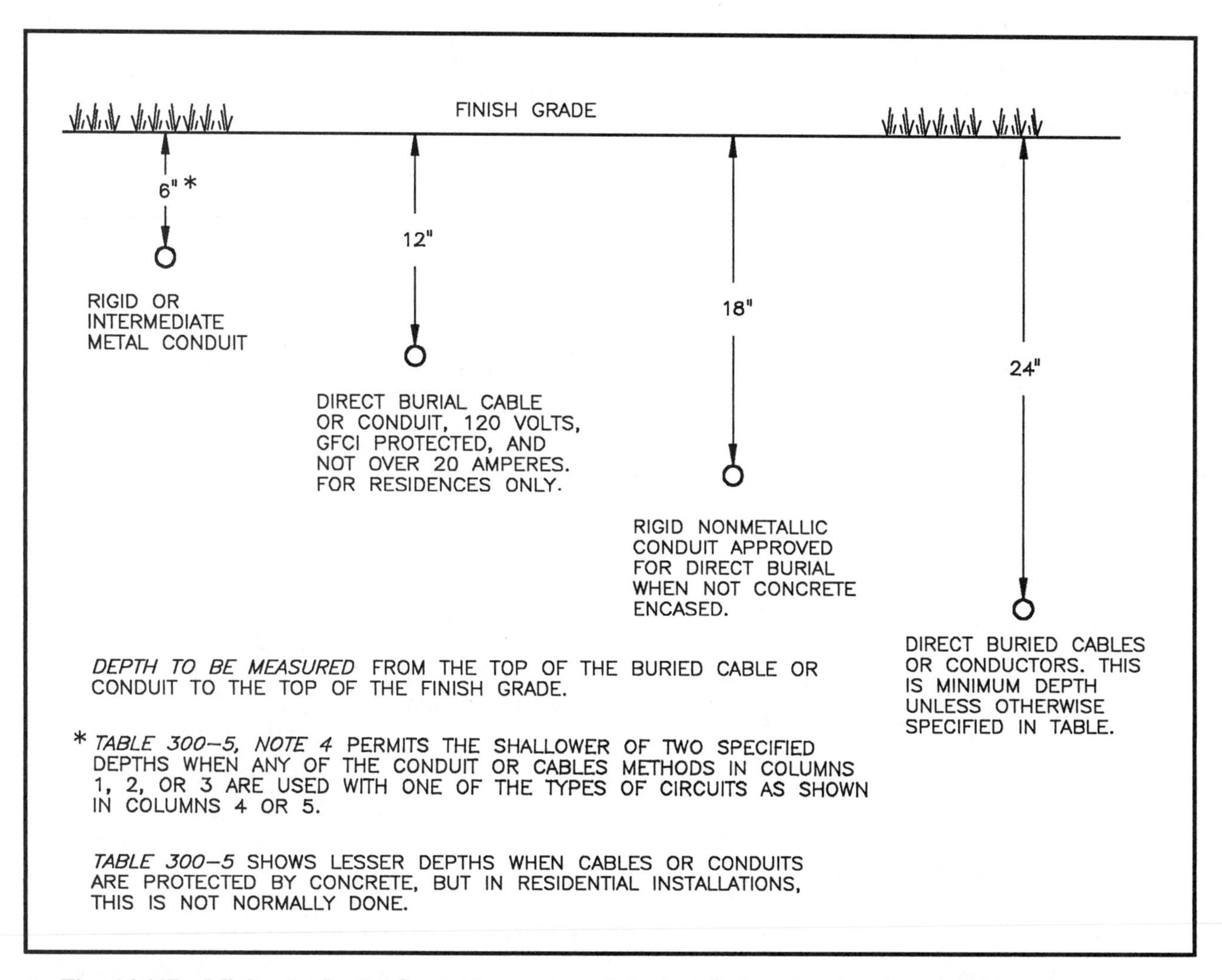

► **Fig. 16-11B Minimum depths for cables and conduits installed underground. See *Table 300-5* (page 229) for depths for other conditions.** ◄

OVERHEAD GARAGE DOOR OPENER Ⓐ E

The overhead garage door opener in this residence is plugged into the receptacle provided for this purpose on the garage ceiling. This particular overhead door opener is rated 5.8 amperes at 120 volts. The actual installation of the overhead door will be done by the carpenter or a specially trained overhead door installer.

Principles of Operation

The overhead door opener contains a motor, gear reduction unit, motor reversing switch, and electric clutch. This unit is preassembled and wired by the manufacturer.

Almost all residential overhead door openers use split-phase, capacitor-start motors in sizes from 1/4 to 1/2 horsepower. The motor size selected depends upon the size (weight) of the door to be raised and lowered. By using springs in addition to the gear reduction unit to counterbalance the weight of the door, a small motor can lift a fairly heavy door.

There are two ways to change the direction of a split-phase motor: (1) reverse the starting winding with respect to the running winding; (2) reverse the running winding with respect to the starting winding. The motor can be run in either direction by properly connecting the starting and running winding leads of the motor to the reversing switch. When the motor shaft is connected to a chain drive or screw drive, the door can be raised or lowered according to the direction in which the motor is rotating.

The electrically operated reversing switch can be controlled in a number of ways. These methods include push-button stations marked "open-close-stop," "up-down-stop," or "open-close," indoor and outdoor weatherproof push buttons, and key-operated stations. Radio-operated controllers are popular because they permit overhead doors to be controlled from an automobile.

Wiring of Door Openers

The wiring of overhead door openers for resi-

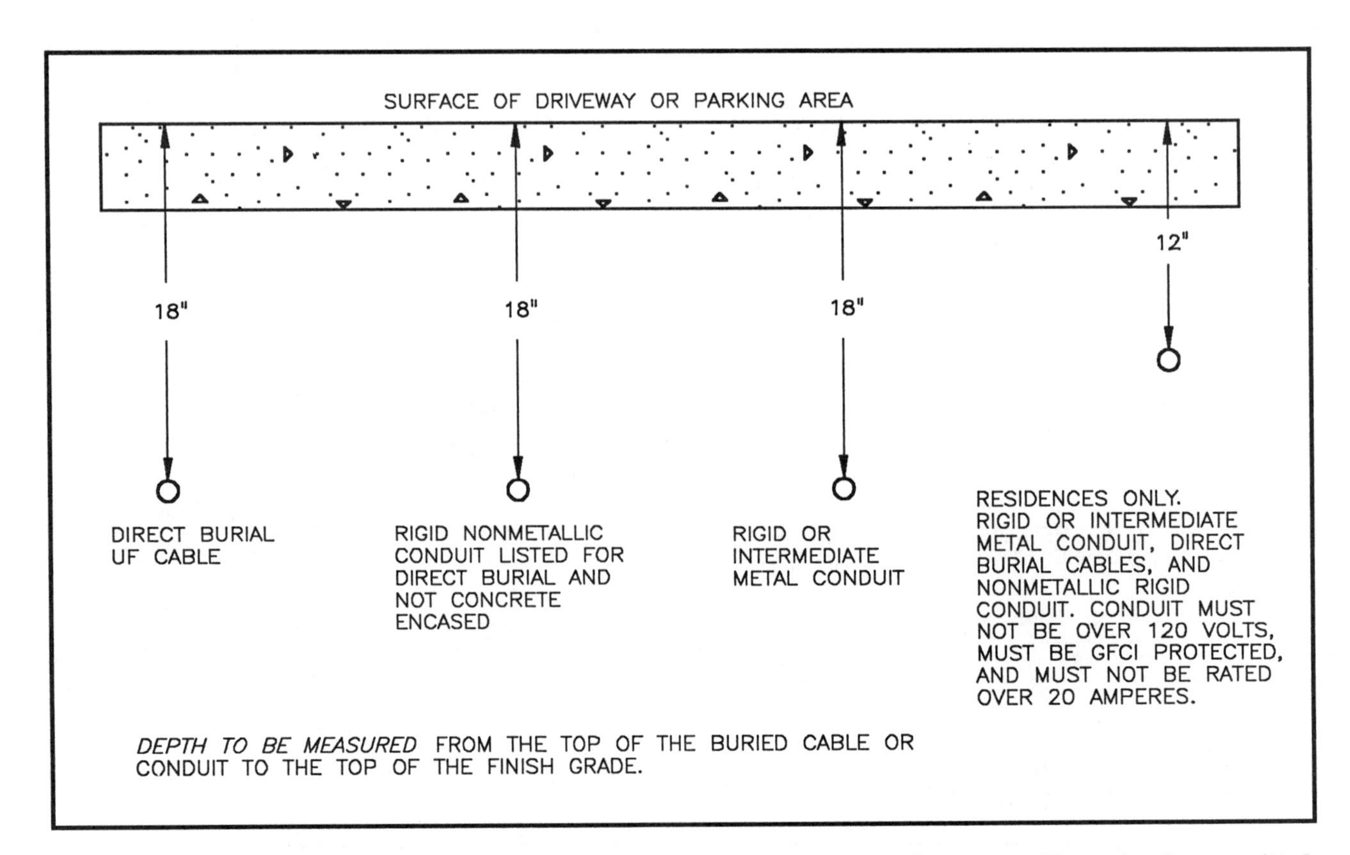

► **Fig. 16-11C Depth requirements under residential driveways and parking areas. These depths permitted for one- and two-family residences.** ◄

dential use is quite simple, since these units are completely prewired by the manufacturer. The electrician must provide a 120-volt circuit close to where the overhead door operators are to be installed. This 120-volt circuit can be wired directly into the overhead door unit; alternatively, a receptacle can be provided into which the cord on the overhead door operator is plugged.

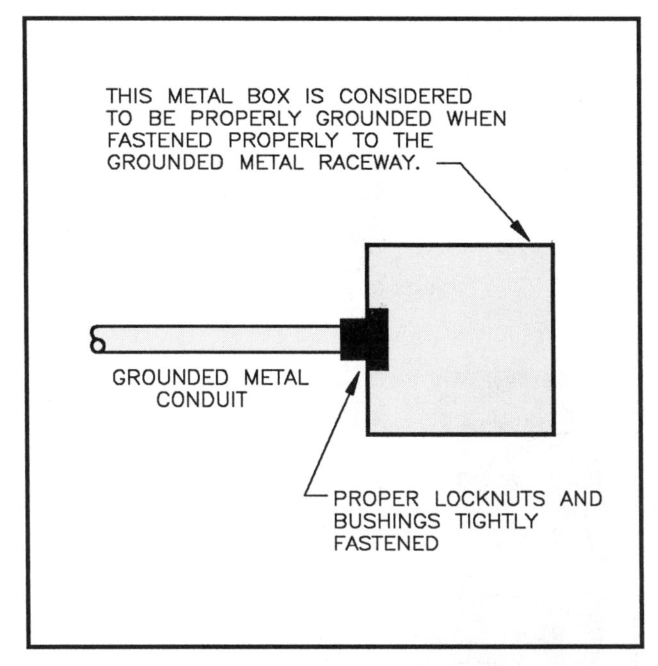

Fig. 16-12 This illustration shows that a box is grounded when properly fastened to the grounded metal raceway. This is acceptable in *Section 250-91(b)*. More and more Code-enforcing authorities and consulting engineers are requiring that a separate equipment grounding conductor be installed in all raceways. This insures effective grounding of the installation. Refer to unit 18.

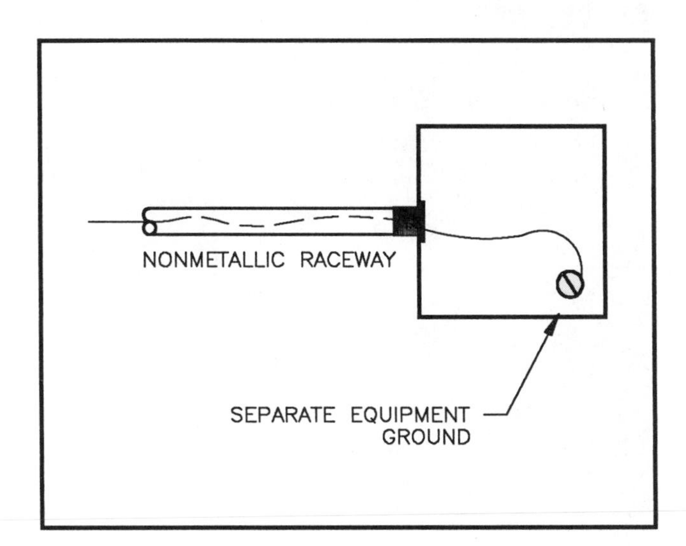

Fig. 16-13 Grounding a metal box with a separate equipment grounding conductor.

The buttons that are pushed to operate the overhead door opener can be placed in any convenient, desirable location. These push buttons are wired with low-voltage wiring, usually 24 volts, figure 16-15. This is the same type of wire used for wiring bells and chimes. From each button, the electrician runs a two-wire, low-voltage cable back to the control unit. Since the residential openers operate on the momentary contact principle for push buttons, a two-wire cable is all that is necessary between the operator and the push buttons. Commercial overhead door openers can require three or more conductors.

It is desirable to install push buttons at each door leading into the garage.

The actual connection of the low-voltage wiring between the controller and push buttons is a parallel circuit connection, figure 16-15.

Overhead door openers can be connected directly by using flexible metal conduit or metal armored cable. The illustration in figure 16-16

Fig. 16-14 Post-type lights.

shows how this can be accomplished. Note that the flexible cable is run between the door opener and an electrical box that has been mounted on the ceiling.

Box-Cover Unit

Instead of merely installing a receptacle outlet on the ceiling near the garage door opener unit, it has been found to be advantageous to provide a box-cover unit, either a switch-fuse type, figure 16-16, or a fuse-receptacle type, figures 16-17 and 16-18. This box-cover unit permits the use of dual-element Type S fuses sized at 110 to 125% of the motor rating.

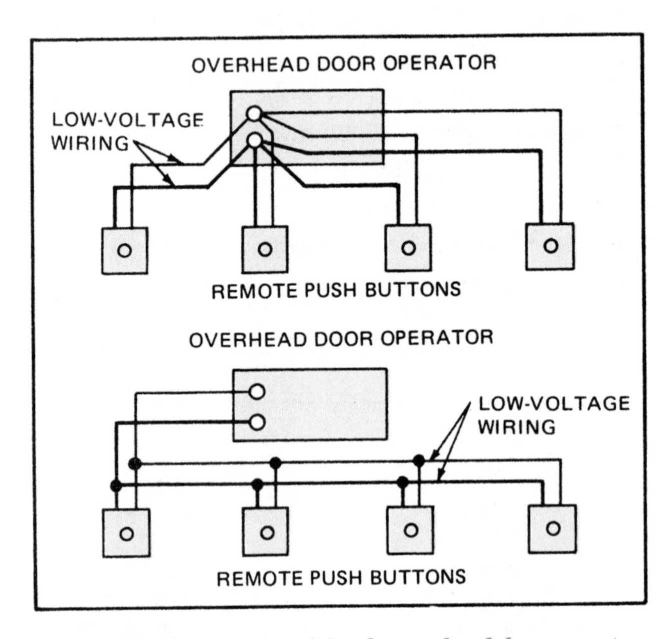

Fig. 16-15 Push-button wiring for overhead door operator.

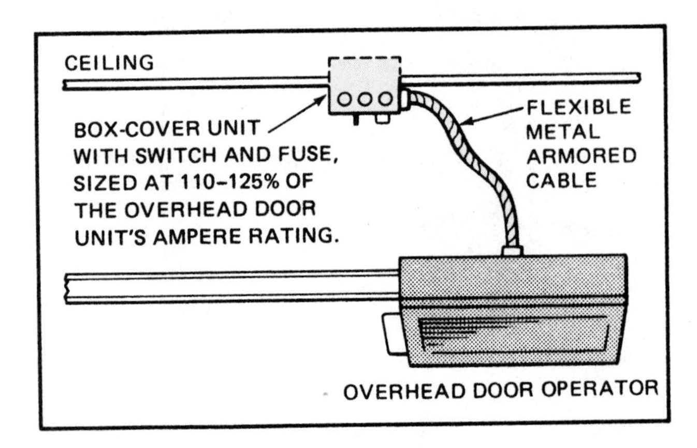

Fig. 16-16 Connections for overhead door operator.

This box-cover unit provides convenient disconnection of power to the overhead door unit for servicing or when the owner is away from home.

Box-cover units are readily available for mounting onto 4-inch octagon boxes, 4-inch square boxes, and handy boxes. Should the need arise to provide for two individual disconnects and individual fusing, then a box-cover unit of the style illustrated in figure 16-18 can be installed.

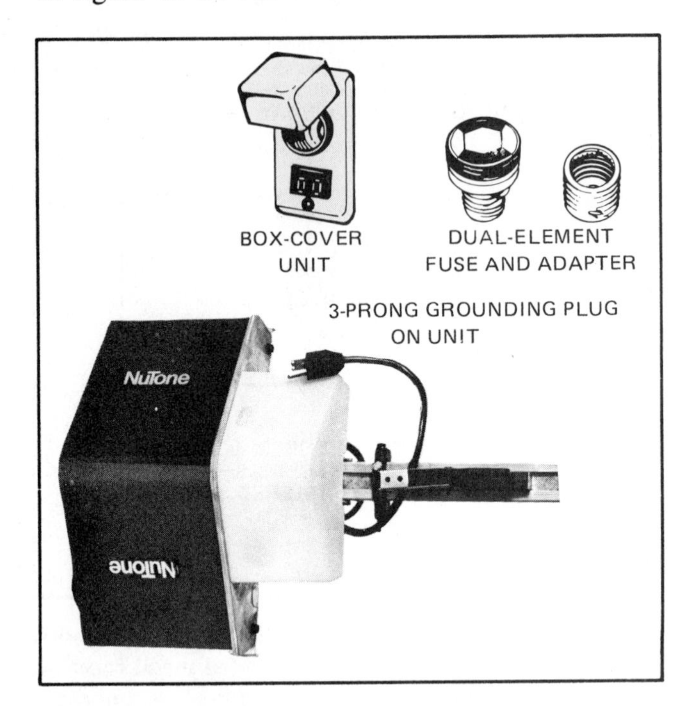

Fig. 16-17 Overhead door opener with three-wire cord connector.

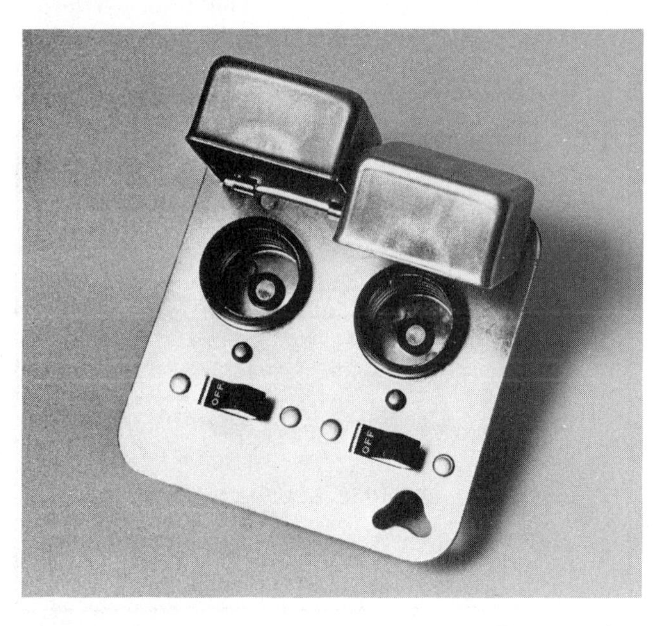

Fig. 16-18 Box-cover unit that provides disconnecting means plus individual overcurrent protection.

Overcurrent Protection

Overhead door operators have integral overload protection, per *Section 430-32(c)(2)*, so the addition of the box-cover fuse device provides added protection, *Section 430-32(c)(1)*. Added overload protection in the form of Type S, dual-element, time-delay fuses installed in the box-cover units is determined by sizing the type S fuse at not over 125% of the full-load current rating of the motor, per *Section 430-32(c)*:

$$5.8 \times 1.25 = 7.25 \text{ amperes}$$

Using 110% sizing, we have

$$5.8 \times 1.10 = 6.38 \text{ amperes}$$

This would give us an ampere rating range to choose from. In the previous instance, a Type S 6 1/4-ampere dual-element fuse would be an excellent choice for backup overload protection of the overhead door operator.

A dual-element Type S fuse installed as shown in figures 16-16 and 16-17 would sense any problem that might arise in the electrical wiring inside the overhead door unit, and would clear the problem before the circuit overcurrent protective device opened, because the ampere rating of the fuse at the overhead door unit is much smaller than the rating of the branch-circuit overcurrent device, figure 16-19. In technical terms this is called *selective coordination*, and is presented in the Commercial Wiring Text.

The receptacle for the overhead door unit does *not* have to be protected by a GFCI, *Section 210-8(a)(2), Exception 1.*

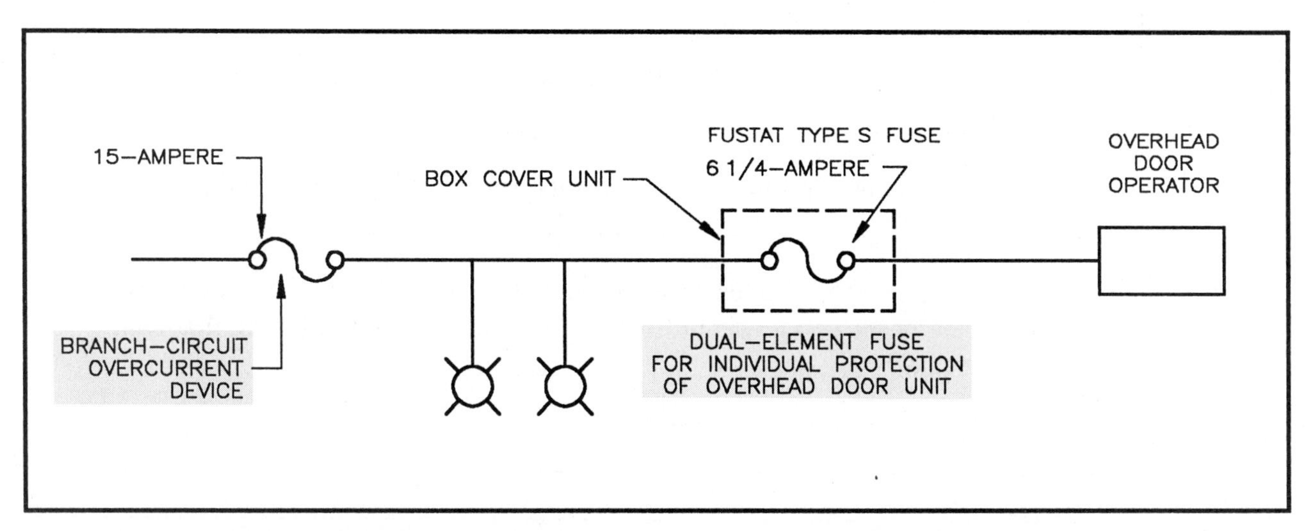

Fig. 16-19 This diagram shows the benefits of installing a small-ampere-rated dual-element time-delay fuse that is sized to provide overload protection for the motor. If an electrical problem occurs at the overhead door opener, only the 6 1/4-ampere fuse will open. The 15-ampere branch circuit is not affected. All other loads connected to the 15-ampere circuit remain energized.

REVIEW

Note: Refer to the Code or plans where necessary.

1. What circuit supplies the garage? ______________________________
2. What is the circuit rating? ______________________________
3. How many GFCI-protected receptacle outlets are connected to the garage circuit?

____

4. a. How many cables enter the wall switch box located next to the side garage door?

   ____

   b. How many circuit conductors enter this box? ______________

   c. How many grounding conductors enter this box? ______________

5. Show calculations on how to select a proper size box for question 4. What kind of box would you use? ______________

   ____

   ____

   ____

   ____

   ____

   ____

   ____

6. a. How many lights are recommended for a one-car garage? ______ for a two-car garage? ______ for a three-car garage? ______

   b. Where are these lights to be located? ______________

   ____

7. From how many points in the garage of this residence are the ceiling lights controlled? ______________

8. GFCI breakers or GFCI receptacles cost approximately $15.00 each. How would *you* arrange the wiring of the garage circuit to make an economical installation that complies with the Code? ______________

   ____

   ____

   ____

   ____

   ____

   ____

9. The total estimated volt-amperes of the garage circuit draws how many amperes? Show your calculations. ______________________________

____

____

____

10. How high from the floor are the switches and receptacles to be mounted? ______

11. What type of cable feeds the post light? ______________________________

►12. All raceways, cables, and direct burial-type conductors require a "cover." Explain the term "cover." ______________________________

____

____

____

13. In the spaces provided, fill in the cover (depth) for the following residential underground installations.

a. Type UF. No other protection. ______________ inches

b. Type UF below driveway. ______________ inches

c. Rigid metal conduit under lawn. ______________ inches

d. Rigid metal conduit under driveway. ______________ inches

e. Electrical metallic tubing between house and detached garage. ______________ inches

f. UF cable under lawn. Circuit is 120 volts, 20 amperes, GFCI protected. ______________ inches

g. Rigid nonmetallic conduit approved for direct burial. No other protection. ______________ inches

h. Rigid nonmetallic conduit. Circuit is 120 volts, 20 amperes, GFCI protected. ______________ inches

i. Rigid conduit passing over solid rock and covered by 2 inches of concrete extending down to rock.◄ ______________ inches

14. What section of the Code prohibits embedding Type UF cable in concrete? ______

____

15. The following is a layout of the garage circuit. Using the suggested cable layout, make a complete wiring diagram of this circuit, using colored pens or pencils to indicate conductor insulation colors.

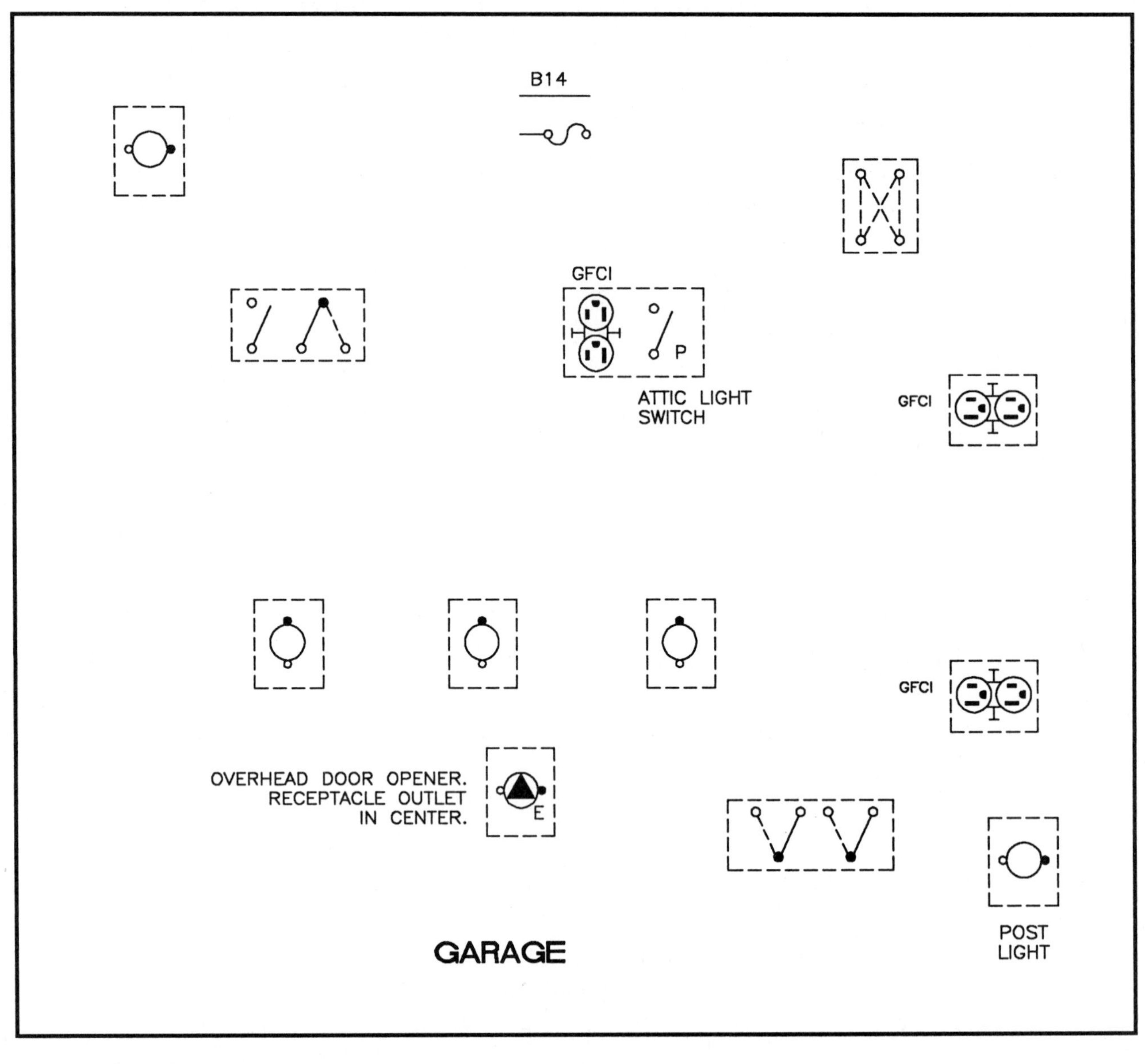

16. What type of motors are generally used for garage door openers? ______________

17. How is the direction of a split-phase motor reversed? ______________________

____

____

18. Motor running overload protection is generally based on what percentage of the full-load running current of the motor? ______________________

19. Show the proper dual-element fuse size for a 1/2-horsepower, 120-volt, single-phase motor to provide running overload protection. See *Table 430-148*.

____

____

____

UNIT 17

Recreation Room

OBJECTIVES

After studying this unit, the student will be able to

- understand three-wire (multiwire) circuits: the advantages and the cautions.
- understand how to install lay-in fixtures.
- calculate watts loss and voltage drop in two-wire and three-wire circuits.
- understand the term *fixture whips*.
- understand the advantages of installing multiwire branch circuits.
- understand problems that can be encountered on multiwire branch circuits as a result of open neutrals.

RECREATION ROOM LIGHTING

The recreation room is well-lighted through the use of six lay-in 2-foot by 4-foot fluorescent fixtures. These fixtures are exactly the same size as two 2′ × 2′ ceiling tiles. The fixtures rest on top of the ceiling tee bars, figure 17-1.

Junction boxes are mounted above the dropped ceiling usually within 1 or 2 feet of the intended fixture location. Four to 6 feet of flexible metal conduit are installed between this junction box and the fixture. These are commonly called *fixture whips*. They contain the correct type and size of conductor suitable for the temperature ratings and load required by the Code for recessed fixtures. See unit 7 for a complete explanation of Code regulations for the installation of recessed fixtures. See figure 17-2.

Fluorescent fixtures of the type shown in figure 17-1 might bear a label stating "Recessed Fluorescent Fixture." The label might also state "Suspended Ceiling Fluorescent Fixture." Although they might look the same, there is a difference.

Recessed fluorescent fixtures are intended for installation in cavities in ceilings and walls and are to be wired according to *Section 410-64* of the *National Electrical Code®*. These fixtures may also be installed in suspended ceilings if they have the necessary mounting hardware.

Suspended fluorescent fixtures are intended only for installation in suspended ceilings where the acoustical tiles, lay-in panels, and suspended grid are not part of the actual building structure.

Although the *National Electrical Code®*, in *Part N*, is entitled "Special Provisions for Flush and Recessed Fixtures," there has been a Formal Interpretation from the National Fire Protection Association FI No. 81-6 that states that fixture installations in suspended ceilings must conform to the requirements of *Part N* of the *NEC®*.

Underwriters Laboratories Standard 1570 covers recessed and suspended ceiling fixtures in detail.

The fluorescent fixtures in the recreation room are four-lamp fixtures having deluxe warm white (WWX) or deluxe cool white (CWX) lamps installed in them. Five of the fixtures are controlled by a single-pole switch located at the bottom of the

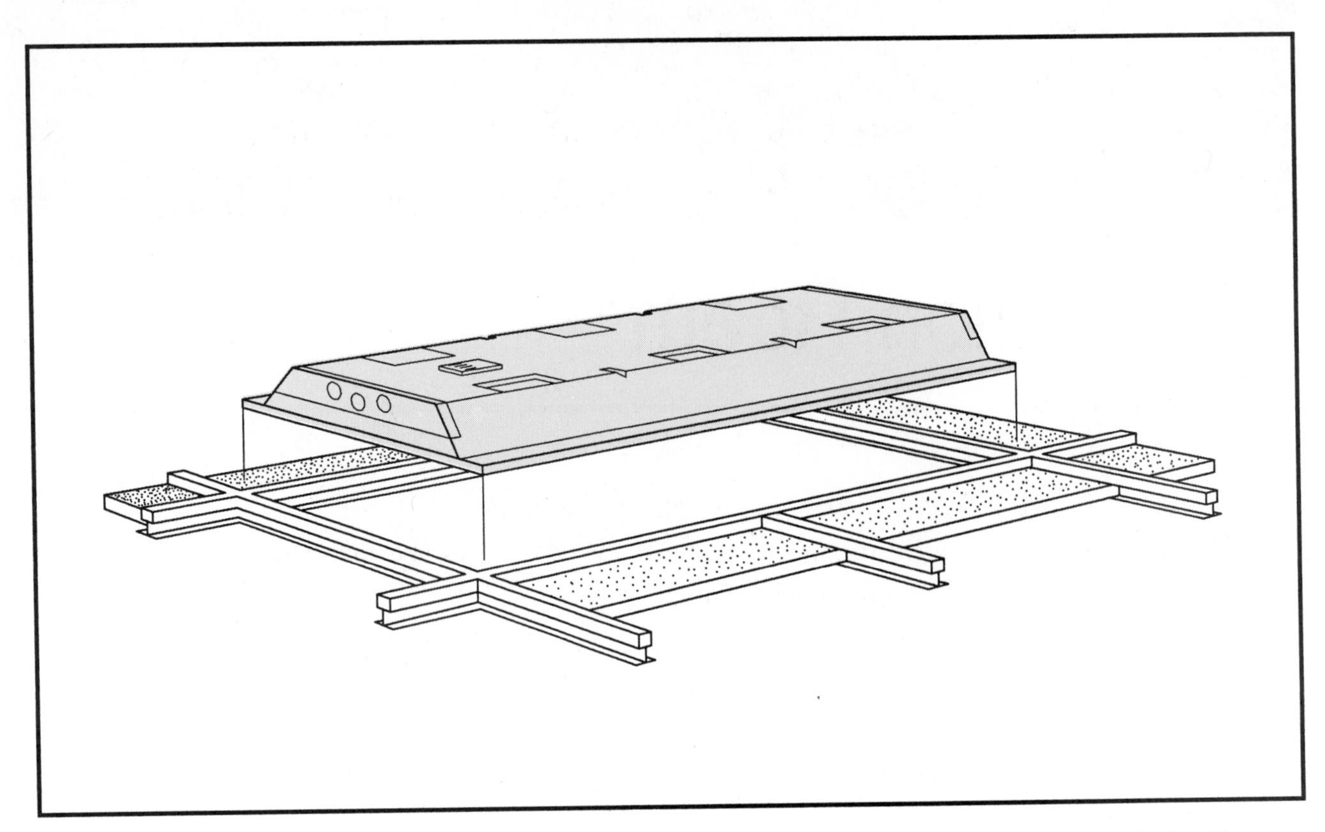

Fig. 17-1 Typical lay-in fluorescent fixture commonly used in conjunction with dropped acoustical ceilings.

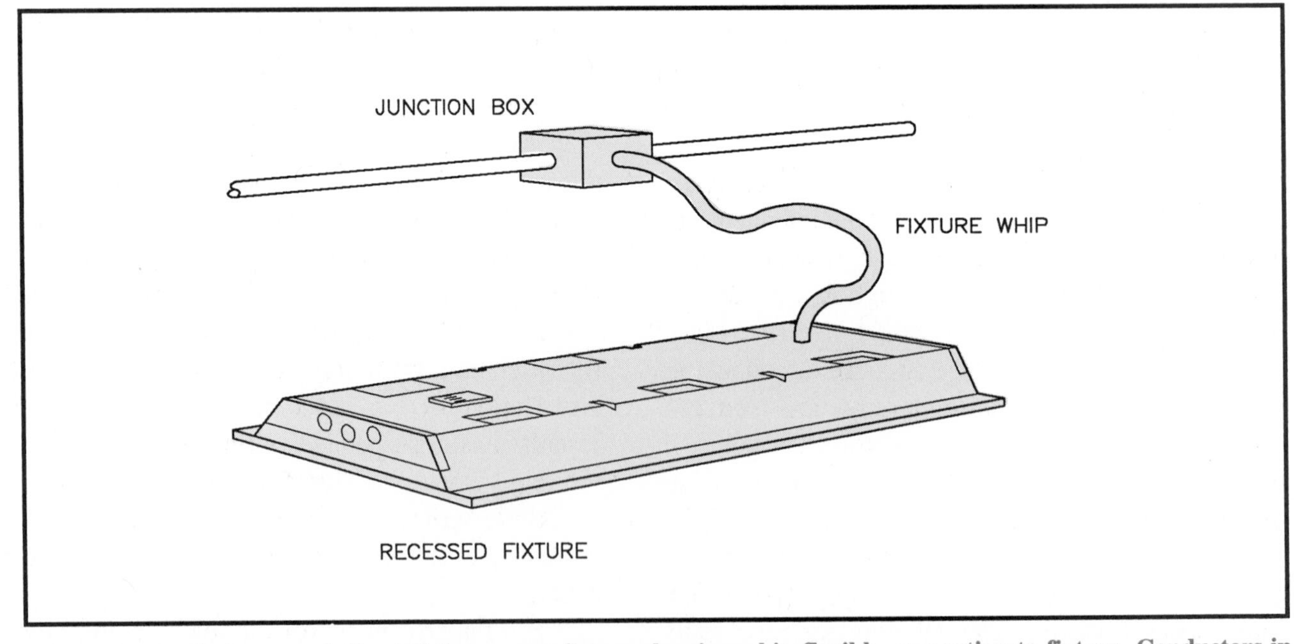

Fig. 17-2 Typical recessed "lay-in" fluorescent fixture showing whip flexible connection to fixture. Conductors in whip must have temperature rating as required on label in fixture. For example, "For Supply Use 90 °C conductors."

stairs. One fixture is controlled by the three-way switches that control the stairwell light fixture hung from the ceiling above the stair landing midway up the stairs.

This was done so that the entire recreation room lighting would not be on continually whenever someone came down the stairs to go into the workshop. This is a practical energy-saving feature.

Circuit B12 feeds the fluorescent fixtures in the recreation room. See cable layout, figure 17-3. Table 17-1 summarizes the outlets and estimated load for the recreation room.

How to Connect Recessed Lay-in Fixtures

Figure 17-2 shows how this is done. A junction box is located above the ceiling near the fixture. Fixture whips are generally made up of 3/8-inch flexible metal conduit, either by the electrician or by a manufacturer that specializes in this type of manufacturing.

Section 350-3, *Exception 3*, permits 3/8-inch flexible metal conduit in lengths not to exceed 6 feet to make the connection between the junction box and the fixture. *Section 410-67(c)* tells us that the junction box shall be at least one foot (305 mm) from the fixture.

Grounding of the fixture is done according to *Section 250-91(b)*, *Exception 1*, through the metal of the 3/8-inch flexible metal conduit when:

1. The flex is not over 6 feet (1.83 m) long.
2. The branch-circuit overcurrent protection does not exceed 20 amperes.
3. The connectors for the flex are listed for grounding purposes. This refers to UL listing.

Flexible metal conduit must be supported within 12 inches (305 mm) of any outlet box, junction box, cabinet or fitting according to *Section 350-4*. A "fixture whip" is exempt from this requirement per *Section 350-4, Exception No. 3.* Yet, many electricians do secure the flex within 12 inches of the outlet box so as not to pull the flex out of the connector while they are making up the electrical connections and installing the fixture in the lay-in ceiling.

DESCRIPTION	QUANTITY	WATTS	VOLT-AMPERES
Circuit Number B11			
Receptacles @ 120 W each	7	840	840
Closet recessed fixtures @ 75 W each	2	150	150
Exhaust fan (2.5 A × 120 V)	1	160	300
TOTALS	10	1150	1290
Circuit Number B12			
Fixture on stairway	1	100	100
Recessed fluorescent fixtures with four 40-W lamps each	6	960	1080
TOTALS	7	1060	1180
Circuit Number B9			
Receptacles @ 120 W each	4	480	480
Wet-bar recessed fixtures @ 75 W each	4	300	300
TOTALS	8	780	780

Table 17-1. Recreation Room: outlets and estimated load (three circuits).

RECEPTACLES AND WET BAR

The circuitry for the recreation room wall receptacles and the wet-bar lighting area introduces a new type of circuit. This is termed a *multiwire circuit* or a *three-wire circuit.*

In many cases the use of a multiwire circuit can save money in that one three-wire circuit will do the job of two two-wire circuits.

Also if the loads are nearly balanced in a three-wire circuit, the neutral conductor carries only the unbalanced current. This results in less voltage drop and watts loss for a three-wire circuit as compared to similar loads connected to separate two-wire circuits.

Figures 17-4 and 17-5 (pages 243 and 244) illustrate the benefits of a multiwire circuit relative to watts loss and voltage drop. The example, for simplicity, is a purely resistive circuit, and loads are exactly equal.

The distance from the load to the source is 50 feet. The conductor size is No. 14 AWG copper. The resistance of each 50-foot length of No. 14 AWG is 0.154 ohm, as calculated from the data in *Table 8* of the *National Electrical Code*®

EXAMPLE 1: TWO 2-WIRE CIRCUITS

Watts loss in each current-carrying conductor is

$$\text{Watts} = I^2R = 10 \times 10 \times 0.154 = 15.4 \text{ watts}$$

Watts loss in all four current-carrying conductors is

$$15.4 \times 4 = 61.6 \text{ watts}$$

Voltage drop in each current-carrying conductor is

$$E_d = IR$$
$$= 10 \times 0.154$$
$$= 1.54 \text{ volts}$$

E_d for both current-carrying conductors in each circuit:

$$2 \times 1.54 = 3.08 \text{ volts}$$

Voltage available at load:

$$120 - 3.08 = 116.92 \text{ volts}$$

EXAMPLE 2: ONE 3-WIRE CIRCUIT

Watts loss in each current-carrying conductor is

$$\text{Watts} = I^2R = 10 \times 10 \times 0.154 = 15.4 \text{ watts}$$

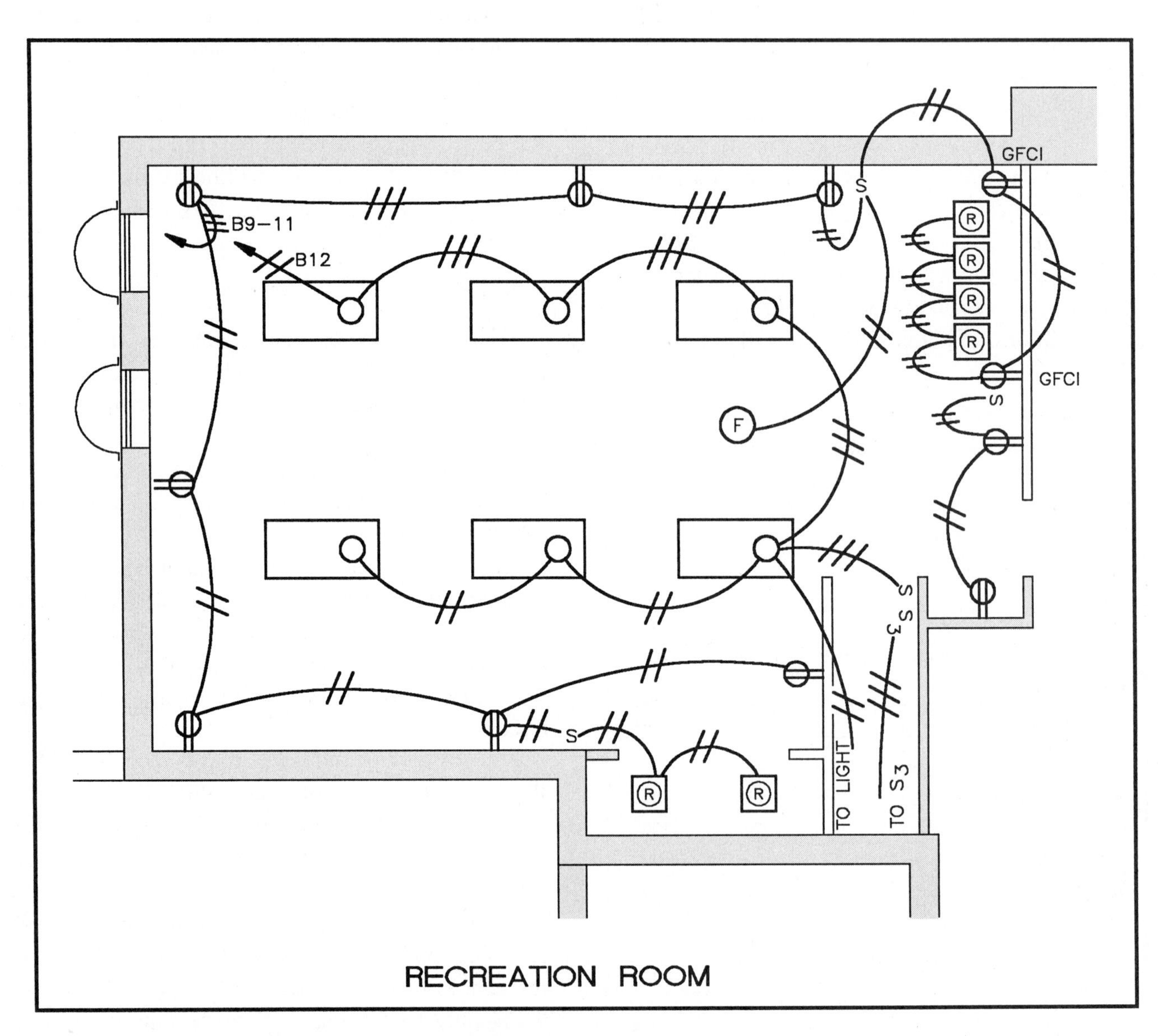

Fig. 17-3 Cable layout for recreation room. The six fluorescent fixtures are connected to Circuit B12. The receptacles located around the room are connected to Circuit B11. The four recessed fixtures over the bar plus the three receptacles located on the bar wall plus the receptacles located to the right of the steps as you come down the steps are connected to Circuit B9. Circuits B9 and B11 are three-wire circuits.

Watts loss in both current-carrying conductors is

$$15.4 \times 2 = 30.8 \text{ watts}$$

Voltage drop in each current-carrying conductor is

$$E_d = IR$$
$$= 10 \times 0.154$$
$$= 1.54 \text{ volts}$$

E_d for both current-carrying conductors is:

$$2 \times 1.54 = 3.08$$

Voltage available at loads:

$$240 - 3.08 = 236.92$$

Voltage available at each load:

$$\frac{236.92}{2} = 118.46 \text{ volts}$$

It is readily apparent that a multiwire circuit can greatly reduce watts loss and voltage drop. This is the same reasoning that is used for the three-wire service-entrance conductors and the three-wire feeder to Panel B in this residence.

In the recreation room, a three-wire cable carrying Circuits B9 and B11 feeds from Panel B to the receptacle closest to the panel. This three-wire cable continues to each receptacle wall box along the same wall. At the third receptacle along the wall, the multiwire circuit splits, where the wet-bar Circuit B9 continues on to feed into the receptacle outlet to the left of the wet bar.

The white conductor of the three-wire cable is common to both the receptacle circuit and the wet-bar circuits. The black conductor feeds the receptacles. The red conductor is spliced straight through the boxes and feeds the wet bar area.

Care must be used when connecting a three-wire circuit to the panel. The black and red conductors *must* be connected to the opposite phases in the panel to prevent heavy overloading of the neutral grounded (white) conductor.

The neutral (grounded) conductor of the three-wire cable carries the unbalanced current. This current is the difference between the current in the black wire and the current in the red wire. For example, if one load is 12 amperes and the other load is 10 amperes, the neutral current is the differ-

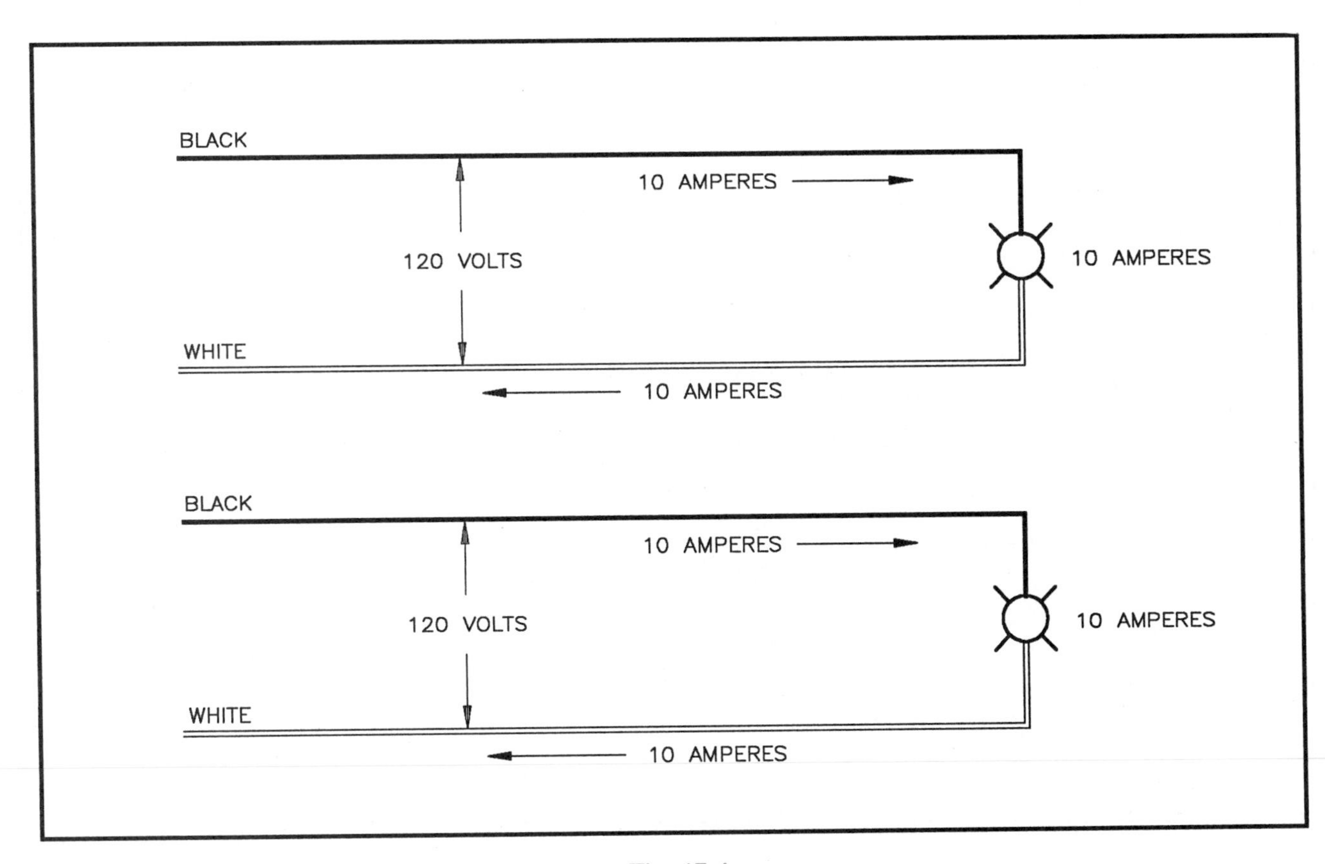

Fig. 17-4

ence between these loads, 2 amperes, figure 17-6.

If the black and red conductors of the three-wire cable are connected to the same phase in Panel A, figure 17-7, the neutral conductor must carry the total current of both the red and black conductors rather than the unbalanced current. As a result, the neutral conductor will be overloaded. All single-phase, 120/240-volt panels are clearly marked to help prevent an error in phase wiring. The electrician must check all panels for the proper wiring diagrams before beginning an installation.

Figure 17-7 shows how an improperly con-

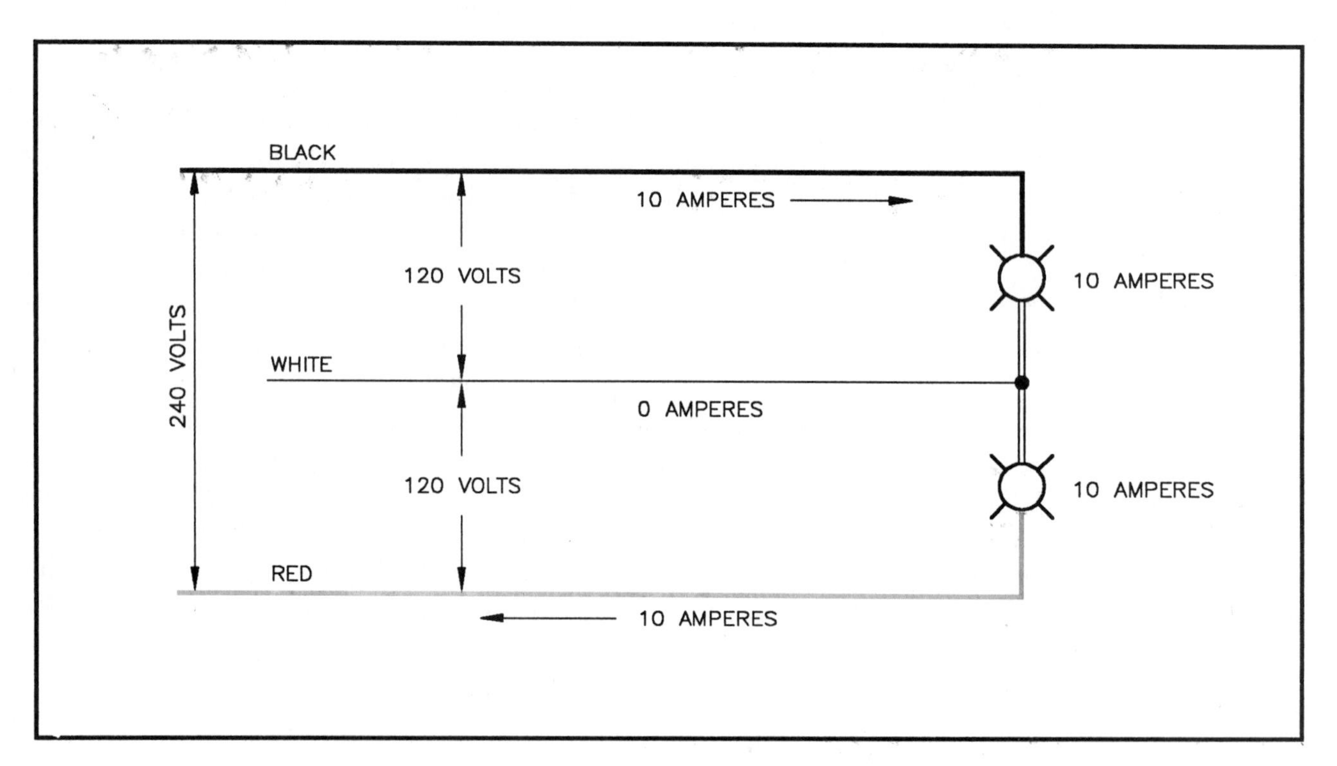

Fig. 17-5

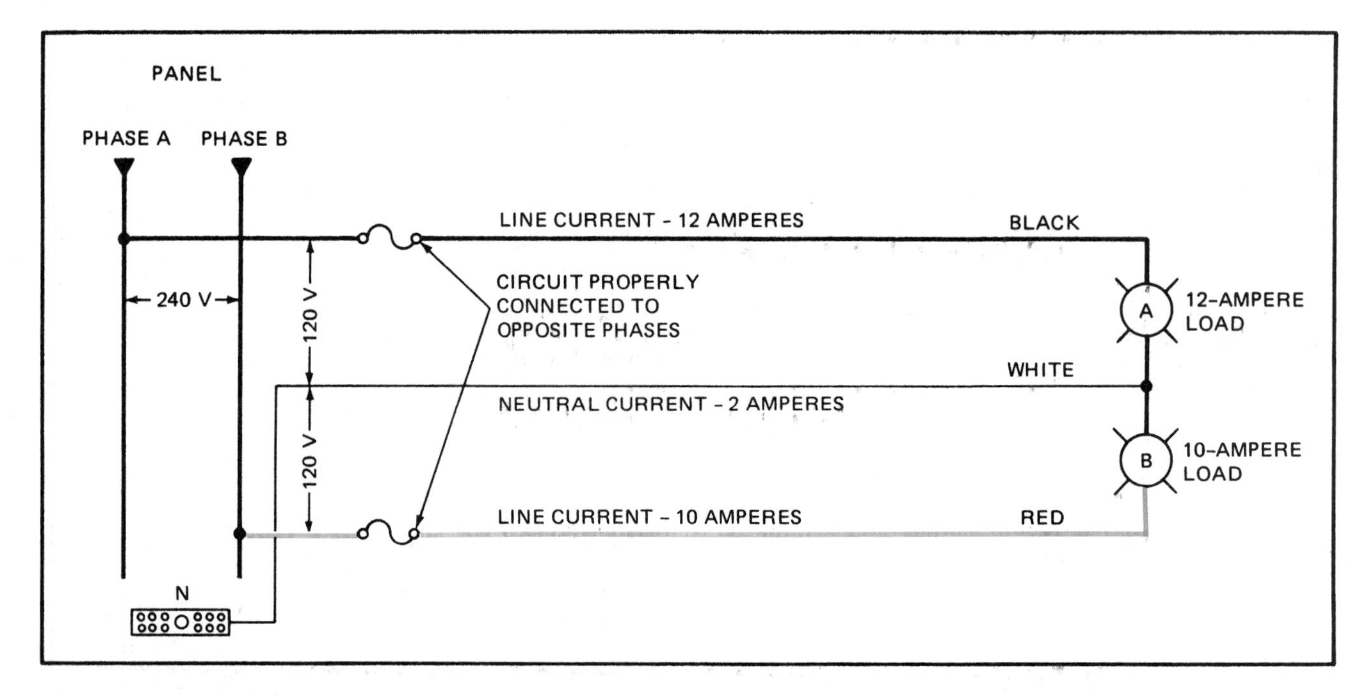

Fig. 17-6 Correct wiring connections for three-wire (multiwire) circuit.

nected three-wire circuit results in an overloaded neutral conductor. If an open neutral occurs on a three-wire circuit, some of the electrical appliances in operation may experience voltages higher than the rated voltage at the instant the neutral opens.

For example, figure 17-8 shows that for an open neutral condition, the voltage across load A decreases and the load across load B increases. If the load on each circuit changes, the voltage on each circuit also changes. According to Ohm's law, the voltage drop across any device in a series circuit is directly proportional to the resistance of that device. In other words, if load B has twice the resistance of load A, then load B will be subjected to twice the voltage of load A for an open neutral condition. To insure the proper connection, care must be used when splicing the conductors.

An example of what can occur should the neutral of a three-wire multiwire circuit open is shown in figure 17-9. Trace the flow of current from phase A through the television set, then through the toaster, then back to phase B, thus completing the circuit. The following simple calculations show why the television set (or stereo or home computer) can be expected to burn up.

$$R_t = 8.45 + 80 + 88.45 \text{ ohms}$$

$$I = \frac{E}{R} = \frac{240}{88.45} = 2.71 \text{ amperes}$$

Voltage appearing across the toaster:

$$IR = 2.71 \times 8.45 = 22.9$$

Voltage appearing across the television:

$$IR = 2.71 \times 80 = 216.8 \text{ volts}$$

This example illustrates the problems that can arise with an open neutral on a three-wire, 120/240-volt multiwire branch circuit.

The same problem can arise when the neutral of the utility company's incoming service-entrance conductors (underground or overhead) opens. The problem is minimized because the neutral of the service is solidly grounded to the water piping system within the building. However, there are cases on record where poor service grounding has resulted in serious and expensive damage to appliances within the home because of an open neutral in the incoming service-entrance conductors. For example, poor service grounding results when relying on a driven iron pipe (that rusts in a short period of time) as the

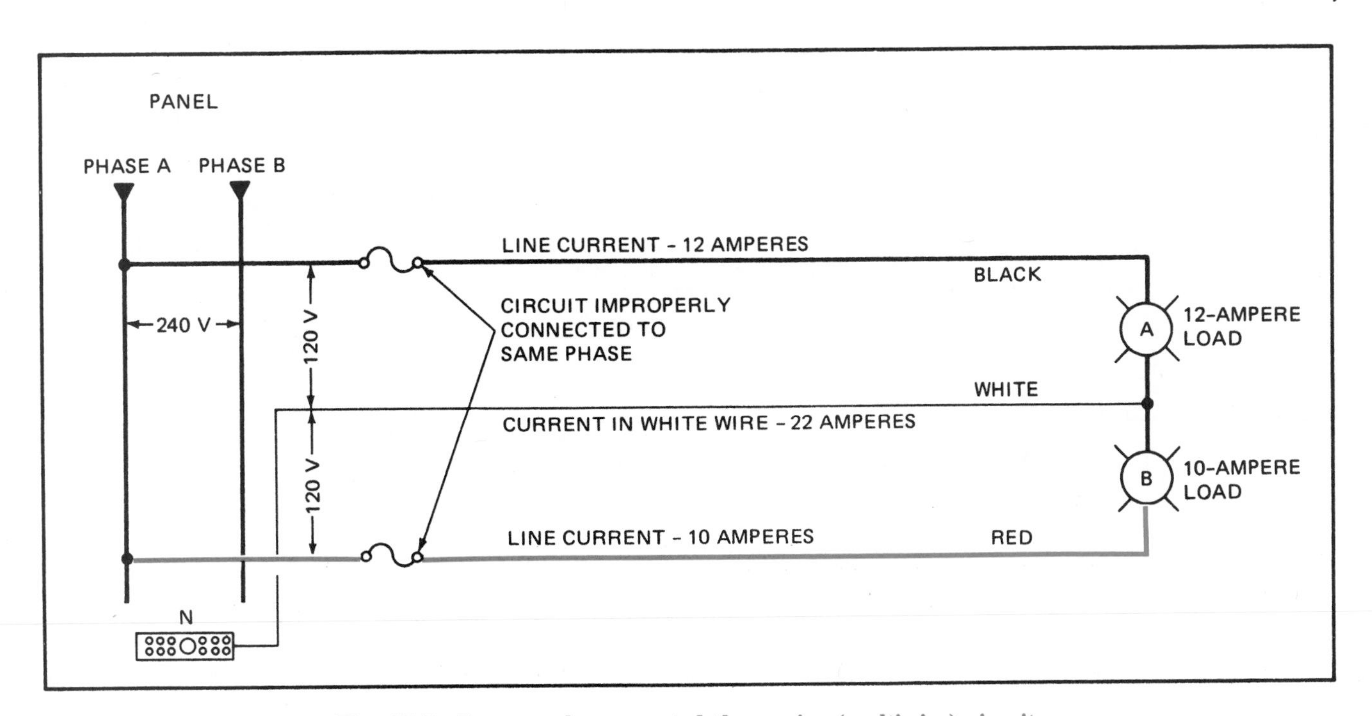

Fig. 17-7 Improperly connected three-wire (multiwire) circuit.

only means of obtaining the service equipment ground.

The integrity of using ground rods only is highly questionable, whether the rod is an iron pipe, a galvanized pipe, or a copper-clad rod. The Code prohibits the use of ground rods as the only means of obtaining a good service-entrance equipment ground.

See unit 27 for details on the grounding and bonding of service-entrance equipment.

In summary, three-wire circuits are used occasionally in residential installations. The two hot conductors of three-wire cable are connected to opposite phases. One white grounded (neutral) conductor is common to both hot conductors.

The two receptacle outlets above the wet bar in the recreation room are GFCI protected as indicated on the plans. However, this is *not* a Code requirement. *Section 210-8(a)(4)* requires GFCI protected receptacles in unfinished basements. This recreation room is finished (paneled). In addition, *Section 210-8(a)(5)* requires GFCI protected receptacles within 6 feet (1.83m) of a *kitchen sink*. Therefore, the small sink located in the recreation room is not governed by *Section 210-8(a)(5)*.

A single-pole switch to the right of the wet bar controls the four recessed fixtures above it. See unit 7 for details pertaining to recessed fixtures.

An exhaust fan of a type similar to the one installed in the laundry is installed in the ceiling of the recreation room to exhaust stale, stagnant, and smoky air from the room.

The basic switch, receptacle, and fixture connections are repetitive of most wiring situations discussed in previous units.

The wall boxes for the receptacles are selected based upon the measurements of the furring strips on the walls. Two-by-two furring strips will require the use of 4-inch square boxes with suitable raised covers. If the walls are furred with 2 × 4s, then possibly sectional device boxes could be used. Select a box that can contain the number of conduc-

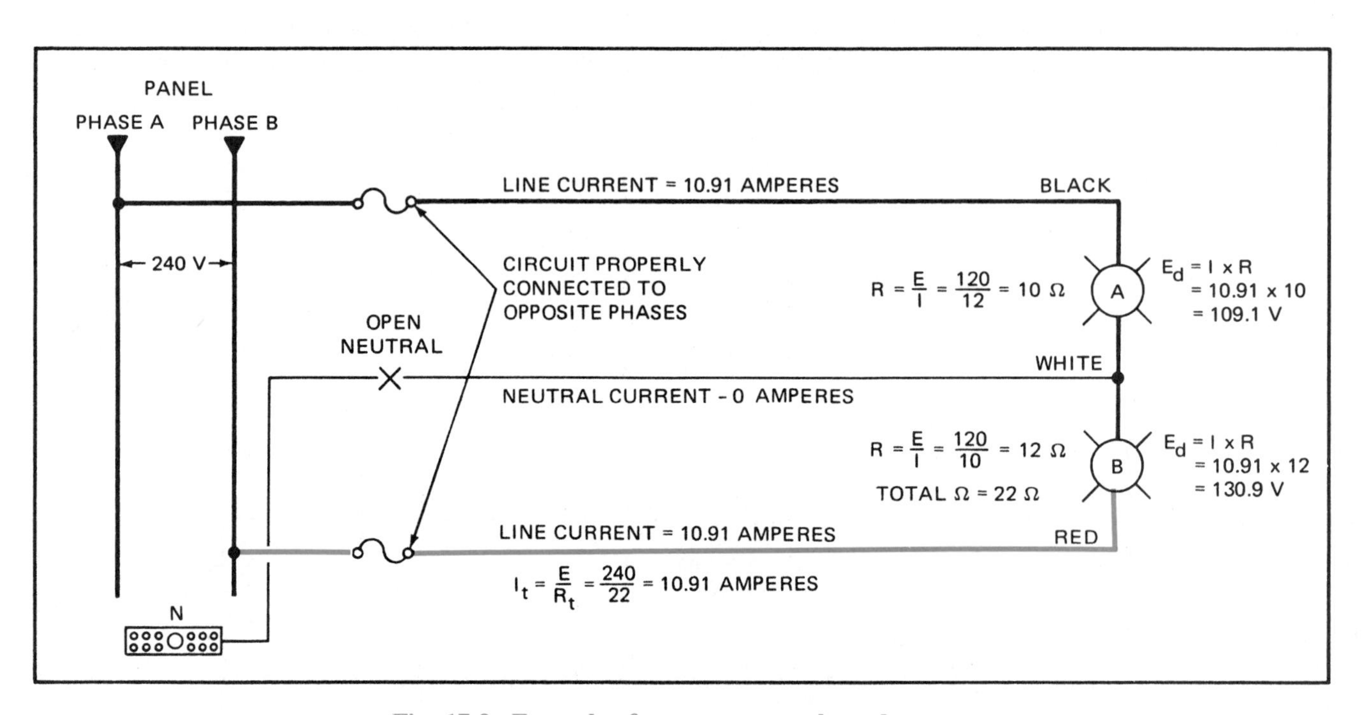

Fig. 17-8 Example of an open neutral conductor.

tors, devices, and clamps to "meet Code," as discussed many times in this text. After all, the installation must be safe.

Probably the biggest factor in deciding what wiring method (cable or electrical metallic tubing) will be used to wire the recreation room is "what size and type of wall furring will be used?" See unit 4 for the in-depth discussion on the mechanical protection required for nonmetallic sheathed cables where the cables will be less than 1-1/4 inch from the edge of the framing members.

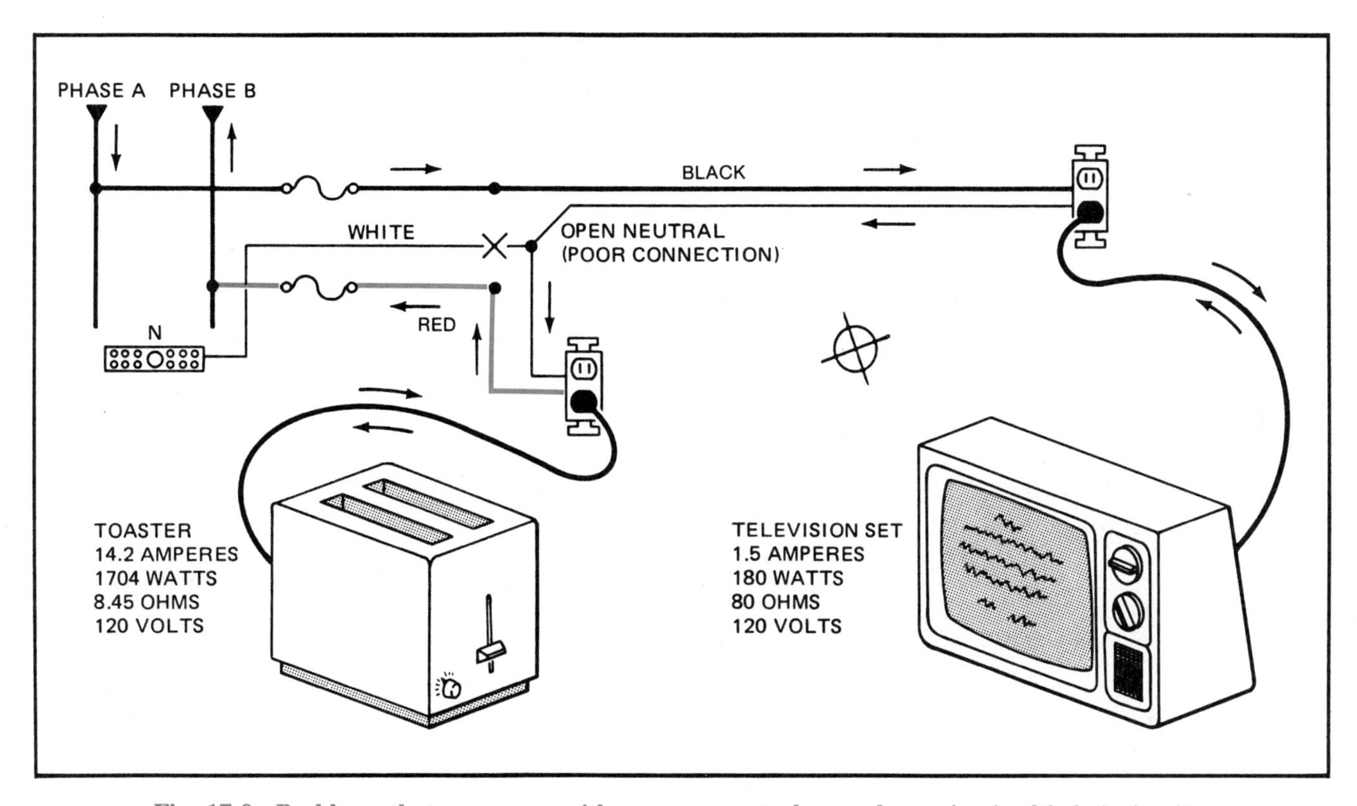

Fig. 17-9 Problems that can occur with an open neutral on a three-wire (multiwire) circuit.

REVIEW

Note: Refer to the Code or plans where necessary.

1. What is the total current draw when all six fluorescent fixtures are turned on?

____

____

____

2. The junction box that will be installed above the dropped ceiling near the fluorescent fixtures closest to the stairway will have _______________ (number of) No. 14 AWG conductors.

3. Why is it important that the hot conductor in a three-wire circuit be properly connected to opposite phases in a panel? _______________________________________

____

4. In the diagram, Load A is rated at 10 amperes, 120 volts. Load B is rated at 5 amperes, 120 volts.

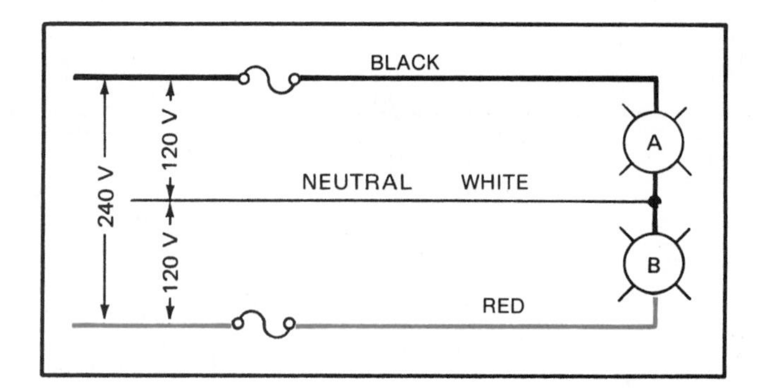

a. When connected to the three-wire circuit as indicated, how much current will flow in the neutral? _______________________________________

b. If the neutral should open, to what voltage would each load be subjected, assuming both loads were operating at the time the neutral opened? Show all calculations.

5. Calculate the watts loss and voltage drop in each conductor in the following circuit.

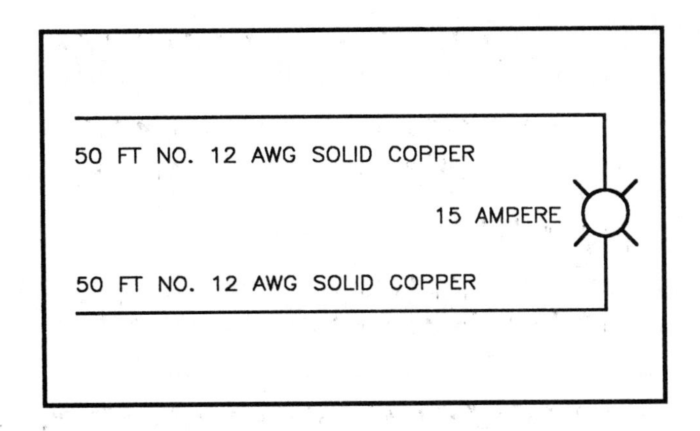

6. Unless specifically designed, all recessed incandescent fixtures must be provided with factory-installed ________________________________

7. If the fluorescent fixtures in the recreation room were to be mounted on the ceiling, what sort of marking would you look for on the label of the fixture? The ceiling is low-density cellulose fiberboard. ____________________________

____

8. What is the current draw of the recessed fixtures above the bar? ____________

____

____

____

9. a. The computed VA load for Circuit B9 is ________________ volt-amperes.
   b. The computed VA load for Circuit B11 is ________________ volt-amperes.
   c. The computed VA load for Circuit B12 is ________________ volt-amperes.
10. Calculate the total current draw for Circuits B9, B11, and B12.

11. Complete the wiring diagram for the recreation room. Follow the suggested cable layout. Use colored pencils or pens to identify the various colors of the conductor insulation.

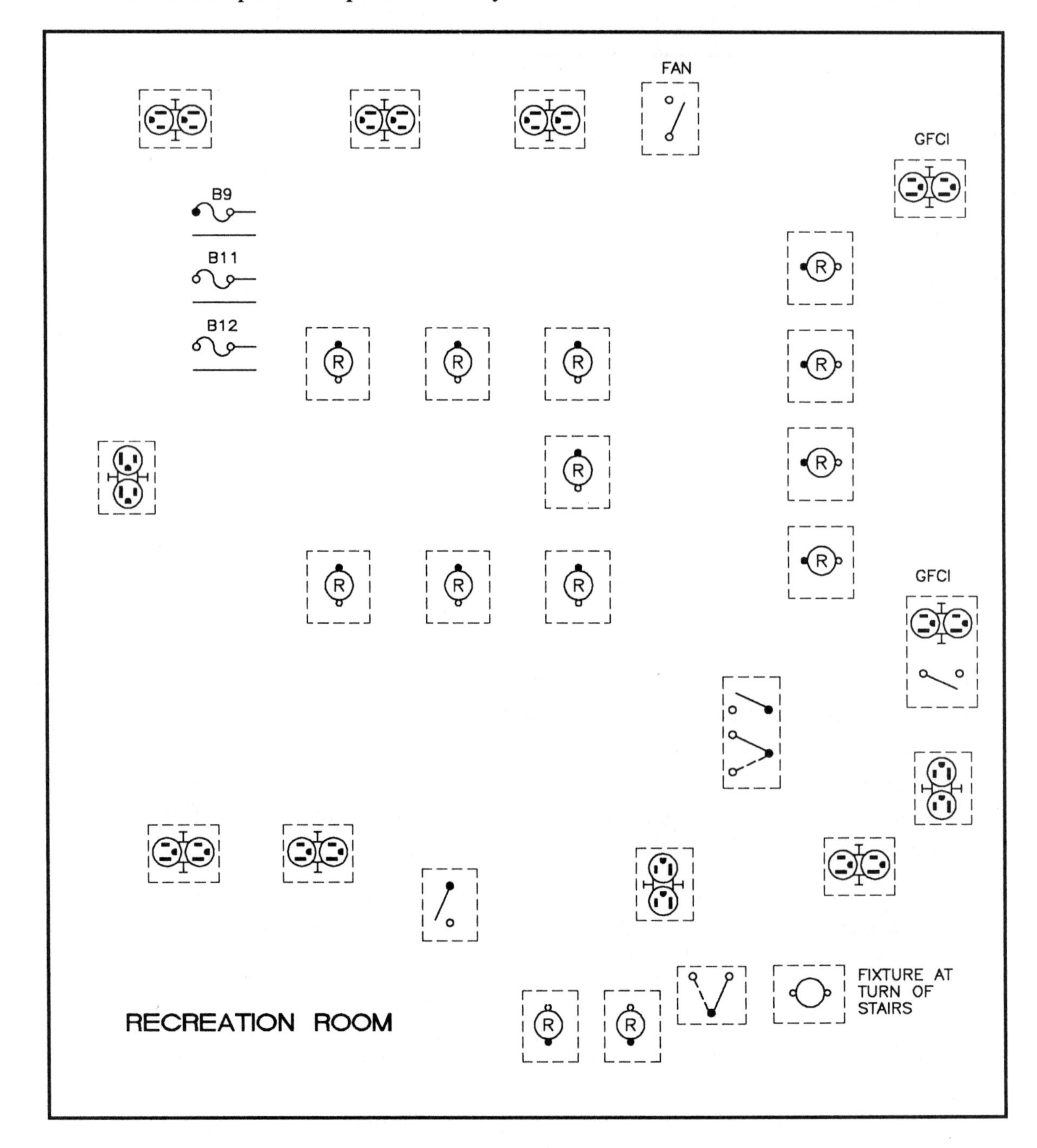

12. a. May a fluorescent fixture that is marked "Recessed Fluorescent Fixture" be installed in a suspended ceiling? ______

b. May a fluorescent fixture that is marked "Suspended Ceiling Fluorescent Fixture" be installed in a recessed cavity of a ceiling? ______

c. What part of the Code covers the installation of recessed fixtures? ______

______

______

d. What part of the Code covers the installation of suspended ceiling fixtures?

______

______

UNIT 18

Lighting Branch Circuit, Receptacle Circuits for Workshop

OBJECTIVES

After studying this unit, the student will be able to

- understand the meaning, use, and installation of multioutlet assemblies.
- understand where GFCI protection is required in basements.
- understand the Code requirements for conduit installation.
- select the proper outlet boxes to use for surface mounting.
- make conduit fill calculations based upon the number of conductors in the conduit.
- make use of derating and correction factors for determining conductor current-carrying capacity.

The workshop area is supplied by more than one circuit:

A13 Separate circuit for freezer

A17 Lighting

A18 Plug-in strip (multioutlet assembly)

A20 Two receptacles on window wall

The wiring method in the workshop is electrical metallic tubing (EMT). Figure 18-1 shows the conduit layout for the workshop.

A single-pole switch at the entry controls all four ceiling porcelain lampholders. However, note on the plans that two of these lampholders have pull-chains, which would allow the homeowner to turn these lampholders on and off as needed, figure 18-2. They would not have to be on all of the time; energy savings is the end result.

In addition to supplying the lighting, Circuit A17 also feeds the smoke detectors, chime transformers, and ceiling exhaust fan. The current draw of the smoke detectors and chime transformer is extremely small. Refer to Table 18-1, Workshop Outlet Count and Estimated Load for Lighting and Table 18-2, Outlet Count and Estimated Load for Three 20-Ampere Circuits.

Smoke detectors are discussed in detail in unit 26. Smoke detectors are *not* to be connected to GFCI-protected circuits.

WORKBENCH LIGHTING

Two two-lamp, 40-watt fluorescent fixtures are mounted above the workbench to reduce shadows over the work area. The electrical plans show that these fixtures are controlled by a single-pole wall

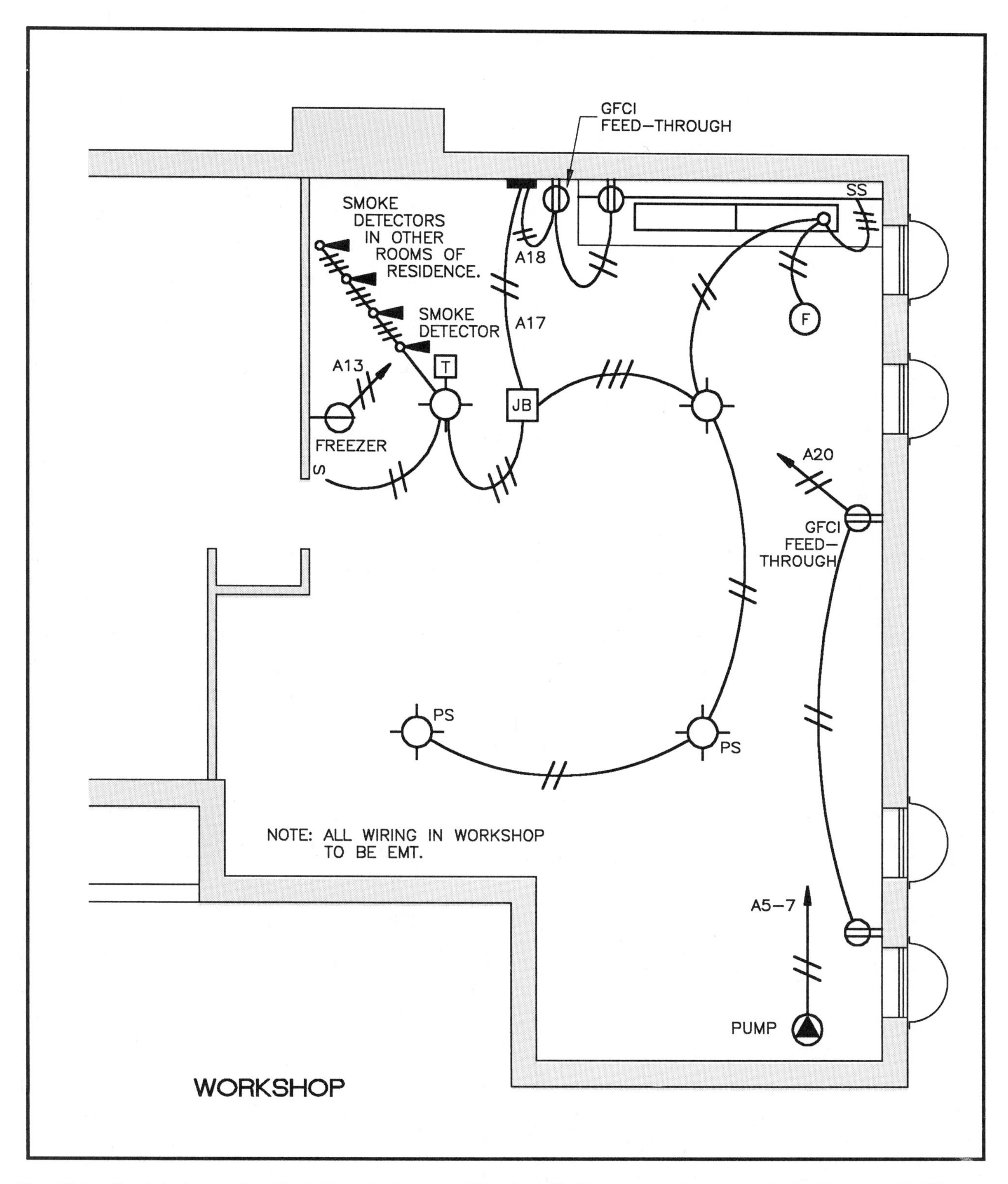

Fig. 18-1 Conduit layout for Workshop. Lighting is Circuit A17. Freezer single receptacle is Circuit A13. The two receptacles on the window wall are connected to Circuit A20. Plug-in strip receptacles and receptacle to the right of Main Panel are connected to Circuit A18. All receptacles are GFCI protected except the single receptacle outlet serving the freezer.

switch. A junction box is mounted immediately above or adjacent to the fluorescent fixture so that the connections can be made readily.

Armored cable, flexible metal conduit, or flexible cord (Type S, SJ, or equivalent) can be used to connect the fixture to the junction box. The fluorescent fixture must be grounded. Thus, any flexible cord used must contain a third conductor for grounding only.

RECEPTACLE OUTLETS

► *Section 210-8(a)(4)* of the *National Electrical Code*® requires that all 125-volt, single-phase, 15- or 20-ampere receptacles installed in unfinished basements and crawl spaces that are at or below ground level must be GFCI protected.

Exceptions to this requirement are:

- for a *single receptacle* that is supplied by a separate dedicated circuit for cord- and plug-connected appliances, such as a refrigerator or freezer. This receptacle must be properly identified. Duplex or triplex receptacles are not permitted under this exception.
- for a *single receptacle* that is installed for a permanently installed sump pump. Duplex or triplex receptacles are not permitted under this exception.
- for the receptacle(s) installed for the laundry equipment as required by *Section 210-52(f)* and *Section 220-4(c)*. ◄

Most large freezers have an alarm or pilot light to indicate when the power to the freezer is lost.

See unit 6 for full discussion of ground fault circuit interrupters.

Section 210-52(g) makes it mandatory to install at least one receptacle outlet in a basement. If the laundry area is in the basement, then a receptacle outlet must also be installed for the laundry equipment (*Section 210-52(f)*) and be within 6 feet (1.83 m) of the intended location of the washer (*Section 210-50(c)*) and must be a 20-ampere circuit.

DESCRIPTION	QUANTITY	WATTS	VOLT-AMPERES
Ceiling lights @ 100 W each	4	400	400
Fluorescent fixtures Two 40-W lamps each	2	160	200
Chime transformer	1	8	10
Exhaust fan	1	80	90
Smoke detectors @ 1/2 W each	4	2	2
TOTALS	12	650	702

Table 18-1. Workshop: outlet count and estimated load. (Lighting load) (Wire with conduit)

DESCRIPTION	QUANTITY	WATTS	VOLT-AMPERES
20-A Circuit, Number A18			
Receptacle next to main panel	1	180	180
A six-receptacle plug-in multioutlet assembly at 1 1/2 A per outlet (180 VA)	1	1080	1080
TOTALS	2	1260	1260
20-A Circuit, Number A20			
Receptacles on window wall @ 180 W each	2	360	360
TOTALS	2	360	360
20-A Circuit, Number A13			
Single receptacle for freezer (*not* GFCI)	1	—	696
TOTALS	1	1620	2316

Table 18-2. Workshop: outlet count and estimated load for three 20-A circuits. (Wire with conduit) (Receptacle load).

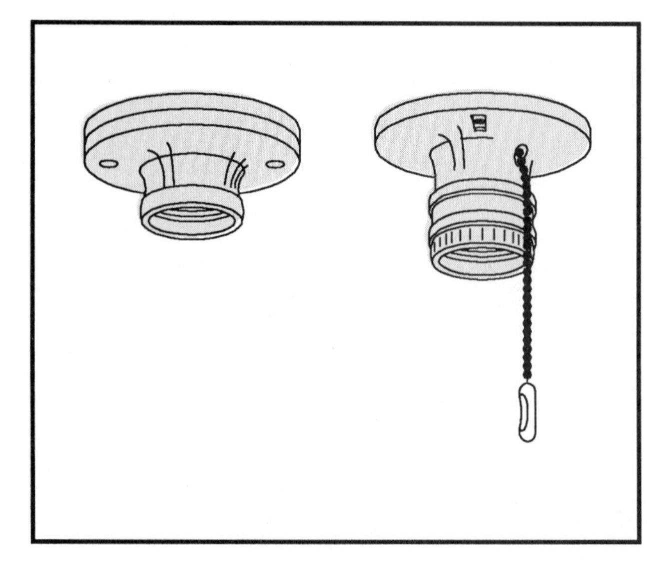

Fig. 18-2 This illustration shows porcelain lampholders of both the keyless and pull-chain types. These lampholders are also available with a receptacle outlet.

The laundry receptacle outlet required in *Section 210-52(f)* is in addition to the basement receptacle required in *Section 210-52(g)*.

The circuit feeding the required laundry receptacle shall have no other outlets connected to it, *Section 220-4(c)*.

When a single receptacle is installed on an individual branch circuit, the receptacle rating must not be less than the rating of the branch circuit, *Section 210-2(b)(1)*.

Typical outlet boxes are shown in figure 18-3.

CABLE INSTALLATION IN BASEMENTS

Unit 4 covers the installation of nonmetallic-sheathed cable (Romex) and armored cable (BX).

Cable assemblies smaller than two No. 6 conductors or three No. 8 conductors must be run through holes bored in joists or on running boards when installed in unfinished basements, *Section 336-12*.

If armored cable is installed in an unfinished basement, then it must follow the building surface or running boards to which it is fastened, *Section 333-11*. An exception to this requirement is armored cable run on the undersides of floor joists in basements where supported at each joist and so located as not to be subject to physical damage, *Section 333-11, Exception No. 2. Note:* The local inspection authority usually interprets the meaning of "subject to physical damage." Similar methods are used to protect cables in exposed areas in basements and cables in attics (see unit 15).

CONDUIT INSTALLATION IN BASEMENTS

Many local electrical codes do not permit cable wiring in unfinished basements other than where absolutely necessary. For example, a cable may be dropped into a basement from a switch at the head of the basement stairs. Transitions between cable and conduit also must be made. Usually this will be a junction box. The electrician must check the local codes before selecting conduit or cable for the installation, to prevent costly wiring errors.

Since local codes vary, the electrician should also check with the local inspection authority that enforces the code. Electrical utility companies can supply additional information on local regulations. The electrician should never assume that the

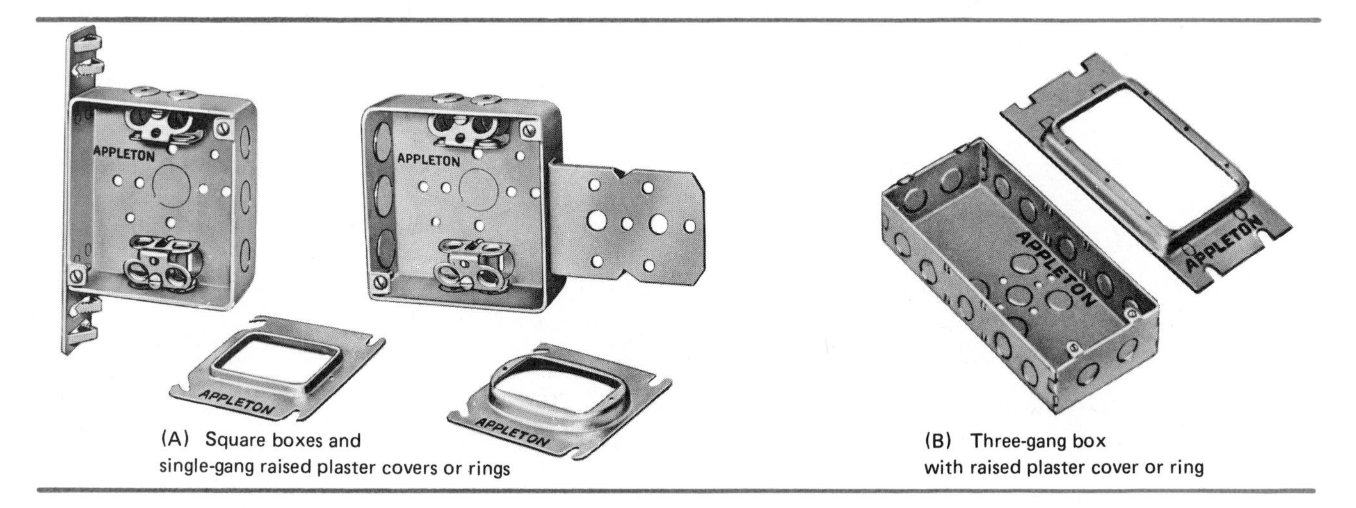

(A) Square boxes and single-gang raised plaster covers or rings

(B) Three-gang box with raised plaster cover or ring

Fig. 18-3 Typical outlet boxes and raised covers.

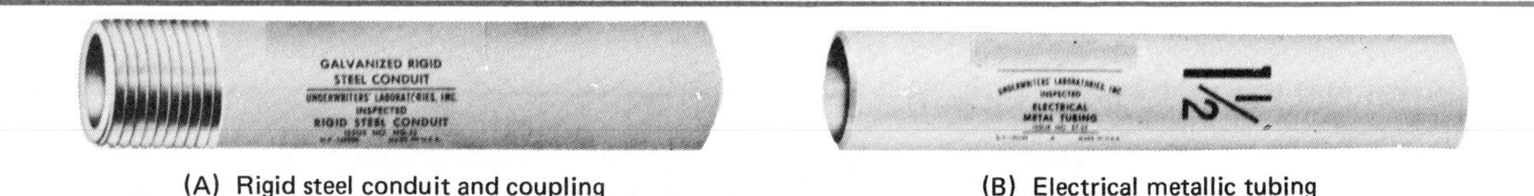

(A) Rigid steel conduit and coupling

(B) Electrical metallic tubing

Fig. 18-4 Types of electrical conduit.

National Electrical Code® is the recognized standard everywhere. Any city or state may pass electrical installation and licensing laws. In many cases these laws are more stringent than the *National Electrical Code*® Types of electrical conduit are shown in figure 18-4.

Outlet boxes may be fastened to masonry walls with lead or plastic anchors, shields, concrete nails, or power-actuated studs.

Conduit may be fastened to masonry surfaces using EMT straps, figure 18-5(A). Cable may be fastened to wood surfaces using staples, figure 18-5(B) or (C).

The conduits on the workshop walls and on the ceiling are exposed to view. The electrician must install the exposed wiring in a neat and skillful manner while complying with the following practices:

- The conduit runs must be straight.
- Bends and offsets must be true.
- Vertical runs down the surfaces of the walls must be plumb.

All conduits and boxes on the workshop ceiling are fastened to the underside of the wood joists, figure 18-6.

If the electrician is able to get to the residence before the basement concrete floor is poured, he might wish to install conduit runs in or under the concrete floor from the main panel to the freezer outlet, the window wall receptacles, and to the water pump disconnect switch location. This can be a cost-saving (labor and material) benefit.

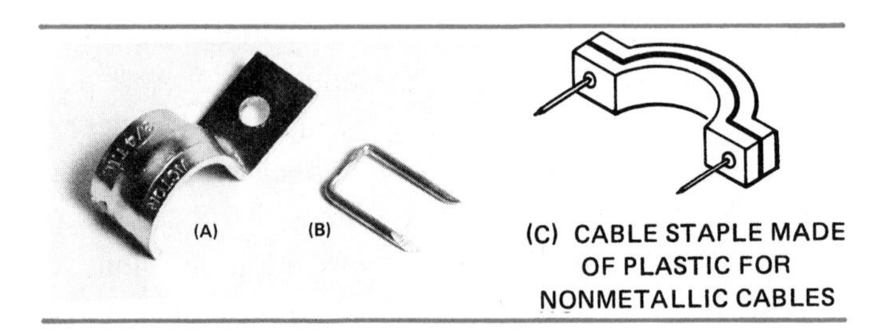

Fig. 18-5 Devices used for attaching conduit and cable to masonry surfaces (A) and wood surfaces (B) or (C).

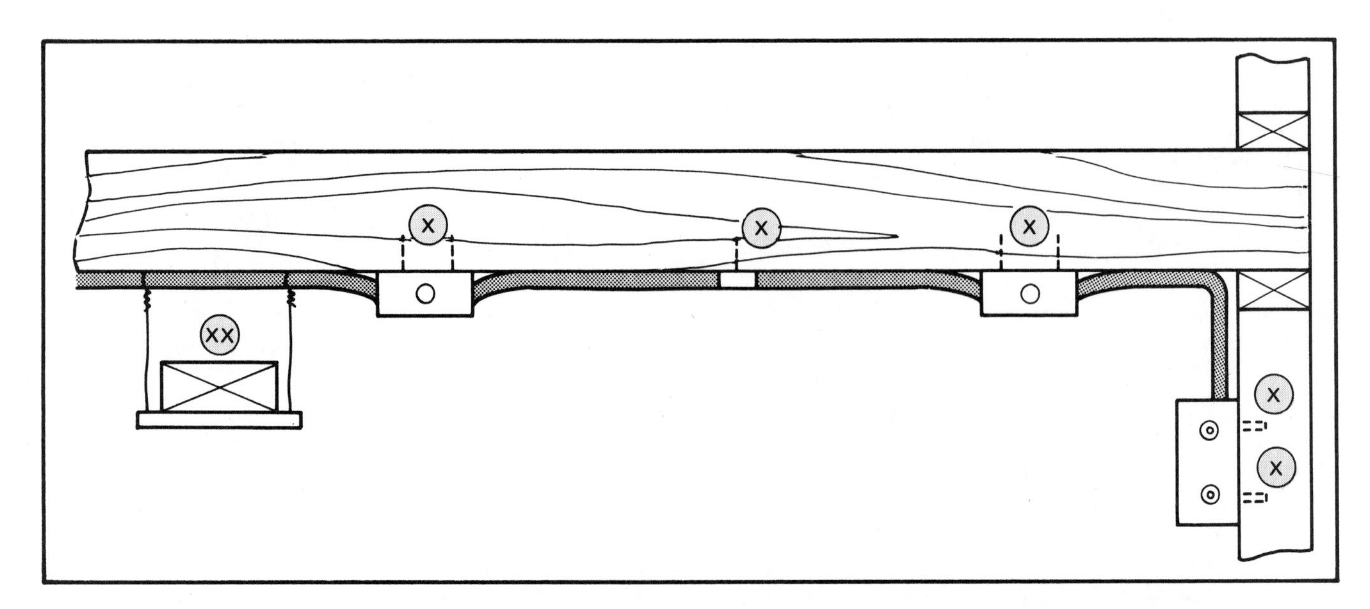

Fig. 18-6 Raceways, boxes, fittings, and cabinets must be securely fastened in place, points X. It is not permissible to hang other raceways, cables, or nonelectrical equipment from an electrical raceway, point XX, *Section 300-11(a), (b)*. Refer to figure 24-20 for exception that allows low-voltage thermostat cables to be fastened to the conduit that feeds a furnace or other similar equipment.

Outlet Boxes for Use in Exposed Installations

For surface wiring such as the workshop, various types and sizes of outlet boxes may be used. For example, a handy box, figure 18-7, could be used for the freezer receptacle. The remaining workshop boxes could be 4-inch square outlet boxes with raised covers, figure 18-8. The box size and type is determined by the maximum number of conductors contained in the box, *Article 370.*

Switch and outlet boxes are available with knockouts of the following sizes: 1/2 inch, 3/4 inch, and 1 inch. The knockout size selected depends upon the size of the conduits entering the box. The conduit size is determined by the maximum percentage of fill of the total cross-sectional area of the conduit by the conductors pulled into the conduit.

Supporting equipment and/or raceways to the joists, walls, and so on, is accomplished with proper nails, screws, anchors, and straps. Be careful when trying to support other equipment from raceways as this is not always permitted. Refer to figure 18-6.

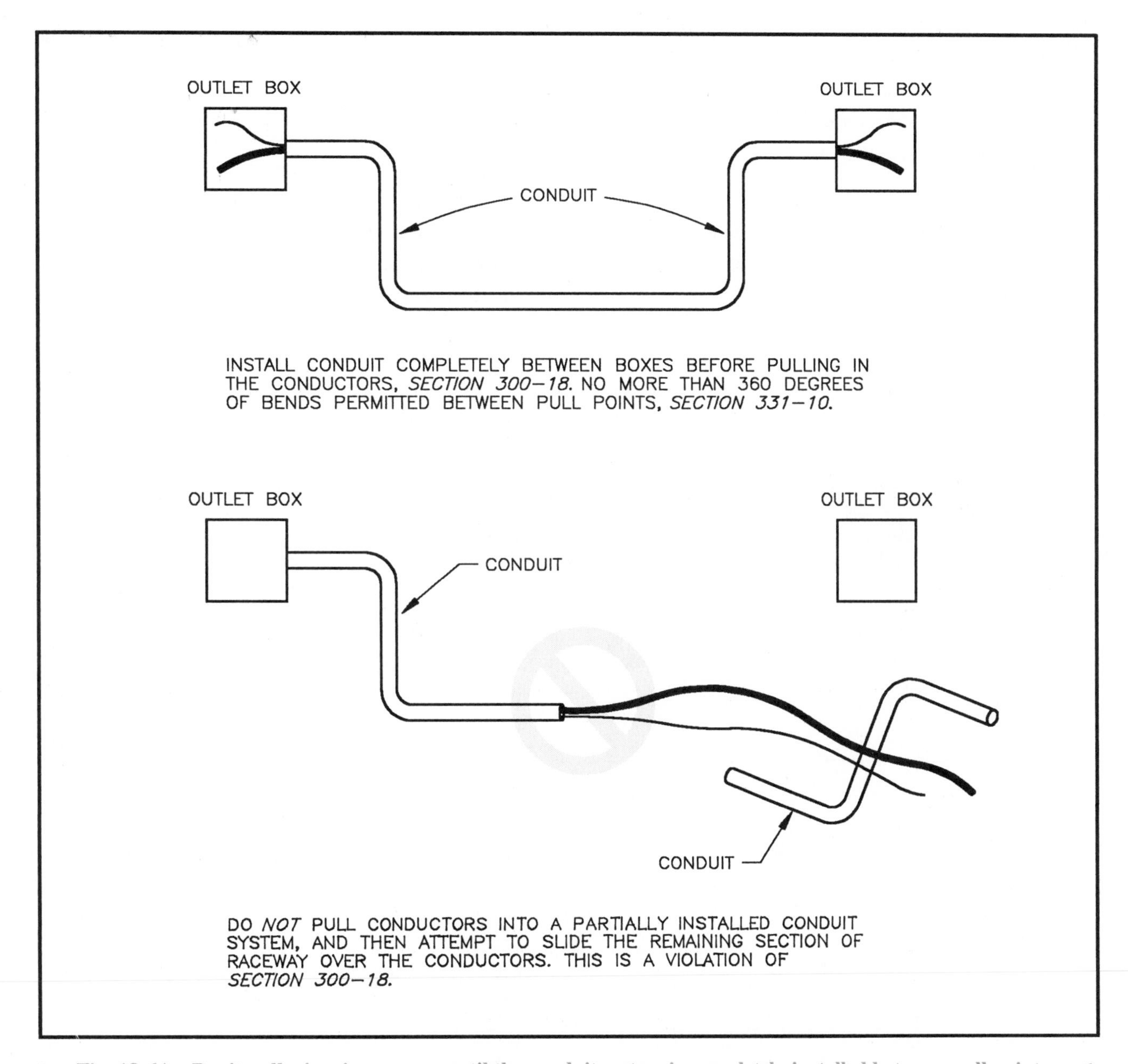

Fig. 18-6A Don't pull wires in raceway until the conduit system is completely installed between pull points.

Conduit Fill Calculations

The electrician must be able to select the proper conduit size to meet various requirements. If the conductors to be installed in a conduit are the same size, then *Tables 3A*, *3B*, and *3C* of *Chapter 9* of the Code can be used to determine the conduit size.

For example, three No. 10 THHN conductors require 1/2-inch conduit, five No. 12 THW conductors require 3/4-inch conduit, and five No. 12 THHN conductors require 1/2-inch conduit.

When the conductors to be used have different sizes, the cross-sectional areas of all conductors must be added. This sum must not exceed the allowable percentage fill of the interior cross-sectional area of the conduit as given in *Table 4* of the Code.

For example, if three No. 6 TW conductors and two No. 8 TW conductors are to be installed in one conduit, the proper conduit size is determined as follows:

1. Find the cross-sectional area of the conductors (using *Table 5*, *Chapter 9* and the dimensions in *Column 5*).

Three No. 6 TW conductors (0.0819 in^2 each)	0.2457 in^2
Two No. 8 TW conductors (0.0471 in^2 each)	0.0942 in^2
Total Area	0.3399 in^2

2. Check *Table 1, Chapter 9* to determine the percentage of fill permitted. When three or more conductors are installed in a conduit, the percentage of fill is 40 percent.
3. Check *Column 5* (40% fill) of *Table 4* to find the conduit size. A 3/4-inch conduit holds up to 0.21 square inch of conductor fill and 1-inch conduit holds up to 0.34 square inch of conductor fill. Therefore, 1-inch conduit is the minimum size that can be used to carry the conductor area determined in step 1.

More and more electrical ordinances (local codes) are requiring that a separate green insulated equipment grounding conductor be installed in every raceway whether rigid or flexible, whether metallic or nonmetallic. This requirement resulted because electrical inspectors were finding loose connectors and couplings where crimp and other

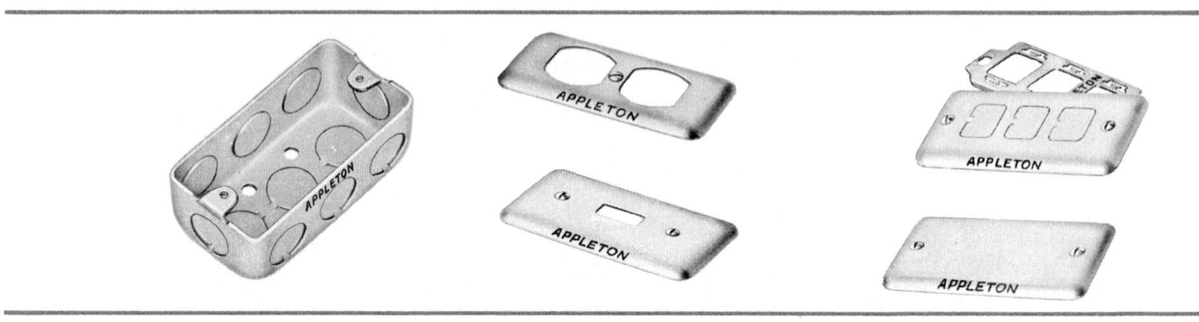

Fig. 18-7 Handy box and covers.

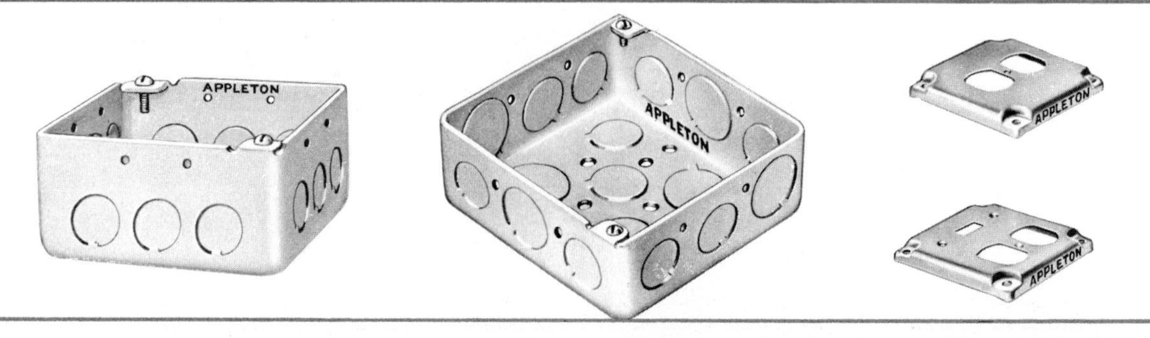

Fig. 18-8 Four-inch square boxes and raised covers.

compression fittings were poorly installed, loose or missing set screws on set-screw-type connectors and couplings, and corrosion. Installing a separate equipment grounding conductor inside the raceway (*not* outside) and sizing it according to *Table 250-95* of the Code will result in an effective ground, as required in *Section 250-51* of the Code.

This separate equipment grounding conductor does take up space, and therefore must be counted when considering conduit fill.

EXAMPLE: A feeder to an electric furnace is fed with two No. 3 THHN copper conductors protected by a set of 100-ampere fuses. This feeder is to be installed in electrical metallic tubing. A separate insulated equipment grounding conductor is to be installed in the raceway. What is the minimum size electrical metallic tubing to be installed?

SOLUTION: According to *Table 250-95*, the minimum size equipment grounding conductor for the 100-ampere feeder is No. 8 AWG copper.

From *Table 5*:

Two No. 3 THHN	0.0995 in²
	0.0995 in²
One No. 8 THHN	0.0373 in²
	0.2363 in²

From *Table 4*:

40% fill 3/4″ EMT = 0.21 in²

40% fill 1″ EMT = 0.34 in²

Therefore, install 1-inch electrical metallic tubing.

► The Code, in *Section 300-18*, states that a conduit system must be completely installed before pulling in the conductors. Refer to figure 18-6A (page 257). ◄

Table 3A. Maximum Number of Conductors in Trade Sizes of Conduit or Tubing
(Based on Table 1, Chapter 9)

Conduit Trade Size (Inches)		½	¾	1	1¼	1½	2	2½	3	3½	4	5	6
Type Letters	**Conductor Size AWG, kcmil**												
TW, XHHW (14 through 8)	14	9	15	25	44	60	99	142					
	12	7	12	19	35	47	78	111	171				
	10	5	9	15	26	36	60	85	131	176			
	8	2	4	7	12	17	28	40	62	84	108		
RHW and RHH (without outer covering), THW	14	6	10	16	29	40	65	93	143	192			
	12	4	8	13	24	32	53	76	117	157			
	10	4	6	11	19	26	43	61	95	127	163		
	8	1	3	5	10	13	22	32	49	66	85	133	
TW,	6	1	2	4	7	10	16	23	36	48	62	97	141
	4	1	1	3	5	7	12	17	27	36	47	73	106
THW,	3	1	1	2	4	6	10	15	23	31	40	63	91
	2	1	1	2	4	5	9	13	20	27	34	54	78
	1		1	1	3	4	6	9	14	19	25	39	57
FEPB (6 through 2), RHW and RHH (without outer covering)	1/0		1	1	2	3	5	8	12	16	21	33	49
	2/0		1	1	1	3	5	7	10	14	18	29	41
	3/0		1	1	1	2	4	6	9	12	15	24	35
	4/0			1	1	1	3	5	7	10	13	20	29
	250			1	1	1	2	4	6	8	10	16	23
	300			1	1	1	2	3	5	7	9	14	20
	350				1	1	1	3	4	6	8	12	18
	400				1	1	1	2	4	5	7	11	16
	500				1	1	1	1	3	4	6	9	14
	600					1	1	1	3	4	5	7	11
	700					1	1	1	2	3	4	7	10
	750					1	1	1	2	3	4	6	9

Note: This table is for concentric stranded conductors only. For cables with compact conductors, the dimensions in Table 5A shall be used.

Table 3B. Maximum Number of Conductors in Trade Sizes of Conduit or Tubing
(Based on Table 1, Chapter 9)

Conduit Trade Size (Inches)		½	¾	1	1¼	1½	2	2½	3	3½	4	5	6
Type Letters	Conductor Size AWG, kcmil												
THWN,	14	13	24	39	69	94	154						
	12	10	18	29	51	70	114	164					
	10	6	11	18	32	44	73	104	160				
	8	3	5	9	16	22	36	51	79	106	136		
THHN,	6	1	4	6	11	15	26	37	57	76	98	154	
FEP (14 through 2),	4	1	2	4	7	9	16	22	35	47	60	94	137
FEPB (14 through 8),	3	1	1	3	6	8	13	19	29	39	51	80	116
PFA (14 through 4/0)	2	1	1	3	5	7	11	16	25	33	43	67	97
PFAH (14 through 4/0)	1		1	1	3	5	8	12	18	25	32	50	72
Z (14 through 4/0)													
XHHW (4 through 500 kcmil)	1/0		1	1	3	4	7	10	15	21	27	42	61
	2/0		1	1	2	3	6	8	13	17	22	35	51
	3/0		1	1	1	3	5	7	11	14	18	29	42
	4/0		1	1	1	2	4	6	9	12	15	24	35
	250			1	1	1	3	4	7	10	12	20	28
	300			1	1	1	3	4	6	8	11	17	24
	350			1	1	1	2	3	5	7	9	15	21
	400				1	1	1	3	5	6	8	13	19
	500				1	1	1	2	4	5	7	11	16
	600				1	1	1	1	3	4	5	9	13
	700					1	1	1	3	4	5	8	11
	750					1	1	1	2	3	4	7	11
XHHW	6	1	3	5	9	13	21	30	47	63	81	128	185
	600				1	1	1	1	3	4	5	9	13
	700					1	1	1	3	4	5	7	11
	750					1	1	1	2	3	4	7	10

Note: This table is for concentric stranded conductors only. For cables with compact conductors, the dimensions in Table 5A shall be used.

Table 3C. Maximum Number of Conductors in Trade Sizes of Conduit or Tubing
(Based on Table 1, Chapter 9)

Conduit Trade Size (Inches)		½	¾	1	1¼	1½	2	2½	3	3½	4	5	6
Type Letters	Conductor Size AWG, kcmil												
	14	3	6	10	18	25	41	58	90	121	155		
	12	3	5	9	15	21	35	50	77	103	132		
RHW,	10	2	4	7	13	18	29	41	64	86	110		
	8	1	2	4	7	9	16	22	35	47	60	94	137
RHH	6	1	1	2	5	6	11	15	24	32	41	64	93
	4	1	1	1	3	5	8	12	18	24	31	50	72
(with	3	1	1	1	3	4	7	10	16	22	28	44	63
outer	2		1	1	3	4	6	9	14	19	24	38	56
covering)	1		1	1	1	3	5	7	11	14	18	29	42
	1/0		1	1	1	2	4	6	9	12	16	25	37
	2/0			1	1	1	3	5	8	11	14	22	32
	3/0			1	1	1	3	4	7	9	12	19	28
	4/0			1	1	1	2	4	6	8	10	16	24
	250				1	1	1	3	5	6	8	13	19
	300				1	1	1	3	4	5	7	11	17
	350				1	1	1	2	4	5	6	10	15
	400				1	1	1	1	3	4	6	9	14
	500				1	1	1	1	3	4	5	8	11
	600					1	1	1	2	3	4	6	9
	700					1	1	1	1	3	3	6	8
	750						1	1	1	3	3	5	8

Note: This table is for concentric stranded conductors only. For cables with compact conductors, the dimensions in Table 5A shall be used.

Table 4. Dimensions and Percent Area of Conduit and of Tubing

Areas of Conduit or Tubing for the Combinations of Wires Permitted in Table 1, Chapter 9.

		Area — Square Inches								
			Not Lead Covered			Lead Covered				
Trade Size	Internal Diameter Inches	Total 100%	2 Cond. 31%	Over 2 Cond. 40%	1 Cond. 53%	1 Cond. 55%	2 Cond. 30%	3 Cond. 40%	4 Cond. 38%	Over 4 Cond. 35%
½	.622	.30	.09	.12	.16	.17	.09	.12	.11	.11
¾	.824	.53	.16	.21	.28	.29	.16	.21	.20	.19
1	1.049	.86	.27	.34	.46	.47	.26	.34	.33	.30
1¼	1.380	1.50	.47	.60	.80	.83	.45	.60	.57	.53
1½	1.610	2.04	.63	.82	1.08	1.12	.61	.82	.78	.71
2	2.067	3.36	1.04	1.34	1.78	1.85	1.01	1.34	1.28	1.18
2½	2.469	4.79	1.48	1.92	2.54	2.63	1.44	1.92	1.82	1.68
3	3.068	7.38	2.29	2.95	3.91	4.06	2.21	2.95	2.80	2.58
3½	3.548	9.90	3.07	3.96	5.25	5.44	2.97	3.96	3.76	3.47
4	4.026	12.72	3.94	5.09	6.74	7.00	3.82	5.09	4.83	4.45
5	5.047	20.00	6.20	8.00	10.60	11.00	6.00	8.00	7.60	7.00
6	6.065	28.89	8.96	11.56	15.31	15.89	8.67	11.56	10.98	10.11

Conduit Bodies (*Section 370-6(c)*)

Conduit bodies are used with conduit installations to provide an easy means to turn corners, terminate conduits, and mount switches and receptacles. They are also used to provide access to conductors, to provide space for splicing (when permitted), and to provide a means for pulling conductors, figure 18-9.

A conduit body:

- must have a cross-sectional area not less than twice the cross-sectional area of the largest conduit to which it is attached, figure 18-10(A). This is a requirement only when the conduit body contains No. 6 AWG conductors or smaller.
- may contain conductors in the amounts as permitted in *Table 1, Chapter 9* of the Code for the size conduit attached to the conduit body.
- with provisions for less than three conduit entries must not contain splices, taps, or devices unless the conduit is marked with its cubic-inch area, figure 18-10(B), and the conduit must be supported "in a rigid and secure manner."
- the conductor fill volume must be properly calculated according to *Table 370-6(b)*, which lists the free space that must be provided for each conductor within the conduit body. If the conduit body is to contain splices, taps, or devices, it must be supported rigidly and securely.

Derating Factors

For more than three current-carrying wires in conduit or cable, refer to figure 18-11.

If a raceway is to contain more than three current-carrying conductors, the student must check *Note 8* to *Tables 310-16* through *310-31*. *Note 8* states that for the conductors listed in the tables, the maximum allowable ampacity must be reduced when more than three current-carrying conductors are installed in a raceway. The reduction factors based on the number of conductors are noted in figure 18-11. When the conduit nipple does not

Table 5. Dimensions of Rubber-Covered and Thermoplastic-Covered Conductors

	Types RFH-2, RH, RHH,*** RHW,*** SF-2		Types TF, THW,† TW		Types TFN, THHN, THWN		Types**** FEP, FEPB, FEPW, TFE, PF, PFA, PFAH, PGF, PTF, Z, ZF, ZFF		Type XHHW, ZW††		Types KF-1, KF-2, KFF-1, KFF-2	
Size AWG kcmil	Approx. Diam. Inches	Approx. Area Sq. In.	Approx. Diam. Inches	Approx. Area Sq. In.	Approx. Diam. Inches	Approx. Area Sq. In.	Approx. Diam. Inches	Approx. Area Sq. Inches	Approx. Diam. Inches	Approx. Area Sq. In.	Approx. Diam. Inches	Approx. Area Sq. In.
Col. 1	Col. 2	Col. 3	Col. 4	Col. 5	Col. 6	Col. 7	Col. 8	Col. 9	Col. 10	Col. 11	Col. 12	Col. 13
18	.146	.0167	.106	.0088	.089	.0062	.081	.0052	...	....	.065	.0033
16	.158	.0196	.118	.0109	.100	.0079	.092	.0066	...	....	.070	.0038
14	30 mils .171	.0230	.131	.0135	.105	.0087	.105 .105	.0087 .0087	...	....	.083	.0054
14	45 mils .204*	.0327*	...	....	...	....	... ...	....	...	....	...	....
14	....	....	.162†	.0206†	...	....	... ...	....	.129	.0131	...	....
12	30 mils .188	.0278	.148	.0172	.122	.0117	.121 .121	.0115 .0115	...	....	.102	.0082
12	45 mils .221*	.0384*	...	....	...	....	... ...	....	...	....	...	....
12	....	....	.179†	.0252†	...	....	... ...	....	.146	.0167	...	....
10	...... .242	.0460	.168	.0222	.153	.0184	.142 .142	.0158 .0158	...	....	.124	.0121
10	....	....	.199†	.0311†	...	....	... ...	....	.166	.0216	...	....
8	...... .328	.0845	.245	.0471	.218	.0373	.206 .186	.0333 .0272	...	....	...	....
8	....	....	.276†	.0598†	...	....	... ...	....	.241	.0456	...	....
6	.397	.1238	.323	.0819	.257	.0519	.244 .302	.0468 .0716	.282	.0625	...	....
4	.452	.1605	.372	.1087	.328	.0845	.292 .350	.0670 .0962	.328	.0845	...	....
3	.481	.1817	.401	.1263	.356	.0995	.320 .378	.0804 .1122	.356	.0995	...	....
2	.513	.2067	.433	.1473	.388	.1182	.352 .410	.0973 .1320	.388	.1182	...	....
1	.588	.2715	.508	.2027	.450	.1590	.420 ...	.1385	.450	.1590	...	....
1/0	.629	.3107	.549	.2367	.491	.1893	.462 ...	.1676	.491	.1893	...	....
2/0	.675	.3578	.595	.2781	.537	.2265	.498 ...	.1948	.537	.2265	...	....
3/0	.727	.4151	.647	.3288	.588	.2715	.560 ...	.2463	.588	.2715	...	....
4/0	.785	.4840	.705	.3904	.646	.3278	.618 ...	.3000	.646	.3278	...	....

Table 5 (Continued)

	Types RFH-2, RH, RHH,*** RHW,*** SF-2		Types TF, THW,† TW		Types TFN, THHN, THWN		Types**** FEP, FEPB, FEPW, TFE, PF, PFA, PFAH, PGF, PTF, Z, ZF, ZFF		Type XHHW, ZW††	
Size AWG kcmil	Approx. Diam. Inches	Approx. Area Sq. In.	Approx. Diam. Inches	Approx. Area Sq. In.	Approx. Diam. Inches	Approx. Area Sq. In.	Approx. Diam. Inches	Approx. Area Sq. Inches	Approx. Diam. Inches	Approx. Area Sq. In.
Col. 1	Col. 2	Col. 3	Col. 4	Col. 5	Col. 6	Col. 7	Col. 8	Col. 9	Col. 10	Col. 11
250	.868	.5917	.788	.4877	.716	.4026	...	....	.716	.4026
300	.933	.6837	.843	.5581	.771	.4669	...	....	.771	.4669
350	.985	.7620	.895	.6291	.822	.5307	...	....	.822	.5307
400	1.032	.8365	.942	.6969	.869	.5931	...	....	.869	.5931
500	1.119	.9834	1.029	.8316	.955	.7163	...	....	.955	.7163
600	1.233	1.1940	1.143	1.0261	1.058	.8791	...	....	1.073	.9043
700	1.304	1.3355	1.214	1.1575	1.129	1.0011	...	....	1.145	1.0297
750	1.339	1.4082	1.249	1.2252	1.163	1.0623	...	....	1.180	1.0936
800	1.372	1.4784	1.282	1.2908	1.196	1.1234	...	....	1.210	1.1499
900	1.435	1.6173	1.345	1.4208	1.259	1.2449	...	....	1.270	1.2668
1000	1.494	1.7530	1.404	1.5482	1.317	1.3623	...	....	1.330	1.3893
1250	1.676	2.2062	1.577	1.9532	...	....	...	....	1.500	1.7671
1500	1.801	2.5475	1.702	2.2751	...	....	...	....	1.620	2.0612
1750	1.916	2.8832	1.817	2.5930	...	....	...	....	1.740	2.3779
2000	2.021	3.2079	1.922	2.9013	...	....	...	....	1.840	2.6590

* The dimensions of Types RHH and RHW.
† Dimensions of THW in sizes No. 14 through No. 8. No. 6 THW and larger is the same dimension as TW.
*** Dimensions of RHH and RHW without outer covering are the same as THW No. 18 through No. 10, solid; No. 8 and larger, stranded.
**** In Columns 8 and 9 the values shown for sizes No. 1 through 4/0 are for TFE and Z only. The right-hand values in Columns 8 and 9 are for FEPB, Z, ZF, and ZFF only.
†† No. 14 through No. 2.

exceed 24 inches, the reduction factor does not apply, figure 18-12.

According to *Note 10*, a neutral conductor carrying only unbalanced currents shall not be counted in determining current-carrying capacities as provided for in *Note 8.* See figure 18-13.

EXAMPLE: What is the correct ampacity for No. 6 THHN when there are six current-carrying wires in the conduit?

$$75 \times 0.80 = 60 \text{ amperes}$$

CORRECTION FACTORS (Due to high temperatures)

When conductors are installed in locations where the temperature is higher than 30 °C (86 °F), the correction factors noted below *Table 310-16* must apply.

For example, consider that four current-carrying No. 3 THWN copper conductors are to be installed in one raceway in an ambient temperature of approximately 90 °F. "Ambient" temperature means "surrounding" temperature. This temperature is common in boiler rooms and is well exceed-

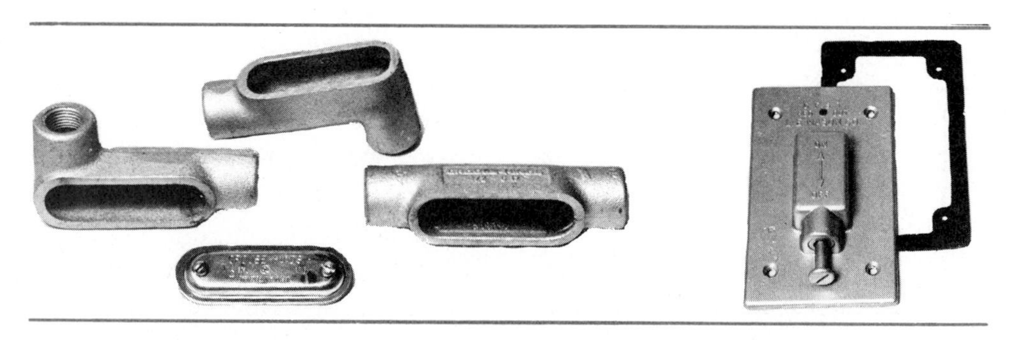

Fig. 18-9 A variety of common conduit bodies and covers.

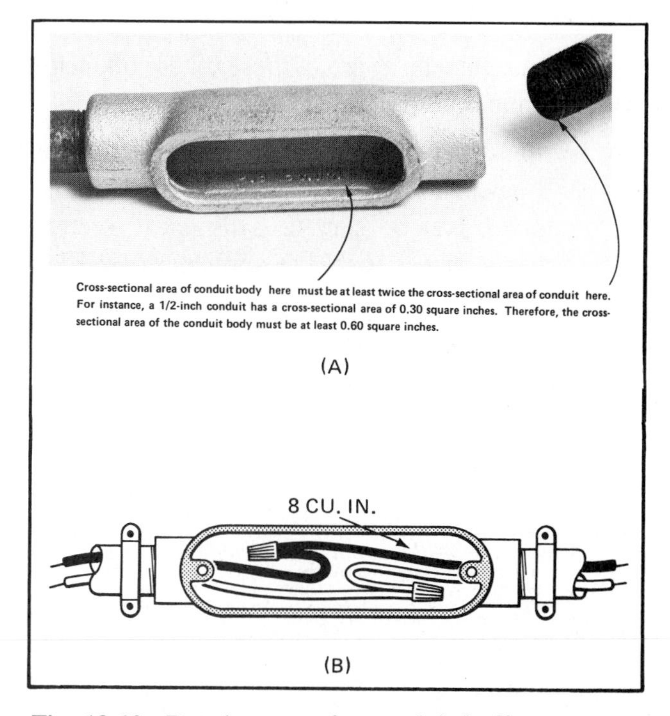

Fig. 18-10 Requirements for conduit bodies.

Number of Conductors	Column A Percent of Values in Tables as Adjusted for Ambient Temperature if Necessary	Number of Conductors	Column B** Percent of Values in Tables as Adjusted for Ambient Temperature if Necessary
4 through 6	80	4 through 6	80
7 through 9	70	7 through 9	70
10 through 24*	70	10 through 20	50
25 through 42*	60	21 through 30	45
43 and above*	50	31 through 40	40
		41 through 60	35

* These factors include the effects of a load diversity of 50 percent.
**No diversity.

(FPN): Column A is based on the following formula:

$A_2 = \sqrt{\frac{0.5N}{E}} \times (A_1)$ where

A_1 = Table ampacity multiplied by factor from Note 8(a)
N = Total number of conductors used to obtain factor from Note 8(a)
E = Desired number of energized conductors
A_2 = Ampacity limit for energized conductors

Fig. 18-11 *Note 8* to *NEC® Table 310-16.*

ed in attics where the sun shines directly onto the roof.

The maximum ampacity for these conductors is determined as follows:

- The ampacity of No. 3 THWN copper conductors from *Table 310-16* = 100 amperes.
- Apply the correction factor for 90°F. This is found in the last column of *Table 310-16* under Ampacity Correction Factors. The first column shows Celsius; the last column shows Fahrenheit.

$$100 \times 0.94 = 94 \text{ amperes}$$

- Apply the derating factor for four current-carrying wires in one raceway.

$$94 \times 0.80 = 75.2 \text{ amperes}$$

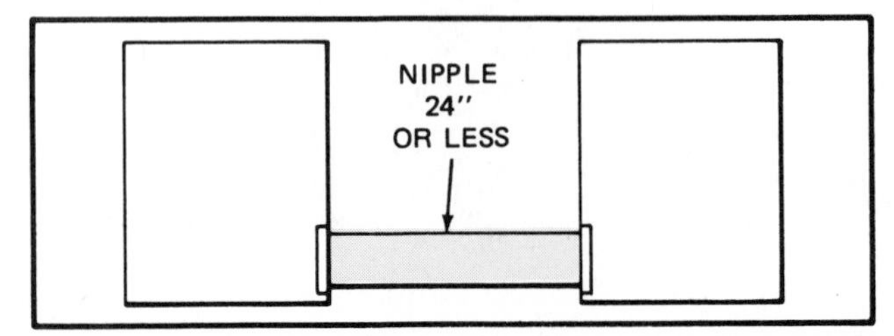

Fig. 18-12 The derating in figure 18-11 for more than three conductors in a raceway is *not* required when the conduit nipple does not exceed 24 inches. See *Note 8, Exception No. 3* to *Table 310-16*, and *Note A3* to *Chapter 9* of the *NEC*®

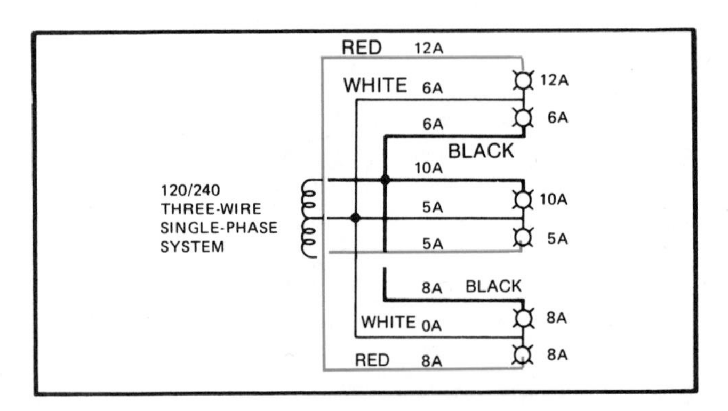

Fig. 18-13 A neutral conductor carrying the unbalanced currents from the other conductors need *not* be counted when determining the ampacity of the conductor, according to *Note 10* to *Table 310-16*. The Code would recognize this example as *two current-carrying conductors*. Thus, in the three multiwire circuits shown, the actual total is nine conductors, but only six current-carrying conductors are considered when the derating factor of figure 18-11 is applied.

- Therefore, 75.2 amperes is the new corrected ampacity for the example given.

Some authorities having jurisdiction in extremely hot areas of the country, such as the southwestern desert climates, will require that service-entrance conductors running up the side of a building or above a roof exposed to direct sunlight be "corrected" per the correction factors below *Table 310-16*. Check this out before proceeding with an installation where this situation might be encountered.

MAXIMUM SIZE OVERCURRENT PROTECTION

An important footnote below *Table 310-16* must be observed when installing No. 14, No. 12, and No. 10 AWG conductors. Notice in the table many obelisks (†) (also called daggers) referring to these sizes of conductors. As can be seen in figure 18-14, the maximum permitted size of the overcurrent protection is less than the ampacities listed in the table.

We have just discussed how to derate because of the number of current-carrying wires in the conduit, and how to apply correction factors because of high temperature. This results in a new ampacity rating for the conductor.

OTHER CODE LIMITATIONS

We must also be aware of these other conductor loading limitations:

- *Section 210-19(a)* — Branch-circuit conductors shall have an ampacity not less than the maximum load to be served. This entire section discusses ranges, cooking appliances, and tap conductors.
- *Section 210-22* — This is the fundamental rule that the total load shall not exceed the branch-circuit rating.

† Unless otherwise specifically permitted elsewhere in this Code, the overcurrent protection for conductor types marked with an obelisk (†) shall not exceed 15 amperes for 14 AWG, 20 amperes for 12 AWG, and 30 amperes for 10 AWG copper; or 15 amperes for 12 AWG and 25 amperes for 10 AWG aluminum and copper-clad aluminum after any correction factors for ambient temperature and number of conductors have been applied.
* For dry and damp locations only. See 75°C column for wet locations.

Fig. 18-14 *Footnote* to *NEC*® *Table 310-16.*

- *Section 210-22(a)* — This section refers to motor loads, *Article 430*. Size conductors for motors at 125% of motor full-load rating.
- *Section 210-22(c)* — For continuous loads, the branch-circuit rating must be not less than the noncontinuous load plus 125% of the continuous load.
- *Section 220-3(a)* — For continuous and noncontinuous loads, the branch-circuit rating must be not less than the noncontinuous load *plus* 125% of the continuous load.
- *Section 220-10(b)* — For continuous and noncontinuous loads, the feeder overcurrent protection rating must be not less than the noncontinuous load *plus* 125% of the continuous load.
- *Section 384-16(c)* — The total loading of any overcurrent device in a panel must not exceed 80% of its rating when the loads will continue for 3 hours or more.
- See *Articles 422 (Appliances)* and *424 (Fixed Electric Space Heating)* for specific loading limitations.

Note: In the above Code references, the term "continuous" is defined as "where the maximum current is expected to continue for 3 hours or more." Loads in homes are generally not considered to be continuous. Many loads in commercial and industrial installations can be considered to be continuous and must be computed as such.

EXAMPLES OF DERATING

To illustrate what has been learned so far about derating and maximum overcurrent protection, suppose that 8 THW copper current-carrying conductors are to be installed in the same conduit. The connected load will be 18 amperes. The size wire to use may be determined as follows:

- No. 14 THW copper wire
  a. The maximum overcurrent protection from figure 18-14 (*footnote* to *Table 310-16*) is 15 amperes.
  b. The ampacity of No. 14 THW copper wire from *Table 310-16* is 20 amperes.
  c. Apply the derating factor (8 current-carrying wires in the raceway) from figure 18-11 (*Note 8* to *Table 310-16*).
  $20 \times 0.70 = 14$ amperes (derated ampacity)
  Thus, No. 14 THW wire is *too small* to supply the 18-ampere load.
- No. 12 THW copper wire
  a. The maximum overcurrent protection from figure 18-14 (*footnote* to *Table 310-16*) is 20 amperes.
  b. The ampacity of No. 12 THW copper wire from *Table 310-16* is 25 amperes.
  c. Apply the derating factor (8 current-carrying wires in the raceway) from figure 18-11 (*Note 8* to *Table 310-16*).
  $25 \times 0.70 = 17.5$ amperes (derated ampacity)
  Thus, No. 12 THW is *too small* to supply the 18-ampere load.
- No. 10 THW copper wire
  a. The maximum overcurrent protection from figure 18-14 (*footnote* to *Table 310-16*) is 30 amperes.
  b. The ampacity of No. 10 THW copper wire from *Table 310-16* is 35 amperes.
  c. Apply the derating factor (8 current-carrying wires in the raceway) from figure 18-11 (*Note 8* to *Table 310-16*).
  $35 \times 0.70 = 24.5$ amperes (derated ampacity)
  Thus, No. 10 THW copper wire is adequate to supply the 18-ampere load.

MULTIOUTLET ASSEMBLY

A multioutlet assembly has been installed above the workbench. For safety reasons, as well as from an adequate wiring viewpoint, it is recommended that any workshop receptacles be connected to a separate circuit. In the event of a malfunction in any power tool commonly used in the home workshop (saw, planer, lathe, or drill), only receptacle Circuit A18 is affected. The lighting in the workshop is not affected by a power outage on the receptacle circuit.

The GFCI feedthrough receptacle to the left of the workbench provides GFCI protection for the entire plug-in strip. See *Section 210-8(a)(4)*. See figure 18-15.

The installation of multioutlet assemblies must conform to the requirements of *Article 353* of the Code.

Load Considerations for Multioutlet Assemblies

Table 220-3(b) gives the general lighting loads in volt-amperes per square foot (0.093 m²) for types of occupancies. The *footnote* to this table mentions that *all* general-use receptacle outlets in one, two, and multioutlet assemblies are to be considered as outlets for general lighting and as such have been included in the volt-amperes per square foot calculations. Therefore, no additional load need be added. This is further confirmed by the last sentence of *Exception No. 1* to *Section 220-3(c)*.

Because the multioutlet assembly, figure 18-16, has been provided in the workshop for the purpose of plugging-in portable tools, a separate Circuit A18 is provided. This circuit has been shown in the service-entrance calculations at 1500 volt-amperes,

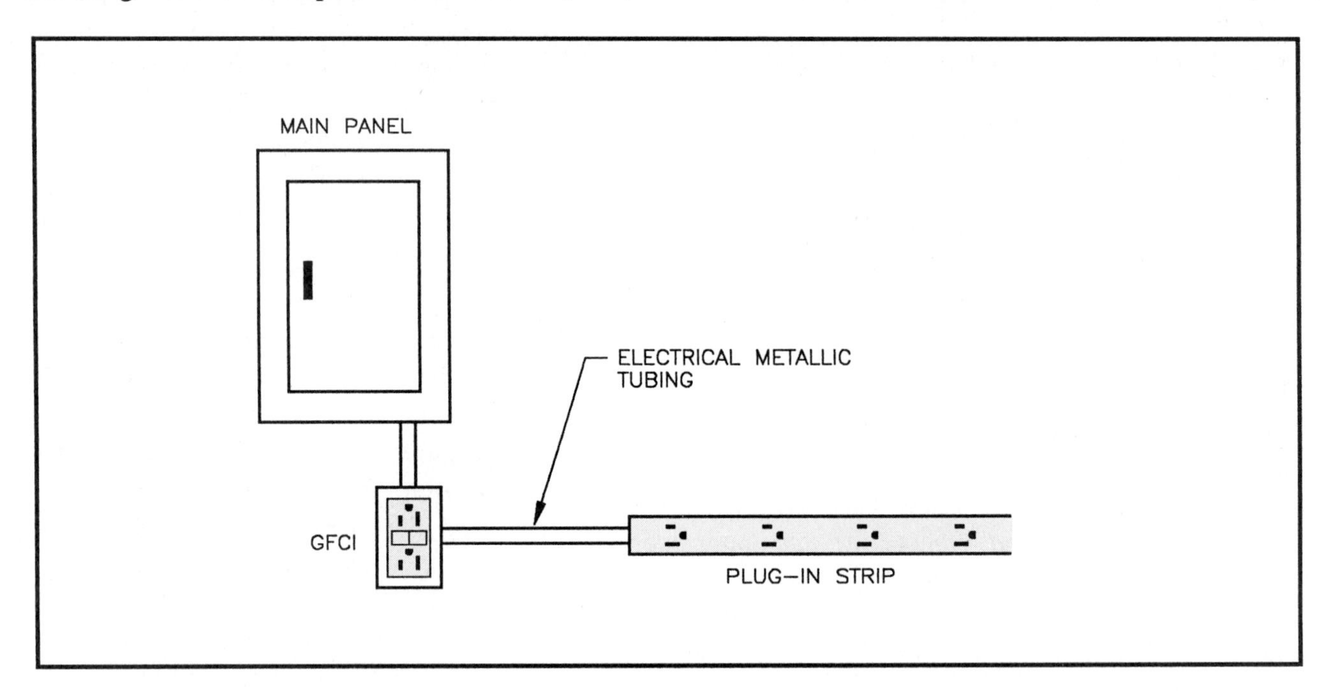

Fig. 18-15 Detail of how the workbench receptacle plug-in strip is connected by feeding through a GFCI feedthrough receptacle located below the Main Panel. This is Circuit A18.

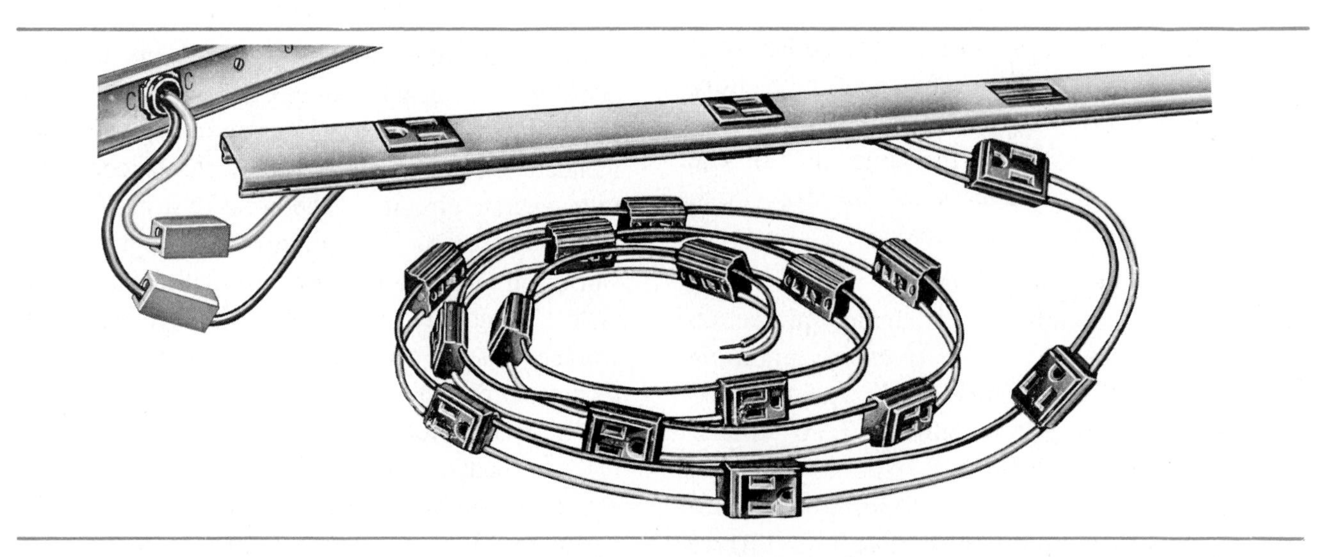

Fig. 18-16 Multioutlet assemblies are permitted to be run through (not within) a dry partition when no outlets are within the partition, and all exposed portions of the assembly can have the cover removed.

similar to the load requirements for small appliance circuits.

Many types of convenience receptacles are available for various sizes of multioutlet assemblies. For example, duplex, single-circuit grounding, split-circuit grounding, and duplex split-circuit receptacles can be used with multioutlet assemblies.

Wiring of the Multioutlet Assembly

The electrician first attaches the metal base of the multioutlet assembly to the wall or baseboard. The receptacles may be wired on the job. If the receptacles are factory prewired, they are snapped into the base, figure 18-17. Covers are then cut to the lengths required to fill the spaces between the receptacles. Manufacturers of these assemblies can supply precut covers to be used when evenly spaced receptacles are installed. These receptacles are usually installed 12 inches (305 mm) or 18 inches (457 mm) apart.

Receptacles can also be installed by snapping them into the cover and then attaching the cover to the metal base.

In general, multioutlet installations are not difficult because of the large number of fittings and accessories available. Connectors, couplings, ground clamps, blank end fittings, elbows for turning corners, and end entrance fittings can all be used to simplify the installation.

Prewired multioutlet assemblies normally are wired with No. 12 AWG conductors and have either 15- or 20-ampere receptacles. The feed for a multioutlet assembly is shown in figure 18-18.

EMPTY CONDUITS

Although not truly part of the workshop wiring, Note 8 to the first floor electrical plans indicates that two empty 1/2-inch EMT raceways are to be installed, running from the workshop to the attic. This is unique, but certainly welcomed when at some later date, additional wiring (telephone, television, etc.) might be required. The empty raceways will provide the necessary route from the basement to the attic, thus eliminating the need to fish through the wall partitions.

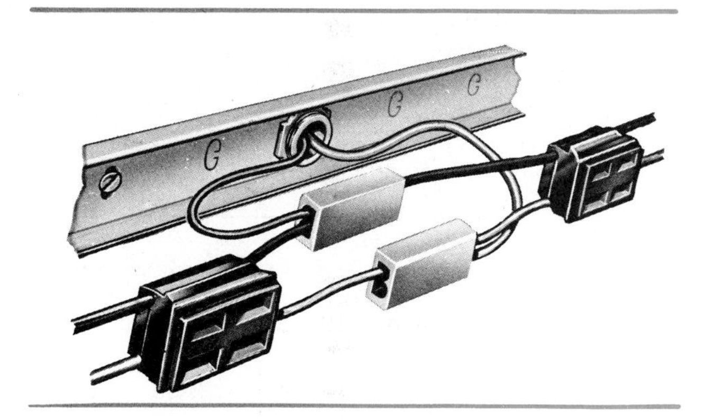

Fig. 18-17 Prewired multioutlet assembly.

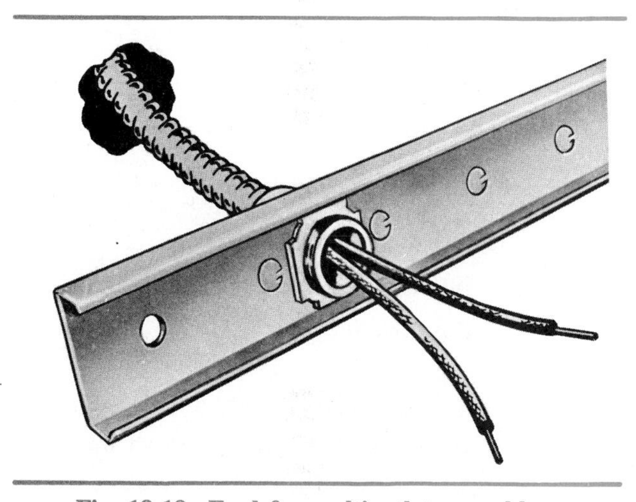

Fig. 18-18 Feed for multioutlet assembly.

REVIEW

1. a. What circuit supplies the workshop lighting? ______________

   b. What circuit supplies the plug-in strip over the workbench? ______________

   c. What circuit supplies the freezer receptacle? ______________

2. Approximately how much 1/2-inch EMT is used for connecting Circuits A13, A17, A18, and A20? ______________

3. How is EMT fastened to masonry? ____

4. What type of lighting fixtures are installed on the workshop ceiling? ___________

5. A check of the plans indicates that two empty 1/2-inch thinwall conduits are installed between the basement and the attic. In your opinion do you feel this is a good idea? Briefly explain your thoughts. ____

6. To what circuit are the smoke detectors connected? ______________________________

7. Is it a Code requirement to connect smoke detectors to a GFCI-protected circuit?

8. When a freezer is plugged into a receptacle, be sure that the receptacle is protected by a GFCI. Is this statement true or false? Explain. ______________________________

9. A sump pump is plugged into a receptacle. This receptacle shall

   a. be GFCI protected

   b. shall not be GFCI protected

   Circle one.

10. When a 120-volt receptacle outlet is provided for the laundry equipment in a basement, what is the *minimum* number of additional receptacle outlets required? What Code reference? ____

11. Derating factors for conductor ampacities must always be taken into consideration when (a) ______________________________ are contained in one raceway. Correction factors are applied when (b) ______________________________.

12. What is the current in the neutral conductor in this circuit at point ⓧ?

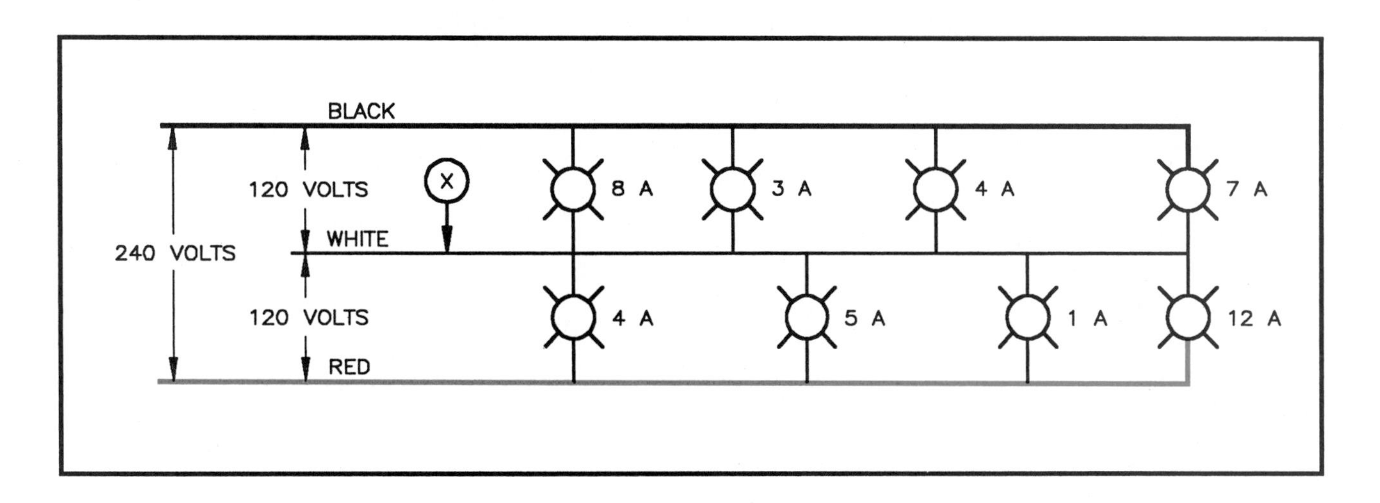

13. When a single receptacle is installed on an individual branch circuit, the receptacle must have a rating ______________ than the rating of the branch circuit.

14. Calculate the total current draw of Circuit A17 if all lighting fixtures and the exhaust fan were turned on.

15. A conduit body must have a cross-sectional area not less than (two) (three) (four) times the cross-sectional area of the (largest) (smallest) conduit to which it is attached. Circle correct answers.

16. A conduit body may contain splices if marked with
    a. the UL logo.
    b. its cubic-inch area.
    c. the size of conduit entries.

    Circle one.

17. What size box would you use where Circuit A17 enters the junction box on the ceiling? ______________

    ____

18. List the proper conduit size for the following. Assume that it is new work and that THHN insulated conductors are used. See *Table 3B*.

    a. 3 No. 14 AWG ______________ f. 4 No. 12 AWG ______________
    b. 4 No. 14 AWG ______________ g. 5 No. 12 AWG ______________
    c. 5 No. 14 AWG ______________ h. 6 No. 12 AWG ______________
    d. 6 No. 14 AWG ______________ i. 4 No. 10 AWG ______________
    e. 3 No. 12 AWG ______________ j. 4 No. 8 AWG ______________

19. According to the Code, what size conduit is required for each of the following combinations of conductors? Assume that these are new installations. Show all calculations.

    a. Three No. 14 AWG THHN, four No. 12 AWG THHN

    b. Two No. 12 AWG TW, three No. 8 AWG TW

    c. Three No. 3 THW, one No. 4 Bare

    d. Three No. 3/0 THHN, two No. 8 THHN

20. a. When more than three current-carrying conductors are installed in one raceway, their allowable ampacities must be reduced according to ________________ to *Table 310-16.*

    b. Four No. 10 THW current-carrying conductors are installed in one conduit. The room temperature will not exceed 86 °F. The connected load is nonmotor, noncontinuous. (1) What is the ampacity of these conductors before derating? (2) What is the maximum overcurrent protection permitted to protect these conductors? (3) What is the ampacity of these conductors before derating? ________________

    ____

    ____

21. What is the minimum number of receptacles required by the Code for basements?

    ____

22. For laundry equipment in basements, what sort of electrical circuitry is required by the Code? ____

____

____

____

____

23. How many receptacles are required to complete the multioutlet assembly above the workbench? ____

24. The following is a layout of the lighting circuits for the workshop. Using the conduit layout in figure 18-1, make a complete wiring diagram. Use colored pencils or pens to indicate conductors.

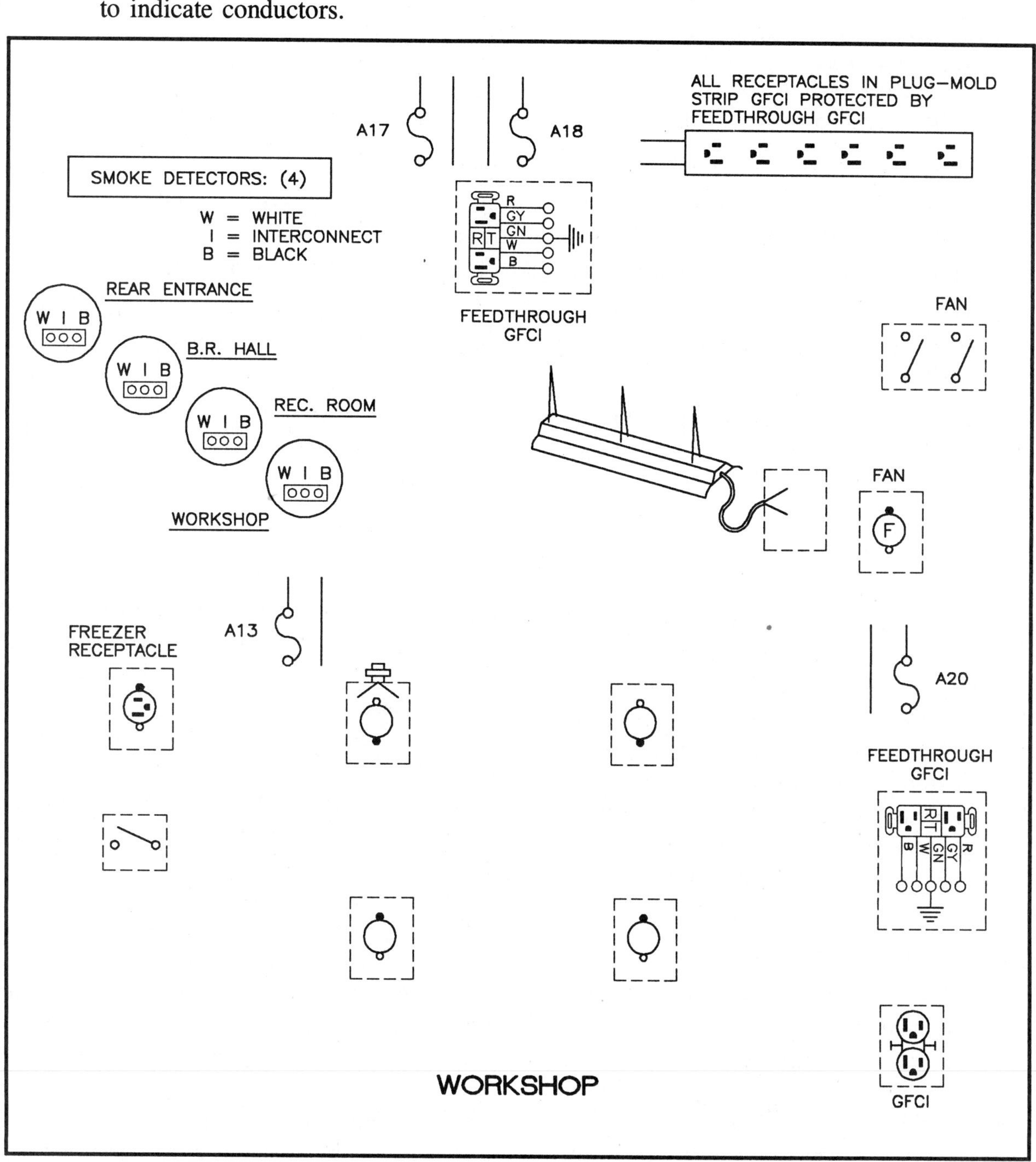

UNIT 19

Special-Purpose Outlets — Water Pump, Water Heater

OBJECTIVES

After studying this unit, the student will be able to

- list the requirements for a deep-well jet pump installation.
- calculate (given the rating of the motor) the conductor size, conduit size, and overcurrent protection required for the pump circuit.
- list some of the electrical circuits used for connecting electric water heaters.
- describe the basic operation of the water heater.
- list the functions of the tank, the heating elements, and the thermostats of the water heater.
- make the proper grounding connections for the water heater.

WATER PUMP CIRCUIT ▲B

All dwellings need a good water supply. City dwellings generally are connected to the city water system. In rural areas, where there is no public water supply, each dwelling has its own water supply system in the form of a well. The residence plans show that a deep-well jet pump is used to pump water from the well to the various plumbing outlets. The outlet for this jet pump is shown by the symbol ▲B. An electric motor drives the jet pump. A circuit must be installed for this system.

JET PUMP OPERATION

Figure 19-1 shows the major parts of a typical jet pump. The pump impeller wheel ① forces water down a drive pipe ② at a high velocity and pressure to a point just above the water level in the well casing. (Some well casings may be driven to a depth of more than 100 feet (30.5 m) before striking water.) Just above the water level, the drive pipe curves sharply upward and enters a larger vertical suction pipe ③. The drive pipe terminates in a small nozzle or jet ④ in the suction pipe. The water emerges from the jet with great force and flows upward through the suction pipe.

Water rises in the suction pipe, drawn up through the tailpipe ⑤ by the action of the jet. The water rises to the pump inlet and passes through the impeller wheel of the pump. Part of the water is forced down through the drive pipe again. The remaining water passes through a check valve and enters the storage tank ⑥.

The tailpipe is submerged in the well water to a depth of about 10 feet (3.05 m). A foot valve ⑦ and strainer ⑧ are provided at the end of the tailpipe. The foot valve prevents water in the pumping equipment from draining back into the well when the pump is not operating.

When the pump is operating, the lower part of

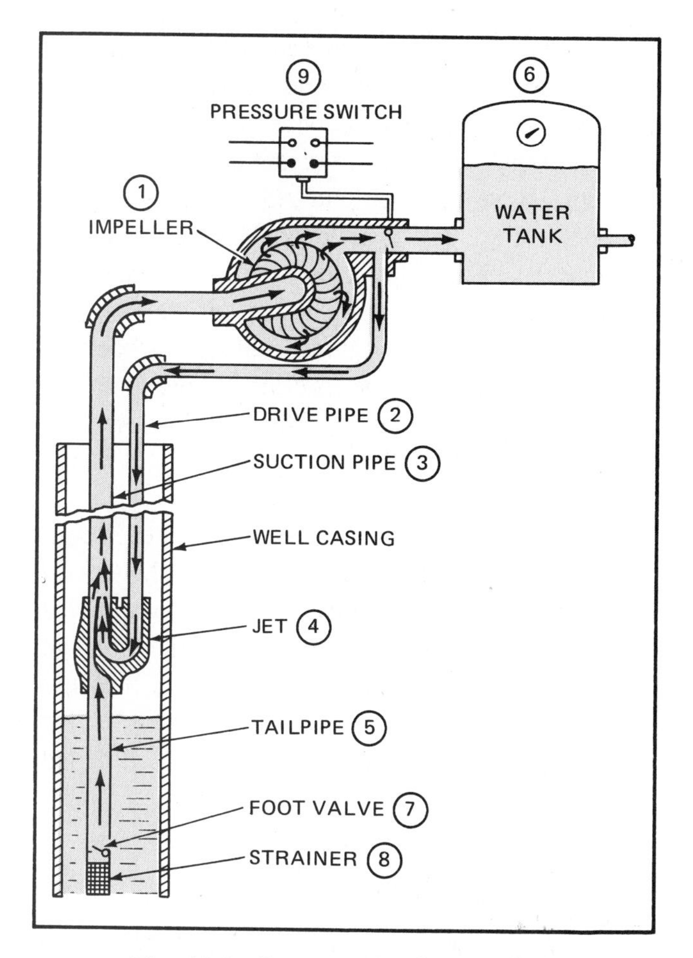

Fig. 19-1 Components of a jet pump.

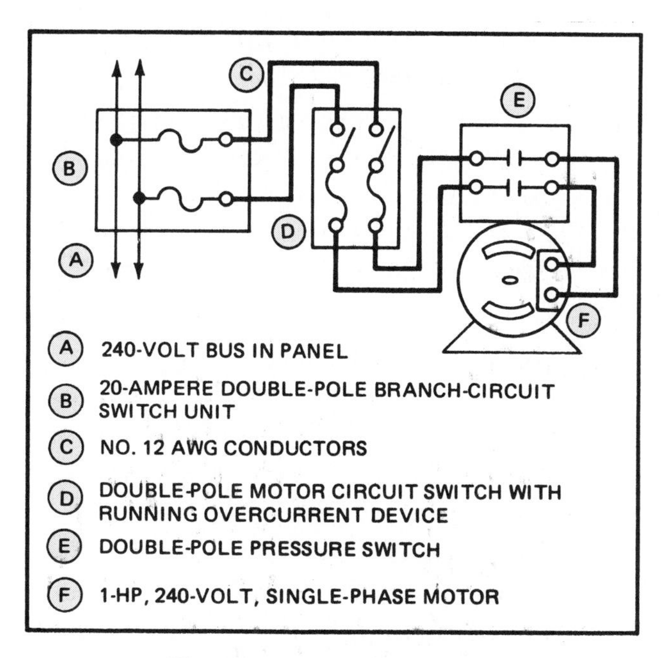

Fig. 19-2 The pump circuit.

the storage tank fills with water and air is trapped in the upper part. As the water rises in the tank, the air is compressed. When the air pressure is 40 pounds per square inch, a pressure switch ⑨ disconnects the pump motor. The pressure switch is adjusted so that as the water is used and the air pressure falls to 20 pounds per square inch, the pump restarts and fills the tank again. One pound (lb) = 0.4536 kilograms (kg). One kilogram (kg) = 2.2046 pounds (lb).

The Pump Motor

A jet pump for residential use may be driven by a 1-horsepower (hp) motor at 3400 r/min. A dual-voltage motor is used so that it can be connected to 120 or 240 volts. For a 1-hp motor, the higher voltage is preferred since the current at this voltage is half the value used by the lower voltage. It is recommended that a single-phase, capacitor-start motor be used for a residential jet pump. Such a motor is designed to produce a high starting torque. In pumping operations, a high starting torque helps to overcome the back pressure within the tank and the weight of the water being lifted from the well.

The Pump Motor Circuit

Figure 19-2 is a diagram of the electrical circuit that operates the jet pump motor and pressure switch. This circuit is taken from electrical Panel B in the utility room. *Table 430-148* indicates that a 1-hp single-phase motor has a rating of 8 amperes at 240 volts. Using this information, the conductor size, conduit size, branch-circuit overcurrent protection, and running overcurrent protection can be determined for the motor.

Conductor Size *(Section 430-22)*

$$1.25 \times 8 = 10 \text{ amperes}$$

Table 310-16 indicates that the minimum conductor size permissible is No. 14 AWG. However, the schedule of special-purpose outlets states that No. 12 AWG conductors are to be used for the pump circuit.

Conduit Size *(Table 3A, Chapter 9)*

Two No. 12 conductors require 1/2-inch conduit.

Branch-circuit Overcurrent Protection
(Section 430-52, Table 430-152)

The jet pump motor is a single-phase motor without code letters. Therefore, *Table 430-152*, line 1, can be used.

The motor's branch-circuit overcurrent device may be sized as follows (*Section 430-52* and *Table 430-152*):

- NONTIME DELAY FUSES
  8 × 3 = 24 amperes (use 25-ampere fuse)
  Maximum size: 400% (4.0)
- DUAL-ELEMENT, TIME-DELAY FUSES
  8 × 1.75 = 14 amperes
  (use 15-ampere fuse)
  Maximum size: 225% (2.25)
- INSTANT TRIP BREAKERS
  8 × 7 = 56 amperes setting
  Maximum setting: 1300% (13)
- INVERSE TIME BREAKERS
  8 × 2.5 = 20-ampere setting
  Maximum setting: 400% for F.L.A. of 100 or less.
  300% for F.L.A. of over 100.

Where the values for branch-circuit protective devices determined by *Table 430-152* do not correspond to the standard sizes or ratings of such devices, the next higher size, rating, or setting shall be permitted, *Section 430-52*.

Running Overcurrent Protection for the Motor
(Section 430-32(c)(1))

Dual-element, time-delay fuses must be rated at not more than 125 percent of the full-load current of the motor to provide running overcurrent protection according to *Section 430-32(c)(1)*. The maximum rating of thermal overload devices is also 125 percent of the full-load current rating of the motor: 1.25 × 8 = 10 amperes.

Wiring for the Pump Motor Circuit

The water pump circuit is run in 1/2-inch conduit from Main Panel A to a disconnect switch and controller that are mounted on the wall next to the pump. The EMT may be run across the ceiling of the workshop or be embedded in the concrete floor. Thermal overload devices (sometimes called *heaters*) are installed in the controller to provide running overcurrent protection. Two No. 12 conductors are connected to the double-pole, 20-ampere branch-circuit overcurrent protective device (in Panel B) to form Circuit A5-7. Dual-element, time-delay Fusetron fuses may be installed in the disconnect switch. These fuses will serve the dual function of providing both branch-circuit overcurrent protection for the pump motor and back-up protection for the thermal overloads protecting the motor. These fuses may be rated at not more than 125 percent of the full-load running current of the motor.

A short length of flexible metal conduit is connected between the load side of the thermal overload switch and the pressure switch on the pump. This pressure switch is usually connected to the motor at the factory. In figure 19-2, a double-pole pressure switch is shown at Ⓔ. When the contacts of the pressure switch are open, all conductors feeding the motor are disconnected.

SUBMERSIBLE PUMP

A submersible pump, figure 19-3, consists of a centrifugal pump driven by an electric motor. The pump and the motor are contained in one housing, submersed below the permanent water level within the well casing. When running, the pump raises the water upward through the piping to the water tank. Proper pressure is maintained in the system by a pressure switch. The disconnect switch, pressure and limit switches, and controller are installed in a logical and convenient location near the water tank.

Some water tanks have a precharged air chamber that contains a vinyl bag that separates the air from the water. This assures that air is not absorbed by the water. The absorption of air causes a slow reduction of the water pressure, and ultimately, the tank will fill completely with water with no room for the air. The initial air charge is always maintained.

Power to the motor is supplied by a cable especially designed for use with submersible pumps. This cable is marked "submersible pump cable," and it is generally supplied with the pump and the pump's various other components. The cable is cut to the proper length to reach between the pump and its controller. When needed, the cable may be spliced according to the manufacturer's specifications.

Submersible water pump cable is "tag-marked" for use within the well casing for wiring deep-well water pumps where the cable is not subject to repetitive handling caused by frequent servicing of the pump units.

Submersible pump cable is *not* designed for direct burial in the ground unless it is marked type USE or type UF. Should it be required to run the pump's circuit underground for any distance, it is necessary to install underground feeder cable (UF), type USE, or a raceway with suitable conductors, and then make up the necessary splices in an approved weatherproof junction box. See figure 19-4.

Figure 19-4 illustrates how to provide proper grounding of a submersible water pump and the well casing. Proper grounding of the well casing and submersible pump motor will minimize or eliminate

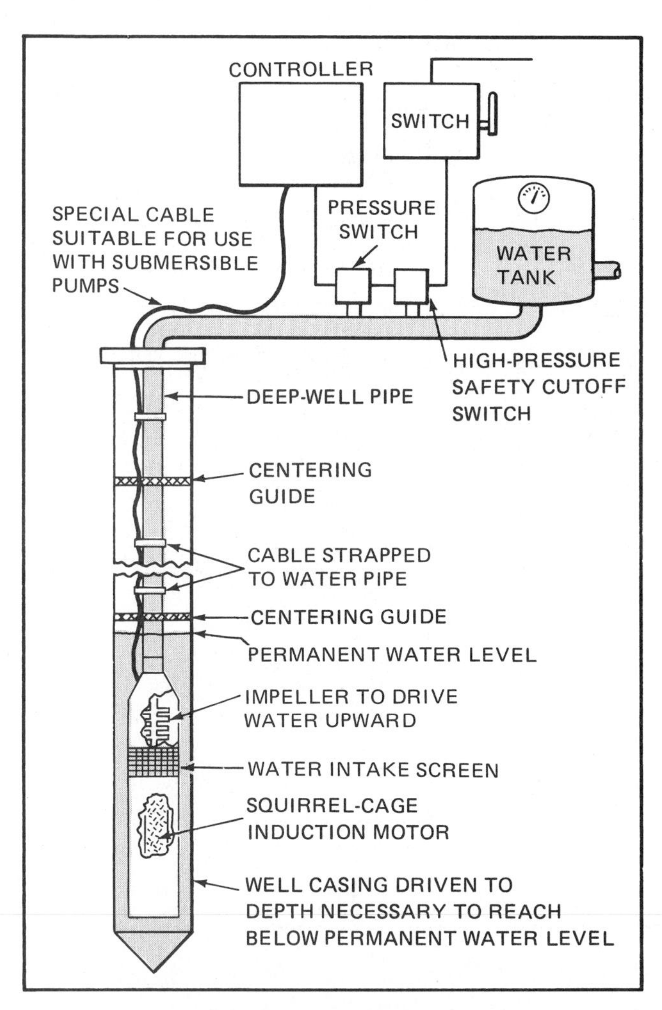

Fig. 19-3 Submersible pump.

stray voltage problems that could occur if the pump motor is not grounded. The subject of stray voltage is discussed in the text *Agricultural Wiring*.

Key code sections that relate to grounding water pumping equipment are:

- ► *250-43(k)* — motor operated water pumps must be grounded ◄
- *250-91(c)* — the earth is not to be used as the sole equipment grounding conductor
- *430-142* — grounding of motors
- *430-144* — grounding of controllers
- *430-145* — acceptable methods of grounding
- ► *547-8(d)* — ground pump and bond casing and pump together ◄

Some electrical codes require that because so much nonmetallic water piping is used today, that a grounding electrode conductor be installed between the neutral bar of the main service disconnecting means and the well casing. This then is the service ground and is sized per *Table 250-94* of the Code. At the casing, the grounding electrode conductor would be attached to a lug termination, or by means of exothermic (Cadweld) welding.

See unit 16 for information about underground wiring.

The controller contains the motor's starting relay, overload protection, starting and running capacitors, lightning arrester, and terminals for making the necessary electrical connections. Thus, there are no moving electrical parts within the pump itself, such as would be found in a typical single-phase, split-phase induction motor that would require a centrifugal starting-switch.

The calculations for sizing the conductors, and the requirements for the disconnect switch, the motor's branch-circuit fuses, and the grounding connections are the same as those for the jet pump motor. The nameplate data and instructions furnished with the pump must be followed.

WATER HEATER CIRCUIT Ⓐ C

All homes require a continuous supply of hot water. To meet this need, one or more automatic water heaters are installed close to the areas in the home having the greatest need for hot water. The proper size and type of water piping is installed to carry the heated water from the water heater to the various plumbing fixtures and to the appliances that require hot water, such as clothes washers and dishwashers.

Modern water heaters have adjustable temperature-regulating controls that maintain the water temperature at the desired setting, generally within a range of 110°–170° F.

For safety reasons, the Code and Underwriters Laboratories require that electric water heaters be equipped with a high-temperature cutoff. This high-temperature control limits the maximum water temperature to approximately 210° F. This control is set at the factory and cannot be changed in the field. According to *Section 422-14*, electric water heaters shall be equipped with a temperature limiting means in addition to the control thermostat. This temperature limiting means must disconnect all ungrounded conductors.

In most residential water heaters, the upper thermostat and high-temperature cutoff are combined into one device.

Pressure/temperature relief valves are usually

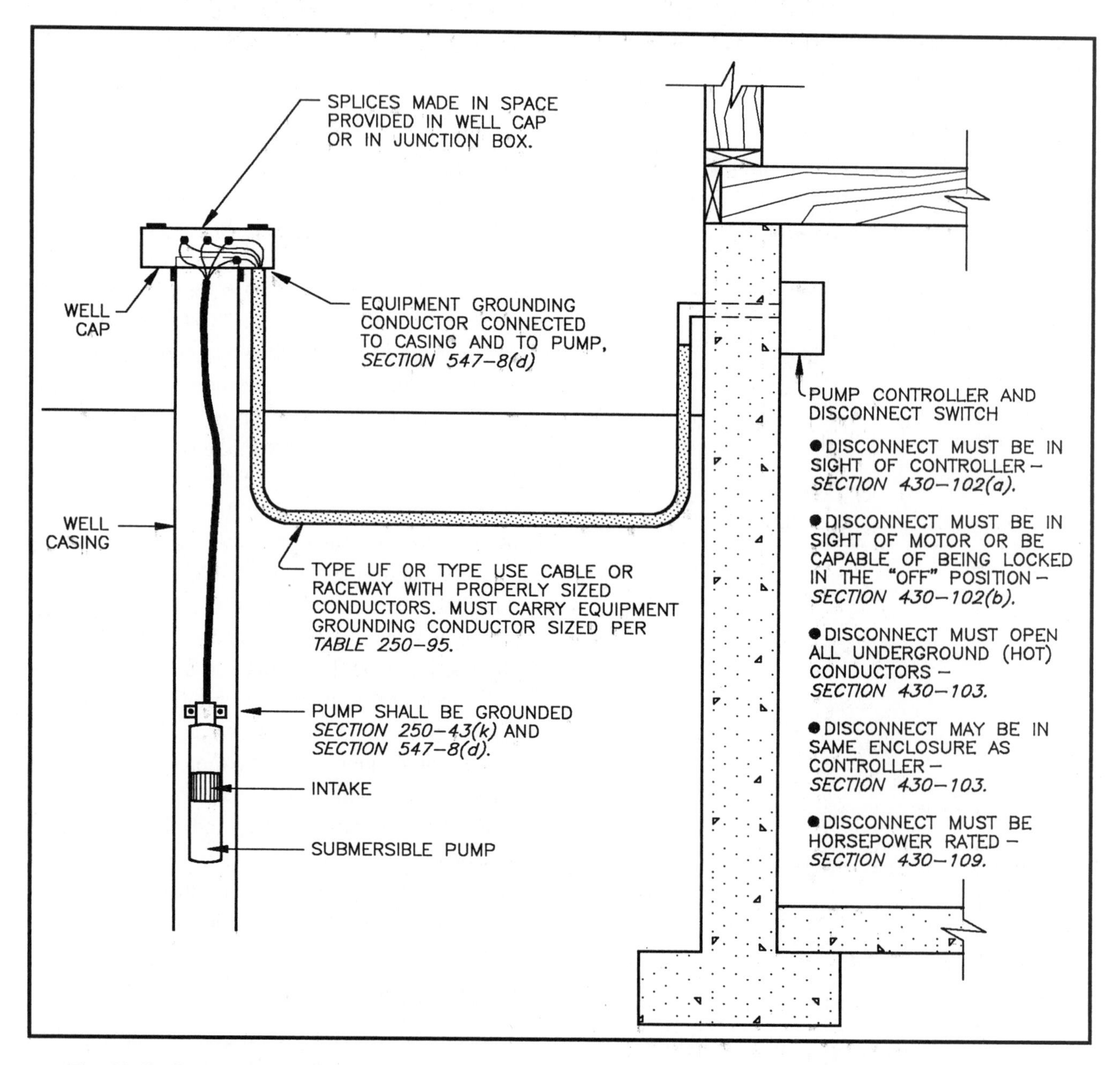

► Fig. 19-4 Grounding requirements for submersible pump. *Section 250-43(k)* and *Section 547-8(d)* require that submersible pumps be grounded. In addition, *Section 547-8(d)* requires that the pump and well casing be bonded together. See text for details. ◄

installed by the plumber into an opening provided and marked for the purpose on the water heater. The plumber will also install tubing from the pressure/temperature relief valve downward to within 6 inches of the floor so that any discharge will exit close to the floor. This reduces the possibility of scalding someone that might be standing nearby. The manufacturer of the water heater can also provide a factory installed pressure/temperature relief valve.

Pressure/temperature relief valves are installed to prevent the water heater tank from rupturing under excessive pressures and temperatures should the adjustable thermostat and the high-temperature cutoff fail to open. This would cause the water to boil (212 °F), would create steam, and would lead to possible rupturing of the tank.

Residential water heaters typically are available in 20-, 30-, 40-, 52-, 66-, and 82-gallon sizes.

To reduce corrosion, most water heaters have glass-lined steel tanks and contain one or two magnesium anodes (rods) permanently submerged in the water. Their purpose is to provide "cathodic protection." Simply stated, corrosion problems are minimized as long as the magnesium anodes are in an active state. Information regarding the replacement of the anode is found by checking the manufacturer's literature that is included with every electric water heater.

HEATING ELEMENTS *(Figure 19-11)*

The wattage ratings of electric water heaters vary greatly, depending upon the size of the heater in gallons, the speed of recovery desired, local electric utility regulations, and local plumbing and building Codes.

Speed of recovery is defined as the time required to bring the water temperature to satisfactory levels in a given length of time. For example, it might be stated that a 3800-watt element water heater can heat 17 gallons of water in one hour, and a 5500-watt element water heater can heat 25 gallons in one hour. The insulation surrounding the tank and the circuit voltage as well as the temperature of the incoming cold water affect the speed of recovery.

Wattage ratings might be 1650, 2000, 2500, 3000, 3800, 4500, or 5500 watts. Other wattage ratings are available.

The heating elements are generally rated 240 volts. Elements are also available with 120-volt ratings and 208-volt ratings. These elements will produce rated wattage output at rated voltage. When operated at less than rated voltage, the wattage output will be reduced. The effect of voltage variation is discussed later in this unit. Most heating elements will burn out if operated at voltages 5% higher than for which they are rated.

Underwriters Laboratories requires that the power (wattage) input must not exceed 105% of the water heater's nameplate rating. All testing is done with a supply voltage equal to the heating element's rated voltage.

To reduce the element burnout situation, some manufacturers will supply water heaters that have heating elements that are actually rated for 250 volts, but will mark the water heater 240 volts with its corresponding wattage at 240 volts. This allows a cushion should higher than normal voltages be experienced.

Heating elements consist of a high resistance nickel-chrome (nichrome) alloy wire which is coiled and embedded (compacted) in magnesium oxide in a copper-clad, or stainless steel, tubular housing. The magnesium oxide is an excellent insulator of electricity, yet it effectively conducts the heat from the nichrome wire to the housing. Construction of heating elements is very much the same as the surface units of an electric range. Terminals are provided on the heating element for the connection of the conductors. A threaded hub allows the element to be screwed securely into the side of the water tank. Water heater resistance heating elements are in direct contact with the water for efficient transfer of heat from the element to the water.

Electric water heaters are available with one or two heating elements, figure 19-5. Single-element water heaters will have the heating element located near the bottom of the tank. Two-element water heaters have one heating element near the bottom of the tank and the other about halfway up the tank.

Thermostats similar to the one illustrated in figure 19-10 (page 281) turn the heating elements on and off, thereby maintaining the desired water temperature. In two-element water heaters, the top thermostat is an interlocking type, sometimes referred to as a snap-over type. The bottom thermostat is a single-pole, on-off type. Both thermostats have

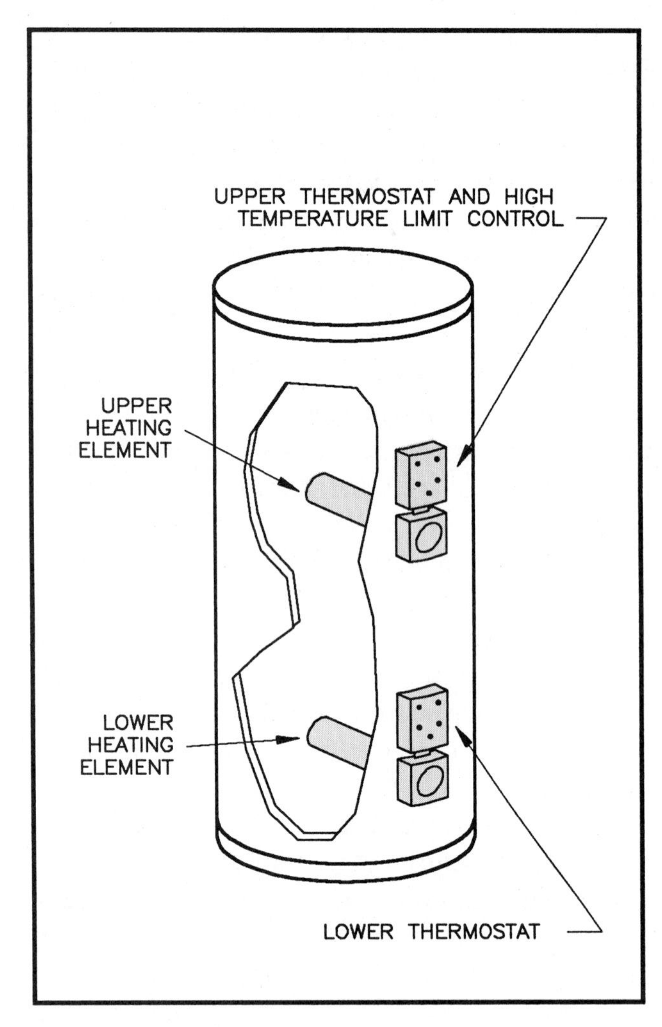

Fig. 19-5 Typical electric water heater showing location of heating elements and thermostats.

silver contacts and have quick make-quick break characteristics to prevent arcing. The various wiring diagrams shown in the figures show that both heating elements cannot be energized at the same time.

SEQUENCE OF OPERATION

There are many ways that water heaters can be connected, usually dependent upon the local power company's regulations, particularly when special low "off-peak" metering is involved. This will be discussed later in this unit. Because there are so many ways to electrically connect these water heaters, only a few types of connections are illustrated. Consult your power company and/or your supplier of electric water heaters to find out what connections are required in your area.

Figures 19-6, 19-7, 19-8, and 19-9 illustrate water heaters that are two-element, non-simultaneous operation type.

Figures 19-6 and 19-7 show the wiring for special low rate "off-peak" metering where electricity to the water heater is "off" during specific hours of the day.

Figures 19-8 and 19-9 show the wiring for separate special low-rate metering, but with the advantage of having 24-hour operation.

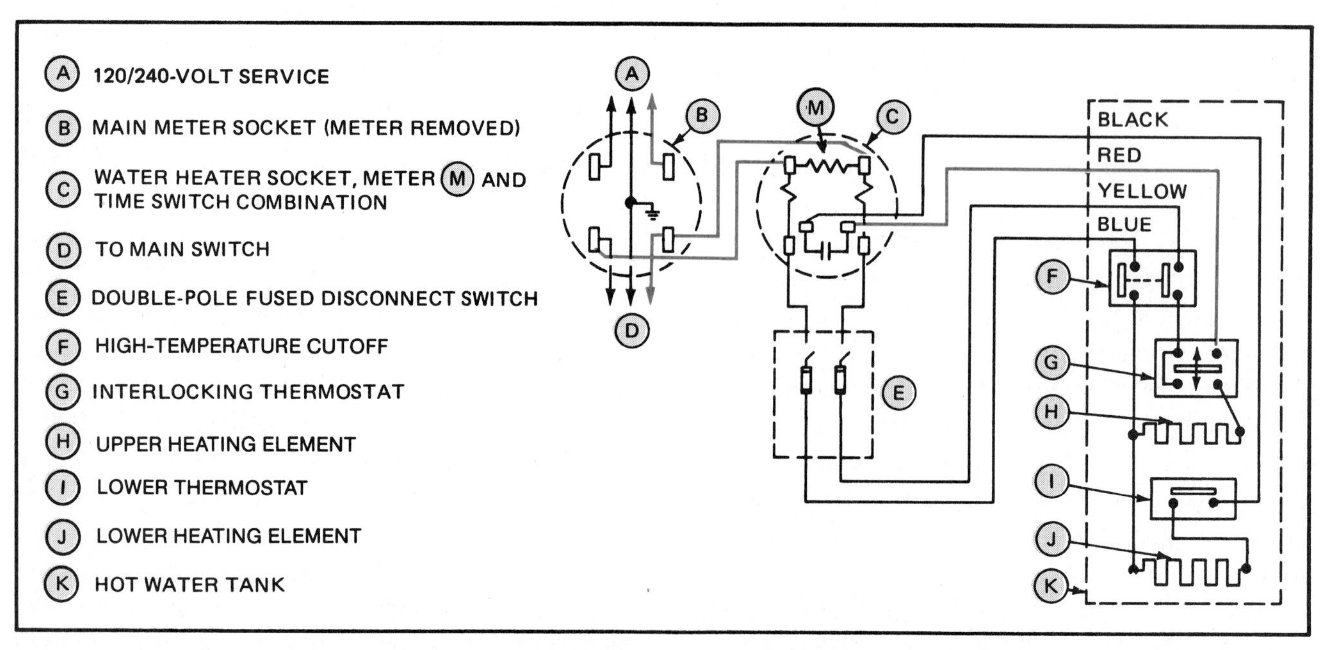

Fig. 19-6 Wiring for typical off-peak water heating installation.

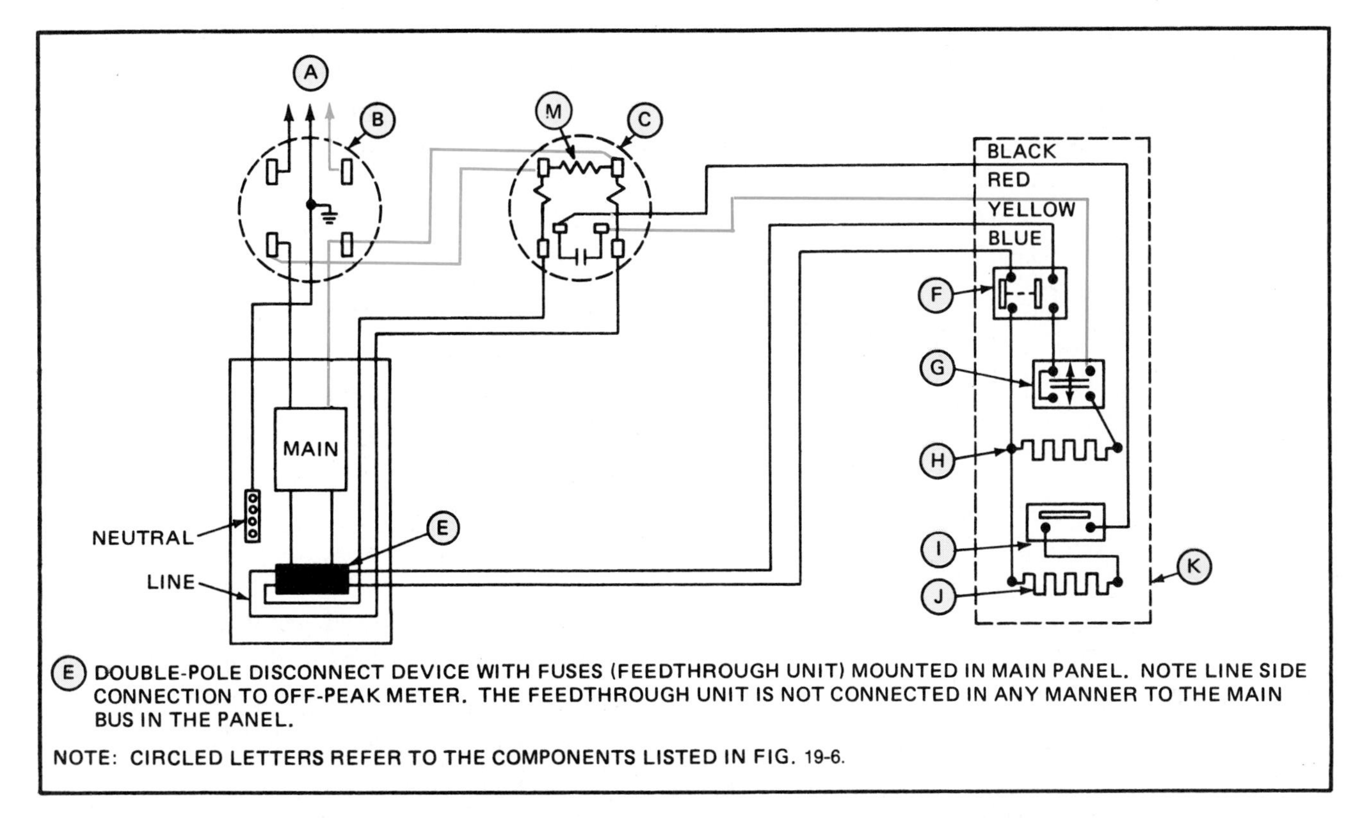

Fig. 19-7 Feedthrough unit protective device for off-peak metering.

When all of the water in the water heater is cold, the upper element heats only the water in the upper half of the tank. The upper element is energized through the lower contacts of the upper thermostat. The lower thermostat also indicates that heat is needed, but heat cannot be supplied because of the open condition of the upper contacts of the upper thermostat.

When the water in the upper half of the tank reaches a temperature of about 150°F., the upper thermostat snaps over. That is, the upper contacts are closed and the lower contacts are opened. The lower element now begins to heat the cold water in the bottom of the tank. When the entire tank is heated to 150°F., the lower thermostat shuts off. Electrical energy is not being used by either element at this point. The heat-insulating jacket around the water heater is designed to keep the heat loss to a minimum. Thus, it takes many hours for the water to cool only a few degrees.

Most of the time, the bottom heating element will keep the water at the desired temperature. Should a large amount of hot water be used, the top element comes on, resulting in a fast recovery of the water in the upper portion of the tank which is where the hot water is drawn from.

METERING AND SEQUENCE OF OPERATION

Electric water heaters are available in a number of styles, such as

- single element.
- double element, simultaneous operation.
- double element, non-simultaneous operation.
- off-peak metering, double-element simultaneous operation.
- off-peak metering, double-element non-simultaneous operation.

There are many ways that a utility company can provide metering and for the electrician to make the connections. Some of the ways include:

1. Connect the water heater to one of the two-pole, 240-volt circuits in the main distribution panel, figure 19-9. This is probably the most common method because the wiring is simple, and additional switches, breakers, time clocks, or meters are not required. This is the connection for the residence in this text. This method provides electricity to the water heater 24 hours of every day.

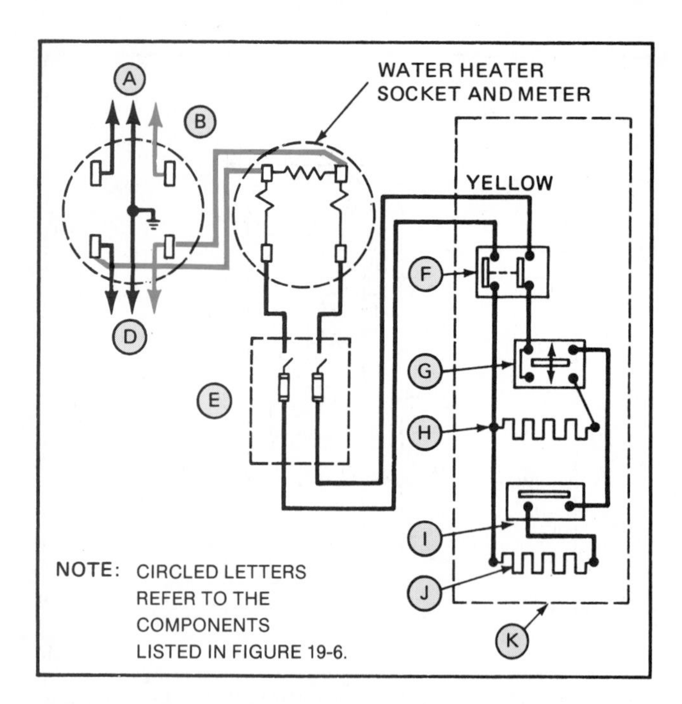

Fig. 19-8 Twenty-four-hour water heater installation.

KILOWATT-HOUR POWER CONSUMPTION OF WATER HEATER IS METERED BY SAME METER USED FOR NORMAL HOUSEHOLD LOADS. ONLY ONE METER IS REQUIRED. POWER RATES ARE USUALLY ON A SLIDING SCALE; THAT IS, THE MORE POWER IN KILOWATT-HOURS USED, THE LOWER THE RATE PER KILOWATT-HOUR.

120/240-VOLT SERVICE

WATER HEATER

YELLOW

BLUE

MAIN HOUSE METER

MAIN

NEUTRAL

F G H I J K

DOUBLE-POLE DISCONNECT DEVICE WITH FUSES CONNECTED TO MAIN BUS IN MAIN PANEL

NOTE: CIRCLED LETTERS REFER TO THE COMPONENTS LISTED IN FIG. 19-6.

Fig. 19-9 Water heater connected to two-pole circuit in Main Panel.

Instead of separate "low-rate" metering that requires an additional watt-hour meter, the homeowner enjoys special rates based upon a sliding scale.

For example, the utility might have the following rate schedule for a single family residence.

METER READING 04-01-90	00532	
METER READING 03-01-90	00007	
TOTAL KILOWATT HOURS USED	00525	
BASIC MONTHLY ENERGY CHARGE		$11.24
1ST. 400 KWH × 0.09208		36.83
NEXT 125 KWH × 0.04627		5.78
AMOUNT DUE		$53.85

State, local, and applicable regulatory taxes would be added to the amount due.

2. Connect the water heater through a separate disconnect switch that is fed through a separate meter. See figure 19-8. In this arrangement, the regular house power is metered at one rate, while the electric water heater is metered at another, usually a lower rate. Note that the water heater meter is connected to the load side of the house meter. Therefore, the reading of the water heater meter is subtracted from the house meter to determine the amount of energy consumed by the water heater and the amount of energy consumed by all other usage in the home. This method provides power to the water heater 24 hours each day as required.

3. Connect the water heater in an "off-peak" metering arrangement, figure 19-6. A separate meter/time switch combination is installed. The utility sets the time switch to be "on" during those periods of time when the demand on the power company is normal. The time clock shuts "off" during those hours deemed peak hours by the utility. The water heater meter is connected to the load side of the regular house meter. The reading of the water heater meter is therefore subtracted from the house meter to determine the amount of energy used by the water heater and the amount of energy used by all other lights and appliances in the house. The water heater rate is lower than the normal

Fig. 19-10 Typical electric water heater controls. The seven (7) terminal control combines water temperature control plus high temperature limit control. This is the type commonly used to control the upper heating element as well as providing high temperature limiting feature. The two (2) terminal control is the type used to control water temperature for the lower heating element. Courtesy THERM-O-DISC, Inc., subsidiary of Emerson Electric Company.

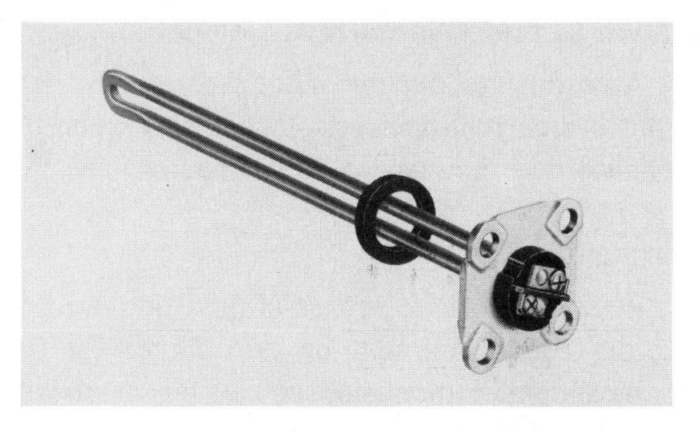

Fig. 19-11 Heating element. Courtesy Edwin L. Wiegand Division, Emerson Electric Company.

rate charged for the general use of electrical energy.

4. Connect the water heater as in figure 19-7. This is a slight variation to what was outlined in #2. In figure 19-7, there is a feedthrough overcurrent device in the panel instead of a separate disconnect switch. The water heater has two elements sequenced for non-simultaneous operation.
5. Connect the water heater circuit as in figure 19-9, but install a time clock in the circuit leading to the water heater. The time clock would be set to be "off" during certain times of the day deemed "peak" hours by the utility. Special incentives are given to the homeowner by the utility for agreeing to such an arrangement.
6. Connect the upper element to the regular source of power, and connect the lower element through the "off-peak" meter. Then, when the time clock of the off-peak meter is "off," the water will be heated through the regular house meter circuit at the regular rate for energy use. This connection will require separate overcurrent devices for each element.
7. Some utilities use radio-controlled switching to control water heater loads. They send a carrier signal through the power line, which is received by the electronic controlling device, thereby turning the power to the water heater "on" and "off" as required by the utility.

WATER HEATER LOAD DEMAND

The following Code sections apply to load demands placed on an electric water heater branch circuit.

Section 422-5(a) requires that an appliance branch circuit shall have a rating not less than the rating marked on the appliance.

Section 422-5(a), Exception No. 2 requires that a continuously loaded appliance must have a branch circuit rating not less than 125% of the marked rating of the appliance. Most inspectors and electricians would not consider residential water heaters to be "continuous duty," as would be the case for most commercial electric water heaters.

Section 422-14(b) applies to all fixed storage water heaters having a capacity of 120 gallons (454.2 L) or less. Such heaters are required to have a branch-circuit rating not less than 125% of the nameplate rating of the water heater.

Section 422-20 requires that the appliance must have a means of disconnect. This would normally be located in the panel (load center), or it could be provided by a separate disconnect switch. Although not normally done, *Section 422-24* of the Code does allow the unit switch that is an integral part of an appliance having an "off" position that disconnects all ungrounded conductors from the source to be considered the required disconnecting means. In many local codes, this is not acceptable. *Section 422-25* further requires that there be another means of disconnect for the appliance, which in the case of a one-family dwelling is the main service disconnect.

Section 422-25 requires that switches and circuit breakers used as the disconnecting means must be of the indicating type, meaning they must clearly show that they are in the "on" or "off" position.

► *Section 422-28(e)* states that for single non-motor-operated electrical appliances, the overcurrent protective device shall not exceed the protective device rating marked on the appliance. If the appliance has no such marking, then the overcurrent device is to be sized as follows:

- If appliance does not exceed 13.3 amperes — 20 amperes
- If appliance draws more than 13.3 amperes — 150% of the appliance rating

For a single non-motor-operated appliance, if the 150% sizing does not result in a standard size overcurrent device rating as listed in *Section 240-6*, then it is permitted to go to the next standard size. This is permitted in *Section 422-28(e), Exception.*

EXAMPLE: A water heater nameplate indicates 4500 watts, 240 volts. What is the maximum size fuse permitted by the Code?

ANSWER:

$$\frac{4500}{240} = 18.75 \text{ amperes}$$

$$18.75 \times 1.5 = 28.125 \text{ amperes}$$

Therefore, the Code permits the use of a 30-ampere fuse. ◄

Section 210-19(a) states that the branch-circuit conductors shall have an ampacity not less than the maximum load to be served.

Section 210-22 states that for any branch-circuit, the total load shall not exceed the rating of the branch-circuit.

The water heater for the residence is connected to Circuit A6-8. This is a 20-ampere, double-pole overcurrent device located in the main panel in the workshop. This circuit is straight 240 volts and does not require a neutral conductor.

Checking the Schedule of Special-purpose Outlets, we find that the electric water heater has two heating elements: a 2000-watt element and a 3000-watt element. Because of the thermostatic devices on the water heater, both elements cannot be energized at the same time. Therefore, the maximum load demand is

$$\text{amperes} = \frac{\text{watts}}{\text{volts}} = \frac{3000}{240} = 12.5 \text{ amperes}$$

For water heaters having two heating elements connected for non-simultaneous operation, the nameplate on the water heater will be marked with the largest element's wattage at rated voltage. For water heaters having two heating elements connected for simultaneous operation, the nameplate on the water heater will be marked with the total wattage of both elements at rated voltage.

Branch-Circuit Overcurrent Device

► According to *Section 422-28(e)*, the branch-circuit overcurrent device is 20 amperes when the appliance does not exceed 13.3 amperes. ◄

Conductor Size

Referring to *Table 310-16* of the Code, we find that the conductors will be No. 12 AWG. The specifications for the residence call for conductors to be Type THHN.

Conduit Size

Table 3A, Chapter 9 of the Code indicates that 1/2-inch conduit (electrical metallic tubing in this residence for all exposed wiring in the basement) is required for two No. 12 THHN conductors. To simplify the installation, a short length (18 to 24 inches . . . 457 to 610 mm) of 1/2-inch flexible metal conduit is attached to the 1/2-inch EMT and is connected to the knockout provided for the purpose on the water heater.

Grounding

The water heater is grounded through the 1/2-inch conduit and the flexible metal conduit.

Code references are *Sections 250-42*, *250-57*, *350-5*, and *422-16*. The subject of grounding was covered in unit 4.

EFFECT OF VOLTAGE VARIATION

As previously stated, the heating elements in the heater in this residence are rated 240 volts. They will operate at a lower voltage with a reduction in wattage. If connected to voltages above their rating, they will have a very short life.

Ohm's law and the wattage formula show how the wattage and current depend upon the applied voltage.

$$R = \frac{E^2}{W} = \frac{240 \times 240}{3000} = 19.2 \text{ ohms}$$

$$I = \frac{E}{R} = \frac{240 \times 240}{19.2} = 12.5 \text{ amperes}$$

If a different voltage is substituted, the wattage and current values change accordingly.

At 220 volts:

$$W = \frac{E^2}{R} = \frac{220 \times 220}{19.2} = 2521 \text{ watts}$$

$$I = \frac{W}{E} = \frac{2521}{220} = 11.46 \text{ amperes}$$

At 230 volts:

$$W = \frac{E^2}{R} = \frac{230 \times 230}{19.2} = 2755 \text{ watts}$$

$$I = \frac{W}{E} = \frac{2755}{230} = 11.97 \text{ amperes}$$

In resistive circuits, current is directly proportional to voltage and can be simply calculated using a ratio and proportion formula. For example, if a 240-volt heating element draws 12.7 amperes at rated voltage, the current draw can be determined at any other applied voltage, say 208 volts.

$$\frac{208}{240} = \frac{X}{12.7}$$

$$\frac{208 \times 12.7}{240} = 11.006 \text{ amperes}$$

To calculate this example if the applied voltage is 120 volts:

$$\frac{120}{240} = \frac{X}{12.7}$$

$$\frac{120 \times 12.7}{240} = 6.35 \text{ amperes}$$

Also, in a resistive circuit, wattage varies as the square of the current. Therefore, when the voltage on a heating element is doubled, the current also doubles and the wattage increases four times. When the voltage is reduced to one-half, the current is halved and the wattage is reduced to one-fourth.

Using the above formulae, it will be found that a 240-volt heating element when connected to 208 volts will have approximately three-quarters of the wattage rating than at rated voltage.

REVIEW

Note: Refer to the Code or the plans where necessary.

WATER PUMP CIRCUIT Ⓐ B

1. Does a jet pump have any electrical moving parts below the ground level? ______
2. Which is larger, the drive pipe or the suction pipe? ______
3. Where is the jet of the pump located? ______
4. What does the impeller wheel move? ______
5. Where does the water flow after leaving the impeller wheel?

   a. ______ b. ______
6. What prevents water from draining back into the pump from the tank? ______

7. What prevents water from draining back into the well from the equipment? ______
____

8. What is compressed in the water storage tank? ______________

9. How is the motor disconnected if pumping is no longer required? ______________
____
____

10. What is a common speed for jet pump motors? ______________

11. Why is a 240-volt motor preferable to a 120-volt motor for use in this residence?
____
____

12. How many amperes does a 1-hp, 240-volt, single-phase motor draw? (See *Table 430-148*.) ______________

13. What size are the conductors used for this circuit? ______________

14. What is the branch-circuit protective device? ______________

15. What furnishes the running protection for the pump motor? ______________
____

16. What is the maximum ampere setting required for running protection of the 1-hp, 240-volt pump motor? ______________

17. Submersible water pumps operate with the electrical motor and actual pump located (Circle correct answer.)
    a. above permanent water level
    b. below permanent water level
    c. half above and half below permanent water level

18. Because the controller contains the motor starting relay and the running and starting capacitors, the motor itself contains ______________

19. What type of pump moves the water upward inside of the deep-well pipe? ______
____

20. Proper pressure of the submersible pump system is maintained by a ______________
____

21. Fill in the data for a 16-ampere electric motor, single-phase, no Code letters.
    a. Branch-circuit protection, nontime-delay fuses.

    Normal size ______________ A Switch size ______________ A

    Maximum size ______________ A Switch size ______________ A

    b. Branch-circuit protection dual-element, time-delay fuses.

    Normal size ______________ A Switch size ______________ A

    Maximum size ______________ A Switch size ______________ A

c. Branch-circuit protection instant-trip breaker.
Normal setting ______________ A
Maximum setting ______________ A

d. Branch-circuit protection — inverse time breaker.
Normal setting ______________ A
Maximum setting ______________ A

e. Branch-circuit conductor size. ______________
Ampacity ______________ A

f. Motor running protection using dual-element time-delay fuses.
Maximum size ______________ A

►22. The *National Electrical Code®* is very specific in its requirement that submersible electric water pump motors be grounded. Where is this specific requirement found in the Code? ______________
______________ ◄

23. Does the *National Electrical Code®* allow submersible pump cable to be buried directly in the ground? ______________
____

24. Must the disconnect switch for a submersible pump be located next to the well?
____
____

WATER HEATER CIRCUIT ⒶC

1. At what temperature is the water in a residential water heater usually maintained?
____

2. Magnesium rods are installed inside the water tank to reduce ______________

3. Is a separate meter required to record the amount of energy used to heat the water?
____

4. What is meant by the phrase "off-peak metering"? ______________
____
____
____
____
____
____

5. What is the most common method of metering water heater loads? ______________
____
____
____
____

6. Two thermostats are generally used in an electric water heater.

   a. What is the location of each thermostat? ______________________________

   ____

   b. What type of thermostat is used at each location? ______________________

   ____

7. a. How many heating elements are provided in the heater in the residence discussed in this text? ____

   b. Are these heating elements allowed to operate at the same time? __________

8. When does the lower heater operate? ______________________________

   ____

9. The Code states that water heaters having a capacity of 120 gallons (454.2 L) or less shall be considered ______________ duty and, as such, the circuit must have a rating of not less than __________% of the rating of the water heater.

10. Why does the storage tank hold the heat so long? ______________________

11. The electric water heater is connected for "limited demand" so that only one heating element can be on at one time.

    a. What size wire is used to connect the water heater? ____________________

    b. What size overcurrent device is used? ______________________________

12. a. If both elements of the water heater are energized at the same time, how much current will they draw? (Assume the elements are rated at 240 volts.)

    b. What size wire is required for the load of both elements? Show calculations.

13. a. How much current do the elements in question 12 draw if connected to 220 volts?

b. What is the wattage value at 220 volts? Show calculations for both.

14. A 240-volt heater is rated at 1500 watts.

a. What is its current rating at rated voltage? ______________________________

b. What current draw would occur if connected to a 120-volt supply?

c. What is the wattage output at 120 volts?

15. For a single, nonmotor-operated electrical appliance rated greater than 13.3 amperes that has no marking that would indicate the size of branch-circuit overcurrent device, what percentage of the nameplate rating is used to determine the maximum size branch-circuit overcurrent device? ______________________

16. A 7,000-watt resistance-type heating appliance is rated 240 volts. What is the maximum size fuse permitted to protect the branch circuit supplying this appliance?

____

UNIT 20

Special-Purpose Outlets for Ranges, Counter-Mounted Cooking Unit $_G$ and Wall-Mounted Oven $_F$

OBJECTIVES

After studying this unit, the student will be able to

- interpret electrical plans to determine special installation requirements for counter-mounted cooking units, wall-mounted ovens, and freestanding ranges.
- compute demand factors for ranges, wall-mounted ovens, and counter-mounted cooking units.
- select proper conductor sizes for wiring installations based on the ratings of appliances.
- ground all appliances properly regardless of the wiring method used.
- describe a seven-heat control and a three-heat control for a dual-element heating unit for a cooktop.
- understand how infinite heat temperature controls operate.
- supply counter-mounted cooking units and wall-mounted ovens by connecting them to one feeder using the "tap" rule.
- understand how to install a feeder to a load center, then divide the feeder to individual circuits to supply the appliances.

The residence being discussed in this text has one built-in oven and one built-in cooktop. Checking the Schedule for Special-Purpose Outlets, we find that the built-in wall-mounted oven is supplied by Circuit B(6–8), and the built-in cooktop is supplied by Circuit b(2–4).

Running separate circuits to individual appliances is probably the most common method used to connect appliances. However, there might be a cost justification for running one larger circuit from the main panel to a junction box located in close proximity to both the oven and the cooktop. From this junction box, "taps" of smaller conductors are made to feed each appliance.

The following text explains the use of a load center and shows how to do the calculations for individual circuits, a single larger circuit with corresponding taps to the individual appliances, and a freestanding range. Which method to use will be governed by local codes and consideration of the cost of installation (time and material).

COUNTER-MOUNTED COOKING UNIT CIRCUIT Ⓖ

Counter-mounted cooking units are available in many styles. Such units may have two, three, or four surface heating elements. The surface heating elements of some cooking units are completely covered by a sheet of high-temperature ceramic. These units have the same controls as standard units and are wired according to the general Code guidelines for cooking units. When the elements of these ceramic-covered units are turned on, the ceramic above each element changes color slightly. Most of these cooking units have a faint design in the ceramic to locate the heating elements. The ceramic is rugged, attractive, and easy to clean.

In the residence in the plans, a separate circuit is provided for a standard, exposed-element, counter-mounted cooking unit. This unit is rated at 7450 watts and 120/240 volts. The ampere rating is:

$$I = \frac{W}{E} = \frac{7450}{240} = 31.04 \text{ amperes}$$

The cooking unit outlet is shown by the symbol Ⓖ.

NEC Table 220-19 shows the demand loads for household cooking appliances. A freestanding range may be supplied by conductors having an ampacity for the demand load shown in the table for the specific range.

However, the demand load for individual wall-mounted oven circuits or individual counter-mounted cooking unit circuits must be calculated at the full nameplate rating (*Note 4* to *Table 220-19*).

Installing the Cooking Unit

Table 310-16 shows that No. 8 THHN conductors may be used as the circuit conductors for the 31.04-ampere cooking unit installation. These circuit conductors are connected to Circuit B2–4. This is a two-pole, 40-ampere circuit in Panel B.

The electrician must obtain the roughing-in dimensions for the counter-mounted cooking unit so that the cable can be brought out of the wall at the proper height. In this way, the cable is hidden from view. To simplify both servicing and installation, a separable connector or a plug and receptacle combination may be used in the supply line, *Section 422-17*.

Most cooktop manufacturers provide a junction box on their units. The electrician can terminate the circuit conductors of the appliance in this box and make the proper connections to the residence wiring. Instead of the junction box, the manufacturer may connect a short length of flexible conduit to the appliance. This conduit contains the conductors to which the electrician will connect the supply conductors. These connections must be made in a junction box furnished by the electrician or the manufacturer of the appliance.

In this residence, the cooktop range is installed in the island. See plans for details. The feed will be a three-conductor nonmetallic-sheathed cable that has three No. 8 AWG copper conductors plus an equipment grounding conductor. The cable runs out of the top of Panel B, across the face of the ceiling joists in the Recreation Room, then upward into the base of the island cabinet. Because the Recreation Room ceiling is a dropped ceiling, the cable is run to a point just below the intended location of the Kitchen island, leaving ample cable length hanging in the basement. Later, when the Kitchen cabinets and the island are installed, and other finishing touches are being done in the Kitchen, the built-in appliances are set into place. Following this, the electrician then completes the final hookup of the built-in appliances.

Note on the plans that a receptacle outlet is to be installed on the side of the island containing the cooktop. This receptacle outlet is supplied by small appliance Circuit B16, figure 12-7. Also study page 42, COUNTERTOPS, and figure 3-6, "Receptacle Locations."

Temperature Effects on Conductors

It is good practice to check the appliance instructions provided by the manufacturer before beginning the installation. The manufacturer may require the supply conductors to have an insulation rated for temperatures higher than the standard insulation rating of 60° C.

Underwriters Laboratories in Standard 858 (Household Electric Ranges) requires appliance manufacturers to attach some form of label or otherwise mark their units to indicate that the supply conductors will be subjected to temperatures higher than 60° C. If such information is not provided, then it is safe to assume that the conductors used may have insulation rated for 60° C. *Table 310-13* gives the conductor insulation temperature limitations and the accepted applications for conductors.

General Wiring Methods

Any of the standard wiring methods may be used to connect counter-mounted cooking units, wall-mounted ovens, freestanding ranges, and clothes dryers. That is, armored cable, nonmetallic-sheathed cable, and conduit may be used. These appliances can also be connected with service-entrance cable, *Article 338* (refer to unit 15).

Local electrical codes can contain information as to restrictions on the use of any of the wiring methods listed.

Grounding Requirements

Section 250-60 of the *National Electrical Code®* requires electric ranges, ovens, cooktop units, and electric clothes dryers to be grounded by any of the following:

a. to the insulated grounded neutral circuit conductor if the grounded neutral conductor is not smaller than No. 10 AWG copper or No. 8 AWG aluminum.

b. to the uninsulated neutral of Type SE service-entrance cable. This is permitted *only* if the branch circuit originates at the service equipment. See figures 20-1 and 20-2.

c. with a separate equipment grounding conductor sized according to *Table 250-95*, run with the circuit conductors in a cable assembly or in a raceway.

d. by means of using grounded metal conduit as the equipment grounding conductor.

e. through properly grounded flexible metal conduit. However, there are limitations on the use of flexible metal conduit as an equipment ground. See *Sections 250-91(b)* and *350-5*. This topic is covered in unit 4.

See unit 15 for additional coverage on the subject of grounding electric ranges, ovens, and cooktop appliances.

Surface Heating Elements for Cooking Units

The heating elements used in surface-type cooking units are manufactured in several steps. First, spiral-wound nichrome resistance wire is impacted in magnesium oxide (a white, chalklike powder). The wire is then encased in a nickel-steel

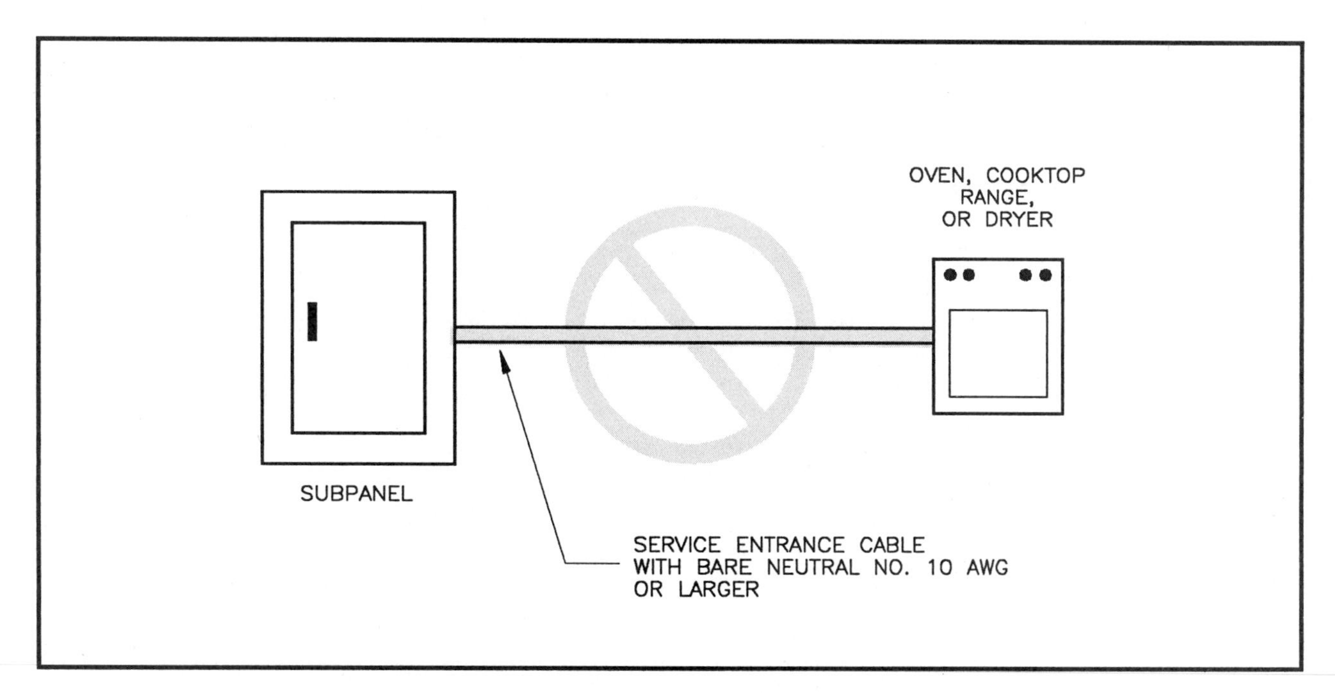

Fig. 20-1 This installation *is not permitted, Section 250-60(c)* and *Section 338-3(b)*, because the service-entrance cable originates at a subpanel.

alloy sheath, flattened under very high pressure, and formed into coils. The flattened surface of the coil makes good contact with the bottoms of cooking utensils. Thus, efficient heat transfer takes place between the elements and the utensils.

Cooktops with a smooth ceramic surface offer the user ease in cleaning, and a neater, more modern appearance. The electric heating elements are in direct contact with the underside of the ceramic top.

Figures 20-3B and 20-4B illustrate typical electric range surface units.

TEMPERATURE CONTROL

Infinite-position Temperature Controls

Modern electric ranges are equipped with "infinite"-position temperature knobs (controls). Older-style heat controls connect the heating elements of a surface unit to 120 volts or 240 volts in series or in parallel to attain a certain number of specific heat levels. Newer switches have a rotating cam that provides an infinite number of heat positions.

Inside these infinite position temperature controls are contacts that open and close repetitively. The circuit to the surface heating element is regulated by the ratio of the time the contacts are in a closed position versus the open position. This is termed *input percentage*.

The contacts open and close as a result of a heater-bimetal device that is an integral part of the rotary control. When the control is turned ON, a separate set of pilot light contacts closes and remains closed, while the contacts for the heating element open and close to attain and maintain the desired temperature of the surface heating unit, figure 20-3A.

The contacts in a temperature control of this type have an expected life of over 250,000 automatic cycles. The knob rotation expected life exceeds 30,000 operations.

Older-style controls, as opposed to the newer infinite-position controls, are still available. One type provides very specific "indents" that snap into place as the knob is rotated to the various heat positions. The same specific heat selections are available with the push-button type controls. Although rarely used on today's electrical appliances, their operation offers an excellent opportunity to explain series parallel circuits, and an opportunity to make electrical calculations of these series, parallel, and series/parallel circuits.

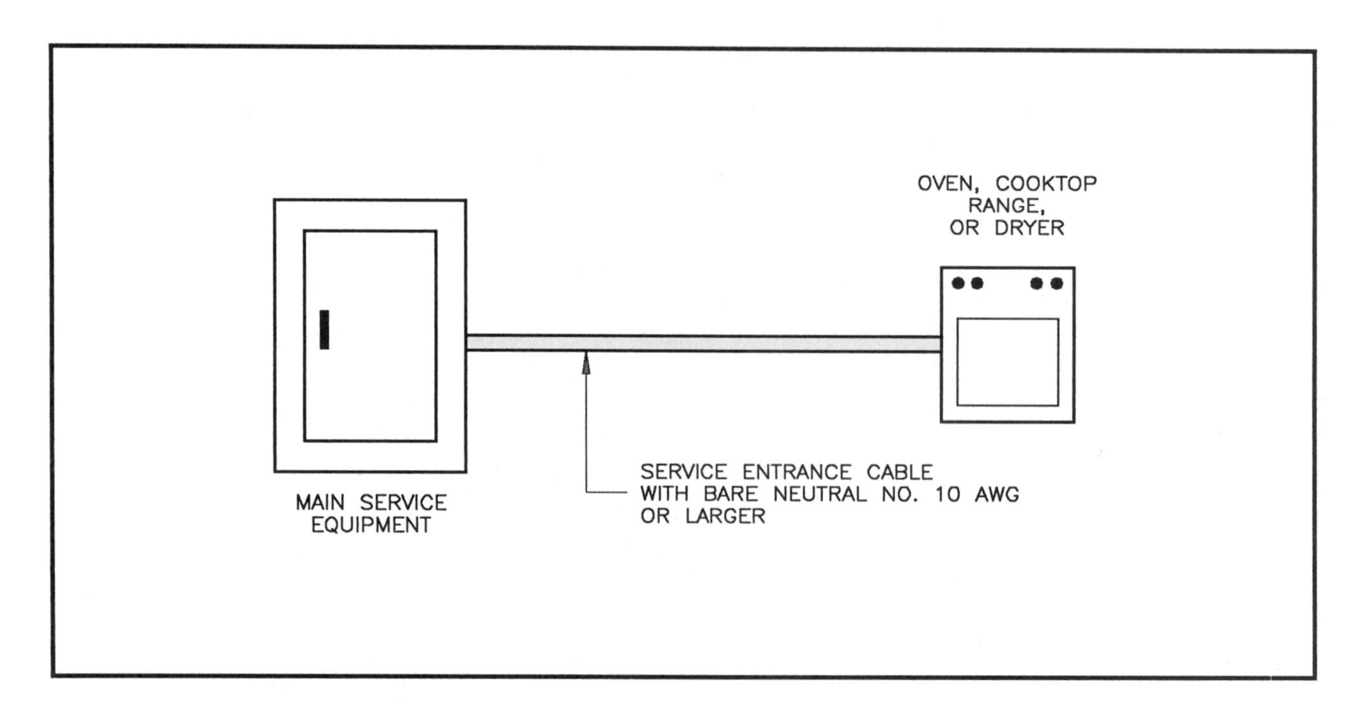

Fig. 20-2 This installation *is permitted*, *Section 250-60(c)* and *Section 338-3(b)*, because the service-entrance cable originates at the main service equipment.

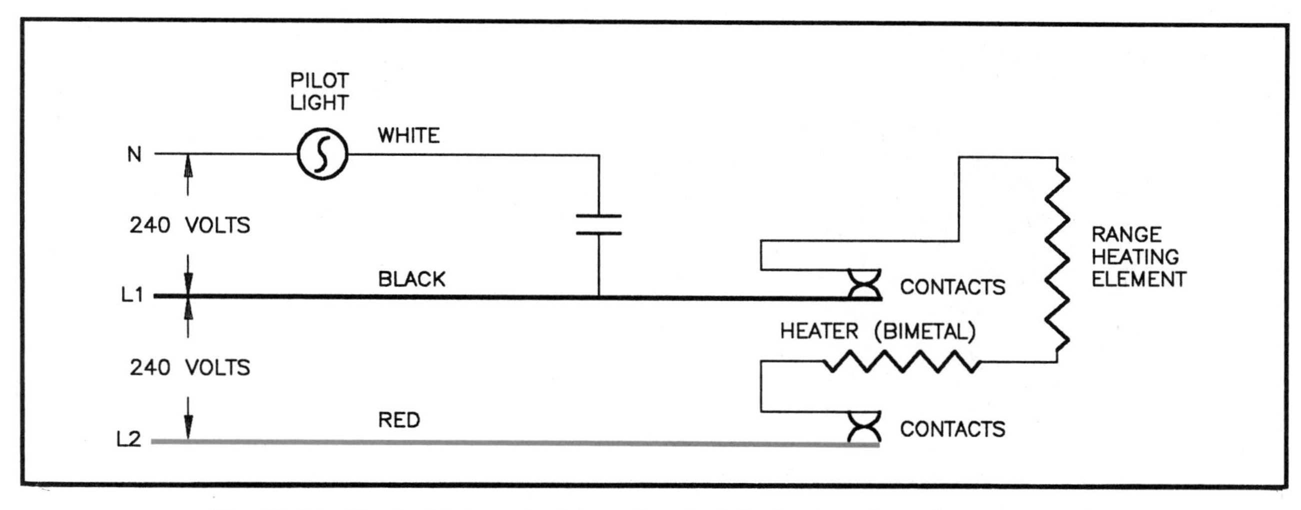

Fig. 20-3A Typical internal wiring of an infinite heat surface element control.

Seven-heat Control for Dual-element Heating Unit

One method of obtaining several levels of heat from one heating coil unit is to use a dual-element unit. Figure 20-4A shows that seven levels of heat can be obtained from such a unit if a seven-point heater switch is used. Note that the two elements of the unit are connected together at the inner end of the steel coil. A return wire (not a resistance wire) is carried back to the terminal block. This arrangement means that the elements can be operated singly, together, or in series.

Figure 20-5 shows the heating element connected with element A as lead 2, element B as lead 1, and the return wire as the common (C). Resistance readings can be taken between these leads as follows:

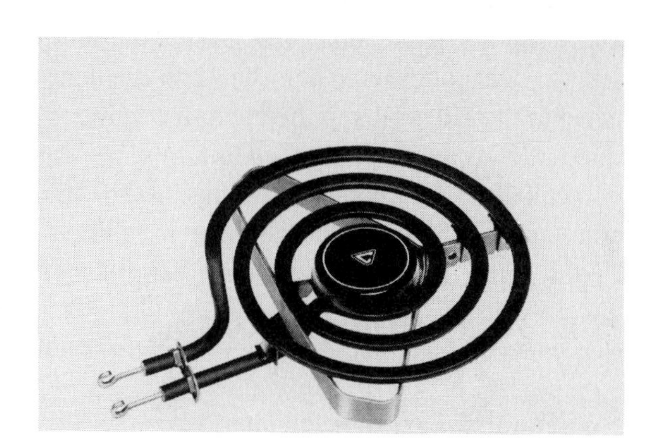

Fig. 20-3B Typical 240-volt electric range surface heating element. This is the type of element used with infinite heat controls. Courtesy of Chromalox, Edwin L. Wiegand Division, Emerson Electric Company.

Test between C and 2	90 ohms (element A)
Test between C and 1 (single operation)	74 ohms (element B)
Test between 2 and 1 (elements in series)	164 ohms (elements A and B in series)
Test between C and 1–2 connected together (elements connected in parallel)	40.6 ohms (elements A and B in parallel)

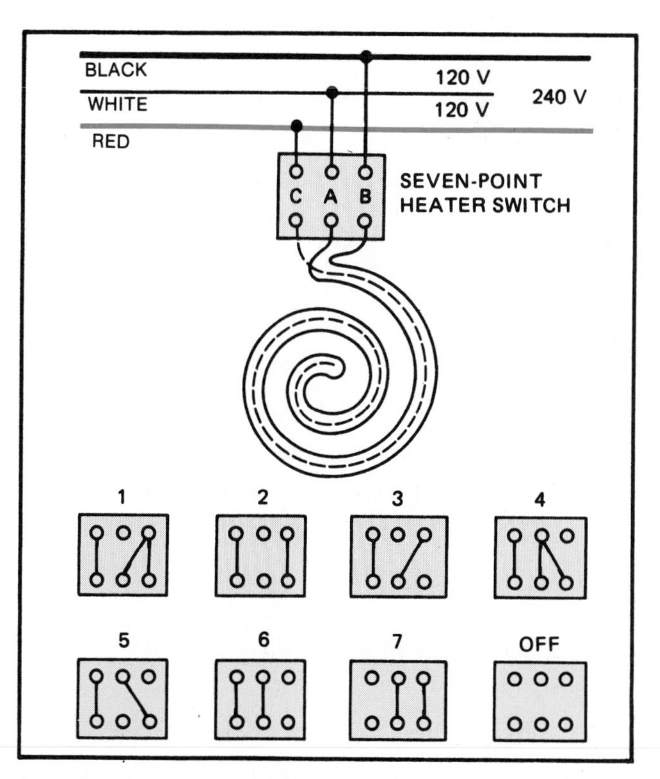

Fig. 20-4A Surface heating element with seven levels of heat.

Fig. 20-4B Typical electric range surface unit of the type used with 7-position controls as discussed in this text. Courtesy of Chromalox, Edwin L. Wiegand Division, Emerson Electric Company.

Element B allows more current to flow than does element A because element B has 74 ohms of resistance as compared to 90 ohms of resistance for element A. As the current increases, the wattage also increases. The eight switch positions shown in figure 20-4A result in the wattages shown in the following list. For each switch position, the formula $W = E^2/R$ is used to find the wattage value.

1. A and B parallel, 240 volts	40.6 ohms	1418 watts
2. B only, 240 volts	74 ohms	778 watts
3. A only, 240 volts	90 ohms	640 watts
4. A and B parallel, 120 volts	40.6 ohms	____ watts*
5. B only, 120 volts	74 ohms	____ watts*
6. A only, 120 volts	90 ohms	____ watts*
7. A and B series, 120 volts	164 ohms	____ watts*
8. Off position	Infinity	0 watts

*These values are omitted because the student is required to calculate the values in the Review section. A sample calculation is completed as follows:

$$W = \frac{E^2}{R} = \frac{240 \times 240}{40.6} = \frac{57600}{40.6} = 1418 \text{ watts}$$

Three-heat Control for Dual-element Heating Unit

A simple, three-heat series-parallel switch is shown in figure 20-6. This switch is similar to the type used on many electric ranges.

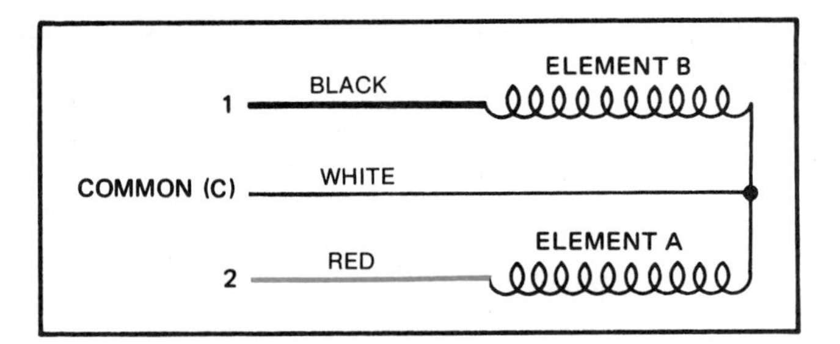

Fig. 20-5 Typical surface unit wiring.

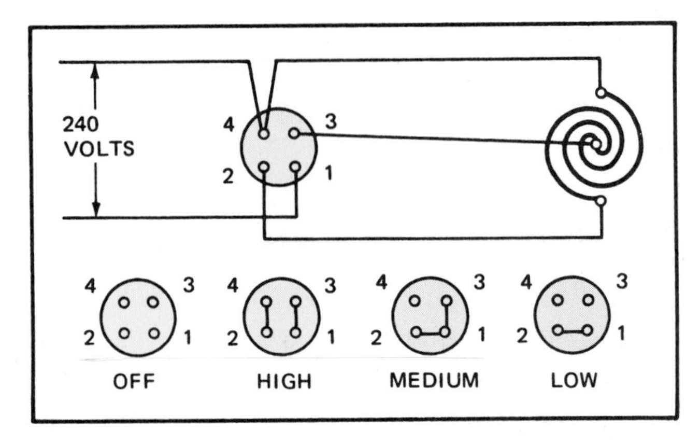

Fig. 20-6 Surface unit with three-heat series-parallel switch.

High-Speed Control

Another type of heating element is the 120-volt flash or high-speed element. When connected to a special control switch, this type of element is connected briefly across 240 volts. In a few seconds, the element is returned automatically to the 120-volt supply for continued operation. A 1250-watt flash unit connected to 240 volts produces 5000 watts of heat in a matter of seconds. This brings the unit to maximum heat quickly. When the flash element is returned to the 120-volt supply, the unit continues to produce 1250 watts, the rated value.

To calculate the maximum wattage at 240 volts, assume that a 1250-watt heating element is rated at 120 volts. This unit has a resistance of:

$$R = \frac{E^2}{W} = \frac{120 \times 120}{1250} = \frac{14400}{1250} = 11.52 \text{ ohms}$$

When this 11.52-ohm element is connected to a 240-volt supply, the wattage is:

$$W = \frac{E^2}{R} = \frac{240 \times 240}{11.52} = \frac{57600}{11.52} = 5000 \text{ watts}$$

Note that the wattage quadruples when the supply voltage is doubled. Conversely, the wattage is reduced to one-fourth its original value when the supply voltage is reduced to one-half its original value.

Heat Generation by Surface Heating Elements

The surface heating elements generate a large amount of radiant heat. For example, a 1000-watt heating unit generates approximately 3412 Btu of heat per hour. Heat is measured in British thermal units (Btu). A *Btu* is defined as the amount of heat required to raise the temperature of one pound of water by one degree Fahrenheit.

WALL-MOUNTED OVEN CIRCUIT Ⓐ$_F$

A separate circuit is provided for the wall-mounted oven. The circuit is connected to Circuit B(6–8), a two-pole, 30-ampere circuit in Panel B. The receptacle for the oven is shown by the symbol Ⓐ$_F$.

The wall-mounted oven installed in the residence is rated at 6.6 kW or 6600 watts at 120/240 volts. The current rating is:

$$I = \frac{W}{E} = \frac{6600}{240} = 27.5 \text{ amperes}$$

Section 422-4(a), Exception No. 3 tells us that branch circuits for household cooking units are to be determined according to *Table 220-19*.

Note 4 to *Table 220-19* states that the branch-circuit load for the wall-mounted oven must be based on the actual nameplate rating of the appliance.

After determining the current rating (I = W/E), refer to *Table 310-16*. We find that for 27.5 amperes, a No. 10 AWG copper TW, THW, or THHN may be used to supply this wall-mounted oven. As previously discussed, for nonmetallic sheathed cable, even though the conductor insulation is 90°C, the ampacity must be determined according to the 60°C column of *Table 310-16*. This is stated in *Section 336-26*.

Caution When Making Connections

Recall in unit 4, under "Conductor Temperature Ratings," that, according to Underwriters Laboratories Standards, when conductor sizes No. 14 through No. 1 AWG are installed, their current-carrying capacity (ampacity) is to be that of the 60° C column of *Table 310-16* unless the terminals are marked with some other temperature rating, such as 75° C or 90° C.

To further emphasize this caution, *Section 336-26* of the Code covering nonmetallic-sheathed cable (Romex) states that even though the insulation is rated 90° C, the ampacity of the conductors is to be based upon the 60° C column of *Table 310-16.* That same statement is found in *Section 333-5*, which covers armored cable (BX).

That is why it is so important to carefully read the appliance manufacturer's installation data for recommended conductor size and type. Check particularly for the temperature rating.

The information previously given on supply conductor connections, conductor temperature limitations, grounding requirements, and general installation procedures for counter-mounted cooking units also applies to wall-mounted ovens.

Self-cleaning Oven

Self-cleaning ovens are popular appliances because they automatically remove cooking spills from the inside of the oven. Lined with high-temperature material, a self-cleaning oven can be set for a cleaning temperature that is much higher than baking and broiling temperatures. When the oven "self-clean" control is turned on, internal temperatures in the oven will approximate 800° F. At this temperature, all residue in the oven, such as grease and drippings, is burned until it is fine ash that can be removed easily. During the self-clean cycle, for safety reasons, a special latching mechanism on the oven door makes it impossible to open the oven door until a predetermined safe temperature is reached. Self-cleaning ovens do not require special wiring and are electrically connected in the same manner as standard electric ovens.

Operation of the Oven

The oven heating elements are similar to the surface elements. The oven elements are mounted in removable frames. A thermostat controls the temperature of these heating units. Any oven temperature can be obtained by setting the thermostat knob to the proper point on the dial. The thermostat controls both the baking element and the broiling element. These elements can be used together to preheat the oven.

A combination clock and timer is used on most ovens. The timer can be preset so that the oven turns on automatically at a preset time, heats for a preset duration, and then turns off.

Microwave Oven

Modern microwave ovens, a result of electronic technology, have introduced many features never before possible. Such features include:

- control of time and temperature.
- programmable multiple-state settings.
- temperature probes for insertion into the food to assure desired temperature.
- ability to "hold" food temperatures for up to one hour following cooking cycle.
- touch-sensitive controls.
- programmable defrost settings, having lower temperatures than those required for actual cooking or baking, so that premature cooking does not occur.

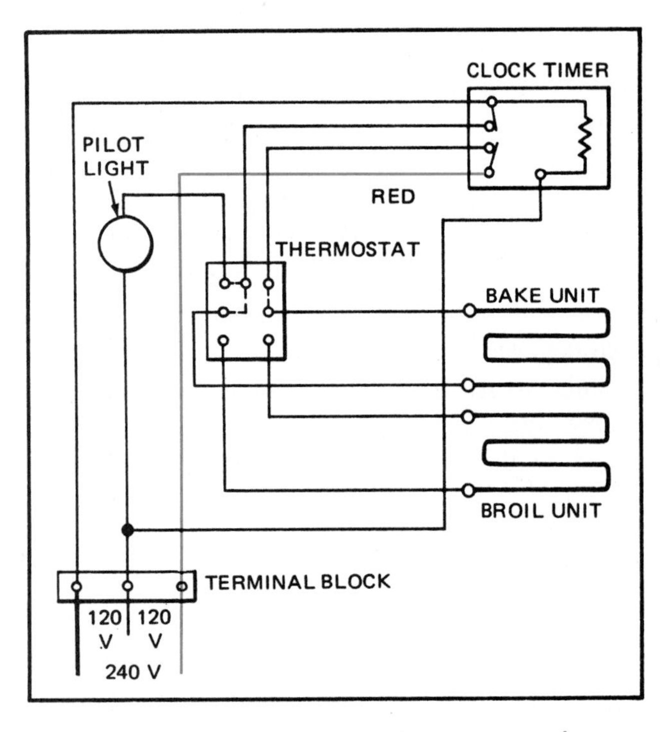

Fig. 20-7 Wiring diagram for an oven unit.

- computer calculations to automatically increase the time of operation when recipe quantities are doubled, tripled, or more.
- delayed starting of oven for up to 12 hours.
- preprogrammed temperature and time for recipes.

The electrician's primary concern is how to calculate the microwave oven load, and how to install the circuit for microwave appliances so as to conform to the Code. The same Code rules apply for both electrical and electronic microwave cooking appliances.

Ovens are insulated with fiberglass or polyurethane foam insulation placed within the oven walls to prevent excessive heat leakage.

Figure 20-7 shows a typical wiring diagram for a standard oven unit.

CIRCUIT *REQUIREMENTS* WHEN MORE THAN ONE WALL-MOUNTED OVEN AND COUNTER-MOUNTED COOKING UNIT ARE SUPPLIED BY ONE CIRCUIT

The Code in *Note 4* to *Table 220-19* states that a counter-mounted cooking unit and not more than two ovens may be connected to one circuit. When only one circuit is used, its capacity must be large enough to serve both appliances.

Because the Code does permit ranges and other household cooking equipment to be connected to a single circuit, certain restrictions must be met. Some of the key restrictions are:

Section 210-19(b)

- the branch-circuit conductors must have an ampacity not less than the branch-circuit rating.
- the branch-circuit conductors' rating must not be less than the load being served.
- the branch-circuit rating must not be less than 40 amperes for ranges of 8 3/4 kW or more.
- for ranges 8 3/4 kW or more, the neutral may be 70% of the branch-circuit rating, but not smaller than No. 10 AWG.
- tap conductors connecting the ranges, ovens, and cooking units to a 50-ampere branch circuit must be rated at least 20 amperes, be ade-

quate for the load, and must not be any longer than necessary for servicing the appliance.

Section 220-19 (Note 4)

- the branch-circuit load for a single wall-mounted oven or a single counter-mounted cooking unit is to be based on the nameplate rating of the appliance.

- if a single counter-mounted cooking unit and not more than two wall-mounted ovens are one branch circuit and are located in the same room, add up the nameplate ratings of the individual appliances, then consider the total to be the same as one range. The calculation for this arrangement is the same as the freestanding range calculation, pages 298–299.

The single circuit method requires that smaller conductors be tapped from the 50-ampere circuit. This means that extra junction boxes, cable connectors, conduit fittings, and wire connectors are used to complete the installation. Thus, the initial cost of installing a 50-ampere circuit may be higher than the cost of installing separate circuits having smaller capacities. Another factor must be considered by the electrician when deciding which method of installing cooking units is more economical. Since the individual overcurrent device for each appliance on an individual circuit has a lower rating, there is greater protection if a short circuit or ground fault occurs on the circuit.

The load demand of a single, freestanding electric range (combined cooking unit and oven) is similar to the load demand of a separate wall-mounted oven and counter-mounted cooking unit. *Note 4* to *Table 220-19* points out special requirements for a branch-circuit load for a counter-mounted cooking unit and not more than two wall-mounted ovens which are all supplied from a single branch circuit and located in the same room. The branch-circuit load for this arrangement must be computed by adding the nameplate ratings on the individual appliances, and treating the total as equivalent to that of one range.

To compute the load demand for the counter-mounted cooking unit and the wall-mounted oven, the nameplate ratings are used.

Counter-mounted cooktop	7450 watts
Wall-mounted oven	6600 watts
Total	14050 watts

The total connected load of 14050 watts at 240 volts is

$$I = \frac{W}{E} = \frac{14050}{240} = 58.5 \text{ amperes}$$

According to *Note 1* of *Table 220-19*, for ranges rated over 12 kW but not over 27 kW, the demand load given in column A of *Table 220-19* must be increased by 5% for each kilowatt or major fraction of a kilowatt in excess of 12 kW.

Assume that the combined load of the counter-mounted cooking unit and the wall-mounted oven is equivalent to the load of a freestanding range rated at 14050 watts. This value exceeds 12000 watts (12 kW) by 2050 watts, or roughly 2 kW.

The demand is determined as follows:

2 kW (kW over 12 kW) × 5% per kW = 10%
8 kW (from Column A of *Table 220-19*) × 0.10 = 0.8 kW
Calculated demand = 8 kW + 0.8 kW = 8.8 kW or = 8800 watts

and,

$$I = \frac{W}{E} = \frac{8800}{240} = 36.7 \text{ amperes}$$

Section 210-23(c) states that fixed cooking appliances may be connected to 40- or 50-ampere circuits. If taps are to be made from 40- or 50-ampere circuits, the taps must be able to carry the load to be served. In no case may the taps be less than 20 amperes, *Section 210-19(b)*.

For example, the 50-ampere circuit shown in figure 20-8 has No. 10 THHN and No. 8 THHN tap conductors. Each of these taps has an ampacity of more than 20 amperes. The branch circuit requires No. 6 Type THHN conductors. *Section 210-19(b), Exception No. 2*, states that the load on the neutral conductor supplying the wall-mounted oven and the counter-mounted cooking unit may be based on 70% of the ampacity of the branch-circuit rating:

$$50 \times 0.7 = 35$$

According to *Table 310-16*, the neutral may be a No. 8 THHN conductor.

USING A LOAD CENTER

In this residence the built-in oven and range are very close to Subpanel B, in which case separate circuits are run to each appliance.

However, there are situations where the Main Panel and the built-in range components are far apart. In these instances, it might make economic sense to install a single subfeeder from the Main Panel to a Load Center, figure 20-9, then create a separate circuit for each appliance.

The maximum power demand for the appliances is obtained from Column A of *Table 220-19*. It was shown that the calculated demand for the 7450-watt cooking unit and the 6600-watt oven is 8.8 kW (36.7 amperes). The two ungrounded conductors are No. 8 THHN wire (40 amperes).

Section 210-19(b) states that the neutral conductor must have a rating of not less than $40 \times 0.70 = 28$ amperes.

Table 310-16 shows that No. 10 THHN wire is the minimum size of copper conductor that can be used for the neutral of this subfeeder circuit.

In figure 20-9, a No. 8 THHN neutral feeder conductor is shown even though the conductor may be reduced to No. 10 THHN, *Section 210-19(b)*. Cable having three conductors of the same size is available.

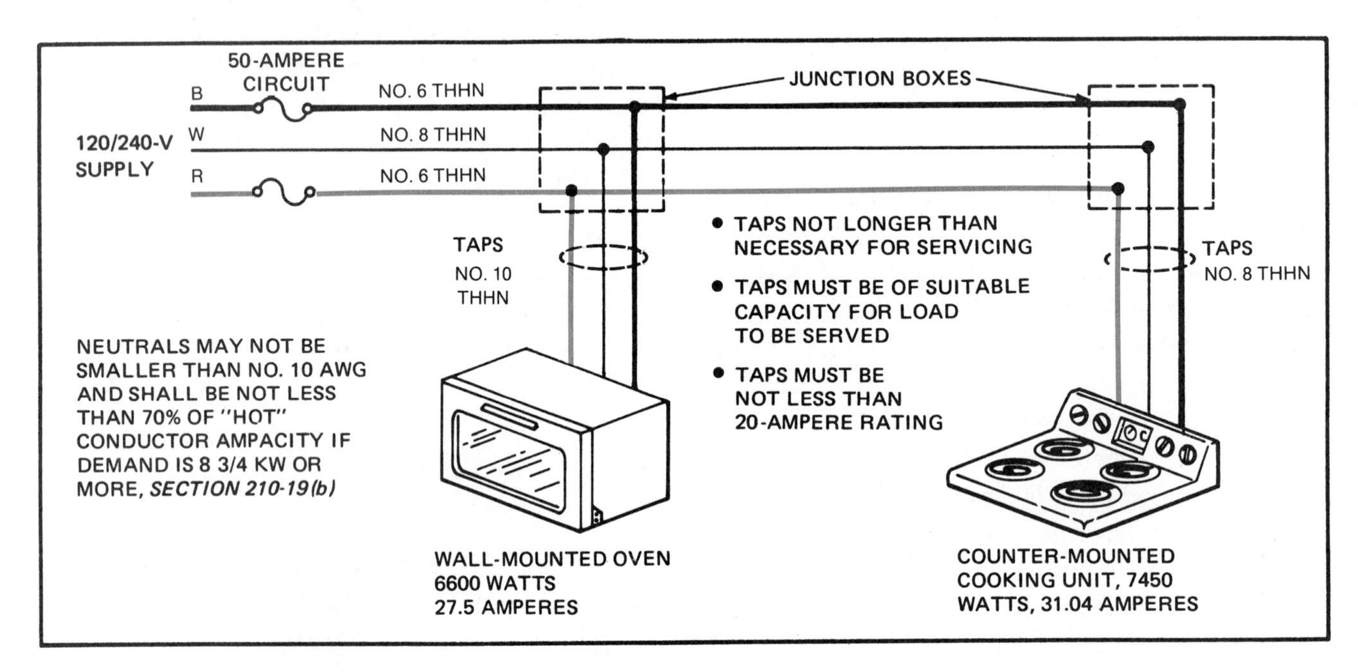

Fig. 20-8 Counter-mounted cooktop and wall-mounted oven connected to one circuit.

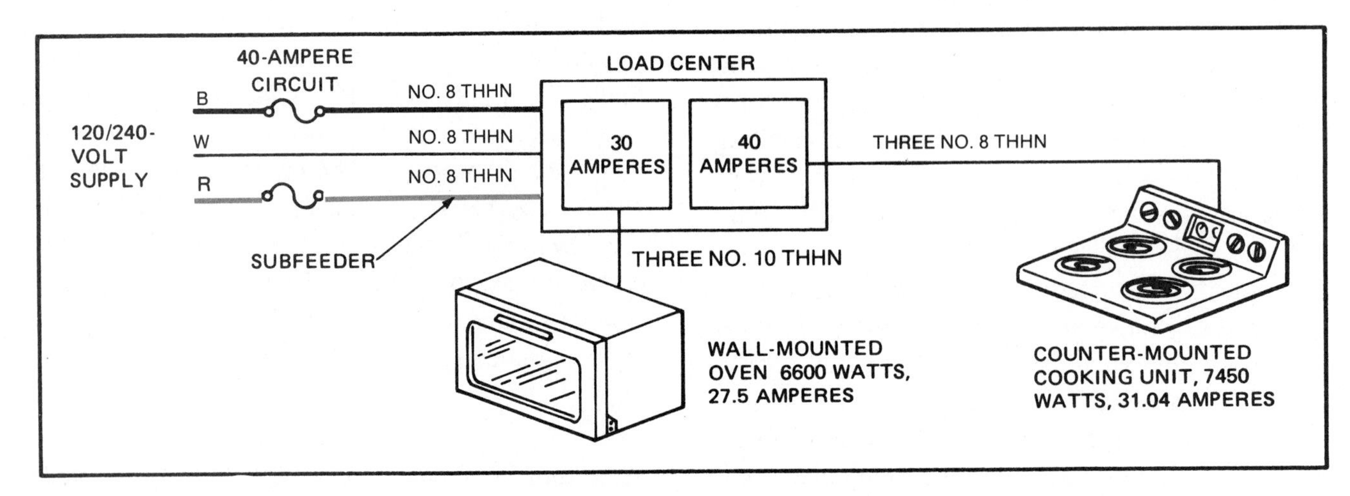

Fig. 20-9 Appliances connected to a load center.

The No. 8 THHN conductors are run from a 40-ampere circuit to the load center. The circuits supplying each appliance are protected at the load center according to their individual current ratings. The fuses used have lower ratings based on the nameplate rating of the appliance (see *Note 4, Table 220-19*).

FREESTANDING RANGE

The demand calculations for a single, freestanding range are simple. For example, a single electric range is to be installed. This range has a rating of 14050 watts (exactly the same as the combined built-in units).

The load for a single range is the same as that for the two built-in units:

1. 14050 watts − 12000 watts = 2050 watts or 2 kW
2. According to *Note 1, Table 220-19*, a 5 percent increase for each kW in excess of 12 kW must be added to the load demand in Column A. 2 kW (kW over 12 kW) × 5% per KW = 10%
3. 8 kW (from Column A) × 0.10 = 0.8 kW
4. Calculated load = 8 kW + 0.8 kW = 8.8 kW
5. In amperes, the calculated load is:

$$I = \frac{W}{E} = \frac{8800}{240} = 36.7 \text{ amperes}$$

No. 8 THHN conductors could be used for the ungrounded conductor. The size of the neutral is based on 70 percent of the current-carrying capacity of the No. 8 THHN conductors, or

$$40 \times 0.70 = 28 \text{ amperes}$$

A No. 10 THHN conductor is to be used for the neutral.

If a range is rated at not over 12 kW, then the maximum power demand is based on Column A of *Table 220-19*. For example, if the range is rated at 11.4 kW, the maximum demand is based on 8 kW:

$$I = \frac{W}{E} = \frac{8000}{240} = 33.3 \text{ amperes}$$

As in the previous example, two No. 8 THHN ungrounded conductors and one No. 10 THHN neutral conductor can be installed for this range. For ranges of 8 3/4 kW or higher rating, the minimum branch-circuit rating shall be 40 amperes, *Section 210-19(b)*.

The disconnecting means for a freestanding range may be a cord-and-plug, the same arrangement as illustrated in figures 18-4 and 18-5 for clothes dryers. For electric ranges, these cords and plugs are generally rated at 50 amperes, figure 20-10.

Fig. 20-10 Typical 50-ampere, three-pole, three-wire, 125/250-volt surface-mount range receptacle.

REVIEW

Note: Refer to the Code or the plans where necessary.

COUNTER-MOUNTED COOKING UNIT CIRCUIT Ⓐ$_G$

1. a. What circuit supplies the counter-mounted cooking unit in this residence? ____
   b. What is the rating of this circuit? ____________________

2. What three methods may be used to connect counter-mounted cooking units? ____

3. Is it permissible to use standard 60° C insulated conductors to connect all counter-mounted cooking units? Why? ____

4. What is the maximum operating temperature (in degrees Celsius) for a
   a. Type TW conductor? ____
   b. Type THW conductor? ____
   c. Type THHN conductor? ____

5. a. May service-entrance cable be used to connect counter-mounted cooking units? ____
   b. What Code section answers question (a)? ____

6. a. What section of the Code applies to grounding a counter-mounted cooking unit? ____
   b. What methods may be used to ground the counter-mounted cooking unit? ____

7. An electric range or clothes dryer may be grounded to the ____ conductor, provided the ____ conductor is not smaller than No. ____ AWG.

8. In the illustration of a typical seven-heat, eight-position switch, figure 20-3A, the wattage values for positions 4, 5, 6, and 7 are omitted. Calculate these values and insert them in the spaces below and on page 294. Show all calculations.

   Position 4

   Position 5

Position 6

Position 7

9. A 120-volt flash or high-speed unit rated at 1000 watts produces ________ watts when connected briefly to the 240-volt source. Show all calculations.

10. When the voltage to an element is doubled, the wattage: (circle one)

    a. increases   b. decreases

11. By how much is the wattage in question 10 increased or decreased? (circle one)

    a. doubled   b. tripled   c. quadrupled   d. halved

12. One kilowatt equals ______________ Btu per hour.

WALL-MOUNTED OVEN CIRCUIT ▲F

1. To what circuit is the wall-mounted oven connected? ______________________

2. An oven is rated at 7.5 kW. This is equal to

    a. __________ watts.

    b. __________ amperes at 240 volts.

3. a. What section of the Code governs the grounding of a wall-mounted oven? ____

    ____

    b. By what methods may wall-mounted ovens be grounded? ______________

    ____

    ____

    ____

    ____

    ____

4. What is the type and rating of the overcurrent device protecting the wall-mounted oven? ____

    ____

    ____

    ____

5. How many feet (meters) of cable are required to connect the oven in the residence?

    ____

6. When connecting a wall-mounted oven and a counter-mounted cooking unit to one feeder, how long are the taps to the individual appliances? ________________

____

____

____

7. The branch-circuit load for a single wall-mounted self-cleaning oven or counter-mounted cooking unit shall be the ______________ rating of the appliance.

8. A 6-kW counter-mounted cooking unit and a 4-kW wall-mounted oven are to be installed in a residence. Calculate the maximum demand according to *Column A, Table 220-19*. Show all calculations.

9. The size of the neutral conductor supplying an electric range may be based on ______ % of the ampacity of the (ungrounded) conductor.

10. a. A freestanding range is rated at 11.8 kW, 240 volts. According to *Column A, Table 220-19*, what is the maximum demand? ________________

    b. What size wire (Type TW) is required?

    c. What size neutral is required? ________________

    ____

11. A double-oven electric range is rated at 18 kW, 240 volts. Calculate the maximum demand according to *Table 220-19*. Show all calculations.

12. a. What size conductors are required for the range in question 11? Show all calculations.

b. What size neutral is required? ______________________________

____

____

13. For ranges of 8 3/4 kW or higher rating, the minimum branch-circuit rating is ______ amperes.

14. When a separate circuit supplies a counter-mounted cooking unit, what Code reference tells us that it is "against Code" to apply the demand factors of *Table 220-19*, and requires us to compute the load based upon the appliance's actual nameplate rating? ______________________________

15. A three-wire No. 10 AWG cable is used to connect a wall-mounted oven. What is the size of the grounding conductor in this cable? ______________________________

16. Is it permitted to ground the frame of the wall-mounted oven in question 15 to the neutral conductor? ______________________________

UNIT 21

Special-Purpose Outlets — Food Waste Disposer Ⓐ$_H$, Dishwasher Ⓐ$_I$

OBJECTIVES

After studying this unit, the student will be able to

- install circuits for kitchen appliances such as a food waste disposer and a dishwasher.
- provide adequate running overcurrent protection for an appliance when such protection is not furnished by the manufacturer of the appliance.
- describe the difference in wiring for a food waste disposer with and without an integral on-off switch.
- determine the maximum power demand of a dishwasher.
- make the proper grounding connections to the appliances.
- describe disconnecting means for kitchen appliances.

In this unit, we will discuss in detail the circuit requirements for food waste disposers and dishwashers as installed in residences. However, since there are many other types of appliances found in homes, the electrician needs to read carefully *Article 422* of the Code. *Article 422* covers all types of appliances in general, branch-circuit requirements, and installation requirements. It also discusses specific types of appliances, etc.

CAUTION: Keep in mind that these 120-volt appliances are *not* permitted to be connected to the small appliance circuits provided for in the kitchen. The small appliance circuits are intended to be used for the numerous types of portable appliances used in the kitchen area.

FOOD WASTE DISPOSER Ⓐ$_H$

The food waste disposer outlet is shown on the plans by the symbol Ⓐ$_H$. The food waste disposer is rated at 7.2 amperes. This appliance is connected to Circuit B19, a separate 20-ampere, 120-volt branch circuit in Panel B. To meet the specifications, No. 12 THHN conductors are used. The con-

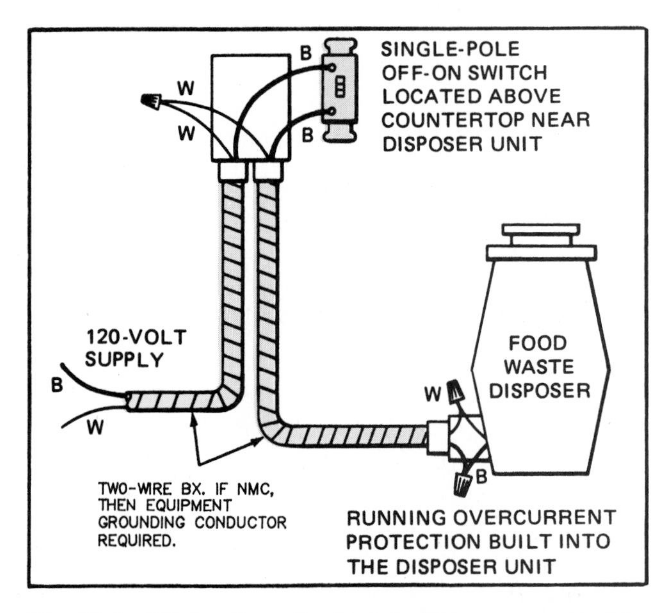

Fig. 21-1 Wiring for a food waste disposer operated by a separate switch located above countertop near the sink.

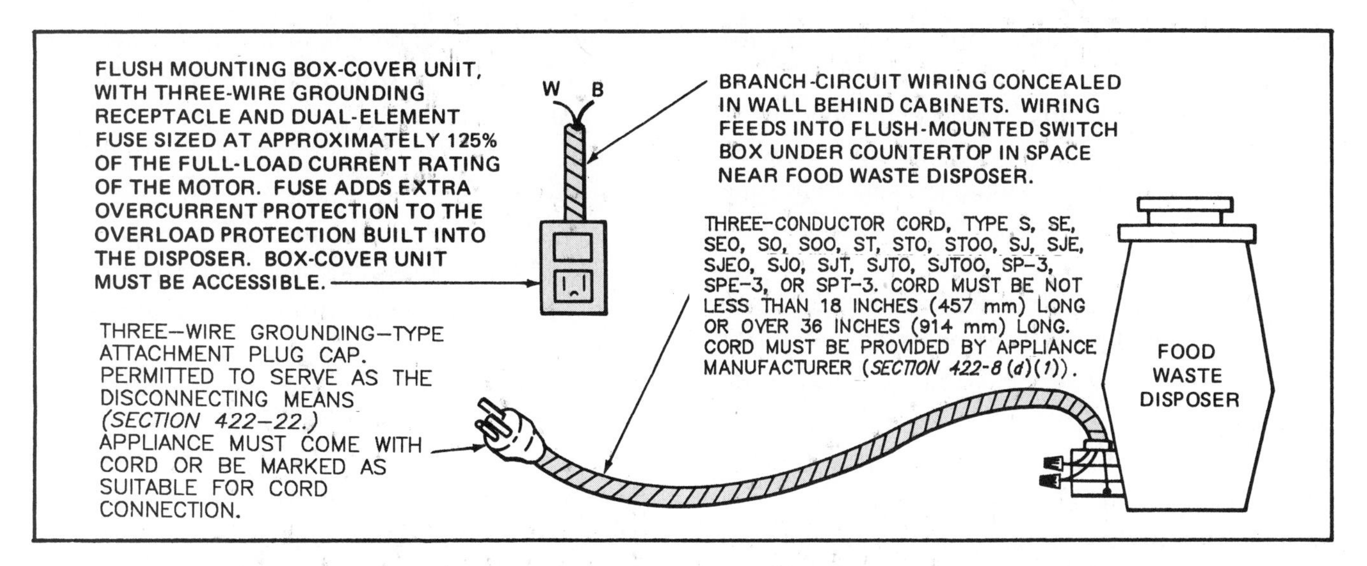

Fig. 21-2 Typical cord connection for a food waste disposer, *Section 422-8(d)(1)*.

ductors terminate in a junction box provided on the disposer.

Overcurrent Protection

The food waste disposer is a motor-operated appliance. Food waste disposers normally are driven by a 1/4- or 1/3-hp, split-phase, 120-volt motor. Therefore, running overcurrent protection must be provided. Most food waste disposer manufacturers install a built-in thermal protector to meet Code requirements. Either a manual reset or an automatic reset thermal protector may be used.

When running overcurrent protection is not built into the disposer, the electrician must install separate protection. Such protection must not exceed 125% of the full-load current rating of the motor. For example, a box-cover unit may be installed under the sink near the food waste disposer. A dual-element fuse of the proper size is then inserted into the box-cover unit. This unit also serves as the disconnecting means.

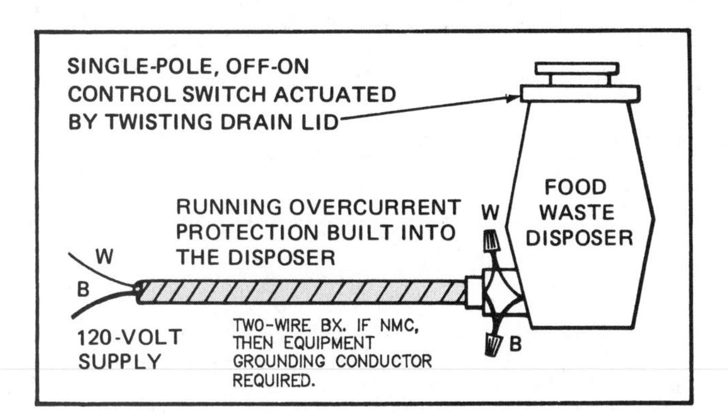

Fig. 21-3 Wiring for a food waste disposer with an integral ON/OFF switch.

Disconnecting Means

All electrical appliances must be provided with some means of disconnecting the appliance, *Article 422*, *Part D*. The disconnecting means may be a separate on-off switch, figure 21-1, a cord connection, figure 21-2, or a branch-circuit switch or circuit breaker, such as that supplying the installation in figure 21-3. See *NEC® Sections 422-21*, *422-24*, *422-25*, and *422-26*.

Some local codes require that food waste disposers, dishwashers, and trash compactors be cord-connected as illustrated in figure 21-2 to make it easier to disconnect the unit, to replace it, to service it, and to reduce noise and vibration. ► The appliance shall be intended or identified for flexible cord connection, *Sections 400-7*, *422-8(c)*, and *422-8(d)*. ◄

Turning the Food Waste Disposer On and Off

Separate On/Off Switch: When a separate ON/OFF switch is used, a simple circuit arrangement can be made by running a two-wire supply cable to a switch box located next to the sink at a convenient location above the countertop. Not only is this convenient, but it positions the switch out of the reach of children, figure 21-1. A second cable is run from the switch box to the junction box of the food waste disposer. A single-pole switch in the switch box provides ON/OFF control of the disposer.

In the kitchen of this residence, the food waste disposer is controlled by a single-pole switch located to the right of the kitchen sink above the countertop, in combination with the light switch and receptacle.

Another possibility, but not as convenient, is to install the switch for the disposer inside the cabinet space directly under the sink near the food waste disposer.

Integral On/Off Switch: A food waste disposer may be equipped with an integral prewired control switch. This integral control starts and stops the disposer when the user twists the drain cover into place. An extra ON/OFF switch is not required with this type of control. To connect the disposer, the electrician runs the supply conductors directly to the junction box on the disposer and makes the proper connections, figure 21-3.

Disposer with Flow Switch in Cold Water Line: Some manufacturers of food waste disposers recommend that a *flow switch* be installed in the cold water line under the sink, figure 21-4. This switch is connected in series with the disposer motor. The flow switch prevents the disposer from operating until a predetermined quantity of water is flowing through the disposer. The cold water helps prevent clogged drains by solidifying any grease in the disposer. Thus, the addition of a flow switch means that the disposer cannot be operated without water.

Figure 21-4 shows one method of installing a food waste disposer with a flow switch in the cold water line.

Grounding

Section 422-16 states that where required by *Article 250*, all exposed, noncurrent-carrying parts are to be grounded in the manner specified in *Article 250*.

The presence of any of the conditions outlined in *Sections 250-42* through *250-45* require that all electrical appliances in the dwelling be grounded. Food waste disposers can be grounded using any of the methods covered in the previous units, such as the metal armor of armored cable or the separate grounding conductor of nonmetallic-sheathed cable, *Section 250-57*. NEVER, NEVER ground a food waste disposer to the grounded circuit conductor!

DISHWASHER Ⓐ$_I$

The dishwasher is supplied by a separate 20-ampere circuit connected to Circuit B19. No. 12 THHN conductors are used to connect the appliance. The dishwasher has a 1/3-hp motor rated at 7.2 amperes and 120 volts. The outlet for the dishwasher is shown by the symbol Ⓐ$_I$. During the drying cycle of the dishwasher, a thermostatically controlled, 1000-watt electric heating element turns on.

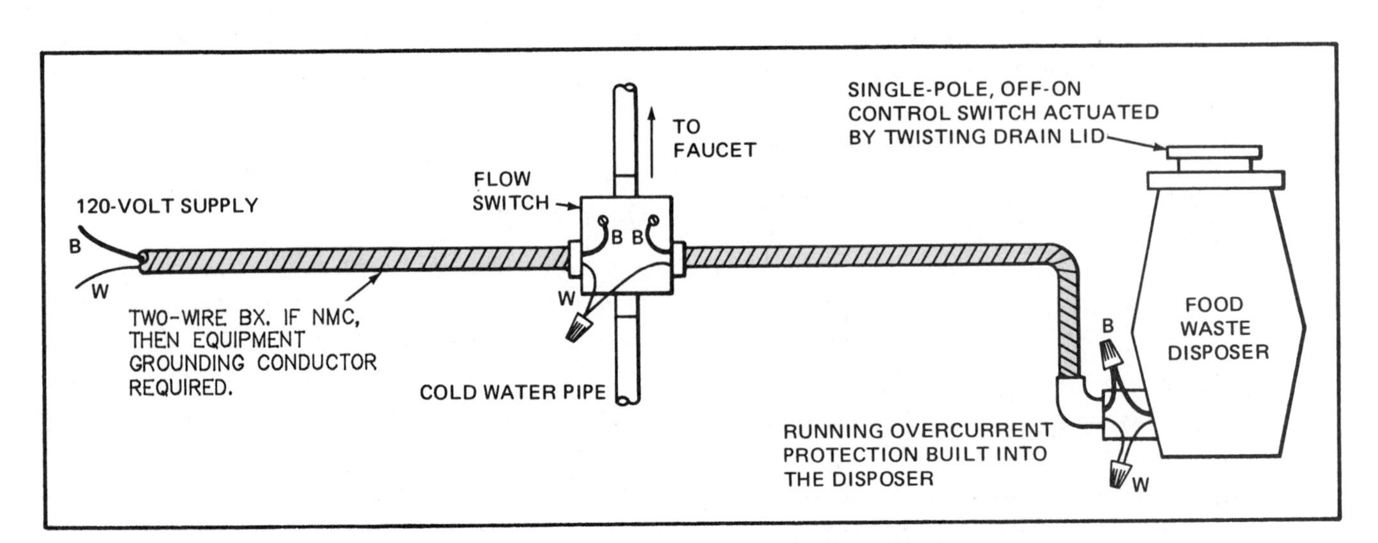

Fig. 21-4 Disposer with a flow switch in the cold water line.

Actual connected load		Circuit calculation
Motor	7.2 A	7.2 A × 1.25 = 9 A
Heater	8.3 A	8.3 A
Total	15.5 A	17.3 A

Use a 20-ampere circuit.

For most dishwashers, the motor does not run during the drying cycle. Thus, the actual maximum demand on the branch circuit would only be the larger of the two loads — the 1000-watt heating element. In some dishwashers, there is a fan or blower to speed up the drying time. In either case, for the dryer example shown, a 20-ampere circuit is more than adequate.

Energy-saving dishwashers are equipped with a built-in booster water heater which allows the homeowner to adjust the regular water heater temperature to a temperature of 120 degrees, or lower, when desired. The booster heater then raises the water temperature from 120 degrees to approximately 155 degrees, the temperature considered necessary to sanitize dishes.

Wiring the Dishwasher

The dishwasher manufacturer normally supplies a terminal or junction box on the appliance. The electrician connects the supply conductors in this junction box. The electrician must verify the dimensions of the dishwasher so that the supply conductors are brought through the wall at the proper point.

The dishwasher is a motor-operated appliance and requires running overcurrent protection. Such protection prevents the motor from burning out if it becomes overloaded or stalled. Normally, the required protection is supplied by the manufacturer as an integral part of the dishwasher motor. If the dishwasher does not have integral protection, the electrician must provide it as part of the installation of the unit. For example, a box-cover unit with appropriate time-delay, dual-element fuses can be installed under the sink. This unit also serves as the disconnecting means according to *Sections 422-21, 422-24*, and *422-26*.

Common Feed for Dishwasher and Food Waste Disposer

There may be instances where the distance from the kitchen sink area to the electrical load center is very long, in which case it may be more economical to run one branch circuit to feed the dishwasher and food waste disposer, instead of two separate circuits. In this case, the electrician must split the circuit in a junction box mounted in the cabinet space below the food waste disposer. If a box-cover unit is used and has the proper fuse sizes, then separate overcurrent protection and separate disconnecting means are available. The connections for this circuit are similar to those for the overhead garage door opener circuit described in figures 16-13 and 16-14.

Both of the appliances (food waste disposer and dishwasher) in this residence are permitted to be supplied with one 20-ampere circuit because the circuit calculations are:

Food waste disposer
$$= 7.2 \times 1.25 = 9 \text{ amperes}$$
Heater in dishwasher
$$= 1000 \div 120 = 8.33 \text{ amperes}$$
Maximum calculated demand = 17.33 amperes

For higher current or wattage rated appliances, it may be necessary to install a 30-ampere branch circuit. Do not connect the garbage disposer or dishwasher to any of the small appliance circuits in the kitchen. To do so would be a violation of the Code. The small appliance circuits are intended to supply cord-connected *portable* appliances only! See unit 12.

Grounding

The dishwasher is required to be grounded. The grounding of appliances is described fully in previous units and in the discussion of the food waste disposer. NEVER, NEVER ground a dishwasher to the grounded circuit conductor!

PORTABLE DISHWASHERS

In addition to built-in dishwashers, portable models are available. These portable units have one hose which is attached to the water faucet and a water drainage hose that hangs in the sink. The obvious location for a portable dishwasher is near the sink. Thus, the dishwasher probably will be plugged into the convenience outlet nearest the sink.

NEC® Section 210-8(a)(5) requires that receptacles installed within 6 feet (1.83 m) of the kitchen sink that serve countertops must be GFCI protected.

Portable dishwashers are supplied with a three-wire cord containing two circuit conductors and one grounding conductor and a three-wire grounding-type attachment plug cap. If the three-wire plug cap is plugged into the three-wire grounding-type receptacle, the dishwasher is adequately grounded. Whenever there is a chance that the user may touch the appliance and a grounded surface at the same time, such as the water pipe or faucet, **the equipment must be grounded to reduce the shock hazard**. The exceptions to *Sections 250-45(c)* and *422-8(d)* permit the double insulation of appliances and portable tools instead of grounding. Double insulation means that the appliance or tool has two separate insulations between the hot conductor and the person using the device. Although refrigerators, cooking units, water heaters, and other large appliances are not double insulated, many portable hand-held appliances and tools have double insulation. All double-insulated tools and appliances must be marked by the manufacturer to indicate this feature.

CORD CONNECTION OF FIXED APPLIANCES

As with the previously discussed, food waste disposer, built-in dishwashers, and trash compactors in residential occupancies are permitted to be cord-connected, figure 21-2.

The Code requirements for dishwashers and trash compactors are very similar to those requirements for food waste disposers, except the cord shall be 3 to 4 feet (0.914 to 1.22 m) long and the receptacle must be located in either the same space of the dishwasher or compactor (behind it) or in the space adjacent to it.

Some electrical inspectors feel that using the cord-connected method of connecting these appliances is better than direct-connection (sometimes referred to as "hard-wired") because servicing of these appliances is generally done by appliance repair people rather than electricians.

► *Section 400-7* covers the uses permitted for flexible cords. Specifically, *Section 400-7(a)(8)* states that the appliance must be fastened in such a manner as to allow ready removal of the appliance for maintenance and repairs. The appliance must be intended or identified for flexible cord connection. The same requirement is found in *Sections 422-8(c)* and *422-8(d)*.

Note that the wording of the Code is that the food waste disposer, dishwasher, or trash compactor must be of the type intended for use in homes and must come provided with the cord. It appears as though the Code does not permit field attachment of the cord to these appliances. ◄

REVIEW

Note: Refer to the Code or the plans where necessary.

FOOD WASTE DISPOSER CIRCUIT Ⓐ$_H$

1. How many amperes does the food waste disposer draw? ____________________
2. a. To what circuit is the food waste disposer connected? ____________________
   b. What size wire is used to connect the food waste disposer? ____________________
3. Means must be provided to disconnect the food waste disposer. The homeowner need not be involved in electrical connections when servicing the disposer if the disconnecting means is ____________________

   ____
4. How is running overcurrent protection provided in most food waste disposer units?

   ____

   ____

5. When running overcurrent protection is not provided by the manufacturer or if additional backup overcurrent protection is wanted, dual-element, time-delay fuses may be installed in a separate box-cover unit. These fuses are sized at not over _____% of the full-load rating of the motor.
6. Why are flow switches sometimes installed on food waste disposers? ____________
7. What Code sections relate to the grounding of appliances? ____________
8. Do the plans show a wall switch for controlling the food waste disposer? ________
9. A separate circuit supplies the food waste disposer in this residence. How many feet (meters) of cable will be required to connect the disposer? ____________

DISHWASHER CIRCUIT ▲I

1. a. To what circuit is the dishwasher in this residence connected? ____________
   b. What size wire is used to connect the dishwasher? ____________
2. The motor on the dishwasher is (circle one)
   a. 1/4 hp b. 1/3 hp c. 1/2 hp
3. The heating element is rated at (circle one)
   a. 750 watts b. 1000 watts c. 1250 watts
4. How many amperes at 120 volts do the following heating elements draw?
   a. 750 watts ____________
   b. 1000 watts ____________
   c. 1250 watts ____________
5. How is the dishwasher in this residence grounded? ____________
6. What type of cord is used on most portable dishwashers? ____________
7. How is a portable dishwasher grounded? ____________
8. What is meant by double insulation? ____________
9. Who is to furnish the dishwasher? ____________
10. What article of the Code specifically addresses electrical appliances? ________

UNIT 22

Special-Purpose Outlets for the Bathroom Ceiling Heat/Vent/Lights, Ⓐ$_J$, Ⓐ$_K$, the Attic Fan Ⓐ$_L$, and the Hydromassage Tub Ⓐ$_A$

OBJECTIVES

After studying this unit, the student will be able to

- explain the operation and switching sequence of the heat/vent/light.
- describe the operation of a humidistat.
- install attic exhaust fans with humidistats, both with and without a relay.
- list the various methods of controlling exhaust fans.
- understand advantages of installing exhaust fans in residences.
- understand the electrical circuit and Code requirements for hydromassage bathtubs.
- understand grounding requirements for electrical equipment.

BATHROOM CEILING HEATER CIRCUITS Ⓐ$_K$ Ⓐ$_J$

Both bathrooms contain a combination heater, light, and exhaust fan installed in the ceiling. The heat/vent/light is such a device and is shown in figure 22-1. The symbols Ⓐ$_K$ and Ⓐ$_J$ represent the outlets for these units.

Each heat/vent/light contains a heating element similar to the surface burners of electric ranges. The appliances also have a single-shaft motor with a blower wheel, a lamp with a diffusing lens, and a means of discharging air to the outside of the dwelling.

The heat/vent/lights specified for this residence are rated 1475 watts at 120 volts. The unit in the Master Bedroom Bathroom Ⓐ$_J$ is connected to Circuit A12, a 20-ampere, 120-volt circuit. In the Front Bedroom Bathroom, the unit is connected to Circuit A11, also a 20-ampere, 120-volt circuit.

The current rating of the 1475-watt unit is

$$I = \frac{W}{E} = \frac{1475}{120} = 12.29 \text{ amperes}$$

Article 424 of the *National Electrical Code*® covers fixed electric space heating. *Section 424-3(b)* requires that the branch-circuit conductors and the overcurrent protective device (fuses or circuit breakers) shall be not less than 125% of the appliances' ampere rating. Thus, 20-ampere, 120-volt branch circuits rather than 15-ampere branch circuits have been chosen to supply the electric ceiling heaters in the bathroom.

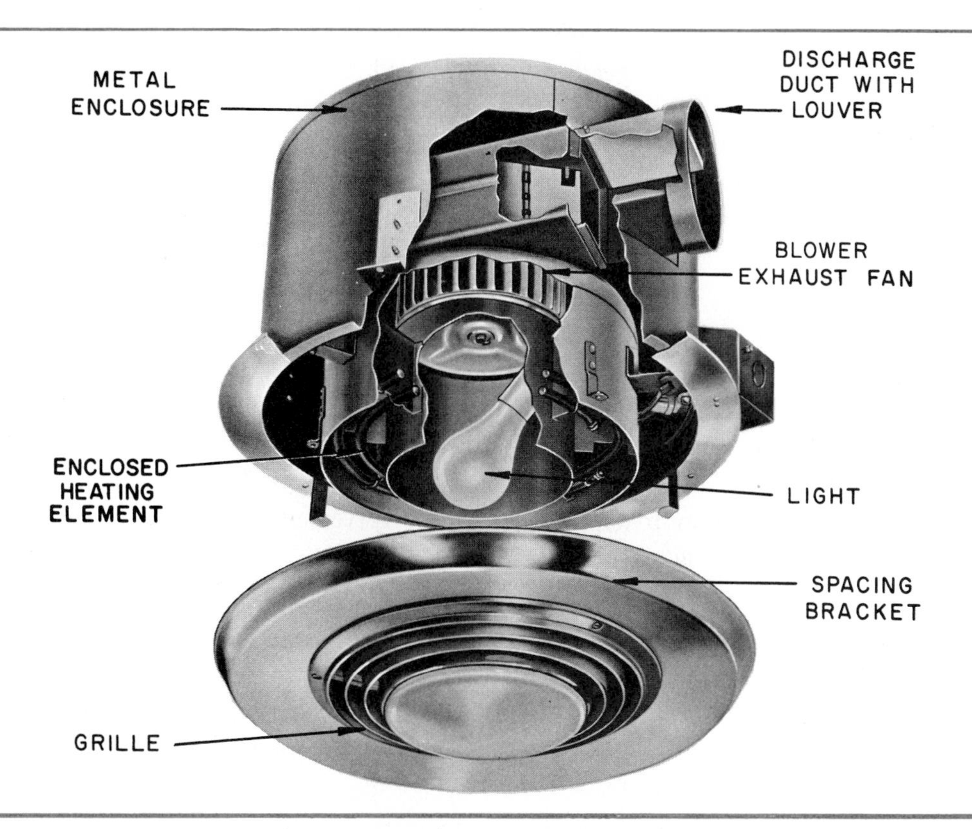

Fig. 22-1 Heat/vent/light. Courtesy of NuTone.

Wiring

In figure 22-2, we see the wiring diagram for the heat/vent/light. The 120-volt line is run to the wall box. A four-wire cable (or two 2-wire cables) runs from the wall box to the unit. The switch location at the wall box could be a 4-inch square, 2 1/8-inch-deep outlet box with a two-gang plaster (drywall) cover suitable for the number of wires, clamps, and wiring devices that will be contained in the wall box. Two 3 1/2-inch-deep device boxes ganged together could also be used.

Grounding the Heat/Vent/Light

Grounding is accomplished by connecting the bare equipment grounding conductor of the nonmetallic-sheathed cable to the grounding terminal or lug provided for this connection on the heat/vent/light unit. If electrical metallic tubing is the wiring method used, then this provides the equipment ground. Do not ground the fixture to the grounded circuit conductor. This is a violation of *Section 250-61(b)* and to do so can result in serious electrical shock hazard. *Section 250-61(b)* states that the grounded circuit conductor *shall not* be used to ground noncurrent-carrying metal parts of equipment *anywhere* beyond the main service disconnect. There are a few exceptions to this, as in the case of

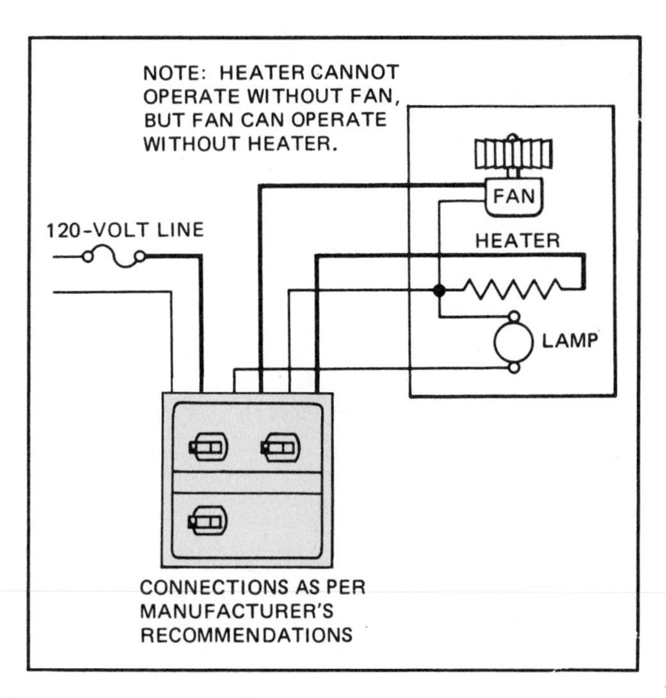

Fig. 22-2 Wiring for heat/vent/light.

frames of ranges, wall-mounted ovens, counter-mounted cooking units, and clothes dryers, *Section 250-60*.

The heat/vent/light is controlled by three switches as shown in figure 22-2:

one switch for the heater

one switch for the light

one switch for the exhaust

Operation of the Unit

When the heater switch is turned on, the heating element begins to give off heat. The heat activates a bimetallic coil attached to a damper section in the housing of the unit. The heat-sensitive coil expands until the damper closes the discharge opening of the exhaust fan. Air is taken in through the outer grille of the unit. The air is blown downward over the heating element. The blower wheel circulates air back into the room area.

If the heater is turned off so that the exhaust fan alone is on, the air is pulled into the unit and is exhausted to the outside of the house by the blower wheel.

ATTIC EXHAUST FAN CIRCUIT Ⓐ L

An exhaust fan is mounted in the hall ceiling between the Master Bedroom and the Front Bedroom, figure 22-3(A). When running, this fan removes hot, stagnant, humid, or smoky air from the dwelling. It will draw in fresh air through open windows. Such an exhaust fan can be used to lower the indoor temperature of the house as much as 10 to 20 degrees.

The exhaust fan in this residence is connected to Circuit A10, a 120-volt, 15-ampere circuit in the main panel utilizing No. 12 THHN wire.

Exhaust fans can also be installed in a gable of a house, figure 22-3(B), or in the roof.

On hot days the air in the attic may reach a temperature of 150° F or more. Thus, it is desirable to provide a means of taking this attic air to the outside. The heat from the attic can radiate through the ceiling into the living areas. As a result, the temperature in the living areas is raised and the air-conditioning load increases. Personal discomfort increases as well. These problems are minimized by properly vented exhaust fans, figure 22-4.

Fan Operation

The exhaust fan installed in this residence has a 1/4-horsepower direct-drive motor. Direct-drive means that the fan blade is attached directly to the shaft of the motor. It is rated 120-volts, 60 Hz (60 cycles per second), 5.8 amperes, 696 VA, and operates at a maximum speed of 1050 r/min. The motor is protected against overload by internal thermal protection. The motor is mounted on rubber cushions for quiet operation.

When the fan is not running, the louver remains in the closed position. When the fan is turned on, the louver opens. When the fan is turned off, the louver automatically closes to prevent the escape of air from the living quarters.

Fan Control

Typical residential exhaust fans can be controlled by a variety of switches.

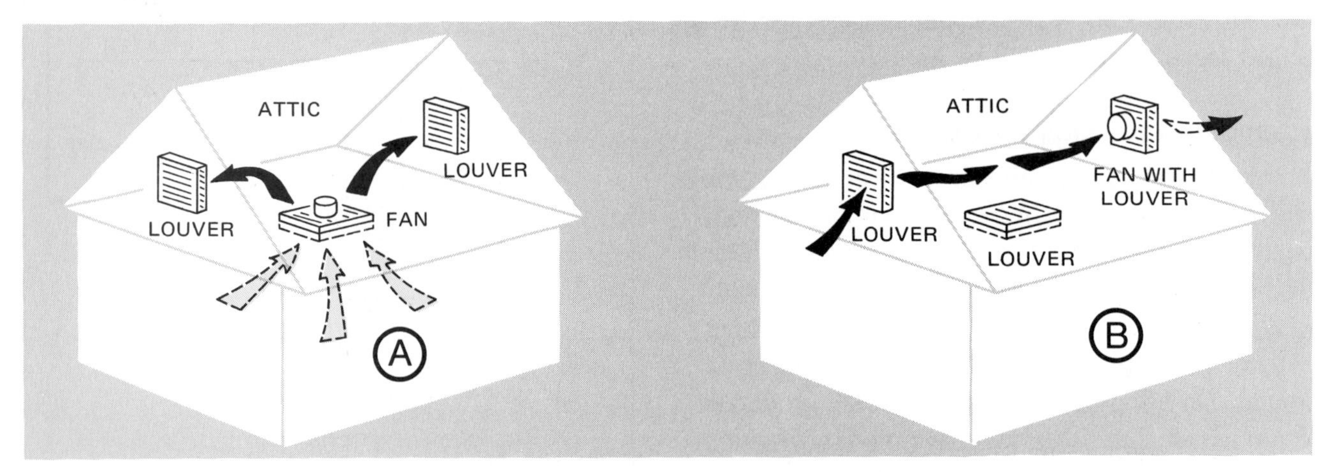

Fig. 22-3 Exhaust fan installation in an attic.

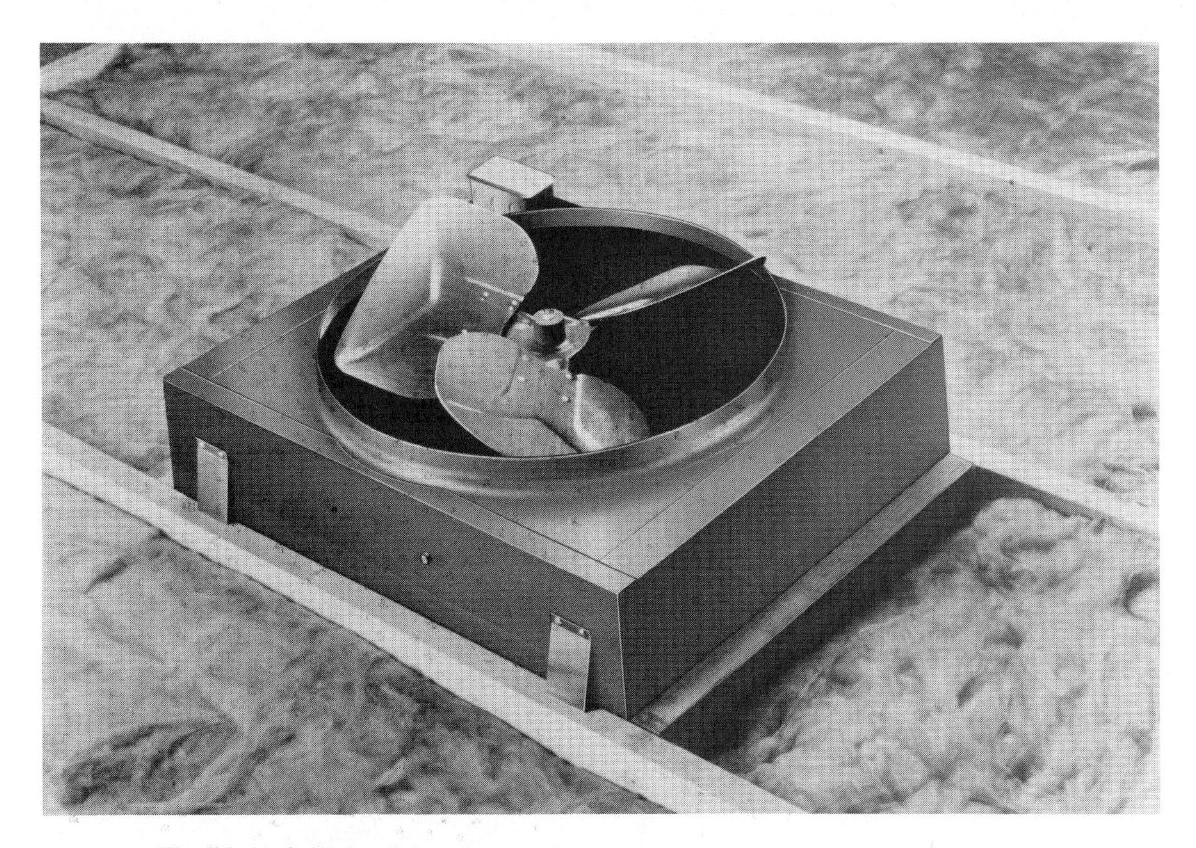

Fig. 22-4 Ceiling exhaust fan as viewed from the attic. Courtesy of NuTone.

Some electricians, architects, and home designers prefer to have fan switches (controls), figures 22-5 and 22-6, mounted 6 feet (1.83 m) above the floor so that they will not be confused with other wall switches. This is a matter of personal preference and should be verified with the owner.

Figure 22-7 shows some of the options available for the control of an exhaust fan.

a. A simple ON/OFF switch. Refer to figure 22-7(A).

b. A speed control switch that allows multiple and/or infinite number of speeds. Refer to figure 22-7(B). (The exhaust fan in this residence is controlled by this type of control.) See also figure 22-5.

c. A timer switch that allows the user to select how long the exhaust fan is to run, up to 12 hours. Refer to figure 22-7(C). See also figure 22-6.

d. A humidity control switch that senses moisture build-ups. Refer to figure 22-7(D). See "Humidity Control" for further details on this type of switch.

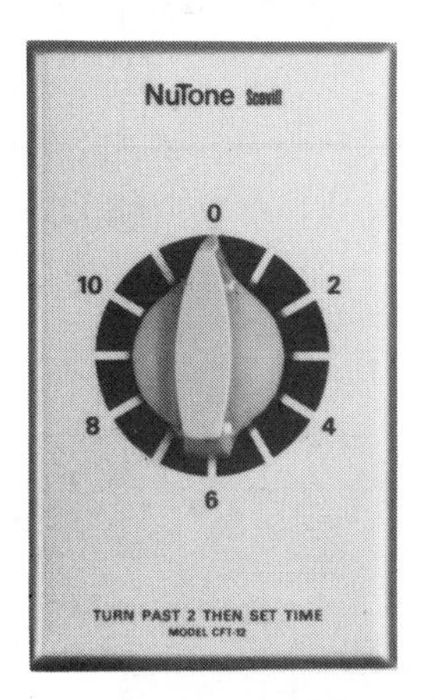

Fig. 22-5 Speed control switches. Courtesy of NuTone.

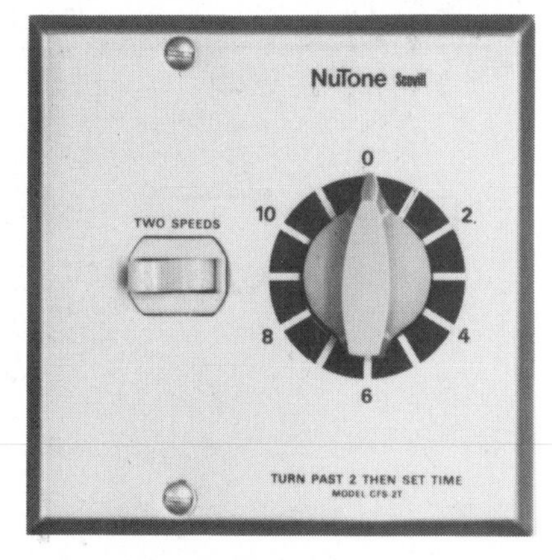

Fig. 22-6 Timer switch. Courtesy of NuTone.

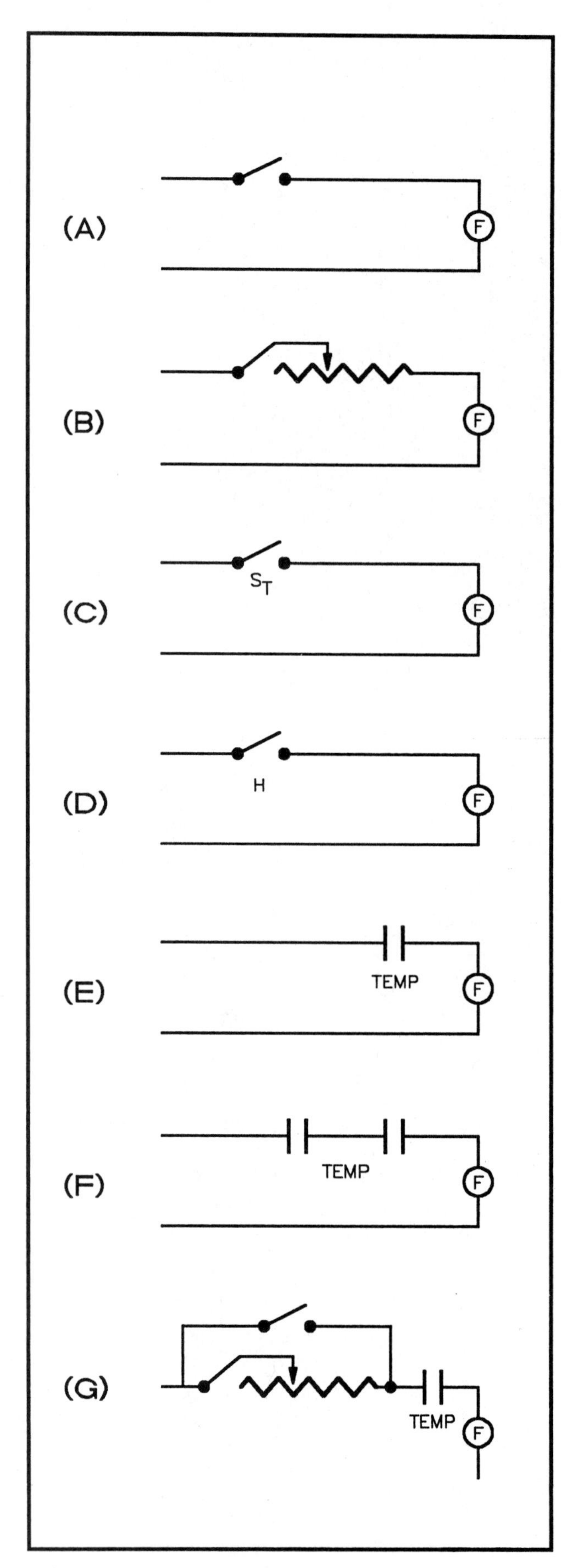

Fig. 22-7 Different options for the control of exhaust fans.

e. Exhaust fans mounted into end gables or roofs of residences are available with an adjustable temperature control on the frame of the fan. This thermostat generally has a start range of 70° to 130° F, and will automatically stop at a temperature 10° below the start setting, thus providing totally automatic ON/OFF control of the exhaust fan. Refer to figure 22-7(E).

f. A high-temperature automatic heat sensor that will shut off the fan motor when the temperature reaches 200° F. This is a safety feature so that the fan will not spread a fire. Connect in series with other control switches. Refer to figure 22-7(F).

g. A combination of controls, this circuit shows an exhaust fan controlled by an infinite-speed switch, a humidity control, and a high-temperature heat sensor. Note that the speed control and the humidity control are connected in parallel so that either can start the fan. The high-temperature heat sensor is in series so that it will shut off the power to the fan when it senses 200° F, even if the speed control or humidity control are in the ON position. Refer to figure 22-7(G).

Overload Protection for the Fan Motor

Several methods can be used to provide running overload protection for the 1/4-hp attic fan motor. For example, an overload device can be built into the motor or a combination overload tripping device and thermal element can be added to the switch assembly. Several manufacturers of motor controls also make switches with overcurrent devices. Such a switch may be installed in any standard flush switch box opening (2″ × 3″). A pilot light may be used to indicate that the fan is running. Thus a two-gang switch box or 1 1/2″ × 4″ square outlet box with a two-gang raised plaster cover may be used for both the switch and the pilot light.

NEC® Section 430-32(c) states that for automatically started motors, the running overload device must not exceed 125% of the full-load running current of the motor. *Table 430-148* shows that the exhaust fan motor has a full-load current rating of 5.8 amperes. Thus the overload device rating is 5.8 × 1.25 = 7.25 amperes.

Type S fuses may be installed for backup protection. These would be installed in the box-cover unit adjacent to the fan. Ratings of 6 1/4, 7, or 8 amperes would be a good choice, since this particular fan motor does have integral overload protection.

Branch-Circuit Short-Circuit Protection for the Fan Motor

For a full-load current rating of 5.8 amperes, *Table 430-152* shows that the rating of the branch circuit overcurrent device shall not exceed 20 amperes if fuses are used ($5.8 \times 3 = 17.4$ amperes; the next standard fuse size is 20 amperes). The overcurrent device rating shall not exceed 15 amperes if circuit breakers are used ($5.8 \times 2.5 = 14.5$ amperes; the next standard circuit breaker size is 15 amperes).

The exhaust fan is connected to Circuit A10, which is a 15-ampere, 120-volt circuit. This circuit does not feed any other loads. The circuit supplies the exhaust fan only.

Time-delay fuses sized at 115 to 125% of the full-load ampere rating of the motor provide both running overload protection and branch-circuit short-circuit protection. If the motor is equipped with inherent, built-in overload protection, then the time-delay fuse in the circuit provides secondary backup protection. Thus, we have double protection against possible motor burnout.

HUMIDITY CONTROL

An electrically heated dwelling can experience a problem with excess humidity due to the "tightness" of the house, because of the care taken in the installation of the insulation and vapor barriers. High humidity is uncomfortable. It promotes the growth of mold and the deterioration of fabrics and floor coverings. In addition, the framing members, wall panels, and plaster or drywall of a dwelling may deteriorate because of the humidity. Insulation must be kept dry or its efficiency decreases. A low humidity level can be maintained by automatically controlling the exhaust fan. One type of automatic control is the *humidistat*. This device starts the exhaust fan when the relative humidity reaches a high level. The fan exhausts air until the relative humidity drops to a comfortable level. The comfort level is about 50% relative humidity. Adjustable settings are from 0 to 90% relative humidity.

The electrician must check the maximum current and voltage ratings of the humidistat before the device is installed. Some humidistats are low-voltage devices and require a relay. Other humidistats are rated at line voltage and can be used to switch the motor directly (a relay is not required). However, a relay must be installed on the line-voltage humidistat if the connected load exceeds the maximum allowable current rating of the humidistat.

The switching mechanism of a humidistat is controlled by a nylon element, figure 22-8. This element is very sensitive to changes in humidity. A bimetallic element cannot be used because it reacts to temperature changes only.

Wiring

The humidistat and relay in the dwelling are installed using two-wire No. 14 AWG cable. The cable runs from Circuit A10 to a 4-inch square, 1 1/2-inch-deep outlet box located in the attic near the fan, figure 22-9(B). A box-cover unit is mounted on this outlet box. The box-cover unit serves as a disconnecting means within sight of the motor as required by *Sections 422-27* and *430-102*. Motor overload protection is provided by the dual-element time-delay fuses installed in the box-cover unit.

Mounted next to the box-cover unit is a relay that can carry the full-load current of the motor. Low-voltage thermostat wire runs from this relay to the humidistat. (The humidistat is located in the hall between the bedrooms.) The thermostat cable is a

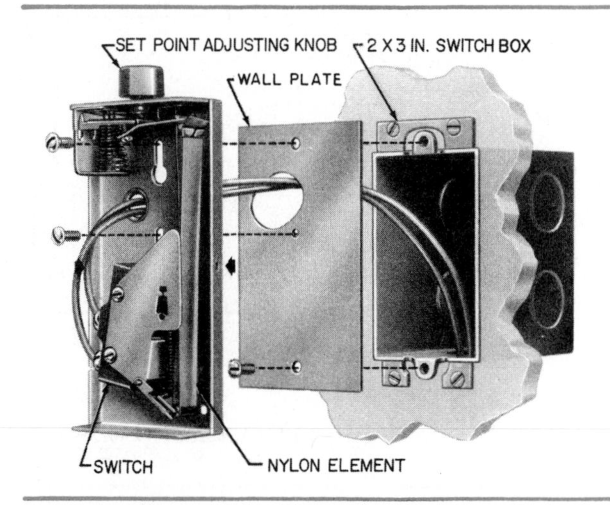

Fig. 22-8 Details of a humidity control used with an exhaust fan.

two-conductor cable since the humidistat has a single-pole, single-throw switching action. The line-voltage side of the relay provides single-pole switching to the motor. Normally the white grounded conductor is not broken.

Although the example in this residence utilizes a relay in order to use low-voltage wiring, it is certainly proper to install line-voltage wiring and use line-voltage thermostats and humidity controls. These are becoming more and more popular, with the switches actually being solid-state speed controllers permitting operation of the fan at an infinite number of speeds between high and low settings.

Grounding

As with most other appliances in this residence, the metal parts of the exhaust fan are to be grounded to the equipment grounding conductor that is part of the cable, or through the metal of the raceway if electrical metallic tubing is the wiring method used. Never ground equipment to the white grounded circuit conductor. This is prohibited in *Section 250-61(b)* of the Code.

APPLIANCE DISCONNECTING MEANS

To clean, adjust, maintain, or repair an appliance, it must be disconnected to prevent personal injury. *Sections 422-20* through *422-27* outline the basic methods for disconnecting fixed, portable, and stationary electrical appliances. Note that each appliance in the dwelling conforms to one or more of the disconnecting methods listed. The more important Code rules for appliance disconnects are as follows:

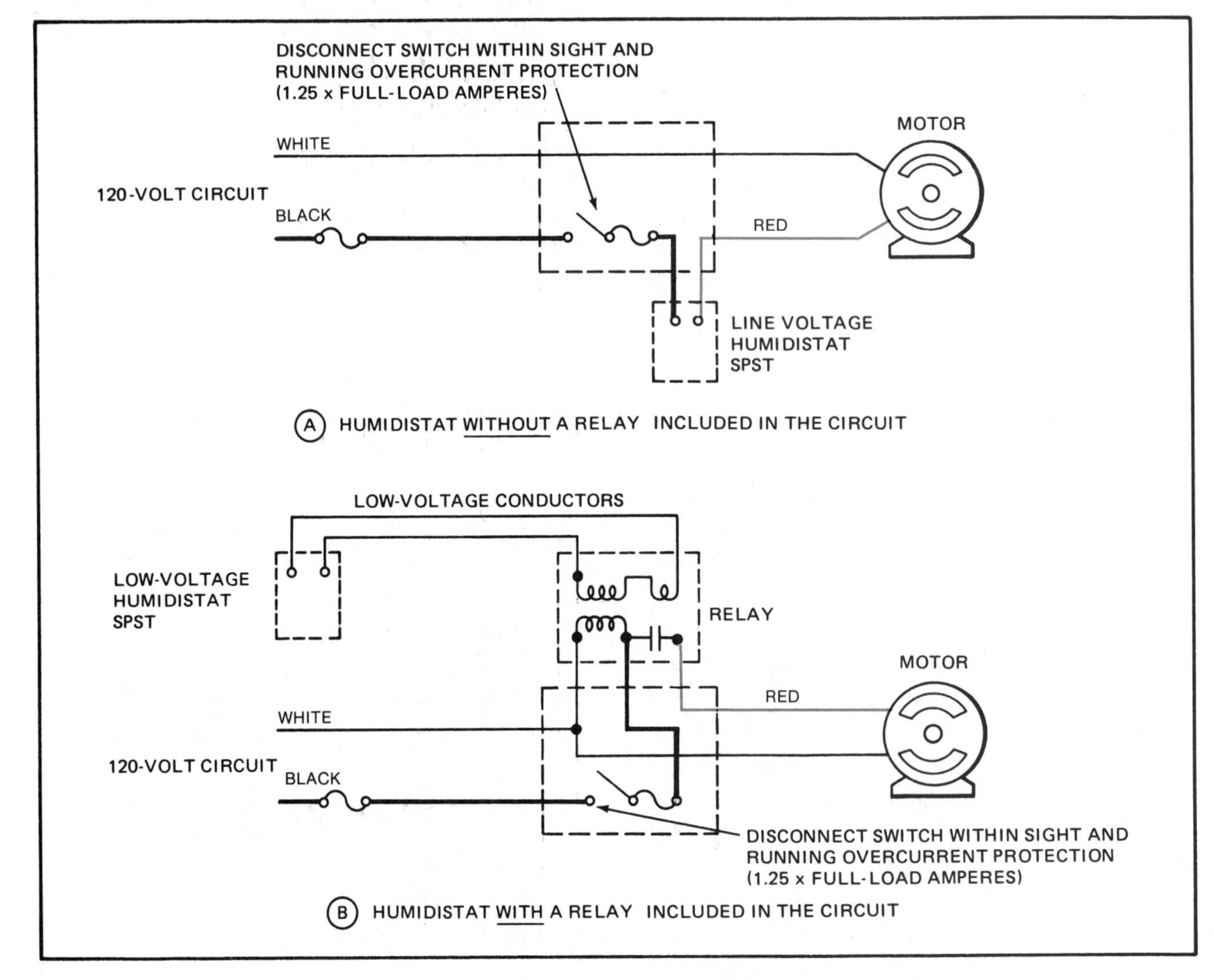

Fig. 22-9 Wiring for the humidistat control.

- each appliance must have a disconnecting means.
- the disconnecting means may be a separate disconnect switch if the appliance is permanently connected.
- the disconnecting means may be the branch-circuit switch or circuit breaker if the appliance is permanently connected and not over 1/8 hp or 300 volt-amperes.
- the disconnecting means may be an attachment plug cap if the appliance is cord connected.
- the disconnecting means may be the unit switch on the appliance only if other means for disconnection are also provided. That is, in a single-family residence, the service disconnect serves as the other means.
- the disconnect must have a positive ON/OFF position.
- the disconnect must be within sight of a motor-driven appliance where the motor is more than 1/8 hp.

Read *Sections 422-20* through *422-27* for the details of appliance disconnect requirements.

HYDROMASSAGE TUB CIRCUIT Ⓐ A

The Master Bedroom is equipped with a hydromassage tub. A hydromassage bathtub is sometimes referred to as a whirlpool bath.

Section 680-4 of the Code defines a hydromassage bathtub as "a permanently installed bathtub equipped with a recirculating piping system, pump and associated equipment. It is designed so that it can accept, circulate, and discharge water upon each use."

In other words, fill — use — drain!

The significant difference between a hydromassage bathtub and a regular bathtub is the recirculating piping system and the electric pump that circulates the water. Both types of tubs are drained completely after each use.

Spas and hot tubs are intended to be filled, then used. They are *not* drained after each use because they have a filtering and heating system.

Another basic difference between a spa and a hot tub is that spas are constructed of manmade material such as fiberglass, acrylics, plastics, or concrete, whereas hot tubs are made of wood, such as redwood, cypress, cedar, oak, or teak.

Electrical Connections

The hydromassage tub is fed with a separate Circuit A9, a 20-ampere, 120-volt circuit using No. 12 AWG THHN conductors.

The Schedule of Special-Purpose Outlets indicates that the hydromassage bathtub in this residence has a 1/2-hp motor that draws 10 amperes.

Section 680-70 requires that the circuit supplying a hydromassage bathtub must be GFCI protected with a Class A type, as discussed in unit 6.

Section 680-71 tells us that lighting fixtures, switches, receptacles, and other electrical equipment located in the same room, and not directly associated with the hydromassage bathtub, shall be installed in accordance with the requirements of *Chapters 1* through *4* of the Code. This section recognizes that a hydromassage tub is used in much the same manner as a regular bathtub, and as such, does not introduce any additional hazards other than those of a regular bathtub.

Spas, hot tubs, and swimming pools do introduce additional electrical shock hazards, and these issues are covered in unit 30.

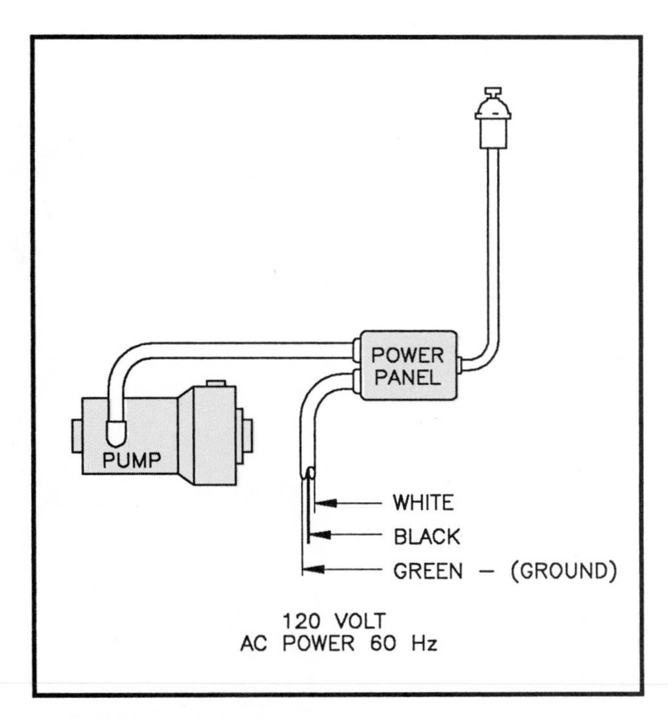

Fig. 22-10 Typical wiring diagram of a hydromassage tub showing the motor, power panel, and electrical supply leads.

The hydromassage bathtub electrical control is prewired by the manufacturer. All that is necessary for the electrician to do is to run the separate 20-ampere, 120-volt GFCI-protected circuit to the end of the tub where the pump and control are located. Generally the manufacturer will supply a 3-foot length of watertight flexible conduit that contains one black, one white, and one green equipment grounding conductor, figure 22-10. Make sure that the equipment grounding conductor of the circuit is properly connected to the green grounding conductor of the hydromassage tub.

DO NOT CONNECT THE GREEN GROUNDING CONDUCTOR TO THE GROUNDED (WHITE) CIRCUIT CONDUCTOR!

Proper electrical connections are made in the junction box where the branch-circuit wiring is to be connected to the hydromassage wires. This junction box, because it contains splices, must be accessible.

The pump and power panel may also need servicing. Access may be from underneath or the end, whichever is convenient for the installation, figure 22-11.

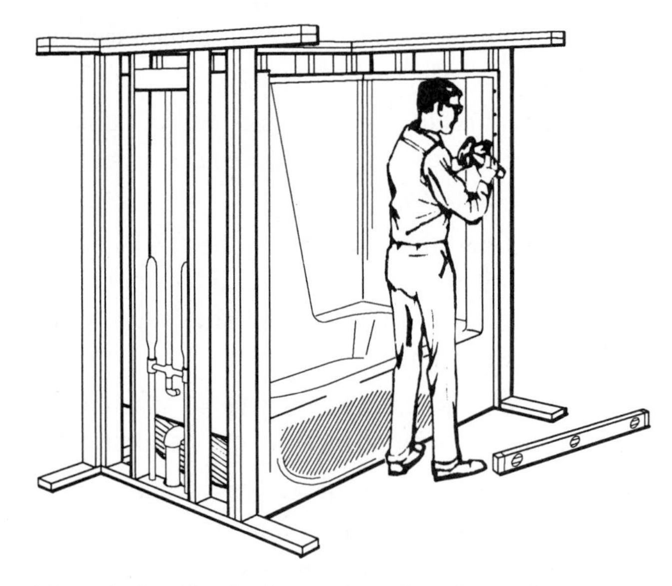

Fig. 22-11 The basic roughing-in of a hydromassage bathtub is similar to that of a regular bathtub. The electrician runs a separate 20-ampere, 120-volt GFCI-protected circuit to the area where the pump and control are located. Check manufacturer's specifications for this data. An access panel from the end or from below is necessary to service the wiring, the pump, and the power panel.

Figure 22-12 is a photograph of a typical hydromassage tub.

Fig. 22-12 A typical hydromassage bathtub, sometimes referred to as a whirlpool. Courtesy of the Kohler Company.

REVIEW

Note: Refer to the Code or the plans where necessary.

BATHROOM CEILING HEATER CIRCUITS Ⓐ$_J$ Ⓐ$_K$

1. What is the wattage rating of the heat/vent/light? ______
2. To what circuits are the heat/vent/lights connected? ______
3. Why is a 4-inch square, 2 1/8-inch deep box with a two-gang raised plaster ring or similar deep box used for the switch assembly for the heat/vent/light? ______
4. a. How many wires are required to connect the control switch and the heat/vent/light? ______

   b. What size wires are used? ______
5. Can the heating element be energized when the fan is not operating? ______
6. Can the fan be turned on without the heating element? ______
7. What device can be used to provide automatic control of the heating element and the fan of the heat/vent/light? ______
8. Where does the air enter the heat/vent/light? ______
9. Where does the air leave this unit? ______
10. Who is to furnish the heat/vent/light? ______
11. For a ceiling heater rated 1200 watts at 120 volts what is the current draw? ______

ATTIC EXHAUST FAN CIRCUIT Ⓐ$_L$

1. What is the purpose of the attic exhaust fan? ______
2. At what voltage does the fan operate? ______
3. What is the horsepower rating of the fan motor? ______
4. Is the fan direct- or belt-driven? ______
5. How is the fan controlled? ______
6. What is the setting of the running overcurrent device? ______
7. What is the rating of the running overcurrent protection if the motor is rated at 10 amperes? ______

8. What is the basic difference between a thermostat and a humidistat? ______________

9. What size conductors are to be used for this circuit? ______________
10. How many feet (meters) of cable are required to complete the wiring for the attic exhaust fan circuit? ______________
11. May the metal frame of the fan be grounded to the grounded circuit conductor? ______________
12. What section of the Code prohibits grounding equipment to a grounding circuit conductor? ______________

HYDROMASSAGE BATHTUB CIRCUIT Ⓐ A

1. What circuit supplies the hydromassage bathtub? ______________
2. Which of the following statements "meets Code"?
   a. The circuit shall have GFCI protection.
   b. The circuit shall not have GFCI protection.
   Circle the correct statement.
3. What conductor size feeds the hydromassage tub? ______________
4. What is the fundamental difference between a hydromassage bathtub and a spa?

5. What sections of the Code reference hydromassage bathtubs? ______________
6. Must the metal parts of the pump and power panel of the hydromassage tub be grounded? ______________

7. Is it permissible to connect the hydromassage tub's green grounding conductor to the branch circuit's grounded (white) conductor? ______________

UNIT 23

Special-Purpose Outlets — Electric Heating $\textcircled{\blacktriangle}_M$, Air Conditioning $\textcircled{\blacktriangle}_N$

OBJECTIVES

After studying this unit, the student will be able to

- list the advantages of electric heating.
- describe the components and operation of electric heating systems (baseboard, cable, furnace).
- describe thermostat control systems for electric heating units.
- install electric heaters with appropriate temperature control according to *National Electrical Code®* rules.
- discuss air conditioners and heat pumps — Code requirements and electrical connections.
- explain how heating and cooling may be connected to the same circuit.

GENERAL DISCUSSION

There are many types of electric heat available for heating homes (e.g., heating cable, unit heaters, boilers, electric furnaces, duct heaters, baseboard heaters, and radiant heating panels). Unit 23 touches upon many of these types. The residence discussed in this text is heated by an electric furnace located in the workshop.

Detailed Code requirements are found in *Article 424* of the *National Electrical Code®, Fixed Electric Space Heating Equipment*.

This text cannot cover in detail the methods used to calculate heat loss and the wattage required to provide a comfortable level of heat in the building. For this residence, the total estimated wattage is 13,000 watts. Depending upon the location of the residence (in the Northeast, Midwest, or South, for example), the heating load will vary.

Electric heating has gained wide acceptance when compared with other types of heating systems. It has a number of advantages over the other types of heating systems. Electric heating is flexible when baseboard heating is used because each room can have its own thermostat. Thus, one room can be kept cool while an adjoining room is warm. This type of zone control for an electric, gas- or oil-fired central heating system is more complex and more expensive.

Electric heating is safer than heating with fuels. The system does not require storage space, tanks, or chimneys. Electric heating is quiet. Electric heat does not add or remove anything from the air. As a result, electric heat is cleaner. This type of heating is considered to be healthier than fuel heating systems which remove oxygen from the air. The only moving part of an electric baseboard heating system is the thermostat. This means that there is a minimum of maintenance.

If an electric heating system is used, adequate insulation must be provided. Proper insulation can keep electric bills to a minimum. Insulation also helps to keep the residence cool during the hot summer months. The cost of extra insulation is offset through the years by the decreased burden on the air-conditioning equipment. Energy conservation measures require the installation of proper and adequate insulation.

TYPES OF ELECTRIC HEATING SYSTEMS

Electric heating units are available in baseboard, wall-mounted, and floor-mounted styles. These units may or may not have built-in thermal overload protection. The type of unit to be installed depends on structural conditions and the purpose for which the room is to be used.

Another method of providing electric heating is to embed resistance-type cables in the plaster of ceilings or between two layers of drywall on the ceiling, *Sections 424-34* through *424-45*.

Electric heat can also be supplied by an electric furnace and a duct system similar to the type used on conventional hot-air central heating systems. The heat is supplied by electric heating elements rather than the burning of fuel. Air conditioning, humidity control, air circulation, electronic air cleaning, and zone control can be provided on an electric furnace system.

► SEPARATE CIRCUIT REQUIRED

Section 422-7 of the Code requires that central heating equipment must be supplied by a separate branch circuit. This includes electric, gas, and oil central furnaces. It also includes heat pumps. The exception to *Section 422-7* is equipment directly associated to central heating equipment, such as humidifiers, electrostatic air cleaners, and so on, which is permitted to be connected to the same circuit.

The logic behind this "separate circuit" Code rule is that there is always a possibility that if the central heating equipment was to be connected to some other circuit, such as a lighting circuit, a fault on that circuit would shut off the power to the heating system. This could cause freezing of water pipes. ◄ Refer to figure 23-1.

Heat pumps are another method of supplying heat.

CONTROL OF ELECTRIC HEATING SYSTEMS

Line-voltage thermostats can be used to control the heat load for most baseboard electric heating systems, figure 23-2. Common ratings for line-voltage thermostats are 2500 watts, 3000 watts, and 5000 watts. (Other ratings are available, however.) The electrician must check the nameplate ratings of the heating unit and the thermostat to insure that the total connected load does not exceed the thermostat rating.

The amperage limit of a thermostat is found using the wattage formula. For example, if a thermostat has a rating of 2500 watts at 120 volts, or 5000 watts at 240 volts, the current value is

$$I = \frac{W}{E} = \frac{2500}{120} = 20.8 \text{ amperes}$$

OR

$$I = \frac{5000}{240} = 20.8 \text{ amperes}$$

The total connected load for this thermostat must not exceed 20.8 amperes.

When the connected load is larger than the rating of a line-voltage thermostat, as in the case of an electric furnace, low-voltage thermostats may be used. In this case a relay must be connected between the thermostat and load. The relay is an integral component of the furnace. The low-voltage contacts of the relay are connected to the thermostat. The line-voltage contacts of the relay are used to switch the actual heater load. Low-voltage thermostat cable is run between the low-voltage terminals of the relay and the thermostat, figure 23-3.

When the connected load is larger than the maximum current rating of a thermostat and relay combination, the thermostat can be used to control a heavy-duty relay or contactor. An example of such a relay is the magnetic switch used for motor controls. The load side of the magnetic switch feeds a distribution panel containing as many 15- or 20-ampere circuits as needed for the connected load. A 40-ampere load may be divided into three

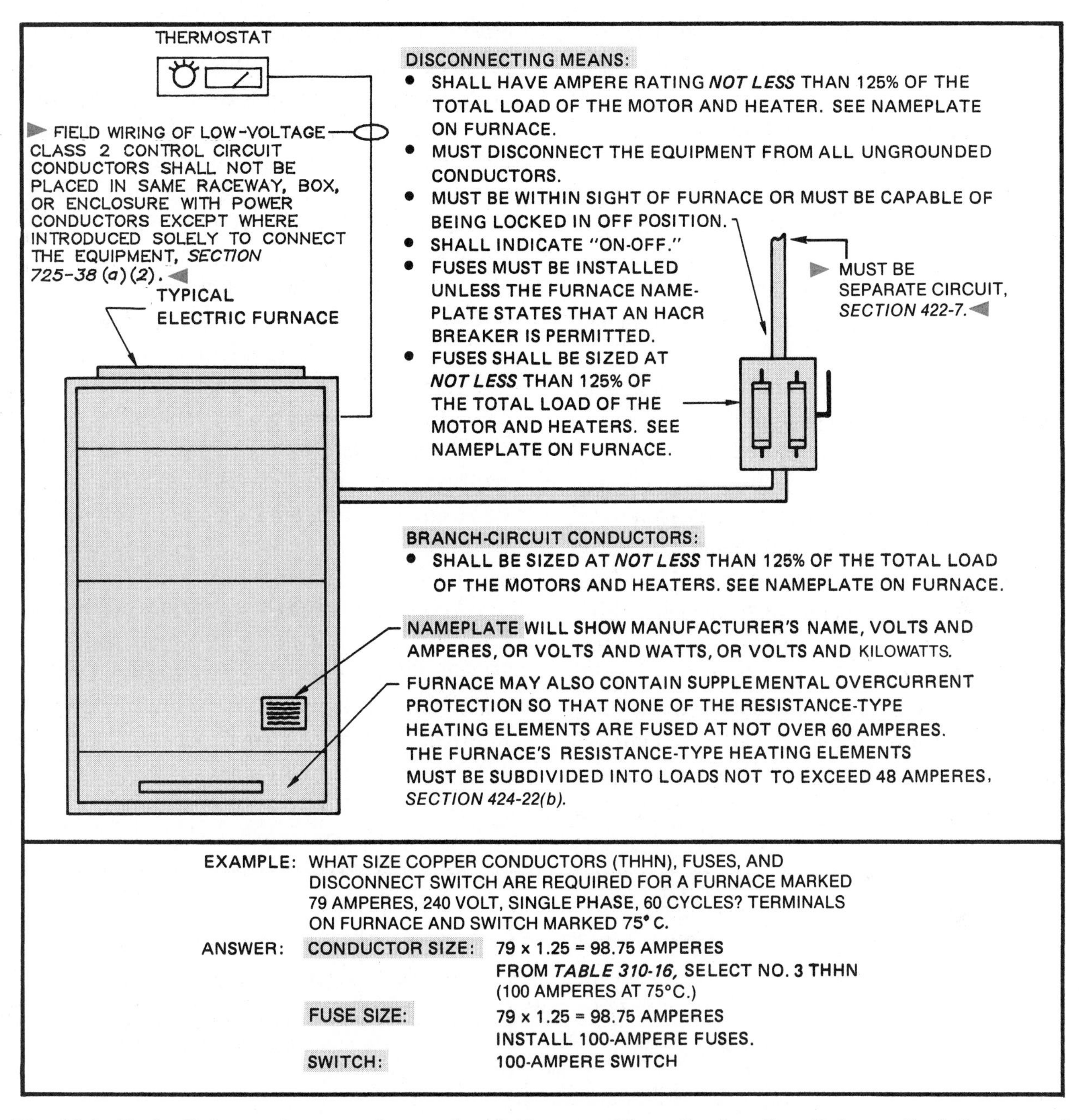

Fig. 23-1 Basic Code requirements for an electric furnace. These "package" units have all of the internal components, such as limit switches, relays, motors, and contactors prewired. The electrician generally need only provide the branch-circuit supply and disconnecting means. See *Article 424* for additional Code information.

Fig. 23-2 Thermostats for electric heating systems.

15-ampere circuits and still be controlled by one thermostat. The proper overcurrent protection is obtained by dividing the 40-ampere circuit into several circuits having lower ratings.

CIRCUIT REQUIREMENTS FOR BASEBOARD UNITS

Figure 23-4 shows typical baseboard electric heating units. These units are rated at 240 volts and are also available in 120-volt ratings.

The wiring for an individual baseboard heating unit or group of units is shown in figures 23-5 and 23-6. A two-wire cable (armored cable or nonmetallic-sheathed cable with ground) would be run from a 240-volt, two-pole circuit in the main panel to the outlet box or switch box installed at the thermostat location. A second two-wire cable runs from the thermostat to the junction box on the heater unit. The proper connections are made in this junction box. Most heating unit manufacturers provide knockouts at the rear and on the bottom of the junction box. The supply conductors can be run through these knockouts. Most baseboard units also have a channel or wiring space running the full length of the unit, usually at the bottom. When two or more heating units are joined together, the conductors are run in this wiring channel. Most manufacturers indicate the type of wire required for these units because of conductor temperature limitations.

A variety of fittings such as internal and external elbows (for turning corners) and blank sections are available from the manufacturer.

Most wall and baseboard heating units are available with built-in thermostats (figure 23-4). The supply cable for such a unit runs from the main panel to the junction box on the unit.

Some heaters have receptacle outlets. Underwriters Laboratories states that when receptacle sections are included with the other components of baseboard heating systems, they must be supplied separately using conventional wiring methods.

Manufacturers of electric baseboard heaters list "blank" sections ranging in length from 2 to 10 feet. These blank sections give the installer the flexibility needed to spread out the heater sections, and to install these blanks where wall receptacle outlets are encountered, figure 23-7.

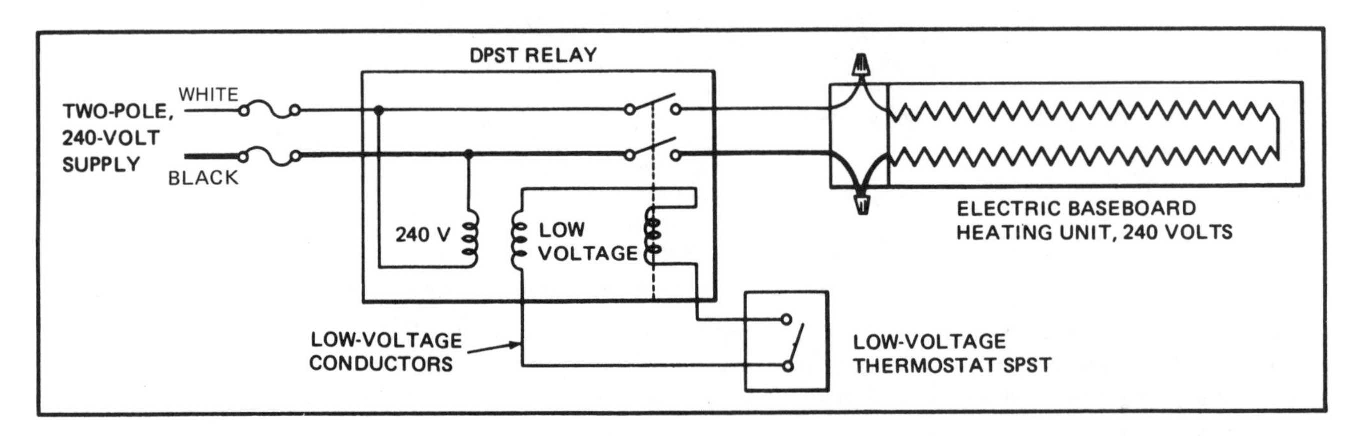

Fig. 23-3 Wiring for baseboard electric heating unit having a low-voltage thermostat and relay.

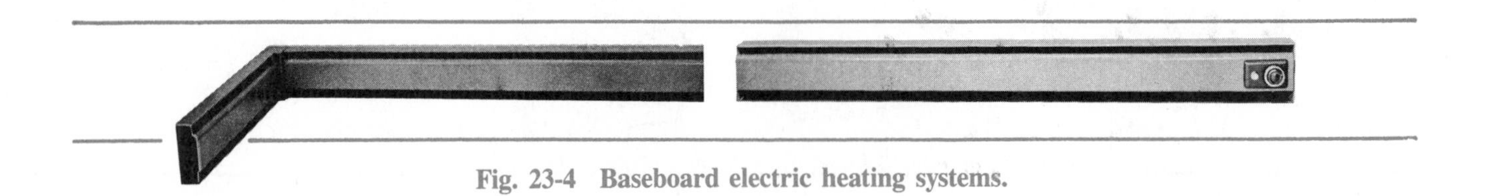

Fig. 23-4 Baseboard electric heating systems.

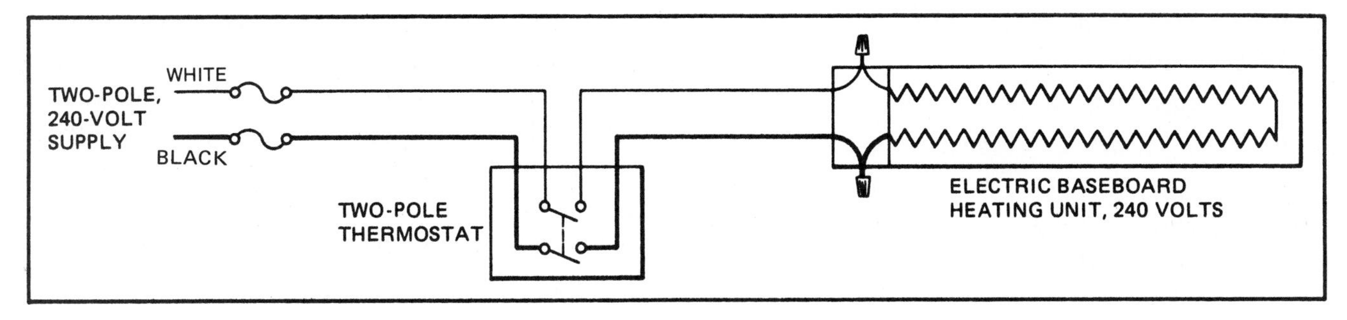

Fig. 23-5 Wiring for a single baseboard electric heating unit.

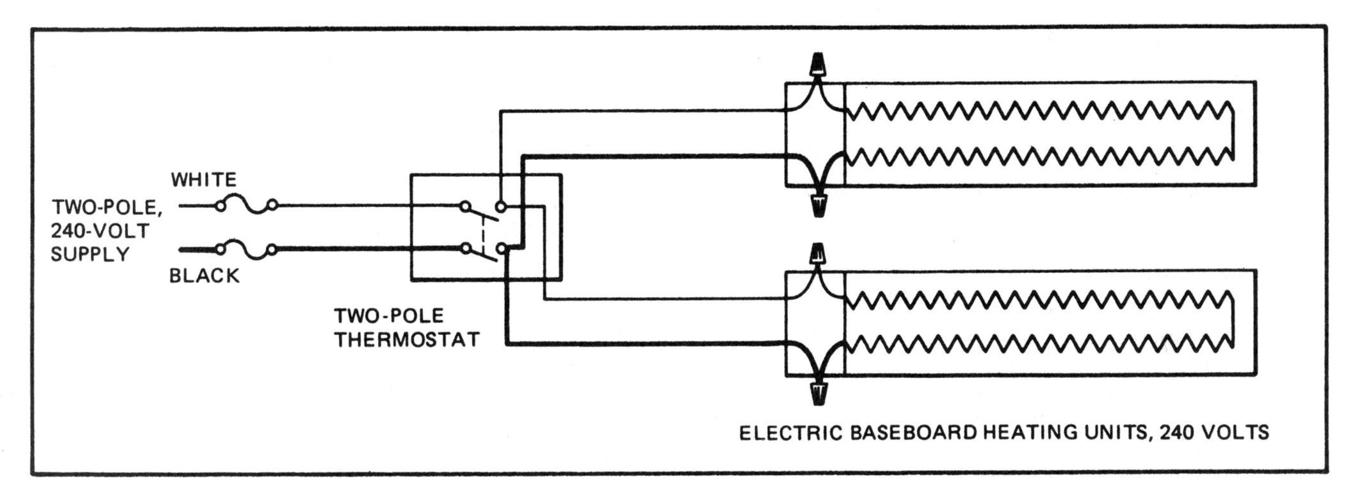

Fig. 23-6 Wiring for baseboard electric heating units at different locations within a room.

LOCATION OF ELECTRIC BASEBOARD HEATERS IN RELATION TO RECEPTACLE OUTLETS

Electric baseboard heaters may not be installed below wall receptacle outlets according to UL standards as referenced in *Section 110-3(b)* of the Code. Instructions furnished with all electric baseboard heaters point this out. Also refer to the *Fine Print Note* following the *Exception* to *Section 210-52(a)*. Refer to figures 23-8A and B. The reasoning for not allowing electric baseboard heaters to be installed below receptacle outlets is the possible fire and shock hazard resulting when a cord hangs over and touches the heated electric baseboard unit.

The electrician must pay close attention to the location of receptacle outlets and the electric baseboard heaters. He must study the plans and specifications carefully.

He may have to fine-tune the location of the receptacle outlets and the electric baseboard heaters, and/or install blank spacer sections, keeping in mind the Code requirements for spacing receptacle outlets and just how much (wattage and length) baseboard heating is required.

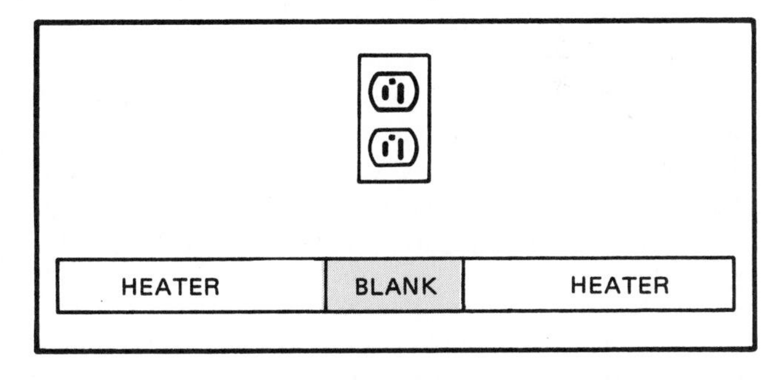

Fig. 23-7 Use of blank baseboard heating sections.

The Code does permit factory-installed or factory-furnished receptacle outlet assemblies as part of permanently installed electric baseboard heaters. These receptacle outlets shall not be connected to the heater circuit, *Section 210-52(a), Exception*. Refer to figure 23-9.

CIRCUIT REQUIREMENTS FOR ELECTRIC FURNACES

In the case of an electric furnace, the source of heat is the resistance heating element(s). The blower motor assembly, filter section, condensate coil, refrigerant line connection, fan motor speed control, high-temperature limit controls, relays, and similar components are much the same as those found in gas furnaces.

In most installations, the supply conductors must be copper conductors having a minimum 90° C temperature rating, such as type THHN. The instructions furnished with the unit will have information relative to supply conductor size and type.

Since proper supply voltage is critical to insure full output of the heating elements, voltage must be maintained at *not less* than 98% of the unit's rated voltage. Another way to state this is that the voltage drop shall not exceed 2% of the unit's rated voltage.

Many electric furnaces furnish the wattage (or kilowatts) output for more than one voltage level. Mathematically, for every 1% drop in voltage, there will be a 2% drop in wattage output. For instance,

Kilowatts	
240 volts	208 volts
12.0	9.0
15.0	11.3
20.0	15.0
25.0	18.8
30.0	22.5

Therefore, the actual voltage at the supply terminals of the electric furnace or baseboard heating unit will determine the true wattage output of the heating elements.

The effect that different voltages have on electric heating elements is covered in unit 20.

The electric heating load used to determine the size of service-entrance conductors and feeders is computed at 100% of the total connected fixed space heating load, *Section 220-15*.

As previously discussed, the electric furnace must be supplied by an individual branch circuit. ► This is a requirement as referenced in *Section 422-7* of the Code. The "separate circuit" requirement pertains to any type of central heating, including oil, gas, electric, and heat pump systems. ◄

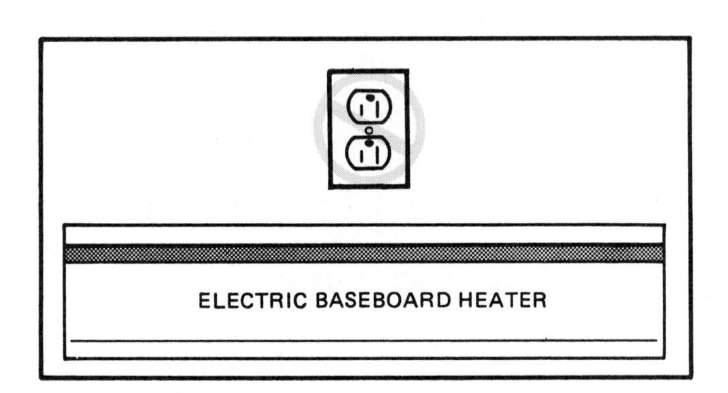

Fig. 23-8A CODE VIOLATION. Baseboard electric heaters shall not be installed below receptacle outlets. This Underwriters Laboratories Standard requirement is part of the instructions furnished by the manufacturer with all electric baseboard heaters. See *NEC® Section 110-3(b)* and FPN to the exception of *Section 210-52(a)*. In this type of installation, cords attached to the receptacle outlet would hang over the heater, creating a fire hazard.

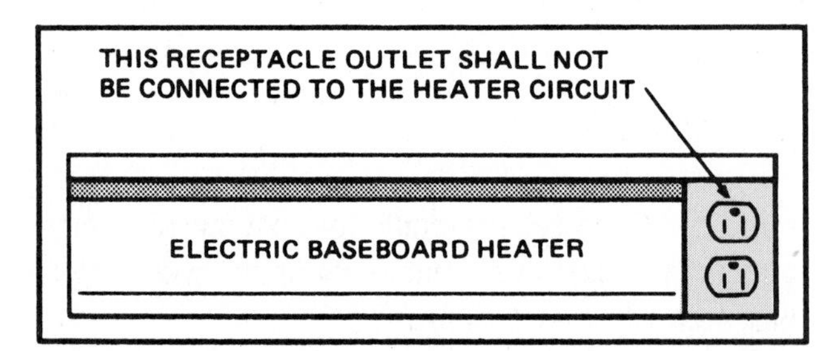

Fig. 23-8B Factory-installed receptacle outlets or receptacle outlet assemblies provided by the manufacturer for use with its electric baseboard heaters may be counted as the required receptacle outlet for the space occupied by a permanently installed heater. See *Exception* to *Section 210-52(a)*.

HEAT PUMPS

A heat pump cools and dehumidifies on hot days and heats on cool days. To provide heat, the pump picks up heat from the outside air. This heat is added to the heat due to the compression of the refrigerant. The resultant heat is used to warm the air flowing through the unit. This warm air is then delivered to the inside of the building. The direction of flow of the refrigerant is reversed for the heating operation.

Heat pumps are available as self-contained packages that provide both cooling and heating. In certain climates, the heat pump alone may not provide enough heat. To overcome this condition, the heat pump may utilize supplementary heating elements contained in the air-handling unit.

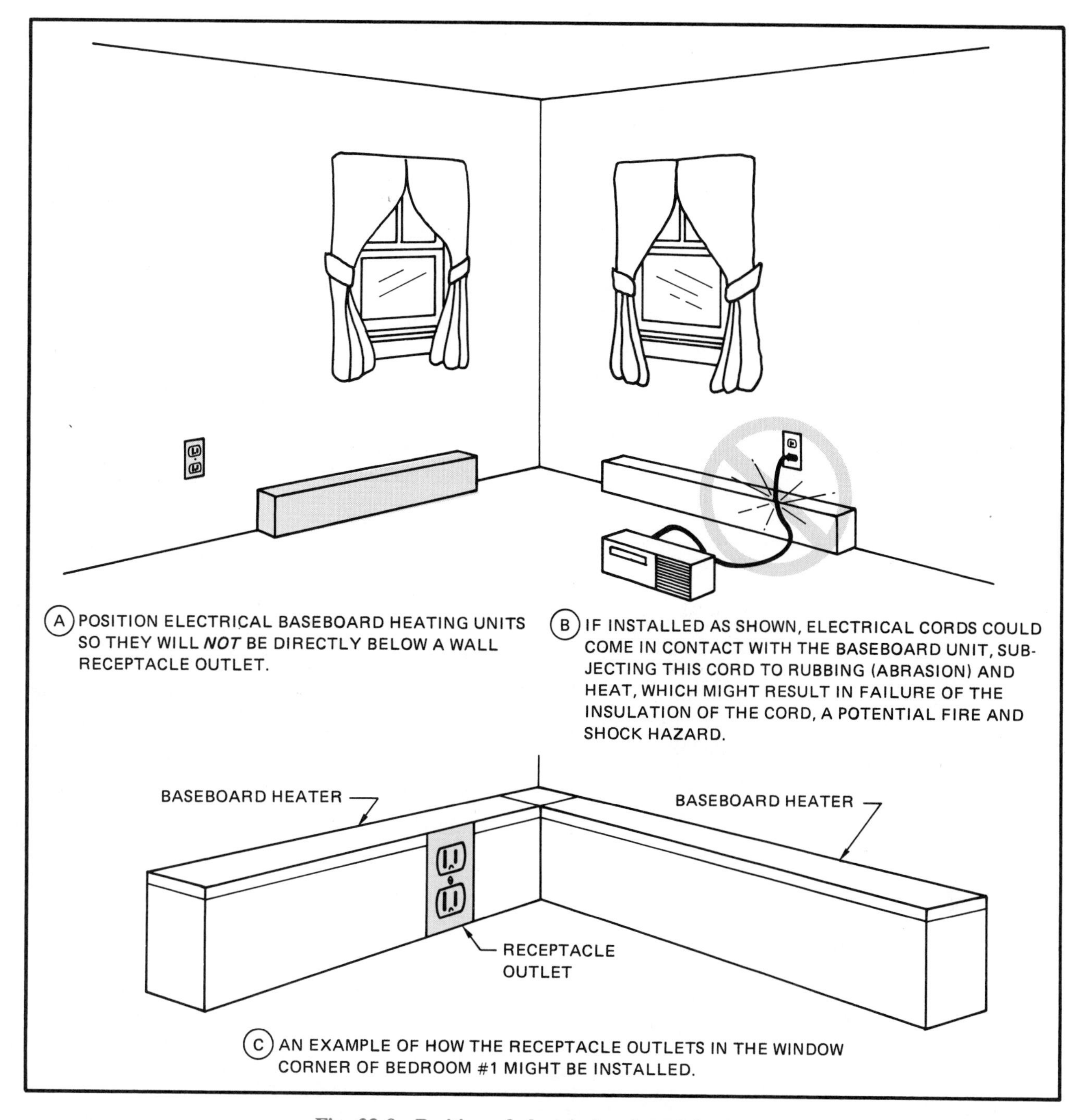

Fig. 23-9 Position of electric baseboard heaters.

GROUNDING

Grounding of electrical baseboard heating units and other electrical equipment and appliances has been covered many times previously in this text and will not be repeated here. To repeat a caution, however, is in order. *Never* ground the equipment to the grounded (white) circuit conductor of the circuit. The only exception to this Code rule is for ranges and dryers, as fully explained in units 15 and 20. See *Sections 250-60* and *250-61(b)*.

MARKING THE CONDUCTORS OF CABLES

Two-wire cable contains one white and one black conductor. Since 240-volt circuits can supply electric baseboard heaters, it would appear that the use of two-wire cable is in violation of the Code. However, *Section 200-7* shows that this is not a Code violation.

According to *Section 200-7*, two-wire cable may be used for the 240-volt heaters if the white conductor is permanently reidentified by paint, colored tape, or other effective means. This step is necessary since it may be assumed that an unmarked white conductor is a grounded conductor having no voltage to ground. Actually, the white wire is connected to a hot phase and has 120 volts to ground. **Thus, a person can be subject to a harmful shock by touching this wire and the grounded baseboard heater (or any other grounded object) at the same time.** Most electrical inspectors accept black paint or tape as a means of changing the color of the white conductor. The white conductor must be made permanently reidentified at the electric heater terminals and at the panels where these cables originate. Reread the section on "Color Identification" in Unit 5.

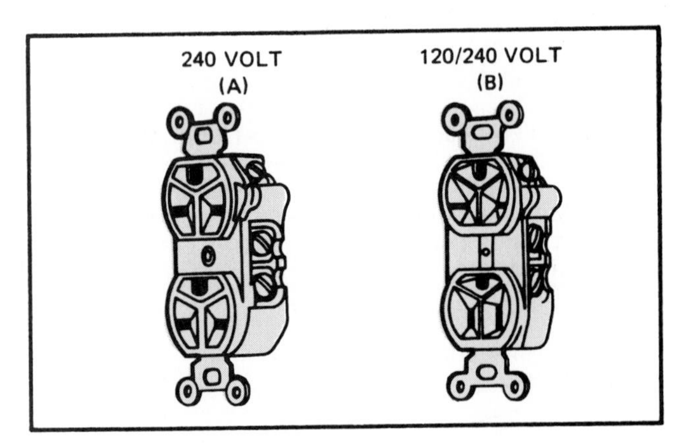

Fig. 23-10 Types of 240-volt receptacles.

ROOM AIR CONDITIONERS

For homes that do not have central air conditioning, window or through-the-wall air conditioners may be installed. These types of room air conditioners are available in both 120-volt and 240-volt ratings. Because room air conditioners are plug-and-cord connected, the receptacle outlet and the circuit capacity must be selected and installed according to applicable Code regulations. The Code rules for air conditioning are found in *Article 440* of the *National Electrical Code*®. The Code requirements for room air-conditioning units are found in *Sections 440-60* through *440-64*.

The basic Code rules for installing these units and their receptacle outlets are as follows:

- the air conditioners must be grounded.
- the air conditioners must be connected using a cord and attachment plug.
- the air-conditioner rating may not exceed 40 amperes at 250 volts, single phase.
- the rating of the branch-circuit overcurrent device must not exceed the branch-circuit conductor rating or the receptacle rating, whichever is less.
- the air-conditioner load shall not exceed 80% of the branch-circuit ampacity if no other loads are served.

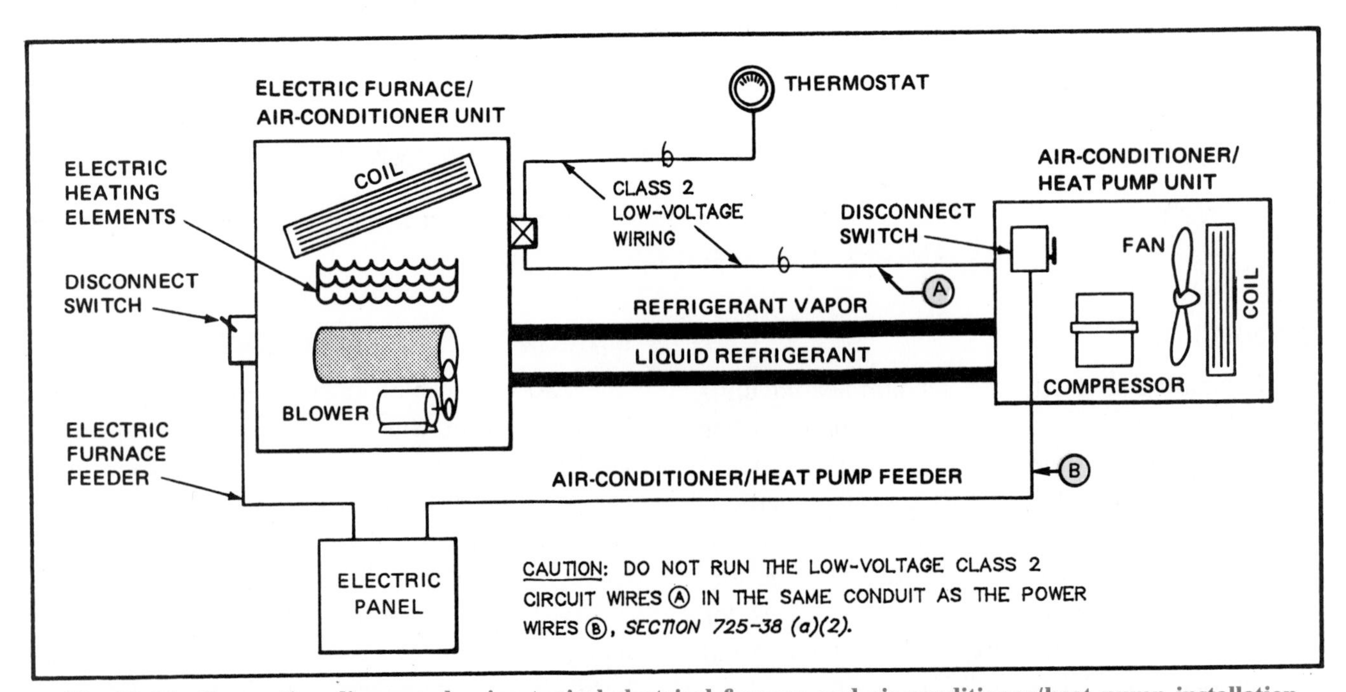

Fig. 23-11 Connection diagram showing typical electrical furnace and air-conditioner/heat pump installation.

- the air-conditioner load shall not exceed 50% of the branch-circuit ampacity if other loads are served.
- the attachment plug cap may serve as the disconnecting means.

RECEPTACLES FOR AIR CONDITIONERS

Figure 23-10 shows two types of receptacles that can be used for 240-volt installations. A combination receptacle is shown in figure 23-10(B). The upper portion of the receptacle is for 120-volt use

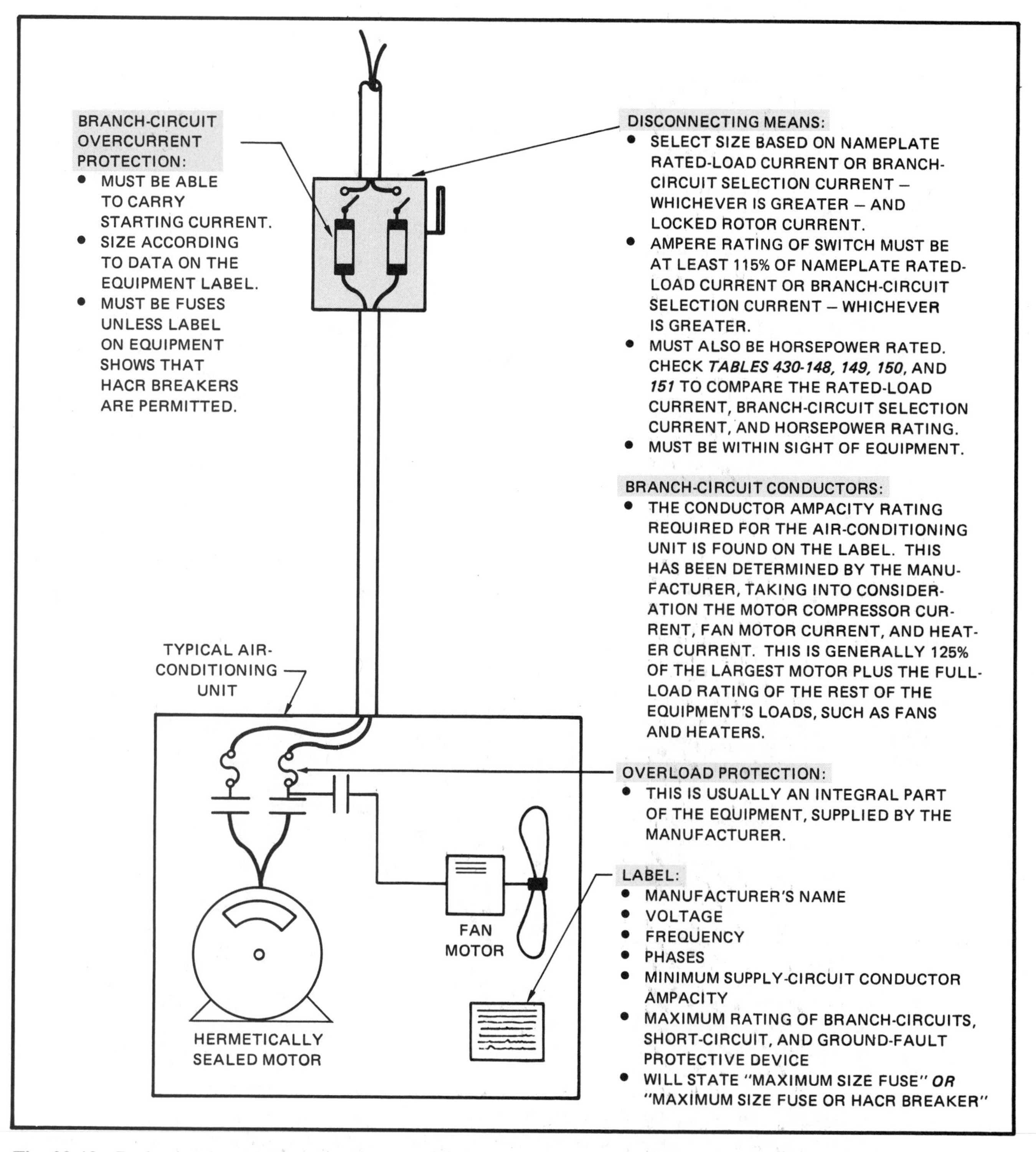

Fig. 23-12 Basic circuit requirements for a typical residential-type air conditioner or heat pump. Reading the label is important in that the manufacturer has determined the minimum conductor size and branch-circuit fuse size. See *Article 440* of the *National Electrical Code®*

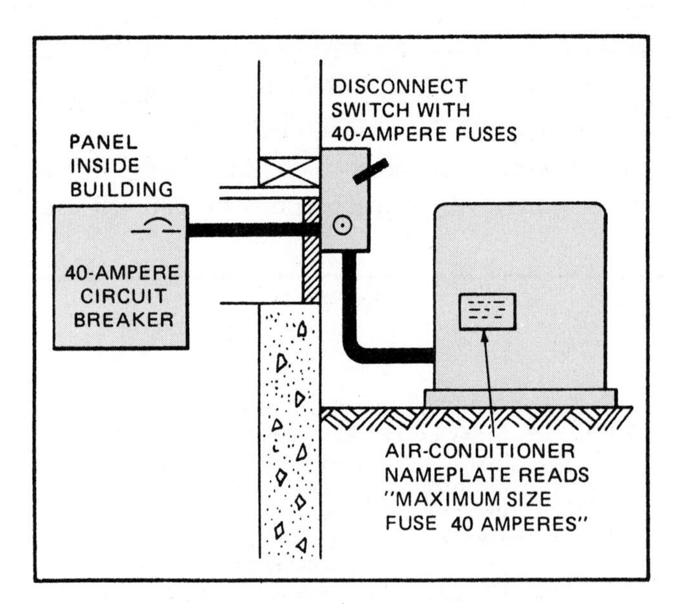

Fig. 23-13 This installation *conforms* to the Code, *Section 440-14*. The disconnect switch is within sight of the unit and contains the 40-ampere fuse called for on the air-conditioner nameplate as the branch-circuit protection.

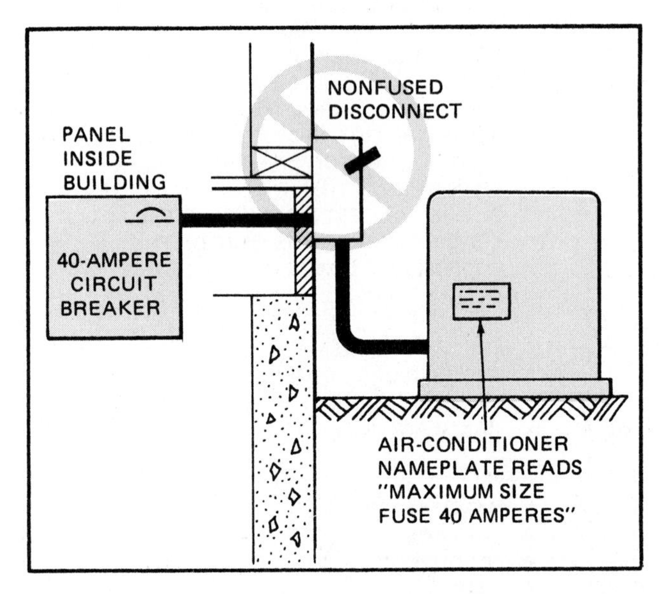

Fig. 23-14 This installation *violates* the Code, *Section 110-3(b)*. Although the disconnect is within sight of the air conditioner, it does not contain fuses. Note that the branch-circuit protection is provided by the 40-ampere circuit breaker inside the building. Note also that the nameplate requires that the branch-circuit protection be 40-ampere fuses maximum. If fused branch-circuit protection were provided at the panel inside the building, the installation would meet Code requirements.

and the lower portion is for 240-volt use. Note that the tandem and grounding slot arrangement for the 240-volt receptacle meets the requirements of *Section 210-7(f)*. This section states that receptacle outlets for different voltage levels must be noninterchangeable with each other.

CENTRAL HEATING AND AIR CONDITIONING

The residence discussed throughout this text has central electric heating and air conditioning consisting of an electric furnace and a central air-conditioning unit.

The wiring for central heating and cooling systems, shown in figure 23-11 (page 328). Note that one feeder runs to the electric furnace and another feeder runs to the air conditioner or heat pump outside the dwelling. Low-voltage wiring is used between the inside and outside units to provide control of the systems. The low-voltage Class 2 circuit wiring must not be run in the same raceway as the power conductors, *Section 725-38(a)(2)*.

SPECIAL TERMINOLOGY

The following terms are unique to air conditioners, heat pumps, and other hermetically sealed motor-compressors.

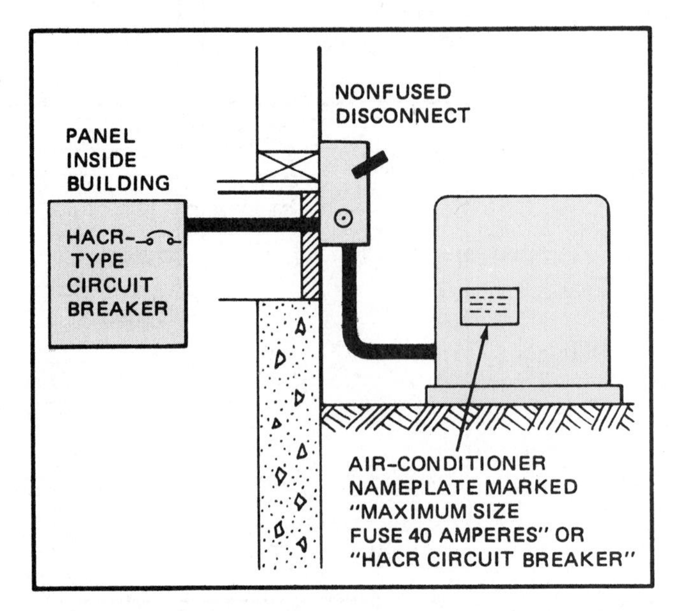

Fig. 23-15 Installation of Type HACR circuit breaker. Permitted only when the label on the breaker and the label on the air conditioner bear the letters "HACR."

Rated-load current is the current drawn when the unit is operating at rated load, at rated voltage, and at rated frequency.

Branch-circuit selection current is the value in amperes to be used instead of the rated-load current for determining conductor sizes, disconnect switch size, controller size, and short-circuit and ground-fault protective device sizes.

Branch-circuit selection current is always equal to or greater than the rated-load current as marked on the equipment.

Refer to figure 23-12 (page 329).

NEC® Section 110-3(b) requires that listed or labeled equipment be used or installed in accordance with any instructions included in the listing or labeling. For example, the nameplate of an air-conditioning unit reads "Maximum Size Fuse 40-amperes." Thus, only 40-ampere fuses can be used. Forty-ampere circuit breakers are not permitted. If the nameplate calls for "Maximum Over-current Protection," then either fuses or circuit breakers may be used.

DISCONNECT TO BE WITHIN SIGHT AND READILY ACCESSIBLE

NEC® Section 440-14 requires that the disconnect for the air conditioner must be within sight of the unit and readily accessible. Figures 23-13 and 23-14 illustrate the requirements of *Section 440-14.* For cord-and-plug-connected appliances, the cord and plug are considered to be the disconnecting means.

Type HACR circuit breakers (2-pole, 15 amperes through 60 amperes, 120/240 volts) are UL listed for group motor application such as found on heating (H), air conditioning (AC), and refrigeration (R). The label on these breakers bears the letters "HACR." The equipment must also be tested and marked "suitable for protection by a Type HACR breaker." Refer to figure 23-15. If the HACR marking is not found on both the circuit breaker and the equipment, then the installation does not "meet Code," unless fuses are installed for the branch-circuit overcurrent protection.

One type of fuse disconnect is shown in figure 23-16.

Fig. 23-16 One type of fuse-type disconnect for air-conditioning equipment. Available in 30-ampere, 240-volt and 60-ampere, 240-volt ratings. Courtesy of Midwest Electric Products, Inc.

The reasoning behind this tough overcurrent protection requirement for air conditioners, heat pumps, and other similar appliances is that one overcurrent device is called upon to protect the entire appliance, including both large wires and motors, as well as small control circuit wires, motors, and relays, figure 23-17.

NONCOINCIDENT LOADS

Loads such as heating and air conditioning are not likely to operate at the same time. As an example, when calculating feeder sizes to an electric furnace and an air conditioner, only the larger load need be considered. See *Section 220-21* of the Code. This is discussed again in unit 29 when service-entrance calculations are presented.

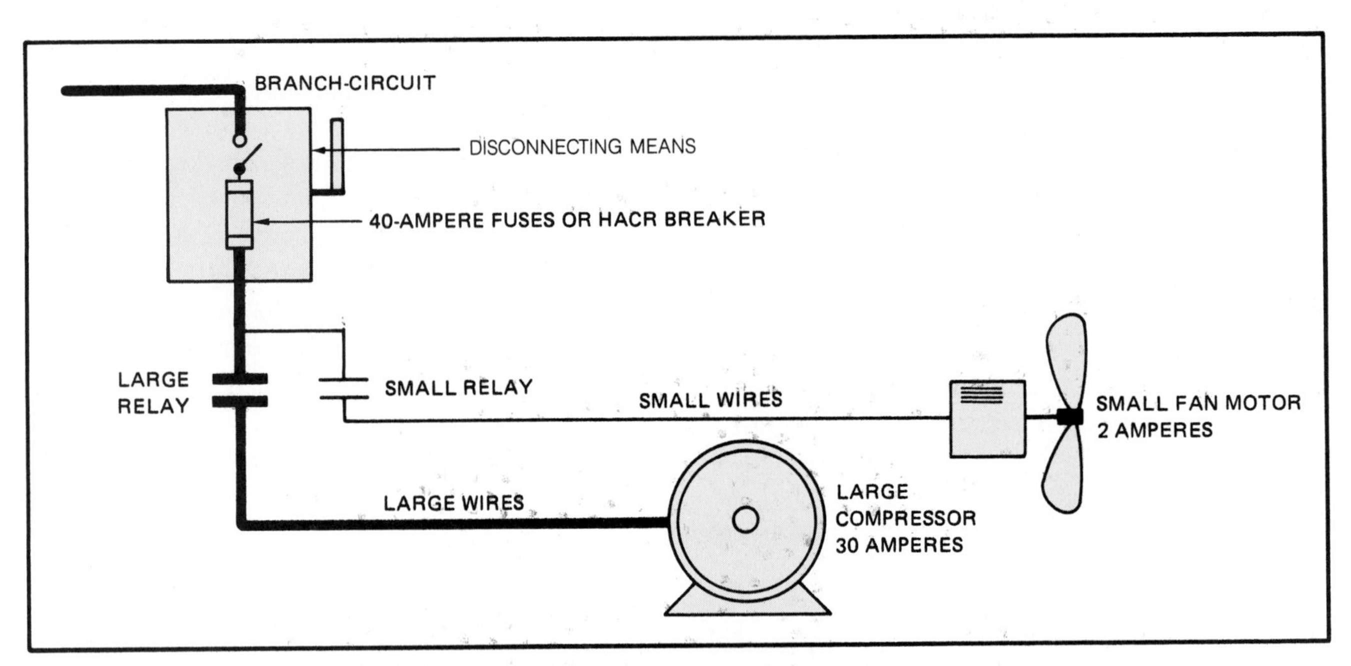

Fig. 23-17 In a typical air-conditioner unit, the branch-circuit protective device is called upon to protect the large components (wire, relay, compressor) and the small components (wire, relay, fan motor).

REVIEW

Note: Refer to the Code or the plans where necessary.

ELECTRIC HEAT

1. a. What is the allowance in watts made for electric heat in this residence? ______

   b. What is the value in amperes of this load? ______

2. What are some of the advantages of electric heating? ______

3. List the different types of electric heating system installations. ______________

____

____

____

____

____

4. There are two basic voltage classifications for thermostats. What are they? ______

____

5. What device is required when the total connected load exceeds the maximum rating of a thermostat? ______________________________

6. The electric heat in this residence is provided by what type of equipment? ______

____

7. At what voltage does the electric furnace operate? ______________________

8. A certain type of control connects electric heating units to a 120-volt supply or a 240-volt supply, depending upon the amount of the temperature drop in a room. These controls are supplied from a 120/240-volt, three-wire, single-phase source. Assuming that this type of device controls a 240-volt, 2000-watt heating unit, what is the wattage produced when the control supplies 120 volts to the heating unit? Show all calculations.

9. What advantages does a 240-volt heating unit have over a 120-volt heating unit?

____

____

____

____

____

____

10. The white wire of a cable may be used to connect to a hot circuit conductor only if ____

____

11. Receptacle outlets furnished as part of a permanently installed electric baseboard heater, when not connected to the heater's branch circuit, (may) (may not) be counted as the required receptacle outlet for the space occupied by the baseboard heater. Electric baseboard heaters (shall) (shall not) be installed beneath wall receptacle outlets. Circle correct answers.

12. The branch circuit supplying a heater must be sized to at least ________% of the heater's rating according to *Section* __________.

13. Compute the current draw of the following electric furnaces. The furnaces are all rated 240 volts.

    a. 7.5 kW

    b. 15 kW

    c. 20 kW

14. For "ballpark" calculations, the wattage output of a 240-volt electric furnace connected to a 208-volt supply will be approximately 75% of the wattage output had the furnace been connected to a 240-volt supply. In question 13, calculate the wattage output of (a), (b), and (c). ____________________

    ____

    ____

    ____

    ____

► 15. A central electric furnace heating system is installed in a home. The circuit supplying this furnace:
    a. may be connected to a circuit that supplies other loads.
    b. shall be connected to a circuit that supplies other loads.
    c. shall be connected to a separate circuit.

16. What section of the Code provides the correct answer to question 15? __________

    ____________________ ◄

AIR CONDITIONING

1. a. When calculating air-conditioner load requirements and electric heating load requirements, is it necessary to add the two loads together to determine the combined load on the system? ____________________

b. Explain the answer to part (a). ______________________________

____

____

2. The total load of an air conditioner shall not exceed what percentage of a separate branch circuit? (circle one)
   a. 75% b. 80% c. 125%
3. The total load of an air conditioner shall not exceed what percentage of a branch circuit that also supplies lighting? (circle one)
   a. 50% b. 75% c. 80%
4. a. Must an air conditioner installed in a window opening be grounded if a person on the ground outside the building can touch the air conditioner? __________

   b. What Code section governs the answer to part (a)? __________
5. A 120-volt air conditioner draws 13 amperes. What size is the circuit to which the air conditioner will be connected? ______________________________
6. What is the Code requirement for receptacles connected to circuits of different voltages and installed in one building? ______________________________
7. When a central air-conditioning unit is installed and the label states "Maximum size fuse 50-amperes," is it permissible to connect the unit to a 50-ampere circuit breaker?

   ____
8. Standard circuit breakers are generally not permitted as the branch-circuit protection for air-conditioning equipment. However, a special situation exists that does permit a specific type of breaker to be used. Explain. ______________________________

   ____

   ____

   ____

   ____

   ____

   ____
9. Briefly, what is the reasoning behind the UL requirement that air conditioners and other multimotor appliances be protected with fuses or special HACR circuit breakers?

   ____

   ____

   ____

   ____

   ____

   ____

   ____
10. What section of the *National Electrical Code*® prohibits the installing of Class 2 control circuit conductors in the same raceway as the power conductors? __________

    ____

UNIT 24

Oil and Gas Heating Systems

OBJECTIVES

After studying this unit, the student will be able to

- understand the basics of gas and oil heating systems.
- understand some of the more important components of gas and oil heating systems.
- interpret basic schematic wiring diagrams.
- explain the principles of the thermocouple and the thermopile in a self-generating system.
- understand and apply the Code requirements for control-circuit wiring, *Article 725, NEC®*

In unit 23, we discussed the circuit requirements for electric heating (furnace, baseboard, ceiling cable). In unit 24, we will discuss typical residential oil and gas heating systems.

Heating a home with gas or oil is possible with:

- Warm air — fan-forced warm air or gravity (hot air rises naturally)
- Hot water — hot water is circulated through the piping system by a circulating pump. The radiators look very much like electric baseboard heating units.

PRINCIPLES OF OPERATION

Most residential gas or oil heating systems operate as follows. A room thermostat is connected to the proper electrical terminals on the furnace or boiler, where the various controls and valves are interconnected in a manner that will provide safe and adequate heat to the home. Some typical wiring diagrams are shown in figures 24-1, 24-2, and 24-3.

Most residential gas and oil heating systems are "packaged units" in which the safety controls, valves, fan controls, limit switches, etc., are integral preassembled, prewired parts of the furnace or boiler.

For the electrician, the "field-wiring" generally consists of:

1. installing and connecting the wire to the thermostat, and
2. installing and connecting the power supply to the furnace or boiler.

Refer to figure 24-4.

MAJOR COMPONENTS

Thermostat, figure 24-5: Responds to temperature changes. Is connected to the controls on the furnace or boiler to turn the unit off and on, thereby maintaining the desired room temperature.

A thermostat can have features such as

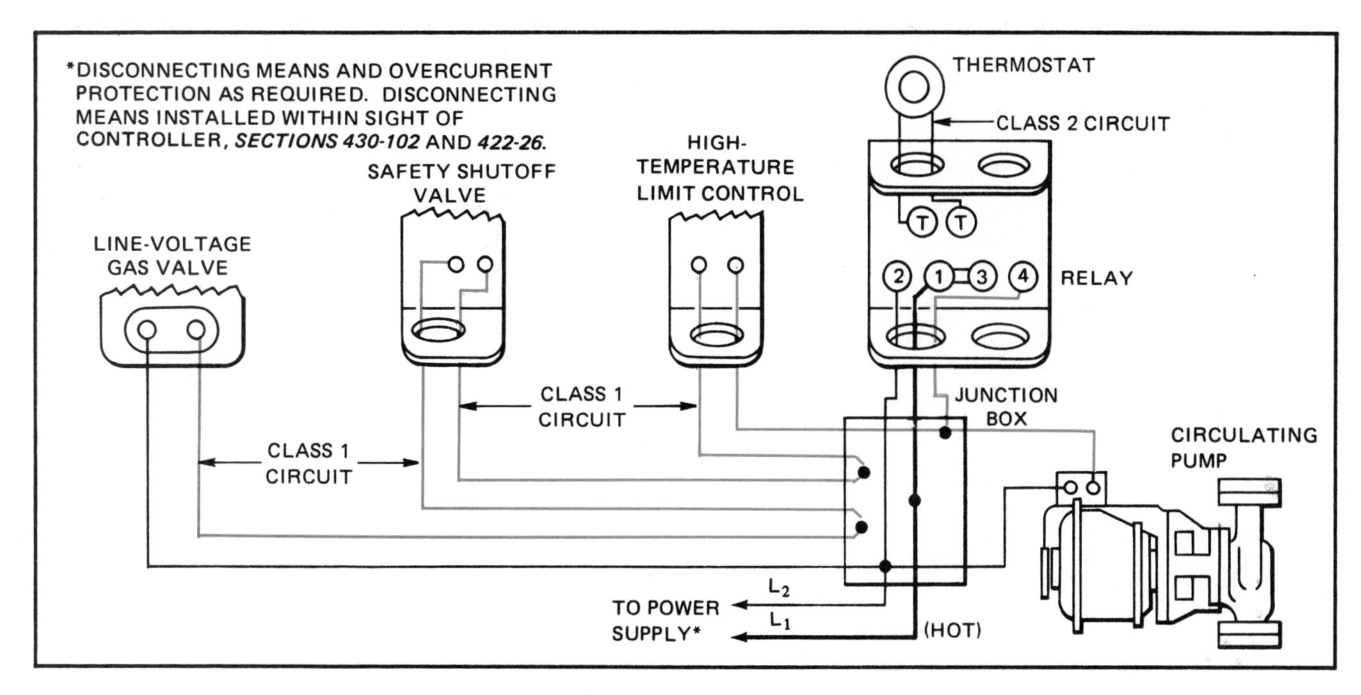

Fig. 24-1 Typical wiring diagram for a gas burner, forced hot water system.

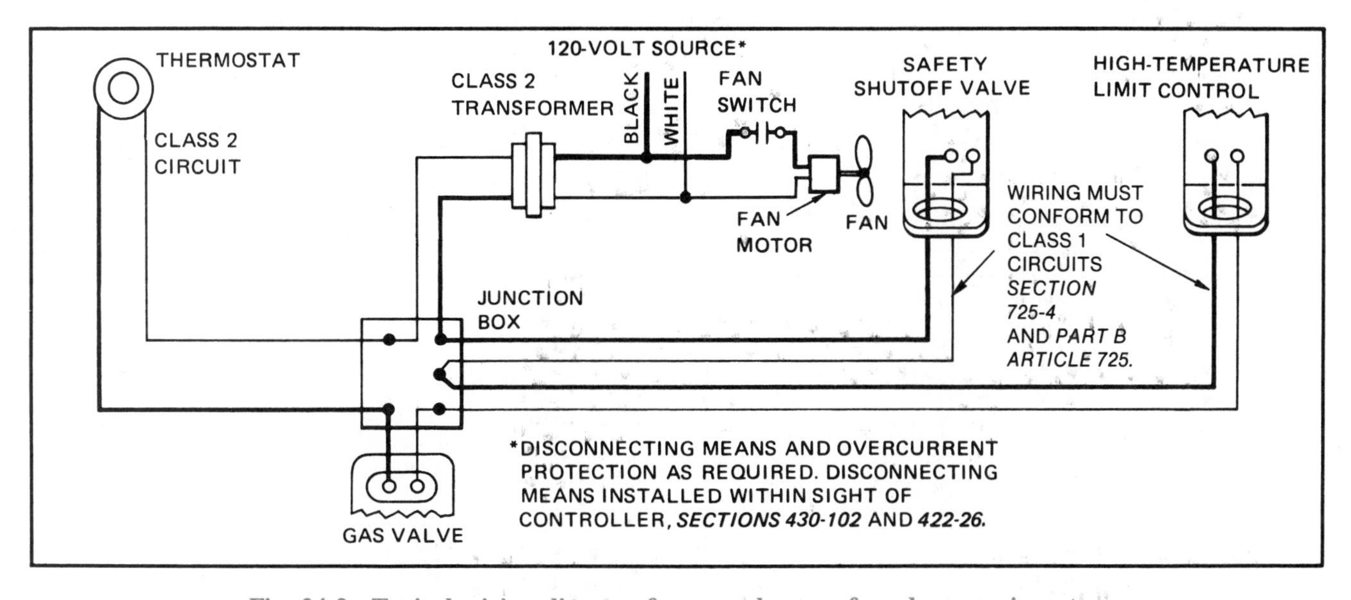

Fig. 24-2 Typical wiring diagram for a gas burner, forced warm air system.

day/night setback, a fan switch for continuous operation of the fan motor, a clock, and a thermometer. Some thermostats control both heating and air-conditioning functions.

For energy savings, thermostats of the types shown in figure 24-5 can maintain reduced temperatures at night and during the day if the home is unoccupied. Some thermostats feature an integral heat anticipator that allows the thermostat to maintain specific comfort requirements to a more narrow range of temperature fluctuations. These heat anticipators are adjustable to match their setting with the current draw of the circuit controlled by the thermostat. Quality thermostats generally are accurate to 1 1/2 degrees.

Primary Control, figure 24-6: Controls the oil burner motor, ignition, and oil valve in response to commands from the thermostat. If the flame fails to

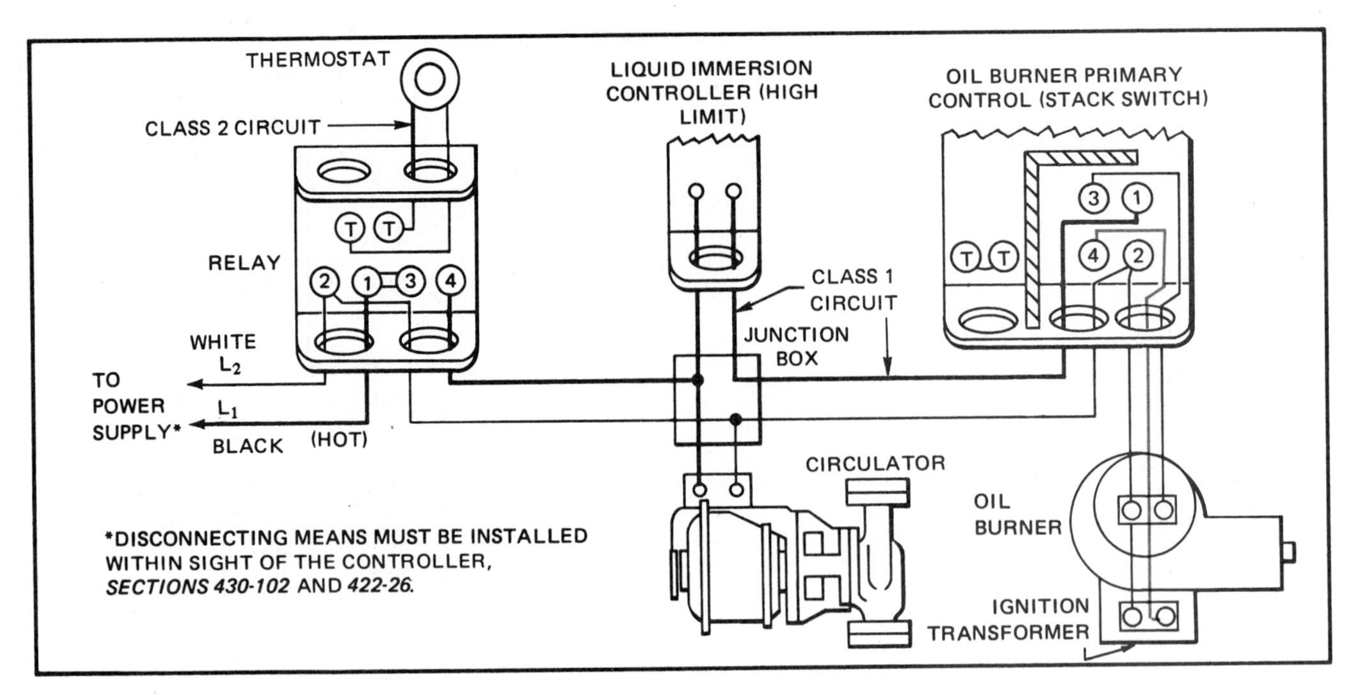

Fig. 24-3 Typical wiring diagram of an oil burner heating installation.

ignite, the controller shuts down the burner. Sometimes referred to as a "stack" switch.

This primary safety control is to an oil burning system what a main gas valve is to a gas burning system.

Primary controllers are available for both constant and intermittent ignition burners. Both types will recycle once if the flame goes out for any reason. If the flame fails to ignite, the controller trips "off" to safety. The control shown in figure 24-6 is designed for stack mounting. Other types of oil burner primary controls are mounted on the oil burner with the flame-sensing device mounted in the blast tube of the burner.

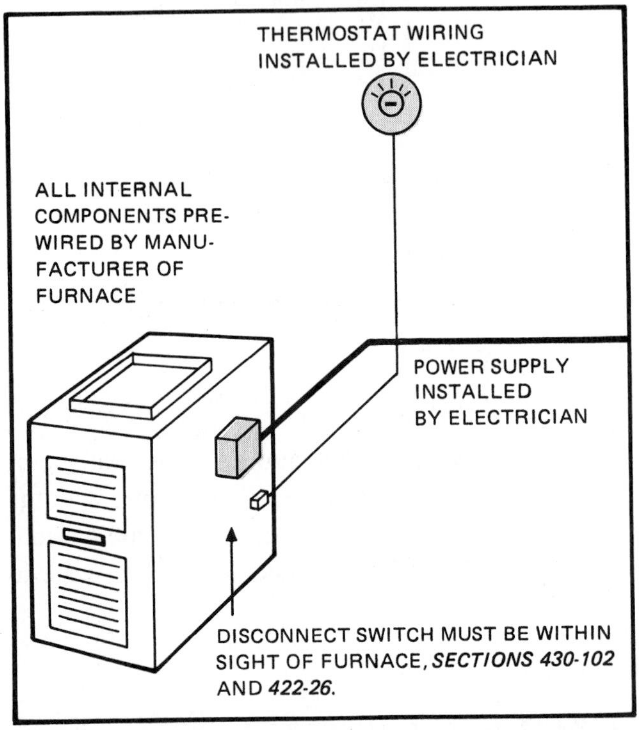

THERMOSTAT WIRING
INSTALLED BY ELECTRICIAN

ALL INTERNAL
COMPONENTS PRE-
WIRED BY MANU-
FACTURER OF
FURNACE

POWER SUPPLY
INSTALLED
BY ELECTRICIAN

DISCONNECT SWITCH MUST BE WITHIN
SIGHT OF FURNACE, *SECTIONS 430-102*
AND *422-26.*

Fig. 24-4 Diagram showing thermostat wiring, power supply wiring, and disconnect switch.

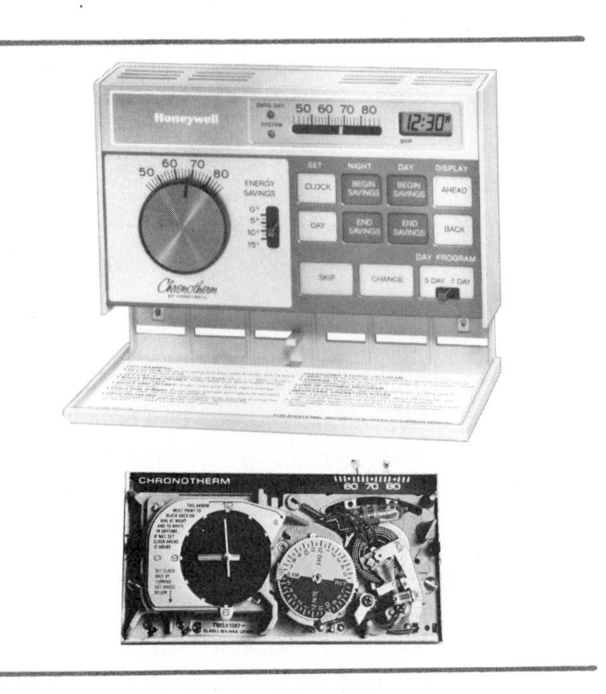

Fig. 24-5 Thermostat.

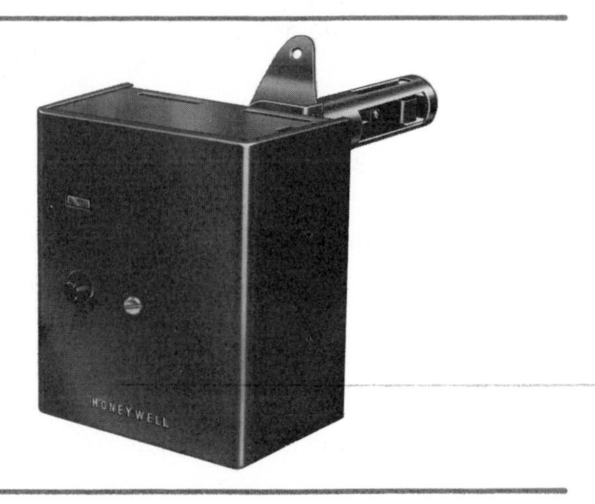

Fig. 24-6 Oil burner primary control.

Fig. 24-7 Switching relay.

Switching Relay, figure 24-7: Is the interconnect between the low-voltage wiring and line-voltage wiring. When a low-voltage thermostat "calls for heat," the relay "pulls in," closing the line-voltage contacts on the relay, which turns on the main power to the furnace or boiler. See figures 24-1 and 24-3. A built-in transformer provides low voltage for the thermostat circuit.

Liquid Immersion Control, figure 24-8: Controls high water and circulating water temperatures. When acting as a "high-limit control," it will shut off the burner if water temperatures in the boiler reach dangerous levels. When acting as a "circulating water temperature control," it makes sure that the water being circulated through the piping system is at the desired temperature, not cold.

An immersion controller must be mounted in the boiler so that the sensing tube is either in direct contact with the water in the boiler or inserted into an immersion well that is in direct contact with the boiler water. The setting of the controller is usually in the range of 180 to 200 degrees Fahrenheit. The setting can be changed by turning the adjusting screw on the face of the control. A stop prevents the temperature from being set any higher than 240° F.

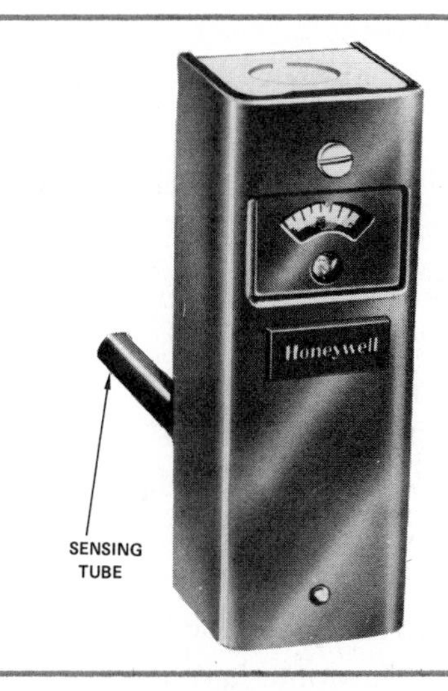

Fig. 24-8 Liquid immersion controller.

Fig. 24-9 Transformer.

Transformer, figure 24-9: Is used where the line-voltage (120-volts) is reduced to low-voltage (24-volts), such as the low-voltage wiring used for thermostat wiring, and control-circuit wiring.

Low-Water Control: Is a device used to sense low water levels in a boiler. Burner cannot come on when water is too low. Similar in appearance to figure 24-8.

Fan Control, figure 24-10: Used to control the fan on forced warm-air systems. It will turn the fan off and on at some predetermined air temperature in the furnace's plenum chamber. Usually set to turn on between 120 and 140 degrees Fahrenheit and to cut off at 90 to 100 degrees Fahrenheit.

Limit Control, figure 24-10: When the air temperature in the plenum of the furnace reaches dangerously high levels (approximately 200° F), the limit control will shut off the burner. The fan, however, will continue to run. When the temperature in the plenum returns to safe levels, the contacts in the limit control will close, allowing the burner to resume normal operation. A limit control may also be called a limit switch. *NOTE*: Fan controllers and limit controllers are available as individual devices and are also available as combination devices.

Fan Motor: The electric motor that forces the warm air through the ducts of a heating system. The fan motor may be single-speed or multi-speed. The fan itself may be belt driven or directly connected to the shaft of the motor. The motor is generally provided with integral overload protection, but, in addition, may be protected with properly sized dual-element fuses installed in the disconnect switch for the furnace.

Water Circulating Pump, figure 24-11: Circulates hot water through the piping system. Is controlled by a liquid immersion control, figure 24-8.

Multizone control can be achieved by providing circulating pumps or zone valves (similar to figure 24-12, only electrically operated). Each circulating pump or zone valve would have its own thermostat, allowing each heating zone to be separately controlled. In multizone systems, the high-limit control is set to maintain the boiler water temperature at a certain value.

Flow Valve, figure 24-12: Prevents the gravity circulation of water through the piping system.

Solid-State Ignition: A feature on some gas furnaces, whereby the need for continuous burning of the pilot light is eliminated. The pilot light comes on only when the themostat calls for heat. This is an energy-saving feature.

Main Gas Valve, figure 24-13: The valve that allows the gas to flow to the main burner of the furnace. It functions as a pressure regulator, a safety shutoff, and as the main gas valve. Should there be a "flame-out," or if the pilot light fails to operate, or in the event of a power failure, this control will shut off the flow of gas. It is the safety control of a gas system, just as a primary control is the safety control of an oil system.

SELF-GENERATING SYSTEM

Several manufacturers of gas burners provide self-generating systems. These systems do not require an outside power supply. Figure 24-14 shows the wiring diagram of a typical self-generating system. Note that the components are connected in series. The small amount of energy required by the gas valve is supplied by a thermopile. Many of these burners are prewired at the factory. Thus, the only wiring at the site is done by

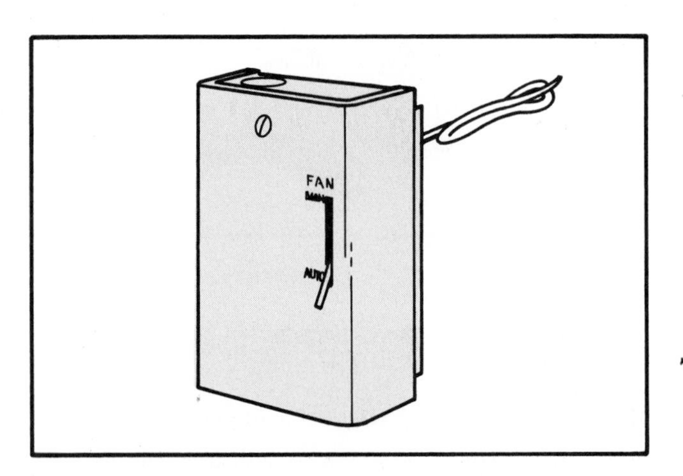

Fig. 24-10 Fan/limit control.

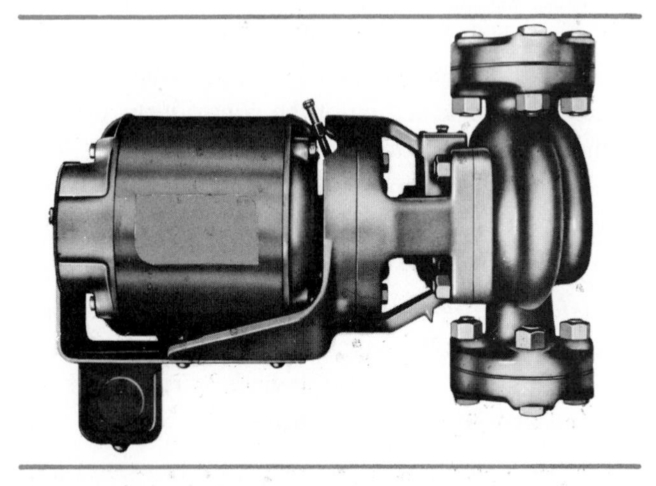

Fig. 24-11 Water circulator.

Fig. 24-12 Flow valve.

the electrician, who runs a low-voltage cable to the thermostat. This system is not affected by a power failure because there is no outside power source.

The self-generating system operates in the millivolt range. A special thermostat is required to function at this low voltage level. All other components of the system are matched to operate at the low voltage.

Thermopile for a Self-generating System

A *thermopile* is a group of thermocouples electrically connected together in series. The following paragraphs describe what a thermocouple is and what it does.

Thermocouple

Whenever two unlike metals are connected to form a circuit, an electric current flows if the junctions of the metals are at different temperatures, figure 24-15. As the temperature difference between the two junctions increases, the electric current increases. Thus, a thermocouple is formed by combining metals such as iron and copper, copper and iron constantan, copper-nickel alloy and chrome-iron alloy, platinum and platinum-rhodium alloy, or chromel and alumel. The types of metals used depend upon the temperatures involved.

In a gas burner system, the source of heat for the thermocouple is the pilot light. The cold junction of the thermocouple remains open and is connected to the pilot safety shutoff gas valve circuit.

A single thermocouple develops a voltage of about 25 to 30 dc millivolts. Thus, the circuit

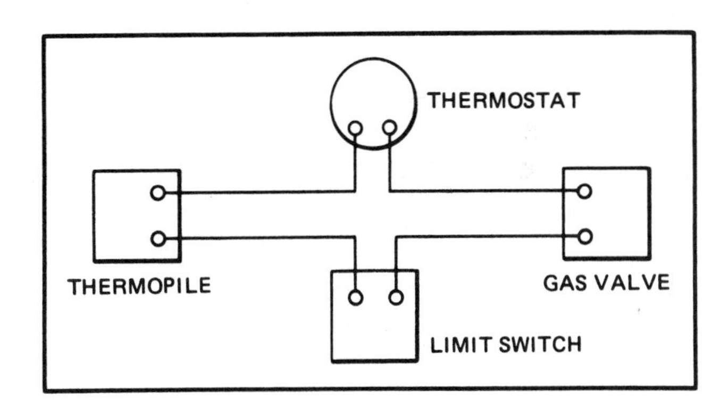

Fig. 24-14 Wiring diagram of a self-generating system.

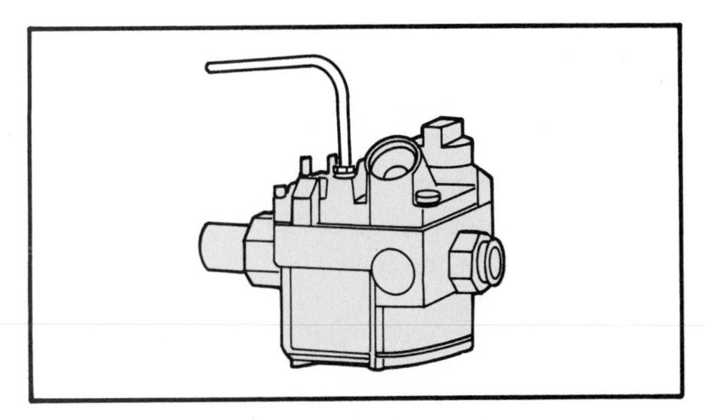

Fig. 24-13 Main gas valve.

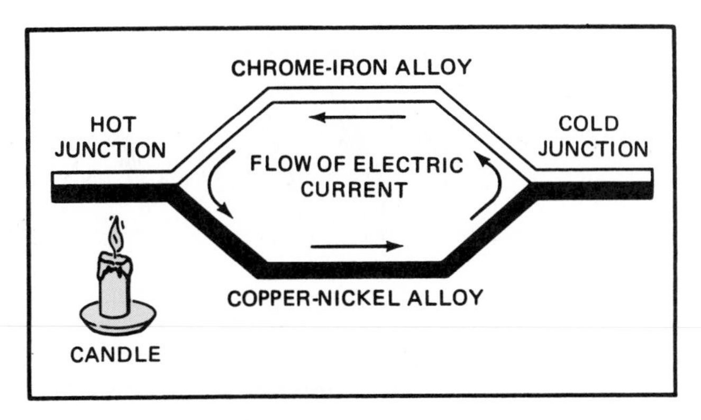

Fig. 24-15 Principle of a thermocouple.

resistance must be kept low. The electrician should follow the manufacturer's recommendations regarding wire size. In addition, it is suggested that any connections other than those at screw terminals be soldered.

Thermopile

When thermocouples are connected in series, a thermopile is formed, figure 24-16. The power output of the thermopile is greater than that of a single thermocouple; typically either 250 or 750 dc millivolts. For both a thermocouple and a thermopile, there must be a temperature difference between the metal junctions. If the junctions marked A in figure 24-16 are heated and the junctions marked B remain cold, the resultant flow of current is shown by the arrows.

SUPPLY-CIRCUIT WIRING

The branch-circuit wiring must be adequate for the power requirements of the motors, fans, relays, pumps, ignition, transformers, and other power-consuming components of the furnace or boiler. The sizing of conductors, overcurrent protection, conduit fill, etc., has already been covered in other units of this text.

► Note that *Section 422-7* of the *National Electrical Code®* requires that any central heating equipment (electric, gas, oil, heat pump) must be supplied by a separate branch circuit. This has been discussed in unit 23, Special-purpose Outlets — Electric Heating, Air Conditioning. ◄

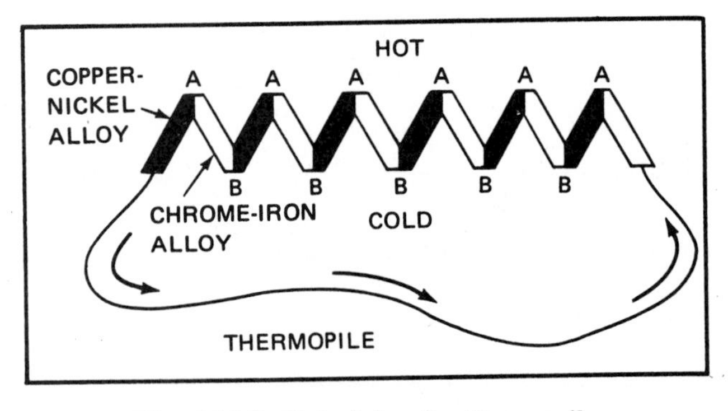

Fig. 24-16 Principle of a thermopile.

CONTROL-CIRCUIT WIRING

Article 725 of the Code defines Class 1, Class 2, and Class 3 remote-control, signaling, and power-limited circuits as circuits that are *not* an integral part of a device or appliance. As soon as the conductors for remote-control, signaling, or power-limited circuits leave the equipment, they are classified as Class 1, Class 2, or Class 3 circuits, figure 24-17. (*Section 725-1.*) There are alternative requirements for Class 1, Class 2, and Class 3 circuits regarding wire sizes, derating factors, overcurrent protection, insulation, wiring methods, and materials that "differentiate" control-circuit wiring from regular (normal) light and power circuits as covered by the Code in Chapters 1 through 4.

In part A of figure 24-17, all of the control circuitry is contained completely within the equipment, and as such, none of the control wiring is governed by *Article 725* of the Code.

In part B of figure 24-17, the control wires leave the equipment. These conductors are classified as Class 1, Class 2, or Class 3, depending upon voltage, current, and other issues discussed in *Article 725*. Motor control circuits may be protected by control fuses (Class CC) in the controller, in which case *Article 725* does not apply. See *Section 430-72* of the Code.

Refer to *Table 725-31(a)* for power limitations of Class 2 and Class 3 circuits.

An example of a Class 1 circuit would be the wiring of figure 24-18 where a failure of the equipment or wiring such as high limit and low water controls would "introduce a direct fire or life hazard" (*Section 725-4*). Because of this potential hazard, *Section 725-18* demands that the wiring be installed in rigid or intermediate metal conduit, rigid nonmetallic tubing, electrical metallic tubing, Type MI or MC cable, or be otherwise suitably protected from physical damage. See *Article 725*, Part B for code requirements applicable to Class 1 circuits.

Class 1 remote-control and signaling circuit voltage must not exceed 600 volts. If the circuit is power limited by the design of the transformer or by separate overcurrent protection, the maximum voltage is 30 volts and 1000 volt-amperes, *Section 725-11(a)*.

The conductors of Class 1 circuits and power

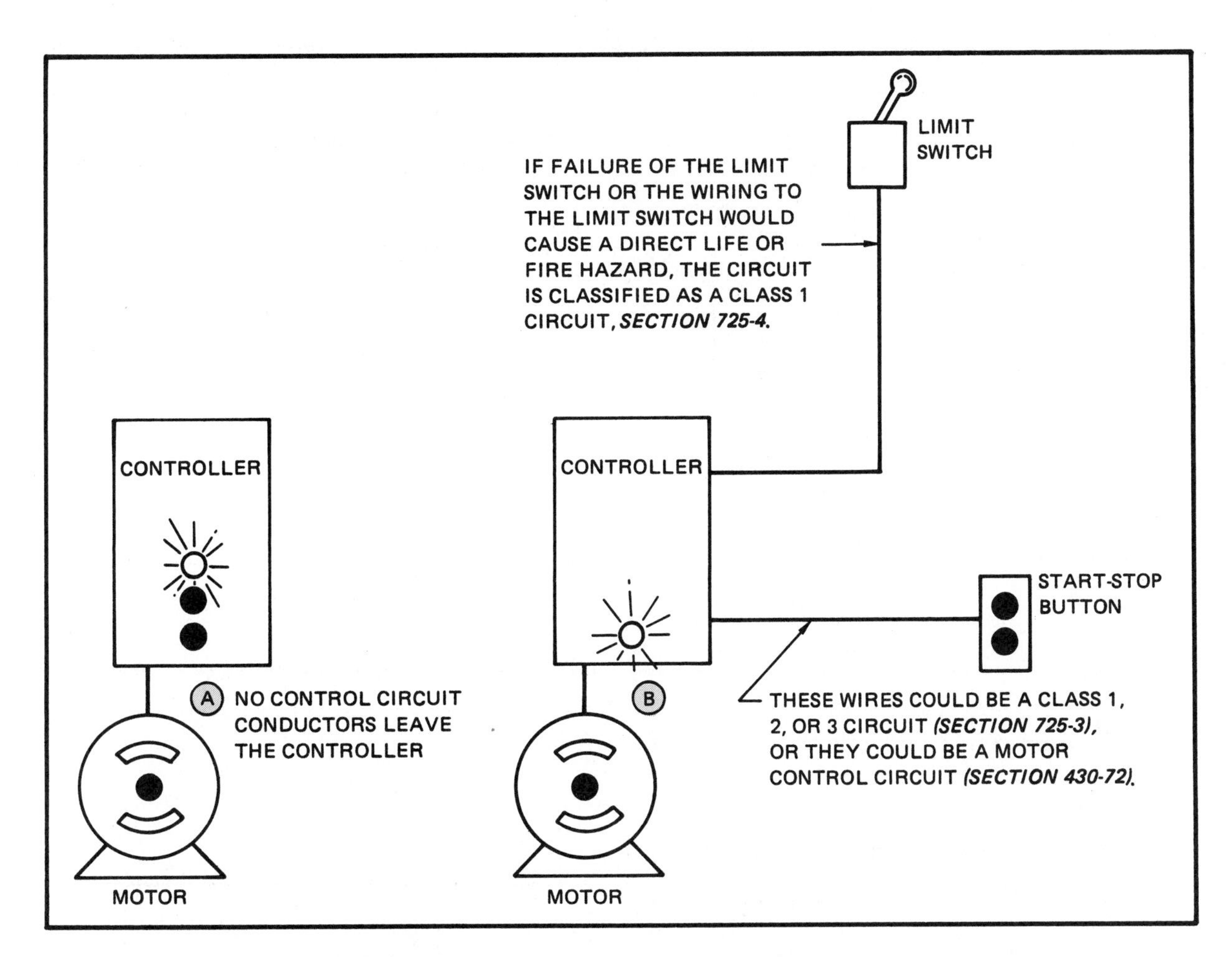

Fig. 24-17 Diagram showing how Class 1, 2, and 3 circuits are categorized.

conductors may be run in the same raceway ONLY if the interconnected equipment is "functionally associated," *Section 725-15*. Most wiring diagrams for residential air-conditioning units indicate that their low-voltage circuitry is Class 2, in which case the field wiring of the power and low-voltage circuits must be run separately. See *Section 725-38(a)(2)* and Figures 24-19A and 24-19B.

An example of a Class 2 circuit is shown in figure 24-18 where the wiring runs to the thermostat. A failure of the thermostat, or a short circuit in the wires running to the thermostat would *not* result in a direct fire or life hazard, since the operation of the high-limit control (or other safety control) would prevent the occurrence of a fire or life hazard.

► Low-voltage conductors (single or multiple) for Class 2 and Class 3 circuits are rated in a "hierarchy based upon their ability to withstand fire." Elaborate tests have been developed to classify these low-voltage conductors.

Table 725-50 shows how these cables must be marked.

Type	Cable Marking
Class 2 Plenum Cable	CL2P
Class 2 Riser Cable	CL2R
Class 2 Cable	CL2
Class 2 Cable, Limited Use	CL2X
Class 3 Plenum Cable	CL3P
Class 3 Riser Cable	CL3R
Class 3 Cable	CL3
Class 3 Cable, Limited Use	CL3X

This list of Class 2 and Class 3 cables is presented in descending order relative to their fire resistance.

Table 725-53 shows which types of cables can be substituted for other types of cables. This table illustrates the substitution list in graphic form.

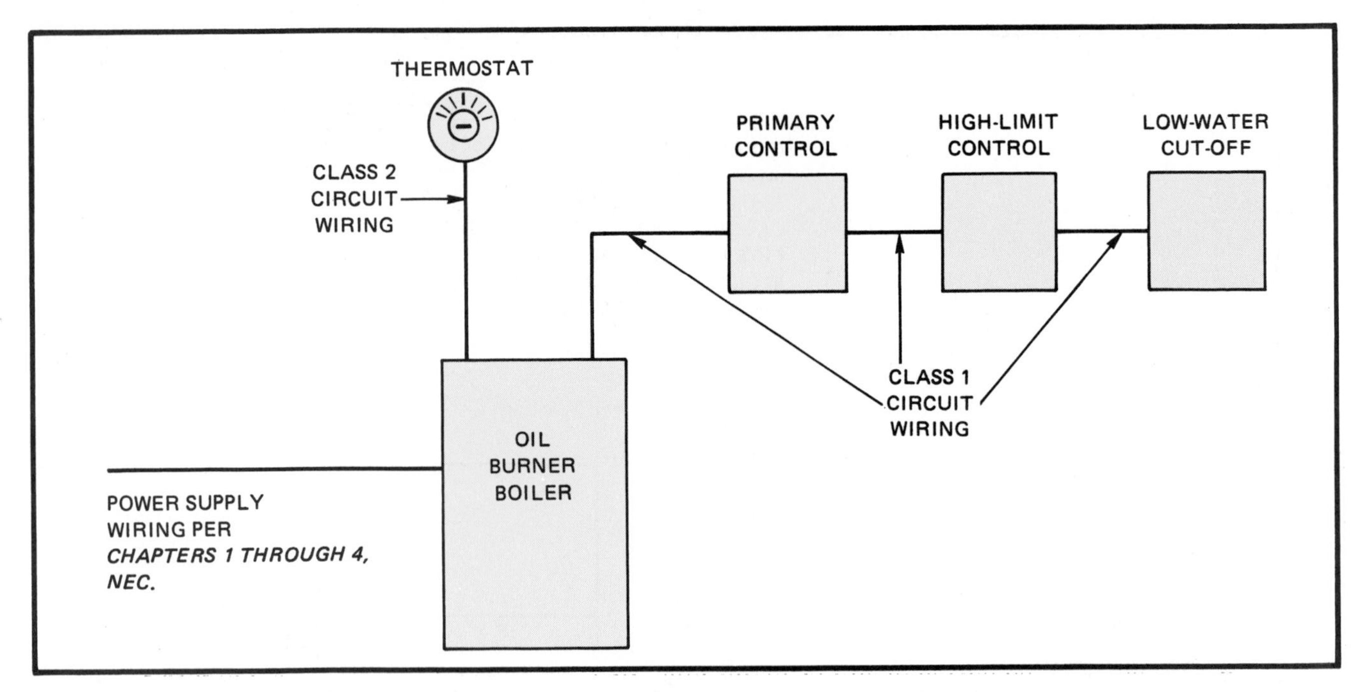

Fig. 24-18 Any failure in Class 1 circuit wiring would introduce a direct fire or life hazard. A failure in the Class 2 circuit wiring would *not* introduce a direct fire or life hazard. See *Section 725-4.*

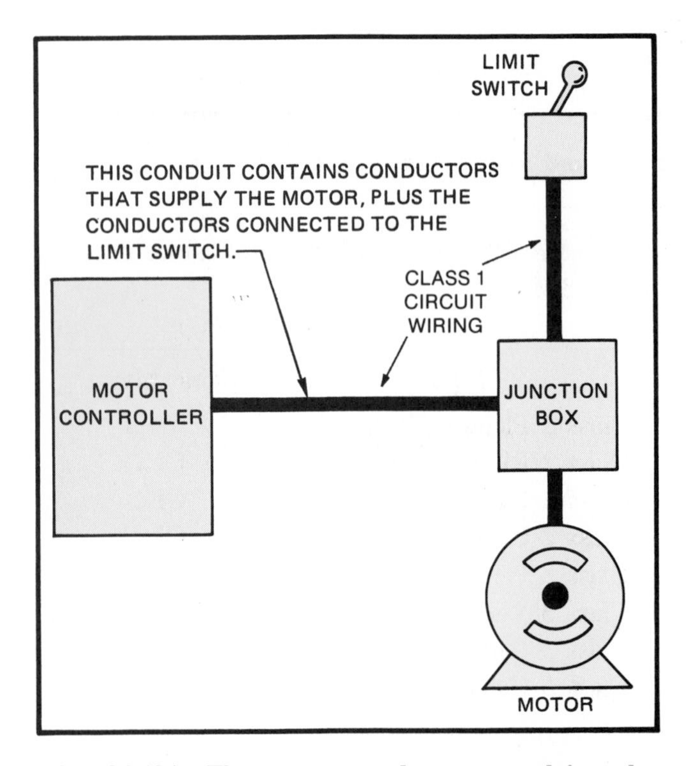

Fig. 24-19A The power conductors supplying the motor and the Class 1 circuit remote-control conductors to the limit switch and controller may be run in the same conduit when the interconnected equipment is "functionally associated." The insulation on the Class 1 circuit conductors must be insulated for the maximum voltage of any other conductors in the raceway. The minimum voltage rating for Class 1 conductors is 600 volts, *Section 725-16.*

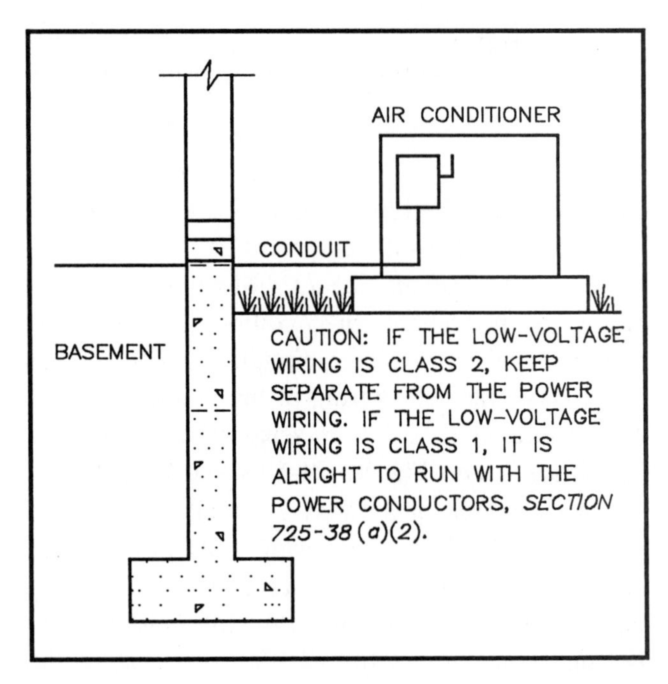

Fig. 24-19B The conduit running out of the basement wall connecting the air-conditioner unit is permitted to contain the power conductors and the Class 1 control circuit conductors because the equipment is "functionally associated." All conductors must be insulated for the maximum voltage of any conductors in the raceway, *Section 725-15.* Class 1 conductors must be insulated for a minimum voltage of 600 volts, according to *Section 725-16.* Standard building wire, such as Type THHN, is rated 600 volts.

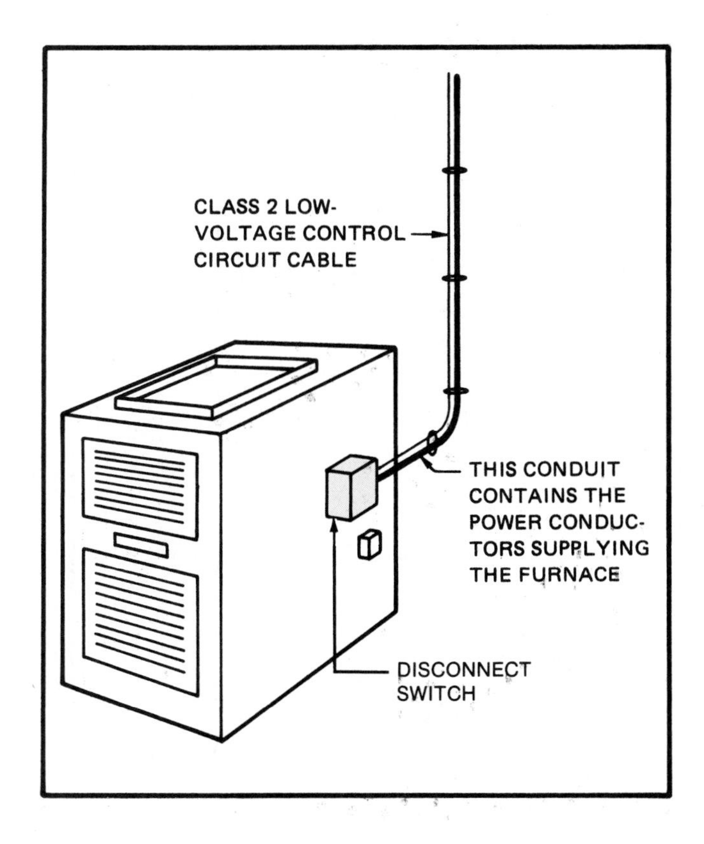

Fig. 24-20 The Code in *Section 300-11(b), Exc. No. 2,* allows Class 2 control circuit conductors (cables) to be supported by the raceway that contains the power conductors supplying electrical equipment such as the furnace shown. *Section 725-38(a)(2)* prohibits installing the low-voltage Class 2 conductor in the same raceway as the power conductors even if the Class 2 conductor has 600-volt insulation.

The different types of Class 2 and Class 3 cables are:

- Type CM .... Communication Wires and Cables
- Type CL2 and CL3 .. Class 2 and Class 3 Remote-Control, Signaling and Power-Limited Cables.
- Type FPL.......... Power-Limited Fire Protective Signaling Cables
- Type MP .... Multipurpose Cables
- Type PLTC .... Power-Limited Tray Cable

The low-cost, low-voltage Class 2 and Class 3 conductors generally installed in residences are the CL2X and CL3X types, *Section 725-51(d)*. ◄

Conductors for Class 2 and Class 3 circuits are *not* permitted to be run in the same raceway, enclosure, box compartment, or fitting with the regular lighting circuits, power circuits, or with Class 1 circuit wiring (*Section 725-38(a)(2)*).

Sometimes it is tempting to cut corners by installing Class 2 conductors in the same raceway as the power conductors. Even if the Class 2 conductors were insulated with 600-volt insulation, the same as the power conductors, this is a Code violation, *Section 725-38(a)(2)*. See figure 24-19B.

It is permitted to fasten a Class 2 low-voltage control cable to a conduit containing the power conductors that supply a furnace or similar equipment (*Section 300-11(b), Exc. No. 2*). Refer to figure 24-20.

Class 2 circuitry is similar to Class 3 circuitry in that for both types of circuits, the power is limited to the values specified in *Tables 725-31(a)* and *725-31(b)*, either (1) "inherently limited power source" where separate overcurrent protection is not required, or (2) by separate overcurrent protection. However, the maximum permitted voltage and current levels are higher in Class 3 circuits.

Conductor sizing, grounding, and overcurrent protection requirements are found in *Article 725* for Class 1, 2, and 3 circuits.

Transformers that are intended to supply Class 2 and Class 3 circuits are listed by Underwriters Laboratories and are marked "Class 2 Transformer" or "Class 3 Transformer."

Figure 24-21 illustrates the definitions of remote control, signaling, and power-limited terminology.

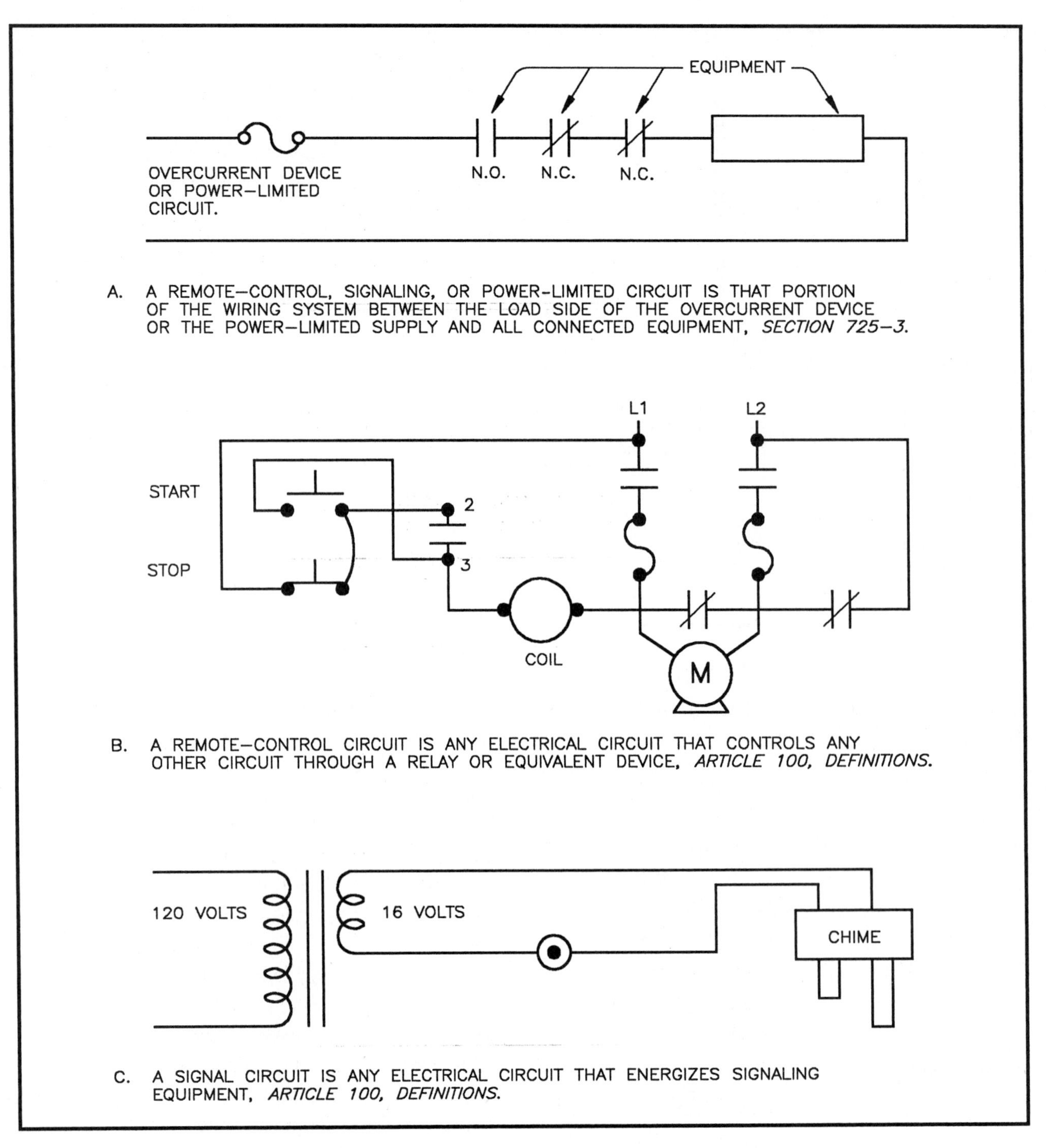

Fig. 24-21 The above diagrams explain a remote-control circuit and a signaling circuit.

REVIEW

Note: Refer to the Code or the plans where necessary.

1. How is the residence in the text heated? ______________________________
2. The Code requires that a disconnecting means must be ____________________ ______________________________ of the furnace or boiler.

3. What device in an oil burning system will shut off the burner in the event of a flameout? ______________________________

4. Name the control that limits dangerous water temperatures in a boiler. ______________________________

5. Name the control that limits the temperature in the plenum of a furnace. ______________________________

6. In the wiring diagram in this unit, are the safety controls connected in series or parallel? ______________________________

7. What is meant by the term "self-generating"? ______________________________

8. Because a self-generating unit is not connected to an outside source of power, the electrician need not use care in the selection of the conductor size or in making electrical connections. Is this statement true or false? Explain. ______________________________

9. Explain what a thermocouple is and how it operates. ______________________________

10. If Class 2 low-voltage conductors are insulated for 32 volts, and the power conductors are insulated for 600 volts, does the Code permit pulling these conductors through the same raceway? ______________________________

11. If Class 2 low-voltage conductors are used on a 24-volt system, but actually insulated for 600 volts (THHN for example) does the Code permit running these conductors together in the same raceway as the 600-volt power conductors? ______________________________

12. Under what conditions may Class 1 conductors and power conductors be installed in the same raceway? ______________________________

UNIT 25

Television, Telephone, and Low-Voltage Signal Systems

OBJECTIVES

After studying this unit, the student will be able to

- install television outlets, antennas, cables, and lead-in wires.
- list CATV installation requirements.
- describe the basic operation of satellite antennas.
- install telephone conductors, outlet boxes, and outlets.
- define what is meant by a signal circuit.
- describe the operation of a two-tone chime and a four-note chime.
- install a chime circuit with one main chime and one or more extension chimes.

TELEVISION (Symbol TV)

Television is a highly technical and complex field. A trained television service person or installer should be consulted before a television system is installed.

General Wiring

According to the plans for the dwelling, television outlets are installed in the following rooms:

Front Bedroom	1
Kitchen/Nook	1
Laundry	1
Living Room	3
Master Bedroom	1
Recreation Room	2
Study/Bedroom	1
Workshop	1

These outlets may be connected in several ways. The method selected depends upon the proposed locations of the outlets. In general, for installation in dwellings, the electrician uses standard sectional switch boxes or 4-inch square, 1 1/2-inch-deep outlet boxes with single-gang raised plaster covers. A box is installed at each point on the system where an outlet is located. Nonmetallic boxes are to be used if the system is to be wired with nonshielded cable, figure 25-1.

Television Installation Methods

One type of television installation is shown in figure 25-2. This is a master amplifier distribution system. The lead-in wire from the antenna is connected to an amplifier placed in an accessible area, such as the basement. Twin-lead, 300-ohm cable runs from the amplifier to each of the outlet boxes. The cable is connected to tap-off units. These units have a 300-ohm input and a 300-ohm output. The tap-off units have terminals or plug-in arrangements so that the 300-ohm, twin-lead cable can be connected between the television outlet and the television receiver.

Fig. 25-1 Nonmetallic boxes and a nonmetallic raised cover.

Amplified systems are generally not necessary in typical residential installations, particularly since the advent of cable television.

A second type of residential television system can also be installed using a *multiset coupler*, figure 25-3. The lead-in wire from the antenna is connected to a specific pair of terminals on the coupler. The cables to each television receiver are connected to other pairs of terminals. Two, three, or four television receivers can be connected to a coupler, depending upon the number of terminal pairs provided. When a coupler is used, an amplifier is not required. A coupler system is less expensive than an amplifier distribution system.

A third method of installing a television system uses shielded 75-ohm coaxial cable. The cable is connected to impedance matching transformers with 75-ohm or 300-ohm outputs. Additional cable is then connected between the matching transformer and the television receiver. Many older television receivers have 300-ohm inputs. Newer models generally have 75-ohm inputs. Some manufacturers recommend that shielded coaxial cable be used to prevent interference and keep the color signal strong. When unshielded lead-in wire is used, distorted color television pictures may result. Both shielded and nonshielded lead-in wire deliver good television reception when properly installed.

Figure 25-4 shows common locations of the television antenna on a residence.

Faceplates are provided for tap-off units to match conventional switch and convenience outlet faceplates. Combination two-gang faceplates also are available for tap-off units, figure 25-5. Both a 120-volt convenience receptacle and a television outlet can be mounted in one outlet box using this combination faceplate.

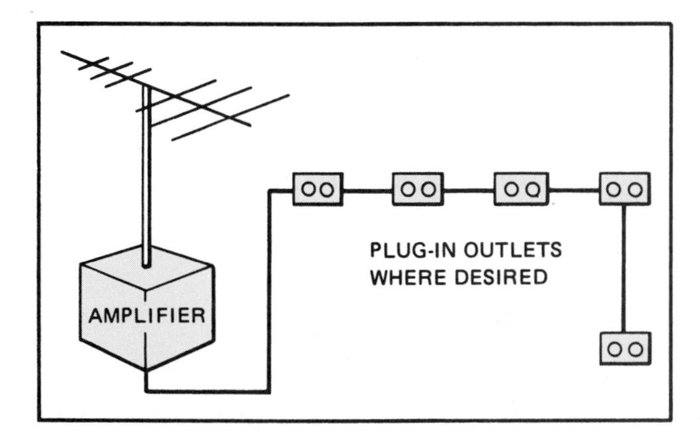

Fig. 25-2 Television master amplifier distribution system.

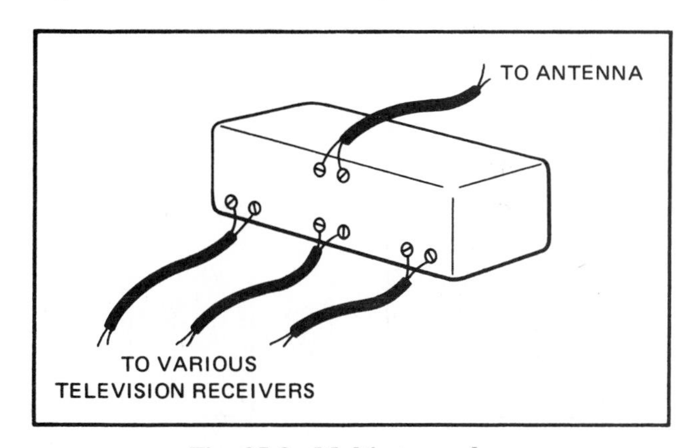

Fig. 25-3 Multiset coupler.

However, a combination installation requires that a metal barrier be installed between the 120-volt section and the television section of the outlet box, *Section 810-18(b)*. This barrier prevents line voltage interference with the television signal and keeps the 120-volt wiring away from the television wiring. In addition, line voltage cables and lead-in wire must be separated to prevent interference when they are run in the same space within the walls and ceilings. The line voltage cables should be fastened to one side of the space and the lead-in cable should be fastened to the other side.

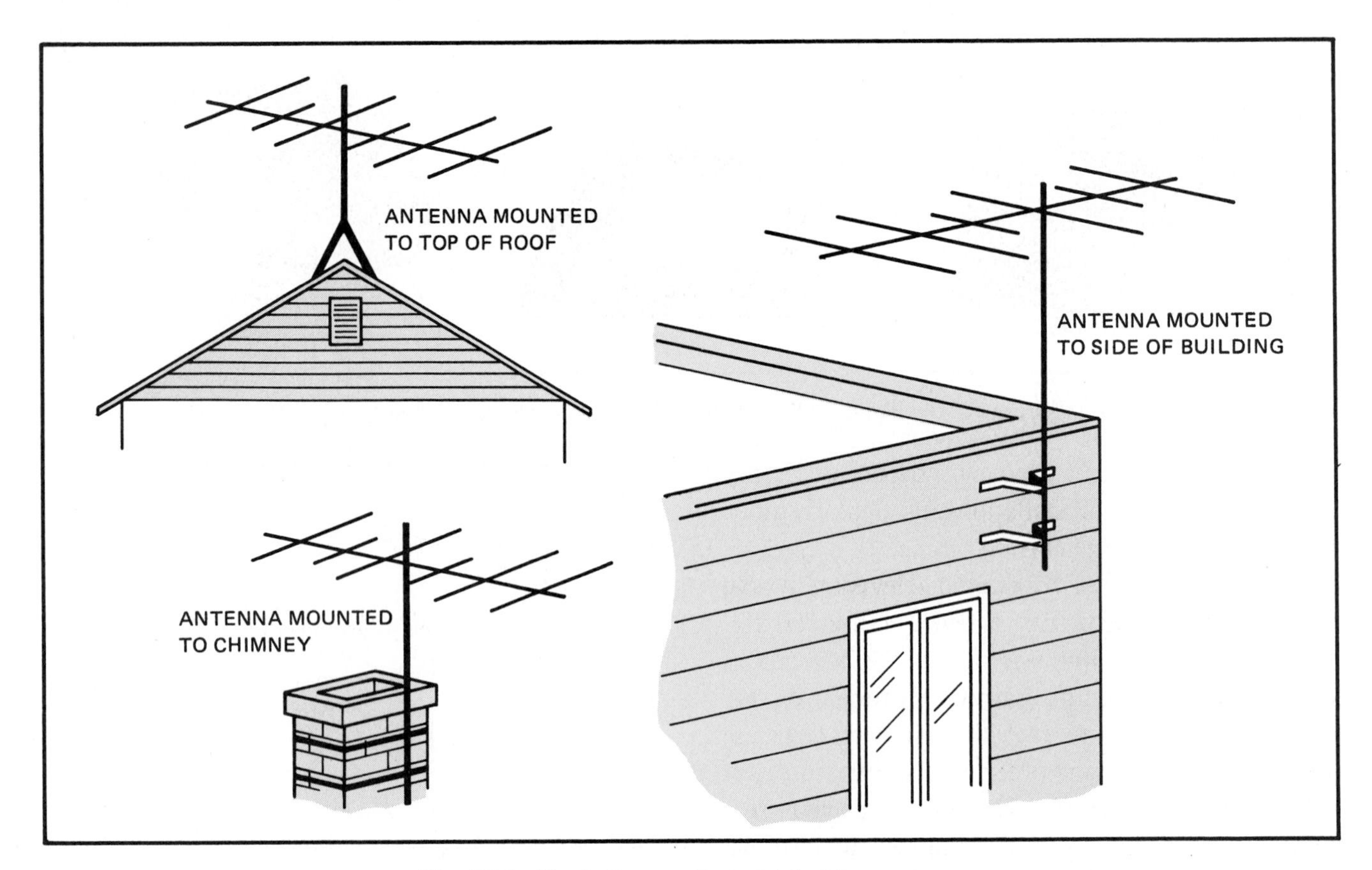

Fig. 25-4 Typical mounting of television antennas.

Fig. 25-5 Typical faceplates.

Code Rules for Cable Television (CATV) (*Article 820*)

Cable television is more correctly called Community Antenna Television, hence the term CATV.

CATV systems are installed both overhead and underground in a community. Then, to supply an individual customer, the CATV company runs coaxial cables through the wall of a residence at some convenient point. Up to this point of entry (*Article 820, Part B*) *and* inside the building (*Article 820, Part E*), the cable company must conform to the requirements of *Article 820* of the Code, plus local codes if applicable. Here in condensed form are some of the key rules.

1. The outer conductive shield of the coaxial cables must be grounded as close to the point of entry as possible, *Section 820-33*.
2. Coaxial cables shall not be run in the same conduits or box with electric light and power conductors, *Section 820-52(2)(b)*.
3. The grounding conductor (*Article 820, Part D*):
   a. Must be insulated.
   b. Must not be smaller than No. 14 copper or other corrosion-resistant conductive material.
   c. May be solid or stranded and must be of copper or other corrosion-resistant conductive material.

d. Must be run in a line as straight as practicable.

e. Shall be guarded from physical damage.

f. Shall be connected to the nearest accessible location on one of the following:

   the building or structure grounding electrode (*Section 250-81*)

   the grounded interior water pipes (*Section 250-80(a)*)

   the metallic power service raceway

   the service equipment enclosure

   the grounding electrode conductor or its metal enclosure

   the grounding conductor

   the grounding electrode of a building's disconnect when these are properly grounded.

g. If *none* of *f* is available, then ground to any of the electrodes per *Section 250-81*, such as metal underground water pipes, metal frame of building, concrete encased electrode, or ground ring.

h. If *none* of *f* or *g* is available, then ground to an effectively grounded metal structure or to any of the electrodes per *Section 250-83*, such as underground gas piping systems, driven rod or pipe, or a metal plate. If grounding to gas piping, make the connection between the gas meter and the street main.

i. *CAUTION:* If any of the preceding results in one grounding electrode for the CATV shield grounding and another grounding electrode for the electrical system, bond the two electrodes together with a bonding jumper no smaller than No. 6 copper or its equal. Grounding and bonding the coaxial cable shield to the same grounding electrode as the main service ground minimizes the possibility that a difference of potential might exist between the shield and other grounded objects, such as a water pipe.

SATELLITE ANTENNAS

A satellite antenna is often referred to as a "dish," figure 25-6. The satellite antenna has a parabolic shape that concentrates and reflects the signal beamed down from one of the many satellites located 22,279 miles above the equator. At the present time, there are 19 commercial satellites in orbit, separated by about 1,838 miles.

Satellite antennas are available in solid or "see-through" types. See-through satellite antennas are constructed of perforated metal. They offer less wind resistance and are lighter in weight than solid dishes.

A transmitter on earth beams an "uplink signal" (5.9 to 6.4 GHz) to the satellite in space. The electronic devices on the satellite reamplify and convert this signal to a "downlink signal" (3.7 to 4.2 GHz), then retransmit the downlink signal back to earth, figure 25-7.

The term *gigahertz* (GHz) means billions of cycles per second.

The term *megahertz* (MHz) means millions of cycles per second.

On earth, the "downlink signal" is picked up by

Fig. 25-6 Satellite antenna. Courtesy of Winegard Company.

the "feedhorn" located on the front of a satellite antenna, figure 25-6.

A "feedhorn" is located at the exact focal point in relationship to the parabolic reflector to gather the maximum amount of signal. The feedhorn funnels the downlink satellite signal into a "low noise amplifier" where the signal is amplified, then fed to a "downconverter" where the signals are converted from high frequency (3.7 to 4.2 GHz) to low frequency (70 MHz) that is proper for the receiver. All of these are mounted in a housing in front of the antenna.

The receiver is located inside the home and is connected to the "downconverter" on the dish with a 75-ohm coaxial cable. The receiver in turn is connected to the television set, figure 25-8.

More and more satellite programmers are scrambling their uplink signals. The home satellite dish owner needs a special descrambling device to be able to decode and view these scrambled channels. Most of the programmers' scrambled signals can be unscrambled by one descrambler.

Motor-driven dishes allow the antenna to be focused from one satellite to another. One manufacturer utilizes a reversible 36-volt dc motor.

Since all satellites in space are in the same orbit, figure 25-9, the controls on the receiver can signal the antenna to move to the "left" for Eastern satellites, or to the "right" for Western satellites.

Depending upon local codes, satellite antennas can be mounted on a post, on a pedestal, or on a roof.

The cables that are installed between the receiver inside the home and the satellite antenna consist of color-coded conductors, making it easy to identify when making the hook-up. The number of conductors necessary depends upon circuit designs of the receiver and the satellite. It is necessary to always follow the manufacturer's installation and wiring instructions.

For residential installations, the cables are generally run underground, either by direct burial or in conduit.

Figure 25-10 shows one method of installing a

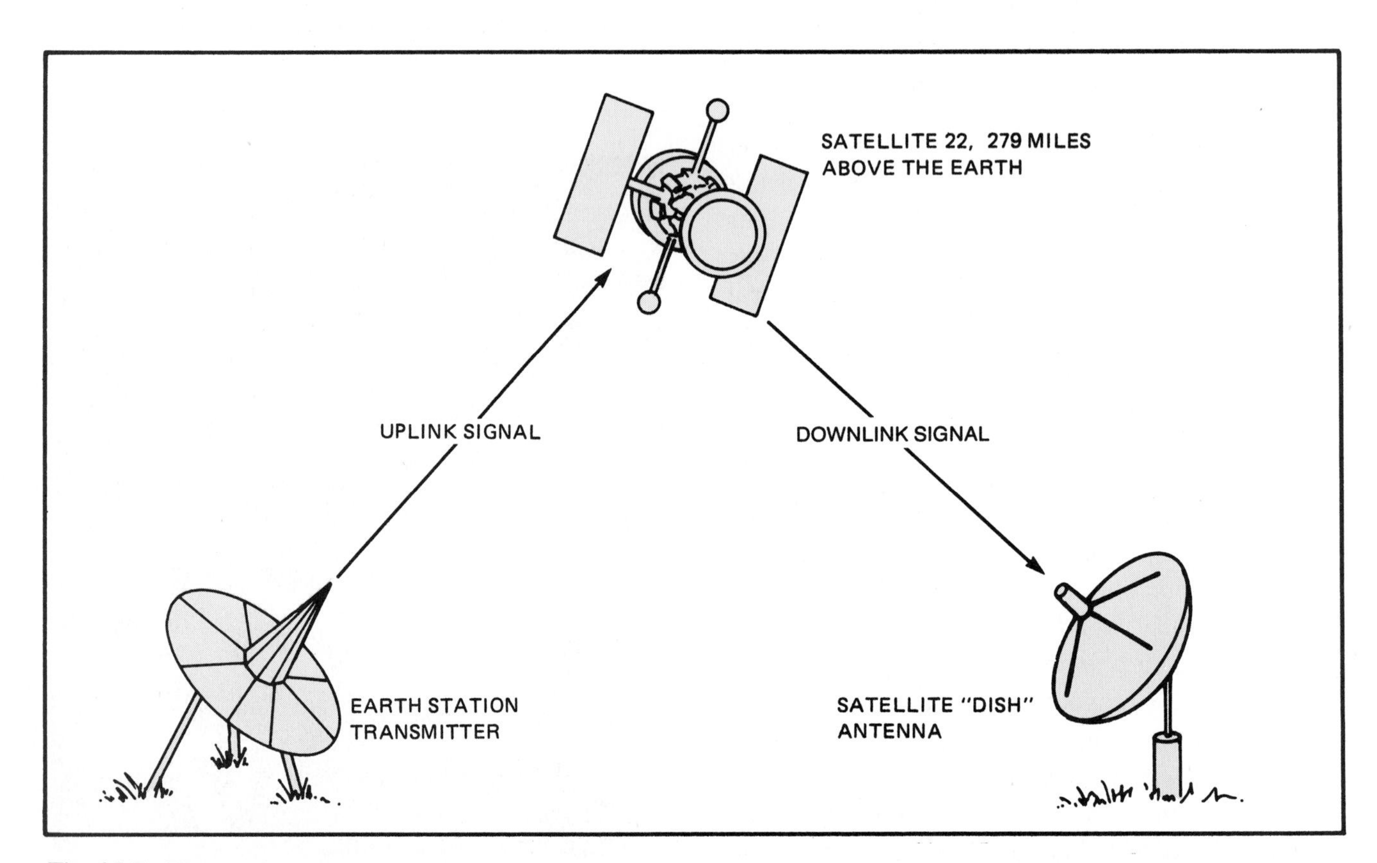

Fig. 25-7 The earth station transmitter beams the signal up to the satellite where the signal is amplified, converted, and beamed back to earth. The power requirements of the satellite are in the range of 5 watts and are provided by the use of solar cells.

satellite antenna to a metal post securely positioned in concrete.

The power consumption of the receiver is small — generally in the range of 150 watts. Since all of the power to the satellite antenna is fed from the receiver, it is necessary only to plug the receiver into a receptacle outlet, in the same manner as connecting a television set, radio, or video cassette recorder.

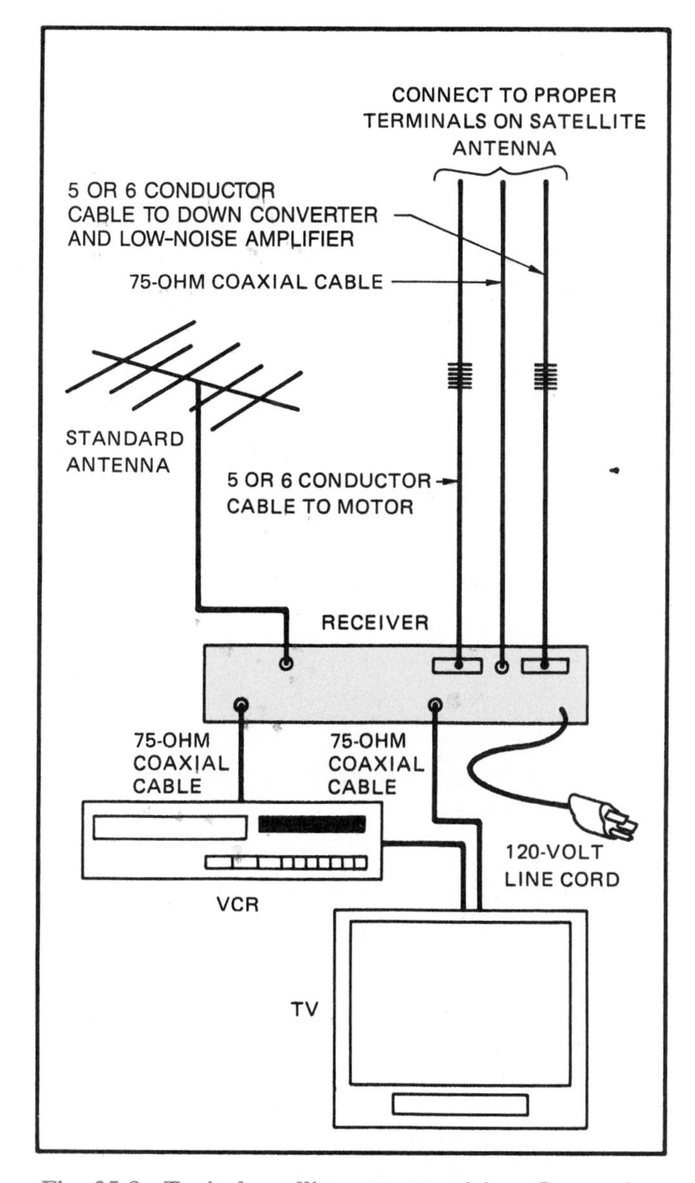

Fig. 25-8 Typical satellite antenna wiring. Connecting to a standard antenna, where this is possible, enables the homeowner to watch one channel while recording another channel — one signal from the satellite and one signal from the standard antenna. It also permits picking up local commercial programming with the receiver turned off.

CODE RULES FOR THE INSTALLATION OF ANTENNAS AND LEAD-IN WIRES

(Article 810)

Home television, video casette recorders (VCRs), compact disc equipment, and AM and FM radios generally come complete with built-in antennas. For those locations in outlying areas on the fringe or out of reach of strong signals, it is quite common to install a separate antenna system.

Either indoor or outdoor antennas may be used with black and white and color televisions. The front of an outdoor antenna is aimed at the television transmitting station. When there is more than one transmitting station and they are located in different directions, a rotor is installed. A rotor turns the antenna on its mast so that it can face in the direction of each transmitter. The rotor is controlled from inside the building. The rotor controller's cord is plugged into a regular 120-volt receptacle to obtain power. A four-wire cable is usually installed between the rotor motor and the control unit. The wiring for a rotor may be installed during the roughing-in stage of construction, running the rotor's four-wire cable into a device wall box, allowing 5 or 6 feet of extra cable, then installing a regular single-gang switch plate when finishing.

Article 810 of the Code covers this subject. Although instructions are supplied with antennas, the following key points of the Code regarding the installation of antennas and lead-in wires should be followed.

1. Antennas and lead-in wires shall be securely supported.
2. Antennas and lead-in wires shall not be attached to the electric service mast.
3. Antennas and lead-in wires shall be kept away from all light and power conductors to avoid accidental contact with the light and power conductors.
4. Antennas and lead-in wires shall not be attached to any poles that carry light and power wires over 250 volts between conductors.
5. Lead-in wires shall be securely attached to the antenna.

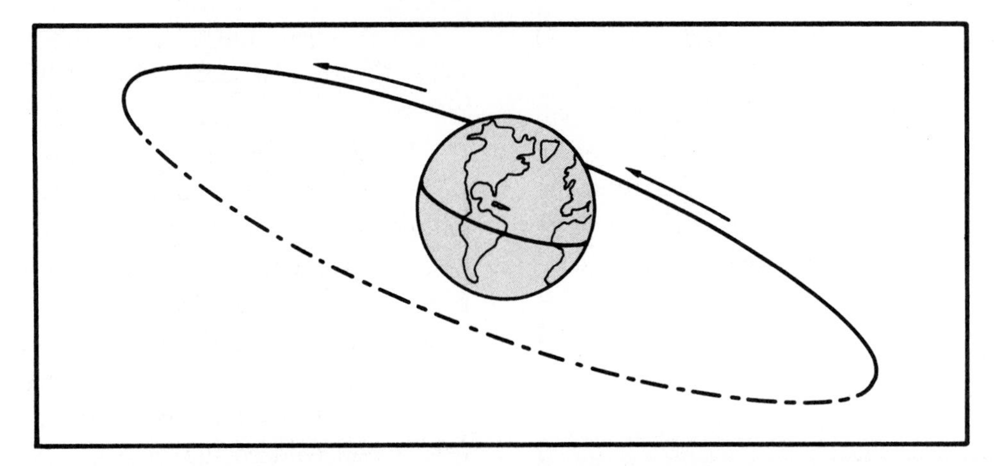

Fig. 25-9 All satellites travel around the earth in the same orbit. They appear to be stationary in space because they are rotating at the same speed at which the earth rotates. This is called "geosynchronous orbit." The satellite receives the uplink signal from earth, amplifies the signal, and transmits it back to earth. The downlink signal is picked up by the satellite antenna.

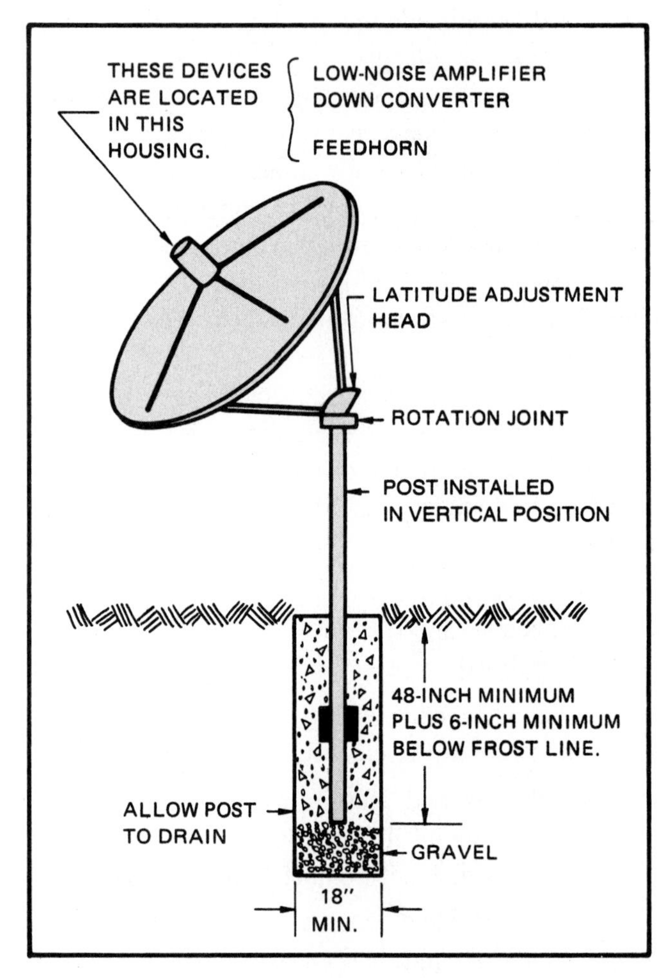

Fig. 25-10 Typical satellite antenna of the type installed in remote areas where reception is poor, or in areas served by CATV, but where the homeowner does not wish to subscribe and pay for the service of the CATV company. Always check with the local inspectors to be sure that all installation code rules are met.

6. Outdoor antennas and lead-in conductors shall not cross over light and power wires.
7. Lead-in conductors shall be kept at least 2 feet (610 mm) away from open light and power conductors.
8. Where practicable, antenna conductors shall not be run under open light and power conductors.
9. On the outside of a building:
   a. Position and fasten lead-in wires so they cannot swing closer than 2 feet (610 mm) to light and power wires having *NOT* over 250 volts between conductors; 10 feet (3.05 m) if *OVER* 250 volts between conductors.
   b. Keep lead-in conductors at least 6 feet (1.83 m) away from a lightning rod system, or bond together according to *Section 250-86.*
10. On the inside of a building:
    a. Keep antenna and lead-in wires at least 2 inches (51 mm) from other open wiring (as in old houses) unless the other wiring is in metal raceway or in metal cable armor.
    b. Keep lead-in wires out of electric boxes unless there is an effective, permanently installed barrier to separate the light and power wires from the lead-in wire.

11. Grounding:
    a. The grounding wire must be copper, aluminum, copper-clad steel, bronze, or similar corrosion-resistant material.
    b. The grounding wire need not be insulated. It must be securely fastened in place; may be attached directly to a surface without the need for insulating supports; shall be protected from physical damage or be large enough to compensate for lack of protection; and shall be run in as straight a line as is practicable.
    c. The grounding conductor shall be connected to the nearest accessible location on: the building or structure grounding electrode (*Section 250-81(a)*); or the grounded interior water pipe (*Section 250-80(a)*); or the metallic power service raceway; or the service equipment enclosure; or the grounding electrode conductor or its metal enclosure.
    d. If none of *c* is available, then ground to any one of the electrodes per *Section 250-81*, such as metal underground water pipe, metal frame of building, concrete-encased electrode, or ground ring.
    e. If neither *c* nor *d* are available, then ground to an effectively grounded metal structure or to any one of the electrodes per *Section 250-83*, such as underground gas piping system, driven rod or pipe, or a metal plate.
    f. The grounding conductor may be run inside or outside of the building.

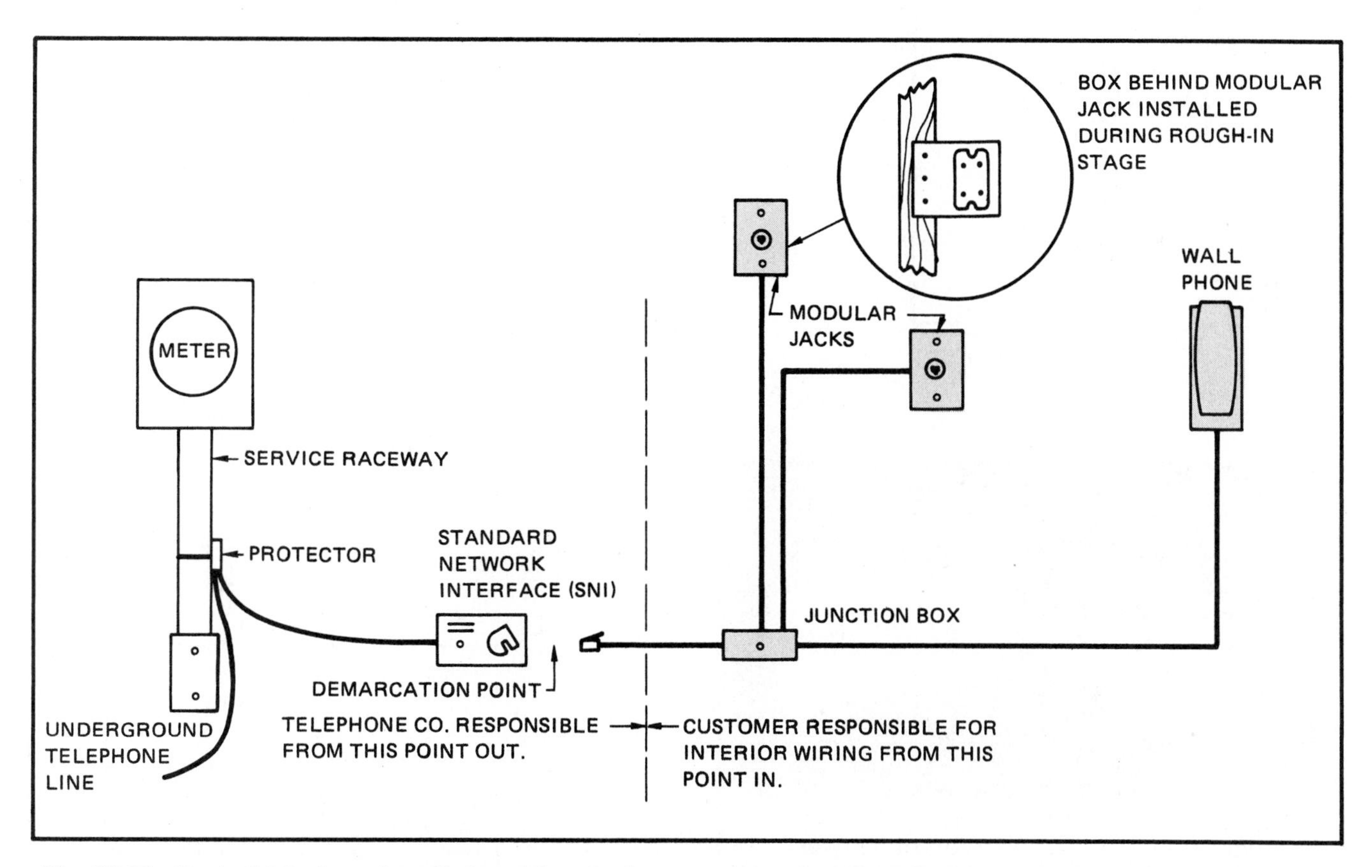

Fig. 25-11 Typical telephone installation. The telephone company installs and connects their underground cable to the protector. In this illustration, the protector is mounted to the service raceway. This establishes the ground that offers protection against hazardous voltages such as a lightning surge on the incoming line. A cable is then run from the protector to a Standard Network Interface (SNI). The electrician or homeowner plugs the interior wiring into the SNI by means of the modular plug. The term *modular* means that most of the phones are plugged-in as opposed to "hard-wired" connections.

g. The grounding conductor shall not be smaller than No. 10 copper or No. 8 aluminum.

h. *CAUTION:* If any of the preceding rules results in one grounding electrode for the antenna and another grounding electrode for the electrical system, bond the two electrodes together with a bonding jumper not smaller than No. 6 copper or its equal.

The objective of grounding and bonding to the same grounding electrode as the main service ground is to reduce the possibility of having a difference of potential between the two systems.

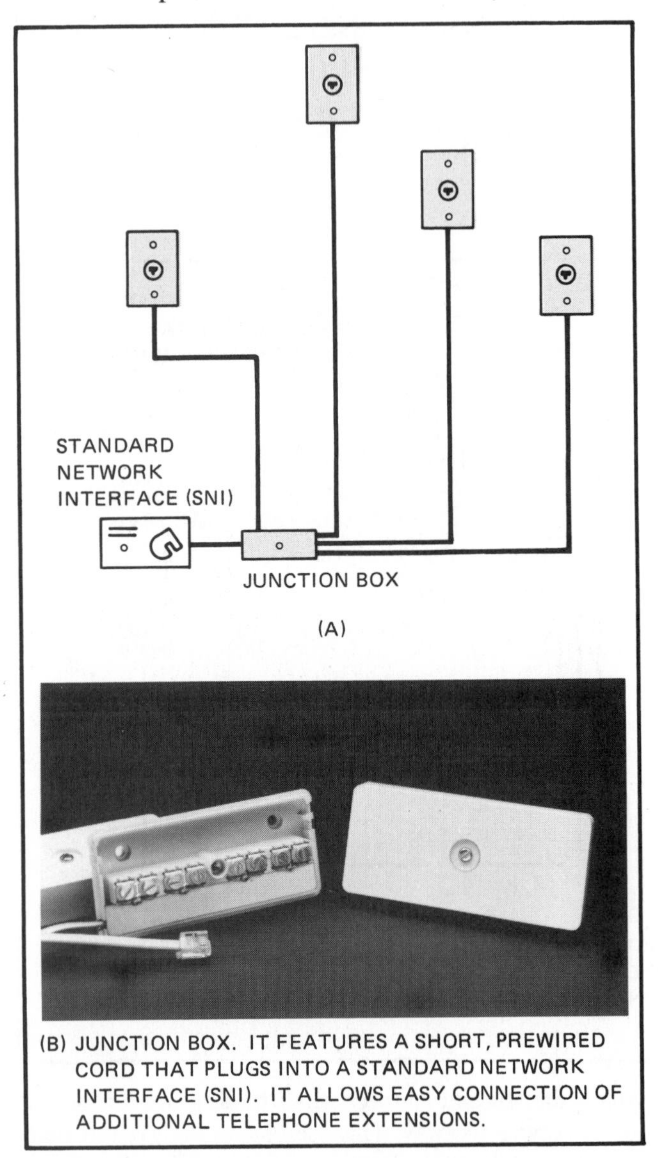

(B) JUNCTION BOX. IT FEATURES A SHORT, PREWIRED CORD THAT PLUGS INTO A STANDARD NETWORK INTERFACE (SNI). IT ALLOWS EASY CONNECTION OF ADDITIONAL TELEPHONE EXTENSIONS.

Fig. 25-12 Individual telephone cables run to each telephone outlet from the telephone company's connection point. Courtesy of KEPtel, Armiger.

TELEPHONE WIRING (Symbol ▶|) (*Article 800*)

Since the deregulation of telephone companies, residential DO-IT-YOURSELF telephone wiring has become quite common. Following is an overview of residential telephone wiring.

The telephone company will install the service line to a residence and terminate at a protector device. The protector protects the system from hazardous voltages. The protector may be mounted either outside or inside the home. Different telephone companies have different rules. Always check with the phone company before starting your installation.

The point where the telephone company ends and the homeowner's interior wiring begins is called the *demarcation point*.

The telephone wiring inside the home may be done by the telephone company, or it may be installed by the homeowner or the electrician. Whatever the case, the rules in *Article 800* of the *National Electrical Code®* apply. Refer to figures 25-11 (page 355), 25-12, and 25-13.

Article 800 of the Code covers telephone wiring as well as fire alarms and burglar alarms.

For the typical residence, the electrician or homeowner will "rough in" the boxes wherever a

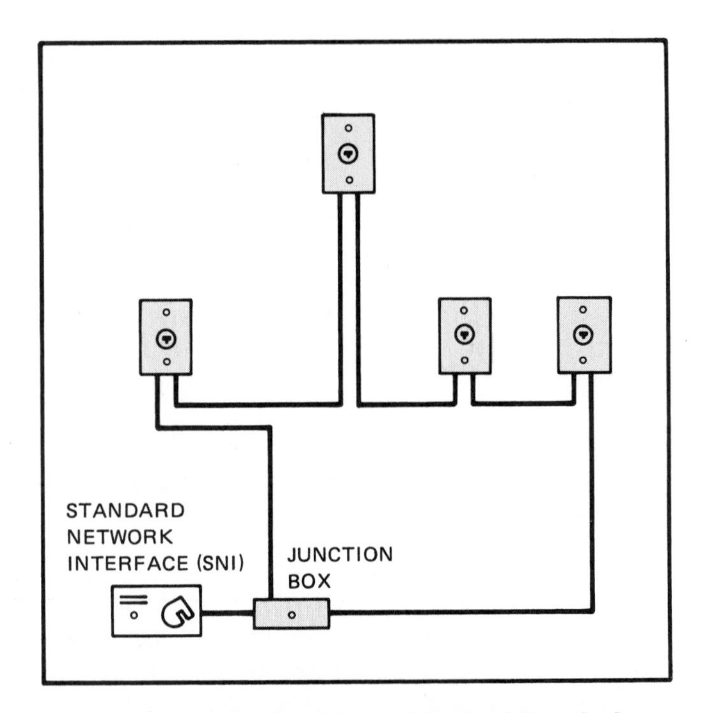

Fig. 25-13 A complete "loop system" of the telephone cable. If something happens to one section of the cable, the circuit can be fed from the other direction.

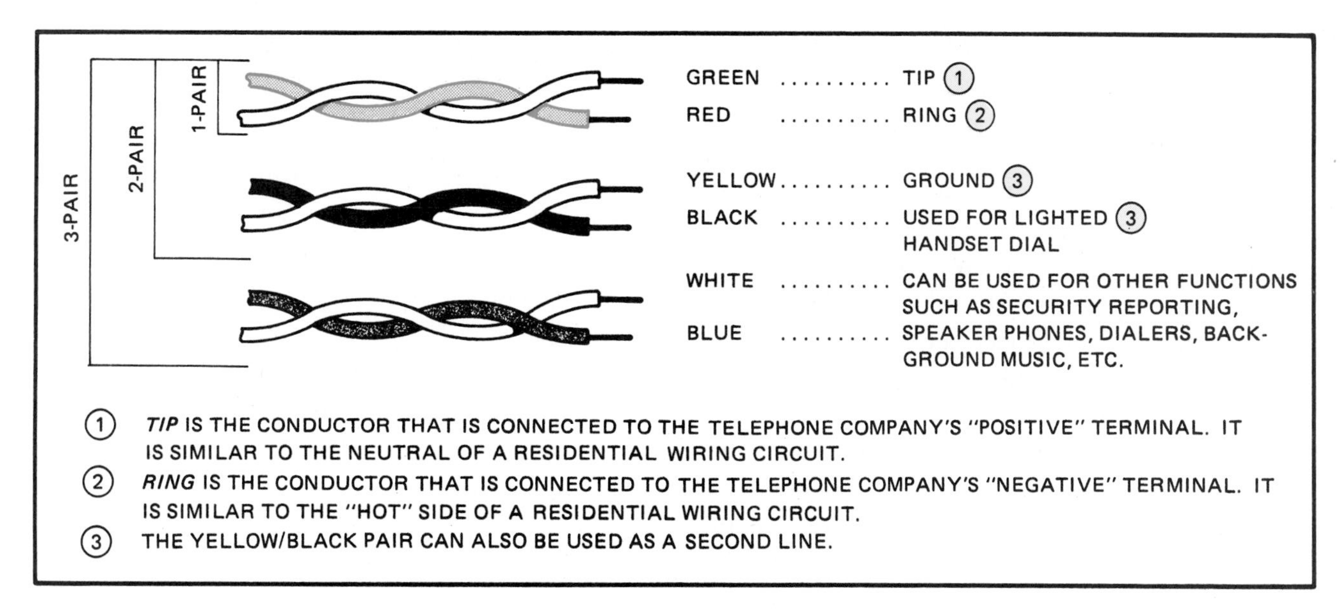

Fig. 25-14 Color coding of telephone cables.

telephone outlet is wanted. The most common interior telephone wiring for a new home is with 4-conductor No. 22 or 24 cable. The outer jacket is usually of a thermoplastic material. This outer jacket most often is a neutral color such as light gray or beige that blends in with decorator colors in the home. This is particularly important in existing homes where the telephone cable may have to be exposed. Cables, mounting boxes, junction boxes, terminal blocks, jacks, adaptors, faceplates, cords, hardware, plugs, etc., are all available through electrical distributors, builders' supply outlets, telephone stores, electronic stores, hardware stores, and similar wholesale and retail outlets. The selection of types of the preceding components is endless.

Telephone Conductors

The color coding of telephone cables is shown in figure 25-14. Figure 25-15 shows some of the many types of telephone cable available.

Cables are available in the following sizes:

2 pair ----- (4 conductors)

3 pair ----- (6 conductors)

6 pair ----- (12 conductors)

12 pair ----- (24 conductors)

25 pair ----- (50 conductors)

50 pair ----- (100 conductors)

Telephone circuits require a separate pair of conductors from the telephone all the way back to the phone company's central switching center. To keep interference that could come from other electrical equipment to a minimum, such as electric motors and fluorescent fixtures, each pair of telephone wires (2-wires) is twisted.

Fig. 25-15 Some of the many types of telephone cable available are illustrated. Note that for ease of installation, the modular plugs and terminals have been attached by the manufacturer of the cables. These cords are available in round and flat configurations, depending upon the number of conductors in the cable.

The recommended circuit lengths are:

No. 24 gauge ----- not over 200 feet

No. 22 gauge ----- not over 250 feet

Many telephone companies suggest that not more than five telephones be connected to one line because there may not be enough power to ring more than five ringers. Most newer telephones indicate on the label or on the carton the telephone REN (Ringer Equivalence Number). The REN refers to the older style electromechanical ringer which is considered to be 1-REN. Electronic ringers use very little power; therefore, more than five phones could be installed on one line. The electrician should check and verify this with the telephone company in his area.

EXAMPLE:

Five 1-REN phones = 5 RENS

Three 1-REN and one 2-REN phone = 5 RENS

Ten 1/2-REN phones = 5 RENS

If you are considering installing more than five telephones in a home, pay attention to the REN numbers.

When several telephones are used in a residence, most telephone companies recommend that at least one of these telephones be permanently installed. This would probably be the wall phone in the kitchen. The remaining phones may remain portable. This means that they can be plugged into any of the phone jacks, figures 25-16 and 25-17. If all of the phones were portable, there might be a time when none of the phones were connected to phone jacks. As a result there would be no audible signal.

The electrical plans show that nine telephone outlets are provided as follows:

Front Bedroom	1
Kitchen/Nook	1
Laundry	1
Living Room	1
Master Bedroom	1
Rear Outdoor Patio	1
Recreation Room	1
Study/Bedroom	1
Workshop	1

Installation of Telephone Conductors, *Sections 800-51* and *800-52*

Telephone conductors:

1. shall be Type CM or Type CMX (for dwellings only), listed for telephone installation as being resistant to the spread of fire.
2. shall be separated by at least 2 inches (51 mm) from light and power conductors unless the light and power conductors are in a raceway, or in nonmetallic-sheathed cable, Type AC cable, or Type UF cable.
3. shall not be placed in any conduit or boxes with electric light and power conductors unless the conductors are separated by a partition.

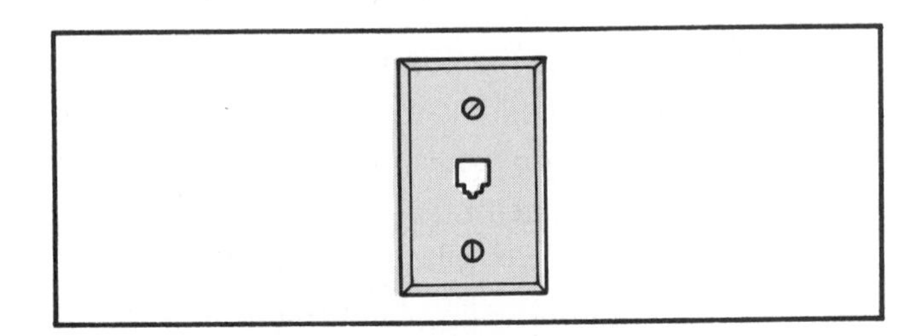

Fig. 25-16 Telephone modular jack.

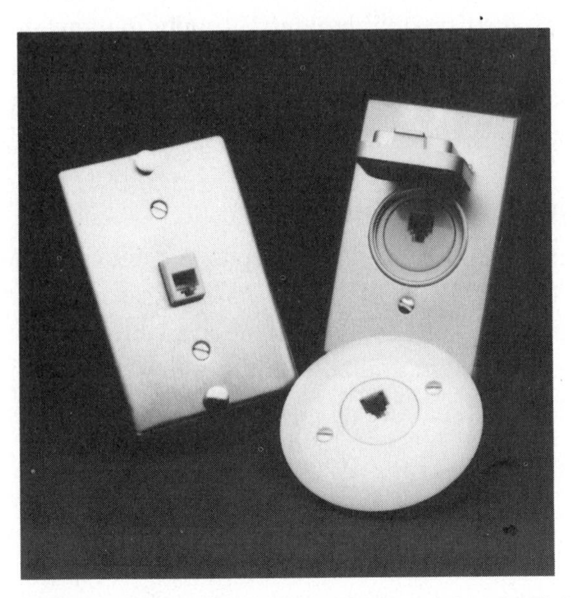

Fig. 25-17 Three styles of wall plates for modular telephone jacks: rectangular stainless steel, weatherproof for outdoor use, and circular. Courtesy of KEPtel, Armiger.

4. may be terminated in either metallic or nonmetallic boxes. Check with the local telephone company or local electrical inspector for specific requirements.
5. should not share the same stud space as the electrical branch-circuit wiring.*
6. should be kept at least 12 inches away from electrical branch-circuit wiring where the telephone cable is run parallel to the branch-circuit wiring. This guards against induction noise.*
7. should not share the same bored holes as electrical wiring, plumbing, or gas pipes.
8. should be kept away from hot water pipes, hot air ducts, and other heat sources that might harm the insulation.*
9. should be attached to the sides of joists and studs with insulated staples, being careful not to crush the cable.*

*These are recommendations from residential telephone manufacturers' installation data.

Conduit

In some communities electricians prefer to install regular device boxes at a telephone outlet location, then "stub" a ½-inch conduit to the basement or attic from this box. Then, at a later date, the telephone cable can be "fished" through the conduit.

Grounding (*Article 800, Part D*)

The telephone company will provide the proper grounding of their incoming cable sheath and protector, generally using wire not smaller than No. 14 AWG copper or equal. Or, they might clamp the protector to the grounded metal service raceway conduit to establish "ground." The grounding requirements are practically the same as for CATV and antennas, which were covered earlier in this unit.

Safety

Open circuit voltage between conductors of an idle pair of telephone conductors can range from 50 to 60 volts dc. The ringing voltage can reach 90 volts ac. Therefore, always work carefully with insulated tools and stay clear of bare terminals and grounded surfaces. Disconnect the interior telephone wiring if work must be done on the circuit, or take the phone off the hook, in which case the dc voltage level will drop and there should be no ac ringing voltage delivered.

SIGNAL SYSTEM (CHIMES)

A signal circuit is described in the *National Electrical Code®* as any electrical circuit that energizes signaling equipment, or one which is used to transmit certain types of signaling messages. Signaling equipment includes such devices as chimes, doorbells, buzzers, code-calling systems, and signal lights.

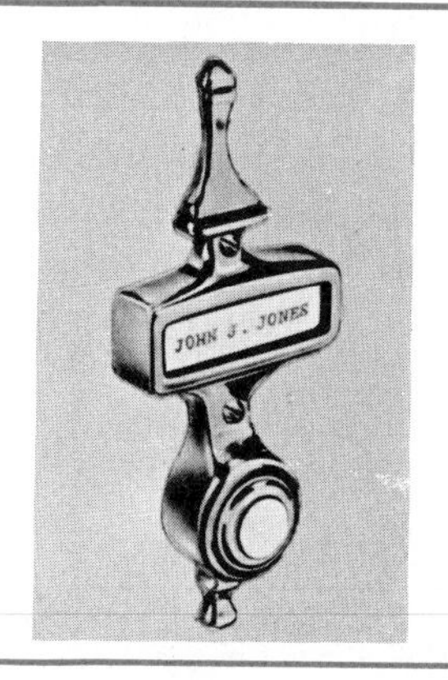

Fig. 25-18 Push buttons for door chimes.

Fig. 25-19 Typical residential door chimes.

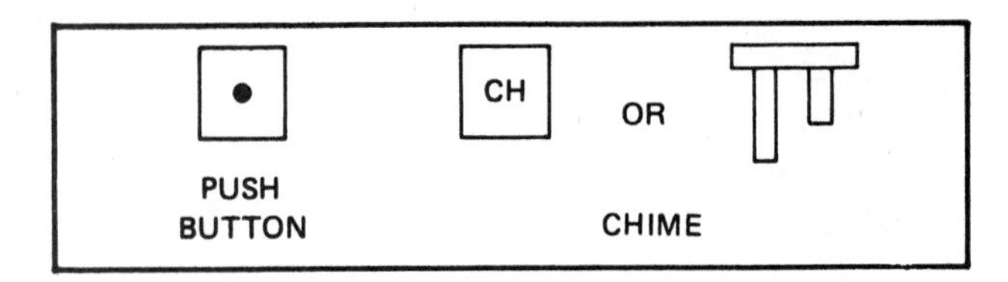

Fig. 25-20 Symbols for chimes and push buttons.

Door Chimes (Symbol CH)

Present-day dwellings often use chimes rather than bells or buzzers to announce that someone is at a door. A musical tone is sounded rather than a harsh ringing or buzzing sound. Chimes are available in single-note, two-note, eight-note (four-tube), and repeater tone styles. In a repeater tone chime, both notes sound as long as the push button is depressed. In an eight-note chime, contacts on a motor-driven cam are arranged in sequence to sound the notes of a simple melody when the chime button is pushed. This type of chime is usually installed in dwellings having three entrances. The chime can be connected so that the eight-note melody sounds for the front door, two notes sound for the side door, and a single note sounds for the rear door. Chimes are also available with clocks and lights.

Electronic chimes may relay their chime tones through the various speakers of an intercom system. When any chime is installed, the manufacturer's instructions must be followed.

The plans show that two chimes are installed in the residence. Two-note chimes are used. Each chime has two solenoids and two iron plungers. When one solenoid is energized, the iron plunger is drawn into the opening of the solenoid. A plastic peg in the end of the plunger strikes one chime tone bar. When the solenoid is deenergized, spring action returns the plunger where it comes to rest against a soft felt pad so that it does not strike the other chime tone bar. Thus, a single chime tone sounds. As the second solenoid is energized, one chime tone bar is struck. When the second solenoid is deenergized, the plunger returns and strikes the second tone bar. A two-tone signal is produced. The plunger then comes to rest between the two tone bars. Generally, two notes indicate front door signaling, and one note indicates rear or side door signaling.

Figure 25-18 (page 359) shows four push-button styles used for chimes. Many other styles are available. Figure 25-19 shows several wall-mounted cover plates for chime units. The symbols used to indicate push buttons and audible signals on the plans are shown in figure 25-20.

Figures 25-21 and 25-22 show one way in which to provide proper backing for chimes. This is obviously done during the rough-in stages of the electrical installation before the walls are closed up.

Transformers

Because of the low-voltage ratings and power limitations of chime transformers, the secondary wiring and the transformer itself are classified as Class 2 circuits.

The transformers used to operate door chimes have a greater capacity than the transformers used with bell and buzzer circuits, figure 25-23. The voltage output of chime transformers ranges between 10 and 24 volts. These transformers are rated from 5 to 30 volt-amperes (watts). Bell transformers have a voltage output range of 6 to 10 volts and a rating of 5 to 20 volt-amperes (watts).

Chime transformers used in dwellings are available with a 16-volt rating. Transformers which give a combination of voltages, such as 4, 8, 12, and 24 volts, also are used.

Underwriters Laboratories lists two types of chime transformers that are normally used for chime wiring. The *inherently limited transformer* is designed to limit the short-circuit current to a maximum of 8 amperes. The *not inherently limited transformer* is rated at 100 volt-amperes or less. This transformer has an overcurrent protective device that limits the voltage and current to the values specified in *Table 725-31*. The open circuit voltage limitation of both types of transformers must not exceed 30 volts. Most transformers suitable for use in dwellings have a built-in thermal overload device. Whenever a short circuit occurs in the bell-wire circuit, the overload device opens and closes repeatedly until the short is cleared.

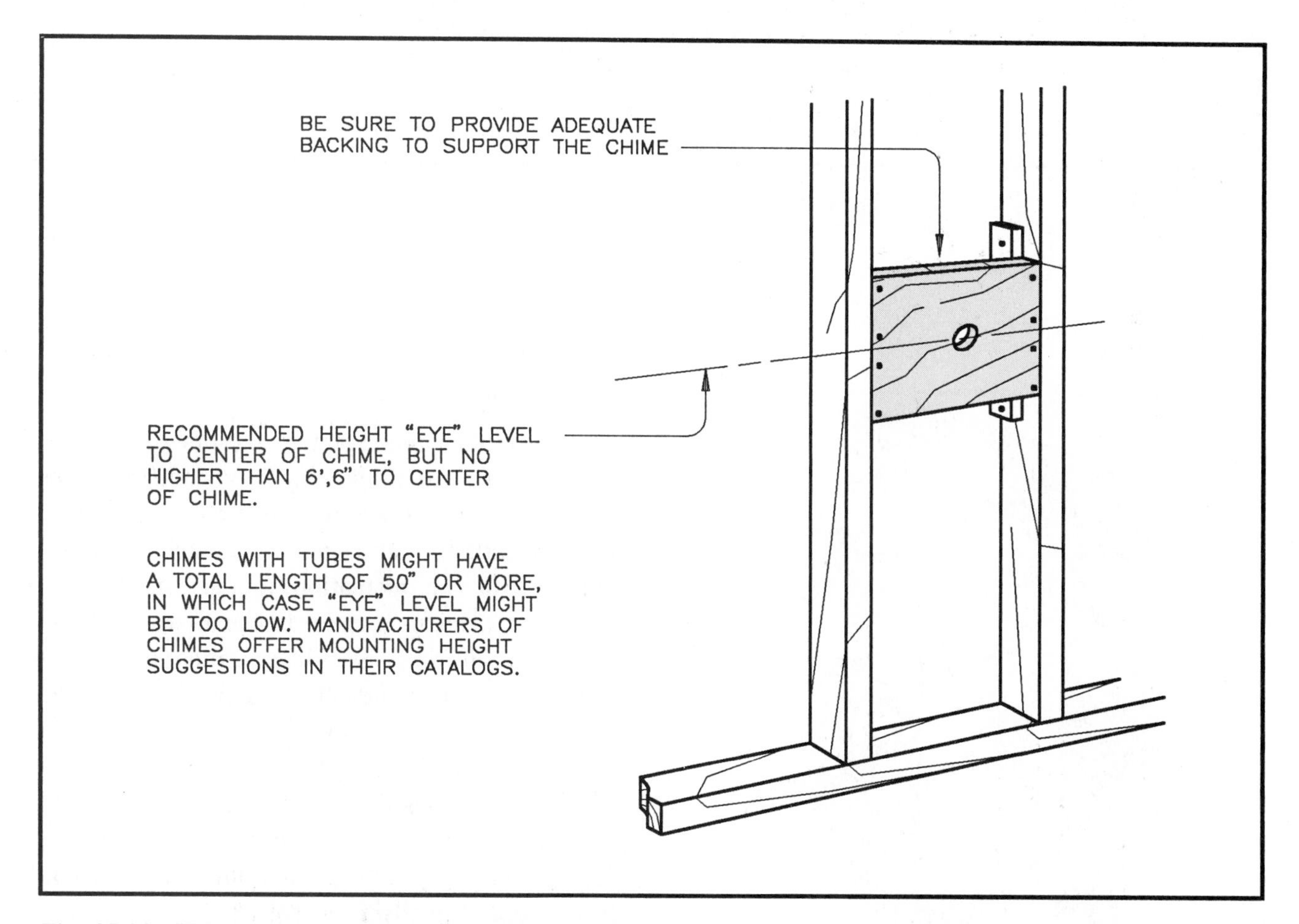

Fig. 25-21 This illustration shows one method of providing support for a surface-mounted chime. This is the rough-in stage, before the walls are closed.

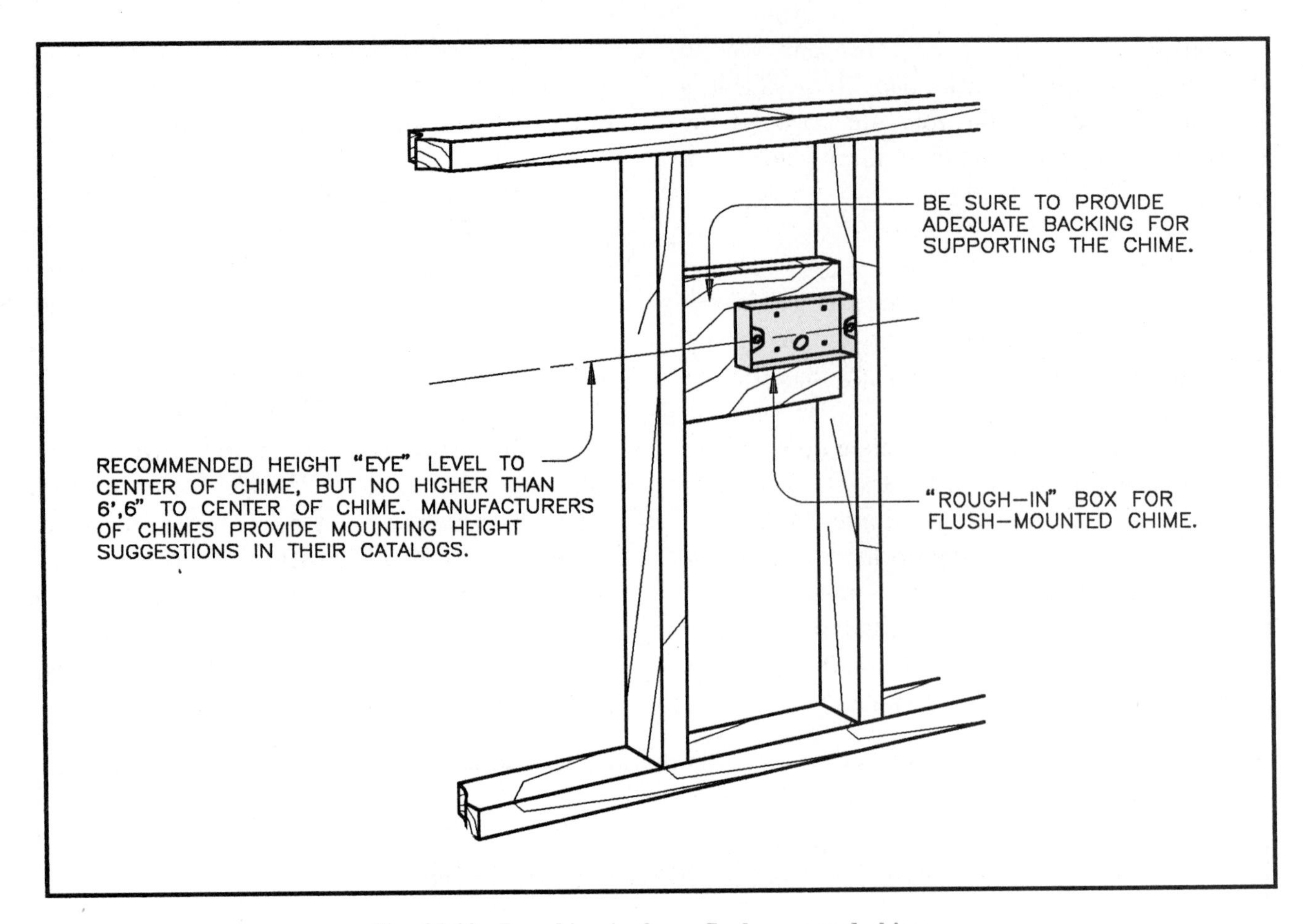

Fig. 25-22 Roughing-in for a flush-mounted chime.

Fig. 25-23 Chime transformer.

Additional Chimes

To extend a chime system to cover a larger area, a second or third chime may be added. In the residence, two chimes are used: one chime is mounted in the front hall and a second (extension) chime is mounted in the recreation room. The extension chime is wired in parallel to the first chime. The wires are run from one chime terminal board to the terminal board of the other chime. The terminals are connected as follows: transformer to transformer, front to front, and rear to rear, figure 25-24.

When chimes are added to a chime circuit, a transformer with a higher wattage rating may be required to energize the greater number of solenoids being used at one time. Also, more wire is used in the circuit and the voltage drop and power loss are greater. This increase is the result of resistance in the wire between the chimes and the transformer.

Section 725-32 does not allow transformers supplying Class 2 systems to be connected in parallel unless listed for interconnection.

If a buzzer (or bell) and a chime are connected to a single transformer and are used at the same time, the transformer will put out a fluctuating voltage. This condition does not allow either the buzzer or the chime to operate properly. The use of a transformer with a larger rating may solve this problem.

The wattage consumption of chimes varies with the manufacturer. Typical ratings are as follows:

TYPE OF CHIME	POWER CONSUMPTION
Standard two-note	10 watts
Repeating chime	10 watts
Internally lighted, two lamps	10 watts
Internally lighted, four lamps	15 watts
Combination chime and clock	15 watts
Motor-driven chime	15 watts
Electronic chime	15 watts

Transformers with ratings of 5, 10, 15, 20, and 30 watts (VA) are available. For a multiple chime installation, wattage ratings for the individual chimes are added. The total value is the minimum transformer rating needed to do the job properly. If there are still technical questions or problems, check the literature supplied by the chime and transformer manufacturers.

Wiring for Chime Installation

Bell Wire and Cable: The wire used for low-voltage bell and chime circuits, and for connecting low-voltage thermostats, is called *bell wire*, *annunciator wire*, or *thermostat wire*.

These wires are generally made of copper and have a thermoplastic insulation suitable for use on 30 volts or less.

Since the current required for bell and chime circuits is rather small, No. 18 AWG conductors are quite often used. If the length of the circuit is very long, a larger conductor is suggested.

Multiconductor cables consist of two, three, or more single wires covered with a single protective insulation. This type of cable is often used for electrical installations because there is less danger of damage to individual wires and it gives a neat appearance to the wiring. The conductors within the cables are color coded to make circuit identification easy.

Bell wire and cable may be fastened directly to surfaces with insulated staples or cleats, or may be installed in the raceways. Be careful not to pierce or crush the wires. The job requirements will determine how the conductors are to be attached. In the residence in the plans, the bell wire is run along the sides of the floor joists in the basement and on the sides of the studs in the walls. The installation of low-voltage wiring is covered in unit 28.

Wiring Circuit: The circuit shown in figure 25-24 is recommended for this chime installation because it provides a hot low-voltage circuit at the front hall

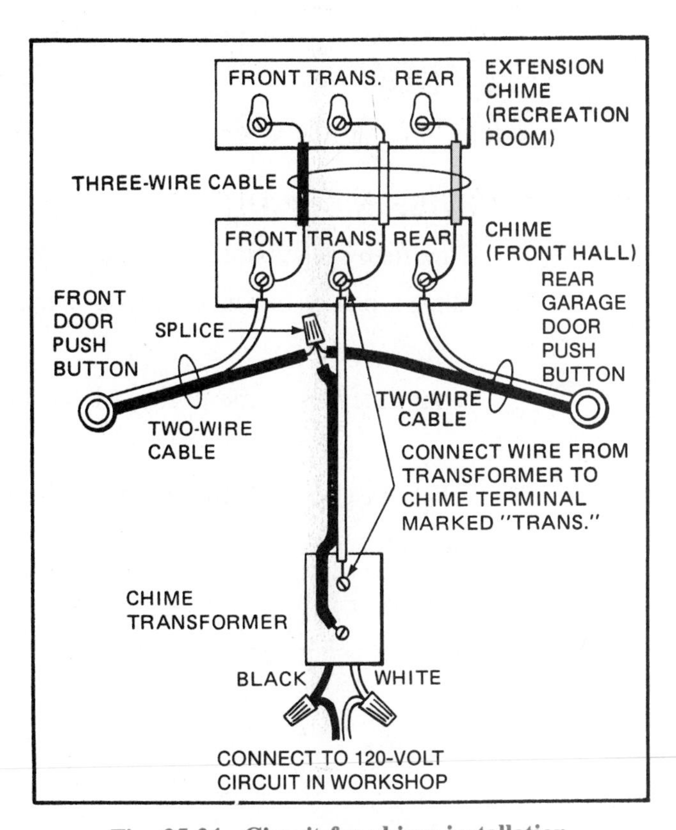

Fig. 25-24 Circuit for chime installation.

location. However, it is not the only way in which these chimes may be connected.

Figure 25-24 shows that a two-wire cable runs from the transformer in the workshop to the front hall chime. A two-wire cable then runs from the chime to both the front and rear door push buttons. A three-wire cable also runs between the front hall chime and the recreation room chime. Because of the hot, low-voltage circuit at the front hall location, a chime with a built-in clock can be used. A four-conductor cable may be run to the extension chime so that the owner may install a clock-chime at this location also.

The plans show that the chime transformer is mounted on one of the ceiling boxes in the utility room. The line voltage connections are easy to make here. Some electricians prefer to mount the chime transformer on the top or side of the distribution panel or load center. The electrician decides where to mount the transformer after considering the factors of convenience, economy, and good wiring practice.

National Electrical Code® (*Article 725*) Rules for Signal Systems

Bell wire with low-voltage insulation must not be installed in the same enclosure or raceway with light or power conductors. Bell wires must be not closer than 2 inches (50.8 mm) to open light or power conductors unless the bell wires are permanently separated from the other conductors by some approved insulation used in addition to the insulation on the wire. Such insulation is provided by the outer jacket on nonmetallic-sheathed cable, UF cable, and armored cable. Furthermore, bell wire with low-voltage insulation may not enter an outlet box or switch box containing light or power conductors unless a metal barrier is used to separate the two types of wiring.

National Electrical Code® requirements for Class 2, low-voltage bell circuits are contained in *Article 725*, *Sections 725-31* through *725-43*.

Remember, low-voltage Class 2 circuits are not permitted to be run in the same raceway as power-circuit conductors.

REVIEW

Note: Refer to the Code or the plans where necessary.

TELEVISION CIRCUIT

1. How many television outlets are installed in this residence? ____________________
2. What types of boxes are recommended when nonshielded lead-in wire is used? __
   ____
3. What determines the design of the faceplates used? ____________________
   ____
4. What must be provided when installing a television outlet and receptacle outlet in one wall box? ____________________
   ____
   ____
   ____
   ____
   ____

5. From a cost standpoint, which system is more economical to install: a master amplifier distribution system or a multiset coupler? Explain the basic differences between these two systems. ______

6. How many wires are in the cable used between a rotor and its controller? ______
7. What precautions should be noted when installing a television antenna and its lead-in wires? ______

8. List the requirements for cable television inside the house. ______

9. What article of the Code references the requirements for cable television antenna installation? ______
10. It is generally understood that grounding and bonding together all metal parts of an electrical system and the metal shield of the cable television cable to the same grounding reference point in a residence will keep both systems at the same voltage level should a surge, such as lightning, occur. Therefore, if the incoming cable to a house that has been installed by the CATV cable company installer has the metal shield grounded to a driven ground rod, does this installation conform to the National Electrical Code? ______
11. The basics of cable television are that a transmitter on earth sends an ______ signal to a ______ where the signal is reamplified and sent back to earth via the ______ signal. This signal is picked up by the ______ on the satellite antenna, which is sometimes referred to as a ______. The feedhorn funnels the signal into a ______ ______ ______, then on to a ______, where the high-frequency signal is converted to a ______-frequency signal suitable for the television receiver.

12. All television satellites rotate above the earth in (the same orbit) (different orbits). Circle correct answer.

13. Television satellites are set in orbit (10,000) (18,000) (22,279) miles above the earth which results in their rotating around the earth at (precisely the same) (different) rotational speed as the earth rotates. This is done so that the satellite "dish" can be focused on a specific satellite (once) (one time each month) (whenever the television set is used). Circle correct answer.

TELEPHONE SYSTEM

1. How many locations are provided for telephones in the residence? ____________

2. At what height are the telephone outlets mounted? ____________________________

____

____

____

____

____

3. Sketch the symbol for a telephone outlet.

4. Is the telephone system regulated by the *National Electrical Code*®? __________

5. a. Who is to furnish the outlet boxes required at each telephone outlet? __________

   b. Who is to furnish the faceplates? _______________________________________

6. Who is to furnish the telephones? ____

7. Who does the actual installation of the telephone equipment? ________________

8. How are the telephone cables concealed in this residence? ___________________

____

____

____

9. The point where the telephone company's cable enters the residence and the interior telephone cable wiring meet is called the ______________________________ point.

10. What are the colors contained in a four-conductor telephone cable assembly and what are they used for? ____

____

____

____

____

____

11. Itemize the Code rules for the installation of telephone cables in a residence. ______

12. If finger contact were made between the red conductor and green conductor at the instant a "ring" occurs, what shock voltage would be felt? ______

SIGNAL SYSTEM

1. What is a signal circuit? ______

2. What style of chime is used in this residence? ______

3. a. How many solenoids are contained in a two-tone chime? ______

   b. What closes the circuit to the solenoid of a chime? ______

4. Explain briefly how two notes are sounded by depressing one push button (when two solenoids are provided). ______

5. a. Sketch the symbol for a push button. ______

   b. Sketch the symbol for a chime. ______

6. a. At what voltage do residence chimes generally operate? ______

   b. How is this voltage obtained? ______

7. What is the maximum volt-ampere rating of transformers supplying Class 2 systems? ______

8. What two types of chime transformers for Class 2 systems are listed by Underwriters laboratories? ______

9. Is the extension chime connected with the front hall chime in series or in parallel? ______

10. How many bell wires terminate at

    a. the transformer? ______

    b. the front hall chime? ______

    c. the extension chime? ______

    d. each push button? ______

11. a. What change in equipment may be necessary when more than one chime is connected to sound at the same time on one circuit? ______________________________

    b. Why? ______________________________

12. What type of insulation is usually found on low-voltage wires? ______________

13. What size wire is installed for signal systems of the type in this residence? ________

14. a. How many wires are run between the front hall chime and the extension chime in the recreation room? ______________________________

    b. How many wires are required to provide a hot low-voltage circuit at the extension chime? ______________________________

15. Why is it recommended that the low-voltage secondary of the transformer be run to the front hall chime location and separate two-wire cables be installed to each push button? ______________________________

16. a. Is it permissible to install low-voltage Class 2 systems in the same raceway or enclosure with light and proper wiring? ______________________________

    b. What Code section covers this? ______________________________

17. a. Where is the transformer in the residence mounted? ______________

    b. To what circuit is the transformer connected? ______________

18. a. How many feet of two-conductor bell wire cable are required? ______________

    b. How many feet of three-conductor bell wire cable are required? ______________

19. How many insulated staples are needed for the bell wire if it is stapled every 2 feet (610 mm)? ______________________________

- DO install a smoke detector on a basement ceiling in close proximity to the stairway to the first floor.
- DO install a detector on a sloped ceiling that has a rise of greater than 1 foot per 8 feet horizontally, at the high end of the room.
- DO place a detector within 3 feet of the peak on a sloped ceiling that has greater than 1 foot rise per 8 feet horizontally.
- DO install smoke and heat detectors at the *top* of an *open* stairway, because heat and smoke travel up.
- DO install in *new* construction, hard-wired detectors directly connected to a 120V ac source. For existing homes, battery-powered units may be installed or hard-wired; 120V ac units may be used.

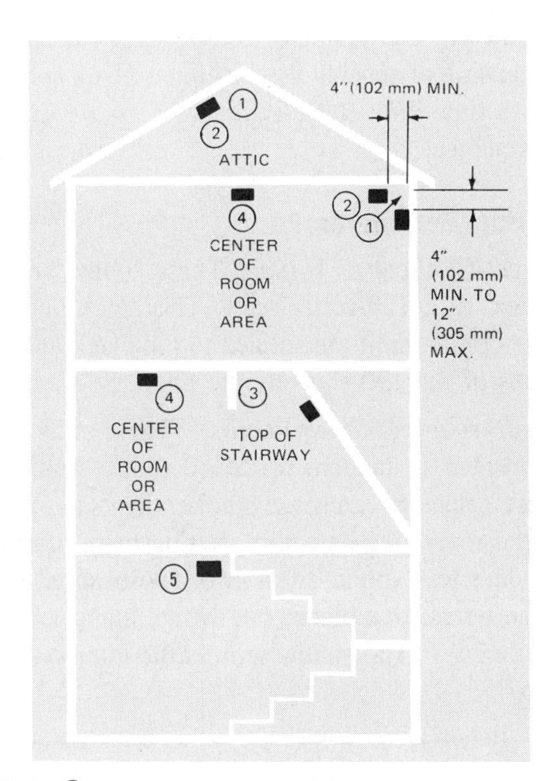

NOTES:
1. Do *not* install detectors in *dead* airspaces.
2. Mount detectors on the bottom edge of joists or beams. The space between these joists and beams is considered to be *dead* airspace.
3. Do *not* mount detectors in *dead* airspace at the top of a stairway if there is a door at the top of the stairway that can be closed. Detectors *should* be mounted at the top of an open stairway because heat and smoke travel upward.
4. Mount detectors in the center of a room or area.
5. Basement smoke detectors must be located in close proximity to the stairway leading to the floor above.

Fig. 26-4 Recommendations for the installation of heat and smoke detectors.

- DO NOT install detectors in the *dead* airspace at the top of a stairway that can be closed off by a door.
- DO NOT place the edge of a detector mounted on the ceiling closer than 4 inches (102 mm) from the wall.
- DO NOT place a wall-mounted detector closer than 4 inches (102 mm) from the ceiling and not farther than 12 inches (305 mm) from the ceiling.
- DO NOT install a detector where the ceiling meets the wall since this is considered dead airspace where smoke and heat may not reach the detector.
- DO NOT connect detectors to wiring that is controlled by a wall switch.
- DO NOT connect detectors to a circuit that is protected by a ground-fault circuit interrupter (GFCI).
- DO NOT install detectors where relative humidity exceeds 85%.
- DO NOT install detectors in bathrooms with showers, laundry areas, or other areas where large amounts of visible water vapor exist.

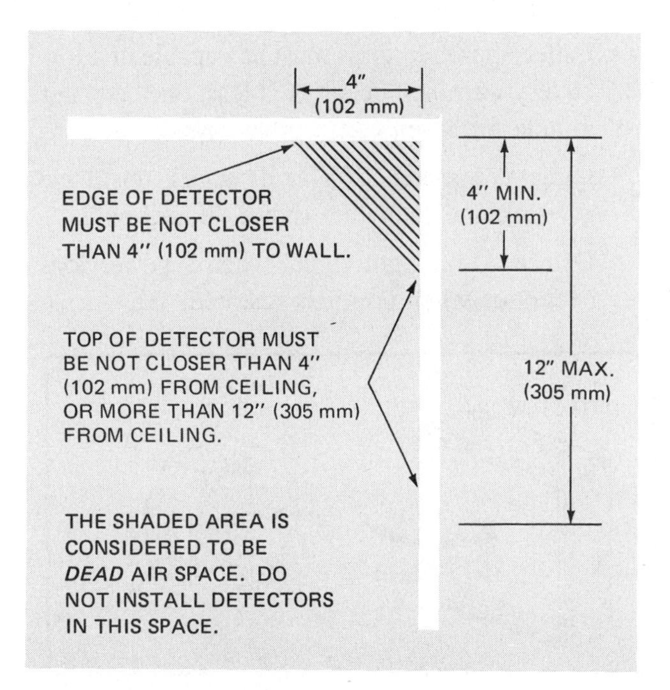

Fig. 26-5 Do not mount detectors in the *dead* airspace where the ceiling meets the wall.

- DO NOT install detectors in front of air ducts, air conditioners, or any high-draft areas where the moving air will keep the smoke or heat from entering the detector.
- DO NOT install detectors in kitchens where the accumulation of household smoke can result in false alarms with certain types of detectors. The photoelectronic type *may* be installed in kitchens, but must *not* be installed directly over the range or cooking appliance.
- DO NOT install where temperature can fall below 32° F or rise above 120° F.
- DO NOT install in garages (vehicle exhaust can set off the detector).
- DO NOT install detectors in airstreams that will pass air originating at the kitchen cooking appliances across the detector. False alarms will result.
- DO NOT install near fluorescent fixtures. Install at least 6 feet (1.8 m) away from a fluorescent fixture.

MANUFACTURERS' REQUIREMENTS

Here are a few of the responsibilities of the manufacturer of smoke and heat detectors (complete data is found in NFPA *Standard No. 74*):

- The power supply must be capable of operating the signal for at least 4 minutes continuously.
- Battery-powered units must be capable of a low battery warning "beep" of at least one beep per minute for seven consecutive days.
- Direct-connected 120V ac detectors must have a visible indicator that shows "power-on."
- Detectors must not signal when a power loss occurs or when power is restored.

FEATURES OF SMOKE AND HEAT DETECTORS

Smoke detectors may contain an indicating light to show that the unit is functioning properly. They also may have a test button which actually simulates smoke. When the button is pushed, it is possible to test the detector's smoke-detecting ability, as well as its circuitry and alarm.

Heat detectors are available that sense a specific *fixed* temperature, such as 135° F or 200° F. Available also are *rate-of-rise* heat detectors that sense rapid changes in temperature (12 degrees to 15 degrees per minute) such as those caused by flash fires.

Fixed and *rate-of-rise* temperature detectors are available as a combination unit. Combination *smoke*, *fixed*, and *rate-of-rise* temperature detectors are also available in one unit.

The spacing of heat detectors shall be as recommended by the manufacturer, because each detector is capable of sensing heat within a given space in a given time limit. NFPA *Standard No. 74* discusses this subject.

Wiring Requirements

Direct-Connected Units: These units are connected to a 120-volt circuit. The black and white wires on the unit are spliced to the black and white wires of the 120-volt circuit, figure 26-6.

Feedthrough (Tandem) Units: These units may be connected in tandem up to 10 units. Any one of these detectors can sense smoke, then send a signal to the remaining detectors, setting them off so that all detectors sound an alarm. These units contain three wires: one black, one white, and one yellow, figure 26-7. The yellow wire is the interconnecting wire.

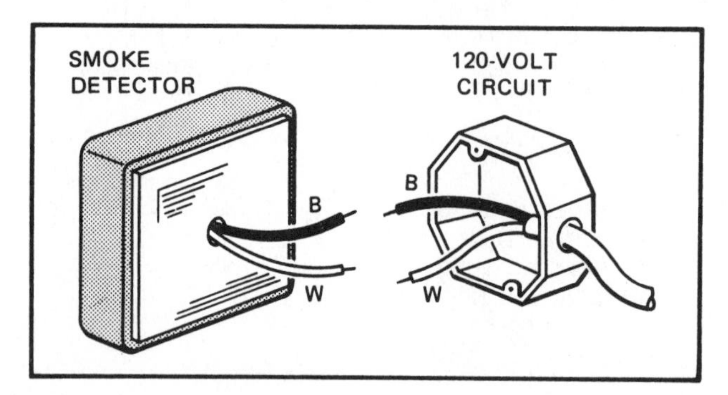

Fig. 26-6 Wiring of direct-connected detector units.

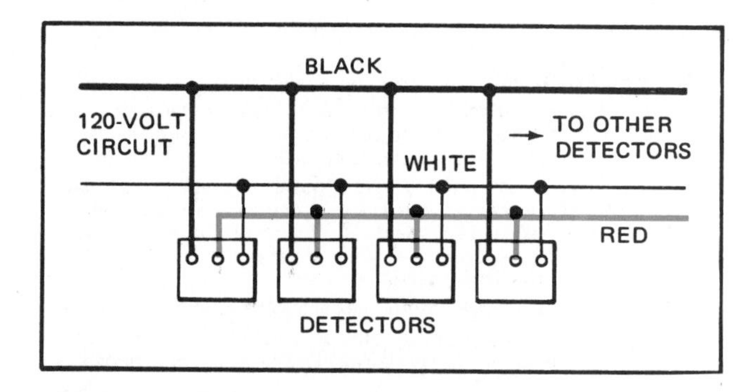

Fig. 26-7 Wiring of feedthrough (tandem) detector units.

Cord-Connected Units: These units operate in the same way as direct-connected units, except that they are plugged into a wall outlet (one that is not controlled by a switch). These units are mounted to the wall on a bracket or screws. A clip on the cord is attached to the wall outlet with the same screw that fastens the faceplate to the outlet, figure 26-8. This prevents the cord from being pulled out, rendering the unit inoperative.

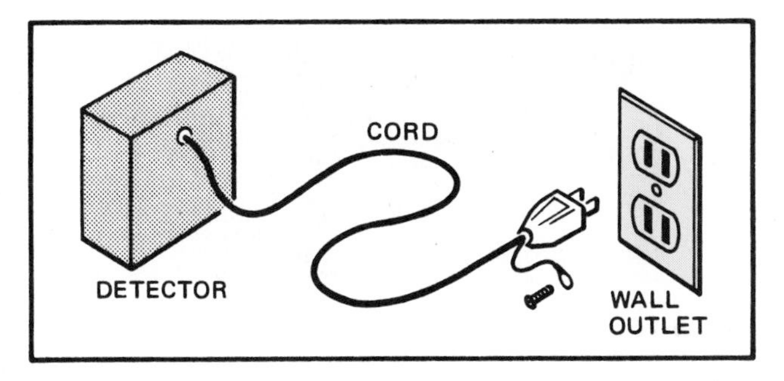

Fig. 26-8 Cord-connected detector units are plugged into wall outlet.

Battery-Operated Units: These units require no wiring; they are simply mounted in the desired locations. Batteries last for about a year. Battery-operated units should be tested periodically, as recommended by the manufacturer.

Always follow the installation requirements and recommendations of the manufacturer of the detecting unit. Refer to *NFPA Standard No. 74* for additional technical data relating to the installation of smoke and heat detectors.

COMBINATION DIRECT/BATTERY/FEED-THROUGH DETECTORS

These units are connected directly to a 120V ac circuit (not GFCI protected) and are equipped with a 9-volt dc battery, figure 26-9. Thus, in the event of a power outage, these detectors will continue to operate on the 9-volt dc battery. The ability to interconnect a number of detectors offers "whole house" protection, because when one unit senses smoke it plus all the other interconnected units will sound an alarm.

Because of the vital importance of smoke and fire detectors to prevent loss of life, it is suggested that the student obtain a copy of NFPA *Standard No. 74.*

Also, read very carefully the application and installation manuals that are prepared by the manufacturer of the smoke and heat detectors.

In this residence we find four smoke detectors connected to Circuit A17. These units are hard-wired directly to the 120-volt ac branch circuit. They are interconnected feedthrough units that will set off signaling in all units should one unit trigger. These interconnected smoke detectors are also powered by a 9-volt dc battery that allows the unit to operate in the event of a power failure in the 120-volt ac supply.

Fig. 26-9 Smoke detector. Operates on 120 volts ac. However, if power fails, the detector continues to operate on the 9-volt battery. Courtesy of BRK Electronics.

SECURITY SYSTEMS

It is beyond the scope of this text to cover each and every type of residential security system. Instead, we will focus on some of the features available from the manufacturers of these systems.

Installed security systems can range from the most simple system to very complex systems. Features and options include intruder detectors for

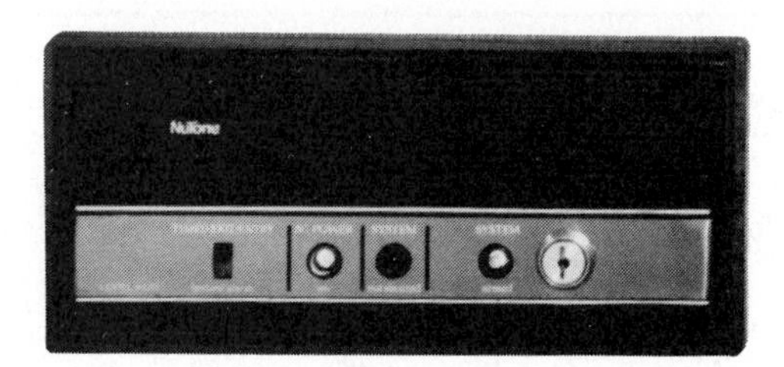

Fig. 26-10 Intruder alarm system. Courtesy of NuTone, Inc.

Fig. 26-11 Infrared detector. Courtesy of NuTone, Inc.

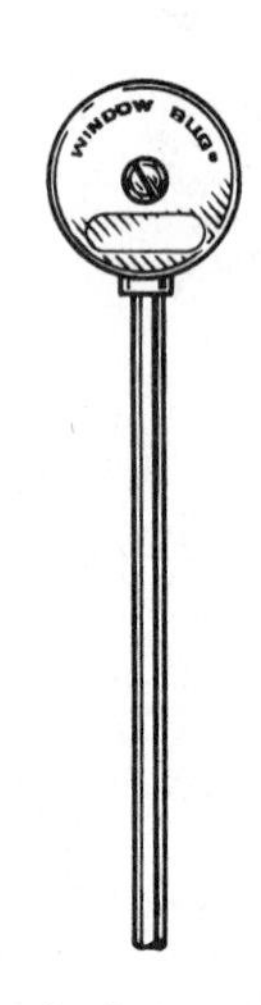

Fig. 26-12 Glass-break detector. Courtesy of NuTone, Inc.

Fig. 26-13 Security alarm keyboard. Courtesy of NuTone, Inc.

Fig. 26-14 Outdoor remote switch. Courtesy of NuTone, Inc.

Fig. 26-15 Indoor remote switch. Courtesy of NuTone, Inc.

Fig. 26-16 Indoor horn alarm. Courtesy of NuTone, Inc.

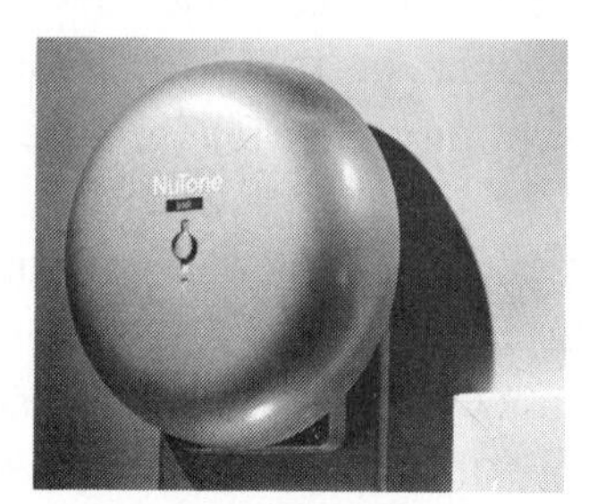

Fig. 26-17 Indoor/outdoor alarm. Courtesy of NuTone, Inc.

Fig. 26-18 Automatic digital communicator. Courtesy of NuTone, Inc.

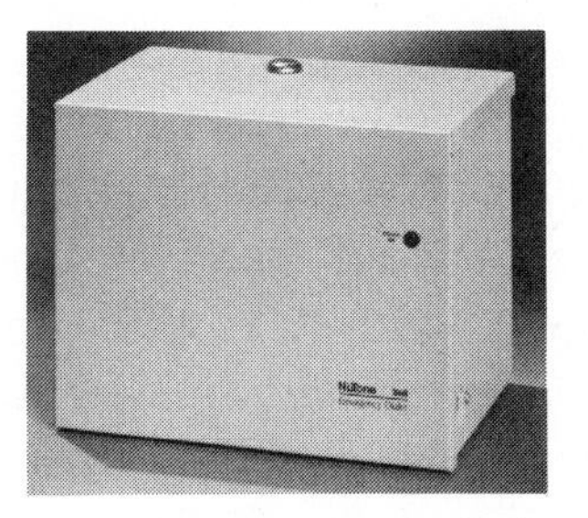

Fig. 26-19 Telephone dialer. Courtesy of NuTone, Inc.

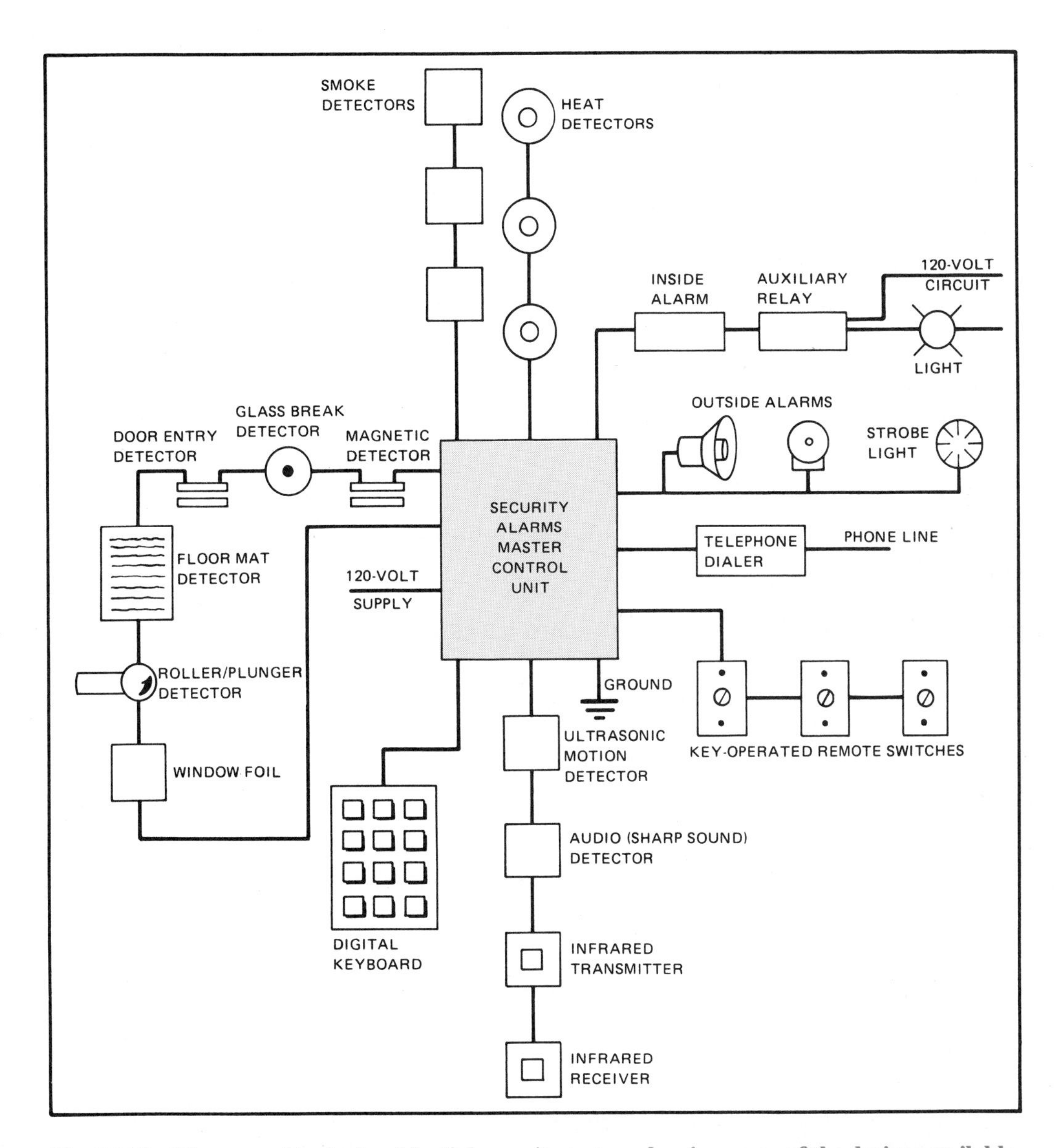

Fig. 26-20 Diagram of typical residential security system showing some of the devices available. Complete wiring and installation instructions are included with these systems. Check local code requirements in addition to following the detailed instructions furnished with the system. Most of the interconnecting conductors are No. 18 AWG.

doors and windows, motion detectors, infrared detectors, and under-the-carpet floor mat detectors for "space" protection, inside and outdoor horns, bells, electronic buzzers, and strobe lights to provide audio (sound) as well as visual detection, and telephone interconnection to preselected telephone numbers such as the police or fire departments, figures 26-10 through 26-19 (pages 374 and 375).

The decision about how complex an individual residential security system should be generally begins with a meeting between the homeowner and the electrician prior to the actual installation. Most electricians will become familiar with a particular manufacturer's selection of systems. Figure 26-20 (page 375) shows the range of devices available for such a system.

These systems are generally on display at lighting fixture display stores which offer the homeowner an opportunity to see a security system "in action."

The wiring of a security system consists of small, easy to install, low-voltage, multiconductor cables made up of No. 18 AWG conductors. The actual installation of these conductors should be done after the regular house wiring is completed to prevent damage to these smaller cables. Usually the wiring can be done at the same time as the chime wiring is being installed. Security system wiring comes under the scope of *Article 725* of the Code.

When wiring detectors such as door entry, glass break, floor mat, and window foil detectors, circuits are electrically connected in series so that if any part of the circuit is opened, the security system will detect the open circuit. These circuits are generally referred to as "closed" or "closed loop." Alarms, horns, and other signaling devices are connected in parallel, since they will all signal at the same time when the security system is set off. Heat detectors and smoke detectors are generally connected in parallel because all of these devices will "close" the circuit to the security master control unit, setting off the alarms.

The instructions furnished with all security systems cover the installation requirements in detail, alerting the installer to Code regulations, clearances, suggested locations, and mounting heights of the systems components.

Always check with the local electrical inspector to determine if there are any special requirements in your locality relative to the installation of security systems.

REVIEW

Note: Refer to the Code or the plans where necessary.

1. What is the name and number of the NFPA standard written about smoke and heat detectors? ______________________

2. Name the two basic types of smoke detectors. ______________________

____

3. Why is it important to mount a smoke or heat detector on the ceiling not closer than 4 inches (102 mm) from a wall? ______________________

____

____

____

4. A basic rule is to install smoke detectors (Circle one)
   a. between the sleeping area and the rest of the house.
   b. at the top of a basement stairway that has a closed door at the top of the stairs.
   c. in a garage that is subject to subzero temperatures.

5. Heat detectors are available in two temperature types. Name them. ____________

____

6. Although NFPA *Standard No. 74* gives many rules for the installation of smoke and heat detectors, always follow the installation recommendations of ____________

____

7. Security systems are usually installed with wire that is much smaller than normal house wires. These wires are generally No. ____ AWG, which is very similar in size to the wiring for chimes.

8. Because the wire used in wiring security systems is rather small and cannot stand rough service and abuse, security wiring should be installed during the rough-in stages of a new house (before) (after) the other power-circuit wiring is completed. Circle correct answer.

9. When more than one smoke detector is installed in a new home, these units must be

   a. connected to separate 120-volt circuits.

   b. connected to one 120-volt circuit.

   c. interconnected so that if one detector is set off, the others would also sound a warning.

   Circle the correct answer(s).

10. It is advisable to connect smoke and heat detectors to circuits that are protected by GFCI devices. TRUE FALSE

11. Because a smoke detector might need servicing or cleaning, be sure to connect it in such a manner that it is controlled by a wall switch. TRUE FALSE

12. Smoke detectors of the type installed in homes must be able to sound a continuous alarm when set off for at least

    a. 4 minutes

    b. 30 minutes

    c. 60 minutes

UNIT 27

Remote-Control Systems — Low Voltage

OBJECTIVES

After studying this unit, the student will be able to

- explain the operation of a low-voltage, remote-control system for lighting circuits.
- interpret the wiring diagrams of various types of low-voltage, remote-control systems.
- install a low-voltage, remote-control system to comply with the requirements of the *National Electrical Code*.®

The general lighting circuits described in this text are wired using standard methods. Another way of installing lighting circuits is to use a remote-control or low-voltage switching system. This system provides very flexible switching.

The remote-control wiring discussed in this unit is one manufacturer's system. In general, it is difficult to mix components of one manufacturer with those of another.

The following text is an overview of how typical remote-control, low-voltage systems operate. There are differences in the form, fit, and function of systems among the many manufacturers of these systems. Electrical wholesalers and distributors usually carry one or two brands of these systems. Once a decision has been reached to install a remote-control, low-voltage system, it is best to stay with one manufacturer's components.

Another type of full remote-control system, plus many added features is covered in unit 31, The Smart House.

A remote-control wiring system uses controlling devices (such as relays) at the equipment to be controlled. An auxiliary means is provided to operate the controlling device. A dial telephone system is one example of a remote-control wiring system. The dial of the telephone operates relays and contactors at a distant point.

A low-voltage, remote-control system can be installed to provide multiswitch control of lighting convenience outlets or small appliances in dwellings. This system is only slightly more expensive than a conventional wiring system. Low-voltage switches are easy to install and the same type of switch is used for light outlets controlled from one or more locations.

The remote-control system for lighting circuits consists of low-voltage, 24-volt relays. These devices control built-in 120-volt contacts from low-voltage controlling switches. The relays are rated at 25 volts ac for use at a nominal voltage of 24 volts. Since a low voltage is used, the shock hazard is less than for a 120-volt system. Low-voltage, remote-control circuits may be installed in two-wire or three-wire cable with No. 18 AWG wire.

This type of installation lends itself to large residences as well as commercial and industrial buildings.

REMOTE-CONTROL SWITCHES

The low-voltage, remote-control system uses a normally open, single-pole, double-throw momentary contact switch. This switch is roughly one-third the size of a standard single-pole switch. It may have three or four terminals or three color-coded lead wires. Figure 27-1 is the wiring diagram of a low-voltage switch. Figure 27-2 shows typical low-voltage switches.

Remote-control switches for lighting systems may be mounted in switch boxes at the switch locations. They can also be mounted on standard raised plaster rings selected for the thickness of the finished walls. The switch boxes or plaster rings must be mounted horizontally for some types of low-voltage switches. The master selector switch requires a horizontally mounted, two-gang raised plaster ring. (The master selector switch is covered later in this section.)

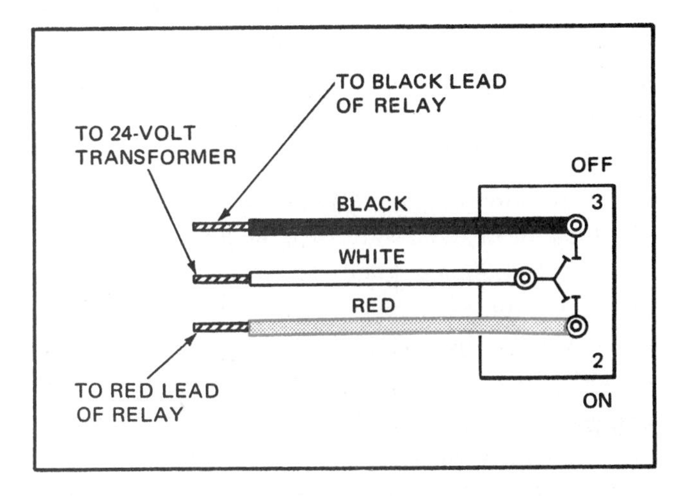

Fig. 27-1 Wiring diagram of a low-voltage switch.

Some local building codes require the use of boxes at the switch locations. The appearance of the installation generally is improved when switch boxes are used. Many devices have pigtail leads that must be spliced to the low-voltage cables. Switch boxes provide the necessary room for splicing. In this way, the splices do not have to be pushed back into the insulation or vapor barrier provided in all electrically heated homes. The use of switch boxes is especially desirable for outside walls.

The white lead of the low-voltage switch is connected to the 24-volt transformer. The red lead of the switch is connected to the red lead of the relay (on) and the black lead is connected to the black lead of the relay (off). (Refer to figure 27-1.)

The master selector switch, figure 27-3, is another type of remote-control switch. This switch can operate from one to twelve circuits. The switch knob is pressed and rotated through the twelve numbers to turn all of the circuits on or off. To turn specific circuits on or off, the knob is turned to the proper number on the dial and is then depressed.

Low-voltage switches may have pilot lights or locator lights to show when the circuit is on. The master selector switch is available with a pilot-lighted directory. This feature allows the user to see immediately which circuits are on or off.

Basically, the master selector switch is a group of individual switches mounted on a single plate. This arrangement makes it easy to select one or all of the individual circuits. When all of the circuits are to be controlled, the master selector switch can sweep all of them. This method is faster and easier than operating each circuit alone. Thus, the master selector switch is convenient for controlling many circuits from a single point. In the residence, for

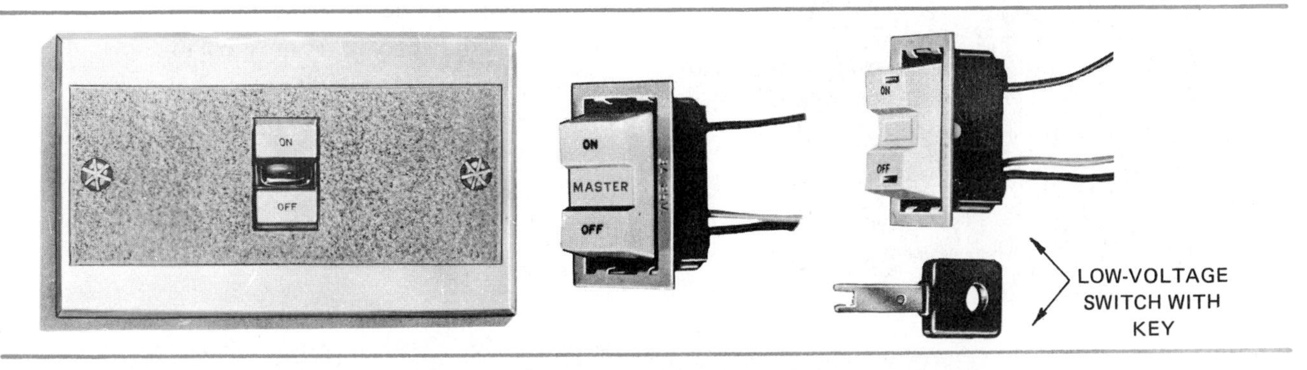

Fig. 27-2 Types of low-voltage switches.

example, a master selector switch can be located in one of the bedrooms. This permits the owner to turn on all of the outdoor lighting to provide some protection against prowlers.

A *motorized master control*, figure 27-4, can provide control for a maximum of 25 circuits. For each touch of the on/off position of a standard low-voltage switch, the motorized master control makes a complete sweep of the 25 positions. Thus, the switch controls the 25 relays connected to these positions.

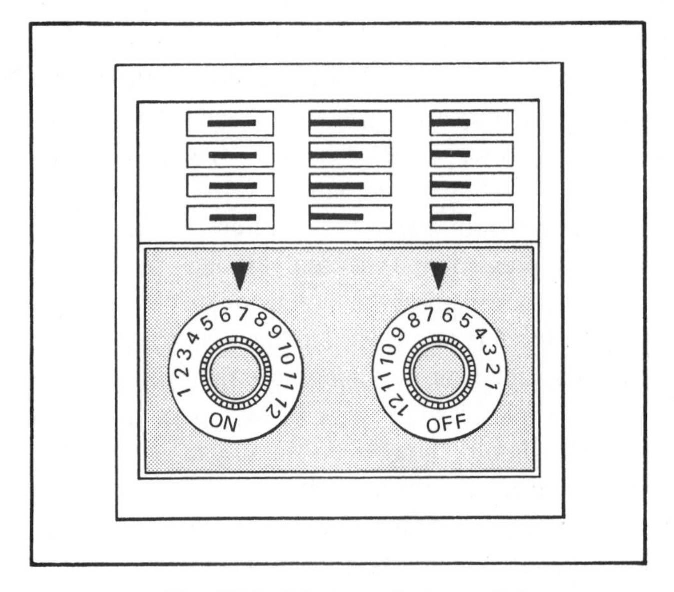

Fig. 27-3 Master selector switch.

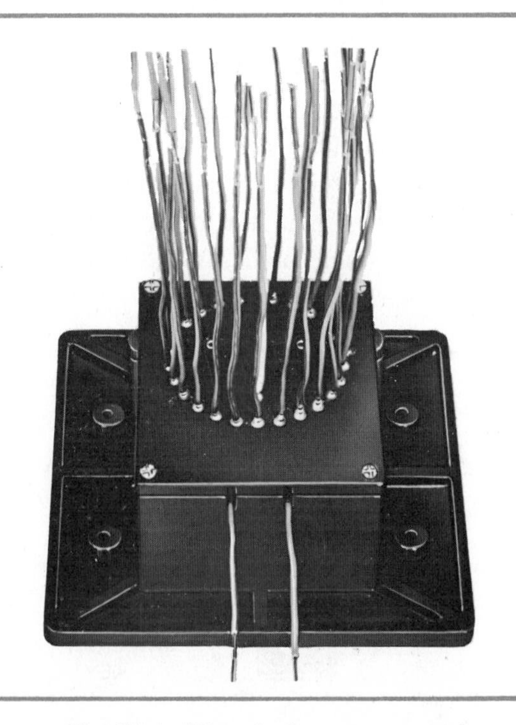

Fig. 27-4 Motorized master control.

LOW-VOLTAGE RELAY

Figure 27-5 shows the wiring of a relay used in low-voltage remote-control systems. This relay operates contacts in the 120-volt lighting circuit. The relay has a split coil design. One coil is used to close the 120-volt circuit. The other coil opens the contactors in the 120-volt circuit. This relay operates on 24 volts. There are two No. 12 AWG black leads for the 120-volt circuits and three No. 20 AWG colored leads for the 24-volt connections.

Basically, the low-voltage relay is a double solenoid. By energizing the coil connected to the red and blue leads, the iron core tries to center itself within the coil. The core moves to the left until it closes the switch contacts. When the power supply to the coil is removed, the switch contacts remain closed. They are held in the on position by a mechanical latching device in the relay. When the coil connected to the black and blue leads is energized, the iron core tries to center itself in this coil. The core now moves to the right to open the switch contacts. The relay is mechanically latched in the off position.

Relay Mounting

The low-voltage relay is small and can be mounted from the inside of a standard outlet box through a 1/2-inch knockout opening. For quiet operation, the relay may be mounted in a 3/4-inch knockout opening using a rubber grommet. When mounted from the inside of the box, the two high-voltage leads of the relay remain inside the outlet box. The low-voltage wires are kept outside the box. The wall of the box serves as a partition between the high and low voltages. The two high-voltage leads inside the box are connected like a standard single-pole switch. The hot or black wire from the 120-volt source goes to one of these leads. The other lead wire from the relay is connected to the lamp terminal. The blue lead of the three low-voltage wires is common to both the off coil and the on coil. The red lead connects internally to the on coil and the black lead connects internally to the off coil.

Plug-in Relay

Plug-in relays are shown in figure 27-6. It is not necessary to splice the line voltage leads for this type of relay. Plug-in relays may be used when a

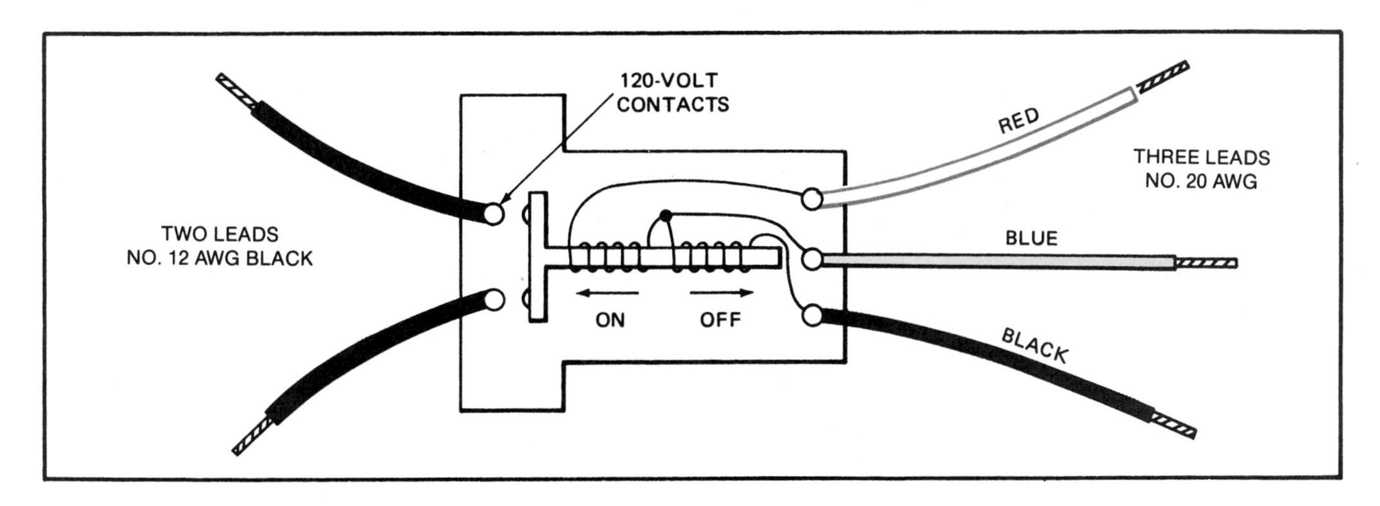

Fig. 27-5 Wiring for a low-voltage relay.

number of relays are to be mounted in one enclosure designed for this purpose.

The relays shown in figure 27-6 are rated for 1 hp, 20 amperes, 125 volts ac/20 amperes, 277 volts ac. The 20-ampere ac rating for both 125 volts and 277 volts means that the relay can be used to control alternating-current loads, including tungsten filament loads and fluorescent loads, up to the full rating of the relay.

TRANSFORMERS

A low-voltage, remote-control system requires just one transformer, figure 27-7. The transformer steps down the lighting circuit voltage from 120 volts to 24 volts for the low-voltage control circuit. The transformer rating is no larger than 100 volt-amperes. When the transformer is overloaded, its output voltage decreases and there is less current output. In other words, the transformer has energy-limiting characteristics which counteract any overload. A specially designed core limits the amount of electrical energy that can be delivered at the secondary or output terminals. Some transformers have a thermal breaker which opens the primary circuit to protect the transformer from overheating. As soon as it cools down, the thermal breaker resets itself and the transformer is automatically reconnected to the line.

Rarely is more than one switch pressed at the same time in a low-voltage, remote-control system. Although 25 or more relays may be connected to the transformer, it is the same as if there were only one. The transformer seems to be underrated for the connected load. Actually, it can control many relays. The transformer manufacturer usually recommends the maximum number of relays that may be operated at one time.

Some manufacturers of remote-control systems suggest that a rectifier be added to the secondary

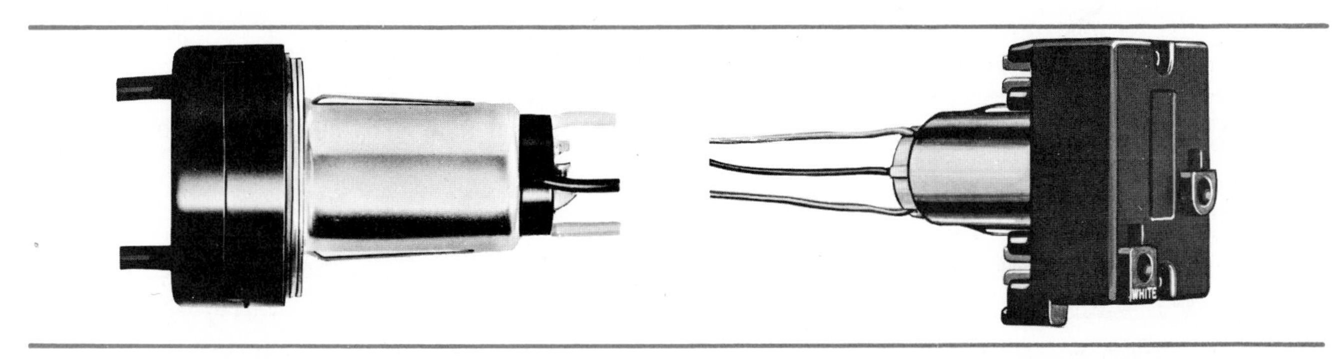

Fig. 27-6 Plug-in relays.

circuit. This device changes the alternating-current supply to direct or pulsating direct current (depending upon whether a half-wave or a full-wave rectifier is used). If a relay is energized for a long time, the alternating-current supply will produce eddy currents inside the laminations. As a result, there is a temperature buildup that may damage the relay. (Recall that this type of relay is meant to provide momentary contact operation only.) Direct current or pulsating direct current does not produce as much heat in this type of relay as does alternating current.

CONDUCTORS

No. 18 AWG conductors are generally used for low-voltage, remote-control systems, figure 27-8. Larger conductors may be required for long runs. A two- or three-conductor cable is used for the installation. Insulation on the individual conductors is generally a thermoplastic material that is color-coded for ease in identification.

The installation of a remote-control system is simplified by using color-coded low-voltage cables. Cables are available in color combinations such as blue-white, red-black, or black-white-red. When these cables are installed correctly in a dwelling, the wires are connected like color to like color. A cable suitable for either overhead or underground installations is available for use in a low-voltage, remote-control system installed outdoors.

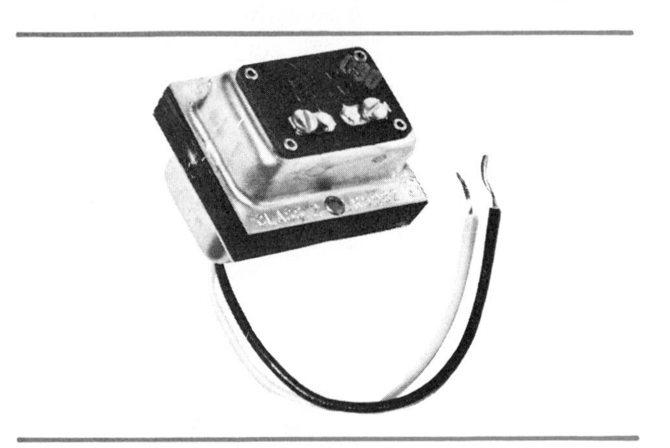

Fig. 27-7 Transformer for remote-control systems.

INSTALLATION PROCEDURE

The installation of a remote-control system in new construction begins with the roughing-in of the 120-volt system. All of the switch legs or switch loops are omitted for the system. In other words, the 120-volt supply conductors are brought directly to each outlet or switch box that will have a switched lighting fixture or switched outlet attached to it. Although the various devices used may be numbered or color coded in different ways, the electrician must remember that only three low-voltage conductors are required between the relay and the switch. In addition, two low-voltage conductors must be run from the transformer to either the relay or the switch to carry the low-voltage supply. These wires are spliced like color to like color. Another method is to assign certain colors to correspond to the numbers on the relays and switches. The wiring is then completed by slipping the relay through the 1/2-inch knockout in the outlet or switch box. The 120-volt connections are made in the standard way.

A relay with a built-in transformer can be used. No. 18/3 cable is run between the relay and the switch. The connections are made like color to like color. The black and white line voltage relay leads are connected to the supply circuit. The red line voltage relay lead is connected to the lamp terminal. (This red lead is actually the switch leg.)

The low-voltage cables are stapled in place in new construction. When existing buildings are rewired, the low-voltage cables may be fished through the walls and ceiling or run behind moldings and baseboards.

Under certain conditions, many or all of the relays should be placed at one location. For example, this is done when a motorized master control is used. Special relay ganging boxes can be used. In

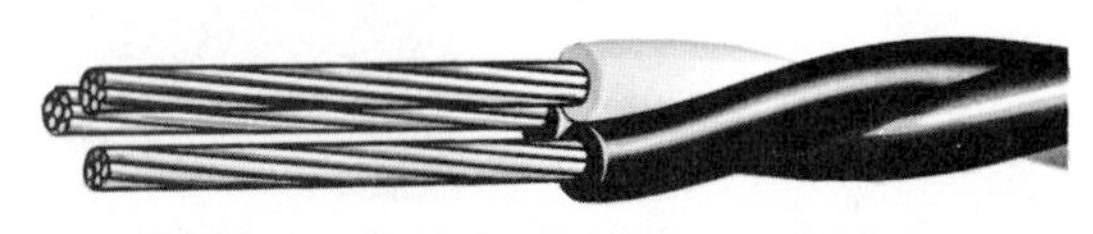

THREE-CONDUCTOR CABLE

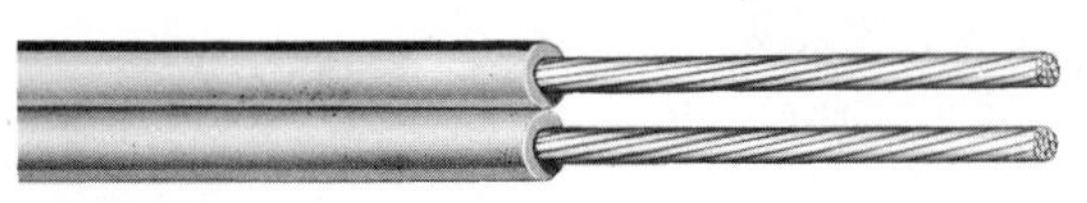

TWO-CONDUCTOR CABLE

Fig. 27-8 Cable for remote-control systems.

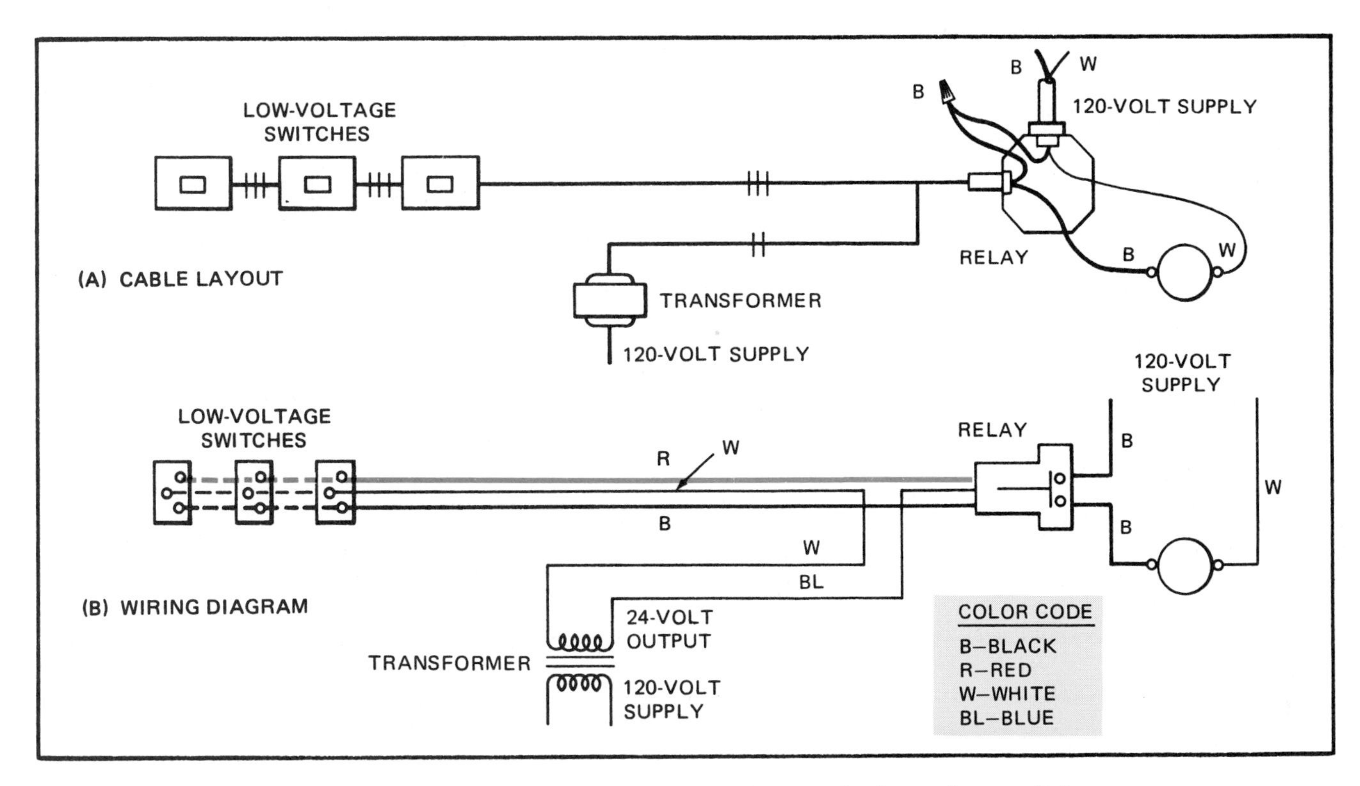

Fig. 27-9 Typical cable layout and wiring diagram for low-voltage switches.

this type of box, all of the relays are inserted through a metal barrier. The line voltage connections are made on one side of this barrier and the low-voltage connections are made on the other side.

TYPICAL WIRING INSTALLATION

One method of installing remote-control, low-voltage switches is shown in figure 27-9. Multipoint control of the light results when the low-voltage, remote-control switches are connected in parallel.

Figure 27-10 shows one method of connecting a master selector control with individual low-voltage switches. The lighting outlets can be placed in different parts of the building and can be connected to operate 120-volt circuits. These outlets are then controlled by their own individual switches or the master selector control. Three lighting circuits are shown but the master selector switch can control up to twelve different relays.

If a master selector switch were installed in the residence, it could control the shrub lights, front entrance light, post light (front), garage lights (front), garage light and post light (rear), garage light (inside), cornice lights, terrace entrance light, and rear porch light. Thus, the entire outside of the residence could be flooded with light instantly, if necessary.

NATIONAL ELECTRICAL CODE®

The low-voltage, remote-control system is not subject to the same Code restrictions as the standard 120-volt system. The low-voltage (24-volt) portion of the remote-control system is a Class 2 system and is regulated by *NEC® Sections 725-31* through *725-42*. UL sets 30 volts RMS as the maximum secondary voltage for Class 2 transformers.

For a remote-control system, all wiring on the supply side of the transformer (120 volts) must conform to the wiring methods given in *Chapter 3* of the *National Electrical Code®* All wiring on the load side of the transformer (24 volts) must conform to *Section 725-38*.

The transformer for a remote-control, low-voltage system is designed so that the power output is limited to meet the requirements of the Code, *Table 725-31(a)*. The power output can also be limited by a combination of the power source and the overcurrent protection, *Section 725-31*. The maximum nameplate rating of the transformer is 100 volt-amperes.

Conductors with thermoplastic insulation are normally used for a low-voltage system. However, the Code does not require thermoplastic insulation, *Section 725-40*.

Conductors for Class 2 systems may not be run

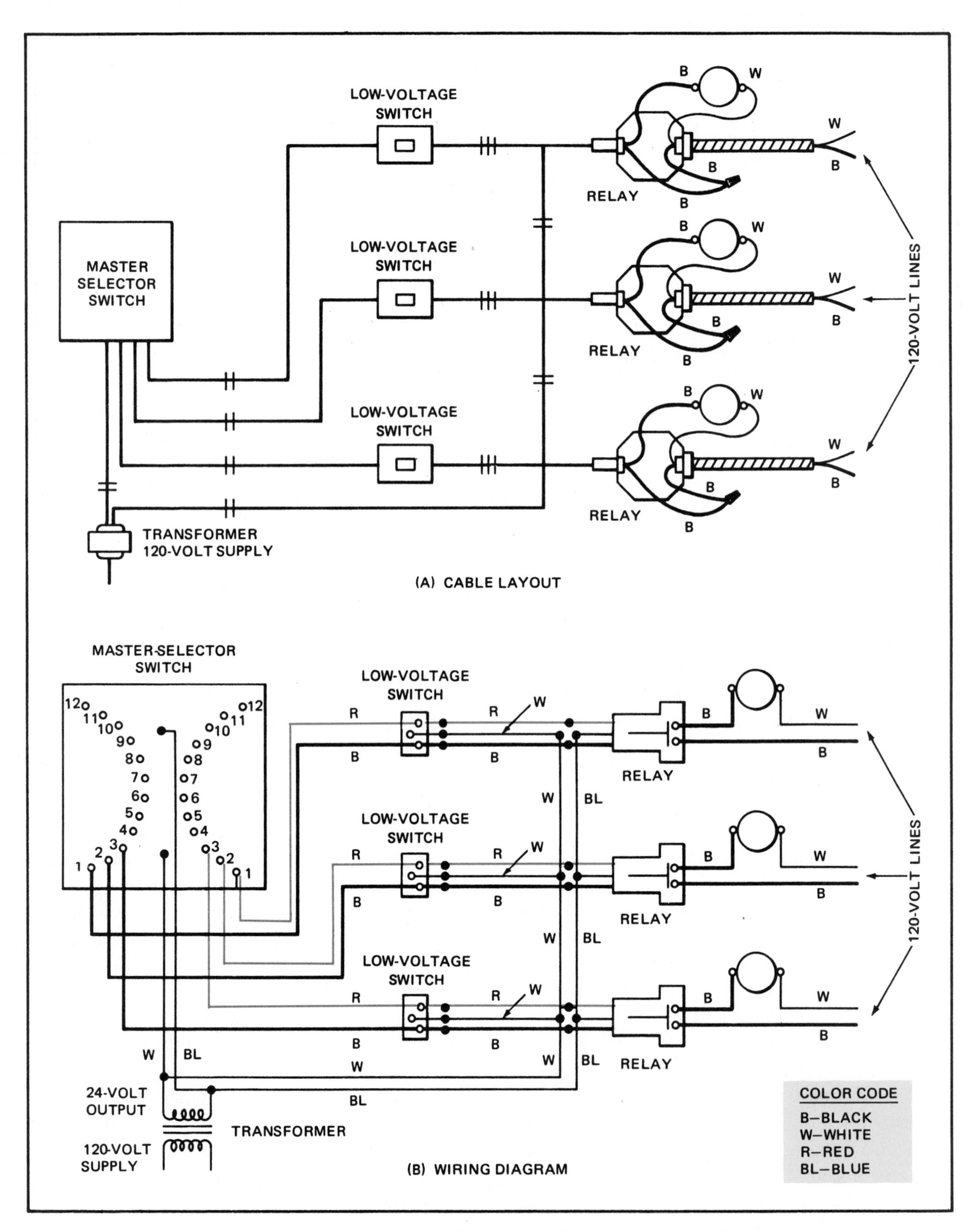

Fig. 27-10 Cable layout and wiring diagrams for a remote-control system.

with regular light and power wiring, *Section 725-38*. Class 2 wiring may enter the same compartment or enclosure as power conductors if separated by a barrier or if introduced solely to connect the equipment, *Section 725-38(a)(2), Exceptions 1 and 2*.

The insulation used on remote-control conductors is rather thin. The electrician must be careful not to damage this insulation. Any damage may result in a short circuit or a relay malfunction. The electrician usually bores holes through the studs and joists of the residence for the small remote-control conductors. They are not pulled through the holes used for the cables for the regular wiring because of the possibility of mechanical damage to the insulation. The electrician must use care when stapling these low-voltage conductors. If the insulation is pinched, a short may occur between the conductors.

The low-voltage system conductors and installation methods are similar to those used to install the chimes in this residence.

REVIEW

Note: Refer to the Code or the plans where necessary.

1. What is the approximate voltage used on low-voltage remote-control systems? ____

2. What are the advantages of a low-voltage system? ____

3. What type of switch is used to control the relays of a low-voltage system? ____

4. Are switch boxes recommended at all low-voltage switch locations? Explain. ____

5. Are three-way and four-way switches used in remote-control circuits? Explain your answer. ____

6. What is a master selector switch? ____

7. What is a motorized master control? ______________________________

____

____

____

____

____

____

8. a. Do the relays connected to a motorized master control operate at the same time? Explain your answer. ______________________________

____

b. Do the relays connected to a master selector switch operate at the same time? Explain your answer. ______________________________

____

9. Explain briefly the operating principle of a low-voltage relay. ______________

____

____

____

____

____

10. What is the electrical rating of relays discussed in this unit? ______________

____

11. What will happen if the common lead from the transformer is connected to the red lead of the relay instead of the blue lead, and the blue and black leads are connected to the switch leads? ______________________________

____

____

____

12. Is it permitted to run low-voltage conductors and line-voltage (120-volt) conductors into the same box? Explain your answer. ______________________________

____

____

____

____

____

13. The maximum nameplate rating of Class 2 transformers is ________ volt-amperes.

14. Since the insulation on low-voltage wire is much thinner than the insulation on regular building wire, the electrician should ______________________________ when installing the conductors to avoid ______________________ the insulation.

15. a. What is the purpose of a rectifier? ______________________________

____

b. Must rectifiers be used on low-voltage systems? ____________________

____

16. What type of conductors are generally used for a low-voltage remote-control system?

____

____

____

____

17. Why are color-coded wires and cable used in low-voltage control systems? _____

____

18. The type of low-voltage wiring discussed in this unit is classified as (Class 1) (Class 2). Circle one.

19. When overcurrent protection is not provided in the secondary circuits of the transformers described in this unit, the transformers must ______________________ their power output.

20. Conductors supplying transformers are governed by what part of the Code? _____

____

21. Complete all line voltage and low-voltage connections in the following diagram. Use colored pencils to indicate conductors. Show low-voltage connections with a dot. Switch No. 1 controls lamps A and C, switch No. 2 controls lamp B, and switch No. 3 controls the bottom of each convenience receptacle. The top of each receptacle is to be hot at all times. The line-voltage wiring is in armored cable.

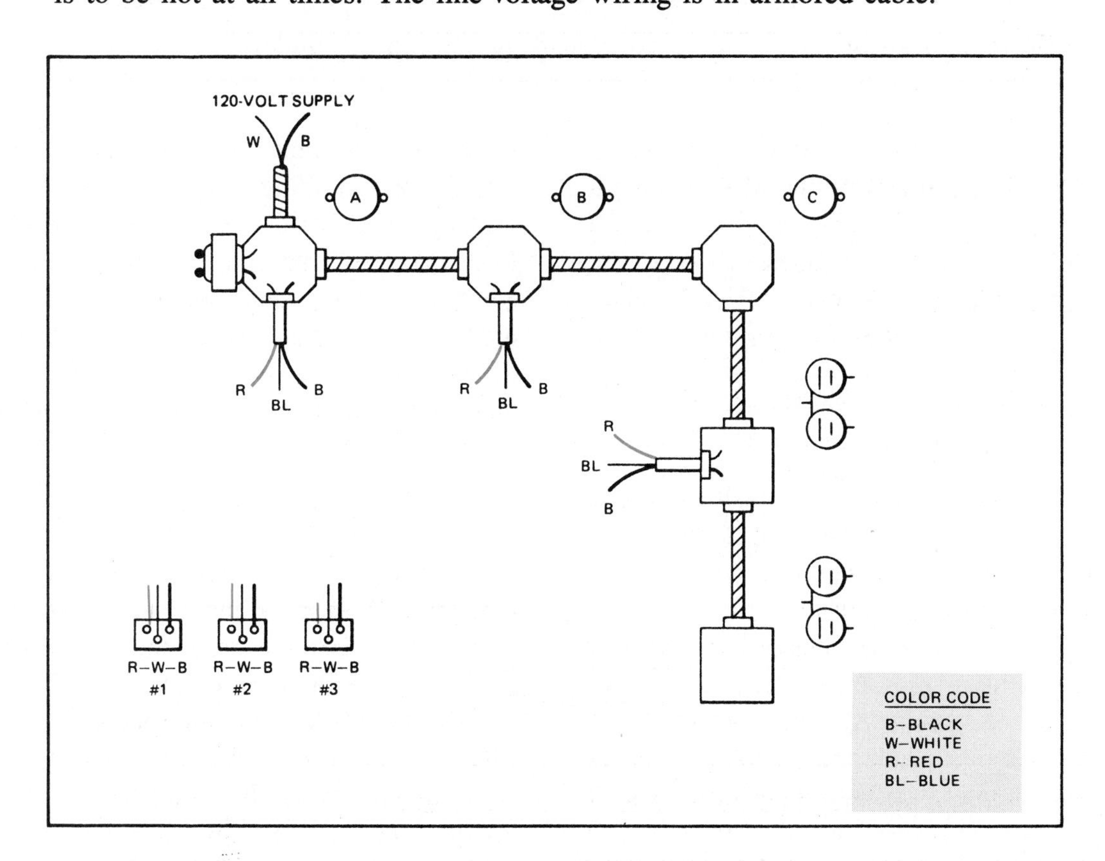

UNIT 28

Service-Entrance Equipment

OBJECTIVES

After studying this unit, the student will be able to

- define electrical service, overhead service, service drop, and underground service.
- list the various Code sections covering the installation of a mast-type overhead service and an underground service.
- discuss the Code requirements for disconnecting the electrical service using a main panel and load centers.
- discuss the grounding of interior alternating-current systems and the bonding of all service-entrance equipment.
- discuss the "UFER" ground method.
- describe the various types of fuses.
- select the proper fuse for a particular installation.
- explain the operation of fuses and circuit breakers.
- explain the term *interrupting rating*.
- determine available short-circuit current using a simple formula.

An electric service is required for all buildings containing an electrical system and receiving electrical energy from a utility company. The *National Electrical Code®* describes the term *service* as the conductors required to deliver energy from the electrical supply source to the wiring system of the premises.

OVERHEAD SERVICE

The Code terms a *service drop* as the overhead service conductors, including splices, if any, which are connected from an outdoor support to the service-entrance conductors at the structure.

The overhead service includes all of the service equipment and installation means from the attachment of the service-drop wires on the outside of the building to the point where the circuits or feeders are tapped to supply specific loads or load centers.

In general, watthour meters are located on the exterior of a building. Local codes may permit the watthour meter to be mounted inside the building. In some cases, the entire service-entrance equipment may be mounted outside the building. This includes the watthour meter and the disconnecting means. Figure 28-1 illustrates the Code terms for the various components of a service entrance.

MAST-TYPE SERVICE ENTRANCE

The mast service, figure 28-1, is a commonly used method of installing a service entrance. The mast service is often used on buildings with low roofs, such as ranch-style dwellings, to insure adequate clearance between the ground and the lowest service conductor.

The service conduit must be run through the

roof as shown in figure 28-1. The conduit is fastened to comply with local code requirements. Methods of fastening this type of service are not covered by the *National Electrical Code*®

Clearance Requirements for Mast Installations

Several factors determine the maximum length of conduit that can be installed between the roof support and the point where the service-drop conductors are attached. These factors are the service-drop length, the system voltage, the roof pitch, and the size and type of conduit (aluminum or steel), figure 28-3.

Service Mast as Support (*Section 230-28*)

The bending force on the conduit increases with an increase in the distance between the roof support and the point where the service-drop conductors are attached. The pulling force of a service drop on a mast service conduit increases as the length of the service drop increases. As the length of the service drop decreases, the pulling force on the mast service conduit decreases.

If extra support is not to be provided, the mast service conduit must be at least 2 inches in diameter. This size prevents the conduit from bending due to the strain of the service-drop conductors. If extra support is provided, it is usually in the form

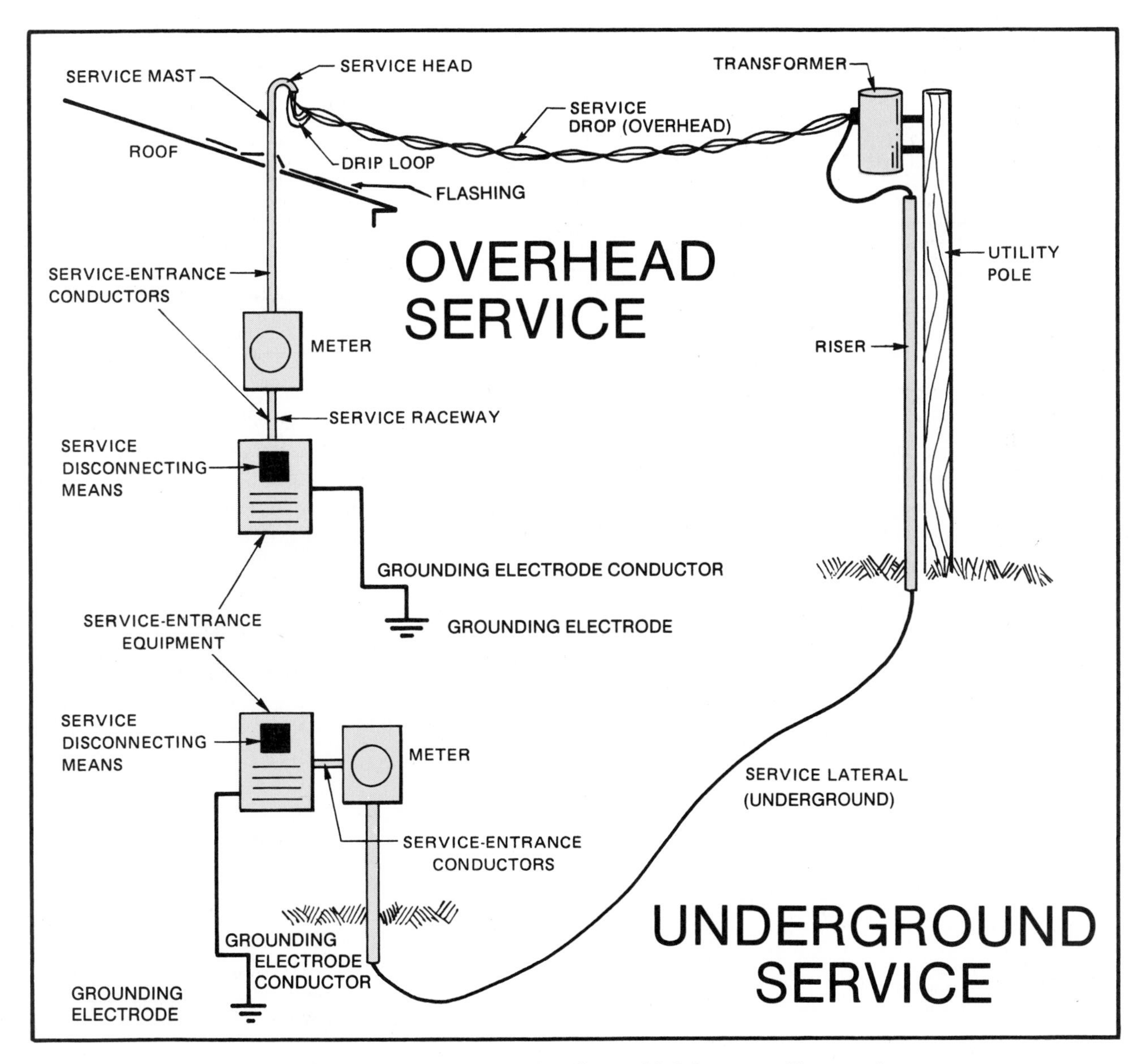

Fig. 28-1 Code terms for services. See figure 28-5 for grounding requirements.

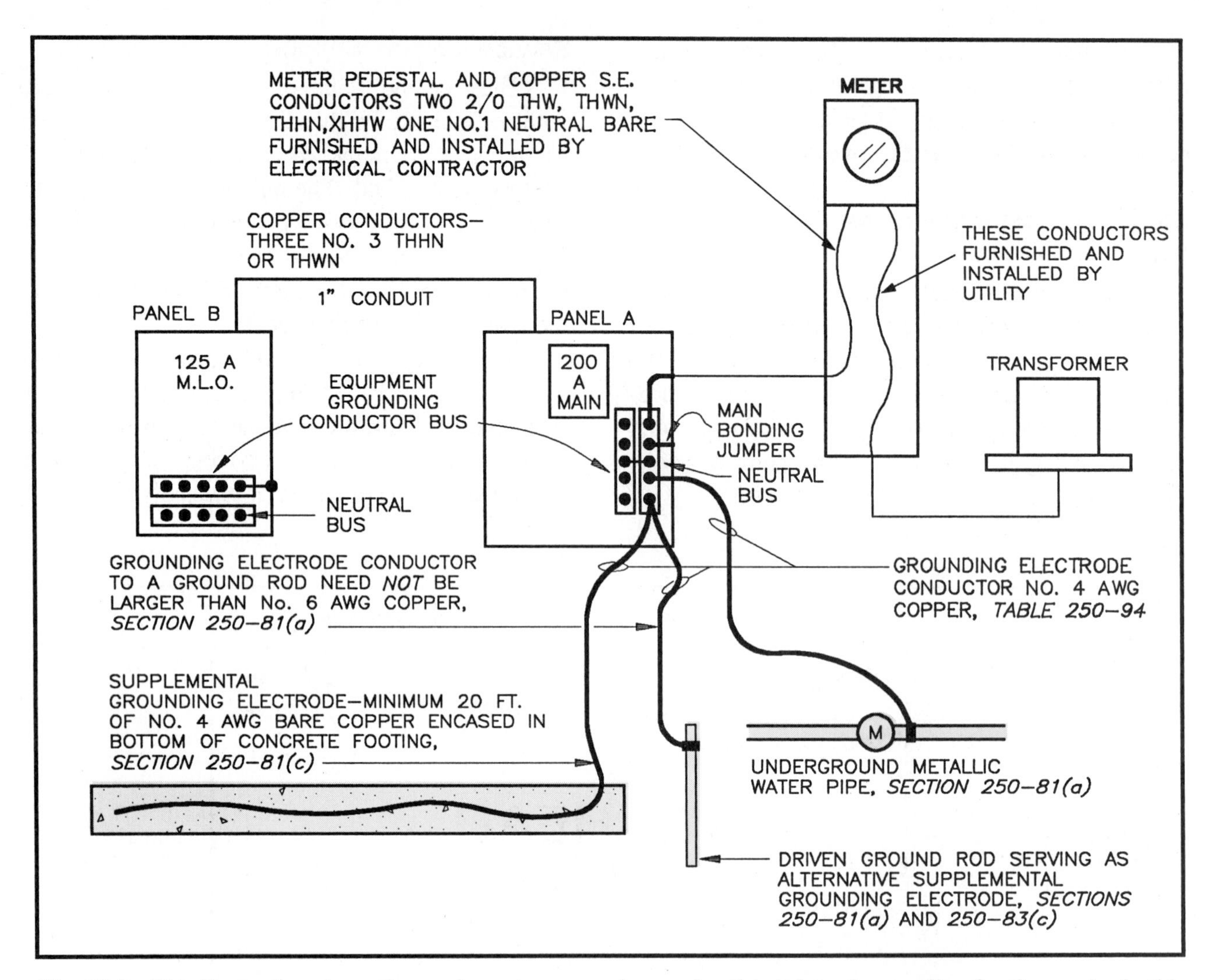

Fig. 28-2 This illustration shows the service-entrance, main panel, subpanel, and grounding for the service in this residence. Note that because the integrity of underground water pipes is questionable, the Code in *Section 250-81(a)* requires a supplemental grounding electrode. Shown are two common types of supplemental grounding electrodes: a ground rod or a concrete-encased grounding electrode. In many instances, the concrete-encased grounding electrode is better than the water pipe ground, particularly since the advent of nonmetallic (PVC) water piping systems.

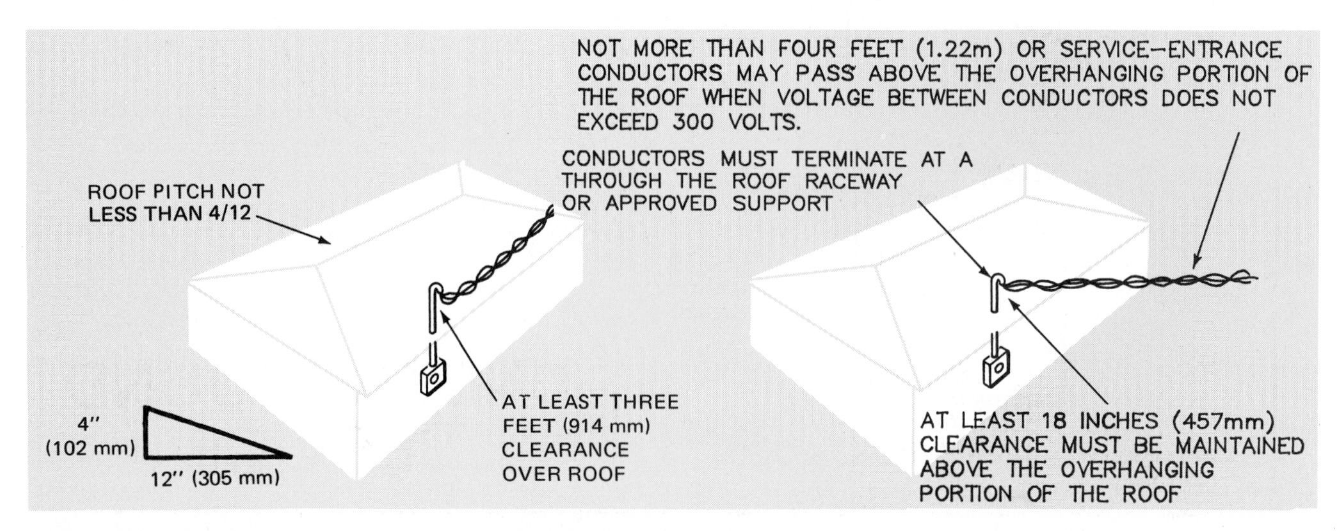

Fig. 28-3 Clearance requirements for service-drop conductors passing over residential roofs, *Section 230-24(a)*, where voltage between conductors does not exceed 300 volts.

of a guy wire attached to the roof rafters by approved fittings.

Consult the utility company and electrical inspection authority for information relating to their specific requirements for clearances, support, and so on, for service masts.

The *NEC*® rules for insulation and clearances apply to the service-drop and service-entrance conductors. For example, the service conductors must be insulated, except where the voltage to ground does not exceed 300 volts. In this case, the grounded neutral conductors are not required to have insulation.

Section 230-24(a) gives clearance allowances for the service drop passing over the roof of a dwelling, figure 28-3.

The installation requirements for a typical service entrance are shown in figures 28-4 and 28-5. Figure 28-4 shows the required clearances above the ground. The wiring connections, grounding requirements and Code references are given in figure 28-5.

UNDERGROUND SERVICE

The underground service means the cable installed underground from the point of connection to the system provided by the utility company.

New residential developments often include underground installations of the high-voltage electrical systems. The conductors in these distribution systems end in the bases of pad-mounted transformers, figure 28-6. These transformers are placed at the rear lot line or in other inconspicuous locations in the development. The transformer and primary high-voltage conductors are installed by, and are the responsibility of, the utility company.

The conductors installed between the pad-mounted transformers and the meter are called *service lateral conductors*. Normally, the electrical utility supplies and installs the service laterals. Figure 28-6 shows a typical underground installation.

The wiring from the external meter to the main service equipment is the same as the wiring for a service connected from overhead lines, figure 28-5. Some local codes may require conduit to be installed underground from the pole to the service-entrance equipment. *NEC*® requirements for underground service are given in *Part D* of *Article 230, Sections 230-30* and *230-31*. The underground conductors must be suitable for direct burial in the earth.

If the electric utility installs the underground

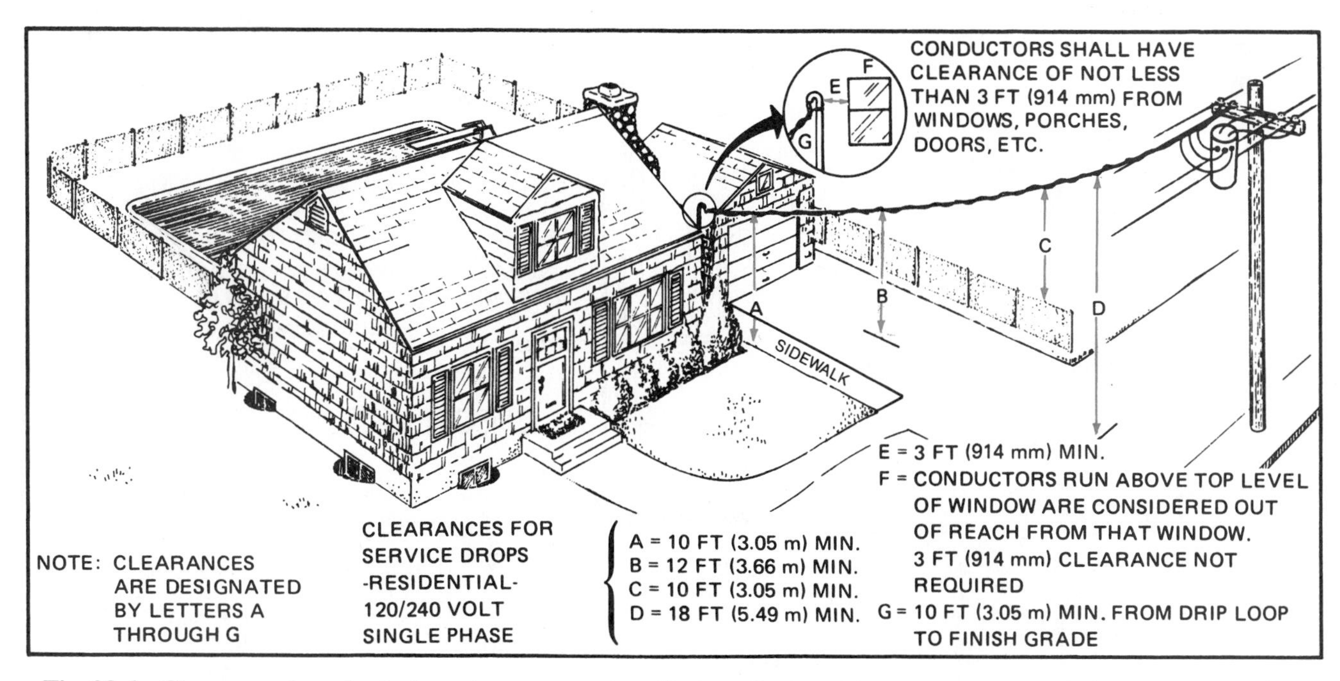

Fig. 28-4 Clearances for a typical service-entrance installation, *Sections 230-24(b)* and *(c)* and *230-26*. (For clearances above swimming pool, see figure 30-6.) Clearances of 10 feet (A) and 12 feet (B) permitted *only* if the service-drop conductors are supported on and cabled together with a grounded base messenger wire and where the voltage does not exceed 150 volts to ground.

service conductors, the work must comply with the rules established by the utility. These rules may not be the same as those given in *NEC® Section 90-2*.

When the underground conductors are installed by the electrician, *Section 230-49* applies. This section deals with the protection of conductors against damage and sealing of underground conduits where they enter a building. *Section 230-49* refers to *Section 300-5*, which covers all situations involving underground wiring.

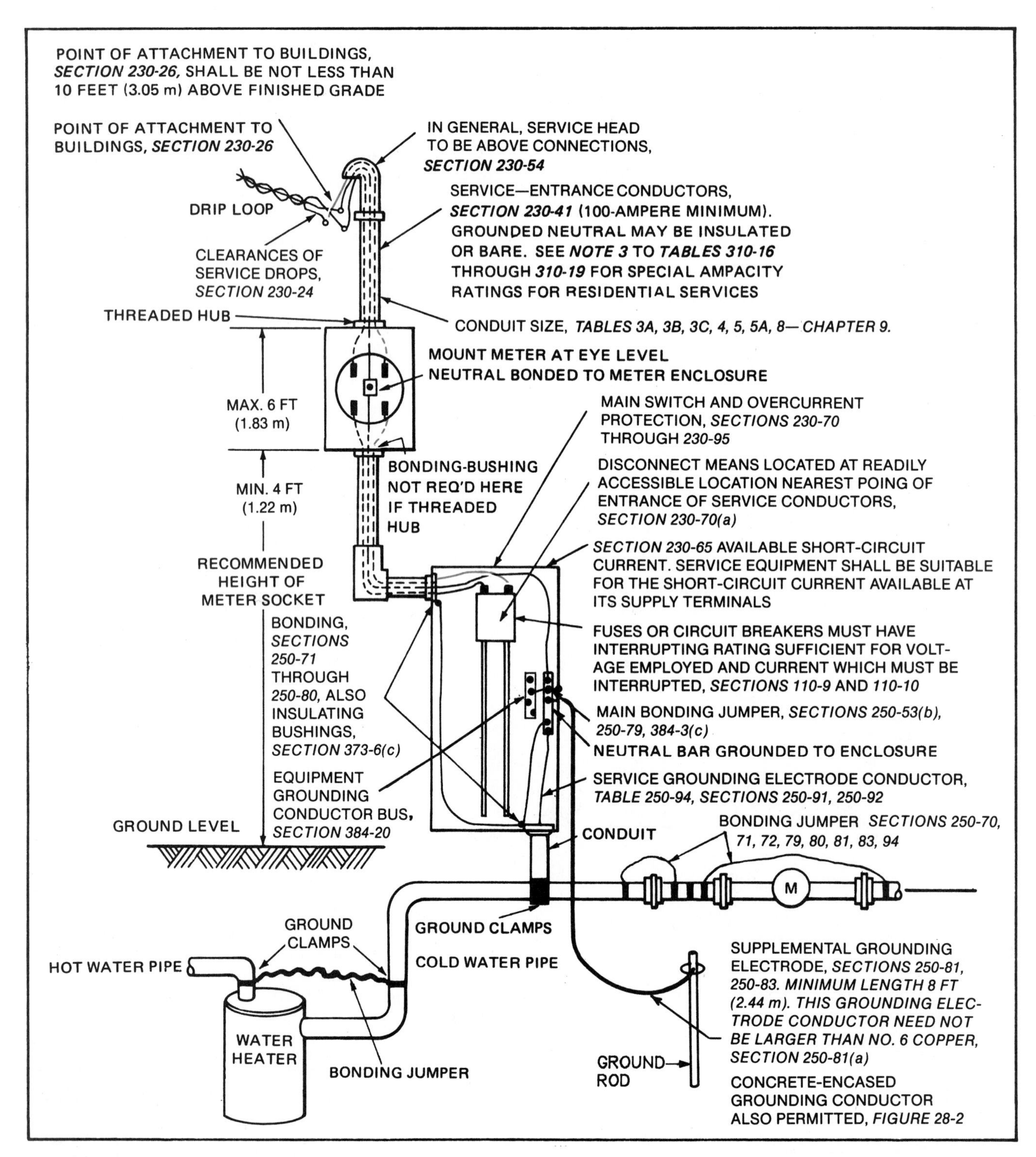

Fig. 28-5 The wiring of a typical service-entrance installation and Code rules for the system grounding.

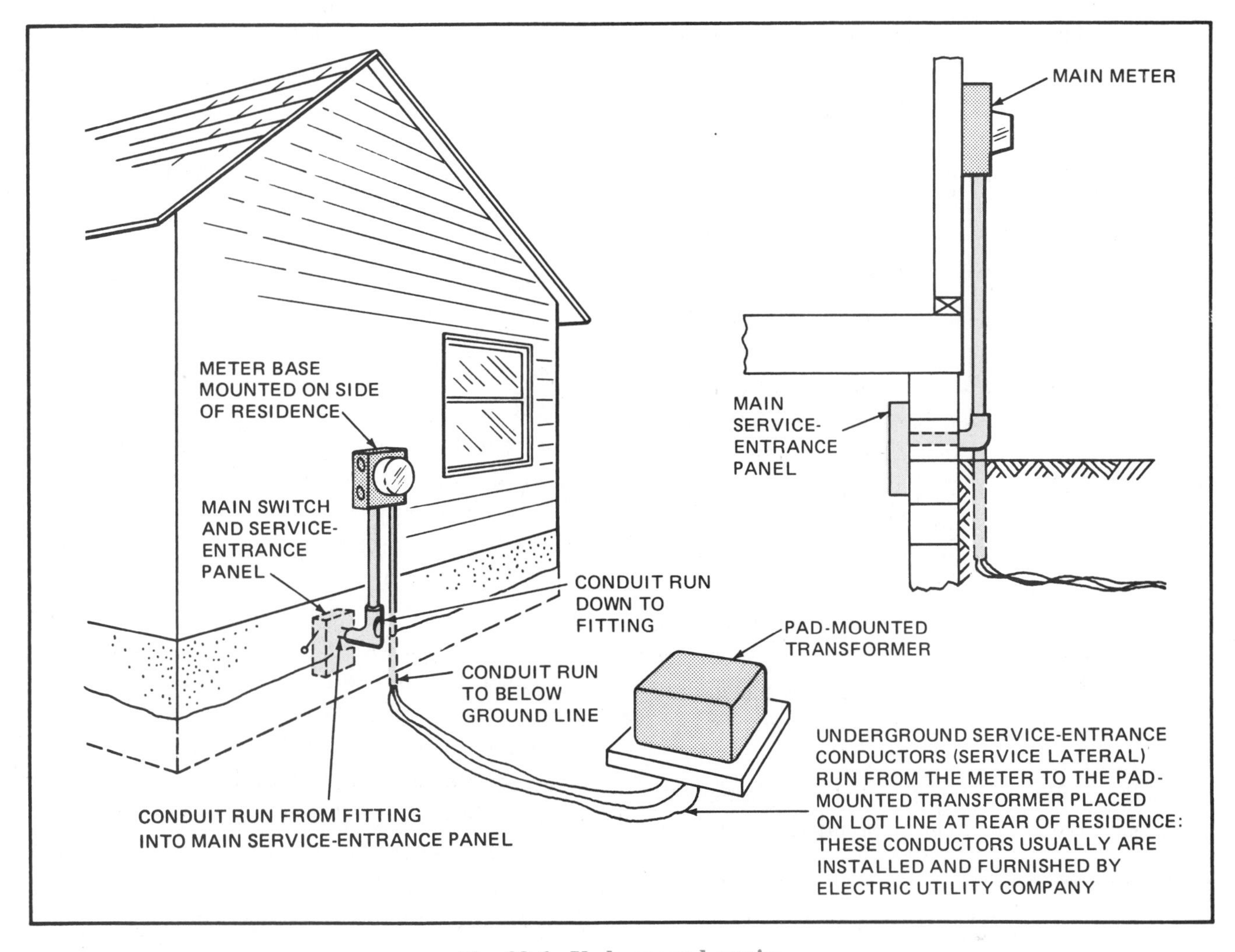

Fig. 28-6 Underground service.

MAIN SERVICE DISCONNECT LOCATION

The main service disconnect means shall be installed at a readily accessible location so that the service-entrance conductors within the building are as short as possible, figure 28-7.

The reason for this rule is that the service-entrance conductors do not have overcurrent protection other than that provided by the utility's transformer fuses. **Should a fault occur on these service-entrance conductors within the building (at the bushing where the conduit enters the main switch, for instance), the arcing could result in a fire.**

The electrical inspector (authority having jurisdiction) must make a judgment as to what is considered to be a readily accessible location nearest the point of entrance of the service-entrance conductors, *NEC® Section 230-70(a)*.

► *Section 384-13* states that all panels must have

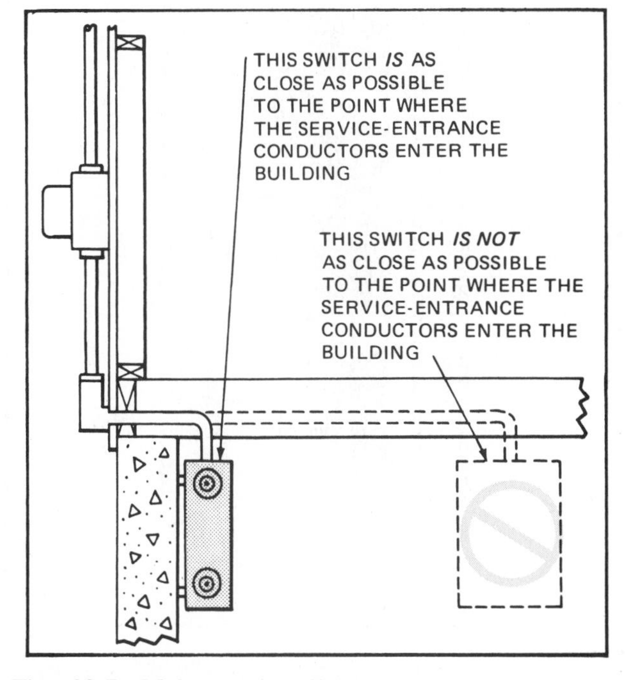

Fig. 28-7 Main service disconnect location, *Section 230-70(a)*.

a "legibly identified" circuit directory indicating what the circuits are for. You can see the circuit directory on the inside of the door of the panel in figure 28-8. The main service panel A circuit directory is shown in figure 28-9. Subpanel B circuit directory is shown in figure 28-11. ◄

Disconnect Means (Panel A)

The requirements for disconnecting the electrical services are covered in *Sections 230-70* through *230-84*. *Section 230-71(a)* requires that the service disconnect means consist of not more than six switches or six circuit breakers mounted in a single enclosure, in a group of separate enclosures, or in or on a switchboard. By complying with *Section 230-71(a)*, all electrical equipment in a building can be disconnected, if necessary, with no more than six hand operations. Some local codes take exception to this rule and state that each service must have a single main disconnect. A panelboard may not contain more than 42 overcurrent devices, *Section 384-15*.

Section 384-16(a) states that a lighting and appliance branch-circuit panelboard shall not have more than two main circuit breakers or two sets of fuses for the panelboard's overcurrent protection and disconnecting means.

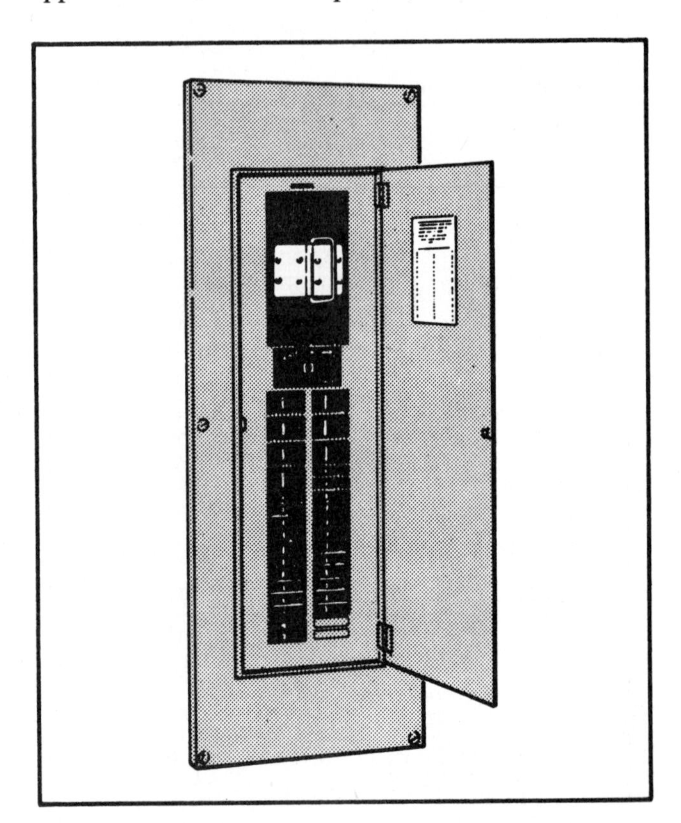

Fig. 28-8 Typical main panel with 200-ampere fusible pullout, suitable for use as service equipment, *Section 230-70(b)*.

The *minimum* rating for a single-family residential service is 100-ampere, three-wire when:

- there are six or more 2-wire branch circuits.
- the initial computed load is 10 kVA or more.

For this residence, Panel A provides a 200-ampere, fusible pullout for the main disconnect. Panel A also has a number of branch-circuit overcurrent devices. These devices are provided for the protection of the many circuits originating from this panel, figures 28-8 and 28-9. This type of panel is listed by Underwriters Laboratories as a load center and as service equipment. The panel meets code requirements in most localities. The overcurrent devices may be fuses or circuit breakers or a combination of both. This is covered later in this unit.

Panel A is located in the workshop. The placement of the main disconnect is determined by the location of the meter. The disconnect is mounted at an accessible point as close as possible to the place where the service conductors enter. The location of the meter depends upon the service drop or the service lateral. The local utility company generally decides where the service drop or service lateral is to be located.

Section 250-61(b) states that a grounded circuit conductor shall not be used for grounding noncurrent-carrying equipment on the load side of the service disconnecting means. Exceptions to this rule are the frames of ranges, wall-mounted ovens, counter-mounted cooking units, and clothes dryers, under the conditions specified by *Section 250-60*.

The panelboard must have an approved means for attaching the equipment grounding conductors when nonmetallic-sheathed cable is used, figure 28-10.

Bonding: At the main service-entrance equipment, the grounded neutral conductor must be bonded to the metal enclosure. For most residential type panels, this main bonding jumper is a bonding screw that is furnished with the panel. This bonding screw is inserted through the neutral bar into a threaded hole in the back of the panel ► itself. This bonding screw must be green in color and must be clearly visible after it is in place, *Section 250-79(b)*. ◄

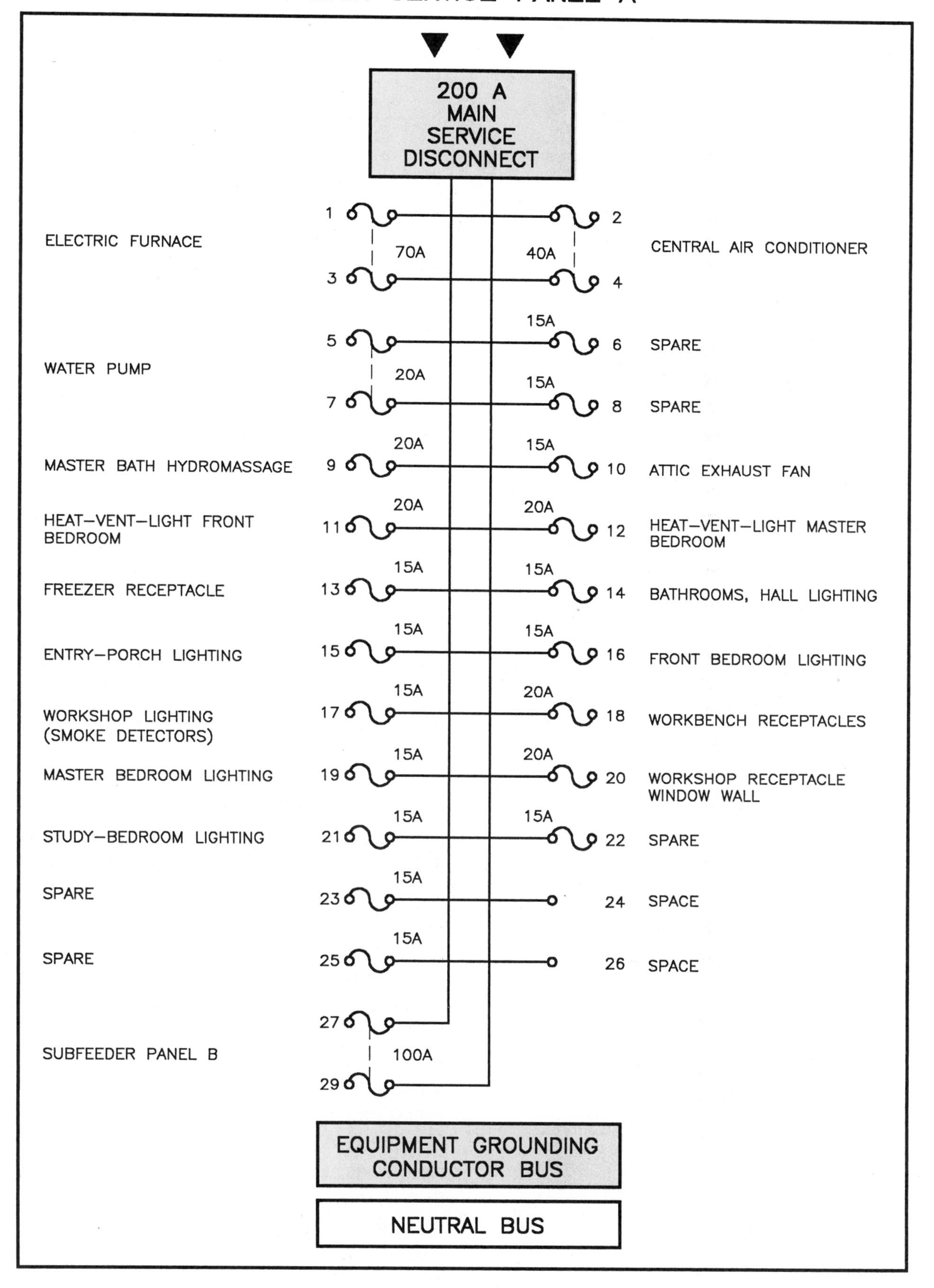

Fig. 28-9 Circuit schedule of main service panel A.

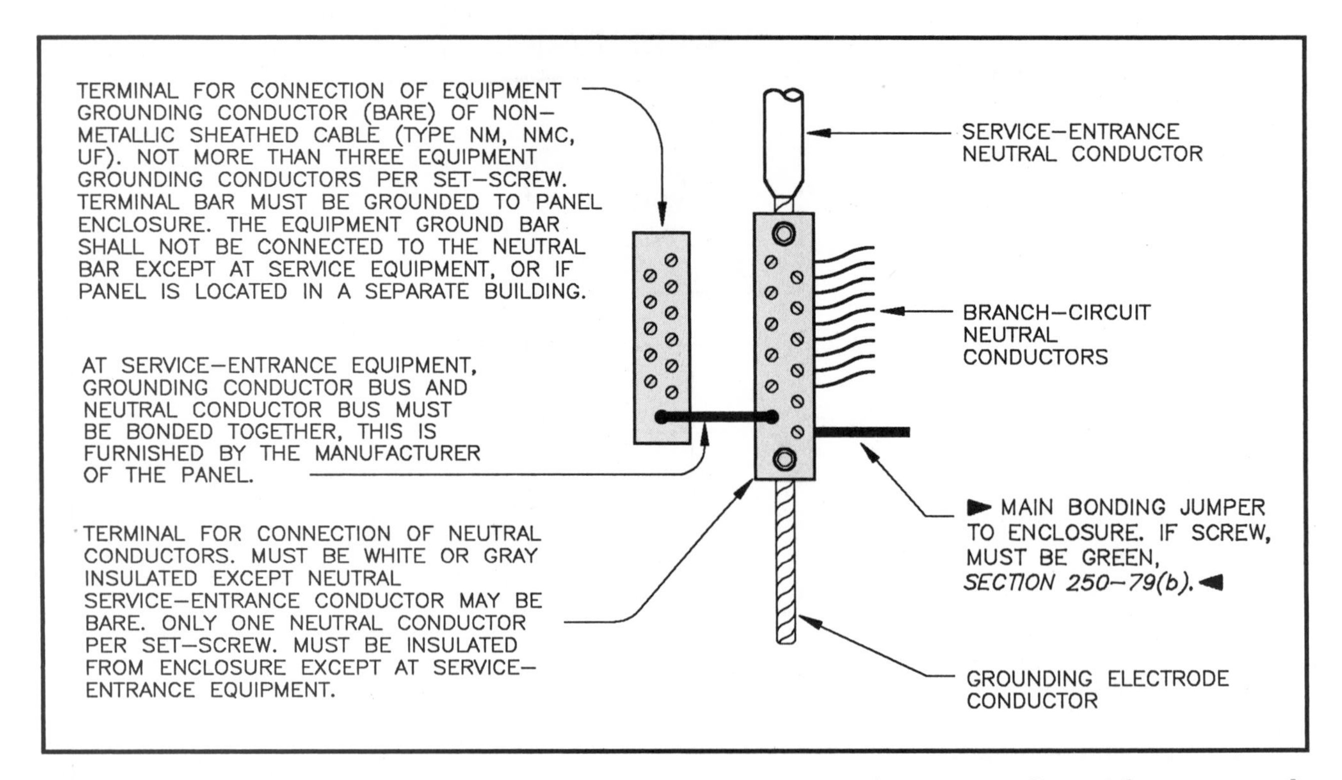

Fig. 28-10 Connections of service neutral, branch-circuit neutral, and equipment grounding conductors at panelboards, *Section 384-20.*

As previously mentioned, *Section 250-61(b)* prohibits connecting the grounded neutral of a system to any equipment on the load side of the main service disconnect. Therefore, the green main-bonding jumper screw furnished with residential-type panels will be inserted at the main service panel *only.* This screw will NOT be installed at subpanels, such as Panel B in the recreation room of the residence.

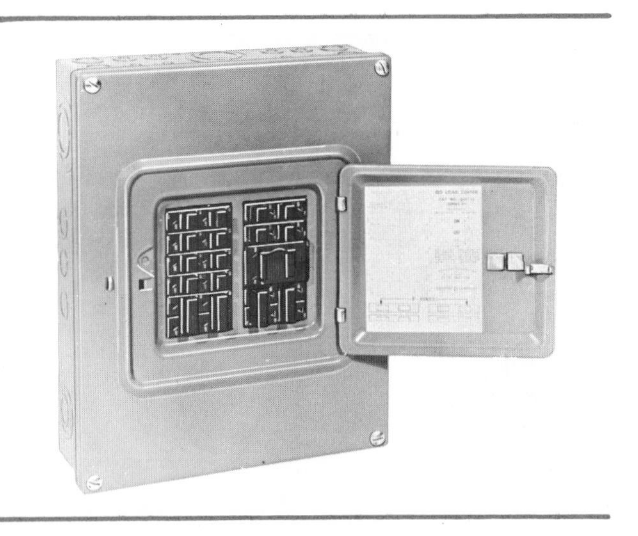

Fig. 28-11 Typical load center of the type installed for Panel B in this residence.

Load Center (Panel B)

When the main service panel is located some distance from those areas having many circuits and/or heavy load concentration, as in the case of the kitchen or laundry of this residence, it is recommended that load centers be installed near these concentrations of load. The individual branch-circuit conductors are run to the load center, not back to the main panel. Thus, the branch-circuit runs are short, and line losses (voltage and wattage) are less than if the circuits had been run all the way back to the main panel.

The cost of material and labor to install the extra load center should be compared to the cost of many branch circuits run back to the main panel to determine if there is a cost or benefit and if the lower line losses justify installing an extra load center.

A typical load center is shown in figure 28-11.

The circuit schedule for Panel B, the load center of the residence, is shown in figure 28-12. Panel B is located in the Recreation Room. It is fed by three No. 3 THHN or THWN conductors run in a 1-inch conduit originating from Panel A. The conductors are protected by a 100-ampere, 240-volt, two-pole overcurrent device in Panel A. The overcurrent protection can be fuses or circuit breakers, or a combination of both. This is discussed later in this unit.

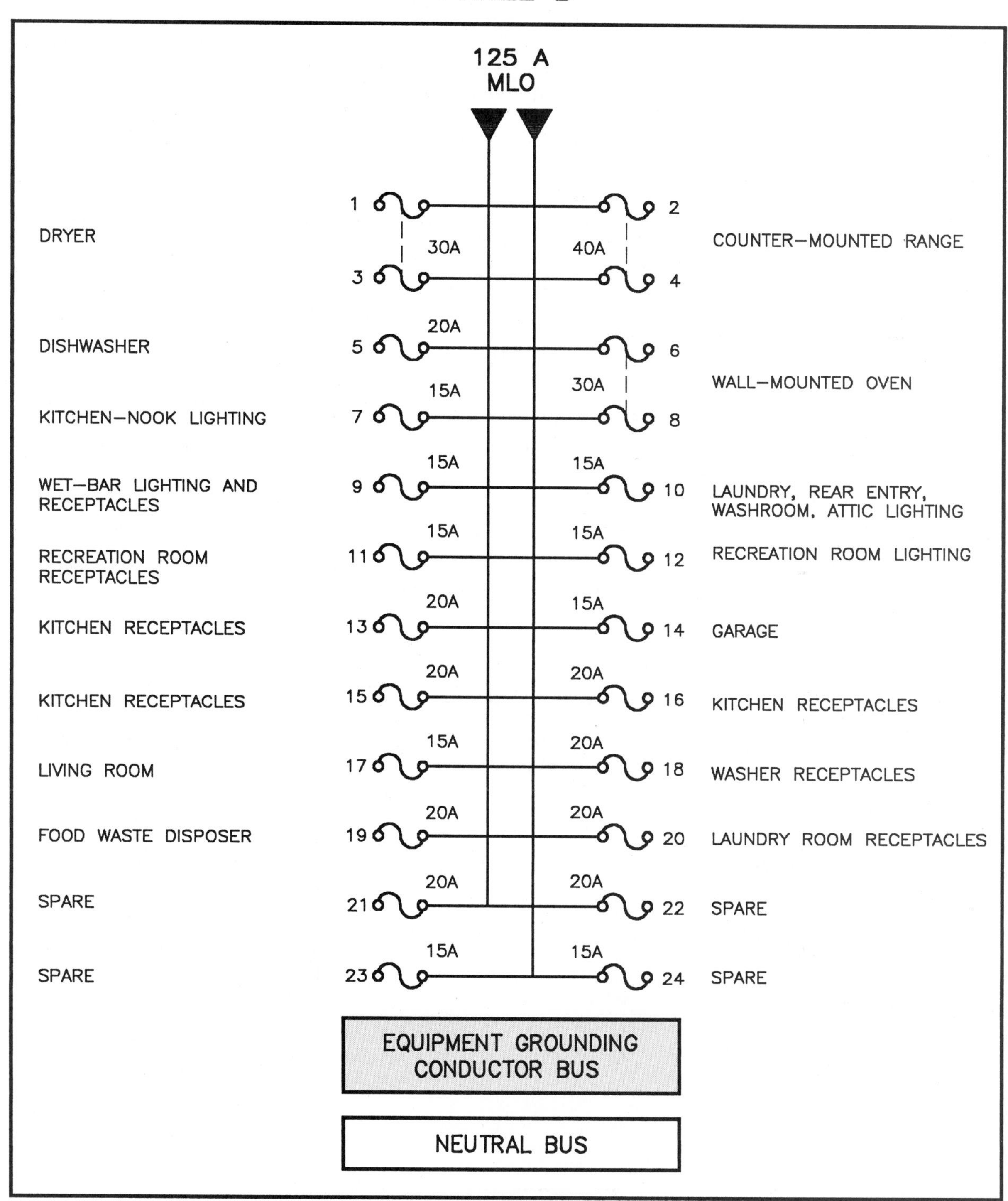

Fig. 28-12 Circuit schedule of subpanel B.

SERVICE-ENTRANCE CONDUIT SIZING

To calculate the proper size conduit that must be installed, the computation is the same as we have illustrated throughout this text relating to conduit fill.

Look up the necessary data in *Tables 4, 5,* and *8* of *Chapter 9* of the Code.

For this residence, we show two possibilities, depending upon the size and type of conductor selected for the service-entrance conductors. As also previously mentioned, a through-the-roof mast service might have to be selected upon mechanical strength instead of conduit fill. See figure 28-13.

METER

Some electric utility companies offer lower rates for water heaters and electric heating units connected to separate meters. The residence in the plans has only one meter. Thus, all lighting, heating, cooking, and water heater loads are registered on one meter.

For a typical overhead service a meter socket, figures 28-14 and 28-6, is mounted at eye level on the outside wall of the house. The service conduit is connected to the socket at a threaded boss (hub) on the top of the socket.

This conduit would be carried down into the

POSSIBILITY ONE		POSSIBILITY TWO	
TWO NO. 2/0 THHN	0.2265 SQ. IN.	TWO NO. 3/0 THHN	0.2715 SQ. IN.
	0.2265 SQ. IN.		0.2715 SQ. IN.
ONE NO. 1 BARE	0.0870 SQ. IN.	ONE NO. 1 BARE	0.1090 SQ. IN.
	0.5400 SQ. IN.		0.6520 SQ. IN.
CONDUIT SIZE	1 1/4 INCH	CONDUIT SIZE	1 1/2 INCH

Fig. 28-13 Examples of conduit fill for the service entrance in this residence. Obtain conductor area from *Table 5.* Use *Table 4* to find conduit fill percentages.

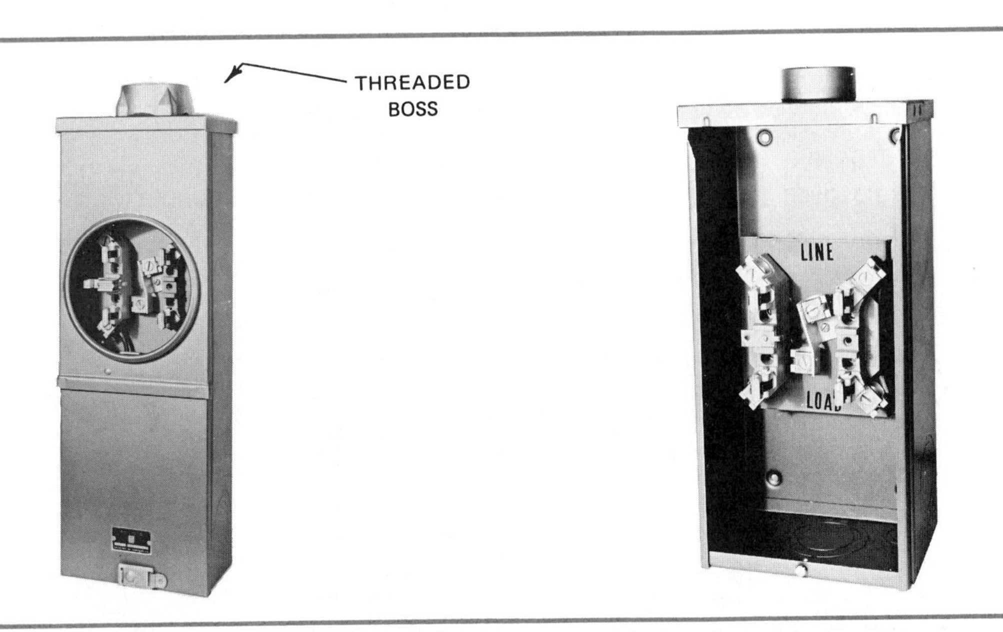

Fig. 28-14 Typical meter sockets.

SECTION 300-5(g) AND *SECTION 300-7* STATE THAT WHEN RACEWAYS PASS THROUGH AREAS HAVING GREAT TEMPERATURE DIFFERENCES, SOME MEANS MUST BE PROVIDED TO PREVENT PASSAGE OF AIR BACK AND FORTH THROUGH THE RACEWAY. NOTE THAT OUTSIDE AIR IS DRAWN IN THROUGH THE CONDUIT WHENEVER A DOOR OPENS. COLD OUTSIDE AIR MEETING WARM INSIDE AIR CAUSES THE CONDENSATION OF MOISTURE. THIS CAN RESULT IN RUSTING AND CORROSION OF VITAL ELECTRICAL COMPONENTS. EQUIPMENT HAVING MOVING PARTS, SUCH AS CIRCUIT BREAKERS, SWITCHES, AND CONTROLLERS, IS ESPECIALLY AFFECTED BY MOISTURE. THE SLUGGISH ACTION OF THE MOVING PARTS IN THIS EQUIPMENT IS UNDESIRABLE.

INSULATION OR OTHER TYPE OF SEALING COMPOUND CAN BE INSERTED AS SHOWN TO PREVENT THE PASSAGE OF AIR.

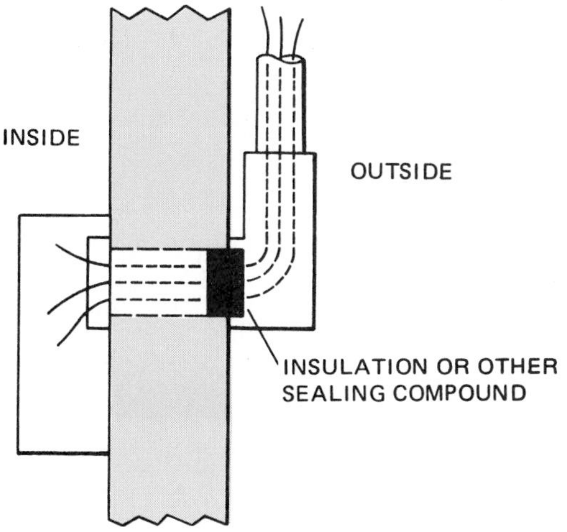

INSIDE

OUTSIDE

INSULATION OR OTHER
SEALING COMPOUND

Fig. 28-15 Installation of conduit through a basement wall.

basement, connecting to the main service equipment as shown in figures 28-6, 28-7, and 28-15. Proper fittings must be used, and the conduit must be sealed where it penetrates the wall, figure 28-15.

Probably more common for residential services is the "pedestal" as shown in figure 28-16. Many utilities furnish and install watthour meters on a pedestal of this type. Although usually mounted on the side of a house, they can be installed on the lot line between residential properties. The utility generally runs the underground lateral service-entrance conductors in a trench from a pad-mount transformer to the line side of the metering portion of the pedestal. The electrician's responsibility is to install the service-entrance conductors from the load side of the meter to the line side of the main service disconnect switch.

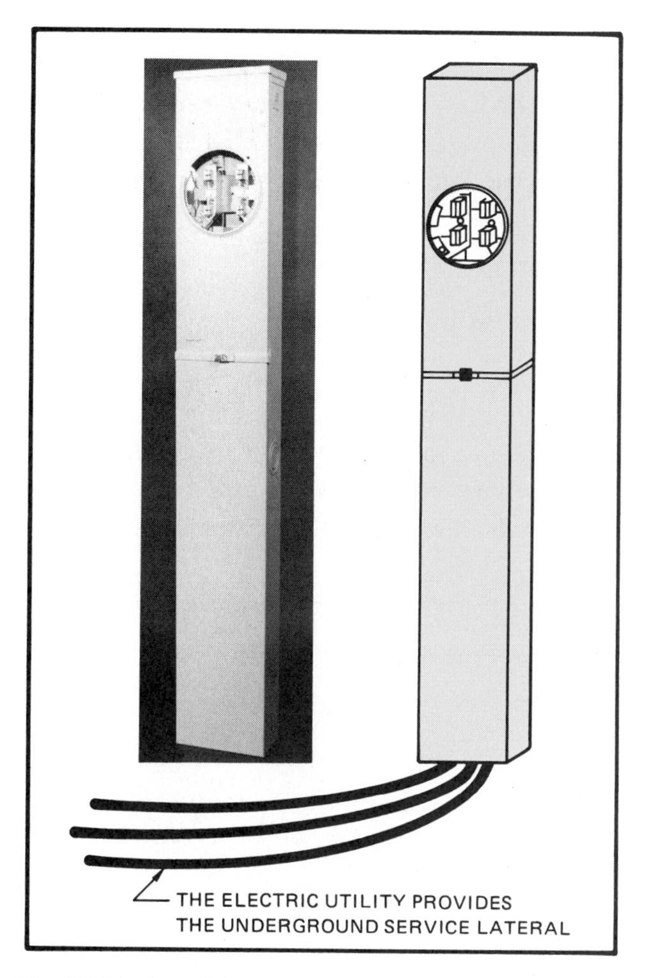

Fig. 28-16 Installing a metal "pedestal" allows for ease of installation of the underground service-entrance conductors by the utility. Courtesy of Millbank Manufacturing Co.

COST OF USING ELECTRICAL ENERGY

A watthour meter is always connected into some part of the service-entrance equipment. In residential metering, a watthour meter would normally be installed as part of the service entrance.

All electrical energy consumed is metered so that the utility can bill the customer on a monthly or bimonthly basis.

The kilowatt (kW) is a convenient unit of electrical power. One thousand watts (W) is equal to one kilowatt (kW). The watthour meter measures both wattage and time. As the dials of the meter turn, the kilowatt-hour (kWh) consumption is continually recorded.

Utility rates are based upon "so many cents per kilowatt-hour."

EXAMPLE: One kWh will light a 100-watt light bulb for 10 hours. One kWh will operate a 1000-watt electric heater for 1 hour. Therefore, if the electric rate is 8 cents per kilowatt-hour, the use of a 100-watt bulb for 10 hours would cost 8 cents. Or a 1000-watt electric heater could be used for 1 hour at a cost of 8 cents.

$$\text{kWh} = \frac{\text{watts} \times \text{hours}}{1000} = \frac{100 \times 10}{1000} = 1 \text{ kWh}$$

The cost of the energy used by an appliance is

$$\text{Cost} = \frac{\text{watts} \times \text{hours used} \times \text{cost per kWh}}{1000}$$

PROBLEM: Find the cost of operating a color television for 8 hours. The set is rated at 175 watts. The electric rate is 9.6 cents per kilowatt-hour.

SOLUTION:

$$\text{Cost} = \frac{175 \times 8 \times 0.096}{1000} = 0.1344 \text{ (13.4 cents)}$$

Assume that a meter presently reads 18672 kWh. The previous meter reading was 17895 kWh. The difference is 777 kWh.

Here is how a typical electric bill might look:

GENERIC ELECTRIC COMPANY

Days of Service	from	to	Due Date
33	01-28-91	03-01-91	03-25-91

Present reading	18672
Last reading	17895
Kilowatt-hours	777
Rate/kWh	0.08
Amount	62.16
State tax	3.23
Now due	65.39

Some utilities apply a fuel adjustment charge that may increase or decrease the electric bill. These charges are on a "per KWh" basis. This fuel adjustment charge enables the electric utility to recover from the consumer extra expenses it might incur for fuel costs used in generating electricity. These charges can vary each time the utility prepares the bill, without it having to apply to the regulatory agency for a rate change.

Many utilities increase their rates during the summer months when people turn on their air conditioners. This air-conditioning load taxes the utility's generating capabilities during the peak summer months. The higher rate structure also gives people an incentive for not setting their thermostats too low, thus saving energy.

To encourage customers to conserve energy, a utility company might offer the following rate structure:

First 500 kWh @ 6.12 cents per kWh

Over 599 kWh @ 10.64 cents per kWh (summer rate)

Over 500 kWh @ 8.64 cents per kWh (winter rate)

GROUNDING — WHY GROUND?

Electrical systems and their conductors are grounded to minimize voltage spikes when lightning strikes, or when other line surges occur. Grounding stabilizes the normal voltage to ground.

Electrical metallic conduits and equipment are grounded so that the equipment voltage to ground is kept to a low value. In this way, the shock hazard is reduced.

Proper grounding means that overcurrent devices can operate faster when responding to ground faults. Effective grounding occurs when a low-impedance (ac resistance) ground path is provided. A low-impedance ground path means that there is a high value of ground-fault current. As the ground-fault current increases, there is an increase in the speed with which a fuse will open or a circuit breaker will trip. Thus, as the overcurrent device senses and opens the circuit faster, less equipment damage results. This is called *inverse time*, which simply stated means that the higher the value of current, the less time it will take to operate the overcurrent device.

The arcing damage to electrical equipment and

conductor insulation is closely related to the value of ampere-squared-seconds (I^2t), where

I = current flowing from phase to ground, or from phase to phase, in amperes

t = time needed by the overcurrent device to open the circuit, in seconds

This expression shows that there will be less equipment and/or conductor damage when the fault current is kept to a low value, and when the time that the fault current is allowed to flow is kept to a minimum.

Grounding electrode conductors and equipment grounding conductors carry an insignificant amount of current under normal conditions. However, when a ground fault occurs, these grounding conductors must be capable of carrying whatever value of fault current might flow for the time it takes the overcurrent protective device to clear the fault.

This would be referred to as the "conductor's short-time withstand rating." *Tables 250-94* and *250-95* of the Code are based upon the fact that copper conductors can withstand

One ampere for 5 seconds for every 30 circular mils.

For example, a No. 8 AWG copper conductor has a short-time 5-second withstand rating of

$$\frac{16\,519 \text{ CM}}{30 \text{ CM}} = 550 \text{ amperes}$$

It can be seen that too much fault current flowing will damage the conductor insulation or even completely melt off the conductor, resulting in loss of the grounding path.

The short-time withstand rating not only applies to grounding conductors but to *all* other circuit conductors as well.

GROUNDING ELECTRODE SYSTEMS, *ARTICLE 250, PART H*

In the grounding electrode system, rather than grounding a single item such as the neutral conductor, the electrician must be concerned with grounding and bonding together an entire system. The term *system* means the service neutral conductor, the grounding electrode, hot and cold water pipes, gas pipes, service-entrance equipment, and jumpers installed around meters. If any of these system parts becomes disconnected or open, the integrity of the grounding system is maintained through other paths. This means that all parts of the system must be tied (bonded) together.

Figure 28-17 and the following steps illustrate what can happen if an entire system is not grounded.

1. A live wire contacts the gas pipe. The bonding jumper Ⓐ is not installed originally.
2. The gas pipe now has 120 volts on it. The pipe is hot.
3. The insulating joint in the gas pipe results in a poor path to ground; assume the resistance is 8 ohms.

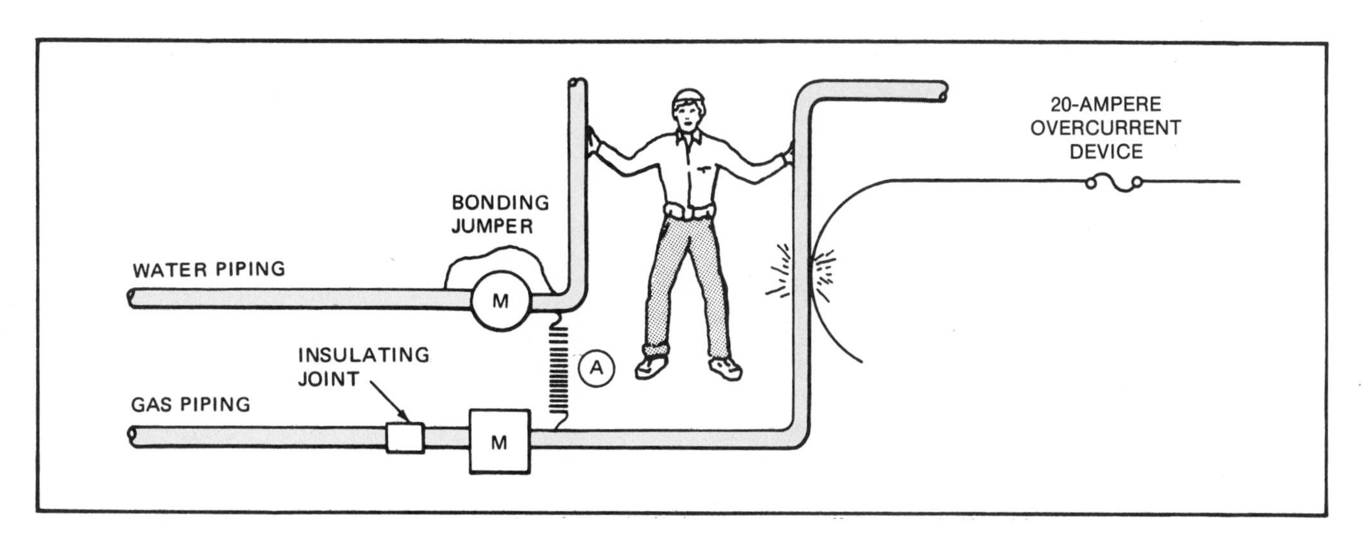

Fig. 28-17 System grounding. See text for explanation.

4. The 20-ampere overcurrent device does not open:

$$I = \frac{E}{R} = \frac{120}{8} = 15 \text{ amperes}$$

5. If a person touches the hot gas pipe and the water pipe at the same time, current flows through the person's body. If the body resistance is 12 000 ohms, the current is:

$$I = \frac{E}{R} = \frac{120}{12000} = 0.01 \text{ ampere}$$

This value of current passing through a human body can cause death.

6. The overcurrent device is now "seeing" 15 + 0.01 = 15.01 amperes; however, it still does not open.
7. If the *system grounding* concept had been used, bonding jumper Ⓐ would have kept the voltage difference between the water pipe and the gas pipe at zero. Thus, the overcurrent device would open. If 10 feet (3.05 m) of No. 4 AWG copper wire is used as the jumper, then the resistance of the jumper is 0.00259 ohm, per *Table 8, Chapter 9, NEC®* The current is

$$I = \frac{E}{R} = \frac{120}{0.00259} = 46332 \text{ amperes}$$

(In an actual system, the impedance of all of the parts of the circuit would be much higher. Thus, a much lower current would result. The value of current, however, would be enough to cause the overcurrent device to open.)

Advantages of System Grounding

To appreciate the concepts of proper grounding, let us review some important Code sections.

Section 250-1 emphasizes what is to be grounded and bonded. There are two *Fine Print Notes* to *Section 250-1* that are extremely important.

(FPN): Systems and circuit conductors are grounded to limit voltages due to lightning, line surges, or unintentional contact with higher-voltage lines, and to stabilize the voltage to ground during normal operation. Systems and circuit conductors are solidly grounded to facilitate overcurrent device operation in case of ground faults.

(FPN): Conductive materials enclosing electrical conductors or equipment or forming part of such equipment are grounded to limit the voltage to ground on these materials and to facilitate overcurrent device operation in case of ground faults. See *Section 110-10.*

Section 250-51 discusses "Effective Grounding Path" (1) to be permanent and continuous, (2) to have capacity to conduct safely *any* fault current likely to be imposed on it, and (3) to have sufficiently low impedance to limit the voltage to ground and to facilitate the operation of the circuit protective device in the circuit.

Section 250-70 is the general Code rule for bonding, stating that "Bonding shall be provided where necessary to assure electrical continuity and the capacity to conduct safely *any* fault current likely to be imposed."

Section 250-75 states in part that the bonding of metal raceways, enclosures, fittings, and so on, that serve as the grounding path "shall be effectively bonded where necessary to assure electrical continuity and the capacity to conduct safely *any* fault current likely to be imposed on them."

Section 240-1, Fine Print Note (FPN) points out that overcurrent protection for conductors and equipment is provided to open the circuit if the current reaches a value that will cause an excessive or dangerous temperature in conductors or conductor insulation. See also *Sections 110-9* and *110-10* for requirements for interrupting rating and protection against fault currents.

The system grounding concept has several advantages.

- The potential voltage differences between the parts of the system are minimized, reducing the shock hazard.
- The impedance of the ground path is minimized. This results in a higher current flow in the event of a ground fault: the lower the impedance, the higher the current flow. This means that the overcurrent device will open faster under fault conditions.

Methods of Grounding

Reviewing figure 28-5 shows that both the metal hot and cold water pipes, the service raceways, the metal enclosures, the service switch, and the neutral conductor are bonded together to form a *grounding electrode system, Section 250-81.*

For discussion purposes regarding this residence, a driven ground rod is the supplemental electrode as required by *Section 250-81(a).* This ground rod supplements the water pipe ground, *Section 250-83(c).*

The size of a grounding electrode conductor that is run to a concrete-encased electrode (sometimes referred to as a UFER ground, and named after the gentleman who developed the concept) need not be larger than No. 4 AWG copper wire, per *Section 250-94, Exception No. 1(b).*

Where a driven rod (*Section 250-83*) is the supplementary grounding electrode, the size of the grounding electrode conductor that is the *sole* connection to the driven rod need not be larger than No. 6 copper or No. 4 aluminum, *Section 250-81(a).*

If the driven ground rod is not selected as the required supplemental grounding electrode, then any one of the following items may be used:

- the metal frame of a building (the residence is constructed of wood).
- at least 20 feet (6.1 m) of steel reinforcing bars, 1/2-inch (12.7-mm) minimum diameter, or at least 20 feet (6.1 m) of bare copper conductor not smaller than No. 4 AWG. Both of these must be encased in concrete at least two-inches (50.8-mm) thick and in direct contact with the earth, such as near the bottom of a foundation or footing.
- at least 20 feet (6.1 m) of bare copper wire encircling the building, having a minimum size of No. 2 AWG, buried directly in the earth at least 2 1/2 feet (762 mm) deep.
- ► metal underground gas piping system — *Section 250-83(a)* prohibits the use of metal underground gas piping system to be used as a grounding electrode.◄
- ground plates — at least 2 square feet ($0.186 \, m^2$).* See *Section 250-83(d).*

  *More than one rod or plate must be installed if the resistance to ground exceeds 25 ohms. See *Section 250-84.*

The requirements for a grounding electrode system are covered in *NEC® Article 250*, parts G, H, J, and K.

There is little doubt that this concept of *grounding electrode systems* gives rise to many interpretations of the Code. The electrician must check with the local code authority to determine the local interpretation. For example, some electrical inspectors may not require a bonding jumper to be installed between the cold and hot water pipes, as shown in figures 28-5 and 28-19. They are of the opinion that an adequate bond is made through the water heater itself, and through the many hot/cold water mixing faucets.

Other electrical inspectors will require that the cold and hot water pipes be bonded together. One reason for this requirement is that some water heaters contain insulating fittings that reduce corrosion caused by electrolysis inside the tank. Another reason given to justify their stand is that while the water heater installed first may contain no insulating fittings, a replacement heater may have insulating fittings. Thus, there will be no bond between the hot and cold water pipes. When there is any doubt as to the bonding requirements, bond the pipes together as shown in figure 28-5.

GROUNDING THE SERVICE WHEN NONMETALLIC WATER PIPE IS USED

Today it is quite common to find the main water supply to a residence installed with plastic (PVC) piping. Even interior water piping systems may be nonmetallic piping. Some building codes prohibit the use of nonmetallic piping. Nevertheless it is important that the electrical system be properly grounded and bonded.

Probably the most popular method to provide for adequate grounding of services where nonmetallic water services are encountered, or where the soil is so dry (sand) that driven ground rods cannot attain a 25 ohms or less low-resistance ground, *Section 250-84*, or where the metallic underground water piping in direct contact with the earth is less than 10 feet, is to install a *concrete-encased electrode*, commonly referred to as a *UFER ground*, figure 28-18.

The UFER ground consists of a bare copper conductor

- not smaller than No. 4 AWG
- not less than 20 feet (6.1 m) long encased by at least 2 inches (50.8 mm) of concrete in the footing or foundation
- in concrete that is in direct contact with the earth
- lying within and near the bottom of the concrete
- solid or stranded

Some soil conditions are so bad relative to relying on the soil for grounding purposes, that the UFER ground becomes the primary grounding electrode, as opposed to the supplemental ground.

SUMMARY — SERVICE-ENTRANCE EQUIPMENT GROUNDING

When grounding service-entrance equipment, figure 28-19, the following Code rules must be observed.

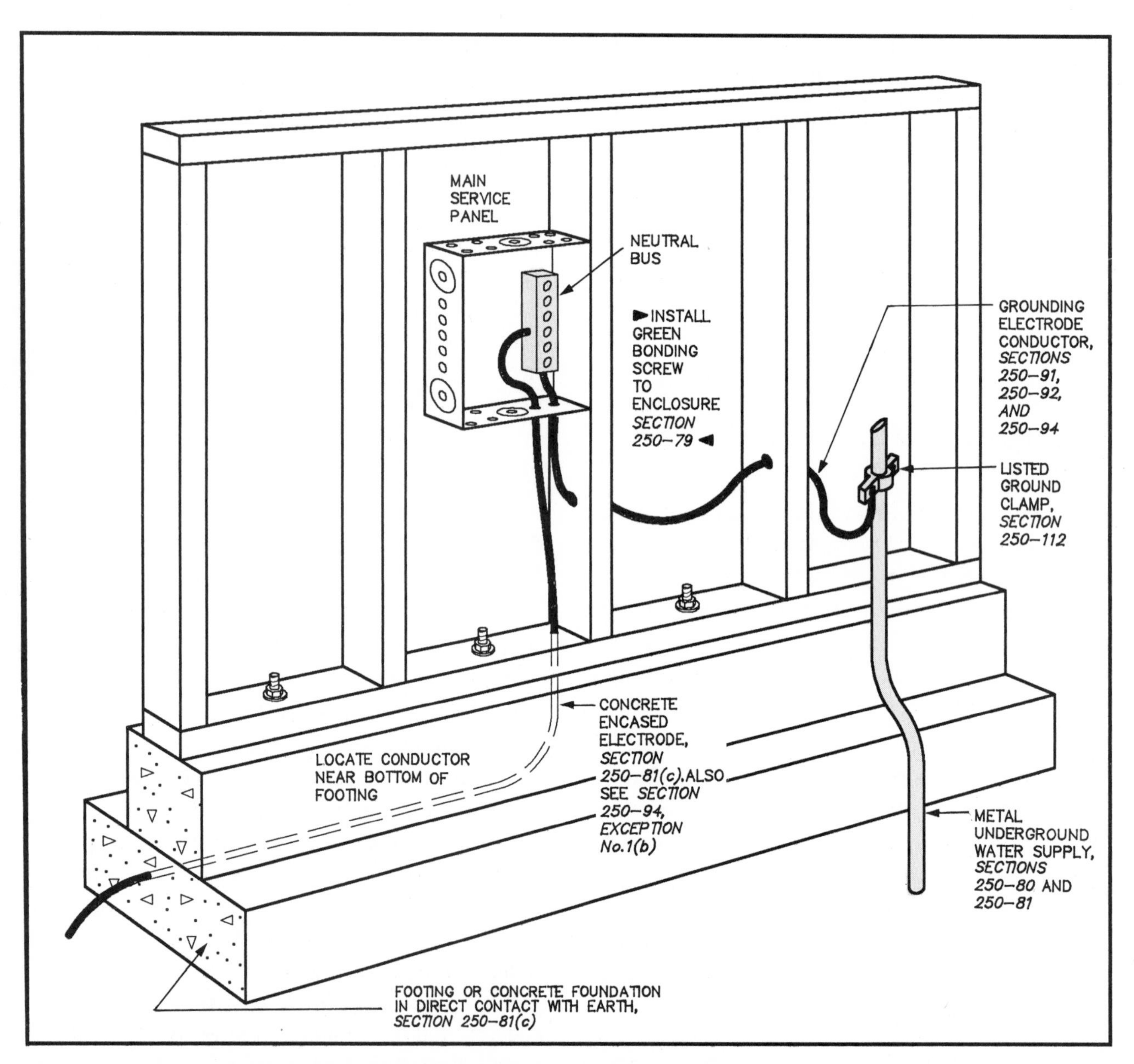

Fig. 28-18 ***Section 250-81*** **of the Code states that if available, an underground metallic water pipe 10 feet or longer in direct contact with the earth shall be used as the grounding electrode. This water pipe ground *must be supplemented* by at least one additional electrode. Here we show a concrete-encased No. 4 AWG bare copper conductor laid near the bottom of the footing, encased by at least 2 inches of concrete. The *minimum* length of the conductor is 20 feet (6.1 m). See *Section 250-81(c)*.**

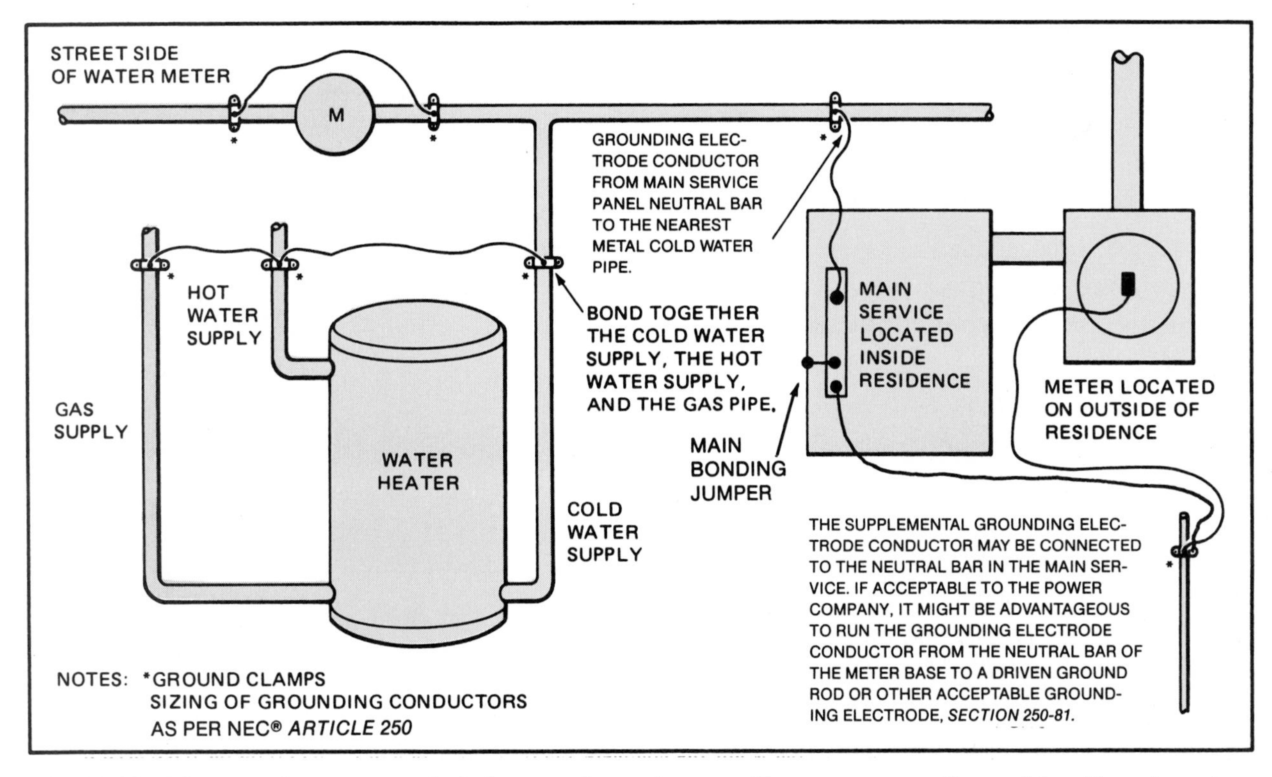

Fig. 28-19 Diagram shows one method that may be used to provide proper grounding and bonding of service-entrance equipment for a typical one-family residence.

- The system must be grounded when the maximum voltage to ground does not exceed 150 volts, *Section 250-5(b)(l)*.
- All grounding schemes shall be installed so that no objectionable currents will flow over the grounding conductors and other grounding paths, *Section 250-21(a)*.
- The ground electrode conductor must be connected to the supply side of the service disconnecting means. It must not be connected to any grounded circuit conductor on the load side of the service disconnect, *Section 250-23(a)*.
- The neutral conductor must be grounded, *Section 250-25*.
- Tie (bond) everything together. See the following section of this unit on bonding. *Sections 250-80(a)* and *(b)*, *250-81*, and *250-71*.
- The grounding electrode conductor used to connect the grounded neutral conductor to the grounding electrode must not be spliced, *Sections 250-53* and *250-91(a)*.
- The grounding electrode conductor is to be sized according to *Table 250-94*.
- The metal hot and cold water piping system shall be bonded to the service equipment enclosure, to the grounded conductor at the service, and to the grounding electrode conductor, *Section 250-80(a)*.
- The grounding electrode conductor must be connected to the metal underground water pipe when 10 feet (3.05 m) long or more, including the well casing, *Section 250-81*.
- In addition to grounding the service equipment to the underground water pipe, an additional electrode must be used, such as a bare conductor in the footing, *Section 250-81(c)*, a grounding ring, *Section 250-81(d)*, a metal underground gas piping system, *Section 250-83(a)*, a rod or pipe electrodes, *Section 250-83(c)*, or plate electrodes, *Section 250-83(d)*.
- The grounding electrode conductor shall be copper, aluminum or copper-clad aluminum, *Section 250-91*.
- The grounding electrode conductor may be solid or stranded, uninsulated, covered or bare, and must not be spliced, *Section 250-91*.

- Bonding shall be provided around all insulating joints or sections of the metal piping system that may be disconnected, *Section 250-112*.
- The connection to the grounding electrode must be accessible, *Section 250-112*.
- The grounding conductor must be connected tightly using the proper lugs, connectors, clamps, or other approved means, *Section 250-115*.

Section 230-65 should be reviewed at this point (this section is described in detail in *Electrical Wiring—Commercial*). This section requires adequate interrupting capacity for services (discussed later in this unit).

This residence is supplied by three No. 2/0 THWN or THHN service-entrance conductors. According to *Table 250-94*, a No. 4 AWG grounding electrode conductor is required. This conductor may be run in conduit or cable armor, or it may be run exposed if it is not to be subjected to severe physical damage. For this residence, a No. 4 AWG armored ground cable is run from the top of Panel A across the Workshop ceiling to the water pipe, where it is terminated in a proper ground clamp.

BONDING

Section 250-71 lists the parts of the service-entrance equipment that must be bonded. *Section 250-72* lists the methods approved for bonding this service equipment. Bonding bushings, figure 28-20, and bonding jumpers are installed on service-entrance equipment to insure a low-impedance path to ground if a fault occurs on any of the service-entrance conductors. Service-entrance conductors

Fig. 28-20 Insulated bonding bushing with grounding lug.

Fig. 28-21 Insulated bushings.

are not fused at the service head. Thus, the short-circuit current on these conductors is limited only by the following: (1) the capacity of the transformer or transformers supplying the service equipment and (2) the distance between the service equipment and the transformers. The short-circuit current can easily reach 20000 amperes or more in dwellings. Fault currents can easily reach 40000 to 50000 amperes or more in apartments, condominiums, and similar dwellings. These installations are usually served by a large-capacity transformer located close to the service-entrance equipment and the metering. **This extremely high fault current produces severe arcing which is a fire hazard**. The use of proper bonding reduces this hazard to some extent.

Fault current calculations are presented later in this unit. *Electrical Wiring—Commercial* covers these calculations in much greater detail.

Section 250-79 states that the main bonding jumpers must be not smaller than the grounding electrode conductor. The grounding lugs on bonding bushings are sized by the trade size of the bushing. The lugs become larger as the size of the bushing increases.

Section 373-6(c) states that if No. 4 AWG or larger conductors are installed in a raceway, an insulating bushing or equivalent must be used, figure 28-21. This bushing protects the wire from shorting or grounding itself as it passes through the metal bushing. Combination bushings can be used. These bushings are metallic (for mechanical strength) and have plastic insulation. When conductors are installed in electrical metallic tubing, they can be protected at the fittings by the use of connectors with insulated throats.

If the conduit bushing is made of insulating material only, as in figure 28-21, then two locknuts must be used, figure 28-22. *Section 373-6(c)*.

Various types of ground clamps are shown in figures 28-23, 28-24, 28-25, and 28-26. These clamps and their actual connection to the grounding electrode must conform to *Section 250-115* of the Code.

BRANCH-CIRCUIT OVERCURRENT PROTECTION

The overcurrent devices commonly used to protect branch circuits in dwellings are fuses and circuit breakers. *Article 240* of the *National Electrical Code®* discusses overcurrent protection.

Plug Fuses, Fuseholders, and Adapters (*Article 240, Part E*)

Fuses are a reliable and economical form of overcurrent protection. *Sections 240-50* through *240-54* give the requirements for plug fuses, fuseholders, and adapters. These requirements include the following:

- the protective devices shall not be used in circuits exceeding 125 volts between conductors. An exception to this rule is for a system having a grounded neutral where no conductor is more than 150 volts to ground. (This is the case for the 120/240-volt system used in the residence in the plans.)
- the fuses shall have ampere ratings of 0 to 30 amperes.
- plug fuses shall have a hexagonal configuration somewhere on the fuse when rated at 15 amperes or less.
- the screw shell of the fuseholder must be connected to the load side of the circuit.
- Edison-base plug fuses may be used only to replace fuses in existing installations where there is no sign of overfusing or tampering.

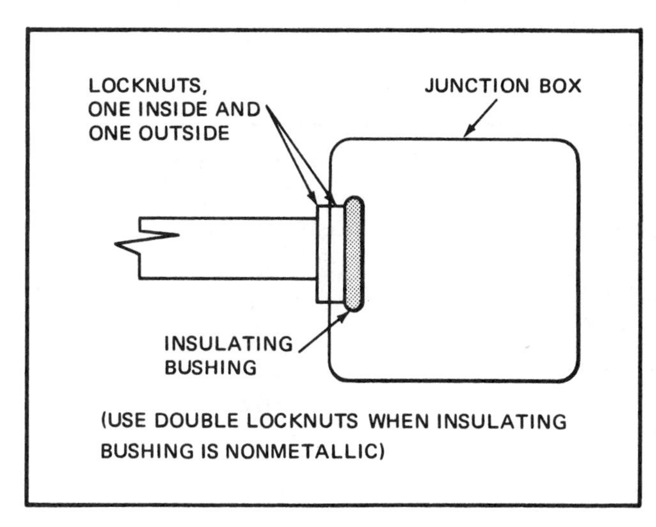

Fig. 28-22 The use of locknuts. See *Section 373-6(c)*.

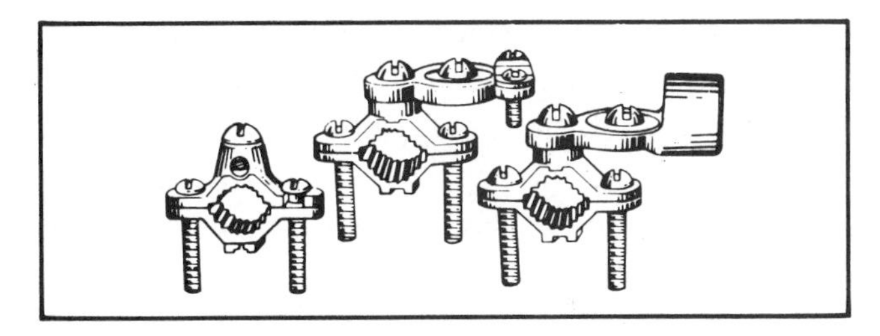

Fig. 28-23 Typical ground clamps used in residential systems.

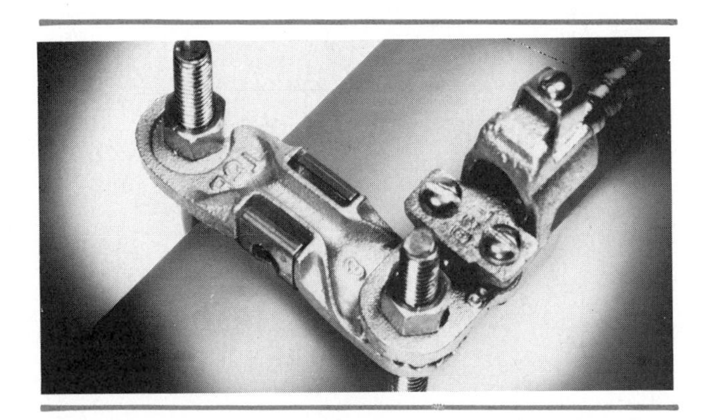

Fig. 28-24 Armored grounding conductor connected with ground clamp to water pipe.

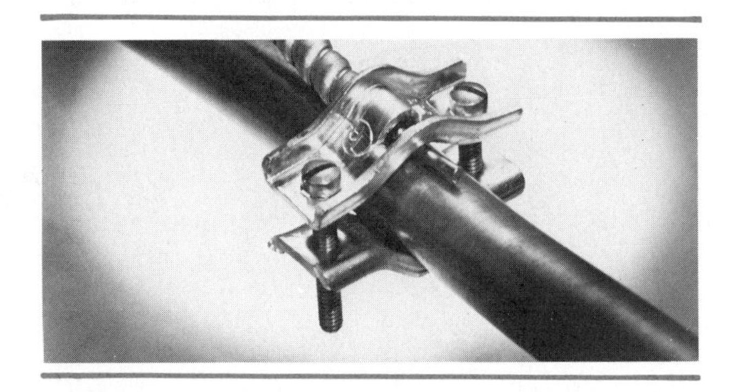

Fig. 28-25 Ground clamp of the type used to bond (jumper) around water meter.

- all new installations shall be in type S fuses.
- Type S fuses are classified at 0 to 15 amperes, 16 to 20 amperes, and 21 to 30 amperes. The reason for this classification is given in the following paragraph.

When the electrician installs fusible equipment, the ampere rating of the various circuits must be determined. Based on this rating, an adapter of the proper size is inserted into the Edison-base fuseholder. The proper Type S fuse is then placed in the adapter. Because of the adapter, the fuseholder is nontamperable and noninterchangeable. For example, assume that a 15-ampere adapter is inserted for the 15-ampere branch circuits in the residence. It is impossible to substitute a Type S fuse with a larger rating without removing the 15-ampere adapter. Another type of fuseholder that can be used is molded into the shape required for the various sizes of Type S fuses.

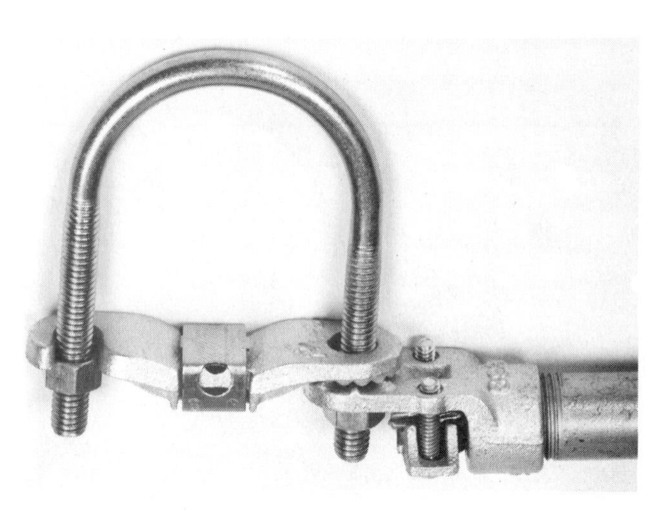

Fig. 28-26 Ground clamp of the type used to attach ground wire to well casings. Courtesy of Thomas & Betts.

Type S fuses and adapters, figure 28-27, are available with ratings in the range from 3/10 of an ampere to 30 amperes. The Type S fuse may have a time-delay feature. If it does, it is known as a dual-element fuse. Momentary overloads do not cause a dual-element fuse to blow. An example of such an overload is the current surge as an electric motor is started. However, a dual-element fuse opens rapidly if there is a heavy overload or a short circuit.

Dual-element Fuse (Cartridge and Plug Type)

The most common type of dual-element cartridge fuse is shown in figure 28-28. These fuses are available in 250-volt and 600-volt sizes with ratings from 0 to 600 amperes, and have an interrupting rating of 200,000 amperes.

A dual-element fuse has two fusible elements connected in series. These elements are known as the *overload element* and the *short-circuit element*. When an excessive current flows, one of the elements opens. The amount of excess current determines which element opens.

The electrical characteristics of these two elements are very different. Thus, a dual-element fuse has a greater range of protection than the single-element fuse. The overload element opens circuits on currents in the low overload range (up to about 500 percent of the fusing rating). The short-circuit element handles only short circuits and heavy overload currents (about 500 percent of the fuse rating).

Overload Element: The overload element opens when excessive heat is developed in the element. This heat may be the result of a loose connection or poor contact in the fuseholder or excessive current.

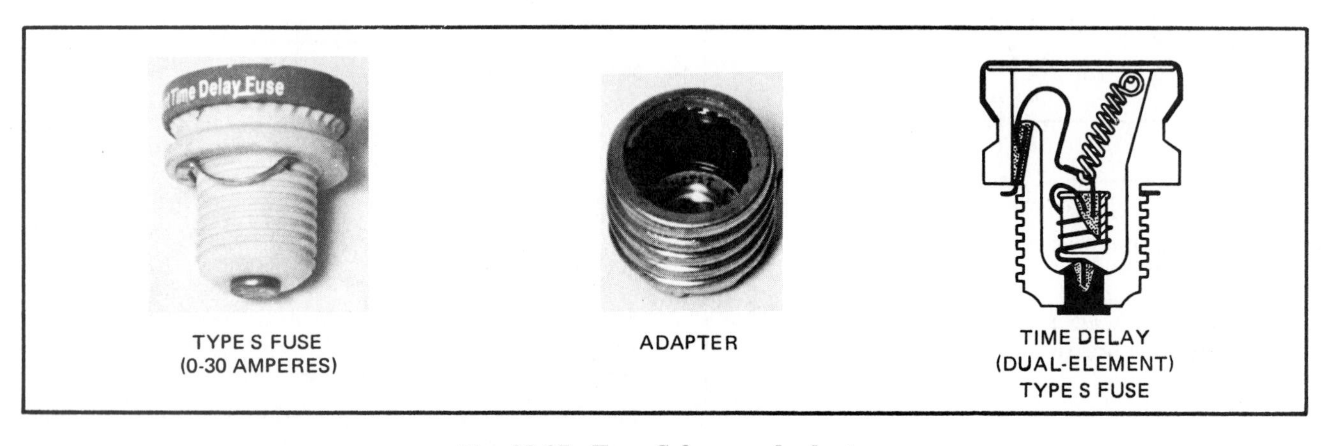

Fig. 28-27 Type S fuses and adapter.

will be visible after the cover of the panel installed).

- Every breaker with an interrupting rating oth than 5000 amperes shall have this ratii marked on the breaker.
- Circuit breakers rated at 120 volts and 277 vol and used for fluorescent loads, shall not l used as switches unless marked "SWD."

Most circuit breakers are ambient temperatu compensated. This means that the tripping point the breaker is not affected by an increase in the su rounding temperature. An ambient-compensat breaker has two elements. One element heats up d to the current passing through it and the heat in t surrounding area. The other element heats because of the surrounding air only. The actions these elements oppose each other. Thus, as the tr ping element tends to lower its tripping po because of external heat, the second elem opposes the tripping element and stabilizes t tripping point. As a result, the current throu the tripping element is the only factor that cau the element to open the circuit. It is a good pract to turn the breaker on and off periodically to "ex cise" its moving parts.

Section 240-6 gives the standard ampere ratii of circuit breakers.

INTERRUPTING RATINGS FOR FUSES AND CIRCUIT BREAKERS

Section 110-9 states that all fuses, circ breakers, and all other electrical devices that br current shall have an interrupting capacity suffici for the voltage employed and for the current wh must be interrupted.

According to *Section 110-10*, all overcurr devices, the total circuit impedance, and the w stand capability of all circuit components (wi contactors, and so on), must be selected so t minimal damage will result in the event of a fa either line-to-line or line-to-ground.

Section 230-65 of the Code states that "serv equipment shall be suitable for the short-cir current available at its supply terminals."

The overcurrent protective device must be a to interrupt the current that may flow under

When the temperature reaches 280° F, a fusible alloy melts and the element opens. Any excessive current produces heat in the element. However, the mass of the element absorbs a great deal of heat before the cutout opens. A small excess of current will cause this element to open if it continues for a long period of time. This characteristic gives the thermal cutout element a large time lag on low overloads (up to 10 seconds at a current of 500 percent of the fuse rating). In addition, it provides very accurate protection for prolonged overloads.

Short-circuit Element: The capacity of the short-circuit element is high enough to prevent it from opening on low overloads. This element is designed to clear the circuit quickly of short circuits or heavy overloads above 500 percent of the fuse ratings.

Dual-element fuses are used on motor and appliance circuits where the time-lag characteristic of the fuse is required. Single-element fuses do not have this time lag. Such fuses blow as soon as an overcurrent condition occurs. The homeowner may be tempted to install a fuse with a higher rating. The fuseholder adapter, however, prevents the use of such a fuse. Thus, the dual-element fuse is recommended for this type of situation.

Three basic types of dual-element fuses are available. The standard plug fuse can be used only to replace blown fuses on existing installations. The Type S fuse is required on all new installations. The cartridge fuse is a dual-element fuse which is available in both ferrule and knife blade styles.

Type SC Fuses

Another style of small dimension cartridge fuse is Type SC, figure 28-29.

Section 240-60(a) recognizes the use of a type of cartridge fuse that is rated at not over 60 amperes for circuits of 300 volts or less to ground. The physical size of this fuse is smaller than that of standard cartridge fuses. The time-delay characteristics of this fuse mean that it can handle harmless current surges or momentary overloads without blowing. However, these fuses open very rapidly under short-circuit conditions.

These fuses prevent the practice of overfusing since they are size limiting for their ampere ratings. For example, a fuseholder designed to accept a 15-ampere, Type SC fuse will not accept a 20-ampere, Type SC fuse. In the same manner, a fuseholder designed to accept a 20-ampere, Type SC fuse will not accept a 30-ampere, Type SC fuse.

Class T Fuses

Another type of physically small fuse is the Class T fuse, figure 28-30. This type of fuse has a high-interrupting rating in sizes from 0 to 600 amperes in both 300-volt and 600-volt ratings. Such fuses are used as the main fuses in a panel having circuit breaker branches, figure 28-31. In this case, the Class T fuses protect the low-interrupting capacity breakers against high-level short-circuit currents.

Most manufacturers list service equipment, metering equipment, and disconnect switches that make use of Class T fuses. This allows the safe installation of this equipment where fault-currents exceed 10,000 amperes, such as on large services, and any service equipment located close to pad-mount transformers, such as are found in

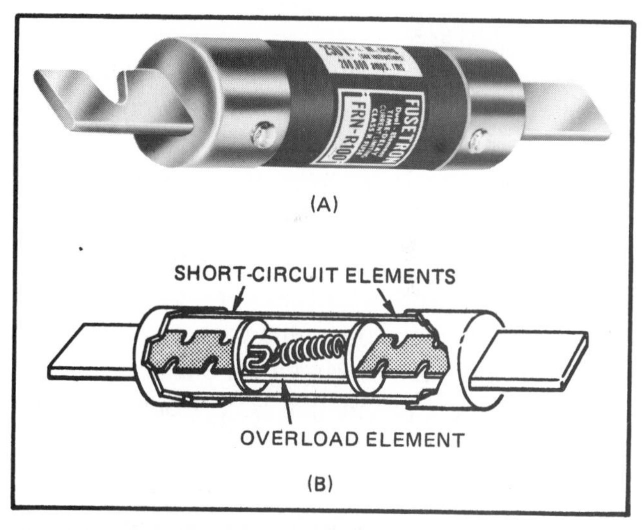

Fig. 28-28 Cartridge-type dual-element fuse (A) is a 250-volt, 100-ampere fuse. The cutaway view in (B) shows the internal parts of the fuse. Photo courtesy of Bussmann Division, Cooper Industries.

Fig. 28-29 Type SC fuses.

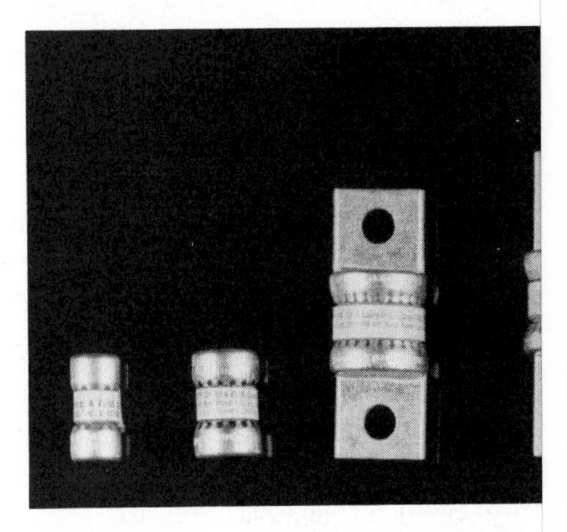

Fig. 28-30 Class T fuses. Illustrated are t
ratings. Courtesy of Bussmann Division,

apartments, condominiums, shopping centers, similar locations.

The standard ratings for fuses are given in *tion 240-6*. These ratings range from 1 amper 6000 amperes.

Circuit Breakers

Installations in dwellings normally use therr magnetic circuit breakers. On a continu overload, a bimetallic element in such a bre moves until it unlatches the inner trip mechanism of the breaker. Momentary s overloads do not cause the element to trip breaker. If the overload is heavy or if there is a s circuit, a magnetic coil in the breaker causes interrupt the branch circuit instantly. (Unit covers the effect of tungsten lamp loads on ci breakers.)

AVAILABLE FAULT CURRENT 20 800 AMPERES

SERVICE EQUIPMENT MUST HAVE SHORT-CIRCUIT RATING EQUAL TO OR GREATER THAN THE AVAILABLE FAULT CURRENT AT SUPPLY TERMINALS

MEETING REQUIREMENTS OF *SECTI*

Fig. 28-31 Fu

- panel must have short-circuit rating adequate for the available fault current.
- for overload conditions and low-level faults on a branch circuit, only the branch breaker will trip off. All other circuits remain energized. System is selective.[1]
- for branch-circuit faults (short circuits or ground faults) that exceed the instant trip setting of the main breaker (usually five times the breaker's ampere rating), both the branch breaker *and* main breaker trip off. For instance, for a branch fault of 700 amperes, a 100-ampere main breaker also trips because the main will trip instantly for faults of 500 (100 × 5) amperes or more. Entire panel is deenergized. System is nonselective.[1]
- check manufacturer's data for time-current characteristic curves and unlatching times of the breakers.

Breaker Main/Breaker Branches (figure 28-34) (Series Connected — Sometimes Called Series Rated)

- must be listed as suitable for use as service equipment.
- must be sized to satisfy the required ampacity as determined by service-entrance calculations and/or local codes.
- must be listed as Series Connected.
- main breaker will have interrupting rating higher than the branch breaker. Example: main C/B has 22000-ampere interrupting rating, whereas branch C/B has 10000-ampere interrupting rating.
- panel must be listed and marked with its maximum fault current rating.
- for branch circuit faults (L-L, L-N, L-G) above the instant trip factory setting (i.e., 200-ampere main breaker with 5 × setting: 200 × 5 = 1000 amperes) both branch and main breakers will trip. System is nonselective.[1]
- for overload conditions and low-level faults on a branch, only the branch breaker will trip. System is selective.[1]
- obtain time-current curves and unlatching time data from the manufacturer.
- the panel's integral rating is attained because when a heavy fault occurs both main and branch breakers open. The two arcs in series add impedance; thus the fault current is reduced to a level less than the 10000 A.I.C. branch breaker.
- ► for series-rated panelboards, where the available fault current exceeds the interrupting rating of the branch circuit breakers, the equipment is marked by the manufacturer, indicating the maximum available fault current that the equipment can safely be connected to, *Section 240-83(c)*.
- when panels containing low interrupting rating circuit breakers are separated from the equipment in which the higher interrupting rating circuit breakers protecting the panel are installed, the system becomes a field-installed "Series-rated System." This panel must be field marked (by the installer electrician), and this marking must be readily visible, *Section 110-22*. This marking calls attention to the fact that the actual available fault current exceeds the interrupting rating of the breakers in the panel, and that the panel is protected by properly selected circuit breakers back at the main equipment.

CAUTION SERIES-RATED SYSTEM

- CAUTION: Closely check the manufacturer's catalog numbers to be sure that the circuit breaker combination you are intending to install "in series" is recognized by Underwriters Laboratories for that purpose. This is extremely important because there are many circuit breakers that have the same physical size and dimensions, yet have different interrupting and voltage ratings, and have not been tested as "series-rated" devices. To install these as a series combination constitutes a violation of *Section 110-3(b)* and could be dangerous. ◄

See manufacturers' catalog numbers to be sure. Also check the UL Yellow Book entitled "Recognized Components Directory."

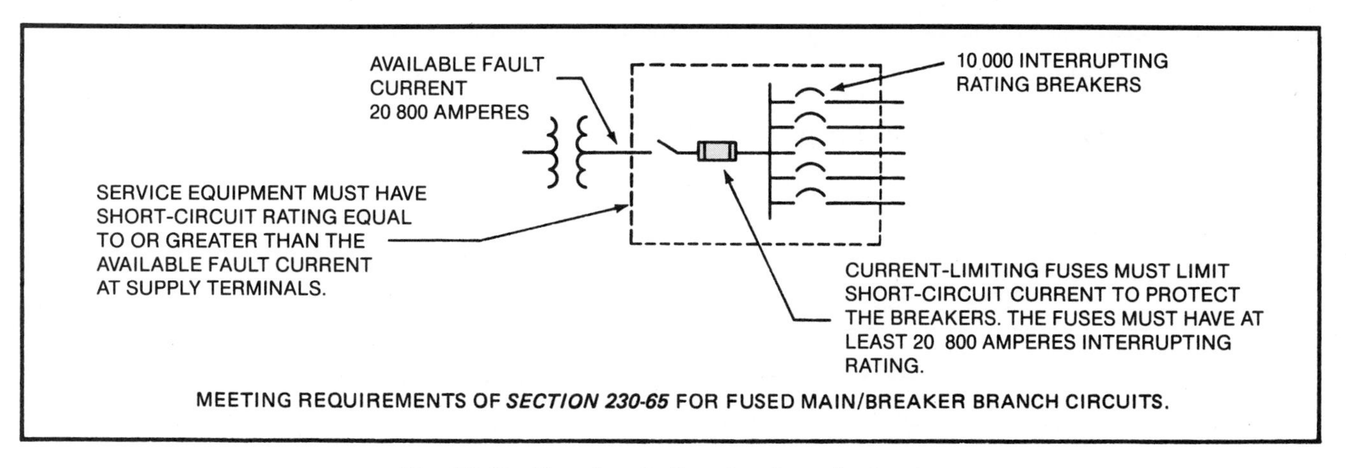

Fig. 28-32 Fused main/breaker branch circuits.

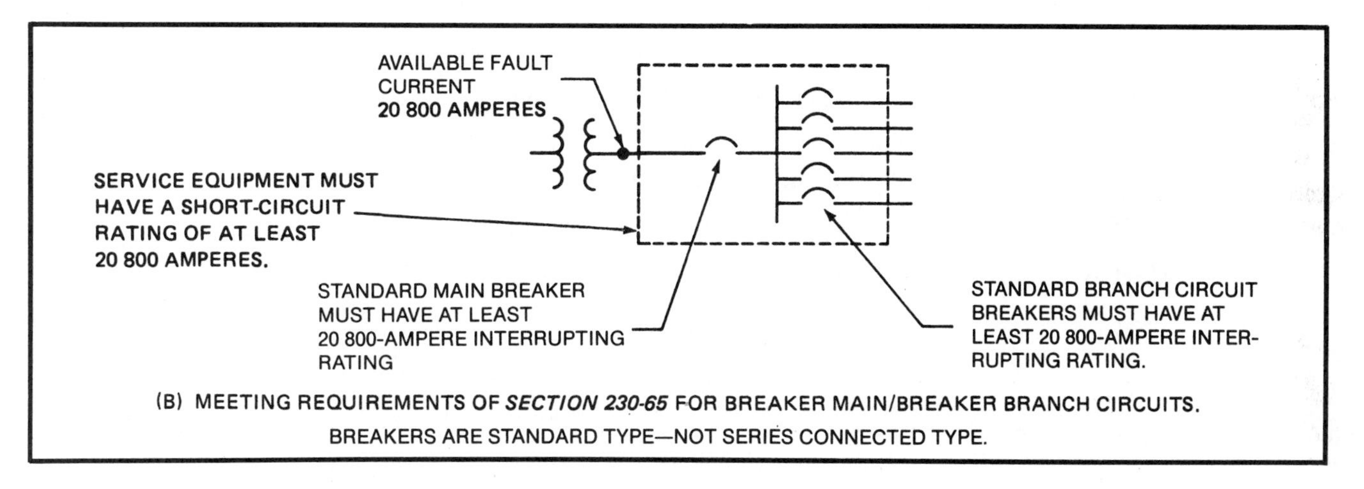

Fig. 28-33 Breaker main/breaker branch circuits.

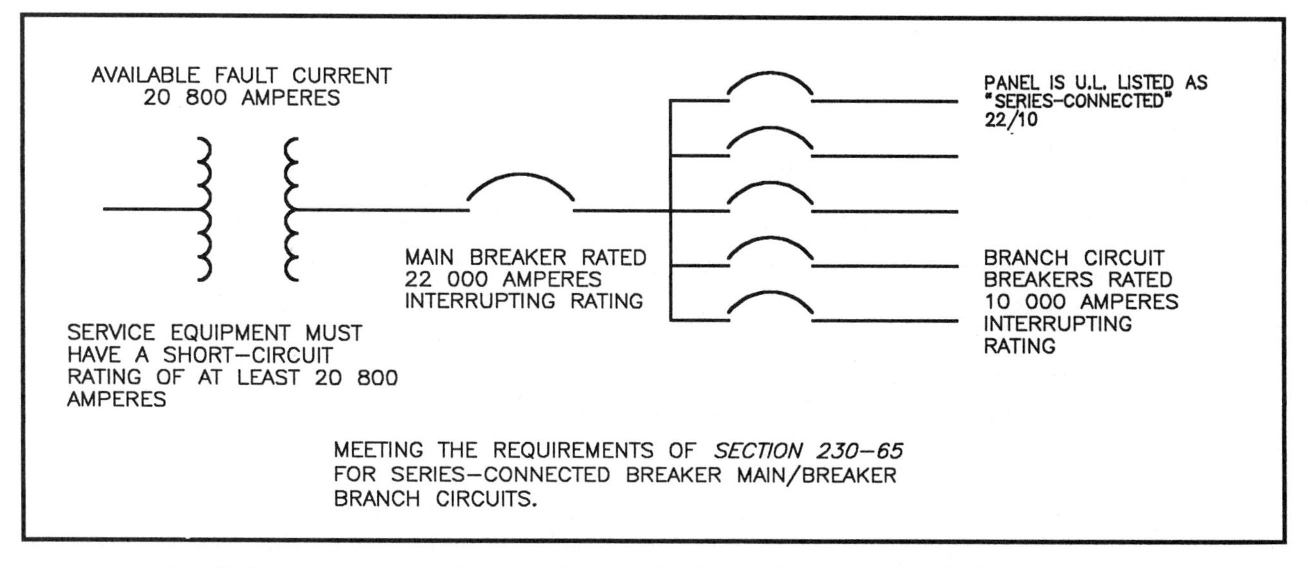

Fig. 28-34 Breaker main/breaker branch circuits (series-connected plastic-case type).

REVIEW

Note: Refer to the Code or the plans where necessary.

1. Where does an overhead service start and end? ______________________________

2. What are service-drop conductors? ______________________________

3. Who is responsible for determining the service location? ______________

4. a. The service head must be located (above) (below) the point where the service-drop conductors are spliced to the service-entrance conductors. Circle one.

   b. What Code section provides the answer to part (a)? ______________

5. What is a mast-type service entrance? ______________________________

6. a. What size and type of conductors are installed for this service? ______________

   b. What size conduit is installed? ______________________________

   c. What size grounding electrode conductor is installed? (not neutral) ________

   d. Is the grounding electrode conductor insulated, armored, or bare? ________

7. How and where is the grounding electrode conductor attached to the water pipe?

8. When a conduit is extended through a roof, must it be guyed? ______________

9. What are the minimum distances or clearances for the following?
   a. Service-drop clearance over private driveway ______
   b. Service-drop clearance over private sidewalks ______
   c. Service-drop clearance over alleys ______
   d. Service-drop clearance over a roof having a roof pitch of not less than 4/12. (Voltage between conductors does not exceed 300 volts.) ______

      ______
   e. Service-drop horizontal clearance from a porch ______
   f. Service-drop clearance from a fence that can be climbed ______
10. What size ungrounded conductors are installed for each of the following residential services? (Use Type THWN copper conductors.) See *Note 3* to *Table 310-16.*
    a. 60-ampere service No. ______ THWN copper
    b. 100-ampere service No. ______ THWN copper
    c. 200-ampere service No. ______ THWN copper
11. What size grounding electrode conductors are installed for the services listed in question 10? (See *Table 250-94.*)
    a. 60-ampere service No. ______ AWG grounding conductor
    b. 100-ampere service No. ______ AWG grounding conductor
    c. 200-ampere service No. ______ AWG grounding conductor
12. What is the recommended height of a meter socket from the ground? ______

    ______

    ______
13. a. May the bare grounded neutral conductor of a service be buried directly in the ground? ______
    b. What section of the Code covers this? ______
14. What exceptions are made regarding the use of bare neutral conductors installed underground? ______

    ______

    ______

    ______

    ______
15. How far must mechanical protection be provided when underground service conductors are carried up a pole? ______
16. a. A method of service disconnect may consist of how many switches or circuit breakers? ______
    b. Must these devices be in one enclosure? ______
    c. What type of main disconnect is provided in this residence? ______

       ______

17. Complete the following table by filling in the columns with the appropriate information.

	CIRCUIT NUMBER	AMPERE RATING	POLES	VOLTS	WIRE SIZE
A. LIVING ROOM RECEPTACLE OUTLETS					
B. WORKBENCH RECEPTACLE OUTLETS					
C. WATER PUMP					
D. ATTIC EXHAUST FAN					
E. KITCHEN LIGHTING					
F. HYDROMASSAGE TUB					
G. ATTIC LIGHTING					
H. COUNTER–MOUNTED COOKING UNIT					
I. ELECTRIC FURNACE					

18. a. What size conductors supply Panel B? ____________

    b. What size conduit? ____________

    c. Is this conduit run in the form of electrical metallic tubing or rigid conduit?

    ____

    d. What size overcurrent device protects the feeders to Panel B? ____________

19. How many electric meters are provided for this residence? ____________

20. a. According to the Code, is it permissible to ground rural service-entrance systems and equipment to driven ground rods only when a metallic water system is available? ____________

    b. What section of the Code applies? ____________

21. What table of the Code lists the sizes of grounding electrode conductors to be used for service entrances of various sizes similar to the type found in this residence?

    ____

22. *Section 250-81* requires that a supplemental ground be provided if the available grounding electrode is: (Circle correct answer.)

    a. water pipe

    b. building steel

    c. concrete encases ground

23. Do the following conductors require mechanical protection?
    a. No. 8 grounding conductor ______
    b. No. 6 grounding conductor ______
    c. No. 4 grounding conductor ______
24. Why is bonding of service-entrance equipment necessary? ______

25. What special types of bushings are required on service entrances? ______

26. When No. 4 AWG conductors or larger are installed in conduit, what additional provision is required on the conduit ends? ______

27. What minimum size copper bonding jumpers must be installed to bond properly the electrical service for the residence discussed in this text? ______

28. a. What is a Type S fuse? ______

    b. Where must Type S fuses be installed? ______

29. a. What is the maximum voltage permitted between conductors when using plug fuses?

    b. May plug fuses (Type S) be installed in a switch that disconnects a 120/240-volt clothes dryer? ______
    c. Give a reason for the answer to (b). ______

30. Will a 20-ampere, Type S fuse fit properly into a 15-ampere adapter? ________

31. What part of a circuit breaker causes the breaker to trip

    a. on an overload? ________ b. on a short circuit? ________

32. What is meant by an ambient-compensated circuit breaker? ________

33. List the standard sizes of circuit breakers up to and including 100 amperes. ________

34. Using the method shown in this unit, what is the approximate short-circuit current available at the terminals of a 50-kVA single-phase transformer rated 120/240 volts? The transformer impedance is 1%.

    a. line-to-line?

    b. line-to-neutral?

35. Where is the service for this residence located? ________

36. a. On what type of wall is Panel A fastened? ________

    b. On what type of wall is Panel B fastened? ________

37. State four possible combinations of service equipment which meet the requirements of *Section 230-65* of the Code.

    a. ________

    b. ________

c. ____

____

____

d. ____

____

____

38. When conduits pass through the wall from outside to inside, the conduit must be ______________________ to prevent air circulation through the conduit.

39. Briefly explain why electrical systems and equipment are grounded. __________

____

____

____

____

____

____

40. What Code section states that all overcurrent devices must have adequate interrupting ratings for the current to be interrupted? ______________________

____

41. All electrical components have some sort of "withstand rating." This rating indicates the ability of the component to withstand fault currents for the time required by the overcurrent device to open the circuit. What Code section refers to withstand ratings with reference to overcurrent protection? ______________________

42. Arcing fault damage is closely related to the value of ______________________

43. In general, systems are grounded so that the maximum voltage to ground does not exceed (Circle one)
    a. 120 volts. b. 150 volts. c. 300 volts.

44. To insure a complete grounding electrode system (Circle one)
    a. everything must be bonded together.
    b. all metal pipes and conduits must be isolated from one another.
    c. the service neutral is grounded to the water pipe only.

45. An electric clothes dryer is rated at 5700 watts. The electric rate is 10.091 cents per kilowatt-hour. The dryer is used continuously for 3 hours. Find the cost of operation, assuming the heating element is on continuously.

46. A heating cable rated at 750 watts is used continuously for 72 hours to prevent snow from freezing in the gutters of the house. The electric rate is 8.907 cents per kilowatt-hour. Find the cost of operation.

47. *Section 230-65* of the Code requires that the service equipment (breakers, fuses, and the panel itself) be rated equal to or greater than ______________________

____

____

48. When the available fault current at the main service equipment exceeds the interrupting rating of the branch-circuit breakers, properly sized main ______________ ______________ ______________ that will limit the current to a value less than the interrupting rating of the branch circuit breakers must be installed.

49. The utility company has provided a letter to the contractor stating that the available fault current at the line side of the main service-entrance equipment in a residence is 17 000 amperes RMS symmetrical, line-to-line. In the space following each statement, write in "Meets Code" or "Violation" of *Section 230-65* of the Code.

    a. Main breaker 10 000-ampere interrupting rating; branch breakers 10 000-ampere interrupting rating. ______________________

    b. Main current-limiting fuse having a 200 000-ampere interrupting rating; branch breakers having 10 000-ampere interrupting rating. ______________________

    ► c. Main breaker has 22 000-ampere interrupting rating; branch breakers have 10 000-ampere interrupting rating. The panel is marked "Series-Connected." ◄

    ____

► 50. While working on the main panel with the panel energized, the electrician inadvertently causes a direct short circuit (line-to-ground) on one of the branch-circuit breakers. The available fault current at the main service equipment is rather high. The panel is labeled "Series-Connected." The main breaker is rated 100 amperes. What will happen? ◄

	TRUE	FALSE
a. Only the branch breaker will trip off.	______	______
b. The branch breaker will not trip off.	______	______
c. The main breaker and the branch breaker will both trip off, resulting in a total power outage.	______	______
d. All of the 120-volt connected loads in the panel will lose power.	______	______
e. All of the 240-volt connected loads and one-half of the 120-volt connected loads will lose power. There will not be a total power outage.	______	______
f. One fuse will open; therefore, all 240-volt connected loads and one-half of the 120-volt connected loads will lose power. There will not be a total power outage.	______	______

51. Repeat question 50 on a main service panel that consists of 100-ampere main current-limiting fuses and breakers for the branch circuits. (Mark TRUE or FALSE.)

a. ______________________ d. ______________________

b. ______________________ e. ______________________

c. ______________________ f. ______________________

52. Here are five commonly used terms in the electrical industry. Enter the letter of the term before its definition.

a. Grounding electrode conductor

b. Main bonding jumper

c. Grounded circuit conductor

d. Equipment grounding conductor

e. Lateral service conductors

______ This is the neutral conductor.

______ This is the term used to define underground service-entrance conductors that run between the meter and the utilities connection.

______ This is the conductor (sometimes a large threaded screw) that connects the neutral bar in the service equipment to the service-entrance enclosure.

______ This is the conductor that runs between the neutral bar in the main service equipment to the grounding electrode (water pipe, ground rod, etc.).

______ This is the bare copper conductor found in nonmetallic-sheathed cable.

UNIT 29

Service-Entrance Calculations

OBJECTIVES

After studying this unit, the student will be able to

- determine the total calculated load of the residence.
- calculate the size of the service entrance, including the size of the neutral conductors.
- understand the Code requirements for services, *Article 230*.
- understand how to read a watthour meter.
- fully understand special Code rules for single-family dwelling service-entrance conductor sizing.
- derate service-entrance conductor if installation is located in extremely hot climate.
- do an optional calculation for computing the required size of service-entrance conductor for residence.

The branch-circuit load values determined in earlier units of this text are now used to illustrate the proper method of determining the size of the service-entrance conductors for the residence. The calculations are based on *National Electrical Code®* requirements. The student must check local and state electrical codes for any variations in requirements that may take precedence over the *National Electrical Code®*

SIZE OF SERVICE-ENTRANCE CONDUCTORS AND SERVICE DISCONNECTING MEANS

Service-entrance conductors, *Section 230-42 (b)(1)(2)*, and the disconnecting means, *Section 230-79(c)*, shall be not smaller than

1. 100 ampere, three-wire, for a single-family dwelling with six or more two-wire circuits
2. 100 ampere, three-wire, for a single-family dwelling with an initial computed load of 10 kVA or more

The electric range used in a dwelling generally has a rating of at least 8 kW, *Table 220-19*. Thus, only 2 kW are available for the remaining appliances, lighting, and special-purpose circuits. There are few, if any, single-family dwellings for which a service smaller than 100 amperes can be installed.

Size and rating of service-entrance conductors are covered in *Section 230-42*. *Section 230-42(a)* states that all the computations shall be done according to *Article 220*. Two methods are permitted by the Code to determine the size of the service-entrance conductors for a dwelling. Method 1 is outlined in *Article 220*, *Parts A* and *B*; method 2 is given in *Article 220*, *Part C*.

Method 1 — *Article 220, Parts A* and *B*

(*NOTE:* The values used in this procedure are taken from calculations in previous units of this text.)

General lighting load (*Sections 220-3b, 220-11*)
3232 sq. ft. @ 3 volt-amperes per sq. ft. = 9696 volt-amperes

Small appliance load (*Section 220-16*)

Kitchen	3	
Laundry	1	
Automatic washer	1	
Workshop	2	(See Note X at end of this calculation)
Total	7	@ 1500 VA per circuit = 10500 volt-amperes

Total general lighting and small appliance load 20196 volt-amperes

Application of demand factors, *Table 220-11*:
3000 volt-amperes @ 100% = 3000 volt-amperes
20196 − 3000 = 17196 @ 35% = 6019 volt-amperes

A Net computed load (less range and "fastened-in-place" appliances) = 9019 volt-amperes

Wall-mounted oven and counter-mounted cooking unit: *Table 220-19, Note 4*:
Wall-mounted oven 7450 volt-amperes
Counter-mounted cooking unit 6600 volt-amperes
Total 14050 volt-amperes (14 kW)

14 kW exceeds 12 kW by 2 kW:
2 kW × 5% = 10% increase
8 kW × 0.10 = 0.8 kW
(*Column A, Table 220-19*)

B 8 + 0.8 = 8.8 kW 8800 volt-amperes

C Net computed load (with cooking unit) = A + B = 17819 volt-amperes

Air conditioner 30 amperes × 240 volts = 7200 volt-amperes

D Electric furnace (*Section 220-15*)
The air-conditioner load is less than the heating load. Therefore, the air-conditioner load need not be included in the service-entrance calculations, *Section 220-21*. 13000 volt-amperes

E Dryer (*Section 220-18*) 5700 volt-amperes

F Freezer 5.8 × 120 = 696 volt-amperes

G Net computed load (lighting, small appliance circuits, range, dryer, freezer) = C + D + E + F = 37215 volt-amperes

There are more than four appliances *in addition* to electric ranges, air conditioners, space heaters, and clothes dryer. A demand factor of 75% may be applied to those appliances "fastened-in-place," *Section 220-17*.

Water heater 3000 volt-amperes
Dishwasher
Motor: 7.2 × 120 = 864 volt-amperes
Heater: 1000 volt-amperes (watts) (maximum demand is 1000 VA because motor and heater do not operate at the same time, *Section 220-21*.) 1000 volt-amperes
Food waste disposer 7.2 × 120 = 864 volt-amperes
Water pump (this is the largest motor; add 25%) 8 × 240 × 1.25 = 2400 volt-amperes
Garage door opener 5.8 × 120 = 696 volt-amperes
Heat-vent-lights (two) 1475 × 2 = 2950 volt-amperes
Attic exhaust fan 5.8 ×120 = 696 volt-amperes
Hydromassage tub 10 × 120 = 1200 volt-amperes

H Total "fastened-in-place" appliance load 12806 volt-amperes
Apply demand factor: 12806 × 0.75 = 9605 volt-amperes

I Lighting, small appliance circuits, ranges, dryer, freezer 37215 volt-amperes

Net computed load (H+I) (lighting, small appliance circuits, ranges, dryer, freezer, plus "fastened-in-place" appliances) 46820 volt-amperes

$$\text{Amperes} = \frac{\text{volt-amperes}}{\text{volts}} = \frac{46820}{240} = 195 \text{ amperes}$$

This could be a No. 3/0 THW, THWN, XHHW, or THHN per *Table 310-16*, or No. 2/0 THW, THWN, XHHW, or THHN per *Note 3* to *Table 310-16*.

NOTE X: The Workshop receptacles have been included as small appliance circuits. These circuits are not true small appliance circuits as defined by the Code. Yet, by including these circuits at 1500 volt-amperes per circuit, the final result provides us with a larger calculated load than if we had included the Workshop receptacles in the general lighting load, or if we had included a load value of 180 volt-amperes per receptacle. We are far over on the safe side with our calculations.

Without compromising Code rules, a quick check of *Table 310-16* shows that for *a full 200-ampere* service, we could install:

250 MCM copper-type TW (ampacity of 215 amperes) (an inspector might permit 4/0 that has an ampacity of 195 amperes)

3/0 copper types THW, THWN, or XHHW (ampacity of 200 amperes)

3/0 copper type THHN (ampacity of 225 amperes) (an inspector might permit 2/0 that has an ampacity of 195 amperes)

► However, without "bending" the Code rules, we find a special ampacity table in *Note 3* to *Table 310-16* that can be used for selecting 120/240-volt, 3-wire, single-phase residential service-entrance conductors and feeder conductors. Referring to this table, we find that No. 2/0 THW, THWN, THHN, or XHHW can be installed for a 200-ampere residential service-entrance.

Table 29-1 replicates *Note 3* to *Table 310-16* in the *NEC*®. The ampacities for a given size of wire are higher than the ampacities listed in *Table 310-16* because of the tremendous diversity of electricity use in homes.

This table may also be used to select conductors for the feeder that runs from main panel A to sub-panel B located in the Recreation Room. ◄

Be careful. Most electrical inspectors will not permit the use of THHN as service-entrance conductors that are installed in the conduit running from the meter socket to the service head because this would be a wet location requiring a conductor having a "W" in its type lettering. See figure 28-5.

Type THHN is acceptable when run between the meter socket and the service entrance panel, figure 28-5.

In certain parts of the country, such as the southwestern desert climates where extremely hot temperatures are common, the "authority having jurisdiction" may require that service-entrance conductors and any other conductors exposed to direct sunlight be corrected (derated) according to the "Ampacity Correction Factors" found below *Table 310-16*. See figure 29-1.

EXAMPLE: If the U.S. weather bureau lists the average summer temperature as 113° F (45° C), then a correction factor of 0.82 must be applied. For instance, a 3/0 XHHW copper conductor per *Table 310-16* is 200 amperes at 86° F (30° C). At 113° F (45° C) the conductor's ampacity is

$$200 \times 0.82 = 164 \text{ amperes}$$

A properly sized conductor capable of carrying 200 amperes safely requires

$$\frac{200}{0.82} = 243.9 \text{ amperes}$$

Therefore, according to *Table 310-16* and the applied correction factor a 250 kcmil Type XHHW copper conductor is required.

Check this out with the electrical inspection department where this situation might be encountered.

Neutral Conductor

Based upon the preceding calculation, the neutral must be able to carry the computed maximum neutral current draw of 93 amperes, per *Section 220-22* of the Code.

According to *Table 310-16*, this could be a

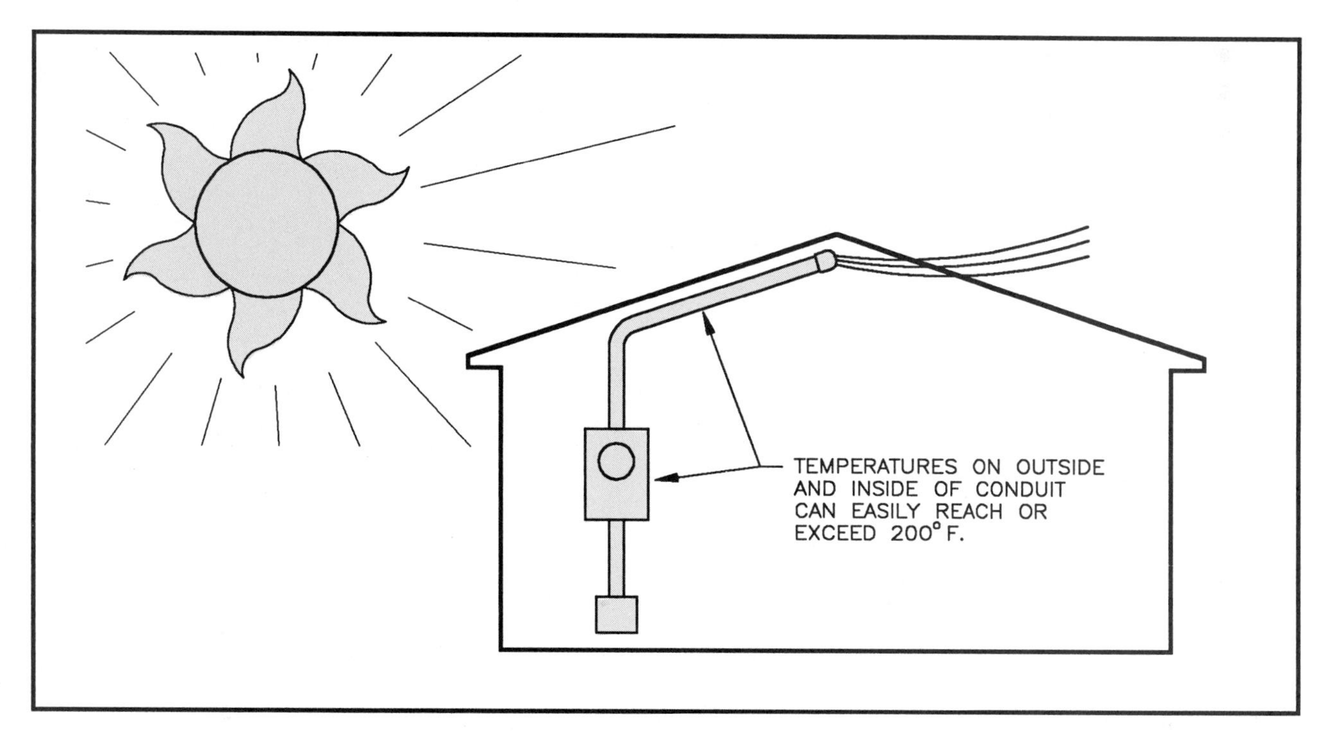

Fig. 29-1 Example of high-temperature location.

No. 3 THW, THHN, THWN, XHHW, or bare copper conductor. Therefore, using the standard calculations for service-entrance conductors, the Code would require two No. 3/0 THW, THWN, or XHHW hot conductors, and one No. 3 THW, THWN, or XHHW or bare neutral conductor. The neutral may be insulated or bare per *Sections 230-30* and *230-40*.

Note 3 following *Table 310-16* of the Code provides special consideration for three-wire, single-phase dwelling service-entrance conductors. Checking this *Note 3*, we find that for a 200-ampere, we could install No. 2/0 THW, THHN, THWN, or XHHW conductors. See *Tables 29-1* and *29-2*.

We can also reduce the neutral conductor size two AWG sizes smaller than the "hot" conductor. This would be a No. 1 AWG (insulated or bare). We proved that the neutral could be as small as a No. 3 per *Section 230-42* and *220-22*. See *Note 3* to *Table 310-16* for reduction of service-entrance neutral conductors.

Therefore, the specifications for this residence call for two No. 2/0 THW, THHN, THWN, or XHHW hot conductors, and one No. 1 AWG neutral conductor.

We mentioned the use of type XHHW conductors as suitable for the service-entrance conductors. Checking *Table 310-16*, we find that XHHW is rated 90° C for dry locations and 75° C when installed in wet locations. Since service-entrance conductors installed underground or where the conductors come out of the weatherhead (service head) are considered to be in wet locations, the ampacity must then be taken from the 75° C column.

Note: The 240-volt loads are not to be included in calculation of neutral conductors.
- Air conditioner
- Electric furnace
- Water pump
- Water heater
- Dishwasher heating element
- Dryer heating elements

General lighting load and small appliance load after applying demand factors	9019 volt-amperes
Range load (oven and counter-mounted unit, *Section 220-22*) 8800 volt-amperes × 0.70 =	6160 volt-amperes
Dryer (motor only) 5.8 × 120 =	696 volt-amperes
Freezer 5.8 × 120 =	696 volt-amperes

A Net computed load	16571 volt-amperes
Four or more "fastened-in-place" appliances:	
Dishwasher (maximum demand, heater only, *Section 220-21*)	1000 volt-amperes
Food waste disposer (*Section 220-14*) 7.2 × 120 × 1.25 =	1080 volt-amperes
Garage door opener 5.8 × 120 =	696 volt-amperes
Heat-vent-lights (two) 1475 × 2 =	2950 volt-amperes
Attic exhaust fan 5.8 ×120 =	696 volt-amperes
Hydromassage tub 10 × 120 =	1200 volt-amperes
Total	7622 volt-amperes
B Apply 75% demand factor: 7622 × 0.75 =	5717 volt-amperes
C Net computed lighting, small appliance circuits, ranges, dryer, freezer, plus "fastened-in-place" appliances: A + B =	22288 volt-amperes

$$\text{Amperes} = \frac{\text{volt-amperes}}{\text{volts}} = \frac{22288}{240} = 93 \text{ amperes}$$

NOTE: These calculations indicate that a No. 3 copper neutral is adequate. Specifications for this residence and *Note 3* to *Table 310-16* indicate that the neutral conductor will be No. 1 copper.

Subpanel B Calculations

General lighting load (*Sections 220-3b, 220-11*) 1420 sq. ft. @ 3 volt-amperes per sq. ft. = (kitchen, living room, laundry, rear entry hall, washroom, recreation room)	4260 volt-amperes
Small appliance load (*Section 220-16*) Kitchen 3; Laundry 1; Automatic washer 1; Total 5 @ 1500 VA per circuit =	7500 volt-amperes
Total general lighting and small appliance load =	11760 volt-amperes
Application of demand factors (*Table 220-11*)	
3000 volt-amperes @ 100% =	3000 volt-amperes
11760 − 3000 = 8760 @ 35% =	3066 volt-amperes
A Net computed load (less range and "fastened-in-place" appliances)	6066 volt-amperes
Wall-mounted oven and counter-mounted range: *Table 220-19, Note 4:*	
Wall-mounted oven	7450 volt-amperes
Counter-mounted range	6600 volt-amperes
Total	14050 volt-amperes

14 kW exceeds 12 kW by 2 kW:
2 kW × 5% = 10% increase
8 kW × 0.10 = 0.8 kW
(*Column A, Table 220-19*)

B 8 + 0.8 = 8.8 kW =	8800 volt-amperes
C Net computed load (with range) = A + B = C	14866 volt-amperes
Dryer (*Section 220-18*)	5700 volt-amperes
Dishwasher Motor: 7.2 × 120 = 864 volt-amperes Heater: 1000 watts (volt-amperes) (maximum demand is 1000 VA because motor and heater do not operate at the same time, *Section 220-21*)	1000 volt-amperes
Food waste disposer 7.2 × 120 =	864 volt-amperes
Garage door opener 5.8 × 120 =	696 volt-amperes
Total "fastened-in-place" appliance load	8260 volt-amperes
Lighting, small appliance circuits, ranges, dryer =	14866 volt-amperes

D Net computed load (lighting, small appliance circuits, ranges, dryer, dishwasher, food waste disposer, garage door opener)	23126 volt-amperes

$$\text{Amperes} = \frac{\text{volt-amperes}}{\text{volts}} = \frac{23126}{240} = 96 \text{ amperes}$$

Checking the specifications and also figure 28-2, we find that three No. 3 THHN conductors supply Panel B.

Method 2 (Optional Calculations — Dwellings) — *Article 220, Part C*

A second method for determining the load for a one-family dwelling is given in *Section 220-30.* This method simplifies the calculations, but may be used only when the service-entrance conductors have an ampacity of 100 amperes or more. In most cases, the service-entrance conductors are smaller than those permitted by *Article 220, Parts A and B.*

Let's take a look at *Part C* of *Article 220.* This is an alternative method of computing service loads and feeder loads. It is referred to as the *optional method.*

Section 220-30(a) tells us that we are permitted to calculate service-entrance conductors and feeder conductors (both phase conductors and the neutral conductor) using *Table 220-30. Section 220-30* addresses single-family dwellings. *Section 220-32* addresses multifamily dwellings.

Section 220-30(b) tells us to

1. include 1500 volt-amperes for each 20-ampere small appliance circuit.
2. include 3 volt-amperes per square foot for lighting and general-use receptacles.
3. include the nameplate rating of appliances:
   - that are fastened in place
   - that are permanently connected
   - that are located to be connected to a specific circuit

   Appliances listed in *Section 220-30(b)(3)* are ranges, wall-mounted ovens, counter-mounted cooking units, clothes dryers, and water heaters.
4. include nameplate ampere rating or kVA for motors and all low-power-factor loads. The intent of the reference to low power factor is to address such loads as low-cost, low-power-factor fluorescent ballasts of the type that might be used in the recessed lay-in fluorescent fixtures in the Recreation Room. It is always recommended that high-power-factor ballasts be installed.

The first sentence of *Table 220-30* tells us to select the largest load from a list of four types of load possibilities. Note that the list of four possibilities references air conditioners, heat pumps, central electric heat (furnaces), and separately controlled electric space heaters, such as electric baseboard heating units.

So let's begin our optional calculation for the residence discussed in this text. The residence has an air conditioner and an electric furnace.

Air conditioner (refer to *Table 220-30(1)*)

$$30 \times 240 = 7200 \text{ volt-amperes}$$

Electric furnace (refer to *Table 220-30(2)*)

$$13000 \times 0.65 = 8450 \text{ volt-amperes}$$

Therefore we will select the electric furnace load for our calculations because it is the largest load. It is also a noncoincidental load, as defined in *Section 220-21.* We can omit the air-conditioner load from our calculations from here on.

We now add up all of the other loads

General lighting load 3232 sq. ft. @ 3 volt-amperes per sq. ft. =	9696 volt-amperes
Small appliance circuits (7) @ 1500 volt-amperes each	10500 volt-amperes
Wall-mounted oven (nameplate rating)	6600 volt-amperes
Counter-mounted cooking unit (nameplate rating)	7450 volt-amperes
Water heater (nameplate rating)	3000 volt-amperes
Clothes dryer (nameplate rating)	5700 volt-amperes
Dishwasher (maximum demand, heater only)	1000 volt-amperes
Food waste disposer 7.2 × 120	864 volt-amperes

Water pump 8 × 240	1920 volt-amperes
Garage door opener 5.8 × 120	696 volt-amperes
Heat-vent-lights (2) 1475 × 2	2950 volt-amperes
Attic exhaust fan 5.8 ×120	696 volt-amperes
Hydromassage tub 10 × 120	1200 volt-amperes
Total other loads	52272 volt-amperes

We can now complete our optional calculation:

Enter electric furnace load 13000 × 0.65 =	8450 volt-amperes
Plus	
first 10 kVA of all other loads at 100%	10000 volt-amperes
Plus	
remainder all other loads at 40%:	
52272 − 10000 = 42272	
42272 × 0.40	16909 volt-amperes
Total	35359 volt-amperes

$$\text{Amperes} = \frac{\text{volt-amperes}}{\text{volts}} = \frac{35359}{240} = 147.3 \text{ amperes}$$

No. 1 THW or THWN conductors or equivalent could be installed for this service, *Table 310-16, Note 3*. However, the residence in the plans is to have a full 200-ampere, 120/240-volt service consisting of two No. 2/0 THW, THHN, THWN, or XHHW phase conductors and one No. 1 bare neutral conductor.

SPECIAL AMPACITY RATINGS FOR SINGLE-PHASE, THREE-WIRE RESIDENTIAL SERVICES AND FEEDERS ONLY (*Note 3, Table 310-16*)		
COPPER CONDUCTOR (AWG) FOR INSULATION OF RH-RHH-RHW-THW-THWN-THHW-THHN-XHHW	ALUMINUM OR COPPER-CLAD ALUMINUM CONDUCTORS (AWG)	SERVICE-ENTRANCE AMPACITY RATING
4	2	100
3	1	110
2	1/0	125
1	2/0	150
1/0	3/0	175
2/0	4/0	200
3/0	250 kcmil	225
4/0	300 kcmil	250
250 kcmil	350 kcmil	300
350 kcmil	500 kcmil	350
400 kcmil	600 kcmil	400

► Table 29-1. The above table shows special ampacity ratings for selecting 120/240 volt, 3-wire, single-phase service-entrance conductors and feeder conductors. This table is for residential installations only. ◄

Special Ampacity Ratings for Single-Phase, Three-Wire Residential Services Only

Note 3 to *Table 310-16* of the Code allows specific conductor sizes for single-phase, three-wire residential services that can result in smaller size service-entrance conductors than regular calculations show. See Table 29-1.

Service-Entrance Conductor Size Table

Table 29-2 has been taken from one major city that prefers to show minimum service-entrance conductor size requirements rather than having the electrical contractor make calculations each and every time he installs a service.

SERVICE-ENTRANCE CONDUCTOR SIZE						
	COPPER			ALUMINUM		
SIZE (AMPERES)	PHASE (HOT) CONDUCTORS	NEUTRAL CONDUCTOR	CONDUIT SIZE (INCHES)	PHASE (HOT) CONDUCTORS	NEUTRAL CONDUCTOR	CONDUIT SIZE (INCHES)
100	#2 AWG	#4 AWG	1 1/2	#1 AWG	#2 AWG	1 1/2
125	#1 AWG	#4 AWG	1 1/2	00 AWG	#1 AWG	1 1/2
150	0 AWG	#4 AWG	1 1/2	000 AWG	0 AWG	2
175	00 AWG	#2 AWG	2	0000 AWG	00 AWG	2
200	000 AWG	#2 AWG	2	250 kcmil	000 AWG	2

Table 29-2. Table of service-entrance conductor sizing used in some cities so that service-entrance calculations do not have to be made for each service-entrance installation.

Table 250-94.
Grounding Electrode Conductor for AC Systems

Size of Largest Service-Entrance Conductor or Equivalent Area for Parallel Conductors		Size of Grounding Electrode Conductor	
Copper	**Aluminum or Copper-Clad Aluminum**	**Copper**	***Aluminum or Copper-Clad Aluminum**
2 or smaller	1/0 or smaller	8	6
1 or 1/0	2/0 or 3/0	6	4
2/0 or 3/0	4/0 or 250 kcmil	4	2
Over 3/0 thru 350 kcmil	Over 250 kcmil thru 500 kcmil	2	1/0
Over 350 kcmil thru 600 kcmil	Over 500 kcmil thru 900 kcmil	1/0	3/0
Over 600 kcmil thru 1100 kcmil	Over 900 kcmil thru 1750 kcmil	2/0	4/0
Over 1100 kcmil	Over 1750 kcmil	3/0	250 kcmil

Where multiple sets of service-entrance conductors are used as permitted in Section 230-40, Exception No. 2, the equivalent size of the largest service-entrance conductor shall be determined by the largest sum of the areas of the corresponding conductors of each set.

Where there are no service-entrance conductors, the grounding electrode conductor size shall be determined by the equivalent size of the largest service-entrance conductor required for the load to be served.

* See installation restrictions in Section 250-92(a).

(FPN): See Section 250-23(b).

Note that the conductor sizes in Table 29-2 are larger than conductor sizes permitted by *Note 3* to *Table 310-16* of the *National Electrical Code*.® But since the Code is considered to be a minimum standard, it is within the realm of local authorities to publish code requirements specifically for their communities.

Main Disconnect

The main disconnect in the residence is a 200-ampere pullout-type panel. Many electric utilities state that the conductors feeding this type of disconnect must be the same size as the ampere rating of the disconnect. Some utilities also state that the neutral conductor cannot be reduced in size because it is not always possible to foresee the type of load that may be connected to the panel. As a result, it is sometimes difficult to conform to all the rules of the electric utility as well as the local electrical code. It is believed, however, that by installing two No. 2/0 THW, THWN, THHN, or XHHW, most local and state electrical code requirements will be satisfied.

Grounding Electrode Conductor

The grounding electrode conductor connects the main service-equipment neutral bar to the grounding electrode. The grounding electrode might be the underground metallic water piping system, a driven ground rod, or a concrete-encased ground. All of this has been discussed in unit 28.

Grounding electrode conductors are sized according to *Table 250-94* of the Code.

As an example, this residence is supplied by No. 2/0 AWG copper service-entrance conductors. Checking *Table 250-94*, we find that the minimum grounding electrode conductor must be a No. 4 AWG copper conductor.

Fig. 29-2 Typical single-phase watthour meter.

READING THE METER

Figure 29-2 shows a typical single-phase watt-hour meter with five dials. From left to right, the dials represent single units, tens, hundreds, thousands, and tens of thousands in kilowatt-hours.

Starting with the first dial, record the last number the pointer has passed. Continue doing this with each dial until the full reading is obtained. The reading on the five-dial meter in figure 29-3 is 18672 kilowatt-hours.

If the meter reads 18975 one month later, figure 29-4, by subtracting the previous reading of 18672 it is found that 303 kilowatt-hours were used during the month.

The number is multiplied by the rate per kilowatt-hour, and the power company bills the consumer for the energy used. The utility may also add a fuel adjustment charge.

Fig. 29-3 The reading of this five-dial meter is 18672 kilowatt-hours.

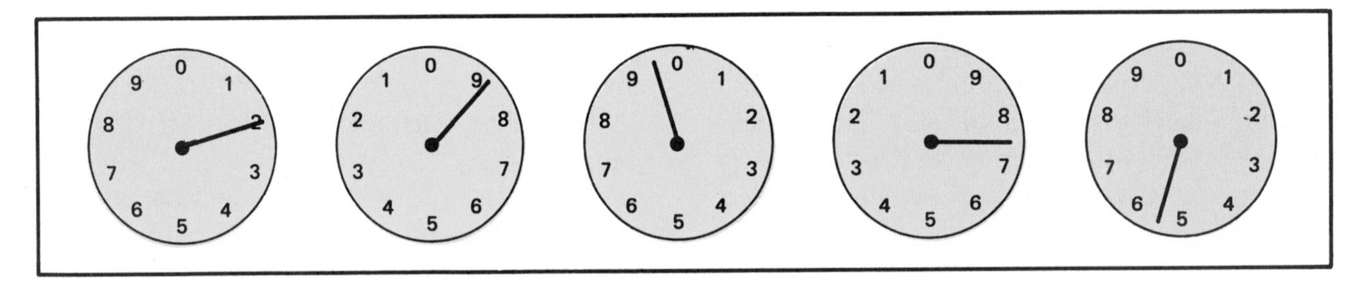

Fig. 29-4 One month later, the meter reads 18975 kilowatt-hours, indicating that 303 kilowatt-hours were used during the month.

REVIEW

Note: Refer to the Code or the plans where necessary.

1. When a service-entrance calculation results in a value of 10 kW or more, what is the minimum size service required by the Code? ____________________

2. a. What is the unit load per square foot for the general lighting load of a residence?

   ____

   b. What are the demand factors for the general lighting load in dwellings? ______

   ____

   ____

   ____

   ____

   ____

3. a. What is the ampere rating of the circuits that are provided for the small appliance loads? ____________

   b. What is the minimum number of small appliance circuits permitted by the Code?

   ____

   c. How many small appliance circuits are included in this residence? ____________

4. Why is the air-conditioning load for this residence omitted in the service calculations?

   ____

5. What demand factor may be applied when four or more fixed appliances are connected to a service, in addition to an electric range, air conditioner, clothes dryer, or space heating equipment? (*Section 220-17*) ____________

6. What load may be used for an electric range rated at not over 12 kW? (*Table 220-19*)

   ____

7. What is the load for an electric range rated at 16 kW? (*Table 220-19*) Show calculations.

8. What is the computed load when fixed electric heating is used in a residence? (*Section 220-15*) ____________

9. On what basis is the neutral conductor of a service entrance determined? ____________

10. Why is it permissible to omit an electric space heater, water heater, and certain other 240-volt equipment when calculating the neutral service-entrance conductor for a residence? ____________

11. a. What section of the Code contains an optional method for determining residential service-entrance loads? ____________

    b. Is this section applicable to a two-family residence? ____________

12. Calculate the minimum size of service-entrance conductors required for a residence containing the following: floor area 24′ × 38′ (7.3 m × 11.6 m); 12-kW electric range; 5-kW dryer consisting of a 4-kW, 240-V heating element, a 120-V motor, a 120-V light (combined motor and light is 1 kW); 2200-W, 120-V sauna heater; 12-kW, 240-V electric heat (six units); 12-A, 240-V air conditioner; 3-kW, 240-V water heater. Determine the sizes of the ungrounded conductors and the neutral conductor. Use type THWN conductors. Be sure to include the small appliance circuits and the laundry circuit. Use Method 1.

    Two No. ____________ THWN ungrounded conductors

    One No. ____________ THWN neutral (or bare neutral if permitted)

    One No. ____________ AWG grounding electrode conductor to water meter

STUDENT CALCULATIONS

13. Read the meter shown. Last month's reading was 22796. How many kilowatt-hours of electricity were used for the current month?

UNIT 30

Swimming Pools, Spas, Hot Tubs

OBJECTIVES

After studying this unit, the student will be able to

- recognize the importance of proper swimming pool wiring with regard to human safety.
- discuss the hazards of electrical shock associated with faulty wiring in, on, or near pools.
- describe the differences between permanently installed pools and pools which are portable or storable.
- understand and apply the basic Code requirements for the wiring of swimming pools, spas, hot tubs, and hydromassage bathtubs.

POOL WIRING (*Article 680 NEC®*)

For easy reference, a detailed drawing of the Code requirements for swimming pool wiring is included in the pocket on the inside of the back cover of this text. You will want to refer to this drawing continually as you study this unit.

Swimming pools, wading pools, hydromassage bathtubs, therapeutic pools, decorative pools, hot tubs, and spas must be wired according to the *National Electrical Code®, Article 680*. To protect people using such pools, specific rules unique to pool wiring have been developed over the years. Extreme care is required when wiring all equipment associated with pools.

ELECTRICAL HAZARDS

A person can suffer an immobilizing or lethal shock in a residential-type pool in either of two ways.

1. An electrical shock can be transmitted to someone in a pool who touches a "live" wire, or the "live" casing or enclosure of an appliance, such as a hair dryer, radio, or extension cord, among others, figure 30-1.
2. In the event that an appliance falls into the pool, an electrical shock can be transmitted to a person in the water, by means of voltage gradients in the water. Refer to figure 30-2 for an illustration of this life-threatening hazard.

As shown in the figure, "rings" of voltage radiate outwardly from the radio to the pool walls. These rings can be likened to the rings that form when a rock is thrown into the water. The voltage rings or gradients range from 120 volts at the radio to zero volts at the pool walls. The pool walls are assumed to be at ground or zero potential. The gradients, in varying degrees, are found in the entire body of water.

Figure 30-2 shows voltage gradients in the pool of 90 volts and 60 volts. (This figure is a simplification of the actual situation in which there are many voltage gradients.) In this case, the voltage differential, 30 volts, is an extremely hazardous value. **The**

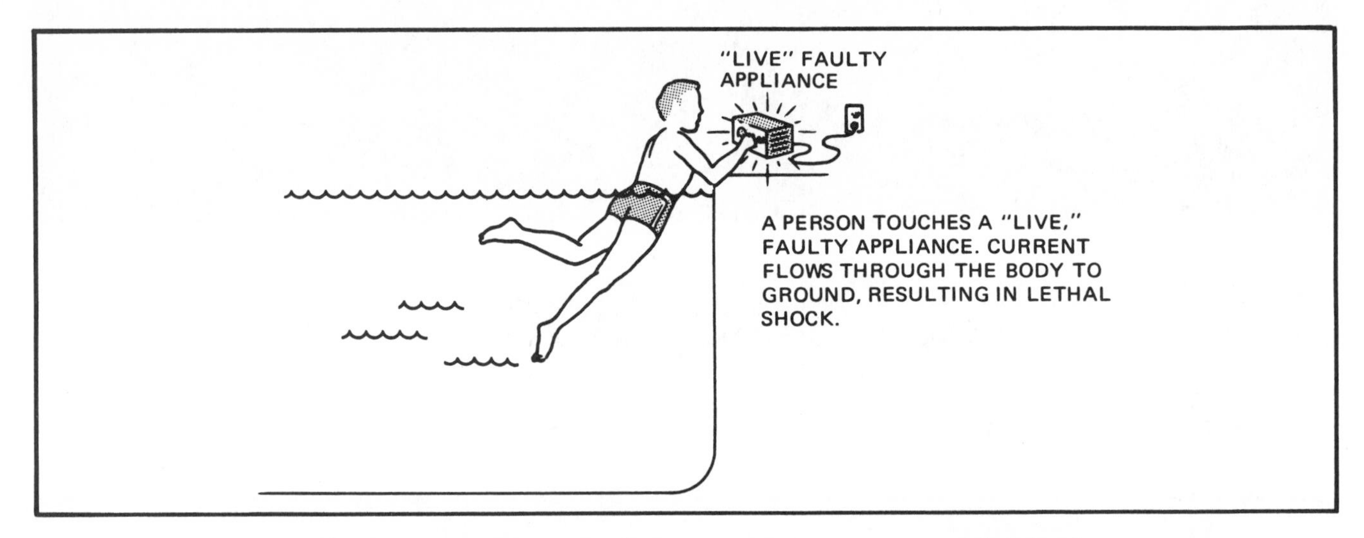

Fig. 30-1 Touching a "live," faulty appliance can cause lethal shock.

person in the pool, who is surrounded by these voltage gradients, is subject to severe shock, immobilization (which can result in drowning), or actual electrocution. Tests conducted over the years have shown that a voltage gradient of 1 1/2 volts per foot can cause paralysis.

Study the fold-out Code Requirements for Swimming Pool Wiring located in the pocket on the inside back cover of this text. You will note that in general, underwater swimming-pool lighting fixtures must be positioned at least 18 inches below the normal water level. The reasoning behind this rule is that when a person is in the water close to an underwater lighting fixture, the relative position of the fixture is well below the person's heart. Fixtures that are permitted to be installed no less than 4 inches below the normal water level have undergone tougher, abnormal impact tests, such as those tests given to cleaning tools or other mechanical objects. The lenses on these fixtures can withstand these abnormal impact tests, whereas the standard underwater fixtures that are required to be 18 inches below the water level are subjected to normal impact tests that duplicate the impact of a person kicking the lens.

The shock hazard to a person in a pool is quite different from that of the normal "touch" shock hazard to a person not submersed in water. The water makes contact with the entire skin surface of the body rather than just at one "touch" point. Also, skin wounds, such as cuts and scratches, reduce the body's resistance to shock to a much lower value than that of the skin alone. Body openings such as ears, nose, and mouth further reduce the body

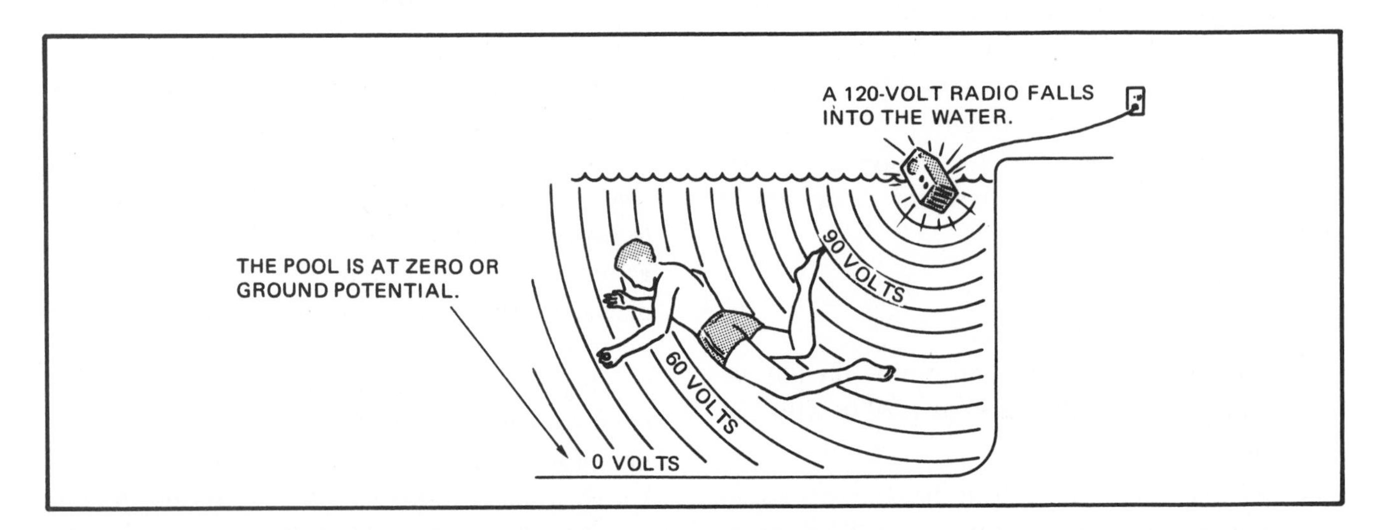

Fig. 30-2 Voltage gradients surrounding a person in the pool can cause severe shock, drowning, or electrocution.

resistance. As Ohm's law states, for a given voltage, the lower the resistance, the higher the current.

Another hazard associated with spas and hot tubs is prolonged immersion in hot water. If the water is too hot and/or the immersion too long lethargy (drowsiness, hyperthermia) can set in. This can increase the risk of drowning. The Underwriters Laboratories Standard #1563 establishes maximum water temperature at 104°F. The suggested maximum time of immersion is generally 15 minutes.

Instructions furnished with spas and hot tubs specify the maximum temperature and time permitted for using the spa or hot tub.

Hot water, in combination with drugs and alcohol, presents a real hazard to life. Caution must be observed at all times.

The figures that follow show the *National Electrical Code®* rules that relate to safety procedures for wiring pools.

Wiring Methods

Throughout *Article 680* of the Code, you will find that the wiring must be installed in rigid metal conduit, intermediate metal conduit, rigid nonmetallic conduit, Type MC cable, or, in some cases, electrical metallic tubing. For wet-niche fixtures, brass or other corrosion-resistant metal must be used.

► *Section 680-25(c), Exception No. 3* allows any type of wiring method recognized in *Chapter 3* of the Code for that portion of the interior wiring of one-family dwelling installations that supplies pool-associated pump motors, provided the wiring method contains an equipment grounding conductor no smaller than No. 12 AWG. ◄

Be sure to read the requirements of *Article 680* closely to determine the proper wiring method to be used for a particular situation.

CODE-DEFINED POOLS

The Code describes a *permanently installed swimming pool* as one that is located in the ground, on the ground, or in a building. The manner of construction is such that the pool cannot be disassembled readily for storage, whether or not it is served by any electrical circuits.

The Code describes a *storable swimming pool* or *wading pool* as one that has a maximum lengthwise dimension of 18 feet (5.49 m), and a maximum wall height of 42 inches (1.07 m). The construction of this type of pool is such that it can be disassembled for storage and reassembled to its original form.

► Pools that have nonmetallic inflatable walls are considered to be storable pools regardless of their dimensions. Storable pools are covered in *Article 680, Part C.* ◄

GROUNDING AND BONDING OF SWIMMING POOLS

Grounding (*Section 680-24*)

Section 680-24 of the Code requires that all of the following items *must* be grounded, as illustrated in figure 30-3.

- wet- and dry-niche lighting fixtures.
- all electrical equipment within 5 feet (1.52 m) of the inside wall of the pool.
- all electrical equipment associated with the recirculating system of the pool.
- junction boxes.
- transformer enclosures.
- ground-fault circuit interrupters.
- panelboards that are not part of the service equipment supply the electrical equipment associated with the pool.

Proper grounding and bonding ensures that all of the metal parts in and around the pool area are at the same ground potential, thus reducing the shock hazard. Proper grounding and bonding practices also facilitate the opening of the overcurrent protective device (fuse or circuit breaker) in the event of a fault in the circuit. Grounding is covered in other units of this text.

Grounding Conductors: Grounding conductors *must*

- be run in the same conduit with the circuit conductors, or be part of an approved flexible cord assembly, as used for the connection of wet-niche underwater lighting fixtures.

- terminate on equipment grounding terminals provided in the junction box, transformer, ground-fault circuit interrupter, subpanel, or other specific equipment.

It is important to note that metal conduit by itself is NOT considered to be an adequate grounding means for equipment grounding in and around pools.

Bonding (*Section 680-22*)

Section 680-22 of the Code requires that all metal parts of a pool installation *must* be bonded together by connecting the parts to a common bonding grid, figure 30-4. The No. 8 or larger solid copper bonding wire need not be connected to any service equipment, a remote panelboard, or any grounding electrode. It need only "tie everything

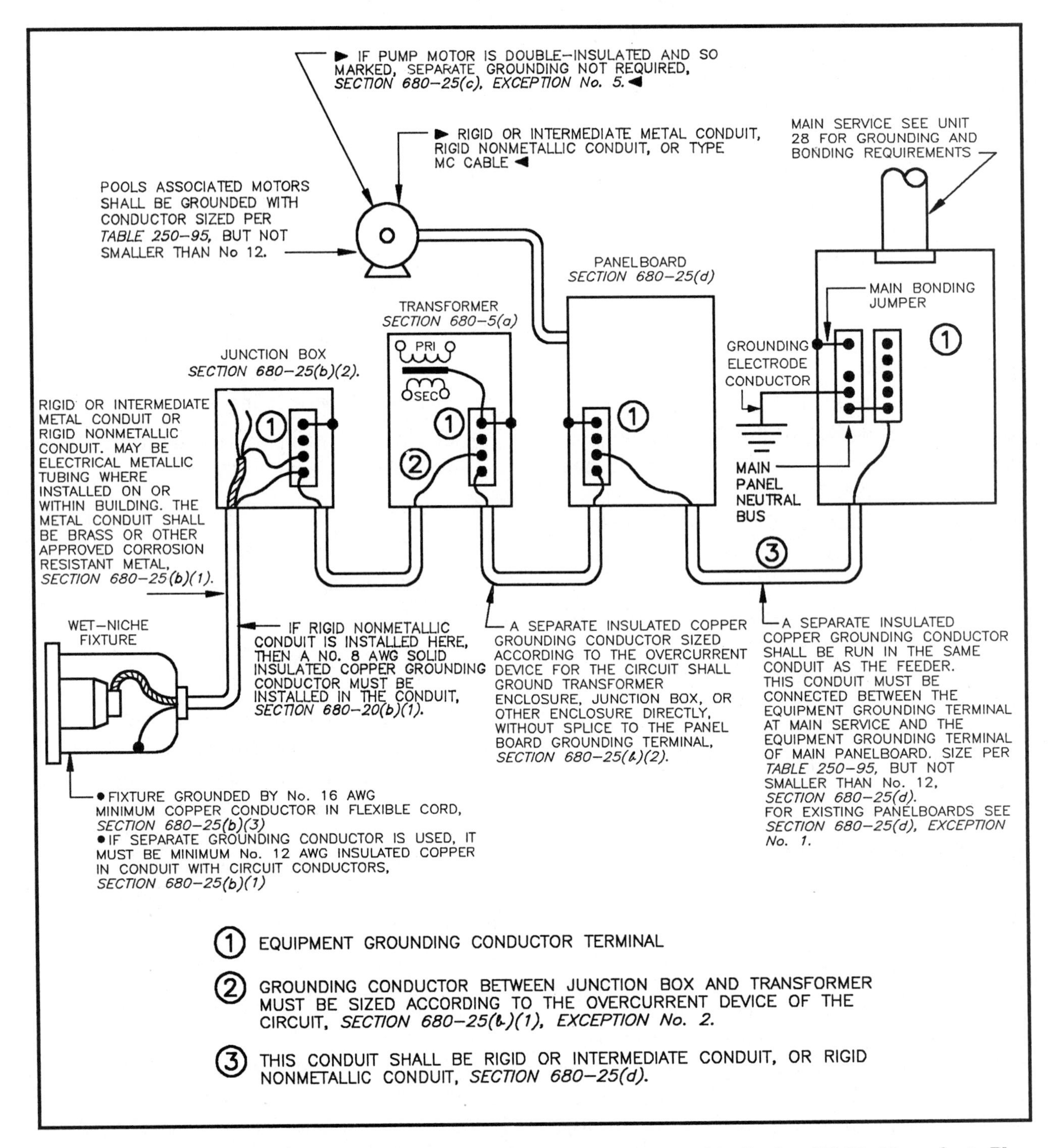

Fig. 30-3 Grounding of important metal parts of a swimming pool as covered in *Section 680-25*. Also refer to Plan 10/10 in the back of the text.

together." This helps keep all metal parts in and around the pool at the same voltage potential, reducing the shock hazard brought about by stray voltages and voltage gradients.

The bonding grid, *Section 680-22(b)*, may be

- the steel reinforcing bars in the concrete, or
- the wall of a bolted or welded metal pool, or
- a solid copper wire no smaller than No. 8.

Bonding Conductors: Bonding conductors

- need *not* be installed in conduit.
- may be connected directly to the equipment that requires bonding, by means of brass, copper, or copper-alloy clamps.

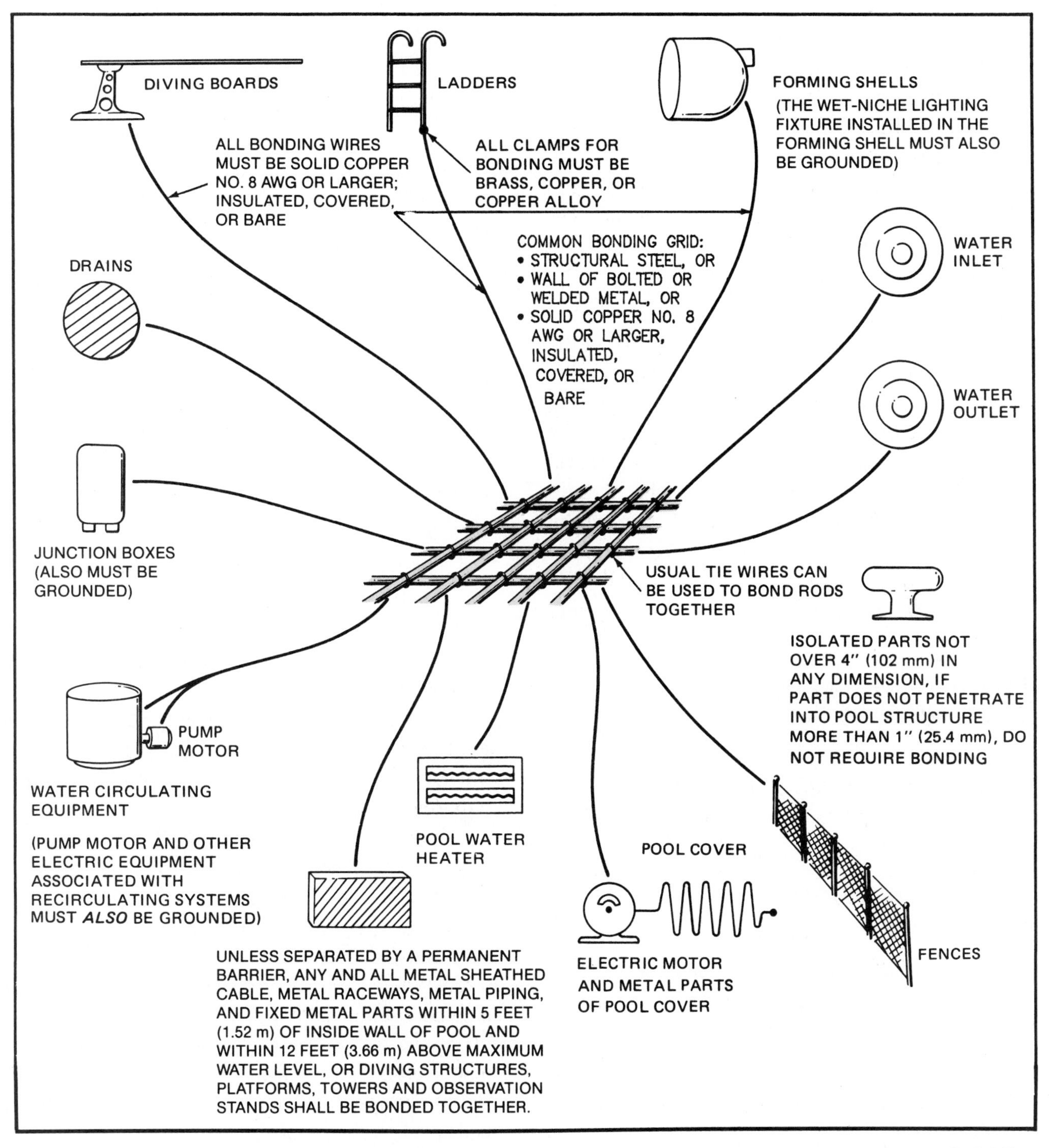

Fig. 30-4 Bonding the metal parts of swimming pool installations, *Section 680-22*, National Electrical Code® In addition to bonding, some of the above items must also be grounded. Refer to *Sections 680-24* and *25* for specifics. Also see Plan 10/10 in the back of the text.

LIGHTING FIXTURES UNDERWATER

Article 680, Part B covers underwater lighting fixtures for permanently installed pools.

There are three types of underwater lighting fixtures:

1. Dry-niche: Dry-niche lighting fixtures are mounted in the side walls of the pool and are designed to relamp from the rear. These fixtures are waterproof.
2. Wet-niche: Wet-niche lighting fixtures are mounted in the side wall of the pool and are designed to be relamped from the front. The supply cord is stored inside the fixture. The cord should be long enough to reach the top of the pool deck for lamp replacement. The forming shell to which the fixture attaches is intended to fill up with pool water.
3. ► No-niche: No-niche lighting fixtures are mounted on a bracket on the inside wall of the pool. The supply conduit terminates on this bracket. The supply cord runs through this conduit to a deck box for connection to the circuit wiring. The extra supply cord is stored in the space behind the no-niche fixture. The cord should be long enough to reach the top of the pool deck for lamp replacement. This type of fixture is mainly installed in retrofit installations and for above-ground pools. ◄

ELECTRIC HEATING OF SWIMMING POOL DECKS
(*Section 680-27*)

The Code requirements for the electric heating units are illustrated in figure 30-5.

Permanently wired radiant heaters:

- shall be suitably guarded.
- shall be securely fastened.
- shall not be located over the pool.
- shall be at least 5 feet (1.52 m) back from the inside edge of the pool.
- shall be mounted at least 12 feet (3.66 m) vertically above pool deck unless approved otherwise.

Unit heaters:

- shall be rigidly mounted on the structure.
- shall be totally enclosed or guarded.
- shall not be located over the pool.
- shall be at least 5 feet (1.52 m) back from the inside edge of the pool.

Radiant heat cables:

- are not permitted to be embedded in or below the pool.

Circuit sizing for electric heaters:

The branch-circuit conductors and the branch-circuit overcurrent protective devices shall be rated not less than 125% of the heater's nameplate rating according to *Section 680-9* of the Code.

► Of the many Code rules covering pools, note in *Section 680-25(c)* that the circuit conductors and the equipment grounding conductor for pool-associated motors must be installed in rigid metal conduit, rigid nonmetallic conduit, or Type MC cable. Electrical metallic tubing (EMT) is permitted where installed on or within buildings. ◄

SPAS AND HOT TUBS
(*Article 680, Part D*)

The basic difference between a spa and a hot tub is that a hot tub is constructed of wood such as redwood, teak, cypress, or oak, whereas a spa is made of plastics, fiberglass, concrete, tile, or other manmade products.

Spas and hot tubs contain electrical equipment for heating and recirculating water. They have no provisions for direct connection to the building plumbing system.

Spas and hot tubs are intended to be filled and used. They are not drained after each use, as in the case of a regular bathtub.

Some spas and hot tubs are furnished with a cord-and-plug set when shipped by the manufacturer. Where these can be converted from cord-and-plug connection to "hard wiring" they are listed and identified as "convertible" spas and hot tubs. The manufacturer must furnish clear instructions as to how the conversion from cord-and-plug set to "hard wiring" is to be done.

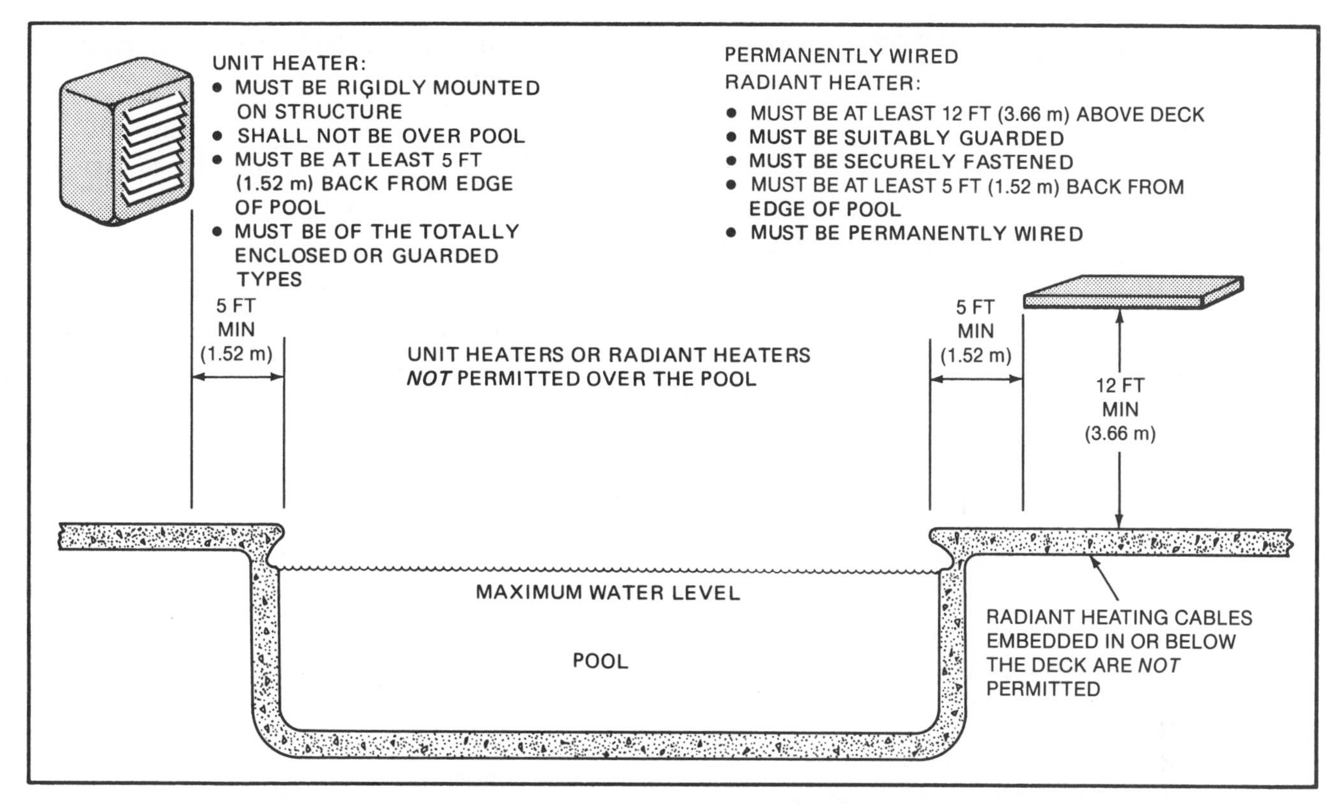

Fig. 30-5 Deck-area electric heating within 20 feet from inside edge of swimming pools, *Sections 680-9* **and** *680-27,* ***National Electrical Code®***

Outdoor Installations, *Section 680-40*

Spas and hot tubs installed outdoors *must* conform to the installation requirements discussed previously for regular swimming pools, *Article 680, Parts A* and *B.*

Exception No. 1 to *Section 680-40* states that the metal bands and hoops that secure the wood staves of a spa or hot tub do not have to be grounded.

Listed packaged units are permitted to have a cord not longer than 15 feet (4.56 m). Such units must be GFCI protected, *Section 680-40, Exception No. 2.*

► Listed packaged units that have a factory installed remote control panel may be connected with liquid-tight flexible conduit not over 3 feet (914 mm) long. ◄

Indoor Installation

Spas and hot tubs installed indoors *must* conform to the following requirements:

- must be connected using the wiring methods covered in *Chapter 3* of the Code.
- may be cord-and-plug connected if rated 20 amperes or less.

Receptacles, *Section 680-41(a)*:

- ► at least one receptacle outlet must be installed at least 5 feet (1.52 m), but not more than 10 feet, (3.05 m) from the inside edge of the spa or hot tub. ◄
- any other receptacles installed on the property must be located at least 5 feet (1.52 m) from the inside edge of the spa or hot tub.
- any 125-volt receptacle located within 10 feet (3.05 m) of the inside walls of the spas or hot tub must be GFCI protected.
- any receptacles that supply power to the spa or hot tub must be GFCI protected no matter how far they are from the spas or hot tub.

Lighting Fixtures, *Section 680-41(b)*:

- all lighting outlets and fixtures located within 5 feet (1.52 m) from the inside wall of the spa or hot tub or directly over the spa or hot tub must be at least 7 1/2 feet (2.29 m) above the maximum water level and must be GFCI protected.

There are a few exceptions to this:

1. GFCI protection is not required if the fixture is 12 feet (3.66 m) or more above maximum water level.
2. Recessed lighting fixtures may be installed less than 7 1/2 feet (2.29 m) above the hot tub or spa if the fixture:
   - is GFCI protected, and
   - ► has a glass or plastic lens, and
   - has a nonmetallic trim suitable for use in damp locations, or
   - has an electrically isolated metal trim suitable for use in damp locations. ◄
3. Surface-mounted fixtures may be installed less than 7 1/2 feet (2.29 m) above the hot tub or spa if the fixture:
   - is GFCI protected, and
   - ► has a plastic or glass globe, and
   - has a nonmetallic body, or
   - has a metallic body that is isolated from contact.
   - is suitable for use in damp locations. ◄

- if any underwater lighting is to be installed, the rules discussed previously for regular swimming pools apply.

Wall Switches, *Section 680-41(c)*:

- wall switches must be located at least 5 feet (1.52 m) from the inside walls of the spa or hot tub.

Grounding, *Section 680-40(f)*:

The requirements for grounding spas and hot tubs are:

- ground all electrical equipment within 5 feet (1.52 m) of the inside wall of the spa or hot tub.
- ground all electrical equipment associated with the circulating system, including the pump motor.
- grounding *must* conform to all of the applicable Code rules of *Article 250*.
- grounding conductor connections *must* be made according to the applicable requirements of *Chapter 3* of the Code.
- if equipment is connected by means of a flexible cord, the grounding conductor must be part of the flexible cord, and must be fastened to a fixed metal part of the equipment.

Bonding, *Sections 680-40(d)* and *(e)*:

The bonding requirements for spas and hot tubs are similar to those for regular swimming pools. Bond together:

- all metal fittings within or attached to the spa or hot tub.
- all metal parts of electrical equipment associated with the circulating system, including pump motors.
- all metal pipes, conduits, and metal surfaces within 5 feet (1.52 m) of the inside edge of the spa or hot tub. This bonding is not required if the materials are separated from the spa or hot tub by a permanent barrier, such as a wall or building.
- all electrical devices and controls *not* associated with the spa or hot tub located less than 5 feet from the inside edge of the spa or hot tub. If located 5 feet (1.52 m) or more from the hot tub or spa, bonding is not required.

Bonding is to be accomplished by means of threaded metal piping and fittings, by metal-to-metal mounting on a common base or frame, or by means of a No. 8 or larger, solid, insulated, covered or bare bonding jumper.

Electric Water Heaters, *Section 680-41(h)*:

- must be listed for the purpose.
- must have their heating elements into loads not over 48 amperes, and protected at not over 60 amperes. This requirement is usually met by the manufacturer of the heater.
- must have branch-circuit conductors rated not less than 125 percent of the total load as indicated on the nameplate of the heater.

HYDROMASSAGE BATHTUBS
(Article 680, Part G)

Hydromassage bathtubs, together with their associated electrical components, must be GFCI protected.

Because a hydromassage bathtub does not constitute any more of a shock hazard than a regular bathtub, the *National Electrical Code*® permits all other wiring (fixtures, switches, receptacles, and other equipment) in the same room but not directly associated with the hydromassage bathtub to be installed according to all the normal Code requirements covering installation of that equipment in bathrooms.

This residence has a hydromassage tub in the Master Bathroom. It is a prewired unit furnished with a 3-foot length of 1/2″ watertight conduit that contains a black, a white, and a green equipment grounding conductor.

The hydromassage tub is powered by a 1/2-hp, 115-volt, 3450-r/min, single-phase motor rated at 10 amperes.

A hydromassage tub and all of its associated equipment generally provided by the manufacturer must be connected to a circuit that has GFCI protection.

The hydromassage tub is connected to a separate 120-volt, 20-ampere Circuit A9. For more details on the circuitry for the hydromassage bathtub see unit 22.

A hydromassage tub is also known as a whirlpool tub or whirlpool bathtub.

Figure 22-12 illustrates one manufacturer's current production-model hydromassage bathtub.

By definition in *Article 680* of the Code, a hydromassage bathtub is intended to be filled (used), then drained after each use, whereas the spa or hot tub is filled with water and not drained after each use.

FOUNTAINS
(Article 680, Part E)

The home discussed in this text does not have a fountain, however, we will discuss two key issues. It is significant to note that self-contained, portable fountains having no dimension over 5 feet (1.52 m) do not have to conform to *Part E* of *Section 680*. When fountains share water with a regular swimming pool, the fountain wiring must conform to pool wiring requirements.

UNDERWRITERS LABORATORIES STANDARDS

The U.L. Standards of interest are:

UL 676 Underwater Lighting Fixtures

UL 943 Ground-Fault Circuit Interrupters

UL 1081 Swimming Pumps, Filters, and Chlorinators

UL 1241 Junction Boxes for Underwater Lighting Fixtures

UL 1261 Electric Water Heaters for Pools and Spas

UL 1563 Electric Hot Tubs, Spas, and Associated Equipment

SUMMARY

The figures in this unit and a plan-sized diagram included in the back of the text presents a detailed overview of the Code requirements for pool wiring.

REVIEW

Note: Refer to the Code or the plans where necessary.

1. The article of the *National Electrical Code*® that covers most of the requirements for wiring of swimming pools is *Article* ________________.

2. Name the two ways in which a person may sustain an electrical shock when in a pool.

____

____

____

____

____

3. The Code in *Article 680* discusses two types of pools. Name and describe each type.

____

____

____

____

4. Use *Section 680-24* of the Code to determine if the following items must be grounded.

	TRUE	FALSE
a. Wet-and-dry-niche lighting fixtures.	______	______
b. Electrical equipment located within 5 feet (1.52 m) of inside edge of pool.	______	______
c. Electrical equipment located within 10 feet (3.05 m) of inside edge of pool.	______	______
d. Recirculating equipment and pumps.	______	______
e. Lighting fixtures installed more than 15 feet (4.56 m) from inside edge of pool.	______	______
f. Junction boxes, transformers, and GFCI enclosures.	______	______
g. Panelboards that supply the electrical equipment for the pool.	______	______
h. Panelboards 20 feet (6.1 m) from the pool that do not supply the electrical equipment for the pool.	______	______

5. Grounding conductors (must) (may) be run in the same conduit as the circuit conductors. Circle correct answer.

6. Grounding conductors (may) (may not) be spliced with wire-nut types of wire connectors. Circle correct answer.

7. The purpose of grounding and bonding is to ______________________________

____

____

____

8. What parts of a pool must be bonded together? ______________________________

____

____

____

____

____

____

9. May electrical wires be run above the pool? Explain. ______________________

10. What is the closest distance that a receptacle may be installed to the inside edge of a pool? ______________________

11. Receptacles located within 15 feet (4.56 m) from the inside edge of a pool must be protected by a ______________________

12. Lighting fixtures installed over a pool must be mounted at least (10 feet) (3.05 m), (12 feet) (3.66 m), (15 feet) (4.56 m), above the maximum water level. Circle correct answer.

13. Grounding conductor terminations in wet-niche metal-forming shells, as well as the conduits entering junction boxes or transformer enclosures where the conduit runs directly to the wet-niche lighting fixture, must be ______________ with a (an) ______________ to prevent corrosion to the terminal and to prevent the passage of air through the conduit, which could result in corrosion.

14. Wet-niche lighting fixtures are accessible (from the inside of the pool) (from a tunnel) (on top of a pole). Circle correct answer.

15. Wet-niche lighting fixtures operating at above 15 volts must be protected by ______________________

16. Dry-niche lighting fixtures are accessible (from a tunnel, or passageway, or deck) (from the inside of the pool) (on top of a pole). Circle correct answer.

17. In general, it is *not* permitted to install conduits under the pool or within 5 feet (1.52 m) measured horizontally from the inside edge of the pool. True or false? Explain. ____

18. Junction boxes, wet-niche lighting fixtures, and transformer and GFCI enclosures have one thing in common. They all (are made of bronze) (have threaded hubs) (must be mounted at least 8 inches, or 203 mm, above the deck). Circle correct answer.

19. Lighting fixtures are permitted to be mounted less than 5 feet (1.52 m) measured horizontally from the inside edge of the pool only if they are (made of plastic) (rigidly fastened to an existing structure) (controlled by a wall switch). Circle correct answer.

20. The Code permits radiant heating cable to be buried in or below the deck of a pool. True or false? ____________________

21. For indoor spas and hot tubs, the following statements are either true or false. Check one.

	TRUE	FALSE
a. Receptacles may be installed within 5 feet (1.52 m) from the edge of the spa or hot tub.	________	________
b. All receptacles within 10 feet (3.05 m) of the spa or hot tub must be GFCI protected.	________	________
c. Any receptacles that supply power to pool equipment must be GFCI protected.	________	________
d. Wall switches must be located at least 10 feet (3.05 m) from the pool.	________	________
e. Lighting fixtures above the pool or within 5 feet (1.52 m) from the inside edge of the pool must be GFCI protected.	________	________

22. Bonding and grounding of electrical equipment in and around spas and hot tubs (are required by the Code) (are not required by the Code) (are decided by the electrician). Circle correct answer.

23. Where a spa or hot tub is installed in an existing bathroom, an existing receptacle outlet within 5 feet (1.52 m) of the tub is permitted to remain, but only if the circuit feeding the receptacle outlet has ____________________ protection.

24. For the circuit supplying a hydromassage (whirlpool) bathtub, be sure that the circuit (does not have GFCI protection that could result in nuisance tripping) (is GFCI protected). Circle one.

25. No-niche fixtures shall not have any exposed metal parts. (T) (F) (Circle correct answer.)

26. The maximum voltage that no-niche fixtures are intended to be connected to is:
    a. 12 volts
    b. 15 volts
    c. 24 volts

27. Swimming pool pump motors do not have to be separately grounded if they are listed and identified as having ____________________

____

____________________.

UNIT 31

Smart-House Wiring

OBJECTIVES

After studying this unit, the student will be able to

- understand the basics of Smart-House wiring.
- understand the requirements of *Article 780, National Electrical Code*®, "Closed-Loop and Programmed Power Distribution."
- understand how present Code rules apply to Smart-House installations.
- understand the special terminology used in conjunction with Smart-House wiring.
- identify symbols unique to Smart-House systems.

INTRODUCTION

Introduced in the 1987 edition of the *National Electrical Code*®, *Article 780* defines and covers "Closed-Loop" and "Programmed Power Distribution." As we discuss in this unit, today's microelectronics technology enables electrical and electronics equipment such as appliances, low-voltage signaling, stereo, cable television, heating and cooling, telephone, security systems, and smoke and heat detectors to be interconnected and controlled through a unique wiring system.

Instead of separate, stand-alone systems for each of the aforementioned items, one closed-loop and programmed power distribution system can be installed that will serve all of them.

One such system, called the *Smart House*, has been developed by The National Association of Home Builders. This organization has developed the concepts, convenience features, safety improvements, and other benefits their system can provide to the homeowner.

Various leading electronic and electrical manufacturers design their products to conform to the specifications of the Smart-House concepts.

TESTING AND LISTING

At the time of this writing, Underwriters Laboratories is working on a standard for Closed-Loop Residential Power Distribution Equipment, which will bring together and coordinate the many, many individual standards into conformance with the new standard.

For example, a lighting fixture must conform to a specific UL standard. An appliance must meet the requirements of another standard. A low-voltage system must meet the requirements of still another standard. The same holds true for television antenna systems, grounding, ground-fault protection, and so on.

Yet, *Article 780* of the *NEC*® permits all of these to be totally integrated into a single unique system that will tie all of these previously individual systems together at a common point, all made possible because of electronics.

CODE REQUIREMENTS

All of the Code rules discussed throughout this text apply to Smart-House installations. *Article 780* of the Code *supplements* the general Code rules.

For example, the dryer circuit conductor sizing is determined by the same method as discussed in unit 15. In other words, an electric clothes dryer will still have its load computed on the basis of 5700 volt-amperes for the purpose of service-entrance and feeder calculations. Yet, when this dryer is connected to and supplied by a closed-loop system, a signal that the dryer has completed its drying cycle could be sent to any and all television sets in the house.

This interchange of information from one appliance to another appliance might be considered an event where one appliance "talks" to another appliance.

THE SMART-HOUSE COMPONENTS

Unique to a closed-loop and programmed power distribution system are the systems components, including three types of hybrid cable which can be used to wire the entire house and all its systems.

Branch Cable

This hybrid cable is a flat ribbon-type cable that contains:

- three No. 14 or No. 12 AWG, 120-volt ac power conductors (hot, neutral, ground)
- two No. 18 AWG, 24-volt dc for the uninterruptible power supply (UPS)
- coaxial cable for television, audio, and other high-speed data transmission
- telephone wires
- low-voltage control wires

All of these individual conductors are enclosed in a tough outer jacket, very similar in appearance to nonmetallic-sheathed cable, figures 31-1 and 31-2.

Branch cables originate at the main *service center* and terminate at outlets called *convenience centers* and *switch/sensor outlets*.

A special tool is used to make all of the connections at the termination points as required. The branch cable runs from convenience center to convenience center in the same way that Type NM cable is run between outlets.

Fixture Cable

This cable delivers 120 volts ac to lighting fixtures and small fixed-in-place appliances such as range hoods and bathroom fans. It also contains

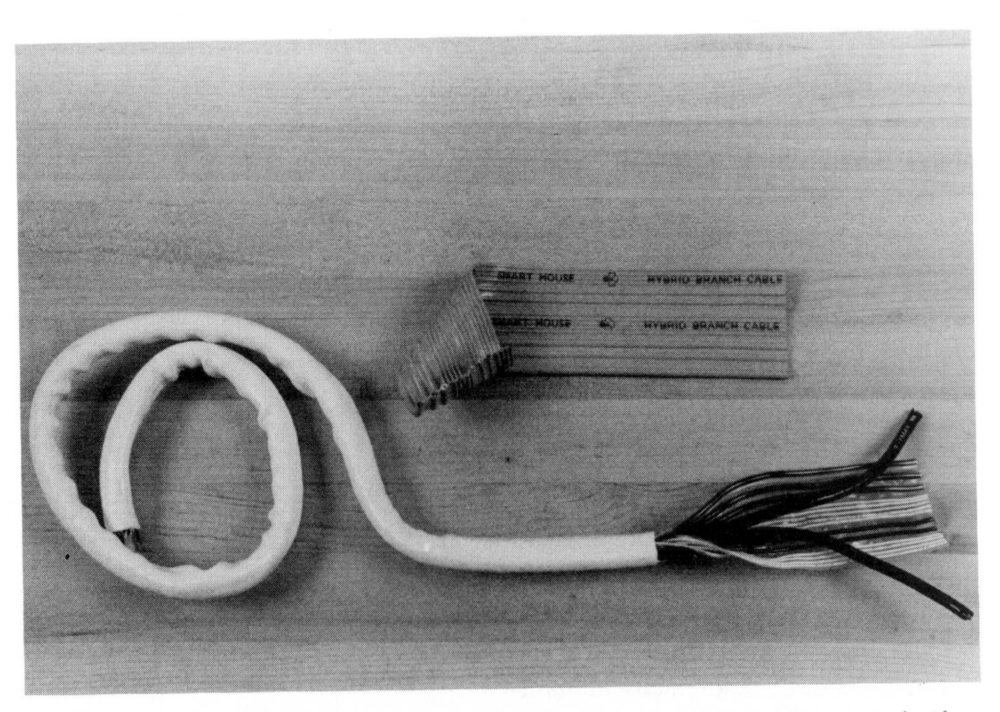

Fig. 31-1 A typical hybrid branch cable that connects all power and communication services to Smart-House convenience centers throughout the home. Courtesy of Smart House, L.P.

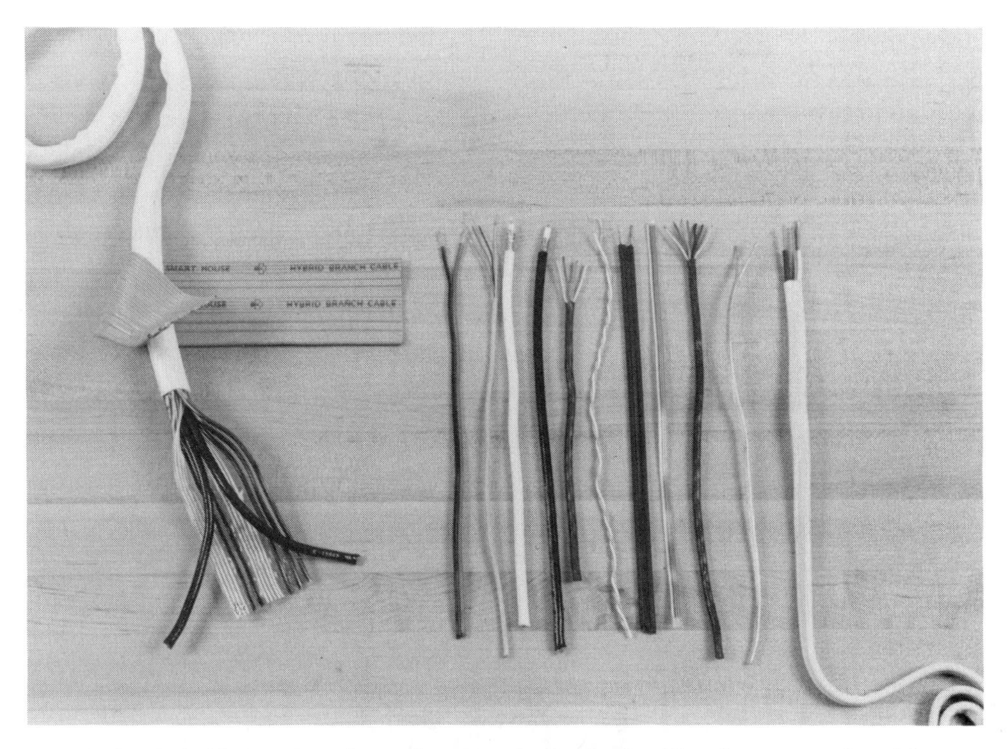

Fig. 31-2 This illustration shows the many individual cables that are replaced by one multiconductor hybrid branch cable. Courtesy of Smart House, L.P.

24-volt dc uninterruptible power supply wires for the control communication that might be desired between that fixture or appliance and the control point. This cable runs between a 120-volt lighting fixture and the convenience center.

Switch/Sensor Cable

This third type of cable carries the 24-volt dc UPS and controls communications between a convenience center and a switch/sensor outlet. System sensors such as smoke detectors, heat detectors, window security systems, electronic wall switches, and so on, install in these outlets. The wall switches, when touched, send a signal back to the system controller, which in turn senses what has been requested and responds accordingly. Is it a request to turn on a light? Is it a request to turn on an appliance? The possibilities are endless.

Service Center

The service center is the brain of the entire installation, figure 31-3. Electric utility power (service entrance), telephone, and television feed into the service center. The service center contains the following:

Load Center: The load center is much the same as a regular load center. It contains the main and branch fuses or circuit breakers for overcurrent protection. It provides transient voltage surge suppression for protection of sensitive electronic equipment and ground-fault protection for all or part of the house wiring. As with all installations, the electrician must provide proper overcurrent protection for the many branch circuits as required by the Code, and must install overcurrent devices that have adequate interrupting ratings according to *Sections 230-65*, *110-9*, and *110-10* of the Code. Part of the load center is a branch-circuit interface where all of the hybrid branch cables terminate. The individual wires (phone, UPS, etc.) are routed to the respective service center components.

Uninterruptible Power Supply (UPS): This is a 24-volt dc system that supplies control system power during a power outage. It allows the system's "memory" to remain intact and also provides sufficient power to ignite gas appliances. It is not designed for supplying load current at 120 or 240 volts to lighting or appliance loads.

System Controller: This is the computer of the system, containing both the hardware and the software. Divided even further, the system controller contains regional and area controllers. Each area controller manages up to two branches, and the regional controller coordinates activities between area controllers and other service center components.

Coaxial Headend: This is the interconnecting point of the system for communication services, such as telephone and cable television. Distribution to the many convenience centers throughout the house starts here.

Telephone Gateway: Here is where the telephone services enter the home. Multiple external phone lines, internal phone lines, and intercom capability are possible by proper connections at the telephone gateway.

Branch Cable Interface: All of the hybrid branch

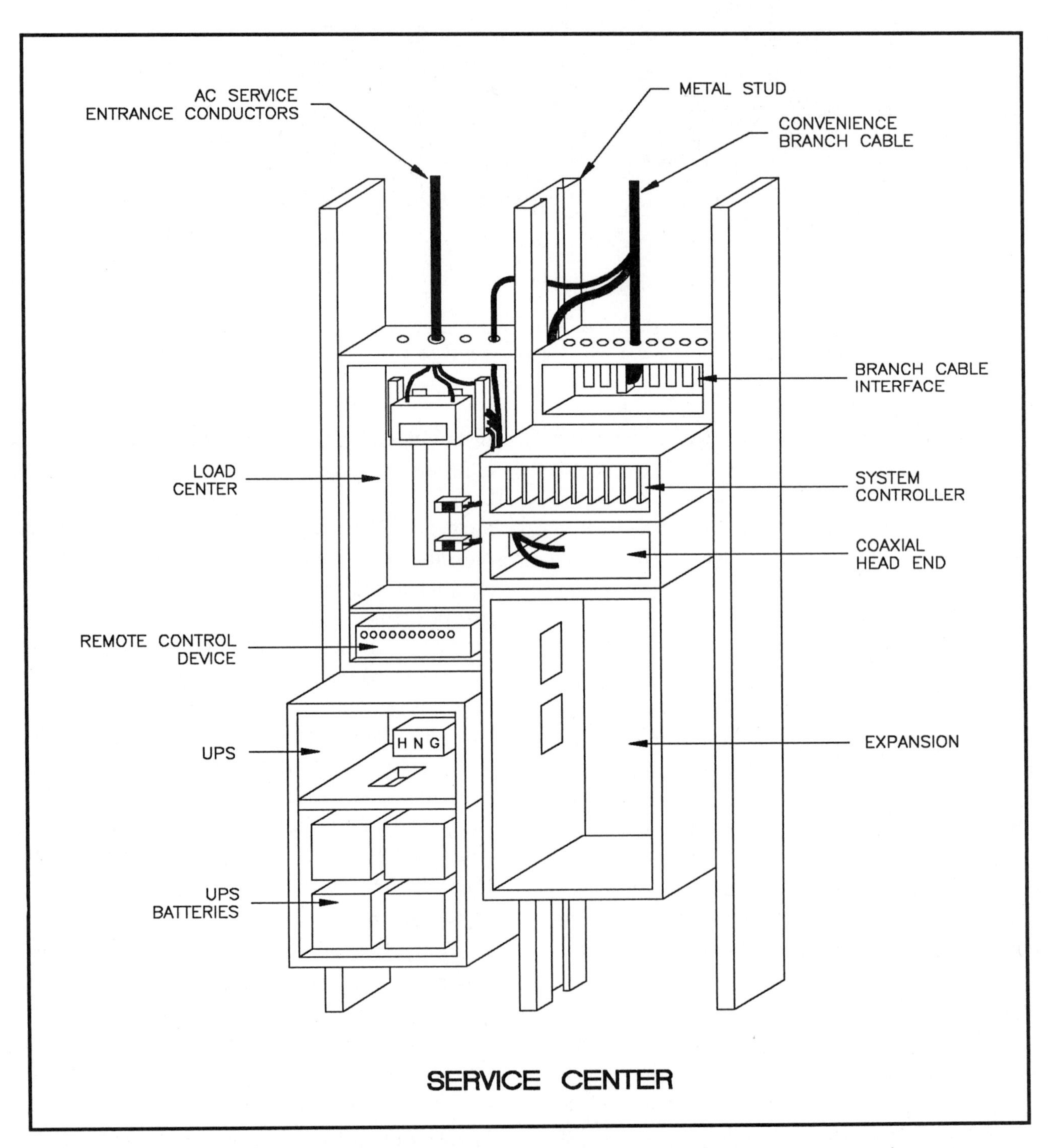

Fig. 31-3 Illustration showing all of the components that make up the Smart-House service center.

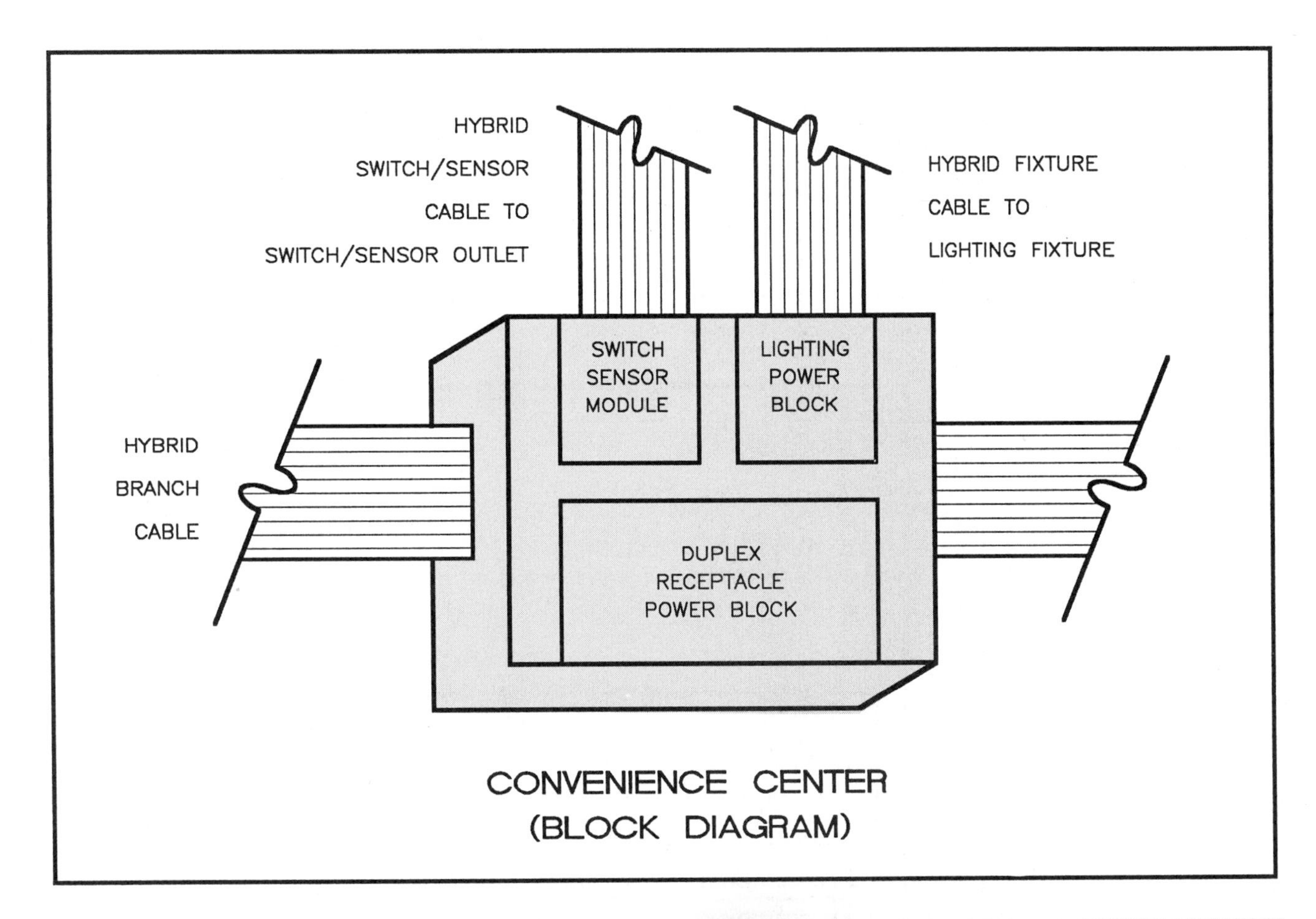

cables terminate here. The individual wires (phone, UPS, etc.) are routed to the proper service-center components.

Remote-Control Device (RCD): As previously mentioned, hybrid cable contains No. 12 or No. 14 AWG conductors for the 120-volt lighting and small appliance circuits. Large appliances such as ranges, dryers, electric furnaces, and so on, are wired using acceptable wiring methods as required by the Code. But when remote control of these dedicated appliances is desired, it is the RCD that permits remote control. An example might be to have the homeowner telephone home and command the thermostat to be turned to a higher temperature prior to returning home.

All of the aforementioned components are 14 inches wide for installation between standard 16-inch-on-center studs.

Convenience Centers (Wall Outlets)

Because the hybrid cable contains many conductors that provide 120-volt circuitry, telephone, television, 24-volt dc, a specially configured wall receptacle outlet is installed. This is called a *convenience center*, figures 31-4A and 31-4B. This special receptacle supplies 120-volt power, television, telephone, 24-volt dc, and control communications, all from one location using special configuration plugs. Inside the convenience center is a power block that incorporates an electronic chip

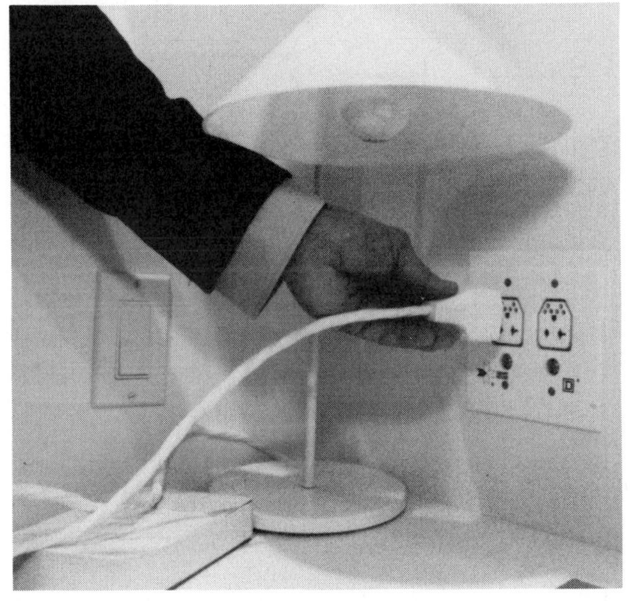

Fig. 31-4A A block diagram of a convenience center.

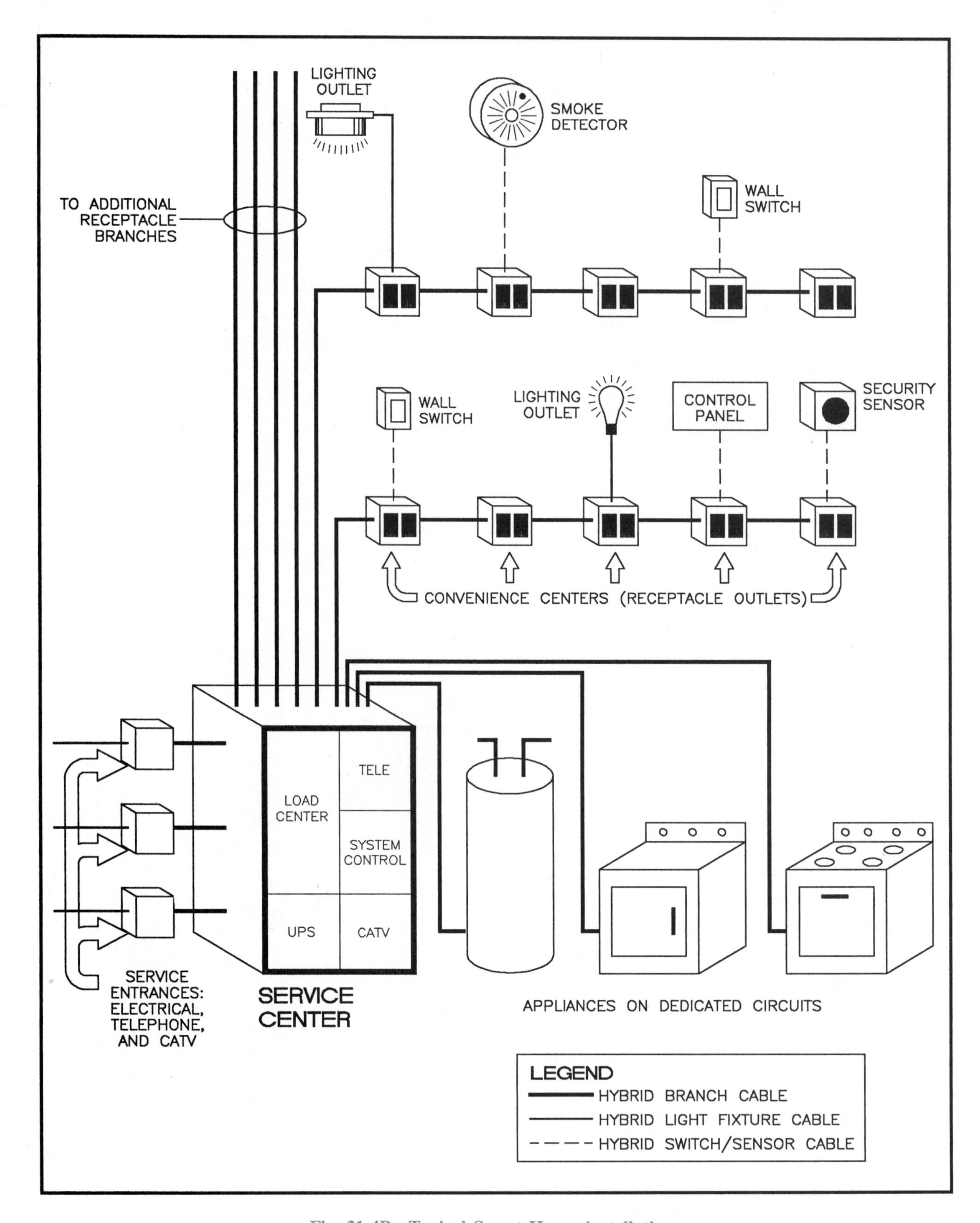

Fig. 31-4B Typical Smart House installation.

(microprocessor) that provides the closed-loop control. There is no power at the convenience center until an appliance is plugged in and a valid request for energy is sent back to the system controller. If a person (baby, child, or adult) inserts something like a paper clip into the receptacle, 120-volt power will not appear because the proper signal is not being sent to the system controller. This is a safety feature of the system.

The male attachment plug cap on the appliance also has the same special configuration of the convenience center outlet. The pins or prongs of the attachment plug cap match up with the proper connections of the convenience center. Older appliances can also be connected to a convenience center outlet. These older appliances receive only 120 volts ac and cannot make use of UPS or control communications.

Power Blocks

In each convenience center are two devices called *power blocks* that communicate with the system controller for remote-control applications or simple switching.

Type 1. Controls and supplies 120-volt power to the duplex receptacle portion of the convenience center (rated 15 or 20 amperes). GFCI protection is one option; non-GFCI protection is another option.

Type 2. Controls 120-volt lighting fixtures or other 120-volt dedicated small appliances that are connected to the convenience center. Dimming and nondimming versions are available.

Switch/Sensor Outlets

A switch sensor outlet is in effect a high-tech device box. System sensors (smoke, heat, intrusion, etc.) and wall switches install in switch/sensor outlets. These devices communicate with the system controller by means of the switch/sensor cable.

Appliances

Appliances for the Smart House are grouped into three categories:

1. Simple — No electronic chips are required in the appliance. The only signal transmitted is the appliance's status. Otherwise the appliance operation is normal.
2. Normal — This appliance contains a chip for communicating digital information back to the system controller.
3. Complex — This appliance contains a microprocessor for complex "messaging" between the system controller and other appliances.

SMART-HOUSE WIRING SYMBOLS

Because a Smart House installation introduces components somewhat new and different from standard electrical installations, a variety of new symbols are required. Figure 31-5 shows these symbols.

Technical training programs relating to the installation of the wiring and systems for the Smart House are available to electrical-joint apprenticeship training committees.

Gas appliances are also available using previously mentioned Smart House control. For instance, a gas appliance is plugged in, and through electronic communication, requests gas energy. The main gas supply shuts off in the event that smoke, fire, or gas leaks are detected.

SUMMARY

The entire Smart-House concept is to provide in one system all of the features presently requiring separate systems. Communication between the switch/sensor outlets, the convenience centers, and the system controller makes this all possible. The homeowner programs the control for any and all lighting outlets, appliance outlets, security systems, and so on. The system can also monitor energy consumption, can detect gas leaks, and do many other tasks.

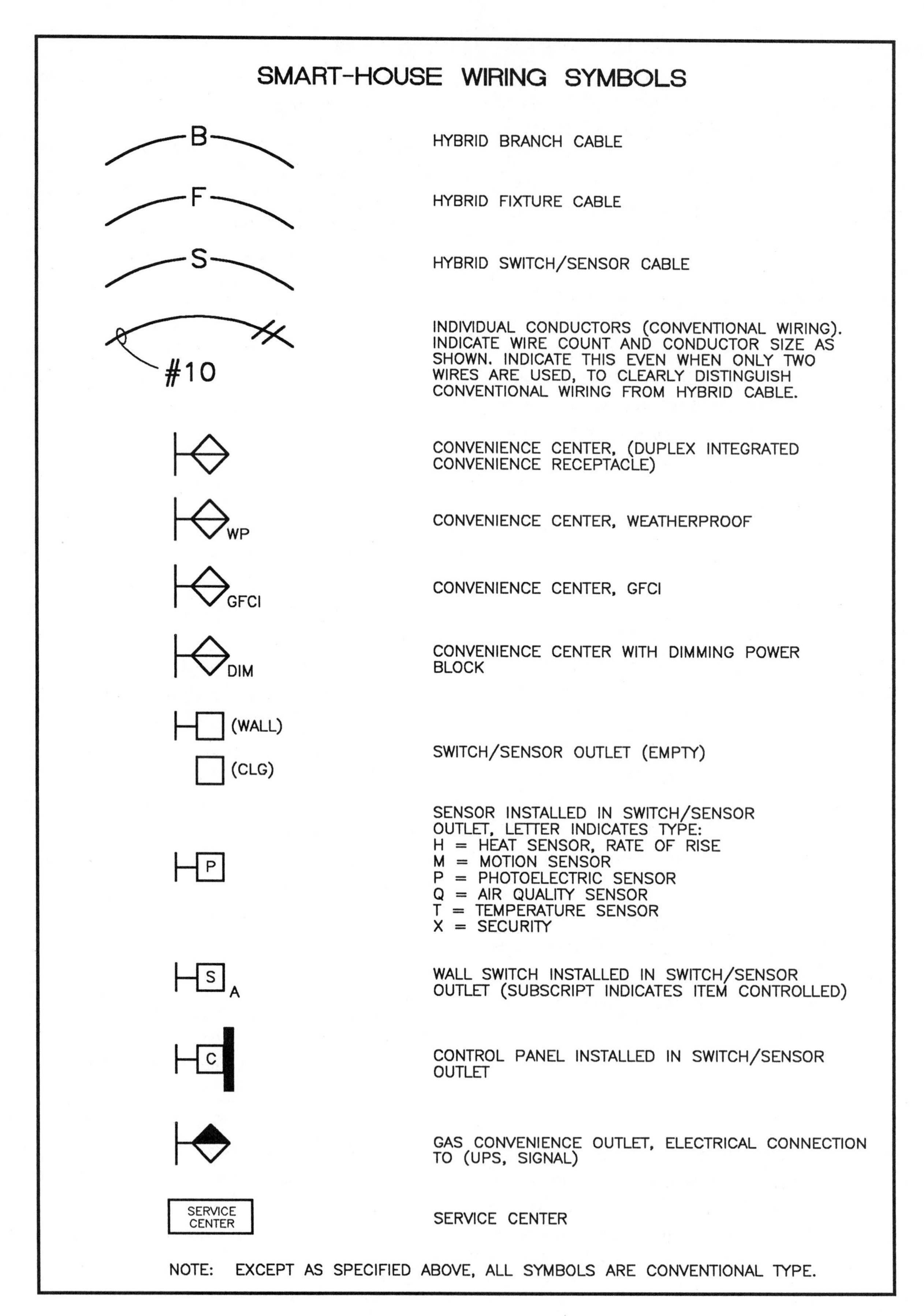

Fig. 31-5 Smart-House wiring symbols.

REVIEW

Note: Refer to the Code or the plans where necessary.

1. *Article 780* of the Code references a rather unique system. What is the name of *Article 780*? ______

2. *Article 780* of the Code is very short. How can a complete residential installation using the Smart-House design be covered in such a short article? ______

3. A special type of cable is used for Smart-House wiring. It contains conductors that supply
   a. ______
   b. ______
   c. ______
   d. ______
   e. ______

4. What are these special cables called? ______

5. Name the three types of hybrid cables used in Smart-House wiring.
   a. ______
   b. ______
   c. ______

6. The "brain" of the Smart House is called the *service center.* Name its main components.
   a. ______
   b. ______
   c. ______
   d. ______
   e. ______
   f. ______
   g. ______

7. The Smart-House wall receptacle outlets are called ______

8. If a child inserted a metal paper clip into the hot slot of a convenience center receptacle and at the same time touched a metal water faucet, would he receive an electrical shock? Why? ______

9. What type of outlets are installed where communication is required to be fed back to a convenience center, then back to the system controller? ______

Specifications for Electrical Work – Single-Family Dwelling

1. **GENERAL:** The "General Clause and Conditions" shall be and are hereby made a part of this division.

2. **SCOPE:** The electrical contractor shall furnish and install a complete electrical system as shown on the drawings and/or in the specifications. Where there is no mention of the responsible party to furnish, install, or wire for a specific item on the electrical drawings, the electrical contractor will be responsible completely for all purchases and labor for a complete operating system for this item.

3. **WORKMANSHIP:** All work shall be executed in a neat and workmanlike manner. All exposed conduits shall be routed parallel or perpendicular to walls and structural members. Junction boxes shall be securely fastened, set true and plumb, and flush with finished surface when wiring method is concealed.

4. **LOCATION OF OUTLETS:** The electrical contractor shall verify location, heights, outlet and switch arrangements, and equipment prior to rough-in. No additions to the contract sum will be permitted for outlets in wrong locations, in conflict with other work, and so on. The owner reserves the right to relocate any device up to 10 feet prior to rough-in, without any charge by the electrical contractor.

5. **CODES:** The electrical installation is to be in accordance with the latest edition of the *National Electrical Code®* all local electrical codes, and the utility company's requirements.

6. **MATERIALS:** All materials shall be new and shall be listed and bear the appropriate label of Underwriters Laboratories, Inc. or another nationally recognized testing laboratory for the specific purpose. The material shall be of the size and type specified on the drawings and/or in the specifications.

7. **WIRING METHOD:** Wiring, unless otherwise specified, shall be nonmetallic-sheathed cable, armored cable, or electrical metallic tubing, adequately sized and installed according to the latest edition of the *National Electrical Code®* and local ordinances.

8. **PERMITS AND INSPECTION FEES:** The electrical contractor shall pay for all permit fees, plan review fees, license fees, inspection fees, and taxes applicable to the electrical installation and shall be included in the base bid as part of this contract.

9. **TEMPORARY WIRING:** The electrical contractor shall furnish and install all temporary wiring for hand-held tools and construction lighting per latest OSHA standards and include all cost in base bid.

10. **WORKSHOP:** Workshop wiring is to be installed in electrical metallic tubing using steel compression gland fittings.

11. **NUMBER OF OUTLETS PER CIRCUIT:** In general, not more than (10) lighting and/or receptacle outlets shall be connected to any one lighting branch circuit. Exceptions may be made in the case of low-current-consuming outlets.

12. **CONDUCTOR SIZE:** General lighting branch circuits shall be No. 14 AWG copper protected by 15-ampere overcurrent devices.

    Small appliance circuits shall be No. 12 AWG copper protected by 20-ampere overcurrent devices.

    All other circuits: wire and overcurrent device as required by the Code.

13. **LOAD BALANCING:** The electrical contractor shall connect all loads, branch circuits, and feeders per Panel Schedule, but shall verify and modify these connections as required to balance connected and computed loads to within 10% variation.

14. **SPARE CONDUITS:** Furnish and install two empty 1/2-inch thinwall (EMT) conduits between workshop and attic for future use.

15. **GUARANTEE OF INSTALLATION:** The electrical contractor shall guarantee all work and materials for a period of one full year after final acceptance by the architect/engineer and owner.

16. **APPLIANCE CONNECTIONS:** The electrical contractor shall furnish all wiring materials and make all final electrical connections for all permanently installed appliances such as, but not limited to, furnace, water heater, water pump, built-in ovens and ranges, food waste disposer, dishwasher, and clothes dryer.

    These appliances are to be furnished by owner.

17. **CHIMES:** Furnish and install two (2) two-tone door chimes where indicated on the plans, complete with two (2) push buttons and suitable chime transformer. Allow $150.00 for above items. Chimes and buttons to be selected by owner.

18. **DIMMERS:** Furnish and install dimmer switches where indicated.

19. **EXHAUST FANS:** Furnish, install, and provide connections for all exhaust fans indicated on the plans, including, but not limited to, ducts, louvers, trims, speed controls, and lamps. Included are recreation room, laundry, rear entry powder room, range hood, and bedroom hall ceiling fan. Allow a sum of $500.00 in base bid for this. This allowance does not include the two bathroom heat/vent/light fixtures.

20. **FIXTURES:** A fixture allowance of $1250.00 shall be included in the electrical contractor's bid. This allowance shall include the furnishing and installation of all surface, recessed, track, strip, pendant, and/or hanging fixtures, complete with lamps. This allowance includes the three bathroom medicine cabinets with lights.

    This allowance does not include the two bathroom ceiling heat/vent/light fixtures. Labor for installation of fixtures shall be included in base bid.

21. **HEAT/VENT/LIGHT CEILING FIXTURES:** Furnish and install two (2) heat/vent/light units where indicated on the plans complete with switch assembly, ducts, louvers, required to perform the heating, venting, and lighting operations as recommended by the manufacturer.

22. **PLUG-IN STRIP:** Where noted in the workshop, furnish and install multioutlet assemblies with outlets 18 inches (457 mm) on center.

23. **SWITCHES, RECEPTACLES, AND FACEPLATES:** All flush switches shall be of the quiet ac-rated toggle type. They shall be mounted 46 inches to center above the finished floor unless otherwise noted.

    Receptacle outlets shall be mounted 12 inches to center above the finished floor unless otherwise noted. All convenience receptacles shall be of the grounding type. Furnish and install where indicated, ground-fault circuit interrupter receptacles to provide ground-fault circuit protection as required by the *National Electrical Code*® All wiring devices are to be provided with ivory handles or faces and shall be trimmed with ivory faceplates except in the kitchen, where chrome-plated steel faceplates shall be used.

    Receptacle outlets, where indicated, shall be of the split-circuit design.

24. **TELEVISION OUTLETS:** Furnish and install 4-inch square, 1 1/2-inch-deep outlet boxes with single-gang raised plaster covers at each television outlet where noted on the plans. Mount at the same height as receptacle outlets. Furnish and install 75-ohm coaxial cable to each television outlet from a point in the workshop near the main service-entrance switch. Allow 6 feet (1.83 mm) of cable in workshop. Furnish and install television plug-in jacks at each location. Faceplates are to match other faceplates in home. All remaining work done by others.

25. **TELEPHONES:** Furnish and install a 3-inch-deep device box or 4-inch square box, 1 1/2 inches deep, with suitable single-gang raised plaster cover at each telephone location, as indicated on the plans.

    Furnish and install four-conductor, No. 18 AWG copper telephone cable to each designated telephone location, terminate in proper modular jack, complete with faceplates.

    Installation shall be according to any and all applicable *National Electrical Code*® and local code regulations.

26. **SERVICE ENTRANCE:** Furnish and install one (1) 200-ampere, 120/240-volt, single-phase, three-wire combination main pullout disconnect complete with 200-ampere Class T fuses in workshop where indicated on the plans. Branch-circuit protection in panel to

incorporate Type S plug fuses, Type SC cartridge fuses, Fusetron® cartridge fuses, or circuit breakers. Panel to have 100,000-ampere interrupting rating.

27. Service-entrance underground lateral conductors to be furnished and installed by utility. Meter equipment (pedestal type) to be furnished by utility and installed by electrical contractor where indicated on plans. Electrical contractor to furnish and install all panels, conduits, fittings, conductors, and other materials required to complete the service-entrance installation from the demarcation point of the utility's equipment to and including the Main Panel.

28. Service-entrance conductors supplied by the electrical contractor shall be two (2) No. 2/0 THW, THHN , THWN, or XHHW phase conductors and one (1) No. 1 bare neutral conductor. Install 1 1/2-inch electrical metallic tubing from Main Panel A to meter pedestal.

29. Bond and ground service-entrance equipment in accordance with latest edition of the *National Electrical Code*,® local, and utility code requirements.

30. **SUBPANEL:** Furnish and install one (1) 20-circuit, 120/240-volt, single-phase, three-wire load center in Recreation Room. Load center to have 125-ampere mains. Feed load center with three (3) No. 3 THHN or THWN conductors or equivalent protected by a 100-ampere two-pole overcurrent device in main panel. Install conductors in 1-inch electrical metallic tubing.

    This panel to be fusible similar to main panel.

31. **CIRCUIT IDENTIFICATION:** All panelboards shall be furnished with typed-card directories with proper designation of the branch-circuit feeder loads and equipment served. The directories shall be located in the panel in a holder for clear viewing.

32. The electrical contractor shall seal and weatherproof all penetrations through foundations, exterior walls, and roofs.

33. Upon completion of the installation, the electrical contractor shall review and check the entire installation, clean equipment and devices, and remove surplus materials and rubbish from the owner's property, leaving the work in neat and clean order and in complete working condition. The electrical contractor shall be responsible for the removal of any cartons, debris, and rubbish for equipment installed by the electrical contractor, including equipment furnished by the owner or others and removed from the carton by the electrical contractor.

34. **SPECIAL-PURPOSE OUTLETS:** Install, provide, and connect all wiring for all special-purpose outlets. Upon completion of the job, all fixtures and appliances shall be operating properly. See Plans and other sections of the Specifications for information as to who is to furnish the fixtures and appliances.

Schedule of Special-Purpose Outlets

SYMBOL	DESCRIPTION	VOLTS	HORSE-POWER	APPLIANCE AMPERE RATING	TOTAL APPLIANCE WATTAGE RATING (OR VA)	CIRCUIT AMPERE RATING	POLES	WIRE SIZE THHN	CIRCUIT NUMBER	COMMENTS
A	Hydromassage tub, master bedroom	120	1/2	10	1200	20	1	12	A9	Connect to Class "A" GFCI. Separate circuit.
B	Water pump	240	1	8	1920	20	2	12	A(5–7)	Run circuit to disconnect switch on wall adjacent to pump; protect with Fusetron dual-element time-delay fuses sized at 125% of motor's F.L.A.
C	Water heater: top element 2000W. Bottom element 3000W.	240	–0–	8.33 12.50 20.83	2000 3000 5000	20	2	12	A(6–8)	Connected for limited demand.
D	Dryer	120/240	120V 1/6 Motor Only	23.75	5700 Total	30	2	10	B(1–3)	Provide flush mounted 30-A dryer receptacle.
E	Overhead garage door opener	120	1/4	5.8	696	15	1	14	B14	Unit comes with 3W cord. Provide box-cover unit (fuse/switch). Install Fustat Type S fuse, 8 amperes. Unit has integral protection. Fustat fuses are additional "back-up" protection. Connect to garage lighting circuit.
F	Wall-mounted oven	120/240	–0–	27.5	6600	30	2	10	B(6–8)	
G	Countertop range	120/240	–0–	31	7450	40	2	8	B(2–4)	
H	Food waste disposer	120	1/3	7.2	864	20	1	12	B19	Controlled by S.P. switch on wall.
I	Dishwasher	120	1/3 Motor Only	Motor 7.20 Htr. 8.33 Total 15.53	1000W 864 1864	20	1	12	B5	
J	Heat/vent/light master B.R. bath	120	–0–	12.3	1475	20	1	12	A12	
K	Heat/vent/light front B.R. bath	120	–0–	12.3	1475	20	1	12	A11	
L	Attic exhaust fan	120	1/4	5.8	696	15	1	14	A10	Run circuit to 4" square box. Locate near fan in attic. Provide box-cover unit. Install Fustat Type S fuse, 8 amperes. Unit has integral protection. Fustat fuses provide additional back-up protection.
M	Electric furnace	240	1/3 Motor	Motor 3.5 Htr. 50.7 Total 54.2	13000	70	2	4	A(1–3)	The overcurrent device and branch-circuit conductors shall not be less than 125% of the total load of the heaters and motor. 54.2 × 1.25 = 67.75 (*Section 424-3(b)*).
N	Air conditioner	240	–0–	30	7200	40	2	8	A(2–4)	Compressor rated load amperes 27.8 Compressor locked rotor amperes 135.0 Condenser fan full load amperes 2.2 Condenser locked rotor amperes 4.5 Branch-circuit dual-element fuse 40.0 Minimum circuit ampacity 37.5
O	Freezer	120	1/4	5.8	696	15	1	14	A13	Install single receptacle outlet. Do not provide GFCI protection.

35. **ALTERNATIVE BID Low-Voltage, Remote-Control System:** The electrical contractor shall submit an alternative bid on the following:

Furnish and install a complete low-voltage remote-control system to accomplish the same results as would be obtained with the conventional switching arrangement as indicated on the electrical plans.

In addition, furnish and install one (1) 12-position master selector switch in Master Bedroom or as directed by the architect or owner. Outlets to be controlled by this switch to be selected by the owner.

In addition, furnish and install two motorized 25-circuit masters. These motor-operated controls shall be controlled from the Front Hall and Master Bedroom, or as indicated by the architect or owner. Connect motor-operated master control in such a manner that each and every switch-controlled lighting outlet and switch-controlled receptacle outlet may be turned off or on from the aforementioned control stations.

All low-voltage wiring to conform to the *National Electrical Code®*.

APPENDIX

USEFUL FORMULAS

TO FIND	SINGLE PHASE	THREE PHASE	DIRECT CURRENT
AMPERES when kVA is known	$\frac{kVA \times 1000}{E}$	$\frac{kVA \times 1000}{E \times 1.73}$	not applicable
AMPERES when horsepower is known	$\frac{hp \times 746}{E \times \%eff. \times pf}$	$\frac{hp \times 746}{E \times 1.73 \times \%eff. \times pf}$	$\frac{hp \times 746}{E \times \%eff.}$
AMPERES when kilowatts are known	$\frac{kW \times 1000}{E \times pf}$	$\frac{kW \times 1000}{E \times 1.73 \times pf}$	$\frac{kW \times 1000}{E}$
KILOWATTS	$\frac{I \times E \times pf}{1000}$	$\frac{I \times E \times 1.73 \times pf}{1000}$	$\frac{I \times E}{1000}$
KILOVOLT AMPERES	$\frac{I \times E}{1000}$	$\frac{I \times E \times 1.73}{1000}$	not applicable
HORSEPOWER	$\frac{I \times E \times \%eff. \times pf}{746}$	$\frac{I \times E \times 1.73 \times \%eff. \times pf}{746}$	$\frac{I \times E \times \%eff.}{746}$
WATTS	$E \times I \times pf$	$E \times I \times 1.73 \times pf$	$E \times I$

I = amperes   E = volts   kW = kilowatts   kVA = kilovolt-amperes

hp = horsepower   % eff. = percent efficiency   pf = power factor

EQUATIONS BASED ON OHM'S LAW:

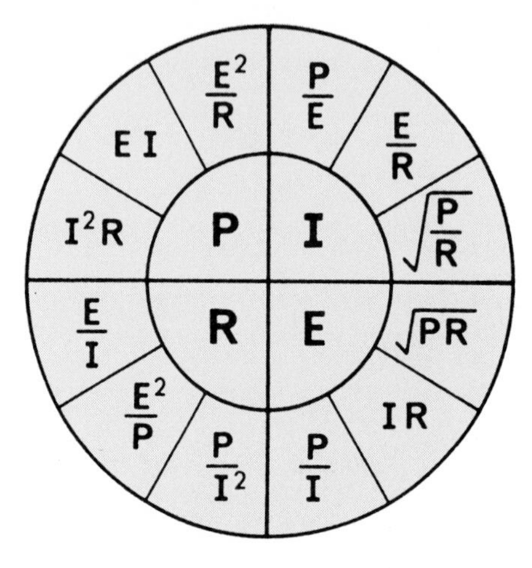

P = POWER, IN WATTS
I = CURRENT, IN AMPERES
R = RESISTANCE, IN OHMS
E = ELECTROMOTIVE FORCE, IN VOLTS

INDEX

Note that italic page numbers indicate that the entry comes from a figure or table.